Lecture Notes in Artificial Intelligence   12033

Subseries of Lecture Notes in Computer Science

More information about this series at http://www.springer.com/series/1244

Ngoc Thanh Nguyen · Kietikul Jearanaitanakij ·
Ali Selamat · Bogdan Trawiński ·
Suphamit Chittayasothorn (Eds.)

Intelligent Information and Database Systems

12th Asian Conference, ACIIDS 2020
Phuket, Thailand, March 23–26, 2020
Proceedings, Part I

Springer

Editors
Ngoc Thanh Nguyen ⓘ
Department of Applied Informatics
Wrocław University of Science
and Technology
Wrocław, Poland

Ali Selamat ⓘ
Faculty of Computer
Science and Information
University Teknologi Malaysia
Kuala Lumpur, Malaysia

Suphamit Chittayasothorn ⓘ
King Mongkut's Institute
of Technology Ladkrabang
Bangkok, Thailand

Kietikul Jearanaitanakij ⓘ
King Mongkut's Institute
of Technology Ladkrabang
Bangkok, Thailand

Bogdan Trawiński ⓘ
Department of Applied Informatics
Wrocław University of Science
and Technology
Wrocław, Poland

ISSN 0302-9743   ISSN 1611-3349   (electronic)
Lecture Notes in Artificial Intelligence
ISBN 978-3-030-41963-9   ISBN 978-3-030-41964-6   (eBook)
https://doi.org/10.1007/978-3-030-41964-6

LNCS Sublibrary: SL7 – Artificial Intelligence

This Springer imprint is published by the registered company Springer Nature Switzerland AG
The registered company address is: Gewerbestrasse 11, 6330 Cham, Switzerland

Preface

ACIIDS 2020 was the 12th event in a series of international scientific conferences on research and applications in the field of intelligent information and database systems. The aim of ACIIDS 2020 was to provide an international forum of research workers with scientific backgrounds on the technology of intelligent information and database systems and its various applications. The ACIIDS 2020 conference was co-organized by King Mongkut's Institute of Technology Ladkrabang (Thailand) and Wrocław University of Science and Technology (Poland) in cooperation with the IEEE SMC Technical Committee on Computational Collective Intelligence, European Research Center for Information Systems (ERCIS), The University of Newcastle (Australia), Yeungnam University (South Korea), Leiden University (The Netherlands), Universiti Teknologi Malaysia (Malaysia), BINUS University (Indonesia), Quang Binh University (Vietnam), and Nguyen Tat Thanh University (Vietnam). It took place in Phuket, Thailand during March 23–26, 2020.

The ACIIDS conference series is already well established. The first two events, ACIIDS 2009 and ACIIDS 2010, took place in Dong Hoi City and Hue City in Vietnam, respectively. The third event, ACIIDS 2011, took place in Daegu (South Korea), followed by the fourth event, ACIIDS 2012, in Kaohsiung (Taiwan). The fifth event, ACIIDS 2013, was held in Kuala Lumpur (Malaysia) while the sixth event, ACIIDS 2014, was held in Bangkok (Thailand). The seventh event, ACIIDS 2015, took place in Bali (Indonesia), followed by the eighth event, ACIIDS 2016, in Da Nang (Vietnam). The ninth event, ACIIDS 2017, was organized in Kanazawa (Japan). The 10th jubilee conference, ACIIDS 2018, was held in Dong Hoi City (Vietnam), followed by the 11th event, ACIIDS 2019, in Yogyakarta (Indonesia).

For this edition of the conference we received 285 papers from 43 countries all over the world. Each paper was peer reviewed by at least two members of the international Program Committee and the international board of reviewers. Only 105 papers with the highest quality were selected for an oral presentation and publication in these two volumes of the ACIIDS 2020 proceedings.

Papers included in these proceedings cover the following topics: knowledge engineering and Semantic Web; natural language processing; decision support and control systems; computer vision techniques; machine learning and data mining; deep learning models; advanced data mining techniques and applications; multiple model approach to machine learning; application of intelligent methods to constrained problems; automated reasoning with applications in intelligent systems; current trends in artificial intelligence; optimization, learning, and decision-making in bioinformatics and bioengineering; computer vision and intelligent systems, data modeling and processing for industry 4.0; intelligent applications of the Internet of Things (IoT) and data analysis technologies; intelligent and contextual systems; intelligent systems and algorithms in information sciences; intelligent supply chains and e-commerce; privacy, security, and trust in artificial intelligence; and interactive analysis of image, video, and motion data in life sciences.

The accepted and presented papers focus on new trends and challenges facing the intelligent information and database systems community. The presenters showed how research work could stimulate novel and innovative applications. We hope that you found these results useful and inspiring for your future research work.

We would like to express our sincere thanks to the honorary chairs for their support: Prof. Suchatvee Suwansawat (President of King Mongkut's Institute of Technology, Ladkrabang, Thailand), Cezary Madryas (Rector of Wrocław University of Science and Technology, Poland), Prof. Moonis Ali (President of the International Society of Applied Intelligence, USA), and Prof. Komsan Maleesee (Dean of Engineering, King Mongkut's Institute of Technology, Ladkrabang, Thailand).

Our special thanks go to the program chairs, special session chairs, organizing chairs, publicity chairs, liaison chairs, and local Organizing Committee for their work towards the conference. We sincerely thank all the members of the international Program Committee for their valuable efforts in the review process, which helped us to guarantee the highest quality of the selected papers for the conference. We cordially thank the organizers and chairs of special sessions who contributed to the success of the conference.

We would like to express our thanks to the keynote speakers for their world-class plenary speeches: Prof. Włodzisław Duch from the Nicolaus Copernicus University (Poland), Prof. Nikola Kasabov from Auckland University of Technology (New Zealand), Prof. Dusit Niyato from Nanyang Technological University (Singapore), and Prof. Geoff Webb from the Monash University Centre for Data Science (Australia).

We cordially thank our main sponsors: King Mongkut's Institute of Technology Ladkrabang (Thailand), Wrocław University of Science and Technology (Poland), IEEE SMC Technical Committee on Computational Collective Intelligence, European Research Center for Information Systems (ERCIS), The University of Newcastle (Australia), Yeungnam University (South Korea), Leiden University (The Netherlands), Universiti Teknologi Malaysia (Malaysia), BINUS University (Indonesia), Quang Binh University (Vietnam), and Nguyen Tat Thanh University (Vietnam). Our special thanks are also due to Springer for publishing the proceedings and sponsoring awards, and to all the other sponsors for their kind support.

We wish to thank the members of the Organizing Committee for their excellent work and the members of the local Organizing Committee for their considerable effort. We cordially thank all the authors, for their valuable contributions, and the other participants of this conference. The conference would not have been possible without their support. Thanks are also due to many experts who contributed to making the event a success.

We would like to extend our heartfelt thanks to Jarosław Gowin, Deputy Prime Minister of the Republic of Poland and Minister of Science and Higher Education, for his support and honorary patronage of the conference.

March 2020

Ngoc Thanh Nguyen
Kietikul Jearanaitanakij
Ali Selamat
Bogdan Trawiński
Suphamit Chittayasothorn

Organization

Honorary Chairs

Suchatvee Suwansawat	President of King Mongkut's Institute of Technology Ladkrabang, Thailand
Cezary Madryas	Rector of Wrocław University of Science and Technology, Poland
Moonis Ali	President of International Society of Applied Intelligence, USA
Komsan Maleesee	Dean of Engineering, King Mongkut's Institute of Technology Ladkrabang, Thailand

General Chairs

Ngoc Thanh Nguyen	Wrocław University of Science and Technology, Poland
Suphamit Chittayasothorn	King Mongkut's Institute of Technology Ladkrabang, Thailand

Program Chairs

Kietikul Jearanaitanakij	King Mongkut's Institute of Technology Ladkrabang, Thailand
Tzung-Pei Hong	National University of Kaohsiung, Taiwan
Ali Selamat	Universiti Teknologi Malaysia, Malaysia
Edward Szczerbicki	The University of Newcastle, Australia
Bogdan Trawiński	Wrocław University of Science and Technology, Poland

Steering Committee

Ngoc Thanh Nguyen (Chair)	Wrocław University of Science and Technology, Poland
Longbing Cao	University of Technology Sydney, Australia
Suphamit Chittayasothorn	King Mongkut's Institute of Technology Ladkrabang, Thailand
Ford Lumban Gaol	Bina Nusantara University, Indonesia
Tu Bao Ho	Japan Advanced Institute of Science and Technology, Japan
Tzung-Pei Hong	National University of Kaohsiung, Taiwan
Dosam Hwang	Yeungnam University, South Korea
Bela Stantic	Griffith University, Australia

Geun-Sik Jo	Inha University, South Korea
Hoai An Le-Thi	University of Lorraine, France
Zygmunt Mazur	Wrocław University of Science and Technology, Poland
Toyoaki Nishida	Kyoto University, Japan
Leszek Rutkowski	Częstochowa University of Technology, Poland
Ali Selamat	Universiti Teknologi Malaysia, Malaysia

Special Session Chairs

Marcin Pietranik	Wroclaw University of Science and Technology, Poland
Chutimet Srinilta	King Mongkut's Institute of Technology Ladkrabang, Thailand
Paweł Sitek	Kielce University of Technology, Poland

Liaison Chairs

Ford Lumban Gaol	Bina Nusantara University, Indonesia
Quang-Thuy Ha	VNU University of Engineering and Technology, Vietnam
Mong-Fong Horng	National Kaohsiung University of Applied Sciences, Taiwan
Dosam Hwang	Yeungnam University, South Korea
Le Minh Nguyen	Japan Advanced Institute of Science and Technology, Japan
Ali Selamat	Universiti Teknologi Malaysia, Malaysia

Organizing Chairs

Wiboon Prompanich	King Mongkut's Institute of Technology Ladkrabang, Thailand
Adrianna Kozierkiewicz	Wrocław University of Science and Technology, Poland
Krystian Wojtkiewicz	Wrocław University of Science and Technology, Poland

Publicity Chairs

Rathachai Chawuthai	King Mongkut's Institute of Technology Ladkrabang, Thailand
Marek Kopel	Wrocław University of Science and Technology, Poland
Marek Krótkiewicz	Wrocław University of Science and Technology, Poland

Webmaster

Marek Kopel   Wroclaw University of Science and Technology,
                                         Poland

Local Organizing Committee

Pakorn Watanachaturaporn   King Mongkut's Institute of Technology Ladkrabang,
                                         Thailand
Sorayut Glomglome   King Mongkut's Institute of Technology Ladkrabang,
                                         Thailand
Watchara Chatwiriya   King Mongkut's Institute of Technology Ladkrabang,
                                         Thailand
Sathaporn Promwong   King Mongkut's Institute of Technology Ladkrabang,
                                         Thailand
Putsadee Pornphol   Phuket Rajabhat University, Thailand
Maciej Huk   Wrocław University of Science and Technology,
                                         Poland
Marcin Jodłowiec   Wrocław University of Science and Technology,
                                         Poland

Keynote Speakers

Włodzisław Duch   Nicolaus Copernicus University, Poland
Nikola Kasabov   Auckland University of Technology, New Zealand
Dusit Niyato   Nanyang Technological University, Singapore
Geoff Webb   Monash University Centre for Data Science, Australia

Special Sessions Organizers

1. *CSHAC 2020: Special Session on Cyber-Physical Systems in Healthcare:*
   Applications and Challenges

Michael Mayo   University of Waikato, New Zealand
Abigail Koay   University of Waikato, New Zealand
Panos Patros   University of Waikato, New Zealand

2. *ADMTA 2020: Special Session on Advanced Data Mining Techniques*
   and Applications

Chun-Hao Chen   Tamkang University, Taiwan
Bay Vo   Ho Chi Minh City University of Technology, Vietnam
Tzung-Pei Hong   National University of Kaohsiung, Taiwan

3. CVIS 2020: Special Session on Computer Vision and Intelligent Systems

Van-Dung Hoang	Quang Binh University, Vietnam
Kang-Hyun Jo	University of Ulsan, South Korea
My-Ha Le	Ho Chi Minh City University of Technology and Education, Vietnam
Van-Huy Pham	Ton Duc Thang University, Vietnam

4. MMAML 2020: Special Session on Multiple Model Approach to Machine Learning

Tomasz Kajdanowicz	Wrocław University of Science and Technology, Poland
Edwin Lughofer	Johannes Kepler University Linz, Austria
Bogdan Trawiński	Wrocław University of Science and Technology, Poland

5. CIV 2020: Special Session on Computational Imaging and Vision

Manish Khare	Dhirubhai Ambani Institute of Information and Communication Technology, India
Prashant Srivastava	NIIT University, India
Om Prakash	Inferigence Quotient, India
Jeonghwan Gwak	Korea National University of Transportation, South Korea

6. ISAIS 2020: Special Session on Intelligent Systems and Algorithms in Information Sciences

Martin Kotyrba	University of Ostrava, Czech Republic
Eva Volna	University of Ostrava, Czech Republic
Ivan Zelinka	VŠB – Technical University of Ostrava, Czech Republic
Pavel Petr	University of Pardubice, Czech Republic

7. IOTAI 2020: Special Session on Internet of Things and Artificial Intelligence for Energy Efficiency-Recent Advances and Future Trends

Mohamed Elhoseny	Mansoura University, Egypt
Mohamed Abdel-Basset	Zagazig University, Egypt

8. DMPI-4.0 2020: Special Session on Data Modelling and Processing for Industry 4.0

Du Haizhou	Shanghai University of Electric Power, China
Wojciech Hunek	Opole University of Technology, Poland
Marek Krótkiewicz	Wrocław University of Science and Technology, Poland
Krystian Wojtkiewicz	Wrocław University of Science and Technology, Poland

9. *ICxS 2020: Special Session on Intelligent and Contextual Systems*

Maciej Huk	Wroclaw University of Science and Technology, Poland
Keun Ho Ryu	Ton Duc Thang University, Vietnam
Goutam Chakraborty	Iwate Prefectural University, Japan
Qiangfu Zhao	University of Aizu, Japan
Chao-Chun Chen	National Cheng Kung University, Taiwan
Rashmi Dutta Baruah	Indian Institute of Technology Guwahati, India

10. *ARAIS 2020: Special Session on Automated Reasoning with Applications in Intelligent Systems*

Jingde Cheng	Saitama University, Japan

11. *ISCEC 2020: Special Session on Intelligent Supply Chains and e-Commerce*

Arkadiusz Kawa	Łukasiewicz Research Network – The Institute of Logistics and Warehousing, Poland
Justyna Światowiec-Szczepańska	Poznań University of Economics and Business, Poland
Bartłomiej Pierański	Poznań University of Economics and Business, Poland

12. *IAIOTDAT 2020: Special Session on Intelligent Applications of Internet of Things and Data Analysis Technologies*

Rung Ching Chen	Chaoyang University of Technology, Taiwan
Yung-Fa Huang	Chaoyang University of Technology, Taiwan
Yu-Huei Cheng	Chaoyang University of Technology, Taiwan

13. *CTAIOLDMBB 2020: Special Session on Current Trends in Artificial Intelligence, Optimization, Learning, and Decision-Making in Bioinformatics and Bioengineering*

Dominik Vilimek	VŠB – Technical University of Ostrava, Czech Republic
Jan Kubíček	VŠB – Technical University of Ostrava, Czech Republic
Marek Penhaker	VŠB – Technical University of Ostrava, Czech Republic
Muhammad Usman Akram	National University of Sciences and Technology Pakistan, Pakistan
Vladimir Juras	Medical University of Vienna, Austria
Bhabani Shankar Prasad Mishra	KIIT University, India
Ondrej Krejcar	University of Hradec Kralove, Czech Republic

14. *SAILS 2020: Special Session on Interactive Analysis of Image, Video and Motion Data in Life Sciences*

Konrad Wojciechowski	Polish-Japanese Academy of Information Technology, Poland
Marek Kulbacki	Polish-Japanese Academy of Information Technology, Poland
Jakub Segen	Polish-Japanese Academy of Information Technology, Poland
Zenon Chaczko	University of Technology Sydney, Australia
Andrzej Przybyszewski	UMass Medical School, USA
Jerzy Nowacki	Polish-Japanese Academy of Information Technology, Poland

15. *AIMCP 2020: Special Session on Application of Intelligent Methods to Constrained Problems*

Jarosław Wikarek	Kielce University of Technology, Poland
Mukund Janardhanan	University of Leicester, UK

16. *PSTrustAI 2020: Special Session on Privacy, Security and Trust in Artificial Intelligence*

Pascal Bouvry	University of Luxembourg, Luxembourg
Matthias R. Brust	University of Luxembourg, Luxembourg
Grégoire Danoy	University of Luxembourg, Luxembourg
El-ghazali Talbi	University of Lille, France

17. *IMSAGRWS 2020: Intelligent Modeling and Simulation Approaches for Games and Real World Systems*

Doina Logofătu	Frankfurt University of Applied Sciences, Germany
Costin Bădică	University of Craiova, Romania
Florin Leon	Gheorghe Asachi Technical University, Romania

International Program Committee

Muhammad Abulaish	South Asian University, India
Waseem Ahmad	Waiariki Institute of Technology, New Zealand
R. S. Ajin	Idukki District Disaster Management Authority, India
Jesus Alcala-Fdez	University of Granada, Spain
Bashar Al-Shboul	University of Jordan, Jordan
Lionel Amodeo	University of Technology of Troyes, France
Toni Anwar	Universiti Teknologi Petronas, Malaysia
Taha Arbaoui	University of Technology of Troyes, France
Mehmet Emin Aydin	University of the West of England, UK
Ahmad Taher Azar	Prince Sultan University, Saudi Arabia
Thomas Bäck	Leiden University, The Netherlands

Amelia Badica	University of Craiova, Romania
Costin Badica	University of Craiova, Romania
Kambiz Badie	ICT Research Institute, Iran
Hassan Badir	École Nationale des Sciences Appliquées de Tanger, Morocco
Zbigniew Banaszak	Warsaw University of Technology, Poland
Dariusz Barbucha	Gdynia Maritime University, Poland
Ramazan Bayindir	Gazi University, Turkey
Maumita Bhattacharya	Charles Sturt University, Australia
Leon Bobrowski	Białystok University of Technology, Poland
Bülent Bolat	Yildiz Technical University, Turkey
Mariusz Boryczka	University of Silesia, Poland
Urszula Boryczka	University of Silesia, Poland
Zouhaier Brahmia	University of Sfax, Tunisia
Stephane Bressan	National University of Singapore, Singapore
Peter Brida	University of Žilina, Slovakia
Andrej Brodnik	University of Ljubljana, Slovenia
Piotr Bródka	Wroclaw University of Science and Technology, Poland
Grażyna Brzykcy	Poznan University of Technology, Poland
Robert Burduk	Wrocław University of Science and Technology, Poland
Aleksander Byrski	AGH University of Science and Technology, Poland
Tru Cao	Ho Chi Minh City University of Technology, Vietnam
Leopoldo Eduardo Cardenas-Barron	Tecnologico de Monterrey, Mexico
Oscar Castillo	Tijuana Institute of Technology, Mexico
Dariusz Ceglarek	WSB University in Poznań, Poland
Stefano A. Cerri	University of Montpellier, France
Zenon Chaczko	University of Technology Sydney, Australia
Altangerel Chagnaa	National University of Mongolia, Mongolia
Somchai Chatvichienchai	University of Nagasaki, Japan
Chun-Hao Chen	Tamkang University, Taiwan
Rung-Ching Chen	Chaoyang University of Technology, Taiwan
Shyi-Ming Chen	National Taiwan University of Science and Technology, Taiwan
Leszek J. Chmielewski	Warsaw University of Life Sciences, Poland
Sung-Bae Cho	Yonsei University, South Korea
Kazimierz Choroś	Wrocław University of Science and Technology, Poland
Kun-Ta Chuang	National Cheng Kung University, Taiwan
Piotr Chynał	Wrocław University of Science and Technology, Poland
Dorian Cojocaru	University of Craiova, Romania
Jose Alfredo Ferreira Costa	Federal University of Rio Grande do Norte (UFRN), Brazil

Ireneusz Czarnowski — Gdynia Maritime University, Poland
Piotr Czekalski — Silesian University of Technology, Poland
Theophile Dagba — University of Abomey-Calavi, Benin
Tien V. Do — Budapest University of Technology and Economics, Hungary
Grzegorz Dobrowolski — AGH University of Science and Technology, Poland
Rafał Doroz — University of Silesia, Poland
Habiba Drias — University of Science and Technology Houari Boumediene, Algeria
Maciej Drwal — Wrocław University of Science and Technology, Poland
Ewa Dudek-Dyduch — AGH University of Science and Technology, Poland
El-Sayed M. El-Alfy — King Fahd University of Petroleum and Minerals, Saudi Arabia
Keiichi Endo — Ehime University, Japan
Sebastian Ernst — AGH University of Science and Technology, Poland
Nadia Essoussi — University of Carthage, Tunisia
Rim Faiz — University of Carthage, Tunisia
Victor Felea — Universitatea Alexandru Ioan Cuza din Iaşi, Romania
Simon Fong — University of Macau, Macau SAR
Dariusz Frejlichowski — West Pomeranian University of Technology, Poland
Blanka Frydrychova Klimova — University of Hradec Králové, Czech Republic
Mohamed Gaber — Birmingham City University, UK
Marina L. Gavrilova — University of Calgary, Canada
Janusz Getta — University of Wollongong, Australia
Daniela Gifu — Universitatea Alexandru Ioan Cuza din Iaşi, Romania
Fethullah Göçer — Galatasaray University, Turkey
Daniela Godoy — ISISTAN Research Institute, Argentina
Gergo Gombos — Eötvös Loránd University, Hungary
Fernando Gomide — University of Campinas, Brazil
Antonio Gonzalez-Pardo — Universidad Autónoma de Madrid, Spain
Janis Grundspenkis — Riga Technical University, Latvia
Claudio Gutierrez — Universidad de Chile, Chile
Quang-Thuy Ha — VNU University of Engineering and Technology, Vietnam
Dawit Haile — Addis Ababa University, Ethiopia
Pei-Yi Hao — National Kaohsiung University of Applied Sciences, Taiwan
Spits Warnars Harco Leslie Hendric — BINUS University, Indonesia
Marcin Hernes — Wrocław University of Economics and Business, Poland
Francisco Herrera — University of Granada, Spain
Koichi Hirata — Kyushu Institute of Technology, Japan

Bogumiła Hnatkowska	Wrocław University of Science and Technology, Poland
Huu Hanh Hoang	Posts and Telecommunications Institute of Technology, Vietnam
Quang Hoang	Hue University of Sciences, Vietnam
Van-Dung Hoang	Quang Binh University, Vietnam
Jaakko Hollmen	Aalto University, Finland
Tzung-Pei Hong	National University of Kaohsiung, Taiwan
Mong-Fong Horng	National Kaohsiung University of Applied Sciences, Taiwan
Yung-Fa Huang	Chaoyang University of Technology, Taiwan
Maciej Huk	Wrocław University of Science and Technology, Poland
Dosam Hwang	Yeungnam University, South Korea
Roliana Ibrahim	Universiti Teknologi Malaysia, Malaysia
Mirjana Ivanovic	University of Novi Sad, Serbia
Sanjay Jain	National University of Singapore, Singapore
Jarosław Jankowski	West Pomeranian University of Technology, Poland
Kietikul Jearanaitanakij	King Mongkut's Institute of Technology Ladkrabang, Thailand
Khalid Jebari	LCS Rabat, Morocco
Janusz Jeżewski	Institute of Medical Technology and Equipment ITAM, Poland
Joanna Jędrzejowicz	University of Gdańsk, Poland
Piotr Jędrzejowicz	Gdynia Maritime University, Poland
Przemysław Juszczuk	University of Economics in Katowice, Poland
Dariusz Kania	Silesian University of Technology, Poland
Nikola Kasabov	Auckland University of Technology, New Zealand
Arkadiusz Kawa	Poznań University of Economics and Business, Poland
Zaheer Khan	University of the West of England, UK
Muhammad Khurram Khan	King Saud University, Saudi Arabia
Marek Kisiel-Dorohinicki	AGH University of Science and Technology, Poland
Attila Kiss	Eötvös Loránd University, Hungary
Jerzy Klamka	Silesian University of Technology, Poland
Frank Klawonn	Ostfalia University of Applied Sciences, Germany
Shinya Kobayashi	Ehime University, Japan
Joanna Kolodziej	Cracow University of Technology, Poland
Grzegorz Kołaczek	Wrocław University of Science and Technology, Poland
Marek Kopel	Wrocław University of Science and Technology, Poland
Józef Korbicz	University of Zielona Gora, Poland
Raymondus Kosala	BINUS University, Indonesia
Leszek Koszałka	Wrocław University of Science and Technology, Poland
Leszek Kotulski	AGH University of Science and Technology, Poland

Jan Kozak	University of Economics in Katowice, Poland
Adrianna Kozierkiewicz	Wrocław University of Science and Technology, Poland
Ondrej Krejcar	University of Hradec Králové, Czech Republic
Dariusz Król	Wrocław University of Science and Technology, Poland
Marek Krótkiewicz	Wrocław University of Science and Technology, Poland
Marzena Kryszkiewicz	Warsaw University of Technology, Poland
Adam Krzyzak	Concordia University, Canada
Jan Kubíček	VSB – Technical University of Ostrava, Czech Republic
Tetsuji Kuboyama	Gakushuin University, Japan
Elżbieta Kukla	Wrocław University of Science and Technology, Poland
Julita Kulbacka	Wrocław Medical University, Poland
Marek Kulbacki	Polish-Japanese Academy of Information Technology, Poland
Kazuhiro Kuwabara	Ritsumeikan University, Japan
Halina Kwaśnicka	Wroclaw University of Science and Technology, Poland
Annabel Latham	Manchester Metropolitan University, UK
Bac Le	VNU University of Science, Vietnam
Kun Chang Lee	Sungkyunkwan University, South Korea
Yue-Shi Lee	Ming Chuan University, Taiwan
Florin Leon	Gheorghe Asachi Technical University of Iasi, Romania
Horst Lichter	RWTH Aachen University, Germany
Igor Litvinchev	Nuevo Leon State University, Mexico
Rey-Long Liu	Tzu Chi University, Taiwan
Doina Logofatu	Frankfurt University of Applied Sciences, Germany
Edwin Lughofer	Johannes Kepler University Linz, Austria
Lech Madeyski	Wrocław University of Science and Technology, Poland
Nezam Mahdavi-Amiri	Sharif University of Technology, Iran
Bernadetta Maleszka	Wrocław University of Science and Technology, Poland
Marcin Maleszka	Wrocław University of Science and Technology, Poland
Yannis Manolopoulos	Open University of Cyprus, Cyprus
Konstantinos Margaritis	University of Macedonia, Greece
Vukosi Marivate	Council for Scientific and Industrial Research, South Africa
Urszula Markowska-Kaczmar	Wrocław University of Science and Technology, Poland

Danilo Pelusi   University of Teramo, Italy
Bernhard Pfahringer   University of Waikato, New Zealand
Bartłomiej Pierański   Poznan University of Economics and Business, Poland
Dariusz Pierzchała   Military University of Technology, Poland
Marcin Pietranik   Wrocław University of Science and Technology,
                                  Poland
Elias Pimenidis   University of the West of England, UK
Jaroslav Pokorný   Charles University in Prague, Czech Republic
Nikolaos Polatidis   University of Brighton, UK
Elvira Popescu   University of Craiova, Romania
Petra Poulova   University of Hradec Králové, Czech Republic
Om Prakash   University of Allahabad, India
Radu-Emil Precup   Politehnica University of Timisoara, Romania
Małgorzata   University of Silesia, Poland
   Przybyła-Kasperek
Paulo Quaresma   Universidade de Evora, Portugal
David Ramsey   Wrocław University of Science and Technology,
                                  Poland
Mohammad Rashedur   North South University, Bangladesh
   Rahman
Ewa Ratajczak-Ropel   Gdynia Maritime University, Poland
Sebastian A. Rios   University of Chile, Chile
Leszek Rutkowski   Częstochowa University of Technology, Poland
Alexander Ryjov   Lomonosov Moscow State University, Russia
Keun Ho Ryu   Chungbuk National University, South Korea
Virgilijus Sakalauskas   Vilnius University, Lithuania
Daniel Sanchez   University of Granada, Spain
Rafał Scherer   Częstochowa University of Technology, Poland
Juergen Schmidhuber   Swiss AI Lab IDSIA, Switzerland
Ali Selamat   Universiti Teknologi Malaysia, Malaysia
Tegjyot Singh Sethi   University of Louisville, USA
Natalya Shakhovska   Lviv Polytechnic National University, Ukraine
Donghwa Shin   Yeungnam University, South Korea
Andrzej Sicmiński   Wrocław University of Science and Technology,
                                  Poland
Dragan Simic   University of Novi Sad, Serbia
Bharat Singh   Universiti Teknology PETRONAS, Malaysia
Paweł Sitek   Kielce University of Technology, Poland
Andrzej Skowron   Warsaw University, Poland
Adam Słowik   Koszalin University of Technology, Poland
Vladimir Sobeslav   University of Hradec Králové, Czech Republic
Kamran Soomro   University of the West of England, UK
Zenon A. Sosnowski   Białystok University of Technology, Poland
Chutimet Srinilta   King Mongkut's Institute of Technology Ladkrabang,
                                  Thailand
Bela Stantic   Griffith University, Australia

Jerzy Stefanowski	Poznań University of Technology, Poland
Stanimir Stoyanov	University of Plovdiv "Paisii Hilendarski", Bulgaria
Ja-Hwung Su	Cheng Shiu University, Taiwan
Libuse Svobodova	University of Hradec Králové, Czech Republic
Tadeusz Szuba	AGH University of Science and Technology, Poland
Julian Szymański	Gdańsk University of Technology, Poland
Krzysztof Ślot	Łódź University of Technology, Poland
Jerzy Świątek	Wrocław University of Science and Technology, Poland
Andrzej Świerniak	Silesian University of Technology, Poland
Ryszard Tadeusiewicz	AGH University of Science and Technology, Poland
Muhammad Atif Tahir	National University of Computing and Emerging Sciences, Pakistan
Yasufumi Takama	Tokyo Metropolitan University, Japan
Maryam Tayefeh Mahmoudi	ICT Research Institute, Iran
Zbigniew Telec	Wrocław University of Science and Technology, Poland
Dilhan Thilakarathne	Vrije Universiteit Amsterdam, The Netherlands
Satoshi Tojo	Japan Advanced Institute of Science and Technology, Japan
Bogdan Trawiński	Wrocław University of Science and Technology, Poland
Trong Hieu Tran	VNU University of Engineering and Technology, Vietnam
Ualsher Tukeyev	Al-Farabi Kazakh National University, Kazakhstan
Olgierd Unold	Wrocław University of Science and Technology, Poland
Natalie Van Der Wal	Vrije Universiteit Amsterdam, The Netherlands
Jorgen Villadsen	Technical University of Denmark, Denmark
Bay Vo	Ho Chi Minh City University of Technology, Vietnam
Gottfried Vossen	ERCIS Münster, Germany
Wahyono Wahyono	Universitas Gadjah Mada, Indonesia
Lipo Wang	Nanyang Technological University, Singapore
Junzo Watada	Waseda University, Japan
Izabela Wierzbowska	Gdynia Maritime University, Poland
Krystian Wojtkiewicz	Wrocław University of Science and Technology, Poland
Michał Woźniak	Wrocław University of Science and Technology, Poland
Krzysztof Wróbel	University of Silesia, Poland
Marian Wysocki	Rzeszow University of Technology, Poland
Farouk Yalaoui	University of Technology of Troyes, France
Xin-She Yang	Middlesex University, UK
Tulay Yildirim	Yildiz Technical University, Turkey
Piotr Zabawa	Cracow University of Technology, Poland

Sławomir Zadrożny	Systems Research Institute, Polish Academy of Sciences, Poland
Drago Zagar	University of Osijek, Croatia
Danuta Zakrzewska	Łódź University of Technology, Poland
Katerina Zdravkova	Ss. Cyril and Methodius University in Skopje, Macedonia
Vesna Zeljkovic	Lincoln University, USA
Aleksander Zgrzywa	Wroclaw University of Science and Technology, Poland
Jianlei Zhang	Nankai University, China
Zhongwei Zhang	University of Southern Queensland, Australia
Maciej Zięba	Wrocław University of Science and Technology, Poland
Adam Ziębiński	Silesian University of Technology, Poland

Program Committees of Special Sessions

Special Session on Cyber-Physical Systems in Healthcare: Applications and Challenges (CSHAC 2020)

Michael Mayo	University of Waikato, New Zealand
Abigail Koay	University of Waikato, New Zealand
Panos Patros	University of Waikato, New Zealand

Special Session on Advanced Data Mining Techniques and Applications (ADMTA 2020)

Tzung-Pei Hong	National University of Kaohsiung, Taiwan
Tran Minh Quang	Ho Chi Minh City University of Technology, Vietnam
Bac Le	VNU University of Science, Vietnam
Bay Vo	Ho Chi Minh City University of Technology, Vietnam
Chun-Hao Chen	Tamkang University, Taiwan
Chun-Wei Lin	Harbin Institute of Technology, China
Wen-Yang Lin	National University of Kaohsiung, Taiwan
Yeong-Chyi Lee	Cheng Shiu University, Taiwan
Le Hoang Son	VNU University of Science, Vietnam
Vo Thi Ngoc Chau	Ho Chi Minh City University of Technology, Vietnam
Van Vo	Ho Chi Minh University of Industry, Vietnam
Ja-Hwung Su	Cheng Shiu University, Taiwan
Ming-Tai Wu	University of Nevada, Las Vegas, USA
Kawuu W. Lin	National Kaohsiung University of Applied Sciences, Taiwan
Tho Le	Ho Chi Minh City University of Technology, Vietnam
Dang Nguyen	Deakin University, Australia
Hau Le	Thuyloi University, Vietnam
Thien-Hoang Van	Ho Chi Minh City University of Technology, Vietnam
Tho Quan	Ho Chi Minh City University of Technology, Vietnam

Ham Nguyen	University of People's Security, Vietnam
Thiet Pham	Ho Chi Minh University of Industry, Vietnam
Nguyen Thi Thuy Loan	Nguyen Tat Thanh University, Vietnam
Mu-En Wu	National Taipei University of Technology, Taiwan
Eric Hsueh-Chan Lu	National Cheng Kung University, Taiwan
Chao-Chun Chen	National Cheng Kung University, Taiwan
Ju-Chin Chen	National Kaohsiung University of Science and Technology, Taiwan

Special Session on Computer Vision and Intelligent Systems (CVIS 2020)

Yoshinori Kuno	Saitama University, Japan
Nobutaka Shimada	Ritsumeikan University, Japan
Muriel Visani	University of La Rochelle, France
Heejun Kang	University of Ulsan, South Korea
Cheolgeun Ha	University of Ulsan, South Korea
Byeongryong Lee	University of Ulsan, South Korea
Youngsoo Suh	University of Ulsan, South Korea
Kang-Hyun Jo	University of Ulsan, South Korea
Hyun-Deok Kang	Ulsan National Institute of Science and Technology, South Korea
Van Mien	University of Exeter, UK
Chi-Mai Luong	University of Science and Technology of Hanoi, Vietnam
Thi-Lan Le	Hanoi University of Science and Technology, Vietnam
Duc-Dung Nguyen	Institute of Information Technology, Vietnam
Thi-Phuong Nghiem	University of Science and Technology of Hanoi, Vietnam
Giang-Son Tran	University of Science and Technology of Hanoi, Vietnam
Hoang-Thai Le	VNU University of Science, Vietnam
Thanh-Hai Tran	Hanoi University of Science and Technology, Vietnam
Anh-Cuong Le	Ton Duc Thang University, Vietnam
My-Ha Le	Ho Chi Minh City University of Technology and Education, Vietnam
The-Anh Pham	Hong Duc University, Vietnam
Van-Huy Pham	Tong Duc Thang University, Vietnam
Van-Dung Hoang	Quang Binh University, Vietnam
Huafeng Qin	Chongqing Technology and Business University, China
Danilo Caceres Hernandez	Universidad Tecnologica de Panama, Panama
Kaushik Deb	Chittagong University of Engineering and Technology, Bangladesh
Joko Hariyono	Civil Service Agency of Yogyakarta, Indonesia
Ing. Reza Pulungan	Universitas Gadjah Mada, Indonesia
Agus Harjoko	Universitas Gadjah Mada, Indonesia

Sri Hartati	Universitas Gadjah Mada, Indonesia
Afiahayati	Universitas Gadjah Mada, Indonesia
Moh. Edi Wibowo	Universitas Gadjah Mada, Indonesia
Wahyono	Universitas Gadjah Mada, Yogyakarta, Indonesia

Special Session on Multiple Model Approach to Machine Learning (MMAML 2020)

Urszula Boryczka	University of Silesia, Poland
Abdelhamid Bouchachia	Bournemouth University, UK
Robert Burduk	Wrocław University of Science and Technology, Poland
Oscar Castillo	Tijuana Institute of Technology, Mexico
Rung-Ching Chen	Chaoyang University of Technology, Taiwan
Suphamit Chittayasothorn	King Mongkut's Institute of Technology Ladkrabang, Thailand
José Alfredo F. Costa	Federal University (UFRN), Brazil
Ireneusz Czarnowski	Gdynia Maritime University, Poland
Fernando Gomide	State University of Campinas, Brazil
Francisco Herrera	University of Granada, Spain
Tzung-Pei Hong	National University of Kaohsiung, Taiwan
Konrad Jackowski	Wrocław University of Science and Technology, Poland
Piotr Jędrzejowicz	Gdynia Maritime University, Poland
Tomasz Kajdanowicz	Wrocław University of Science and Technology, Poland
Yong Seog Kim	Utah State University, USA
Bartosz Krawczyk	Virginia Commonwealth University, USA
Kun Chang Lee	Sungkyunkwan University, South Korea
Edwin Lughofer	Johannes Kepler University Linz, Austria
Hector Quintian	University of Salamanca, Spain
Andrzej Siemiński	Wrocław University of Science and Technology, Poland
Dragan Simic	University of Novi Sad, Serbia
Adam Słowik	Koszalin University of Technology, Poland
Zbigniew Telec	Wrocław University of Science and Technology, Poland
Bogdan Trawiński	Wrocław University of Science and Technology, Poland
Olgierd Unold	Wrocław University of Science and Technology, Poland
Michał Woźniak	Wrocław University of Science and Technology, Poland
Zhongwei Zhang	University of Southern Queensland, Australia
Zhi-Hua Zhou	Nanjing University, China

Special Session on Computational Imaging and Vision (CIV 2020)

Ishwar Sethi	Oakland University, USA
Moongu Jeon	Gwangju Institute of Science and Technology, South Korea
Jong-In Song	Gwangju Institute of Science and Technology, South Korea
Taek Lyul Song	Hangyang University, South Korea
Ba-Ngu Vo	Curtin University, Australia
Ba-Tuong Vo	Curtin University, Australia
Du Yong Kim	Curtin University, Australia
Benlian Xu	Changshu Institute of Technology, China
Peiyi Zhu	Changshu Institute of Technology, China
Mingli Lu	Changshu Institute of Technology, China
Weifeng Liu	Hangzhou Danzi University, China
Ashish Khare	University of Allahabad, India
Moonsoo Kang	Chosun University, South Korea
Goo-Rak Kwon	Chosun University, South Korea
Sang Woong Lee	Gachon University, South Korea
U. S. Tiwary	IIIT Allahabad, India
Ekkarat Boonchieng	Chiang Mai University, Thailand
Jeong-Seon Park	Chonnam National University, South Korea
Unsang Park	Sogang University, South Korea
R. Z. Khan	Aligarh Muslim University, India
Suman Mitra	DA-IICT, India
Bakul Gohel	DA-IICT, India
Sathya Narayanan	NTU, Singapore

Special Session on Intelligent Systems and Algorithms in Information Sciences (ISAIS 2020)

Martin Kotyrba	University of Ostrava, Czech Republic
Eva Volna	University of Ostrava, Czech Republic
Ivan Zelinka	VŠB – Technical University of Ostrava, Czech Republic
Hashim Habiballa	Institute for Research and Applications of Fuzzy Modeling, Czech Republic
Alexej Kolcun	Institute of Geonics, ASCR, Czech Republic
Roman Senkerik	Tomas Bata University in Zlin, Czech Republic
Zuzana Kominkova Oplatkova	Tomas Bata University in Zlin, Czech Republic
Katerina Kostolanyova	University of Ostrava, Czech Republic
Antonin Jancarik	Charles University in Prague, Czech Republic
Petr Dolezel	University of Pardubice, Czech Republic
Igor Kostal	The University of Economics in Bratislava, Slovakia
Eva Kurekova	Slovak University of Technology in Bratislava, Slovakia

Leszek Cedro	Kielce University of Technology, Poland
Dagmar Janacova	Tomas Bata University in Zlin, Czech Republic
Martin Halaj	Slovak University of Technology in Bratislava, Slovakia
Radomil Matousek	Brno University of Technology, Czech Republic
Roman Jasek	Tomas Bata University in Zlin, Czech Republic
Petr Dostal	Brno University of Technology, Czech Republic
Jiri Pospichal	The University of Ss. Cyril and Methodius (UCM), Slovakia
Vladimir Bradac	University of Ostrava, Czech Republic
Petr Pavel	University of Pardubice, Czech Republic
Jan Capek	University of Pardubice, Czech Republic

Special Session on Internet of Things and Artificial Intelligence for Energy Efficiency-Recent Advances and Future Trends (IOTAI 2020)

Xiaohui Yuan	University of North Texas, USA
Andino Maseleno	Universiti Tenaga Nasional, Malaysia
Amit Kumar Singh	National Institute of Technology Patna, India
Valentina E. Balas	Aurel Vlaicu University of Arad, Romania

Special Session on Data Modelling and Processing for Industry 4.0 (DMPI-4.0 2020)

Jörg Becker	Westfälische Wilhelms-Universität, Germany
Rafał Cupek	Silesian University of Technology, Poland
Helena Dudycz	Wroclaw University of Economics and Business, Poland
Marcin Fojcik	Western Norway University of Applied Sciences, Norway
Du Haizhou	Shanghai University of Electric Power, China
Marcin Hernes	Wroclaw University of Economics and Business, Poland
Wojciech Hunek	Opole University of Technology, Poland
Marek Krótkiewicz	Wrocław University of Science and Technology, Poland
Florin Leon	Technical University Asachi of Iasi, Romania
Jing Li	Shanghai University of Electric Power, China
Jacek Piskorowski	West Pomeranian University of Technology Szczecin, Polska
Khouloud Salameh	American University of Ras Al Khaimah, UAE
Predrag Stanimirović	University of Nis, Serbia
Krystian Wojtkiewicz	Wrocław University of Science and Technology, Poland
Feifei Xu	Shanghai University of Electric Power, China

Special Session on Intelligent and Contextual Systems (ICxS 2020)

Adriana Albu	Polytechnic University of Timisoara, Romania
Basabi Chakraborty	Iwate Prefectural University, Japan
Chao-Chun Chen	National Cheng Kung University, Taiwan
Dariusz Frejlichowski	West Pomeranian University of Technology Szczecin, Poland
Diganta Goswami	Indian Institute of Technology Guwahati, India
Erdenebileg Batbaatar	Chungbuk National University, South Korea
Goutam Chakraborty	Iwate Prefectural University, Japan
Ha Manh Tran	Ho Chi Minh City International University, Vietnam
Hong Vu Nguyen	Ton Duc Thang University, Vietnam
Hideyuki Takahashi	Tohoku Gakuin University, Japan
Intisar Chowdhury	University of Aizu, Japan
Jerzy Świątek	Wroclaw University of Science and Technology, Poland
Józef Korbicz	University of Zielona Gora, Poland
Keun Ho Ryu	Chungbuk National University, South Korea
Khanindra Pathak	Indian Institute of Technology Kharagpur, India
Kilho Shin	Gakashuin University, Japan
Maciej Huk	Wroclaw University of Science and Technology, Poland
Marcin Fojcik	Western Norway University of Applied Sciences, Norway
Masafumi Matsuhara	Iwate Prefectural University, Japan
Min-Hsiung Hung	Chinese Culture University, Taiwan
Miroslava Mikusova	University of Žilina, Slovakia
Musa Ibrahim	Chungbuk National University, South Korea
Nguyen Khang Pham	Can Tho University, Vietnam
Plamen Angelov	Lancaster University, UK
Qiangfu Zhao	University of Aizu, Japan
Quan Thanh Tho	Ho Chi Minh City University of Technology, Vietnam
Rafal Palak	Wroclaw University of Science and Technology, Poland
Rashmi Dutta Baruah	Indian Institute of Technology Guwahati, India
Senthilmurugan Subbiah	Indian Institute of Technology Guwahati, India
Sonali Chouhan	Indian Institute of Technology Guwahati, India
Takako Hashimoto	Chiba University of Commerce, Japan
Tetsuji Kuboyama	Gakushuin University, Japan
Tetsuo Kinoshita	RIEC, Tohoku University, Japan
Thai-Nghe Nguyen	Can Tho University, Vietnam
Zhenni Li	University of Aizu, Japan

Special Session on Automated Reasoning with Applications in Intelligent Systems (ARAIS 2020)

Yuichi Goto	Saitama University, Japan
Shinsuke Nara	Muraoka Design Laboratory, Japan
Hongbiao Gao	North China Electric Power University, China
Kazunori Wagatsuma	CIJ Solutions, Japan
Yuan Zhou	Minjiang Teachers College, China

Special Session on Intelligent Supply Chains and e-Commerce (ISCEC 2020)

Carlos Andres Romano	Polytechnic University of Valencia, Spain
Costin Badica	University of Craiova, Romania
Davor Dujak	University of Osijek, Croatia
Waldemar Koczkodaj	Laurentian University, Canada
Miklós Krész	InnoRenew, Slovenia
Paweł Pawlewski	Poznan University of Technology, Poland
Paulina Golińska-Dawson	Poznan University of Economics and Business, Poland
Adam Koliński	Łukasiewicz Research Network – The Institute of Logistics and Warehousing, Poland
Marcin Anholcer	Poznan University of Economics and Business, Poland

Special Session on Intelligent Applications of Internet of Things and Data Analysis Technologies (IAIOTDAT 2020)

Goutam Chakraborty	Iwate Prefectural University, Japan
Bin Dai	University of Technology Xiamen, China
Qiangfu Zhao	University of Aizu, Japan
David C. Chou	Eastern Michigan University, USA
Chin-Feng Lee	Chaoyang University of Technology, Taiwan
Lijuan Liu	University of Technology Xiamen, China
Kien A. Hua	Central Florida University, USA
Long-Sheng Chen	Chaoyang University of Technology, Taiwan
Xin Zhu	University of Aizu, Japan
David Wei	Fordham University, USA
Qun Jin	Waseda University, Japan
Jacek M. Zurada	University of Louisville, USA
Tsung-Chih Hsiao	Huaoiao University, China
Tzu-Chuen Lu	Chaoyang University of Technology, Taiwan
Nitasha Hasteer	Amity University Uttar Pradesh, India
Chuan-Bi Lin	Chaoyang University of Technology, Taiwan
Cliff Zou	Central Florida University, USA
Hendry	Satya Wacana Christian University, Indonesia

Special Session on Current Trends in Artificial Intelligence, Optimization, Learning, and Decision-Making in Bioinformatics and Bioengineering (CTAIOLDMBB 2020)

Sajid Gul Khawaja	National University of Sciences and Technology, Pakistan
Tehmina Khalil	Mirpur University of Sciences and Technology, Pakistan
Arslan Shaukat	National University of Sciences and Technology, Pakistan
Ani Liza Asmawi	International Islamic University, Malaysia
Martin Augustynek	VŠB – Technical University of Ostrava, Czech Republic
Martin Cerny	VŠB – Technical University of Ostrava, Czech Republic
Klara Fiedorova	VŠB – Technical University of Ostrava, Czech Republic
Habibollah Harun	Universiti Teknologi Malaysia, Malaysia
Lim Kok Cheng	Universiti Tenaga Nasional, Malaysia
Roliana Ibrahim	Universiti Teknologi Malaysia, Malaysia
Jafreezal Jaafar	Universiti Teknologi Petronas, Malaysia
Vladimir Kasik	VŠB – Technical University of Ostrava, Czech Republic
Ondrej Krejcar	University of Hradec Kralove, Czech Republic
Jan Kubíček	VŠB – Technical University of Ostrava, Czech Republic
Kamil Kuca	University of Hradec Kralove, Czech Republic
Petra Maresova	University of Hradec Kralove, Czech Republic
Daniel Barvík	VŠB – Technical University of Ostrava, Czech Republic
David Oczka	VŠB – Technical University of Ostrava, Czech Republic
Dominik Vilimek	VŠB – Technical University of Ostrava, Czech Republic
Sigeru Omatu	Osaka Institute of Technology, Japan
Marek Penhaker	VŠB – Technical University of Ostrava, Czech Republic
Lukas Peter	VŠB – Technical University of Ostrava, Czech Republic
Alice Krestanova	VŠB – Technical University of Ostrava, Czech Republic
Chawalsak Phetchanchai	Suan Dusit University, Thailand
Antonino Proto	VŠB – Technical University of Ostrava, Czech Republic
Naomie Salim	Universiti Teknologi Malaysia, Malaysia
Ali Selamat	Universiti Teknologi Malaysia, Malaysia

Imam Much Subroto	Universiti Islam Sultan Agung, Indonesia
Lau Sian Lun	Sunway University, Malaysia
Takeru Yokoi	Tokyo Metropolitan International Institute of Technology, Japan
Hazli Mohamed Zabil	Universiti Tenaga Nasional, Malaysia
Satchidananda Dehuri	Fakir Mohanh University, India
Pradeep Kumar Mallick	KIIT University, India
Subhashree Mishra	KIIT University, India
Cem Deniz	NYU Langone, USA
P. V. Rao	VBIT Hydrabad, India
Tathagata Bandyopadhyay	KIIT University, India

Special Session on Interactive Analysis of Image, Video and Motion Data in Life Sciences (SAILS 2020)

Artur Bąk	Polish-Japanese Academy of Information Technology, Poland
Grzegorz Borowik	Warsaw University of Technology, Poland
Wayne Brookes	University of Technology Sydney, Australia
Leszek Chmielewski	Warsaw University of Life Sciences, Poland
Zenon Chaczko	University of Technology Sydney, Australia
David Davis	University of Technology Sydney, Australia
Aldona Barbara Drabik	Polish-Japanese Academy of Information Technology, Poland
Marcin Fojcik	Western Norway University of Applied Sciences, Norway
Carlo Giampietro	University of Technology Sydney, Australia
Katarzyna Musial-Gabrys	University of Technology Sydney, Australia
Tomasz Górski	Polish Naval Academy, Poland
Adam Gudyś	Silesian University of Technology, Poland
Doan Hoang	University of Technology Sydney, Australia
Celina Imielińska	Vesalius Technologies LLC, USA
Frank Jiang	University of Technology Sydney, Australia
Henryk Josiński	Silesian University of Technology, Poland
Anup Kale	University of Technology Sydney, Australia
Sunil Mysore Kempegowda	University of Technology Sydney, Australia
Ryszard Klempous	Wroclaw University of Technology, Poland
Ryszard Kozera	The University of Life Sciences - SGGW, Poland
Julita Kulbacka	Wroclaw Medical University, Poland
Marek Kulbacki	Polish-Japanese Academy of Information Technology, Poland
Aleksander Nawrat	Silesian University of Technology, Poland
Jerzy Paweł Nowacki	Polish-Japanese Academy of Information Technology, Poland
Eric Petajan	LiveClips LLC, USA
Andrzej Polański	Silesian University of Technology, Poland

Andrzej Przybyszewski	UMass Medical School Worcester, USA
Joanna Rossowska	Polish Academy of Sciences, Poland
Jakub Segen	Gest3D LLC, USA
Aleksander Sieroń	Medical University of Silesia, Poland
Carmen Paz Suarez Araujo	University of Las Palmas, Spain
José Juan Santana Rodríguez	University of Las Palmas, Spain
Adam Świtoński	Silesian University of Technology, Poland
Agnieszka Szczęsna	Silesian University of Technology, Poland
David Tien	Charles Sturt University, Australia
Konrad Wojciechowski	Polish-Japanese Academy of Information Technology, Poland
Robin Braun	University of Technology Sydney, Australia

Special Session on Application of Intelligent Methods to Constrained Problems (AIMCP 2020)

Peter Nielsen	Aalborg University, Denmark
Paweł Sitek	Kielce University of Technology, Poland
Antoni Ligęza	AGH University of Science and Technology, Poland
Sławomir Kłos	University of Zielona Góra, Poland
Grzegorz Bocewicz	Koszalin University of Technology, Poland
Izabela E. Nielsen	Aalborg University, Denmark
Zbigniew Banaszak	Koszalin University of Technology, Poland
Małgorzata Jasiulewicz-Kaczmarek	Poznan University of Technology, Poland
Robert Wójcik	Wrocław University of Science and Technology, Poland
Arkadiusz Gola	Lublin University of Technology, Poland
Marina Marinelli	University of Leicester, UK
Masood Ashraf	Aligarh Muslim University, India
Ali Turkyilmaz	Nazarbayev University, Kazakhstan
Chandima Ratnayake	University of Stavanger, Norway
Marek Magdziak	Rzeszów University of Technology, Poland

Special Session on Privacy, Security and Trust in Artificial Intelligence (PSTrustAI 2020)

M. Ilhan Akbas	Embry-Riddle Aeronautical University, USA
Christoph Benzmüller	Freie Universität Berlin, Germany
Roland Bouffanais	Singapore University of Technology and Design, Singapore
Bernabe Dorronsoro	University of Cadiz, Spain
Rastko Selmic	Concordia University, Canada
Ronaldo Menezes	University of Exciter, UK
Apivadee Piyatumrong	NECTEC, Thailand
Khurum Nazir Junejo	Ibex CX, Pakistan

Daniel Stolfi	University of Luxembourg, Luxembourg
Juan Luis Jiménez Laredo	Normandy University, France
Kittichai Lavangnananda	King Mongkut's University of Technology Thonburi, Thailand
Jun Pang	University of Luxembourg, Luxembourg
Marco Rocchetto	ALES, United Technologies Research Center, Italy
Jundong Chen	Dickinson State University, USA
Emmanuel Kieffer	University of Luxembourg, Luxembourg
Fang-Jing Wu	Technical University Dortmund, Germany
Hannes Frey	University Koblenz-Landau, Germany
Umer Wasim	University of Luxembourg, Luxembourg
Christian M. Adriano	University of Potsdam, Germany

Intelligent Modeling and Simulation Approaches for Games and Real World Systems (IMSAGRWS 2020)

Alabbas Alhaj Ali	Frankfurt University of Applied Sciences, Germany
Costin Bădică	University of Craiova, Romania
Petru Cașcaval	Gheorghe Asachi Technical University, Romania
Gia Thuan Lam	Vietnamese-German University, Vietnam
Florin Leon	Gheorghe Asachi Technical University, Romania
Doina Logofătu	Frankfurt University of Applied Sciences, Germany
Fitore Muharemi	Frankfurt University of Applied Sciences, Germany
Minh Nguyen	Frankfurt University of Applied Sciences, Germany
Julian Szymański	Gdańsk University of Technology, Poland
Paweł Sitek	Kielce University of Technology, Poland
Daniel Stamate	University of London, UK

Contents – Part I

Deep Learning Models

Advanced Data Mining Techniques and Applications

Multiple Model Approach to Machine Learning

Contents – Part II

Computer Vision and Intelligent Systems

Data Modelling and Processing for Industry 4.0

Intelligent Applications of Internet of Things and Data Analysis Technologies

Intelligent and Contextual Systems

Intelligent Systems and Algorithms in Information Sciences

Intelligent Supply Chains and e-Commerce

Privacy, Security and Trust in Artificial Intelligence

Interactive Analysis of Image, Video and Motion Data in Life Sciences

Knowledge Engineering and Semantic Web

Toward Blockchain-Assisted Gamified Crowdsourcing for Knowledge Refinement

Helun Bu^(✉) and Kazuhiro Kuwabara

College of Information Science and Engineering,
Ritsumeikan University, Kusatsu, Shiga 525-8577, Japan
is0385pr@ed.ritsumei.ac.jp

Abstract. This paper reports an application of blockchains for knowledge refinement. Constructing a high-quality knowledge base is crucial for building an intelligent system. One promising approach to this task is to make use of "the wisdom of the crowd," commonly performed through crowdsourcing. To give users proper incentives, gamification could be introduced into crowdsourcing so that users are given rewards according to their contribution. In such a case, it is important to ensure transparency of the rewards system. In this paper, we consider a refinement process of the knowledge base of our word retrieval assistant system. In this knowledge base, each piece of knowledge is represented as a triple. To validate triples acquired from various sources, we introduce yes/no quizzes. Only the triples voted "yes" by a sufficient number of users are incorporated into the main knowledge base. Users are given rewards based on their contribution to this validation process. We describe how a blockchain can be used to ensure transparency of the process, and we present some simulation results of the knowledge refinement process.

Keywords: Blockchain · Knowledge refinement · Gamified crowdsourcing

1 Introduction

This paper describes an application of blockchains for a knowledge refinement process. Constructing a high-quality knowledge base is important for an intelligent system. Knowledge refinement is necessary to increase the value of the knowledge base. Several approaches to refining knowledge represented as knowledge graphs have been reported [13].

One promising approach is to harness the power of many users, for example, through crowdsourcing [6]. The inherent problem in crowdsourcing is maintaining a high quality output. It is important to motivate users (or workers) to produce robust results.

Gamification is one way to keep users motivated in crowdsourcing [12]. Gamification is the process of introducing game-like elements to a non-gaming context. Games with a purpose (GWAP) is an example of gamification, where intended tasks are executed as by-products of playing games [2]. GWAP is also applied to

© Springer Nature Switzerland AG 2020
N. T. Nguyen et al. (Eds.): ACIIDS 2020, LNAI 12033, pp. 3–14, 2020.
https://doi.org/10.1007/978-3-030-41964-6_1

refine knowledge graphs [14]. In the ESP game, which is a prominent example of GWAP, two users who do not communicate with each other are asked to label an image. Points are given to a user who puts the same label as a paired user [1]. By devising proper game rules, an incentive can be given to users to input correct answers.

Introducing a point system is expected to motivate users to earn more points and complete more tasks. In such situations, correctly calculating points is important. In particular, when points cannot be calculated at the time users are playing the game or completing the task and are calculated later according to the user's past contribution, it is necessary to ensure that the users' records that are used for points calculation are not altered in any way. This point calculation process needs to be transparent so that any user can examine its basis. For this purpose, we utilize blockchains, which is a distributed ledger technology that has been proposed as the basis of cryptocurrency [3]. Blockchains allow the data to be stored and shared over the network, with a guarantee of being free from tampering.

In this paper, we focus on the example of the knowledge base used in our word retrieval assistant system [9]. This system is intended to support people with word-finding difficulties. Through a series of questions and answers, the system tries to guess what the person wants to express but cannot find a name for. The knowledge base is used to formulate questions.

For this system, we accumulate knowledge contents from various sources such as scraping websites or obtaining inputs from human users. Because of the nature of this system, its knowledge content should cover topics that often appear in everyday conversation. They are generally not related to specialized domain knowledge; rather, they are about things related to daily life. Thus, the participation of many casual users to construct the knowledge base would be effective.

However, when many casual users contribute to knowledge content, the quality of the knowledge may become an issue. To ensure high quality knowledge, we first store newly acquired pieces of knowledge into a temporary knowledge base, and only move the validated knowledge into the main knowledge base.

In this validation process, we employ a concept of gamified crowdsourcing similar to the one explored in [8]. More specifically, we make simple yes/no quizzes from the contents of the temporary knowledge base, and we present these to users. When enough votes for agreeing with the content are accumulated, the corresponding knowledge content is judged to be valid, and is incorporated into the main knowledge base. As an incentive to users, we award points based on a user's past inputs. If they contribute to the validation process, they are given bonus points as rewards. Blockchain technology is utilized to record the users' input and ensure the transparency of the reward calculation.

The remainder of the paper is organized as follows. The next section describes some related work focusing on applications of blockchains for handling knowledge, and Sect. 3 presents the knowledge refinement process. Section 4 describes the prototype implementation, and Sect. 5 reports simulation experiments and discusses their results. Section 6 concludes.

2   Related Work

Blockchains were developed as an underlying technology for cryptocurrency and were intended for use as the public ledger for transactions on a network. Blockchains are now applied not only to cryptocurrency, but also to other areas such as health care, data provenance, and mobile communication networks [11]. One notable application area is the Internet of Things (IoT), where blockchains are utilized to construct a *knowledge market* in which IoT systems that perform artificial intelligence (AI) tasks at the edge of networks can exchange knowledge in a peer-to-peer fashion [10]. For knowledge management in enterprises, *Knowledge Blockchain* was proposed [5] to audit knowledge evolution and provide proof of provenance of knowledge.

A blockchain is also used as a decentralized database where data from participants are stored transparently. Knowledge graphs represented in the Resource Description Framework (RDF) are stored using a blockchain technology, called GraphChain [15]. In addition, blockchains are applied to the decentralized construction of knowledge graphs [16]. In this system, company-level domain knowledge about employee's skills is constructed from the participation of employees in the company. This system also introduces a voting scheme and a reward mechanism for employees who contribute to the knowledge construction. As another example, AUDABLOK was proposed as a software framework to allow citizens to participate in refining open data [4]. In AUDABLOK, blockchains are utilized to audit users' contributions and provide rewards to users.

In this paper, we focus on the validation process of knowledge contents by casual users and conduct simulation experiments to examine the method's characteristics.

3   Knowledge Refinement Process

3.1   Target Knowledge Base

As an example knowledge base, we chose the knowledge content used in our word retrieval assistant system [9]. A typical problem for people with aphasia is word-finding difficulty; they have a clear image of what they want to say, but cannot recall a proper word to express it. This is similar to the situation when you visit a foreign country and you do not know how to say in local language.

For this kind of difficulty, a human caregiver called a conversation partner often asks a series of questions, such as *Is it food?* or *Is its color red?* Through their responses to the questions, the conversation partner extracts the name of the thing the person with aphasia wants to express.

The word retrieval assistant system aims to provide a similar function to a human conversation partner, but using a computer. The system contains a knowledge base about relevant things and produces an appropriate question to ask the user. According to the user's reply, the next question to ask is determined.

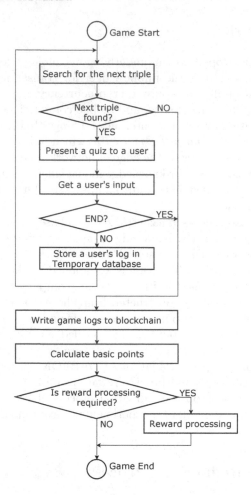

Fig. 1. Flowchart for a game session.

In the early prototype system we developed, knowledge is represented as a triple of *subject, predicate,* and *object* as in the RDF. For example, the fact that *the color of the apple is* red is represented as (`<apple>`, `<color>`, `<red>`).

3.2   Knowledge Acquisition and Refinement

It is important to acquire enough knowledge for the word retrieval assistant system to work well. Techniques to acquire new knowledge include system developers constructing knowledge content manually from scratch or letting a user input the correct word when it is not produced by the system [7].

In addition, we may extract knowledge from data available on the internet, such as Wikipedia. Alternatively, we may use a framework of crowdsourcing to elicit inputs from many casual users. To facilitate such a process, gamification

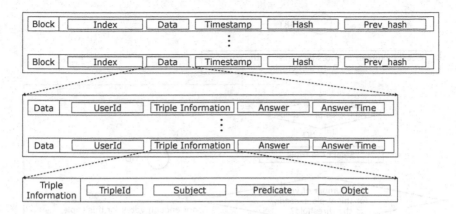

Fig. 2. Data stored in blockchain.

concepts can be applied. For example, when we need data for a triple, we may present a form that consists of three items, one or two of which are blank so that a user can fill them in.

One of the problems with data acquired using such methods is that they may not be correct. To assure high quality of knowledge content, we validate each triple's correctness before they are used. The acquired knowledge content is first stored in the temporary knowledge base. Only the knowledge content validated by users is moved into the main knowledge base.

To validate the contents of the temporary knowledge base, we employ a yes/no quiz. For example, suppose the triple (<apple>, <color>, <red>) is in the temporary knowledge base. A yes/no quiz is presented to a user asking if the color of the apple is red is correct or not. The user may answer YES, NO, or DON'T KNOW. If a certain number of YES votes compared with NO votes are obtained, the triple is considered to be true and is moved into the main knowledge base.

This process is formulated into a kind of game. After a user starts a game, a yes/no quiz sentence is presented with the possible choices: YES, NO, DON'T KNOW, and END. If the user answers END, the game session ends. Otherwise, a user's reply is recorded in the game server, and the next quiz is presented with the same possible choices (Fig. 1).

When one game session is finished, the ending processing is performed. This includes storing logs of the game session into blockchain by creating a new block (Fig. 2). The game history consists of triples that correspond to quizzes that were presented to the user and the user's responses to them. It also includes the date and time that the user played the game.

Furthermore, basic points are given to the user for each of their answers. Finally, possible bonus rewards are calculated, as explained in the next subsection.

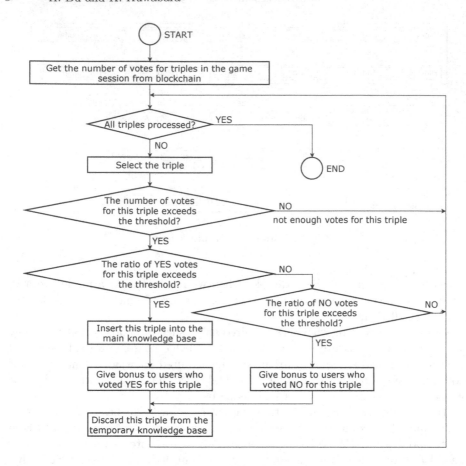

Fig. 3. Reward processing.

3.3   Reward Processing

For triples that are used in the game session, we check whether enough votes (YES or NO) from the game playing are obtained for the triple (Fig. 3). We set a certain threshold S, and check whether a triple has accumulated more than S votes. The data are obtained from the blockchain so that transparency of data is ensured. If the triple has accumulated more than S votes and the ratio of YES votes over the total number of votes for the triple is greater than a threshold α, then the triple is judged to be correct. The triple is removed from the temporary knowledge base and moved into the main knowledge base. Similarly, if the ratio of NO votes over the total number of votes for the triple is greater than a threshold β, the triple is judged to be incorrect and is removed from the temporal knowledge base.

When a triple is judged to be correct, the users who voted YES for this triple are given bonus points. Similarly, when a triple is judged to be incorrect,

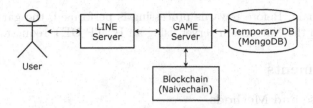

Fig. 4. Prototype configuration.

(1) During a game session   (2) After a game session

Fig. 5. Screenshots of a prototype chat system.

the users who voted NO for this triple are given bonus points. This check is done for all the triples that appeared in the game session.

4   Implementation

Figure 4 shows the configuration of our early prototype system, which is constructed as a chat system. We implemented it using LINE messaging service[1], a popular chat service in Japan and other countries. A user can access the system which is implemented as a chatbot, via a smartphone so that it can be used even in small periods of free time. When the user starts talking to the chatbot, the chatbot presents a quiz that displays the sentence generated from a target triple and buttons (YES, NO, DON'T KNOW) to input the user's response along with a button to end a game (Fig. 5(1)). When the game ends, the summary of the game results are shown (Fig. 5(2)).

We used the implementation of Naivechain[2] for blockchain. A MongoDB server is used to store the temporary data generated during a game session, and also for simulation experiments explained in the next section. Each time a game session ends, a list of the user's inputs during the game session is sent to the Naivechain server using an HTTP POST request, and the session data are added

---

[1] https://line.me/en/.
[2] https://github.com/lhartikk/naivechain.

to the blockchain. Before rewards processing is performed, the games' logs are retrieved from the Naivechain server using an HTTP GET request.

5   Experiments

5.1   Purpose and Methods

To examine the characteristics of the proposed method with some variations of the game design, we conducted simulation experiments rather than probabilistic analysis. In addition, as our early prototype with the particular implementation of a blockchain has performance issues in terms of scalability, the simulations were performed using MongoDB as the data store.

We assumed M triples and N virtual users, and let virtual users participates in the games. We examined whether the triples can be properly validated and how much rewards virtual users receive.

More specifically, the conditions of the simulation are as follows:

- The number of triples (M) was set to 1000, and the number of virtual users (N) was set to 1000.
- User i ($1 \leq i \leq N$) answers correctly with probability $p_c(i)$, where $p_c(i)$ is uniformly distributed over $[0.6, 1.0]$. Thus, the average is 0.8. In addition, a user is assumed not to answer with DON'T KNOW.
- All the triples are assumed to be true. Thus, user i answers with YES with probability $p_c(i)$. Otherwise, user i answers NO.
- A user is selected from a set of N users in sequence, and the selected user plays a game. The user is assumed to answer k quizzes in one game session. The value of k is set randomly from the range of $[3, 7]$. The user is given 1 point per quiz as a basic point.
- As one game session finishes, we check whether there are newly validated triples. The threshold for judging a triple to be correct or not (S) is set to 10, and the threshold ratio for judging the triple to be correct (α) is set to 0.8. That is, a triple that accumulates 10 votes, among which more than 8 votes are for YES, is considered correct. The users who voted YES for the triple are given 100 bonus points. After triples to be validated are determined and bonus points are calculated, the next user is selected and another game is started.
- One simulation run is terminated when there are no more triples to be presented to any user. Note that the same triple will not be presented to the same user twice. Once a triple is removed from the temporary knowledge base, it is not used for a game.

Triple Selection Method. We tested three variations in selecting the next triple to use for a yes/no quiz. When the next triple is searched for, the triples that have already been used for the target users are removed first. Then, from the remaining triples, we used the following methods to select the next triple to present.

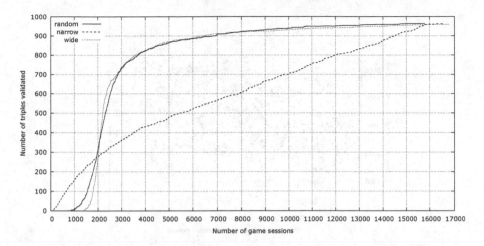

Fig. 6. Changes in the number of triples validated and the number of game sessions in one simulation run.

- *random*: A triple is selected randomly; this is the baseline method.
- *narrow*: A triple that has been presented to users more often is given higher priority. This method effectively focuses on some triples to validate. More specifically, each triple is given a weight equivalent to 10 times the number of times it has been used as a question to users.
- *wide*: A triple that has been presented to users less often is given higher priority. This method effectively broadens the range of target triples. More specifically, each triple is given a weight equivalent to 10 times the maximum number of times any triple has been used as a question minus the number of times the particular triple has been used as a question.

User Reliability. We also ran another set of simulations under different conditions to account for user reliability. Each user's reliability is calculated based on their contribution to the validation of triples. The threshold S for determining whether a triple is valid is set according to the sum of the reliability of users who voted for the triple so that fewer users are required to validate a triple when users with high reliability vote for it. This process is expected to result in fewer games required for validation.

For the simulation condition, the reliability of user i, R_i is set as $R_i = \min(0.002 \times b_i + 1, 3)$ where b_i denotes user i's bonus points. Thus, the range of R_i is $1 \leq R_i \leq 3$, meaning that a user of high reliability counts as, at most, three users.

5.2   Results and Discussion

Simulations were run five times without considering user reliability. Across the simulation runs, the average number of validated triples for each method was

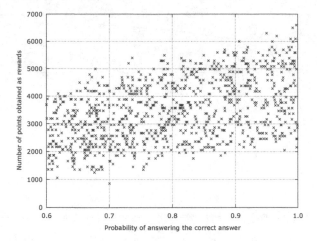

Fig. 7. Relationship between probability of answering a correct answer and points obtained.

967 (*random* method), 966.2 (*narrow* method) and 964 (*wide* method). Generally, this kind of yes/no quizzes can validate enough triples.

The different triple selection methods impacted how the number of triples that are were validated through the games changes, as shown in Fig. 6. As seen in this chart, after a certain number of games, the number of validated triples rapidly increases and then saturates for the *random* and *wide* methods. Among the approaches, the *wide* method is slow to validate triples, but the number of validated triples gradually increases compared with the *random* method.

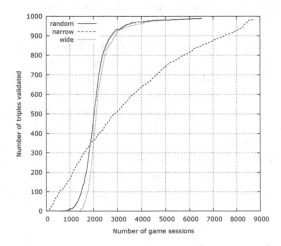

Fig. 8. Changes in the number of triples validated and the number of game sessions (with user reliability considered).

Although the *narrow* method shows a steady increase in the number of validated triples, it is soon surpassed by the other methods.

We also plotted the number of points user i obtained against the user's probability of giving the correct answer $(p_c(i))$ in one simulation run with the *random* method (Fig. 7). The correlation coefficient between was 0.45, indicating a moderate positive relationship. Generally, a user who has a higher probability of answering a correct answer tends to obtain more points as expected.

In addition, the simulation results with user reliability taken into consideration are shown in Fig. 8. As seen in this chart, the number of game sessions required to validate triples were decreased by introducing user reliability. With the parameter settings in the simulation runs, there was no adverse effect such as erroneously validating triples. It is a future task to confirm that introduction of user reliability does not cause any adverse effect in a more realistic situation.

6   Conclusion and Future Work

This paper described the use of blockchain in knowledge refinement. We adopted the concept of gamified crowdsourcing and used blockchains to ensure transparency of the user reward calculation. Our simulation experiments indicate that the proposed approach has the potential to be used as a method of knowledge refinement.

In this paper, we focused on the process of validating the knowledge content already in the knowledge base. We plan to extend the application of blockchains to the process of gathering the knowledge content.

Furthermore, we conducted only simulation experiments with virtual users. Blockchains were introduced to ensure the transparency of the user reward calculation so as not to hinder the users' incentives. Future work should evaluate the proposed method not only in terms of performance scalability in real-world environments, but also in terms of human users' subjective impressions.

Acknowledgements. This work was partially supported by JSPS KAKENHI Grant Number 18K11451.

References

1. von Ahn, L., Dabbish, L.: Labeling images with a computer game. In: Proceedings of the SIGCHI Conference on Human Factors in Computing Systems, CHI 2004, pp. 319–326. ACM, New York, April 2004. https://doi.org/10.1145/985692.985733
2. von Ahn, L., Dabbish, L.: Designing games with a purpose. Commun. ACM **51**(8), 58–67 (2008). https://doi.org/10.1145/1378704.1378719
3. Antonopoulos, A.M.: Mastering Bitcoin: Programming the Open Blockchain. O'Reilly Media, Sebastopol (2017)
4. Emaldi, M., Zabaleta, K., López-de Ipiña, D.: AUDABLOK: engaging citizens in open data refinement through blockchain. Proceedings **31**(1), 52 (2019). https://doi.org/10.3390/proceedings2019031052

5. Fill, H.G., Häerer, F.: Knowledge blockchains: applying blockchain technologies to enterprise modeling. In: Proceedings of the 51st Hawaii International Conference on System Sciences (2018). https://doi.org/10.24251/HICSS.2018.509
6. Howe, J.: Crowdsourcing: Why the Power of the Crowd Is Driving the Future of Business. Crown Business, New York (2009)
7. Iwamae, T., Kuwabara, K., Huang, H.H.: Toward gamified knowledge contents refinement - case study of a conversation partner agent. In: Proceedings of the 9th International Conference on Agents and Artificial Intelligence. ICAART 2017, vol. 1, pp. 302–307 (2017)
8. Kurita, D., Roengsamut, B., Kuwabara, K., Huang, H.H.: Simulating gamified crowdsourcing of knowledge base refinement: effects of game rule design. J. Inf. Telecommun. **2**(4), 374–391 (2018). https://doi.org/10.1080/24751839.2017.1401259
9. Kuwabara, K., Iwamae, T., Wada, Y., Huang, H.-H., Takenaka, K.: Toward a conversation partner agent for people with aphasia: assisting word retrieval. In: Czarnowski, I., Caballero, A.M., Howlett, R.J., Jain, L.C. (eds.) Intelligent Decision Technologies 2016. SIST, vol. 56, pp. 203–213. Springer, Cham (2016). https://doi.org/10.1007/978-3-319-39630-9_17
10. Lin, X., Li, J., Wu, J., Liang, H., Yang, W.: Making knowledge tradable in edge-AI enabled IoT: a consortium blockchain-based efficient and incentive approach. IEEE Trans. Industr. Inf. **15**(12), 1 (2019). https://doi.org/10.1109/TII.2019.2917307
11. Lu, Y.: The blockchain: state-of-the-art and research challenges. J. Ind. Inf. Integr. **15**, 80–90 (2019). https://doi.org/10.1016/j.jii.2019.04.002
12. Morschheuser, B., Hamari, J., Koivisto, J., Maedche, A.: Gamified crowdsourcing: conceptualization, literature review, and future agenda. Int. J. Hum. Comput. Stud. **106**(Supplement C), 26–43 (2017). https://doi.org/10.1016/j.ijhcs.2017.04.005
13. Paulheim, H.: Knowledge graph refinement: a survey of approaches and evaluation methods. Semant. Web **8**(3), 489–508 (2017)
14. Re Calegari, G., Fiano, A., Celino, I.: A framework to build games with a purpose for linked data refinement. In: Vrandečić, D., et al. (eds.) The Semantic Web - ISWC 2018, pp. 154–169. Springer, Cham (2018)
15. Sopek, M., Gradzki, P., Kosowski, W., Kuziski, D., Trójczak, R., Trypuz, R.: GraphChain: a distributed database with explicit semantics and chained RDF graphs. In: Companion Proceedings of the The Web Conference 2018, WWW 2018, pp. 1171–1178. International World Wide Web Conferences Steering Committee, Republic and Canton of Geneva, Switzerland (2018). https://doi.org/10.1145/3184558.3191554
16. Wang, S., Huang, C., Li, J., Yuan, Y., Wang, F.: Decentralized construction of knowledge graphs for deep recommender systems based on blockchain-powered smart contracts. IEEE Access **7**, 136951–136961 (2019). https://doi.org/10.1109/ACCESS.2019.2942338

Predicting Research Collaboration Trends Based on the Similarity of Publications and Relationship of Scientists

Tuong Tri Nguyen[1](✉), Ngoc Thanh Nguyen[2], Dinh Tuyen Hoang[3,4],
and Van Cuong Tran[4]

[1] University of Education, Hue University, Hue, Vietnam
tuongtringuyen@gmail.com
[2] Faculty of Computer Science and Management,
Wroclaw University of Science and Technology, Wroclaw, Poland
Ngoc-Thanh.Nguyen@pwr.edu.pl
[3] Department of Computer Engineering,
Yeungnam University, Gyeongsan, South Korea
hoangdinhtuyen@gmail.com
[4] Quang Binh University, Dong Hoi, Vietnam
vancuongqbuni@gmail.com

Abstract. Nowadays, collaboration is indispensable in solving increasingly complex problems. In the academic context, research collaboration influences many aspects of research problems approached. The research collaboration is beneficial for scientists, especially early-career scientists, to determine potential successful collaborations. Predicting the trend of collaboration is an important step in improving the quality of research collaboration between scientists. In this study, we propose a method for predicting research collaboration trends by taking into account the research similarity and the relationship between scientists. The research similarity is computed by considering the author's profiles. The co-author graph is built to explore new collaborators based on the connections weigh between scientists. We are currently in the process of developing a real system and our system shows promising results in predicting the potential success collaborators.

Keywords: Research collaboration · Collaboration · Research trend

1 Introduction

The fast-growing nature of scholarly data has become a bottleneck in the management and utilization of scholarly information. Scholarly data embodies information such as authors, citations, papers, co-authors, academic networks and so on. Several scholarly data resources due exist that aid scholars in getting easy access to useful technical data analysis, which includes but not limited to ResearchGate, CiteSeer, the DBLP Computer Science Bibliography, and Google

ⓒ Springer Nature Switzerland AG 2020
N. T. Nguyen et al. (Eds.): ACIIDS 2020, LNAI 12033, pp. 15–24, 2020.
https://doi.org/10.1007/978-3-030-41964-6_2

Scholar. Gaining access to scholarly data has proven to yield some fruits and for instance, forecasting the impact of a particular paper, understanding citation relationship which can improve the scientific ranking of a researcher. This, thus, indicates that scholarly data is not only useful in academia, but it acts as a guiding factor in the growth and advancement of scientific tools. Gaining access to valuable and suitable data has become a complex and challenging task.

Accessing the relevant materials required has become successful with the use of recommender systems and techniques [18]. The recommendation systems have often been applied in many areas such as movie recommendations, Amazon product recommendations [4, 10, 19], contend that recommending collaborators is a crucial task in the academic domain. According to Minkov et al. [15], without collaboration among scientists, one may not be able to climb the scientific ladder. Prior studies in this area revealed that cooperation amongst scientists is becoming more and more useful, which is a crucial step in becoming successful. Statistically, there have been over 89 million publications dating from 1900 to 2015. According to Dong et al. [7], there has been a sharp increase in the size of author's list of publications as well as international cooperation rates, which also show a 25-fold rise during the past 116 years.

Building a recommendation system which can assist scientist in related field is the main priority in scientific research, and hence, collaborators recommendation method has gained significant research interest in research years [1, 2, 20]. The research collaboration trend is needed in the 4.0 technology era. It considers working together in the research process towards a common goal of discovering new scientific knowledge and coming up with new initiatives [9]. Scientific outcomes are usually defined by the progress of finding suitable collaborators especially students or young scientists who are still known in the research domain with little experience. Besides, prior studies had shown that scientists with a well-established network of collaborators usually more research than those without such well-connect network [5].

The issue of finding notable collaborators is necessarily an essential task in the scientific domain. Solving the problem of finding a worthwhile collaborator, some scholars developed a collaborator recommender systems [2, 9]. A social network for academic can be represented as a weighted directed graph of nodes (such as scientists or groups) which has unique relationships (for instance, friendship and co-authorship, as shown in Fig. 3). The connections between authors based on some academic background such as co-author information, research topics, academic events attended, and so on. Finding new collaborators in academic social networking is often a challenging task with the vast data. Some collaborative research techniques have been proposed and bibliographic systems built as well which include the DBLP, CiteSeerX, and ResearchGate. Nevertheless, the previous studies faced some limitations in terms of consistency, in particular they cannot adequately address the entire academic network. Therefore, this leads to a lack of consistent methods of modeling effective academic networks [2, 9]. Besides, connections between it do not examine to reach the accurate data sets with different types of information modeled individually (Fig. 2).

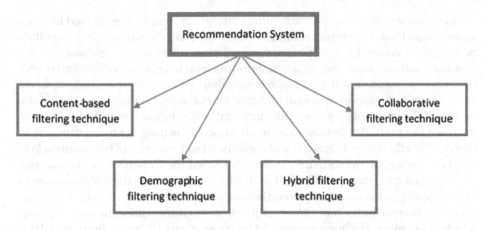

Fig. 1. Taxonomy of recommender system

The rest of this paper is organized as follows. In the next section, related work on recommendation systems concentrating on research collaboration is investigated briefly. In Sect. 3, the research collaboration network model is presented. The Proposed Method are shown in Sect. 4. Lastly, conclusions and future work are presented in Sect. 5.

2 Related Work

Solving the research problem is a challenging and complex task; hence, there is need for scientists to cooperate with others. How scientist collaborates with others is an essential factor in understanding the quality of their results. In recent years, there is a growing interest in the issue of how to effectively recommend a collaborator. There are several approaches to the study of collaboration: Karim Alinami and colleagues [3] proposed a recommender system to aid authors searching for a co-author to collaborate with. Their proposal is based on the supposition that research could have a brilliant idea but not necessarily an expert in that field, therefore, they will need to find a partner who is an expert in that field to co-work with to guarantee the success of the research. Using the offline dataset of ACM and online data from Elsevier API, they propose a prototype system that can assist researchers in getting the best match for their future scientific publication by simply using keywords and the area of expertise. The work of Zhao et al. [23] seeks to investigate the importance of finding a research collaborator when seeking a research fund. The study is driven by the fact that research is a complex and expensive task that requires funding and finding the right applicant becomes a challenging task. They, however, propose a method by using the social media recommendation system to assist in this task. Similar to this is the work of Yang et al. [22] who applied of Nearest Neighbor Based Personal Rank Algorithm for Collaborator recommendation was championed by using historical cooperation of the target user, they could recommend collaborators based on random walk algorithm using social media.

In summary, the collaborator recommendation system can be divided into two approaches. First, the system recommends the best collaborators who have collaborated with the target scientist to reinforce existing collaborations. Second, the system finds out the most potential collaborators who have never collaborated with the target scientist in order to make new collaborations. Several previous works concentrated on proposing a search engine for collaboration by using the combination of Cosine similarity, Jaccard similarity, and relation strength similarity to measure the similarity between the publication of authors. Also, keywords from their publications were extracted and computed the frequency of the common keywords to determine the similarity value. They used the CollabSeer system as well as both author similarity and lexical similarity to recommend the collaborators. Li et al. [13] examined how to propose the cooperation of authors in a co-authoring network. Several indicators are considered by co-author information on paper, which determines the importance of the connection. By analyzing the DBLP dataset, the author was able to compute the collaboration strength between scientists. They found out the top K scientists of the list of potential collaborators in order to recommend to the target scientist [14].

3 Research Collaboration Network Model

Let's consider that a research collaboration can be defined as the co-working of scientists to achieve the final goal that is producing novel scientific knowledge or methods or application. According to [9], research collaborative scientists are included as follows: (a) those included in the list of authors engaged in publication; (b) those who have an important role for one or more of the main elements of the research; (c) those who have cooperated on a research project for long periods of time, or for a large portion of the project, or regular or significant contributors. In this paper, to compute the relationship between scientists for predicting the research collaboration trends, some basic definitions can be used.

Definition 1. *[9]*
   Let O_S be a set of scientists, which is described as follows:
   $O_S =< Name, Impact, Field, Affiliation, Position, CoAuthor >$
   Let O_P be the set of papers which are written by scientists O_{S_P}, $(O_{S_P} \subseteq O_S)$;
   Let O_V be the set of publication that scientists have published their papers.
   Let O_I be the set of institution;
   Let O be the set of scientific objects and O can be one of four types as the following:

$$O = O_S \cup O_P \cup O_V \cup O_I$$

Definition 2. *[9] The research collaborations are represented by two functions as follows:*
   $R_h : O \times O \rightarrow [0, 1];$
   $R_f : O \times O \rightarrow [0, 1];$

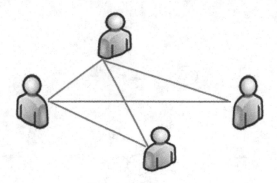

Fig. 2. A research collaboration network

Where R_h is a function related to historical research collaboration among scientists; R_f is a function related to research collaboration among scientists in the future. Therefore, the greater the value of the R_f, the greater the chance of cooperation between the scientists in the future (Fig. 3).

Axiom 1 (Nonnegative) *[9]* $\forall o_1, o_2 \in O_S$:
$R_h(o_1, o_2) \geq 0$ *and* $R_h(o_1, o_1) = 1$;
$R_f(o_1, o_2) \geq 0$ *and* $R_f(o_1, o_1) = 1$;

Axiom 2 (Asymmetric) $\forall o_1, o_2 \in O_S \land o_1 \neq o_2 \land o_1^P \neq o_2^P$:
$R_h(o_1, o_2) \neq R_h(o_1, o_2)$
$R_f(o_1, o_2) \neq R_f(o_1, o_2)$

Axiom 3 (Transitive) $\forall o_1, o_2, o_3, o_4 \in O_S \land o_1 \neq o_2 \neq o_3 \neq o_4$:
(1)$R_h(o_1, o_2) \approx R_h(o_1, o_3)$ *or* $R_h(o_2, o_1) \approx R_h(o_3, o_1)$
$=> R_f(o_3, o_2) > 0, R_f(o_2, o_3) > 0$
(2) $R_h(o_1, o_2) \approx R_h(o_1, o_3)$ *and* $R_h(o_4, o_2) \approx R_h(o_4, o_3)$
$=> R_f(o_1, o_4) > 0, R_f(o_4, o_1) > 0$

4   Predicting Methodology

A scientist's work is a collection of all publications, including journals, conferences, finished projects. The relationship between the two researchers is based on the constraints that have been done together (with the same affiliation, published the same paper, working on the same topic, project, published book, etc.) or through any middle relationship.

4.1   Network Connecting Scientists

This section discusses how to build a network to connect the scientists. Let $O_S = \{S_1, S_2, ..., S_N\}$ be a set of scientists. First, the study focused on computing the constraints that have been done together, including the number of co-authors and then computing the research similarity.

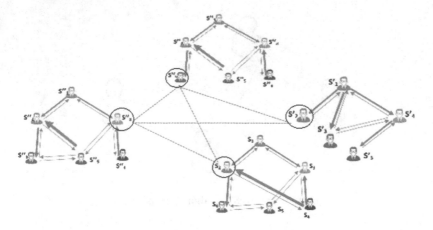

Fig. 3. A research collaboration network of scientists

Research Collaboration Network. The quantitative relationship is represented on an edge between two scientists S_i and S_j be an unequal relationship, it depends on the number of co-author publications and the number of individual publications.

Definition 3. *The weight of $w(S_i, S_j)$ is defined to measure the research collaboration between scientists S_i and S_j as follows:*

$$w(S_i, S_j) = \frac{|P_{S_i} \cap P_{S_j}|}{|P_{S_i}|} \tag{1}$$

Research Similarity (RS). Research topics similarity of scientists are expressed in their publications. The score of two scientists with a similar research topic can be computed by formula (1).

Definition 4. *The function $RS(S_1, S_2)$ is defined to measure the research similarity between scientists S_1 and S_2 as follows:*

$$RS(S_1, S_2) = \sum_{p_j \in P_{S_2}, p_i \in P_{S_1}} \frac{similarity(p_i, p_j)}{(|P_{S_1}| \times |P_{S_2}|)} \tag{2}$$

where $|P_{S_1}|$, $|P_{S_2}|$ indicate the number of all publications by scientists S_1 and S_2. Function $similarity(p_i, p_j)$ returns the content similarity between the abstracts of papers p_i and p_j. The greater is the value of $RS(S_1, S_2)$, the higher is the research related between scientists. Here, the feature keyword is extracted from their publications in DBLP to measure content similarity. To calculate the value of the function $similarity(p_i, p_j)$, we consider using Doc2Vec model [12] to process of calculating the similarity.

4.2   Proposed Method

The details of the proposed method are presented in the following subsections.

(1) Firstly, to build a network connecting scientists, we extract the information from DBLP, and compute the weight of $w(S_i, S_j)$ to measure the research collaboration between scientists S_i and S_j for determining the edge from S_i to S_j or not. The basic of this issue based on three axiom 1, 2, 3 and the result as shown in Fig. 3.

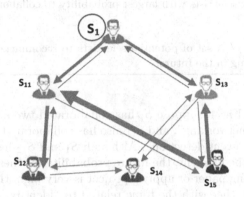

Fig. 4. The network of scientist S_1

(2) The axiom 3, as a standard to decide the list potential scientists with each considering scientist.

(3) By computing the research similarity between scientists and using a threshold for each considered scientist (S_1), we have the result as shown in Fig. 4.

(4) The list scientists with strong research collaboration with S_1 will be showed as the result of predicting the research collaboration trends.

4.3   Predicting Collaboration Algorithm

This section introduces about the algorithm for computing the probability to collaborate with target scientist S_i from his network (Algorithm 1). For example as shown in Fig. 4, we try to find a set of potential scientists to recommend to target scientist S_i for collaborating in the future.

The algorithm works based on databases collected from DBLP and Scient-Direct sources, based on extracting information of authors (including published publications and co-author information and research topics they have done and in progress) to build directed graphs. To predict the collaboration between two scientists in the future, we considered each pair of scientists who have no connected directly but they are in a sub-graph of a scientist (group) to propose future collaboration if the research similarity is greater than a given threshold.

Algorithm 1. Predicting collaboration in the future with S_i

Input: - Homogeneous data;
        - Target scientist S_i;
Output: Top k potential scientists;

1: Preprocessing data;
2: Creating a directed graph related to target scientist S_i;
3: **for** each scientist S_j in the graph with $w(S_i, S_j) = 0$ **do**
4:    Calculating the probability to collaborate with target scientist S_i;
5:    Sorting the list;
6:    Catching top k scientists with largest probability to collaborate with target scientist S_i;
7: **end for**;
8: **return** $R_f(S_i)$; // A set of potential scientists to recommend to target scientist S_i for collaborating in the future.

For example, in Fig. 4, suppose S_1 has collaborated two articles with S_{11} on the topic of "social networking", and S_{11} also has collaborated some articles with S_{12} on the topic of "event detection". Although S_1 and S_{12} have no cooperation with each other at the moment, the future probability for these two scientists to collaborate on writing paper or apply a project is very high, they are completely capable to work together with the topic related to "identifying events based on social networks". With illustrating from Fig. 4, by applying Algorithm 1, we can illustrate the result as shown in Fig. 5.

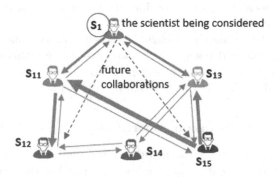

Fig. 5. The future collaborations of S_1

5 Conclusion and Future Work

Predicting research collaboration trends based on scientist's profiles and their publish is a hot issue in academia nowadays. We focused on exploiting data to extract information and to compute the similarity between authors (scientists). We already published some research results in the previous paper [9], therefore, in this paper, we have focused our research on evaluating a number of similar measures and proposed a model to find the potential relationship to predict the trend of research cooperation of scientist that hasn't yet tested the results.

In the next studies, we will complete the system and compare with other systems. Moreover, we will analyze other issues such as scientific impact evaluation, isolate scientists. Some young scientists have never collaborated with anyone in the past, it is the isolated vertex of the graph. We can handle by considering extra features such as being the same affiliation, the collaboration between affiliations. Our system operates based on databases collected from DBLP and ScientDirect sources on the basis of extracting the information of the authors to build a directed graph. The process of determining the weight of each edge for which each node is an author. We consider pairs of authors who had never been connected but who were in a sub-graph of a certain scientist to propose future collaborations. The system will provide a list of potential scientists to recommend to a user (as the scientist being considered) as shown in Fig. 5, and the user can establish information and remove that incorrect information for the system to move towards a more complete recommendation system.

Acknowledgment. This study is funded by Research Project No. DHH2018-03-109 of Hue University, Vietnam.

References

1. Abramo, G., D'Angelo, C.A., Di Costa, F.: Research collaboration and productivity: is there correlation? High. Educ. **57**(2), 155–171 (2009)
2. Ahuja, G.: Collaboration networks, structural holes, and innovation: a longitudinal study. Adm. Sci. Q. **45**(3), 425–455 (2000)
3. Alinani, K., Wang, G., Alinani, A., Narejo, D.H.: Who should be my co-author? Recommender system to suggest a list of collaborators. In: 2017 IEEE International Symposium on Parallel and Distributed Processing with Applications and 2017 IEEE International Conference on Ubiquitous Computing and Communications (ISPA/IUCC), pp. 1427–1433. IEEE (2017)
4. Burke, R., Felfernig, A., Göker, M.H.: Recommender systems: an overview. AI Mag. **32**(3), 13–18 (2011)
5. Chen, H.H., Gou, L., Zhang, X.L., Giles, C.L.: Discovering missing links in networks using vertex similarity measures. In: Proceedings of the 27th Annual ACM Symposium on Applied Computing, pp. 138–143. ACM (2012)
6. Deng, H., King, I., Lyu, M.R.: Formal models for expert finding on DBLP bibliography data. In: Proceedings of the 8th IEEE International Conference on Data Mining, ICDM 2008, Pisa, Italy, 15–19 December 2008, pp. 163–172 (2008)
7. Dong, Y., Ma, H., Shen, Z., Wang, K.: A century of science: globalization of scientific collaborations, citations, and innovations. arXiv preprint arXiv:1704.05150 (2017)
8. Han, S., He, D., Brusilovsky, P., Yue, Z.: Coauthor prediction for junior researchers. In: Proceedings of Social Computing, Behavioral-Cultural Modeling and Prediction - 6th International Conference, SBP 2013, Washington, DC, USA, 2–5 April 2013, pp. 274–283 (2013)

9. Hoang, D.T., Nguyen, N.T., Tran, V.C., Hwang, D.: Research collaboration model in academic social networks. Enterp. Inf. Syst. **13**(7–8), 1023–1045 (2019)
10. Hornick, M., Tamayo, P.: Extending recommender systems for disjoint user/item sets: the conference recommendation problem. IEEE Trans. Knowl. Data Eng. **24**(8), 1478–1490 (2012)
11. Klink, S., Reuther, P., Weber, A., Walter, B., Ley, M.: Analysing social networks within bibliographical data. In: Proceedings of the 17th International Conference on Database and Expert Systems Applications, DEXA 2006, Kraków, Poland, 4–8 September, pp. 234–243 (2006)
12. Le, Q.V., Mikolov, T.: Distributed representations of sentences and documents. arXiv preprint arXiv:1405.4053 (2014)
13. Li, J., Xia, F., Wang, W., Chen, Z., Asabere, N.Y., Jiang, H.: ACRec: a co-authorship based random walk model for academic collaboration recommendation. In: Proceedings of the Companion Publication of the 23rd International Conference on World Wide Web Companion, pp. 1209–1214. ACM (2014)
14. Luong, N.T., Nguyen, T.T., Hwang, D., Lee, C.H., Jung, J.J.: Similarity-based complex publication network analytics for recommending potential collaborations. J. Univ. Comput. Sci. **21**(6), 871–889 (2015)
15. Minkov, E., Charrow, B., Ledlie, J., Teller, S., Jaakkola, T.: Collaborative future event recommendation. In: Proceedings of the 19th ACM International Conference on Information and Knowledge Management, pp. 819–828. ACM (2010)
16. Newman, M.E.: Scientific collaboration networks. I. network construction and fundamental results. Phys. Rev. E **64**(1), 016131 (2001)
17. Nguyen, T.T., Hwang, D., Jung, J.J.: Social tagging analytics for processing unlabeled resources: a case study on non-geotagged photos. In: Camacho, D., Braubach, L., Venticinque, S., Badica, C. (eds.) Intelligent Distributed Computing VIII. SCI, vol. 570, pp. 357–367. Springer, Cham (2015). https://doi.org/10.1007/978-3-319-10422-5_37
18. Resnick, P., Varian, H.R.: Recommender systems. Commun. ACM **40**(3), 56–59 (1997)
19. Ricci, F., Rokach, L., Shapira, B.: Introduction to recommender systems handbook. In: Ricci, F., Rokach, L., Shapira, B., Kantor, P.B. (eds.) Recommender Systems Handbook, pp. 1–35. Springer, Boston (2011). https://doi.org/10.1007/978-0-387-85820-3_1
20. Subramanyam, K.: Bibliometric studies of research collaboration: a review. J. Inf. Sci. **6**(1), 33–38 (1983)
21. Sun, Y., Barber, R., Gupta, M., Aggarwal, C.C., Han, J.: Co-author relationship prediction in heterogeneous bibliographic networks. In: International Conference on Advances in Social Networks Analysis and Mining, ASONAM 2011, Kaohsiung, Taiwan, 25–27 July 2011, pp. 121–128 (2011)
22. Yang, C., Liu, T., Liu, L., Chen, X.: A nearest neighbor based personal rank algorithm for collaborator recommendation. In: 2018 15th International Conference on Service Systems and Service Management (ICSSSM), pp. 1–5. IEEE (2018)
23. Zhao, J., Dong, K., Yu, J.: Recommending funding collaborators with scholar social networks. In: 2014 International Conference on Data Science and Advanced Analytics (DSAA), pp. 122–127. IEEE (2014)

Assessing the Influence of Conflict Profile Properties on the Quality of Consensus

Adrianna Kozierkiewicz⬛, Marcin Pietranik[(⊠)]⬛, and Mateusz Sitarczyk⬛

Faculty of Computer Science and Management, Wroclaw University of Science
and Technology, Wybrzeze Wyspianskiego 27, 50-370 Wroclaw, Poland
{adrianna.kozierkiewicz,marcin.pietranik,mateusz.sitarczyk}@pwr.edu.pl

Abstract. Asserting a high quality of data integration results frequently
involves broadening a number of merged data sources. But does more
always mean more? In this paper we apply a consensus theory, originating from the collective intelligence field, and investigate which parameters describing a collective affects the quality of its consensus, which
can be treated as an output of the data integration, most prominently.
Eventually, we identified, either analytically or experimentally, adjusting which properties of the conflict profile (input data) asserts exceeding
expected integration quality. In other words-which properties have the
biggest influence and which are insignificant.

Keywords: Consensus theory · Collective intelligence · Knowledge
management · Data integration

1 Introduction

In recent years, available data have become more and more dispersed. Also, the
amount of data that can be meaningful (and therefore, valuable) is reaching
unparalleled levels. Developing a method of data integration, which allows effective merging a representation of such heterogenous and diverse data, into their
unified version, while asserting its maximal quality, has become one of the most
researched areas of computer science. A variety of different data integration tools
can be found in the literature.

No matter which approach is chosen, for a large set of data, such unification
can be characterized by a very high computational complexity. This entails a
long time required to perform such a procedure. However, to obtain a good
representation of the input data it is necessary to feed the integration algorithm
with all the available data. The natural approach is adding more data sources,
which intuitively should increase the quality of their integration. But does more
always mean more?

We adopt a consensus theory [12] as a foundation of our work. Originating from the collective intelligence field, it is based on an assumption, that the
integrated input data are formally describable using a mathematical representation called a conflict profile. Such representation is defined as a finite set of

N. T. Nguyen et al. (Eds.): ACIIDS 2020, LNAI 12033, pp. 25–36, 2020.
https://doi.org/10.1007/978-3-030-41964-6_3

different perspectives on the same topic. These perspectives can be expressed in different ways e.g. as trees, tuples or as simple vectors of values describing some object from a domain of interest. The output of the integration of the whole profile (which is further called a consensus) must meet a set of formally defined requirements. Thanks to such an approach, we can obtain good results (which was proved in [13]) of the data integration, which quality can be quantitatively measured due to the accepted formal restrictions.

In this paper, we would like to investigate and analyze the properties of the aforementioned conflict profile. In other words, we would like to check, how adjusting a variety of them, can lead to reaching some minimal, expected quality of the data integration.

Building our research on solid, mathematical foundations, the main task can be defined as follows: *For an assumed minimal and acceptable consensus quality, one should determine to adjust which properties, describing a conflict profile, asserts exceeding the accepted quality's threshold.* In the following, we provide an in-depth analysis of the conflict profile and identify (either analytically, if applicable, or experimentally) which properties of the conflict profile have the biggest influence on the consensus quality and which are insignificant. All researches are conducted for the assumed data representation, defined in subsequent sections.

The remaining sections of this paper are structured as follows. In the next part, an overview of related works is given. In Sect. 3 we provide base definitions that are used in this paper. Section 4 describes our approach to the presented research goal. It is split into two subsections devoted to two different approaches to measuring the quality of the consensus. Section 5 contains a short summary and sheds some light on our upcoming research plans.

2   Related Works

The topic covered in this article is closely related to the area of collective intelligence [17] and reducing the size of the integrated data while preserving its quality is not a completely new idea. [16] explores the performance of a collective in the light of the collective-size and expertise transferability. The authors claim that the collective's size is related to the level of a task's difficulty. This relationship is the strongest for tasks in a medium difficulty range. This research points out that the size of a collective cannot be chosen in separation with a domain of interest in which a collective is expected to perform.

A committee selection scenario is investigated in [5], where a voting rule that captures a positive correlation (synergy) between candidates has been presented. The higher synergy the better the outcome, but this synergy is not correlated with the size of a committee. A similar approach is presented in [4] where authors conducted experiments to discover factors (including cardinality) that allow groups of people to behave intelligently.

[11] examines collective decision making in relation to a performance metric of the collective decision quality. Conducted experiments show the existence of a relationship between both individual and collective intelligence, and furthermore

collective intelligence and collective decision quality. This implicates an influence size of the collective may have on its overall performance.

In [14] authors covered the topic of an automated decision-making process using machine learning methods. However, such methods must be trained before-hand and the opened question concerning the cardinality of the training data remains open. A similar consideration can be found in [10] where authors focus on a comparative analysis of quality and the popularity of articles in Wikipedia. Proposed solutions can be successfully used during the conflict-resolution phase of the data integration procedure. [6] contains a description of a framework for reactive, goal-directed navigation based on analyzing data coming from a mobile sensor network for a path planning of an autonomous mobile robot. The drone may communicate only with a few mobile sensing nodes and based on such data generates a belief map of a sequence of way-points, leading to a possible goal.

In [2] authors investigate an iterative and a collaborative ontology building, proposing that a trust-based consensus can support an efficient conflict resolu-tion among different viewpoints of participants of the process. Presented ideas are further developed in [3] where they are applied in a collaborative video anno-tation using a consensus-based social network.

Research similar to our work may be found in [15]. However, the authors approach the topic by analyzing how a diversity of the collective influences the effectiveness of the collective's performance. Also, a comparison of performance measures is presented.

We investigated a topic close to the one covered in this article in our previous publication [8], where a framework, which allows assessing the quality of the data integration output based solely on the analysis of its input is presented. Now we want to focus on properties of such input, which may serve as a natural extension of our previous research.

3 Basic Notions

Let the finite, nonempty set of a universe of objects be denoted as U. The set U can be considered as descriptions of elements of a certain domain of interest. By the symbol 2^U we denote the powerset of U, which is the set of all subsets of U. Let $\Pi_b(U)$ be the set of all nonempty subsets with repetitions (that cardinality is equal b) of the set U, for $b \in N$, where $\Pi(U) = \bigcup_{b \in N} \Pi_b(U)$ is the set of all nonempty subsets with repetitions of the universe U. Each element that belongs to $\Pi(U)$ is called a knowledge profile (a profile) [12]. $X \in \Pi(U)$ could be understood as a knowledge of a collective and each $x \in X$ as the knowledge of a collective member. The macrostructure of the set U is defined in the following way [12]:

Definition 1. The macrostructure of the set U is a distance function $\delta : U \times U \rightarrow [0, 1]$, which satisfies the following conditions:

1. $\forall_{v,u \in U}, \delta(v, u) = 0 \Leftrightarrow v = u$
2. $\forall_{v,u \in U}, \delta(v, u) = \delta(u, v)$

In this paper, we consider only measures where the transitive condition is not satisfied, thus in Definition 1, we assume the lack of the triangular inequality. This condition is too strong for many practical situations [1]. Therefore, the pair (U, δ) is called a distance space because there is no need to be a metric space.

For the assumed distance space, the consensus choice problem requires establishing the consensus choice function.

Definition 2. By a consensus choice function in the space (U, δ) we mean a function:

$$C : \Pi(U) \to 2^U \tag{1}$$

The set $C(X)$ is the representation of $X \in \Pi(U)$. In many real situations, each profile X can have many representations. Each $c \in C(X)$ we call a consensus of a profile X, which represents a consistent knowledge state of an assumed collective. If we consider the problem of data integration, the profile could be interpreted as input data sources. By the consensus, we call the final, consistent, merged data.

In [12,13] authors presented 10 postulates for the consensus choice functions: reliability, unanimity, simplification, quasi-unanimity, consistency, Condorcet consistency, general consistency, proportion, 1-optimality, 2-optimality. The last two postulates: 1-optimality and 2-optimality play the most important role in solving the consensus choice problem. 1-optimality postulate requires the consensus to be as near as possible to elements of the profile and could be recognized as the best representation of the profile. Postulate 2-optimality allows determining the most 'fair' consensus.

Let us assume, that two auxiliary functions are defined as follows:

- $\delta^1(x, X) = \sum_{y \in X} \delta(x, y)$
- $\delta^2(x, X) = \sum_{y \in X} (\delta(x, y))^2$

Postulates 1-optimality and 2-optimality are formally defined as follows [12,13]:

Definition 3. Let $X \in \Pi(U)$ be a profile and C a consensus choice function. We say that C satisfies a 1-optimality postulate if:

$$(x \in C(X) \Rightarrow (\delta^1(x, X) = \min_{y \in U} \delta^1(y, X)) \tag{2}$$

Also we say that C satisfies a 2-optimality postulate if:

$$(x \in C(X) \Rightarrow (\delta^2(x, X) = \min_{y \in U} \delta^2(y, X)) \tag{3}$$

In many situations, it is not possible to determine a "good", reliable consensus that represents the profile in the best way. That is why, before determining a consensus we need to check that the profile is susceptible to a consensus:

Definition 4. For a given space (U, δ) a profile $X \in \Pi(U)$ is susceptible to a consensus in relation to postulate 1-optimality (for $i = 1$) or 2-optimality (for $i = 2$) iff:

$$\hat{\delta}^i(X) \geq \hat{\delta}^i_{min}(X) \qquad (4)$$

where: $\hat{\delta}^i(X) = \dfrac{\sum\limits_{x,y \in X} \delta^i(x,y)}{n(n+1)}$, $\hat{\delta}^i_x(X) = \dfrac{\sum\limits_{y \in X} \delta^i(x,y)}{n}$, $\hat{\delta}^i_{min}(X) = \min\limits_{x \in U} \hat{\delta}^i_x(X)$, $n = card(X), i = \{1, 2\}$.

For evaluation of the determined consensus the quality measure is defined in the following way [13]:

Definition 5. Let $X \in \Pi(U)$ and $x \in C(X)$. The quality of a consensus x in a profile X we call the following value:

$$Q^i(x, X) = 1 - \frac{\delta^i(x, X)}{card(X)} \qquad (5)$$

where: $i \in \{1, 2\}$.

If the consensus x in the profile X satisfies the criterion for 1-optimality, then we calculate $Q^1(x, X)$, otherwise $Q^2(x, X)$. It is obvious that we want to find a consensus that maximizes the quality measure because we expect the most reliable representations of input data sources. However, in many real situations, we can accept the lower level of quality, if it limits the costs of the consensus determination. Thus, in our research, we would like to consider how to select a profile to ensure the assumed level of consensus quality.

In this paper we assume that the profile X consists of n binary vectors of the length equal to m. Thus, a distance space (U, δ) is composed from: $U = \{u_1, u_2, ...\}$ where elements of the universe are binary vectors and $\delta(w, v) = \sum\limits_{j=1}^{m} |w^j - v^j|$ for such $w, v \in U$ that $w = (w^1, w^2, ..., w^m), v = (v^1, v^2, ..., v^m)$, $v^q, w^q \in \{0, 1\}, q \in \{1, ..., m\}$. The profile is defined as: $X = \{a_1, a_2, ..., a_n\} \in \Pi(U)$, where: $a_i = (a_{i1}, a_{i2}, ..., a_{im}), i \in \{1, ..., n\}$.

For such defined distance space (U, δ), the one-level method of determining the consensus satisfying 1-optimality criterion, is conducted in using the following steps [12]:

Algorithm 1. 1-optimality consensus determination method

Require: $X = \{a_1, a_2, ..., a_n\}, a_i = (a_{i1}, a_{i2}, ..., a_{im}), i \in \{1, ..., n\}$
Ensure: x^*
1: **for all** $j := 1$ to m **do**
2:     $f_j := 0$;
3:     **for all** $i := 1$ to n **do**
4:         **if** $a_{ij} = 1$ **then**
5:             $f_j := f_j + 1$;
6:         **end if**
7:     **end for**
8: **end for**
9: **for all** $j := 1$ to m **do**
10:     **if** $f_j \geq \frac{n}{2}$ **then**
11:         $x_j^* := 1$;
12:     **else**
13:         $x_j^* := 0$;
14:     **end if**
15: **end for**
16: **return** x^*

A heuristic algorithm [12] determining the 2-optimality consensus is presented below:

Algorithm 2. 2-optimality consensus determination method

Require: $X = \{a_1, a_2, ..., a_n\}$
Ensure: x^*
1: Use Algorithm 1 to determine a consensus x^*;
2: $md := \delta^2(x^*, X)$;
3: **for all** $j := 1$ to m **do**
4:     $x_j^* := x_j^* \oplus 1$;
5:     **if** $\delta^2(x^*, X) < md$ **then**
6:         $md := \delta^2(x^*, X)$;
7:     **else**
8:         $x_j^* := x_j^* \oplus 1$;
9:     **end if**
10: **end for**
11: **return** x^*

---

4   Consensus Quality Analysis

4.1   1-Optimality Criterion

In this subsection, some properties about the impact of a profile's content on the consensus quality are proved analytically.

Theorem 1. *For an assumed distance space (U, δ) the consensus satisfy-ing the 1-optimality postulate has the quality at least:* $Q^1(x, X) = 1 - \frac{(\sum\limits_{j \in D} \sum\limits_{i=1}^{n} a_{ij} + \sum\limits_{j \in C} (n - \sum\limits_{i=1}^{n} a_{ij}))}{m*n}$ *where:* $C = \{j : \sum\limits_{i=1}^{n} a_{ij} \geq \frac{n}{2}\}, D = \{j : \sum\limits_{i=1}^{n} a_{ij} < \frac{n}{2}\}$ *and n is the cardinality of X.*

Proof. The proof of Theorem 1 follows from an Algorithm 2. The consensus sat-isfying the 1-optimality postulate is determined using the fact that if for a fixed $j = 1, ..., m$ $\sum\limits_{i=1}^{n} a_{ij} \leq \frac{n}{2}$ (more than half of the elements are zeros) then $u^{j*} = 0$ and $u^{j*} = 1$ otherwise, where $u*$ is the desired consensus. ∎

Based on Theorem 1, we can notice that the quality of the consensus depends on the profile's content. Thus, we formulate theorem which gives us a clue how to choose a profile to ensure an assumed level of the consensus quality.

Theorem 2. *For an assumed distance space (U, δ), where a ratio of occurrences of ones and zeros in the profile is equal $\frac{p}{q}$ (where $p + q = 1$), then to ensure the assumed level of the consensus quality Q_f (where $Q_f \in [\frac{1}{2}, 1]$), one of the following properties must be true:*

$- p \approx \frac{1 - \sqrt{2*Q_f - 1}}{2}$

$- p \approx \frac{1 + \sqrt{2*Q_f - 1}}{2}.$

Proof. Based on Theorem 1, we know that: $Q^1(x, X) = Q_f = 1 - \frac{(\sum\limits_{j \in D} \sum\limits_{i=1}^{n} a_{ij} + \sum\limits_{j \in C} (n - \sum\limits_{i=1}^{n} a_{ij}))}{m*n}$ where: n is the cardinality of X and m is the length of a binary vector. Additionally, from the assumptions we know that in the whole profile we have $m * n * p$ occurrences of ones and $m * n * q$ occur-rences of zeros. Therefore, $\sum\limits_{j \in D} \sum\limits_{i=1}^{n} a_{ij}$ expresses the number of occurrences of ones in columns from the set D and is approximately equal $card(D) * n * p$. Similarly, $\sum\limits_{j \in C} (n - \sum\limits_{i=1}^{n} a_{ij})$ expresses the number of occurrences of zeros in columns from the set C and is approximately equal $card(C) * n * q$. Thus, we obtain: $Q_f \approx 1 - \frac{card(D)*n*p + card(C)*n*q}{m*n}$. After transformations, we have: $Q_f \approx 1 - \frac{card(D)*p + card(C)*(1-p)}{m} \Rightarrow Q_f \approx 1 - \frac{(m - card(C))*p + card(C)*(1-p)}{m}$.

The set C contains a column from the profile, where the number of occur-rences of zeros is higher than the number of occurrences of ones and set D is defined in the opposite way. Thus, $\frac{card(C)}{card(D)}$ can be approximated by a ratio $\frac{p}{q}$. Moreover, we know that $card(D) + card(C) = m$. From both equations we cal-culate that $card(C) = p*m$. Thus, we obtain: $Q_f \approx 1 - \frac{(m - p*m)*p + p*m*(1-p)}{m} \Rightarrow Q_f \approx 1 - \frac{m*p - p^2*m + p*m - p^2 m}{m} \Rightarrow Q_f \approx 1 - \frac{m(-2*p^2 + 2*p)}{m} \Rightarrow Q_f \approx 1 + 2*p^2 - 2*p$. This quadratic equation has two roots: $p \approx \frac{1 - \sqrt{2*Q_f - 1}}{2}$ and $p \approx \frac{1 + \sqrt{2*Q_f - 1}}{2}$. ∎

Table 1. Experts opinions about flu' symptoms

No. of expert	Fever	Stomach ache	Sneeze	Cough
Expert 1	1	0	1	1
Expert 2	1	0	1	1
Expert 3	1	1	1	1
Expert 4	1	0	1	1
Expert 5	1	1	1	1

Example 1. Let us assume that we ask five experts about observed symptoms of flu such as fever, stomach ache, sneeze, cough. We want to ensure, that the final diagnosis determined based on experts' opinions will be trustworthy. Let us suppose that we want to ensure the quality of the consensus equal to or better than $Q_f = \frac{3}{4}$. Based on Theorem 2 we calculate that $p \approx 0.15$ and $q \approx 0.85$ (or otherwise, the number of ones should be greater than zeros). After some simplification of the calculation, we obtain a ratio of occurrences of ones and zeros equal $\frac{3}{17}$ or ones and zeros equal $\frac{17}{3}$. It means that the final diagnosis will be reliable if all experts' answers should not agree only in 3 cases. Table 1 presents the example experts' answers which satisfy the assumed conditions, where "1" means that symptoms occur and "0" otherwise. It is easy to calculate (Algorithm 1) that the consensus satisfying the 1-postulate is equal (1, 0, 1, 1). It could be interpreted that if a patient has a cough, sneeze, and fever he suffers from influenza. Based on Eq. (5) it is easy to show that the quality of the consensus determined based on collected answers is better than $\frac{3}{4}$ and is equal $\frac{17}{20}$.

Theorem 2 sheds new light on a consensus determination problem. Many pieces of research try to verify the impact of the profile's cardinality on the final consensus. However, after our analysis, we can conclude that getting a consensus of a given quality depends only on the content of this profile. If it is more consistent, then we can select the profile's size as small as possible, only paying attention to the consensus susceptibility. For the assumed distance space, the profile is susceptible to the consensus if their cardinality is an odd number [9]. Our conclusions are equivalent to the intuition. It is not important how many incoherent input data are processed, it is not possible to obtain their consistent and reliable representation.

4.2   2-Optimality Criterion

This part of our paper is devoted to the 2-optimality consensus, which is an NP-complete problem [7]. Thus, for the demonstration of its properties, we have used an experimental methodology.

The steps of the conducted procedure are as follows. In the beginning, we randomly generated an initial profile with a given very small size and using considered distribution. The distribution determined the probability of occurrence

Table 2. Results of the experiment for different profile's distributions

Quality	50/50				60/40				70/30			
	Avg	Mode	Mode share %	Max	Avg	Mode	Mode share %	Max	Avg	Mode	Mode share %	Max
0.7	3	3	100	3	3	3	100	3	3	3	100	3
0.75	3	3	100	3	3	3	100	3	3	3	100	3
0.8	3	3	100	3	3	3	100	3	3	3	100	3
0.85	3	3	100	3	3	3	100	3	3	3	100	3
0.9	453	3	81	9	253	3	84	11	3	3	97	9
0.95	3851	5001	77	3	3851	5001	77	5	1852	3	51	7

1 or 0 at each position in vectors. It was denoted as $\frac{p}{q}$ which represents the fact that there was p * 100% chance for drawing 1 and q * 100% for drawing 0. For a generated profile we have used the Algorithm 2 to determine a consensus and compute the quality of achieved results using the formula 5. If the quality was satisfying (greater or equal to the assumed level) the sufficient size of the profile was estimated and the evaluation was finished. Otherwise, the profile was regenerated with a bigger size (to assert the susceptibility for the 1-optimality consensus it was increased by 2) and previous steps were repeated until the expected quality was achieved or the maximal size of the profile was reached.

In our experiments, we accepted some assumptions. The size of vectors in the profile was set to $m = 10$, the initial size of the profile was set to $n_{init} = 3$ and the maximal threshold was set to $n_{max} = 5001$. During our examinations, we noticed, that further extension of the profile after reaching some limit did not improve the quality, only extended the time and cost of the performance. We analyzed six levels of quality Q_f: 0.7, 0.75, 0.8, 0.85, 0.9, 0.95, and the following distributions: 50/50, 60/40, 70/30, 80/20 and 90/10. For each set of parameters, the evaluation of the aforementioned method was repeated 100 times.

To analyze the obtained results (which can be found in Tables 2 and 3) we used the following measures:

– Mode of profile size and its percentage share in the results
– Average profile size
– Maximal profile size in a situation when it was determined (the threshold was not achieved)

We started our analysis with a mode of profile's size. It is a measure, which describes the most often value in the obtained results. In significantly more situations (especially for smaller levels of the quality), experiments showed, that a profile consisting of only 3 elements is sufficient to achieve the given quality. Moreover, the percentage share of modes was in all situations greater than 50% and for quality levels from 0.7 to 0.85, it occurred in 100% of repeats. It is very important for real situations because we would like to choose as little members of a collective as possible, for both economical and availability reasons.

The analysis of an average showed, that for profiles which are not cohesive, their sizes in many situations were too big or the threshold was not achieved.

Table 3. Results of the experiment for different profile's distributions (continuation)

Quality	80/20				80/20			
	Avg	Mode	Mode share %	Max	Avg	Mode	Mode share %	Max
0.7	3	3	100	3	3	3	100	3
0.75	3	3	100	3	3	3	100	3
0.8	3	3	100	3	3	3	100	3
0.85	3	3	100	3	3	3	100	3
0.9	3	3	99	5	3	3	100	3
0.95	3	3	77	25	3	3	97	5

This conclusion can be emphasized by comparing the average with the maximal profile size after filtering situations when the maximal size was reached. In each situation, the maximal value was less than 30, whereas the average takes value even to 3851. It is caused by the fact that for distributions 50/50, 60/40 and 70/30, vectors representing answers, could be diverse, which makes it difficult to determine a high-quality consensus. However, in real situations, where we take into account experts, their answers come from rather more cohesive distributions (like 80/20 or 90/10), therefore their small groups are sufficient, which was confirmed by our experiments.

In the last step of our considerations, we statistically analyzed the size of profiles at the quality level Q_f equal 0.9 and the significance level $\alpha = 0.05$. In many everyday situations, it is the acceptable level, taking into the account also cost of acquiring collective's members. We merged results from all distributions and obtained one sample. In the beginning, we used the Shapiro-Wilk test to check if data comes from the normal distribution. Because $p - value$ was near 0 and $W = 0.1505$, we rejected the null hypothesis, stating that the sample comes from the normal distribution. Next, we used the one-sided Wilcoxon signed-rank test. The null hypothesis claimed that the sample comes from the distribution with a median equal to 4. The alternative hypothesis stated, that the sample comes from a distribution with a median lower than 4. The obtained $p - value = 5.1063e - 63$ allowed us to accept the alternative hypothesis.

The conducted analyses show, that in most situations, the collective size equal to 3 is sufficient to achieve the high-quality consensus, especially when answers are cohesive. It confirms the fact, that very often, for a jury or committee only 3–5 members are chosen, because they can make credible decisions and the cost of their canvass is not high. The results of the experiment also confirmed our consideration from Sect. 4.1 that to assert the assumed quality of the consensus, the size of the profile is not crucial, but only the consistency of the profile is.

5   Summary and Future Works

In this paper, we accepted the consensus theory as a solid, mathematical foundation for performing data integration. Such an approach allowed us to quantitatively measure the quality of the collected results. For an assumed minimal and acceptable consensus quality, we determined adjusting which properties, describing a conflict profile, asserts exceeding the accepted quality's threshold. We provided an in-depth analysis of the conflict profile and identified (either analytically, if applicable, or experimentally) which of its properties have the biggest influence on its consensus quality and which are insignificant.

The provided Theorem 2 sheds new light on a consensus determination problem. It allows us to conclude that achieving a consensus of a given quality depends only on the content of the input profile. It is frequently said in the literature that the profile's cardinality has the biggest impact. However, after our analysis and experimental verification, we can draw the opposite conclusion. For the assumed distance space, if the profile is susceptible to the consensus, only its consistency contributes to the quality of its integration.

In the future, we want to investigate different conflict profile structures (extending them beyond simple binary vectors). Also, we will perform more experiments verifying, whether yes or not, it is not important how much incoherent input data is processed and it is not possible to obtain their consistent and reliable representation.

References

1. Bogart, K.P.: Preference structures I: distances between transitive preference relations. J. Math. Sociol. **3**(1), 49–67 (1973)
2. Duong, T.H., Nguyen, N.T., Nguyen, D.C., Nguyen, T.P.T., Selamat, A.: Trust-based consensus for collaborative ontology building. Cybern. Syst. **45**(2), 146–164 (2014)
3. Duong, T.H., Nguyen, N.T., Truong, H.B., Nguyen, V.H.: A collaborative algorithm for semantic video annotation using a consensus-based social network analysis. Expert Syst. Appl. **42**(1), 246–258 (2015)
4. Green, B.: Testing and quantifying collective intelligence. In: Proceedings of the Collective Intelligence Conference (2015)
5. Izsak, R.: Working together: committee selection and the supermodular degree. In: Sukthankar, G., Rodriguez-Aguilar, J.A. (eds.) AAMAS 2017. LNCS (LNAI), vol. 10642, pp. 103–115. Springer, Cham (2017). https://doi.org/10.1007/978-3-319-71682-4_7
6. Jha, D.K., Chattopadhyay, P., Sarkar, S., Ray, A.: Path planning in GPS-denied environments via collective intelligence of distributed sensor networks. Int. J. Control **89**(5), 984–999 (2016)
7. Kemeny, J.G.: Mathematics without numbers. Daedalus **88**, 577–591 (1959)
8. Kozierkiewicz, A.: The analysis of expert opinions' consensus quality. Inf. Fusion **34**, 80–86 (2017)
9. Kozierkiewicz, A.: Analysis of susceptibility to the consensus for a few representations of collective knowledge. Int. J. Software Eng. Knowl. Eng. **24**(5), 759–775 (2014)
10. Lewoniewski, W., Węcel, K., Abramowicz, W.: Relative quality and popularity evaluation of multilingual wikipedia articles. Informatics **4**(4), 43 (2017)

11. McHugh, K.A., Yammarino, F.J., Dionne, S.D., Serban, A., Sayama, H., Chatterjee, S.: Collective decision making, leadership, and collective intelligence: tests with agent-based simulations and a field study. Leadersh. Quart. **27**(2), 218–241 (2016)
12. Nguyen, N.T.: Consensus Choice Methods and their Application to Solving Conflicts in Distributed Systems. Wroclaw University of Technology Press, Wroclaw (2002). (in Polish)
13. Nguyen, N.T.: Advanced Methods for Inconsistent Knowledge Management. Springer, London (2008). https://doi.org/10.1007/978-1-84628-889-0
14. Pretorius, A., Parry, D.A.: Human decision making and artificial intelligence: a comparison in the domain of sports prediction. In: Proceedings of the Annual Conference of the South African Institute of Computer Scientists and Information Technologists, p. 32. ACM (2016)
15. Nguyen, V.D., Merayo, M.G.: Intelligent collective: some issues with collective cardinality. J. Inf. Telecommun. **1**(2), 127–140 (2017)
16. Wagner, C., Suh, A.: The wisdom of crowds: impact of collective size and expertise transfer on collective performance. In: 47th Hawaii International Conference on System Sciences (HICSS), pp. 594–603. IEEE (2014)
17. Woolley, A.W., Aggarwal, I., Malone, T.W.: Collective intelligence and group performance. Curr. Dir. Psychol. Sci. **24**(6), 420–424 (2015)

OWL RL to Framework for Ontological Knowledge Integration Preliminary Transformation

Bogumiła Hnatkowska[iD], Adrianna Kozierkiewicz[iD], and Marcin Pietranik[(✉)][iD]

Wroclaw University of Science and Technology, Wyb. Wyspiańskiego 27,
50-370 Wrocław, Poland
{Bogumila.Hnatkowska,Adrianna.Kozierkiewicz,
Marcin.Pietranik}@pwr.edu.pl

Abstract. FOKI is a formally defined framework, proposed by authors, which addresses storing, processing, and integrating ontologies. Its model is based on a mathematical apparatus but lacks a concrete syntax. These features make difficult to use standardized benchmark datasets, usually expressed in OWL2, during experimental verification of FOKI's validity. To enable a practical usage of FOKI, a set of bidirectional transformation rules (defined at the abstract syntax level) between the OWL2 RL and the framework is needed. However, due to major differences in base assumptions it is impossible to provide a straightforward translation between FOKI and OWL. Therefore, the aim of the paper is to identify which elements of OWL syntax can be transformed into FOKI formalism (on its current state of development) and which of these rules are bi-directional. The defined rules are illustrated with some overall examples. The paper also provides a short discussion about different approaches to transformation definitions.

Keywords: FOKI · OWL2 · Transformation · Migration · Ontology integration framework

1 Introduction

Ontology integration is a widely discussed topic. It puts attention on a seemingly simple problem of merging a set of ontologies into one ontology. This task eventually results in a unified ontology, which contains all of the knowledge taken from the input sources. Informally speaking, it can be understood as selecting elements of input ontologies which express the same parts of a modeled universe of discourse. Then, such elements should be merged into a single, unified element. Any non-conflicting elements that differentiate input ontologies should be preserved as-is. Although simple to understand, ontology integration is a difficult task in terms of its semantic and computational complexity. The main reason is the fact that any appearing semantic conflicts or inconsistencies should be excluded or resolved.

It is easy to find in the literature a variety of different solutions that address the ontology integration problem. Their common feature is being built on top of OWL2

© Springer Nature Switzerland AG 2020
N. T. Nguyen et al. (Eds.): ACIIDS 2020, LNAI 12033, pp. 37–48, 2020.
https://doi.org/10.1007/978-3-030-41964-6_4

representation and its syntaxes. It entails a potential shortcoming – the ontology integration tools mentioned above are tightly coupled with OWL syntaxes. Therefore, they are separated from formal foundations of the problem at hand. For example, a common approach to designating similar elements in ontologies is calculating similarities between informal comments from `rdfs:comment` tag available in OWL. The problem is that such comments are not defined in the base, mathematical definitions of ontologies. A broad explanation of the motivation behind our research can be found in [13].

The remarks presented above brought to our attention the necessity of providing a Framework for Ontological Knowledge Integration (FOKI), which is a formally grounded tool, with a sound theoretical background and strong ontology definitions. Its foundation is a notion of attributes' semantics, which gives explicit meanings to attributes when they are included in different concepts. For example, the same attribute *address* may carry different meanings while included in the *Home* concept and completely different when incorporated in the *PersonalWebsite* concept. Overcoming the aforementioned drawbacks made possible to provide (in our previous publications [10–12]) several applications that showed the usefulness of our approach in both the task of ontology integration and ontology alignment. However, it the FOKI framework has one important disadvantage. It cannot be easily expressed using the OWL, which makes it challenging to apply in some practical applications. In particular, it is impossible to directly use benchmark ontologies provided by the Ontology Alignment Evaluation Initiative (OAEI), which are a state-of-the-art dataset used to test the usefulness of many ontology-related applications. It would be beneficial to use them to verify algorithms developed based on our framework.

OWL RL is a sublanguage (also referred to as profile) of OWL2 which we chose because it asserts reasoning in polynomial time with respect to the size of the ontology and is an extension of Description Logic based OWL-DL. It brought to our attention the necessity of translating ontologies expressed using OWL-RL into the FOKI framework. Providing a set of rules that could be used to transform an ontology defined in OWL RL to our framework can become invaluable in any upcoming research focused on ontologies.

Preparing such transformations implies a several difficulties to be addressed. First of all, OWL-RL assumes that the modeled domain of discourse is an open world, where FOKI is based on an assumption of a closed-world domain. This entails that OWL ontology inferred from FOKI formalism may be too strict. For example, OWL provides a mechanism to express that two concepts are equivalent, which implies that their set of instances must be disjoint. In FOKI it is easy to check whether or not two sets of instances are disjoint, however such outcome does not necessarily mean that inferred concept equivalency was intended by ontology developer.

Second major difficulty that needs addressing comes from the assumption that in FOKI both attributes and relations are annotated with logic sentences that assign them some explicitly intended semantics. For example, let us suppose the *level* attribute. If we consider this attribute in a context of the *Lecture* concept, we think about a difficulty of educational material. However, the attribute *level* included in the *Apartment* concept can express the apartment's floor within a building. Such formal tool provides good expressiveness. However, from the practical point of view, it drastically increases the difficulty of maintaining the FOKI ontology which was broadly discussed in [7].

The aforementioned problems bring us to the conclusion that on the current state of FOKI's development it is impossible to provide a ruleset that would cover the whole OWL-RL syntax. Therefore, the main goal of the paper is to identify two things: *(i)* which elements of OWL syntax can be clearly transformed into FOKI formalisms, *(ii)* and which of these transformation rules are one- or bi-directional (as explained in the previous paragraphs). This may become a sound foundation for upcoming research in extending FOKI framework into a formalism that is fully translatable from/to OWL. It would be beneficial to take advantage of both the expressiveness of OWL and mathematical formalism of FOKI.

The article is structured as follows. Section 2 provides fundamental definitions of the FOKI framework, on top of which transformation rules are built. An overview of related works is given in Sect. 3. A set of transformation rules from OWL to our framework is given in Sect. 4. Section 5 demonstrates selected transformation rules with illustrative examples. The article ends with a summary and an overview of our future research directions given in Sect. 6.

2 Basic Notions

The FOKI framework is based on a mathematical apparatus that evolved throughout several of our publications [9–13]. It is based on a notion of a real world defined as a pair (A, V), where A is a finite set of attributes that can be used to describe objects, and V is a set of their valuations (domains) where: $V = \bigcup_{a \in A} V_a$. By V_a we call a domain of a particular attribute a. The defined pair (A, V) allows to present a quintuple which is called by us as an (A, V)-based ontology:

$$O = \left(C, H, R^C, I, R^I \right) \tag{1}$$

where: C is a finite set of concepts, H defines generalization relationships between concepts, R^C is a finite set of relations between concepts, I denotes a finite set of instances, R^I is a finite set of relations between concepts' instances.

During the development of an ontology firstly we need to define a set of concepts C. Each concept $c \in C$ is defined as:

$$c = \left(id^c, A^C, V^C, I^C \right) \tag{2}$$

where: id^c is an identifier of the concept c, A^C is a set of its attributes, such that $A^C \subseteq A$, V^C is a set of attributes domains (formally: $V^C = \bigcup_{a \in A^c} V_a$), I^C is a set of c instances. For short, we write $a \in c$ to denote that an attribute a belongs to concept's c set of attributes A^C.

Attributes from the set A do not have any semantics. They become interpretable only when they are a part of a selected concept. Let D_A be a set containing atomic descriptions of attributes. Subsequently, we define a sub-language of the sentence calculus built from elements of D_A and logic operators of conjunction, disjunction and negation. We denote it as L_S^A and it is used for description of semantics of attributes. Formally, we can define a function which assigns a logic sentence from L_S^A to attributes within particular concept:

$$S_A : A \times C \to L_S^A \tag{3}$$

Based on such an approach each attribute being part of different concepts can express different meanings. Let us suppose an attribute *class.* If we consider this attribute in a context of a concept *Student,* we probably think about an educational level. However, attribute *class* included in a concept *Ticket* can express a standard of services.

Based on the previously defined function S_A, we can formally define how attributes included in concepts relate to each other. We distinguish three relations between attributes: *equivalency* (denoted as $\equiv$), *generalization* (denoted as $\leftarrow$) and *contradiction* (denoted as $\sim$):

- Two attributes $a \in A^{c_1}, b \in A^{c_2}$ are semantically equivalent $a \equiv b$ if the formula: $S_A(a, c_1) \Leftrightarrow S_A(b, c_2)$ is a tautology for any two $c_1 \in C_1, c_2 \in C_2$.
- The attribute $a \in A^{c_1}$ in concept $c_1 \in C_1$ is more general than the attribute A^{c_2} in concept $c_2 \in C_2$ $a \leftarrow b$ if the formula $S_A(b, c_2) \Rightarrow S_A(a, c_1)$ is a tautology.
- Two attributes $a \in A^{c_1}, b \in A^{c_2}$ such that $c_1 \in C_1, c_2 \in C_2$, are semantically contradicting $a \sim b$ if the formula $\sim (S_A(a, c_1) \wedge S_A(b, c_2))$ is a tautology.

For a given a concept c, we define its instances from the set I^C as a tuple:

$$i = (id^i, v_c^i) \tag{4}$$

where: id^i is an instance identifier, v_c^i is a function interpreted as a tuple of type A^c with a signature: $v_c^i : A^c \to V^c$. For short we can write $i \in c$ which means that the instance i belongs to the concept c. A set of instances is defined as:

$$I = \bigcup_{c \in C} \{id^i | \left(id^i, v_c^i\right) \in I^c\} \tag{5}$$

By $Ins(c)$ we define a set of identifiers of instances assigned to the concept c:

$$Ins(c) = \{id^i | \left(id^i, v_c^i\right) \in I^c\} \tag{6}$$

To simplify, a function $Ins^{-1} : I \to 2^C$ generates a set of concepts to which an instance with some identifier belongs:

$$Ins^{-1}(i) = \{c | c \in C \bigwedge i \in c\} \tag{7}$$

The last part of the ontology are relations between concepts and instances. R^C is a finite set of relations between concepts, $R^C = \{r_1^C, r_2^C, \ldots, r_n^C\}, n \in N$, such that every $r_i^C \in R^C$ $i \in [1, n]$ is a subset of Cartesian product $r_i^C \subseteq C \times C$.

By analogy to D_A, we define a set D_R containing atomic descriptions of relations. Subsequently, we define a sub-language of the sentence calculus built from elements of DR and logic operators of conjunction, disjunction, and negation. We denote it as

$$S_R : R^C \to L_S^R \tag{8}$$

As a consequence, we can define formal criteria for relationships between relations:

- Two relations $r_1, r_2 \in R^C$ are equivalent $r_1 \equiv r_2$ if the formula: $S_R(r_1) \Leftrightarrow S_R(r_2)$ is a tautology
- Two relations $r_2 \in R^C$ is more general than relation $r_1 \in R^C$ $r_2 \leftarrow r_1$ if the formula: $S_R(r_1) \Rightarrow S_R(r_2)$ is a tautology
- Two relations $r_1, r_2 \in R^C$ are contradicting $r_1 \sim r_2$ if the formula $\sim (S_R(r_1) \wedge S_R(r_2))$ is a tautology

A special type of relation is a hierarchy of concepts $H \subset C \times C$. A pair of concepts (c_1, c_2) may be included in H stating that c_1 is more general than c_2 (denoted as $c_2 \leftarrow c_1$) only if the following criteria are all true:

- $|A^{c_1}| \geq |A^{c_2}|$
- $\forall a_2 \in A^{c_2} \exists a_1 \in A^{c_1} : (a_1 \equiv a_2) \bigvee (a_2 \leftarrow a_1)$
- $Ins(c_1) \subseteq Ins(c_2)$

The relations between instances are denoted as $R^I = \{r_1^I, r_2^I, \ldots, r_n^I\}$, $n \in N$. Every relation from the set R^C has a complementary relation from the set R^I, thus $|R^C| = |R^I|$. In other words, a relation $r_j^C \in R^C$ describes potential connections that may occur between instances of concepts from the set C but $r_j^I \in R^I$ describes what is connected where $j \in [1, n]$. Formal criteria describing such an approach are defined below:

- $r_1, r_2 \subseteq \bigcup_{(c_1,c_2) \in r_j^C} (Ins(c_1) \times Ins(c_2))$
- $(i_1, i_2) \in r_j^I \Rightarrow \exists (c_1, c_2) \in r_j^C : \left(c_1 \in Ins^{-1}(i_1)\right) \bigwedge \left(c_2 \in Ins^{-1}(i_2)\right)$ which means that two instances can be connected by some relation only if there is a relation connecting concepts they belong to
- $(i_1, i_2) \in r_j^I \Rightarrow \sim \exists r_k^I \in R^I : \left((i_1, i_2) \in r_k^I\right) \bigwedge \left(r_j^C \sim r_k^C\right)$ which means that two instances cannot be connected by two contradicting relations.
- $(i_1, i_2) \in r_j^I \bigwedge \exists r_k^I \in R^I : r_k^C \leftarrow r_j^C \Rightarrow (i_1, i_2) \in r_k^I$ if two instances are in a relation and there exists a more general one, then they are also connected by it.

3 Related Works

Ontology transformation – similarly to program transformation – can be defined as the act of changing one ontology into another [19]. The languages of transformed and resulting ontologies are called the source and target languages, respectively. Transformation can be one or bi-directional, lossy or preserving the whole input semantics.

To transform one language to another, one has to define a set of transformation rules and combine them in an algorithm (if ordering of rules matters). These rules can be defined at two levels (see Fig. 1).

At the first level, a specific concrete syntax of the source language is the subject of transformation rules. They manipulate the input content directly, e.g., with the use of

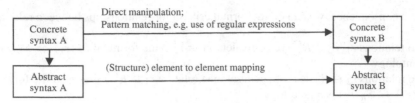

Fig. 1. Possible approaches to ontology transformations.

pattern matching mechanisms like regular expressions. The part which fits the pattern is replaced with a new content written in the target concrete syntax (e.g. [9, 21]). That group includes, e.g., XSLT transformations which can also be applied for OWL2. At the second level, abstract syntaxes (or meta-models) are the subject of interest. Transformation rules map one structural element of the input to one or more structural elements of the output (e.g. [1–6, 14, 15, 18]). This approach assumes that there exist proper tools for vertical translations between a concrete syntax and abstract syntax (tree) in both directions. Definition of transformation rules at that level makes the mapping independent of particular representations what is especially useful for multi-concrete-syntax solutions.

Transformation rules are defined either informally, in natural language (e.g. [1, 3–6, 16]), semi-formally (e.g., with the use of structured tables [14, 15] or patterns with placeholders [9]), or formally, with the use of mathematical apparatus [2, 17]. The formal approach reduces the risk of misinterpretation but could be hard to understand, so it is typically extended with simple explanations. As FOKI is defined in a formal manner that would be also our preference.

Transformation rules can be written in declarative (e.g. implementation part in [18]) or operational manner [4]. Sometimes a hybrid method is used, e.g. [8]. In the cases when the ordering of transformation rules matters, their ordering is provided [17].

Ontologies, especially expressed in OWL2, are subjects of transformations to/from different notations, including other ontology formalisms [17], UML [4, 5, 15, 17, 18], programming languages [3, 6], databases' schemas [1, 2, 14] or business vocabularies and rules [8, 9].

OWL2 uses many concrete syntaxes, e.g., functional, Manchester, XML, RDF Turtle, RDF/XML but all of them share one abstract syntax [20] defined as a sequence of annotations, axioms, and facts.

Mappings between OWL2 concrete syntaxes refer to the same structure in contrast to those for which the structures of the source and target formalisms are different. Exemplary translation (at the first level) between Manchester and functional style is given in [21]. In [22], there is a semi-formal mapping defined at the abstract syntax level between the structural OWL2 specification and RDF graph.

In the paper, we deal with migration between the meta-models of OWL2 and FOKI. As there is no concrete syntax for FOKI, the set of transformation rules have to be defined at the structural level. To avoid misunderstandings, the rules are defined formally, in a declarative manner by the reference to the proper FOKI constructs. The transformation is bi-directional and partially lossy. However, the semantics of OWL is preserved in one-way transformation.

4 Mappings

In this section a set of transformation rules from OWL-RL to the FOKI framework is provided. Every rule is built by a reference to an abstract syntax element of OWL-RL, and an appropriate mathematical formalism of FOKI framework (obviously if such exists and is applicable). Moreover, every rule is labeled with a direction in which the transformation rule can be applied – in both directions or in one. If needed it is further explained in a justification of the rule.

Many of the rules are only one directional from FOKI to OWL, which entails the necessity of extending FOKI with some kind of formal mechanism to enable a bidirectional transformation. This issue also concerns lack for rules for OWL elements *ObjectComplementOf, ObjectIntersectionOf* and *ObjectSomeValuesFrom*. These constructs take advantage of OWL open-world foundations making possible to create "virtual" definitions, which are not explicitly given, but are built from other definitions (Table 1).

Table 1. OWL-RL to FOKI framework transformation rules

No.	OWL RL	Direction	FOKI Framework for Ontological Knowledge Integration
1	*Class*	⇔	$c \in C$
	Classes from OWL correspond to concepts, which are elements of the set C		
2	*ClassAssertion*	⇔	$i \in c$ where: $i \in I$; $c \in C$
	ClassAssertion represents the fact that some individual in OWL is an instance of a specific class. The same meaning, in our framework, is denoted as being a member of the set c of concept instances or (for short) as being a member of the particular concept c		
3	*Datatype*	⇔	$V_a \in V$
	Datatype in OWL defines *DatatypeProperties* ranges. In the FOKI framework, it corresponds to attributes' domains (valid valuations) taken from the set V containing all possible domains for the selected universe of discourse		
4	*DatatypeProperty*	⇔	$a \in A$
	DatatypeProperty represents an attribute. These are defined as elements of the set A for the selected universe of discourse		
5	*DataPropertyAssertion*	⇔	$v_c^i : A^c \rightarrow V^c$; for given $c \in C$ and $i \in c$
	DataPropertyAssertion in OWL is used to define values that *DatatypeProperties* acquired in particular instances. This situation is expressed as the output of the function v_c^i that belongs to every instance i of the c concept. The function returns a vector containing values of attributes assigned to that concept		
6	*DataPropertyDomain*	⇔	$a \in A^c$; $c \in C$
	DataPropertyDomain in OWL assigns *DatatypeProperties* to classes. In the FOKI framework, it is defined that some attribute a is a member of a set of attributes of some concept c		
7	*DataPropertyRange*	⇔	$V_a \in V$
	DataPropertyRange defines assignment of set of valid valuation to a *DatatypeProperty*. In our framework it is expressed as elements of V, which contains all possible domains of attributes		
8	*DataSomeValuesFrom*	⇔	$V_a \in V$
	In the FOKI framework it is not possible in all cases to define a situation in which instances can acquire only some selected values from the attributes' domains. Due to the fact that there are no restrictions on elements of V this rule may cover both general domains as in Rule 7 and some detailed domains, e.g. $V_a = \{1, 2, 3\}$ or $V_a = \{x : x \in N \wedge x < 20\}$		

(continued)

Table 1. (*continued*)

No.	OWL RL	Direction	FOKI Framework for Ontological Knowledge Integration
9	*DifferentIndividuals*	$\Leftarrow$	id^i

In OWL *DifferentIndividuals* denotes that several individuals are all different from each other. In the FOKI framework instances have unique identifiers which can be easily used to distinguish their identities. The obvious distinctiveness of instance identifiers entails using *DifferentIndividuals* could be more frequent than intended

| 10 | *DisjointDataProperties* | $\Leftarrow$ | $V_a \cap V_b = \emptyset$ where: $V_a, V_b \in V$ |

DisjointDataProperties in OWL represents the fact that two (or more) *DataProperties* must have disjoint set of literals for each individual. In the proposed framework it is denoted as disjoint set of attributes domains. This rule is an example of a close-world/open-world dichotomy addressed in Sect. 1. This rule will give more strict results when used to transform ontologies from FOKI to OWL due to the fact that using *DisjointDataProperties* in OWL does not entail that two domains of attributes need to be strictly disjoint

| 11 | *DisjointObjectProperties* | $\Leftarrow$ | $r_1^C \not\equiv r_2^C$ where: $r_1^C, r_2^C \in R^C$ and $r_1^I \cap r_2^I = \emptyset$ |

In OWL *DisjointObjectProperties* states that two object properties cannot simultaneously connect two individuals. In FOKI it can be expressed as a disjoint set of instances relations. However, similarly to earlier rules, one cannot assume that the inferred OWL element *DisjointObjectProperties* was intended by ontology developer

| 12 | *DisjointWith* | $\Leftarrow$ | $c_1 \not\equiv c_2 \not\equiv \ldots \not\equiv c_n$ where
$c_1, c_2, \ldots, c_n \in C \leftrightarrow I^{c1} \cap I^{c2} = \emptyset \wedge \ldots \wedge I^{c_{n-1}} \cap I^{c_n} = \emptyset$ |

In OWL *DisjointWith* denotes that some group of classes cannot share any individuals. In the FOKI framework it can be defined as mutually exclusive sets of concept's instances, which be definition from Eq. 5 is finite

| 13 | *EquivalentClass* | $\Leftarrow$ | $c_1 \equiv c_2$ where: $c_1, c_2 \in C \leftrightarrow I^{c1} = I^{c2}$ |

EquivalentClass in OWL states that two classes share the same set of assigned individuals. In the FOKI framework it can be denoted as equality of sets of concepts' instances

| 14 | *EquivalentDataProperties* | $\Leftarrow$ | $a \in A^{c1}, b \in A^{c2}; c_1, c_2 \in C$ and $S_A(a, c_1) \Leftrightarrow S_A(b, c_2)$ is a tautology |

EquivalentDataProperties in OWL is used to explicitly denote that two *DatatypeProperties* are equivalent. In the FOKI framework it can be achieved using attributes' semantics defined in Eq. 3. The key difference is that in OWL such *DatatypeProperties* are always equivalent, where in the FOKI framework the equivalency may appear only between two attributes that are assigned to concepts (due to the fact that according to Eq. 3 they don't possess any semantics otherwise). Using this rule implies that logic annotations of attributes are provided, which from practical point of view may be difficult. Issues related to this topic are broadly discussed in our other publication [7]

| 15 | *EquivalentObjectProperty* | $\Leftarrow$ | $r_1 \in R_1^C, r_2 \in R_2^C : r_1 \equiv r_2 S_R(r_1) \Leftrightarrow S_R(r_2)$
is a tautology |

In OWL *EquivalentObjectProperty* is used to state that two *ObjectProperties* are semantically equivalent to each other. Such restriction can be easily described in the FOKI framework using relations' semantics from Eq. 8, but inferring OWL constructs may be too strict in terms of ontology developer intention. Moreover, some difficulties related to providing semantic annotation of relations can appear, by analogy to semantic annotations of attributes

| 16 | *InverseOf* | $\Leftarrow$ | $r_1, r_2 \in R^C, \forall (c_1, c_2) \in r_1 \exists! (c_2, c_1) \in r_2 \wedge S_R(r_1) = S_R(r_2)$ |

InverseOf in OWL can be used to denote that two *ObjectProperties* are their "mirror images". For example, the "is_parent" *ObjectProperty* is an inversion of the "is_child" *ObjectProperty*. In the FOKI framework it can be expressed as a restriction put on the set of concepts describing two invert relations

17	*ObjectComplementOf*	–	–
18	*ObjectIntersectionOf*	–	–
19	*ObjectPropertyAssertion*	$\Leftrightarrow$	$(i_1, i_2) \in r_1^I; (c_1, c_2) \in r_1^C$ where: $i_1 \in c_1; i_2 \in c_2$

ObjectPropertyAssertion in OWL is used to connect two individuals via some *ObjectProperty*. A complementary definition can be expressed in FOKI by including a pair of instances to the appropriate set describing a particular instance relation

(*continued*)

Table 1. (*continued*)

No.	OWL RL	Direction	FOKI Framework for Ontological Knowledge Integration
20	*ObjectProperty*	⇔	$r_1^C \in R^C, r_1^I \in R^I$

An *ObjectProperty* in OWL is a relation between concepts in FOKI framework. Therefore, a particular relation must belong to the set of all relations between concepts along with a set of instances connected by such relation that connects concepts they belong to

| 21 | *ObjectPropertyDomain* | ⇔ | $(c_1, x) \in r_1$ where: $c_1, x \in C, r_1 \in R^C$ for any x |

ObjectPropertyDomain is used to describe a domain of an *ObjectProperty* (informally speaking, its "left side"). In FOKI framework it can be defined as a restriction put on the first element of a pair of concepts included in a concept relation. Please note that this rule can only be applied in conjunction with the subsequent Rule 20 which addresses "right side" of a relation

| 22 | *ObjectPropertyRange* | ⇔ | $(x, c_1) \in r_1$ where: $c_1, x \in C, r_1 \in R^C$ for any x |

ObjectPropertyDomain is used to describe a range of an *ObjectProperty* (informally speaking, its "right side"). In FOKI framework it can be defined as a restriction put on the second element of a pair of concepts included in a concept relation

| 23 | *ObjectSomeValuesFrom* | – | – |
| 24 | *SubClassOf* | ⇔ | $(c_1, c_2) \in H$ where: $c_1, c_2 \in C$ |

SubClassOf is used in OWL to define a taxonomy of classes. In the FOKI framework it is defined as a membership of a pair of concepts in the set H describing their hierarchy

| 25 | *SubObjectPropertyOf* | ⇐ | $r_1 \Leftarrow r_2$ where: $r_1, r_2 \in R^C$ |

SubObjectPropertyOf in OWL defines a hierarchy of *ObjectProperties*, in order to express that some *ObjectProperty* is less specific than the other. In the FOKI framework such issue is inferable using relations semantics from Eq. 8

| 26 | *SymmetricProperty* | ⇐ | $\forall (c_1, c_2) \in r_1^C \exists (c_2, c_1) \in r_1^C \wedge$
$\forall (i_1, i_2) \in r_1^I \exists (i_2, i_1) \in r_1^I;$ where $c_1, c_2 \in C,$
$r_1^C \in R^C, r_1^I \in R^I, i_1 \in c_1, i_2 \in c_2$ |

A *SymmetricProperty* in OWL is a restriction that can be used to force that an *ObjectProperty* must be symmetric in terms of connected classes. In the FOKI framework it is easily definable as a symmetry restriction put on a set describing a particular relation on a concept level, and on a its complementary set of instances' relations

5 Example

Defined translation rules from OWL-RL to FOKI are illustrated with a simple example. Main concepts, their hierarchy as well as relationships are represented by a class diagram – see Fig. 1.

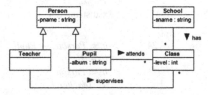

Fig. 2. Graphical representation of an exemplary ontology

Table 2. OWL-RL to FOKI transformation example

Ontology axiom	Rule#	FOKI element	Comments
Declaration(Class(:Class)), Declaration(Class(:Person)), Declaration(Class(:Pupil)), Declaration(Class(:School)), Declaration(Class(:Teacher))	1	$c = (Class,\ \emptyset, \emptyset, \emptyset)$ $c = (Person,\ \emptyset, \emptyset, \emptyset)$ $c = (Pupil,\ \emptyset, \emptyset, \emptyset)$ $c = (School,\ \emptyset, \emptyset, \emptyset)$ $c = (Teacher,\ \emptyset, \emptyset, \emptyset)$	Sets of attributes, their domains and instances are initially empty
Declaration(DataProperty(:album)) Declaration(DataProperty(:level)) Declaration(DataProperty(:pname)) Declaration(DataProperty(:sname))	4	$\{album\} \in A$ $\{level\} \in A$ $\{pname \in A$ $\{sname\} \in A$	Attributes *album, level, pname, sname* belong to the set of all attributes A
Declaration(ObjectProperty(:attends)) Declaration(ObjectProperty(:has)) Declaration(ObjectProperty(:supervises))	20	$attends \in R^C$ $has \in R^C$ $supervises \in R^C$	Relations *attend, has, supervises* belong to the finite set of relations between concepts
ObjectPropertyDomain(:attends :Pupil) ObjectPropertyDomain(:has :School) ObjectPropertyDomain(:supervises :Teacher)	21	$attends \subseteq Pupil \times$ $\_attends \in R^C$ $has \subseteq School \times \_has$ $\in R^C$ $supervises \subseteq Teacher$ $\times \_supervises \in R^C$	Definition of the *attends/has/supervises* relationship with *Pupil/School/Teacher* domain on the left
ObjectPropertyRange(:attends :Class) ObjectPropertyRange(:has :Class) ObjectPropertyRange(:supervises :Class)	22	$attends \subseteq \_ \times Class$ $attends \in R^C$ $has \subseteq \_ \times Class\ has$ $\in R^C$ $supervises \subseteq \_ \times$ $Class\ supervises$ $\in R^C$	Definition of the *attend/has/supervises* relationship with *Class* domain on the right
DataPropertyDomain(:album :Pupil) DataPropertyDomain(:level :Class) DataPropertyDomain(:pname :Person) DataPropertyDomain(:sname :School)	6	$\{album\} \in A^{Pupil}$ $\{level\} \in A^{Class}$ $\{pname\} \in A^{Person}$ $\{sname\} \in A^{School}$	Attribute *album, level, pname, sname* belong to attributes of the *Pupil, Class, Person, School* class, respectively
DataPropertyRange(:album sd:string) DataPropertyRange(:level xsd:int) DataPropertyRange(:pnamexsd: string) DataPropertyRange(:sname xsd:string)	7	$V_{album} =$ $xsd\!:\!string \in V$ $V_{level} = xsd\!:\!int \in V$ $V_{pname} =$ $xsd\!:\!string \in V$ $V_{sname} =$ $xsd\!:\!string \in V$	xsd:string is the range for *album, pname, sname* attributes and xsd:int is the range for *level* attribute, where xsd:string and xsd:int denote a set of all possible strings or integers, respectively
SubClassOf(:Pupil :Person) SubClassOf(:Teacher :Person)	24	$h = Person \leftarrow Pupil$ $h \in H$ $h = Person$ $\leftarrow Teacher$ $h \in H$	Generalization relationship is defined

The ontology was prepared in Protégé tool and saved in OWL functional syntax. In that tool all properties have to inherit either from *topObjectProperty* or from *topDataProperty* what is not reflected in FOKI. Therefore, these axioms were skipped in the translation process.

As FOKI does not have a concrete syntax defined, an informal notation is used to represent results of transformation rules (see Table 2).

6 Summary

The W3C Web Ontology Language (OWL) is a Semantic Web language designed to represent an ontology. It is the most popular syntax to express a rich and complex knowledge about objects taken from the real world and relations between them. It has proved so useful, that in 2004 the Ontology Alignment Evaluation Initiative (OAEI) started coordinating an international initiative to evaluate, compare and improve the tools for ontology mapping and alignment. In order to do so, OAEI provides standardized benchmark datasets in OWL syntax.

In our previous work [13], we proposed a mathematical model of an ontology which was the basis for a Framework for Ontological Knowledge Integration (FOKI) [10–12]. However, for reliable verification of FOKI we should conduct experiments using the aforementioned, well-known tests datasets provided by OAEI in OWL standard. For this purpose, we need a transformation of OWL2 into FOKI and vice-versa. However, as discussed in the paper, on the current state of development it is impossible to provide a set of bi-directional rules that would fully cover both FOKI and OWL.

Therefore, in this work we identified which elements of OWL-RL are translatable into FOKI constructs and vice-versa. For elements that can be transformed, we proposed rules which allow migrating them into the FOKI framework. The downside of our approach is allowing a situation where the transformation from FOKI to OWL may result in OWL axioms that were not indented. For example, a rule concerning *DisjointObjectProperties* is included when concepts' relations are mutually exclusive. However, for a particular FOKI ontology, such a situation may be only accidental. That way, the OWL axiom may be too restrictive.

In the future, we plan to extend the FOKI framework with proper formal tools that would allow a full, bidirectional transformation. That would form a basis for tool development for ontology integration using FOKI.

References

1. Afzal, H., Waqas, M., Naz, T.: OWLMap: fully automatic mapping of ontology into relational database schema. Int. J. Adv. Comput. Sci. Appl. (IJACSA) **7**(11), 7–15 (2017)
2. Andon, P.I., Reznichenko, V.A., Chistyakova, I.S.: Mapping of description logic to the relational data model. Cybern. Syst. Anal. **53**(6), 963–977 (2017)
3. Athanasiadis, I.N., Villa, F., Rizzoli, A.E.: Ontologies, JavaBeans and relational databases for enabling semantic programming. In: 31st Annual International Computer Software and Applications Conference (COMPSAC 2007), Beijing, China (2007)
4. Hajjamy, U.E., Alaoui, K., Alaoui, L., Bahaj, M.: Mapping UML to OWL ontology. J. Theor. Appl. Inf. Technol. **90**(1), 126–143 (2016)

5. Hnatkowska, B.: Automatic SUMO to UML translation. e-Informatica Softw. Eng. J. **10**(1), 51–67 (2016)
6. Hnatkowska, B., Woroniecki, P.: Transformation of OWL2 property axioms to Groovy. In: Tjoa, A.M., Bellatreche, L., Biffl, S., van Leeuwen, J., Wiedermann, J. (eds.) SOFSEM 2018. LNCS, vol. 10706, pp. 269–282. Springer, Cham (2018). https://doi.org/10.1007/978-3-319-73117-9_19
7. Hnatkowska, B., Kozierkiewicz, A., Pietranik, M.: Semi-automatic definition of attributes' semantics for the purpose of ontology integration. Special Issue on Breakthroughs on Cross-Cutting Data Management, Data Analytics and Applied Data Science Information Systems Frontiers (submitted)
8. Karpovic, J., Nemuraite, L., Stankeviciene, M.: Requirements for semantic business vocabularies and rules for transforming them into consistent OWL2 ontologies. In: Skersys, T., Butleris, R., Butkiene, R. (eds.) ICIST 2012. CCIS, vol. 319, pp. 420–435. Springer, Heidelberg (2012). https://doi.org/10.1007/978-3-642-33308-8_35
9. Kendall, E., Linehan, M.H.: Mapping SBVR to OWL2, Technical report, IBM Research report, RC25363 (2013)
10. Kozierkiewicz, A., Pietranik, M.: The knowledge increase estimation framework for ontology integration on the concept level. J. Intell. Fuzzy Syst. **32**(2), 1161–1172 (2017)
11. Kozierkiewicz-Hetmańska, A., Pietranik, M., Hnatkowska, B.: The knowledge increase estimation framework for ontology integration on the instance level. In: Nguyen, N.T., Tojo, S., Nguyen, L.M., Trawiński, B. (eds.) ACIIDS 2017. LNCS (LNAI), vol. 10191, pp. 3–12. Springer, Cham (2017). https://doi.org/10.1007/978-3-319-54472-4_1
12. Kozierkiewicz-Hetmanska, A., Pietranik, M.: The knowledge increase estimation framework for integration of ontology instances' relations. In: Lupeikiene, A., Vasilecas, O., Dzemyda, G. (eds.) DB&IS 2018. CCIS, vol. 838, pp. 172–186. Springer, Cham (2018). https://doi.org/10.1007/978-3-319-97571-9_15
13. Pietranik, M., Nguyen, N.T.: A multi-attribute based framework for ontology aligning. Neurocomputing **146**, 276–290 (2014). https://doi.org/10.1016/j.neucom.2014.03.067
14. Jiménez-Ruiz, E., et al.: BOOTOX: practical mapping of RDBs to OWL 2. In: Arenas, M., et al. (eds.) ISWC 2015. LNCS, vol. 9367, pp. 113–132. Springer, Cham (2015). https://doi.org/10.1007/978-3-319-25010-6_7
15. Sadowska, M., Huzar, Z.: Representation of UML class diagrams in OWL 2 on the background of domain ontologies. e-Informatica Softw. Eng. J. **13**(1), 63–103 (2019)
16. Tirmizi, S.H., et al.: Mapping between the OBO and OWL ontology languages. J. Biomed. Semant. **2**(Suppl 1), S3 (2011). https://doi.org/10.1186/2041-1480-2-S1-S3
17. Xu, Z., Ni, Y., He, W., et al.: Automatic extraction of OWL ontologies from UML class diagrams: a semantics-preserving approach. World Wide Web **15**, 517 (2012). https://doi.org/10.1007/s11280-011-0147-z
18. Zedlitz, J., Jörke, J., Luttenberger, N.: From UML to OWL 2. In: Lukose, D., Ahmad, A.R., Suliman, A. (eds.) KTW 2011. CCIS, vol. 295, pp. 154–163. Springer, Heidelberg (2012). https://doi.org/10.1007/978-3-642-32826-8_16
19. http://www.program-transformation.org/Transform/ProgramTransformation. Accessed 20 Oct 2019
20. https://www.w3.org/TR/2004/REC-owl-semantics-20040210/syntax.html. Accessed 20 Oct 2019
21. https://www.w3.org/TR/owl2-manchester-syntax/. Accessed 20 Oct 2019
22. https://www.w3.org/TR/2012/REC-owl2-mapping-to-rdf-20121211/. Accessed 20 Oct 2019

Natural Language Processing

Cached Embedding with Random Selection: Optimization Technique to Improve Training Speed of Character-Aware Embedding

Yaofei Yang[2], Hua-Ping Zhang[1(✉)], Linfang Wu[3], Xin Liu[4],
and Yangsen Zhang[2]

[1] Beijing Institute of Technology, Beijing, China
kevinzhang@bit.edu.cn
[2] Beijing Information Science and Technology University, Beijing, China
yangyaofei@gmail.com, zhangyangsen@bistu.edu.cn
[3] Hebei University of Science and Technology, Shijiazhuang, China
linfangwu0112@163.com
[4] Beijing Institute of Information Technology, Beijing, China
jfz97@163.com

Abstract. Embedding is widely used in most natural language processing. e.g., neural machine translation, text classification, text abstraction and sentiment analysis etc. Word-based embedding is faster and character-based embedding performs better. In this paper, we explore a way to combine these two embeddings to bridge the gap between word-based and character-based embedding in speed and performance. In the experiments and analysis of Hybrid Embedding, we found it's difficult to make these two different embeddings generate the same embedding vector, but we still obtain a comparable result. According to the results of analysis, we explore a form of character-based embedding called Cached Embedding that can achieve almost the same performance and reduce the extra training time by almost half compared to character-based embedding.

Keywords: Cached Embedding · Word embedding · Char-aware embedding · Time reduction · Training speed · Linguist · Natural language processing

1 Introduction and Background

In natural language processing (NLP) task, projecting text from a sparsity one-hot space to a smaller density space, which is called embedding, is required. There are two main reasons that, in most tasks, words would be treated as the one-hot item and not a character. First, vanilla word-based embedding is a matrix that can quickly achieve the embedding and is easy to train. Second, processing at word level can result in a shorter sentence length, which saves computation time and computer memory.

© Springer Nature Switzerland AG 2020
N. T. Nguyen et al. (Eds.): ACIIDS 2020, LNAI 12033, pp. 51–62, 2020.
https://doi.org/10.1007/978-3-030-41964-6_5

However, vanilla word-based embedding has its disadvantages. The first is the vocabulary problem. It is not a good idea to maintain a fixed-size vocabulary and some words must be *out-of-vocabulary* words. Second, there must be meanings at the character level. Although we can use a larger corpus to train the model to obtain a proper embedding meaning, character-based embedding is a better way to obtain the meaning inside the word.

In neural machine translation, Chung et al. [2] showed that a character-based model can improve the 2 BLEU [10] score. Cherry et al. [1] showed that a pure character-based model is roughly 8 times slower than the *Byte-Pair-Encoding* (BPE) [11] model, which is subword-based. To obtain the character-level meaning, Sennrich et al. [11] and Kudo [8] broke words apart to expose the charac ter-level meaning. In another approach, Kim et al. [7] used a *convolutional neural network* (CNN) to extract character-level n-gram features and encode a word representation that they called character-aware embedding. In our experiments, this character-aware embedding out-performed vanilla word embedding by +10 perplexity points in the Wikitext-2 dataset, but doubled the training time.

In this paper, we seek a way to reduce the training time of this model by combining vanilla word-based and character-aware embedding using random selection to devise a comparable model in both performance and training speed. We employ a sample model that combines character-aware and vanilla word embedding using random selection in training time. Because the output of character-aware embedding is a word representation that is same as in word embedding, this model is called *Hybrid Embedding*.

Upon further analyzing the Hybrid Embedding model, we found that it does not allow character and vanilla word embedding achieve the same or similar result with the same word. Also interesting is that the Hybrid Embedding model achieves the same performance when using only vanilla word embedding, only character-aware embedding, or both.

According to these analyses, we employed a new embedding mode called *Cached Embedding* that allowed vanilla word embedding be a cache of character-aware embedding and that used the cache randomly. This Cached Embedding realized better performance and consumed less time while achieving nearly the same performance compared with character-aware embedding, while reducing the extra training time by half compared with character-aware embedding.

2   LSTM Language Model

Since it is difficult to evaluate embedding itself, in this paper, we apply embeddings to a vanilla *Long Short Term Memory* (LSTM) [5] language model. The perplexity of language model using different embeddings can be a fair way to compare their performance.

2.1   Language Model

We chose a classic two-layer LSTM language model following Zaremba et al. [13]. The structure of this model is shown in Fig. 1.

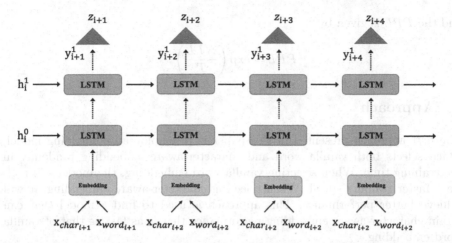

Fig. 1. Structure of LSTM language model. h_i^j is the hidden state of the jth layer and i timestamp, which will be the input of the jth layer and $i+1$ timestamp. The embedding model can be any embedding model that outputs a word representation, and it can take input at the character level, word level, or both. The dotted lines are applied with dropout and the solid lines are not.

The vanilla LSTM language model has three parts: an embedding Layer, stacked LSTM layer, and fully connection project layer.

The inputs of the model are character-level data, $X_{char} = [x_{c_1}, x_{c_2}, \ldots, x_{c_n}]$, and word-level data, $X_{word} = [x_{w_1}, x_{w_2}, \ldots, x_{w_n}]$, where n is the sentence length, x_{c_i} a sequence of characters, and x_{w_i} a word index number in the word vocabulary. $x_{c_i} = [c_1, c_2 \ldots, c_m]$, where c_j is a character index number in the character vocabulary. The embedding layer uses one of these two data or both depending on the type of embedding.

The process in one timestamp of this language model can be represented as follows:

$$e_i = Embed(x_{c_i}, x_{w_i})$$
$$y_i^1 = LSTMs(h_{i-1}^0, h_{i_1}^1, e_i)$$
$$z_i = Proj(y_i^1).$$

The *Embed* is an embedding model, *LSTMs* a two-layers stacked LSTM model, and *Proj* a fully connected network to predict the next word.

2.2   Perplexity

Perplexity (PPL) is a standard evaluation criterion in language modeling, and is defined under the negative log-likelihood (NLL) of a sequence $[w_1, \ldots, w_n]$:

$$NLL = -\sum_{i=1}^{n} \log \Pr(w_i | w_{1:i-1}),$$

and the PPL is given by

$$PPL = \exp\left(\frac{NLL}{T}\right).$$

3 Approach

The first approach presented in this paper is to employ an embedding model, which selects both vanilla word and character-aware embedding randomly in the training time. When selecting vanilla word embedding, the model will realize a faster training speed; When selecting character-aware embedding, it will achieve better performance. This approach is used to find a model that can attain character-aware embedding performance that is as fast as that of vanilla word embedding.

3.1 Character-Aware Embedding

Character-aware embedding is a special character-level embedding that encodes a character-based word sequence to a one-word representation. The output of character-aware embedding has the same form as word embedding, so it is easy to combine this embedding with other word embeddings. The state-of-the-art character-aware embedding is Kim et al. [7]'s character CNN embedding, which uses a CNN to acquire character-level n-gram features.

The details of this model follow.

Let $\mathcal{C}$ be the set of all of the characters, d be the dimensionality of character embeddings and $\mathbf{Q} \in \mathbb{R}^{d \times |C|}$ be the character embedding matrix. Supposing the word sequence is $x_c = [c_1, c_2, \ldots, c_n]$, where $c_i \in \mathcal{C}$ and n is the length of word w, and the character representation C_i of c_i is the ith column of $\mathbf{Q}$, then the process of the character encoding is as follows:

$$W = Highways(F, layers)$$
$$F = concat([f_1, f_2, \ldots, f_k])$$
$$f_i = pooling_{max}(h_i)$$
$$h_i = tanh(cnn_{1d}([C_1, C_2, \ldots, C_n], kernel_i, channel_i) + b_i)$$

where cnn_{1d} is a one-dimensional CNN in the length direction, of which the kernel size is $kernel_i$ and the output channel is $channel_i$. $pooling_{max}$ selects the maximum number in the last dimension. There are k different CNNs that can be employed in this model. $concat$ means concatenate a list of matrixes in the last dimension and it will concatenate k different CNN outputs. Finally, this is achieved through $Highways$, which is a multi-layer highway network [12], and $layers$ is the number of layers,

The performance of this model compared with a language model is shown in Table 1, and the results indicate that this model out-performs vanilla embedding by +10 points in PPL, but almost doubles the training time.

3.2 Hybrid Embedding

Model Details. The details of the first approach, Hybrid Embedding, are very straightforward and shown in Fig. 2.

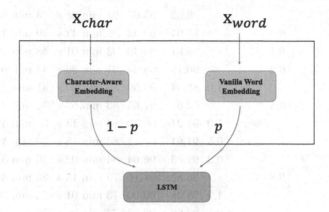

Fig. 2. Hybrid Embedding

Two different embeddings are used in this Hybrid Embedding, namely, vanilla word embedding and Kim et al. [7]'s CNN embedding. In training time, this Hybrid Embedding model will select one of these two embeddings randomly. The selected embedding will obtain the input and send its output to the LSTM. The probability of selecting vanilla embedding p is denoted **word-rate**.

Selection with Decay. In Zhang et al. [14]'s paper, it was shown to be difficult to train the model using a random sampling in the beginning of training, and so in that paper a method called "Sampling with Decay" was employed to well train the model. Inspired by Sampling with Decay, we employed a similar method called "Selection with Decay". The equation used to calculate word-rate, for which e is the training epoch and λ a hyper-parameter, is

$$rate_{word} = rate_{final} * \frac{e^{\lambda e} - 1}{e^{\lambda e}}.$$

Following this equation, we can obtain the word-rate decay from 1 to $rate_{final}$, and λ will control the speed of decay. The word-rate will be zero at the beginning of the training, which is good for character-aware model training. As the training goes on, the word-rate will increase, which will make selection more frequent and accelerate training.

Table 1. Hybrid Embedding results

Model	Selection[a]		PPL		Time[b]	
	Word-rate	λ	Test	Valid	Total	Plus
Word-LSTM			99.69	105.60	47 min 31 s	
Kim-CNN			88.62	93.67	91 min 28 s	43 min 57 s
Hybrid	0.0		89.01	94.44	96 min 43 s	49 min 12 s
	0.1		89.61	94.40	93 min 01 s	48 min 18 s
	0.2		90.13	95.24	91 min 36 s	44 min 05 s
		1.0	89.94	94.69	89 min 20 s	41 min 49 s
	0.5		92.61	97.65	83 min 53 s	36 min 22 s
		1.0	91.21	96.50	84 min 13 s	36 min 42 s
		0.5	91.61	97.13	84 min 15 s	36 min 44 s
		0.3	91.03	96.64	84 min 02 s	36 min 31 s
	0.8		96.80	101.10	73 min 15 s	25 min 44 s
		1.0	94.95	100.00	73 min 01 s	25 min 30 s
		0.5	93.98	99.63	74 min 00 s	26 min 29 s
		0.3	93.54	98.42	75 min 10 s	27 min 39 s
		0.1	91.76	96.62	77 min 17 s	29 min 48 s
	0.9		98.05	102.90	70 min 21 s	22 min 50 s
		1.0	97.96	103.50	70 min 25 s	22 min 54 s
		0.5	95.77	100.90	72 min 55 s	25 min 24 s
		0.3	94.87	99.66	73 min 54 s	26 min 23 s
		0.1	91.87	96.67	76 min 34 s	29 min 03 s

[a] These two parameters are described in Sect. 3.2.
[b] This is the time for training the model; the total is the total time, i.e., 30 epochs for every model. Plus sign denotes the extra time compared with the Word-LSTM.

Results and Analysis. We trained the Hybrid Embedding model with the vanilla LSTM language model described in Sect. 2, and the results for Wikitext-2 datasets are shown in Table 1, illustrating out focus on both training time and performance. In Table 1, the training time of the character-aware model in 30 epochs is **44** min longer than that of the vanilla word model. The proposed Hybrid Embedding model reduces this time to **29** min (**−34%**), a loss of only **3.6%** in performance compared with the Kim-CNN.

In another dimension, the method of Selection with Decay works well, which can increase performance from **91.87** to **98.05** while only using an extra **6** min at a word-rate of 0.9. This performance is better than that achieved with a word-rate of 0.5, which achieved **92.61** PPL and took nearly **84** min without using Selection with Decay.

Another purpose of this model is to hybridize both vanilla word and character-aware embedding. Therefore, we compared the outputs of these two embeddings for the same word by calculating the cosine similarity and length. Furthermore, we used the embedding in a vanilla word language model to train the character-aware embedding. The results are shown in Table 2.

Table 2. Analysis of Hybrid Embedding

Method	Metrics	Means[c]	Variance[d]
Hybrid-0.5	Cosine[a]	0.4802	0.0827
	Divide[b]	1.86	1.124
Pre-train	Cosine	0.8464	0.0820
	Divide	0.8925	0.0412

[a] Cosine denotes cosine similarity.
[b] "Divide" is the L2-norm of the word representation of the character-aware division of the vanilla word.
[c] The mean is the mean of all of the words in the dataset.
[d] The variance of all of the words in the dataset.

The results in Table 2 show that in the Hybrid Embedding model the vanilla word and character-aware embedding have different lengths and different directions. It is difficult to allow them have the same embedding, even if using pre-trained vanilla word embedding to train the character-aware embedding. This proves that the embeddings are in difference sub-spaces. Although we cannot combine these two embeddings, we still obtain a comparable result.

Further analysis is necessary. In the previous experiments, we used full vanilla word embedding in Hybrid Embedding to obtain inferences unless the word is an out-of-vocabulary word. In further experiments, we tested these models in three ways: full word, fully character-aware, and a mixture of these two. The results are presented in Table 3. We found that these three methods of obtaining inferences achieve almost the same perplexities, showing that the LSTM language model learned two different word representations. Moreover, the embedding model be trained well not using all of the datasets. Based on the analyses of Hybrid Embedding, we employed a new method, Cached Embedding, which is described in the next Sect. 3.3.

3.3   Cached Embedding

Model Details. From the analysis in Sect. 3.2, two facts emerge. First, it is difficult to mix vanilla word and character-aware embedding, and, second, it is unnecessary to use all of the datasets to train the embedding. We employ a new embedding model called *Cached Embedding* based on these conditions.

Table 3. Different inference methods

Method	0.0	0.1	0.2	0.5	0.8	0.9
Char	94.40	94.08	95.25	97.58	100.1	102.7
Mixture	94.41	94.14	95.16	97.54	100.1	102.8
Word	94.40	94.13	95.18	97.66	100.1	102.8

The structure of Cached Embedding is shown in Fig. 3. Similar to Hybrid Embedding, there are two embeddings in Cached Embedding that are selected by probability p, which is called the **cache-rate** during the training period. The difference is that the vanilla word embedding becomes the cache of the character-aware embedding, which is called the **embedding cache** during the training period. When character-aware embedding is selected, the data will update the embedding cache . The rest of the unselected data will quickly obtain the embedding representations from the embedding cache. The cache-rate controls the proportion of unselected data.

We trained the embedding with only part of the data because all the data are not necessary for training, and, in addition, we cannot hybridize it simultaneously with training. Treating vanilla word embedding as a cache is a simple way to harmlessly combine these two embeddings.

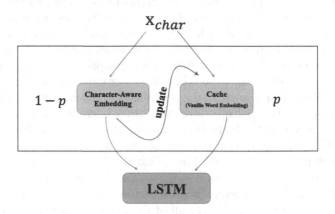

Fig. 3. Cached Embedding

Results. We trained Cached Embedding using the aforementioned LSTM language model, and the results are shown in Table 4, from which it can be seen that the proposed Cached Embedding model reduces the time to **22** min (**−49%**) while losing only **1.5%** in performance. In addition, compare with Hybrid Embedding with the same hyper-parameters, the time consumption and perplexities are better. To be specific, a cache-rate of 0.9 with λ **0.1** obtains a total time of **70** min and a PPL of **89.85** and a word-rate of **0.9** with λ **0.1** obtains **76** min and a PPL of **91.87**. The method Selection with Decay also works well.

Table 4. Cached Embedding results

Model	Selection[a]		PPL		Time[b]	
	Cache-rate	λ	Test	Valid	Total	Plus
Word-LSTM			99.69	105.60	47 min 31 s	
Kim-CNN			88.62	93.67	91 min 28 s	43 min 57 s
Cached	0.0		89.01	94.44	93 min 37 s	46 min 06 s
	0.2		89.37	94.14	84 min 50 s	37 min 19 s
	0.5		90.65	95.68	75 min 07 s	27 min 36 s
		1.0	90.64	95.48	75 min 32 s	28 min 01 s
		0.5	89.94	94.79	76 min 22 s	28 min 51 s
		0.3	89.93	94.50	76 min 38 s	29 min 07 s
	0.8		94.57	99.96	65 min 12 s	17 min 41 s
		1.0	91.67	96.78	66 min 30 s	18 min 59 s
		0.5	91.64	96.63	67 min 00 s	19 min 29 s
		0.3	90.79	95.97	67 min 39 s	20 min 08 s
		0.1	89.68	94.67	71 min 25 s	23 min 54 s
	0.9	1.0	93.93	99.11	64 min 46 s	17 min 15 s
		0.5	92.55	97.99	65 min 49 s	18 min 18 s
		0.3	91.97	96.46	65 min 30 s	17 min 59 s
		0.1	89.85	95.19	70 min 12 s	22 min 41 s
Hybrid	0.0		89.01	94.44	96 min 43 s	49 min 12 s
	0.5	0.3	91.03	96.64	84 min 02 s	36 min 31 s
	0.8	0.1	91.76	96.62	77 min 17 s	29 min 46 s
	0.9	0.1	91.87	96.67	76 min 34 s	29 min 03 s

[a] These two parameters are described in Sect. 3.3.
[b] The time for training the model; the total is the total time, i.e., 30 epochs for every model. Plus sign denotes the extra time compared with the Word-LSTM

4 Experimental Setup

Model Setup. For the character-aware embedding, we followed Kim et al. [7]'s large hyper-parameter settings; that is, the character embedding size is 15, and seven different CNNs are used following the equation $f = [\min\{200, 50 \cdot k\}]$, where k is kernel size and f is filter size. For the highway network, the size of the layer is 2 and the activation function is $ReLU$ [9].

Because the output of character-aware embedding is 1,300, vanilla embedding uses the same dimension size. The dimension of a two-layer LSTM is set to 512, which is slightly different from that used in Kim et al. [7].

Optimization. The model was trained by truncated backpropagation through time [3] and the backpropagation was truncated to 35 time steps. We let stochastic gradient descent (SGD) [6] be the optimizer and set the initial learning rate to 20 and halved it if the perplexity did not decrease by more than 1.0 on the validation set after an epoch. We used a batch size of 20 and trained all of the models with 30 epochs. The parameters in this model were randomly initialized over a uniform distribution with support in the interval $[-0.05, 0.05]$.

For regularization, the dropout [4] was used in the model with probability 0.5. Following Zaremba et al. [13]'s dropout recipe, the dropout was not used in the time direction of the LSTM. In case of gradient explosion, the gradient was clipped to 0.25 with a L_2-norm if it exceeded 5.

Datasets. The Wikitext-2 dataset was used to train, validate, and test. Instead of using a proper processed version, we used the raw dataset to meet our needs at the character level. This raw dataset was processed to two parallel data at both word and character levels. At word level, the vocabulary size was set to 33,276 following the common word version of the Wikitext-2 dataset. At character level, all of the characters were in the vocabulary.

Reproducible Things. For all of the experiments reported in this paper, we used a machine with one NVIDIA 1080TI graphical processing unit (GPU) and one AMD Threadripper 1900X Central Processing Unit (CPU) with 16 GB of memory. We also tested the proposed model using another machine with eight NVIDIA 1080TI GPUs. The time is slightly longer with eight GPUs that with one GPU, and the reason may be the number of PIC-E channels. The one-GPU machine fully granted 16 PCI-E channels for its single GPU, but the eight-GPU machine does not have a sufficient number of PCI-E channels for each GPU.

These machine details are relevant because all of the experiments were very sensitive to machine performance. Although the time consumption will be proportionable, to be reproducible for every number this information is important for interpreting the results of this paper and for other interested researchers.

5 Conclusions

The character-level model in natural language processing exhibits better performance in most cases, but time consumption is one stumbling block to using it to compare results with word and subword levels. Our propose approach is to find a way to reduce the computational complexity of the character-level model with an acceptable performance loss.

In this paper, we began with a character-aware embedding, which is easier to combine with vanilla word-level embedding. To reduce the training time and combine the vanilla word and character-aware embedding, we proposed a Hybrid Embedding with random selection. While we do achieve am increase in training speed and acceptable performance loss, the combination of these two embeddings fails according to subsequent analysis. In addition, Cached Embedding with a random selection model was employed and further analysis conducted.

The experimental results show that the Cached Embedding model can achieve better performance while consuming less time. In the meantime, the method denoted "Selection with Decay" can significantly enhance the performance.

Our proposed model, i.e., **Cached Embedding**, and our preferred training method, **Selection with Decay**, can be applied in any word-based system without any further modification. With a simple replacement, the applied mode can achieve almost the same performance and reduce by half the extra time required compared with pure character-aware embedding.

Acknowledgements. This work was supported by National Science Foundation of China (Grant No. 61772075), National Science Foundation of China (Grant No. 61772081), Scientific Research Project of Beijing Educational Committee (Grant No. KM201711232022), Beijing Municipal Education Committee (Grant No. SZ20171123228), Beijing Institute of Computer Technology and Application (Grant by Extensible Knowledge Graph Construction Technique Project).

References

1. Cherry, C.A., Foster, G., Bapna, A., Firat, O., Macherey, W.: Revisiting character-based neural machine translation with capacity and compression. In: Empirical Methods in Natural Language Processing (2018)
2. Chung, J., Cho, K., Bengio, Y.: A character-level decoder without explicit segmentation for neural machine translation. In: Proceedings of the 54th Annual Meeting of the Association for Computational Linguistics, Berlin, Germany (volume 1: Long Papers), pp. 1693–1703. Association for Computational Linguistics (2016)
3. Graves, A.: Generating sequences with recurrent neural networks. arXiv:1308.0850 [cs], August 2013
4. Hinton, G.E., Srivastava, N., Krizhevsky, A., Sutskever, I., Salakhutdinov, R.R.: Improving neural networks by preventing co-adaptation of feature detectors. arXiv:1207.0580 [cs]. July 2012
5. Hochreiter, S., Schmidhuber, J.: Long short-term memory. Neural Comput. **9**(8), 1735–1780 (1997)
6. Kiefer, J., Wolfowitz, J.: Stochastic estimation of the maximum of a regression function. Ann. Math. Stat. **23**(3), 462–466 (1952)
7. Kim, Y., Jernite, Y., Sontag, D., Rush, A.M.: Character-aware neural language models. In: Proceedings of the Thirtieth AAAI Conference on Artificial Intelligence, AAAI 2016, pp. 2741–2749. AAAI Press (2016)
8. Kudo, T.: Subword regularization: improving neural network translation models with multiple subword candidates. In: Proceedings of the 56th Annual Meeting of the Association for Computational Linguistics, Melbourne, Australia (vol. 1: Long Papers), pp. 66–75. Association for Computational Linguistics (2018)
9. Nair, V., Hinton, G.E.: Rectified linear units improve restricted Boltzmann machines. In: Proceedings of the 27th International Conference on International Conference on Machine Learning, ICML 2010, pp. 807–814. Omnipress, USA (2010)
10. Papineni, K., Roukos, S., Ward, T., Zhu, W.J.: BLEU: a method for automatic evaluation of machine translation. In: Proceedings of the 40th Annual Meeting of the Association for Computational Linguistics (2002)

11. Sennrich, R., Haddow, B., Birch, A.: Neural machine translation of rare words with subword units. In: Proceedings of the 54th Annual Meeting of the Association for Computational Linguistics, Berlin, Germany (vol. 1: Long Papers), pp. 1715–1725. Association for Computational Linguistics, August 2016
12. Srivastava, R.K., Greff, K., Schmidhuber, J.: Highway networks. arXiv:1505.00387 [cs], May 2015
13. Zaremba, W., Sutskever, I., Vinyals, O.: Recurrent neural network regularization. arXiv:1409.2329 [cs], September 2014
14. Zhang, W., Feng, Y., Meng, F., You, D., Liu, Q.: Bridging the gap between training and inference for neural machine translation. In: Proceedings of the 57th Annual Meeting of the Association for Computational Linguistics, pp. 4334–4343, July 2019

A Comparative Study of Pretrained Language Models on Thai Social Text Categorization

Thanapapas Horsuwan[iD], Kasidis Kanwatchara[iD], Peerapon Vateekul$^{(\boxtimes)}$[iD], and Boonserm Kijsirikul$^{(\boxtimes)}$[iD]

Department of Computer Engineering, Faculty of Engineering,
Chulalongkorn University, Bangkok, Thailand
thanapapas.h@gmail.com, kanwatchara.k@gmail.com,
{peerapon.v,boonserm.k}@chula.ac.th

Abstract. The ever-growing volume of data of user-generated content on social media provides a nearly unlimited corpus of unlabeled data even in languages where resources are scarce. In this paper, we demonstrate that state-of-the-art results on two Thai social text categorization tasks can be realized by pretraining a language model on a large noisy Thai social media corpus of over 1.26 billion tokens and later fine-tuned on the downstream classification tasks. Due to the linguistically noisy and domain-specific nature of the content, our unique data preprocessing steps designed for Thai social media were utilized to ease the training comprehension of the model. We compared four modern language models: ULMFiT, ELMo with biLSTM, OpenAI GPT, and BERT. We systematically compared the models across different dimensions including speed of pretraining and fine-tuning, perplexity, downstream classification benchmarks, and performance in limited pretraining data.

Keywords: Language model · Pretraining · Thai social media · Comparative study · Data preprocessing

1 Introduction

Social networks are active platforms rich with a quickly accessible climate of opinion and community sentiment regarding various trending topics. The growth of the online lifestyle is observed by the bustling active communication on social media platforms. Opinion-oriented information gathering systems aim to extract insights on different topics, which have numerous applications from businesses to social sciences. Nevertheless, existing NLP researches on utilizing these abounding noisy user-generated content have been limited despite its potential value.

First introduced in [7], pretrained language models (LMs) have been a topic of interest in the NLP community. This interest has been coupled with works reporting state-of-the-art results on a diverse set of tasks in NLP. In light of the notable benefits of transfer learning, we chose to compare four renowned LMs: ULMFiT [9], ELMo with biLSTM [14], OpenAI GPT [16], and BERT [8].

© Springer Nature Switzerland AG 2020
N. T. Nguyen et al. (Eds.): ACIIDS 2020, LNAI 12033, pp. 63–75, 2020.
https://doi.org/10.1007/978-3-030-41964-6_6

To the best of our knowledge, our work is the first comparative study conducted on pretrained LMs in Thai language. Our LMs were trained in a three-stage process as per suggested in [9]: LM pretraining, LM fine-tuning, and classifier fine-tuning. The goal of unsupervised pretraining is to find a good initialization point to capture the various general meaningful aspects of a language. Befitting Thai language with resource scarcity, we expect that pretraining user-generated content would serve as a solid basis for transfer learning to downstream tasks.

Pantip is the largest Thai internet forum with a huge active community where a diverse range of topics are discussed. The variability of surplus examples from Pantip covers the basic linguistic syntax of Thai language while maintaining the colloquial and noisy nature of online user-generated content. In this paper, we investigate and compare the capability of each LM to capture the relevant features of a domain-specific language via pretraining copious unlabeled data from user-generated content.

The main contributions of this paper are the following:

- We developed unique data preprocessing techniques for Thai social media.
- We pretrained ULMFiT, ELMo, GPT, and BERT on a noisy Thai social media corpus much larger than the existing Thai Wikipedia Dump.
- We compared the language models across different dimensions including speed of pretraining and fine-tuning, perplexity, downstream classification benchmarks, and performance in limited pretraining data.
- Our pretrained models and code can be obtained upon request to the corresponding authors.

This paper is organized as follows. Our data preprocessing techniques are explained in Sect. 2 and the LMs used for pretraining are briefly described in Sect. 3. The datasets used in this paper are described in Sect. 4 and Sect. 5 explains our hyperparameters and evaluation metrics. The results are reported in Sect. 6 and finally concluded in Sect. 7.

2   Our Data Preprocessing for Thai Social Media

Data preprocessing is one of the most important phases in improving the learning comprehension of the LMs. If much irrelevant and redundant information introduces unwanted noise in the training corpus, it is difficult for the models to discover knowledge during the training phase. This is especially true for unfiltered data from user-generated content on social media, where it requires specific methods of data preprocessing unique to the domain.

The Thai webboard Pantip allows members to freely create threads as long as it conforms to a list of actively regulated etiquette. The colloquial nature of the data posted allows for huge amounts of noise to be introduced in the data, such as ASCII arts, language corruption ภาษาวิบัติ irregular spacing, misspelling, character repetition, and spams [10]. The unpredictable noise in the data substantially increases the vagueness of word boundary, which already is a problem in formal Thai language [4]. Additionally, Thai word segmentation is dependent on context. A famous example is the compound word ตากลม, which can be either split into [ตา][กลม] or [ตาก][ลม]. Both are grammatically correct when used within their corresponding context. To ease the impact of the issues, the data preprocessing approaches we employed are as follows:

1. **Length Filtering.** To select meaningful threads to the LM, threads with a title and with a body of more than 100 characters were selected.
2. **Language Filtering.** An n-gram-based text categorization library langdetect[1] was used to filter out the threads that are not labeled as Thai language.
3. **General Preprocessing before Tokenization.** Inspired by [3,9], the techniques include fixing HTML tags, removing duplicate spaces and newlines, removing empty brackets, and adding spaces around '/' and '#'. In addition, character order in Thai language may be typed in a different sequence but visually rendered in the same way. This is due to the fact that vowels, diphthongs, tonal marks, and diacritics may sometimes be physically located above or below the base glyph–allowing different sequential orders to appear visually equivalent. Thus, normalizing the character order is required for the machine to understand the seemingly similar tokens.
4. **Customized Preprocessing before Tokenization.** We also developed and customized techniques suitable for Thai social media. Last character repetition is a common behavior of Thai people analogous to prolonging the vowel sounds of a word in spoken language to emphasize certain emotions. We truncate the word and follow it by a special token. pyThaiNLP [15] adopted a similar technique but we implemented minor modifications of space addition following the token for better tokenization results. Likewise, a special token is used for word repetitions similar to [9] preprocessing technique, which at the time this technique has not been widely used in Thai language preprocessing. Since Thai is a language without word boundaries, our algorithm recognizes words as any character sequence of more than 2 characters with more than 2 repetitions of that sequence. All types of repetitions are truncated to 5 as it provides no higher emotional impact and to limit the vocabulary size.

In addition, we also propose 2 new preprocessing methods: a special token for any numeric strings and a special token for laughing expressions.

We replaced all strings related to numbers with a special token: general numbers, masked and unmasked phone numbers, Thai numbers, date and time, masked prices, and numbers of special forms. Although differentiating the numbers provide some semantic value, the sparsity of the information would most likely make these numbers tail out of vocabulary (OOV) tokens. We believe that this preprocessing method would allow the language models to more generally understand how numbers are used in text.

In an online environment, Thai people often express laughter in written language with an onomatopoeia, utilizing the repetition of '5' followed by an optional '+'. This is due to the fact that the Thai pronunciation of '5' is 'ha'. We replaced all tokens with more than 3 consecutive '5' and an optional '+' with a special laugh token. Although this may have a minor effect on actual

---

[1] github.com/fedelopez77/langdetect.

numbers, this onomatopoeia is very commonly used in Thai online context and it is important for the model to learn this special token. An example is provided in Table 1 for clarification.

Table 1. An example of our preprocessing method. [CREP] and [LAUGH] are special tokens used for character repetition and laughing respectively.

	Before	After
Thai	ฉันชอบมันมากกกก555555+	ฉันชอบมันมาก [CREP] 4 [LAUGH]
Translated	I like it a lotttt hahahahaha[a]	I like it a lot [CREP] 4 [LAUGH]

[a] '5' is pronounced 'ha' in Thai

5. **Tokenization.** We used the pyThaiNLP [15] default tokenizer, which is a dictionary-based Maximum Matching with Thai Character Cluster. However, we created our own aggregated dictionary for tokenization to improve the tokenization accuracy for colloquial user-generated content. The dictionary[2] is compiled from various sources of data, including general words, abbreviations, transliterations, named entities, and self-annotated Thai slangs and commonly used corrupted language. This includes word variants like ฮะ ฮาฟ ฮับ ฮัฟ ค้ราบ ค้าฟ ค้าบ ดับ ครัช which are all word variants of the suffix to indicate formality ครับ. The vocabulary is built from the most common 80k tokens.

6. **General Preprocessing after Tokenization.** Following [9] and [3], some general preprocessing techniques after tokenization were used. This includes ungrouping the emoji's from text, and to lowercase all English words.

7. **Spelling Correction.** In an effort to reduce the number of unnecessary tokens sprouting from incorrectly spelled words, we compiled a list of commonly misspelled word mappings aggregated from various sources. We corrected and standardized the vocabulary used. This is an important task due to the free and lax nature of the corpus, where a single word may be represented in different variants or misspelled and abbreviated into various tokens. Note that not all replacements can be made due to the collision of actual vocabularies and the limited comprehensiveness of the list.

3   Pretrained Language Models in Our Study

3.1   Universal Language Model Fine-Tuning (ULMFiT)

A single model architecture that is used for both LM pretraining and downstream fine-tuning was first introduced in ULMFiT [9]. This allows the weights learnt during pretraining to be reused instead of constructing a new task-specific

---

[2] The dictionary is referenced at our GitHub https://github.com/Knight-H/thai-lm.

model. Howard and Ruder suggested that LM overfits to small datasets and suffers catastrophic forgetting when directly fine-tuned to a classifier. Hence, the ULMFiT approach was proposed to attempt to effectively fine-tune the AWD-LSTM [11] model. ULMFiT is a 3-stage training method consisting of LM pretraining, LM fine-tuning, and classifier fine-tuning. They also proposed novel techniques such as discriminative fine-tuning, gradual unfreezing, and slanted triangular learning rates for stable fine-tuning.

3.2 Embeddings from Language Models (ELMo)

Traditional monolithic word embeddings such as word2vec [12] and GloVe [13] fails to model context-dependent meanings of a word. Hence, ELMo [14] produces contextualized word embeddings by utilizing a pretrained biLM as a fixed feature extractor and incorporate its embedding representation as features into another task-specific model for downstream tasks. The authors suggested that combining the internal states of the LSTM layers allows for rich contextualized word representations on top of the original context-independent word embeddings.

3.3 Generative Pretrained Transformer (GPT)

Sequential computation models used in sequence transduction problems [5,6,17] forbid parallelization in the training examples. The transformer [18] is the first transduction model based solely on self-attention to draw global dependencies between input and output, eliminating the use of recurrence and convolutions. OpenAI introduced GPT [16] by extending the idea to multi-layer transformer decoder for language modeling. Additionally, LM fine-tuning and classifier fine-tuning are done simultaneously by using LM as an auxiliary objective. The authors suggested that this improves the generalization of the supervised model and accelerates convergence.

3.4 Bidirectional Encoder Representations from Transformers (BERT)

ULMFiT [9] and GPT [16] use a unidirectional forward architecture while ELMo [14] uses a shallow concatenation of independently trained forward and backward LMs. With criticism on the standard unidirectional LMs as suboptimal by severely restricting the power of pretrained representations, BERT [8] was proposed as a multi-layer transformer encoder designed to pretrain deep bidirectional representations by jointly conditioning on both left and right context in all layers. Since the standard autoregressive LM pretraining method is not suitable for bidirectional contexts, BERT is trained on masked language modeling (MLM) and next sentence prediction (NSP) tasks. MLM masks 15% of the input sequences at random and the task is to predict those masked tokens, requiring more pretraining steps for the model to converge. The output of the special first token is used to compute a standard softmax for classification tasks.

4   Dataset

4.1   Pretraining Dataset

To collect our Thai social media corpus data, we extracted non-sensitive information from all threads from Pantip.com since 1st January 2013 up until 9th February 2019 using our implementation of the Scrapy Framework [2]. A total of $8,150,965$ threads were extracted. As discussed in Sect. 2, data preprocessing techniques are applied to the corpus. Length filtering and language filtering filtered down the threads to $5,524,831$ and $5,487,568$ respectively. After preprocessing, tokenization, and postprocessing the data, we divided our pretraining dataset into 3 parts: $5,087,568$ threads for training, $200,000$ threads for validation, and $200,000$ threads for testing. The train dataset, validation dataset, and test dataset has a total of $1,262,302,083$ tokens, $4,701,322$ tokens, and $4,588,245$ tokens respectively. By comparison, our pretrain dataset is more than 31 times larger than the Thai Wikipedia Dump with respect number of tokens, which is only on the order of 40M tokens for the training set.

4.2   Benchmarking Dataset

Two Thai social text classification tasks were chosen to benchmark the models for extrinsic model evaluation as shown in Table 2. Since both are originally Kaggle competitions, the Kaggle evaluation server will be used for benchmarking.

Wongnai Challenge: Rating Review Prediction. First initiated as a Kaggle competition, the Wongnai Challenge is to create a multi-class classification sentiment prediction model from textual reviews. As an emerging online platform in Thailand, Wongnai holds a large user base of over 2 million registered users with a surplus of user-written reviews accompanied by a rating score ranging from 1 to 5 stars. This is challenging due to the varying user standards, corresponding to shifting weighted importance of each sentiment in mixed reviews.

Wisesight Sentiment Analysis. The Wisesight Sentiment Analysis is a private Kaggle competition where the task is to perform a multi-class classification on 4 categories: positive, negative, neutral, and question. Wisesight, a social data analytics service provider, provides data from various social media sources with various topics on current internet trends. It should be noted that the topics and the source of the data are much more diverse than that of Wongnai.

5   Experimental Setup

5.1   Implementation Details

ULMFiT. We used the same model hyperparameters as the popular Thai GitHub repository *thai2fit* [3]: the base model is a 4-layer AWD-LSTM with $1,550$ hidden activation units per layer and an embedding size of 400. A BPTT batch size of 70 was used. We applied dropout of 0.25 to output layers, 0.1 to RNN layers, 0.2 to input embedding layers, 0.02 to embedding layers, and weight dropout of 0.15 to the RNN hidden-to-hidden matrix.

Table 2. Datasets, tasks, number of classes, train and test examples, and the average example length measured in tokens. The OOV rate is measured with respect to the original vocabulary of the pretraining corpus.

Dataset	Task	Classes	Train	Test	OOV	Average length
Wongnai	Sentiment classification	5	40k	6.2k	0.710%	126 ± 124
Wisesight	Sentiment classification	4	26.7k	3.9k	2.685%	27 ± 44

ELMo. We used the same biLM architecture from the original implementation [14] with all default hyperparameters, where the LM is a 2-layer biLSTM with 4096 units and 512 dimension projections with another static character-based representations layer with convolutional filters. For both downstream tasks, a 3-layer biLSTM was used with 256 hidden units as the task-specific model.

GPT. Default configurations of [16] were used. The resulting model has 12 layers of transformer each with 12 self-attention heads and 768-dimensional states. We used learnt position embeddings and a maximum sequence length of 256 tokens.

BERT. We used the publicly available $BERT_{BASE}$ unnormalized multilingual cased model, which has a hidden size of 768, 12 self-attention heads, and 12 transformer blocks. Note that the $BERT_{BASE}$ was chosen to have identical hyperparameters as GPT for comparative purposes.

5.2 Evaluation Metrics

A total of 4 tasks were evaluated: the proposed data preprocessing technique in Sect. 2, LM pretraining, LM fine-tuning, and classifier fine-tuning. We chose to benchmark on the easiness to train each model (speed and number of epochs), the intrinsic evaluations (perplexity), and the extrinsic evaluations (downstream classification tasks). In addition, an ablation study of limited corpus data is compared to see the performance of each model in smaller data scenarios.

Data Preprocessing. To benchmark the quality of our unique data preprocessing techniques for Thai social media corpus, we sampled a thread from each dataset and request expert Thai native speakers to help tokenize the samples. At the time of writing, there is no standard corpus for benchmarking the task of colloquial Thai word segmentation. Each character in the thread is labeled as 1 (beginning of word) or 0 (intra-word character). The precision, recall, and F1 score is calculated based on the performance of segmenting each character, where true positives are the correctly segmented beginning of word. The default pyThaiNLP tokenizer [15] Maximum Matching (newmm) is compared between with and without our data preprocessing methods. Unfortunately, labeling tokenization dataset in Thai language requires large amount of effort. Therefore, more extensive experiments will be conducted in the future.

Language Model Pretraining. Pretraining a language model is the most expensive process in the transfer learning workflow. This task is generally performed only once before fine-tuning on a target task. With minimal hyperparameter tuning, we evaluated the pretraining process on: (1) the speed of training in each epoch and (2) the intrinsic perplexity value. Although with the ambiguity that comes with intrinsic metrics, perplexity is one of the traditional methods in LM evaluation. It measures the confidence of the model on the observed sequence via exponentiation of the cross-entropy loss, where cross-entropy loss is defined as the negative sum of the mean LM log-likelihood. Note that this definition applies to different levels of granularity. Due to resource constraints, each model was pretrained for a fixed number of epochs. An NVIDIA P6000 is used to pretrain each model, and the appropriate batch size was selected such that it maximizes the GPU VRAM of 24 GB. The models were trained for 3 epochs and the best performing model was selected. However, since BERT trains using MLM and is able to learn just 15% of the corpus during 1 epoch, we decided to train for the standard 1 million steps [8] (equivalent to around 6.5 epochs).

Language Model Fine-Tuning. Each model was benchmarked on the number of epochs used and the total time until convergence. This process aims to learn the slight differences in data distribution of the target corpus. The models overfit easily due to the modest size of the corpus, thus each LM was fine-tuned until early stopping.

Classifier Fine-Tuning. In this paper, we reported each downstream task performance following the metric used in each Kaggle competition. Wongnai Rating Review Challenge and Wisesight Sentiment Analysis both use classification accuracy for evaluation, which is calculated by the proportion of correctly classified samples out of all the samples. Kaggle ranks the competitors' final standings with the private score, hence this will be used as the benchmark.

6   Results

In this section, we first report the results of our unique preprocessing methods, followed by the results of pretraining the data. We then compare the results of ULMFiT, ELMo, GPT, and BERT with the previous state-of-the-art models in the Thai NLP research community from the Kaggle competition benchmarks.

6.1   Data Preprocessing

Results are shown in Table 3, where our preprocessing method allows the default pyThaiNLP maximum matching (MM) tokenizer to more precisely segment noisy social media data. This is due to the lower false positive tokens segmented by the noisiness of the data, where most of the spams and repetitions are preprocessed correctly. With more comprehensive vocabulary, it allows the tokenizer to segment short colloquial words more accurately. Note that this does not account for the supposed increased comprehension of the models from standardizing the data.

Table 3. Tokenization precision, recall, and F1-score

Tokenizer	Precision	Recall	F1-score
MM+Our preprocessing	95.83%	**98.65%**	**97.22%**
MM	**96.04%**	97.39%	96.71%

6.2   Language Model Pretraining

From Table 4, AWD-LSTM with ULMFiT requires the least amount of time per epoch and the least total time, 100 h and 33 h respectively. Due to resource scheduling limitations, ELMo is trained with 2 P6000 GPUs, making the total time and the time per epoch much lower than the supposed value. With character-level convolutions and character-based operations, ELMo training time should be the longest amongst all the LMs. Transformer-based models require time around more than 1.5x of ULMFiT.

Table 4. Model pretraining time. t_{epoch} is the time used per epoch.

Model	t_{epoch}
ULMFiT	**33 h**
biLM(ELMo) (2 GPU)	52 h
GPT $seq_{max} = 256$	55 h
BERT $seq_{max} = 256$	49 h

The training loss and perplexity are shown in Table 5. BERT has the lowest word-level cross-entropy loss with 15.3857 MLM perplexity. This is expected due to the difference of the MLM prediction task with fully visible beginning and ending context, providing more contextual information to predict the masked word as compared with traditional forward and backward models. In the domain of traditional autoregressive models, GPT has a lower perplexity than ULMFiT. ELMo is not compared to other models due to prediction granularity difference and is reported as is.

6.3   Language Model Fine-Tuning

All the language models are fine-tuned with the target corpus until they give the best result with respect to the validation loss. An NVIDIA P6000 is used for each model and the time required is presented in Table 6. Transformer-based models are shown to overfit quicker than LSTM-based models.

Table 5. Training loss and perplexity after pretraining

Model	Loss	Perplexity
ULMFiT	3.5281	34.0603
GPT $seq_{max} = 256$	3.1735	23.8913
BERT MLM $seq_{max} = 256$	**2.7334**	**15.3857**
biLM(ELMo) (Character-Level)	1.7140	5.5512

Table 6. Language model fine-tuning time. t_{total} is the total time used and t_{epoch} is the time used per epoch.

Model	Wisesight			Wongnai		
	#Epoch	t_{total}	t_{epoch}	#Epoch	t_{total}	t_{epoch}
ULMFiT	11	**11 min**	**1 min**	11	99 min	**9 min**
biLM(ELMo)	5	25 min	5 min	2	64 min	32 min
GPT $seq_{max} = 256$	3	57 min	19 min	3	90 min	30 min
BERT $seq_{max} = 256$	3	36 min	12 min	2	**38 min**	19 min

6.4   Classifier Fine-Tuning

The results of the downstream classification tasks are shown in Table 7. BERT with our pretraining data outperforms all existing models on the private set of Wongnai and Wisesight and obtains 0.9% and 3.2% respective absolute accuracy improvement over the state-of-the-art. Absolute accuracy improvements on all models and tasks are obtained when pretrained with our Thai Social Media data instead of the Thai Wiki Dump.

6.5   Limited Pretraining Corpus

We also investigated the performance of the models in the scenario where the pretraining corpus is limited. This result reflects the learning ability of the models in a language where training data is scarce. We randomly sampled a total of 40M tokens (equivalent to around 234K threads) from the dataset used in our previous experiments. ULMFiT, ELMo, and GPT are trained for 3 epochs while BERT is trained for 30k steps (equivalent to approximately 6.5 epochs on this data). Table 8 shows that ULMFiT and GPT perform considerably well. On the other hand, adding ELMo to LSTM input shows little improvement. This means that ELMo requires a larger corpus to be effective. Although BERT performs well on the Wisesight dataset, it has a drop in performance on Wongnai dataset.

Table 7. Classifier fine-tuning results. Our models are compared to other models: the baseline that predicts the most frequent label, the latest Kaggle competition winner, and public github repositories. The public leaderboard and private leaderboard are calculated with approximately 30% and 70% of the test data respectively.

Model	Wisesight (Acc.)		Wongnai (Acc.)	
	Private	Public	Private	Public
Baseline	0.5809	0.6044	0.4785	0.4785
Kaggle best	0.7597	0.7532	0.5914	0.5814
fastText [3]	0.6131	0.6314	0.5145	0.5109
LinearSVC [3]	–	–	0.5022	0.4976
Logistic regression [3]	0.7499	0.7278	–	–
Thai Wiki Dump Pretraining				
ULMFiT [3]	0.7419	0.7126	0.5931	0.6032
ULMFiT Semi-supervised [3]	0.7597	0.7337	–	–
BERT $seq_{max} = 128$ [1]	–	–	0.5661	0.5706
Ours (Thai Social Media Pretraining)				
ULMFiT	0.7586	0.7346	0.6203	**0.6409**
biLSTM	0.6366	0.6213	0.4773	0.4946
ELMo+biLSTM	0.6866	0.6450	0.5310	0.5226
GPT $seq_{max} = 256$	0.7669	**0.7540**	0.6088	0.6145
BERT $seq_{max} = 256$	**0.7691**	0.7439	**0.6251**	0.6231

Table 8. Limited pretraining corpus results. The public and private scores are calculated with approximately 30% and 70% of the test data respectively.

Model	Wisesight (Acc.)		Wongnai (Acc.)	
	Private	Public	Private	Public
ULMFiT	0.7358	0.7143	0.5984	**0.6290**
biLSTM	0.6366	0.6213	0.4773	0.4946
biLSTM + ELMo	0.6489	0.6095	0.4879	0.4753
GPT $seq_{max} = 256$	0.6931	0.7075	**0.6111**	0.6102
BERT $seq_{max} = 256$	**0.7467**	**0.7244**	0.5650	0.5516

7   Conclusion

Our work shows that by using our unique data preprocessing methods and our pretraining social media data, we can improve the performance of the LMs in the downstream tasks. The improvement of all models from pretraining data of the same domain suggests that pretraining data has a significant impact on

LM performance. Moreover, the possibility for LM pretraining on a noisy corpus shows the ability of the models to learn in spite of the quality of the data.

Results-wise, BERT is the best performing model with respect to classification accuracy. It can achieve state-of-the-art results on both of the benchmarking downstream tasks. However, it has unstable performance on downstream tasks when pretrained on a small corpus and uses a lot of pretraining time. If speed and ease of training are the main considerations, we recommend using AWD-LSTM with ULMFiT due to its speed of pretraining and fine-tuning, while the results are still on par with transformer-based models. Although OpenAI GPT shows promising results with acceptable pretraining speed, it is overshadowed by other models in both aspects. Finally, although ELMo shows significant improvements when compared with the baseline biLSTM, it places a dependency on designing a powerful task-specific model to achieve good performance.

Acknowledgements. In the making of the paper, the authors would like to acknowledge Mr. Can Udomcharoenchaikit for his continuous and insightful research suggestions until the completion of this paper.

References

1. Bert-th. (2019). https://github.com/ThAIKeras/bert
2. Scrapy. (2019). https://github.com/scrapy/scrapy
3. thai2fit. (2019). https://github.com/cstorm125/thai2fit
4. Aroonmanakun, W.: Thoughts on word and sentence segmentation in Thai (2007)
5. Bahdanau, D., Cho, K., Bengio, Y.: Neural machine translation by jointly learning to align and translate. arXiv preprint arXiv:1409.0473 (2014)
6. Cho, K., et al.: Learning phrase representations using RNN encoder-decoder for statistical machine translation. arXiv e-prints arXiv:1406.1078, June 2014
7. Dai, A.M., Le, Q.V.: Semi-supervised sequence learning. arXiv e-prints arXiv:1511.01432, November 2015
8. Devlin, J., Chang, M.W., Lee, K., Toutanova, K.: BERT: pre-training of deep bidirectional transformers for language understanding. arXiv e-prints arXiv:1810.04805, October 2018
9. Howard, J., Ruder, S.: Universal language model fine-tuning for text classification. arXiv e-prints arXiv:1801.06146, January 2018
10. Lertpiya, A., et al.: A preliminary study on fundamental Thai NLP tasks for user-generated web content. In: 2018 International Joint Symposium on Artificial Intelligence and Natural Language Processing (iSAI-NLP), pp. 1–8, November 2018. https://doi.org/10.1109/iSAI-NLP.2018.8692946
11. Merity, S., Shirish Keskar, N., Socher, R.: Regularizing and Optimizing LSTM Language Models. arXiv e-prints arXiv:1708.02182, August 2017
12. Mikolov, T., Chen, K., Corrado, G., Dean, J.: Efficient estimation of word representations in vector space. arXiv e-prints arXiv:1301.3781, January 2013
13. Pennington, J., Socher, R., Manning, C.D.: Glove: global vectors for word representation. In: Empirical Methods in Natural Language Processing (EMNLP), pp. 1532–1543 (2014)
14. Peters, M.E., et al.: Deep contextualized word representations. arXiv e-prints arXiv:1802.05365, February 2018

15. Pythainlp 2.0. (2019). https://github.com/PyThaiNLP/pythainlp
16. Radford, A., Narasimhan, K., Salimans, T., Sutskever, I.: Improving language understanding by generative pre-training (2018)
17. Sutskever, I., Vinyals, O., Le, Q.V.: Sequence to sequence learning with neural networks. arXiv e-prints arXiv:1409.3215, September 2014
18. Vaswani, A., et al.: Attention is all you need. arXiv e-prints arXiv:1706.03762, June 2017

Comparative Analysis of Deep Learning Models for Myanmar Text Classification

Myat Sapal Phyu(✉) [iD] and Khin Thandar Nwet(✉) [iD]

Faculty of Computer Science, University of Information Technology, Yangon, Myanmar
{myatsapalphyu,khinthandarnwet}@uit.edu.mm

Abstract. Text classification is one of the major research areas for Natural Language Processing (NLP). Long Short Term Memory (LSTM), Convolutional Neural Networks (CNN), and their combination models have been applied in many NLP tasks. This paper presents a joint CNN with no max-polling layer and Bidirectional LSTM to fulfill the requirements of each model. The proposed model takes advantage of CNN to extract features and Bi-LSTM to capture long term contextual information from past and future contexts. The proposed model is compared with CNN, Bi-LSTM, RNN, and CNN-LSTM models with pre-trained word embedding on five article datasets in Myanmar language.

Keywords: Text classification · Myanmar language · Deep learning · Pre-trained word embedding · CNN · RNN · CNN-RNN · CNN-LSTM · Bi-LSTM

1 Introduction

In the age of information, people are wasting a lot of time finding their interesting information. Consequently, it is crucial to effectively and quickly extract the most relevant information from a wide range of information. Text classification can negotiate with these problems. It is one of Natural Language Processing (NLP)'s main research areas. Text classification is the arrangement of text into their respective categories such as spam filtering, articles, sentiment analysis, posts, and hate speech identification. Recently, the use of word embedding with a deep learning method has attracted considerable interest in text classification due to their ability to capture semantic relationships of words [2, 6, 8]. Words are considered as basic unit in most of the NLP for implementing continuous word vector representation. This paper focuses in particular on the Myanmar text classification. There is no rule to determine word boundaries for Myanmar language. Since Myanmar language is rich in morphology, it is difficult to learn good representation of words because many word types seldom occur in the training corpus. In order to classify Myanmar text by means of deep learning models, several steps are taken to pre-process Myanmar text such as extracting massive amounts of Myanmar text, removing unnecessary characters, determining words boundaries and converting words into word vectors that keep the context information. Grave et al. [3] published pre-trained word vectors for two hundred forty-six languages trained on common crawl and Wikipedia. They proposed bag-of-character n-grams based on skip-gram that could capture sub-word

© Springer Nature Switzerland AG 2020
N. T. Nguyen et al. (Eds.): ACIIDS 2020, LNAI 12033, pp. 76–85, 2020.
https://doi.org/10.1007/978-3-030-41964-6_7

information to enrich word vectors. The pre-trained sub-word vectors for two hundred seventy-five languages were also released by Heinzerling et al. [5]. Their works are very helpful in resource-scarce languages and can be applied to specific NLP tasks by transferring learning. This paper applies the deep learning models for text classification and pre-trained word embedding trained on Wikipedia for the construction of embedding matrix.

The next sections are as follow, Section 2 addresses the related work of the text classification for both the English and Myanmar languages. Section 3 discusses the pre-processing steps before an embedding layer. Section 4 explains the proposed model. Section 5 explains the experimental section containing the dataset collection, comparison models and experiment results and the paper is concluded in the Sect. 6.

2 Related Work

Conneau et al. [2] have proposed very deep convolutional neural networks (VDCNN) that use twenty nine layers of convolution. VDCNN operated directly on character-level and performance is measured by using eight datasets. Joulin et al. [6] developed a text classification system that is efficient and simple and is denoted as fastText. This model's accuracy is similar to other deep learning classifiers, but using a regular multicore CPU, it takes less than ten minutes for training more than one billion words. Song et al. [13] introduced a context-LSTM-CNN model to use LSTM-based long-range dependencies and used the convolution layer and max-pooling layer to extract local features at specific points. Lai et al. [10] applied bi-directional RNN to capture meaning and max-pooling to capture key components in texts. Kim [8] showed that the use of a single convolution layer in the simple CNN and proposed variations of the CNN models CNN-rand, CNN-static, CNN-non-static, and CNN-multichannel. These models were experimented on seven publicly available datasets and improved the state-of-the-art methods on four out of seven datasets. Zhang et al. [15] compared character-level convolutional networks with word-level ConvNets and RNN for text classification in the English language. In Myanmar language text classification, we also investigated previous research work, such as news classification, spam filtering, and sentiment analysis. Aye et al. [1] improved the accuracy of prediction on informal Myanmar text by considering objective and intensifier words for Myanmar's food and restaurant text reviews. Khine et al. [7] showed the comparison of Naïve Bayes and k-Nearest Neighbors (KNN) algorithms for Myanmar news classification. The experiment showed that KNN is higher in recall and accuracy than Naïve Bayes on 1,200 documents datasets with four categories. Yu et al. [14] developed a corpus annotated with sentiment polarity for Myanmar news. The N-gram model is used to choose features and the Naïve Bayes algorithm is to identify emotions. Kyaw et al. [9] constructed a spam filtering corpus and proposed a Naïve Bayes-based learning algorithm for spam or harm classification. According to the literature review, some deep learning models were improved and explored for Automatic Speech Recognition as in [12]. Most of the Myanmar text classification tasks are performed in lexicon-based and approaches because the challenge of text classification in Myanmar language is the need for huge resources to train in deep learning models. Using pre-trained word vectors can address such resource-requiring problems. In our previous work [12], we performed the comparative analysis of CNN and RNN both on syllable and word level by using three pre-trained vectors and also collected and annotate six Myanmar articles datasets. We use the pre-trained vector that is trained on the skip-gram model in the embedding layer. This paper presents a joint CNN and Bi-LSTM model and compares with most of the baseline deep learning models and their combination models for Myanmar text classification on five datasets.

3 Pre-processing

Pre-processing steps is cruel for Myanmar language because of its nature. Firstly, we extract sentences from text documents. Pre-processing steps contain removing the non-Myanmar character, punctuation marks, and numbers. As this work focus on Myanmar text classification, we remove non-Myanmar characters that do not contain in the Unicode range between [U1000-U104F]. The numbers [U1040-U1049] and the punctuation marks [U104A-U104B] are also removed. Myanmar language has rule to determine the boundary of words. In this work, the BPE tokenizer[1] is used to define the word boundary. Algorithm 1 and 2 show the step by step procedure of preprocessing task. Algorithm 1 shows the step-by-step process to remove the unnecessary characters from the text dataset. Table 1 shows the sample of pre-processing steps for sample input text " ဆေးဝါးလိုအပ်သူများအတွက် အခမဲ့Medicine Box ပဲခူးဆေးရုံကြီး၌စတင်ထားရှိ။". In this sample text, non-Myanmar characters "Medicine Box" and punctuation marks " ။" are removed and the remaining text string " ဆေးဝါးလိုအပ်သူများအတွက်   အခမဲ့ ပဲခူးဆေးရုံကြီး၌စတင်ထားရှိ" is segmented as " ဆေးဝါး_လိုအပ်_သူများအတွက်_အခမဲ့_ပဲခူးဆေးရုံကြီး၌_စတင်_ထားရှိ" by the tokenizer.

Table 1. Pre-processing steps for sample input text

Input Text	ဆေးဝါးလိုအပ်သူများအတွက် အခမဲ့ Medicine Box ပဲခူးဆေးရုံကြီး၌ စတင်ထားရှိ။
Word Segmentation	ဆေးဝါး_လိုအပ်_သူများအတွက်_အခမဲ့_ပဲခူး ဆေးရုံကြီး၌_စတင်_ထားရှိ

---

[1] https://github.com/bheinzerling/bpemb.

Algorithm 1: Removing Unnecessary Characters

Input	: Raw text document
Result	: Text documents D without unnecessary characters
Initialization	: Character c_i, Character code cc_i,

Myanmar Unicode, MU = [u1000-u104f],
Myanmar Digit, MD = [u1040-u104b],
Punctuation Marks, PM = [u104a-u104b], Text String T extracted from
D; i = {0,1,...,n}, i = 0

foreach $c_i \in T$ **do**
　if $cc_i \notin MU$ **then**
　|　**remove** c_i
　|　i++
　end
　if $cc_i \in MD$ **then**
　|　**remove** c_i
　|　i++
　end
　if $cc_i \in PM$ **then**
　|　**remove** c_i
　|　i++
　end
end

3.1 Pre-trained Vector

In this work, we use the pre-trained vector trained on the fastText Skip-gram model[2]. The number of word vectors in this pre-trained vector is 91,497 and the dimension is 300. The pre-trained vectors file is used as vocabulary to convert words into word vectors. Algorithm 2 shows the conversion of segmented words to embedding matrix. Figure 1 shows the step-by-step process before the embedding layer. Table 2 the sample result of embedding matrix for each segmented word.

Table 2. Sample result of embedding matrix for each segmented word

Segmented Word	Sample Embedding Matrix (300 Dimension)
ဆေးဝါး	0.963450 1.260167 -0.309332 0.324999 ……….....0.454490
လို့အဓိ	-0.303027 0.528527 -0.898522 -0.326885 ……….0.060485
သူများအတွက်	-1.601050 -0.07043 0.098535 -0.363543 ………... 0.178379
အခမဲ့	-0.720556 0.595658 -0.711430 0.183147 ……….0.5865110
ပဲခူး	0.800076 0.450307 -0.803149 -0.686676 ……….-0.876465
ဆေးရှုကြိုးဒဲ့	-1.635552 -0.716141 -0.154168 -0.289999 ……….0.643253
စတင်	0.058164 0.033736 -0.705663 0.482859………....0.5694375
ထားရှိ	-0.803120 -0.225860 1.441452 1.634341………... 0.587558

---

[2] https://fasttext.cc/docs/en/pretrained-vectors.html.

Algorithm 2: Word Embedding Matrix

Input   : text documents D without unnecessary characters
Result   : Word embedding matrix for embedding layer
Initialization : Words w_i,
                        where, i = {0,1,...,n}
                        Word embedding matrix, embmatrix[i],
                        Text String T extracted from D, Vocabulary in
                        Pretrained Vectors, V,
                        i = 0,
  segment T into wi by BPE tokenizer
    remove duplicate wi;
    foreach $w_i \in V$ **do**
      if $w_i \in V$ **then**
          embmatrix[i] = vector(w_i)
          i++
      end
    end
  end

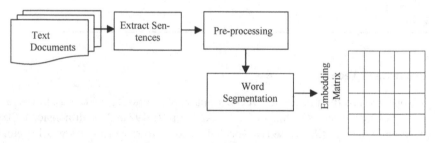

Fig. 1. Pre-processing steps before the embedding layer

4 Model

A joint CNN-Bi-LSTM model is illustrated in Fig. 2. It is basically composed of the following layers.

Embedding Layer: After pre-processing steps, the segmented words are matched with the vocabulary in the pre-trained word vector that is trained on the skip-gram model. Each word in the vocabulary attaches with their corresponding vectors and it can catch context information.

Convolution Layer: Convolution layer performs the convolution process with stride size 1 by using the ReLU activation function $f(x) = \max(x, 0)$. The convolution layer is used to extract features from the embedding matrix and discard the pooling layer because it only captures the most important information and lost the context information.

Bi-LSTM Layer: Bi-LSTM layer is applied as an alternative of pooling layer to capture long term semantic information from both past and future contexts.

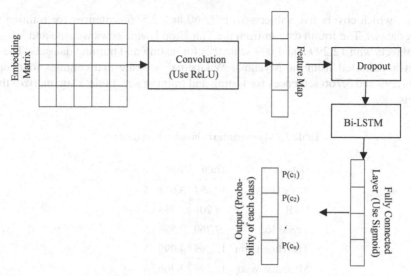

Fig. 2. A joint CNN and Bi-LSTM model

Fully Connected Layer: Fully connected layer used sigmoid activation function to calculate the probabilities of each class. The sigmoid function of $p(c_n)$ is

$$p(c_n) = \frac{1}{1 + e^{-c_n}} \tag{1}$$

The probability of a class does not depend on all other classes' probabilities. It can handle the multi-label problem. Binary cross-entropy is used as a loss function and Adam optimization function with 0.5 dropouts and 16 batch size on 10 epochs are set as hyper-parameters. In addition, the bias and kernel regularizer set ($l2 = 0.01$) in output layer for reducing overfitting problem.

5 Experiment

5.1 Datasets

Empirical exploration is conducted in Myanmar language on five news datasets. These datasets are collected from five daily news websites [12]. Text data are converted into Unicode font by using rabbit converter[3]. Each line represented a sentence annotated with corresponding label. Text data are and shuffled and split into 75% and 25% for training and testing datasets. The first dataset is collected from the 7Day Daily[4] website with 10,884 and 3,628 sentences for train and test sets. The second dataset is collected from the DVB[5] website and it includes five subjects, with 8,201 and 2,733 sentences for the training and testing dataset. The third dataset is collected from The Voice[6] news

---

[3] https://www.rabbit-converter.org/.
[4] http://7daydaily.com/.
[5] http://burmese.dvb.no/.
[6] http://thevoicemyanmar.com/.

website, which covers five subjects with 7,660 and 2,586 sentences for training and testing dataset. The fourth dataset from the Thit Htoo Lwin[7] news website and includes five subjects with 12,299 and 4,099 sentences for testing and training datasets. The last dataset is collected from the Myanmar Wikipedia[8] website that contains four topics with 11,299 and 3,766 sentences for testing and training set. Table 3 summarizes these datasets.

Table 3. Myanmar text classification datasets

Dataset	Train	Test	Classes
7Day Daily	10,884	3,628	5
DVB	8,201	2,733	5
The Voice	7,760	2,586	5
Thit Htoo Lwin	12,299	4,099	5
Myanmar Wiki	11,299	3,766	4

5.2 Comparison Models

In this work, we performed the comparative analysis of a joint CNN and Bi-LSTM model with CNN, RNN, Bi-LSTM, CNN-LSTM models.

Convolutional Neural Networks (CNN): It is an artificial neural network feed-forward, most widely used for visual image analysis. This model has recently achieved significant success in the tasks of text classification. It has three basic components, convolution, pooling, and fully connected layer. ReLU activation functions $f(x) = \max(x, 0)$ is used in convolution layer and it can have several layers of convolution. The pooling layer extracts the most important features. The pooling layer mostly applies Max-pooling. The fully connected layer is the model's output layer and it predicts the class of the input sentences. The fully connected layer commonly uses the Softmax function. Softmax function of $f(x)_i$ is $\frac{e^{x_i}}{\sum_j^C e^{x_j}}$, the probability of a class depends on the probabilities of all other classes.

Recurrent Neural Networks (RNN): RNN is a generalization of feedforward neural networks with the distinction that it has an internal memory that keeps information to persist. It performs the same function for all input data by learning from the previous data. RNN produces the output y_t as in Eq. (2).

$$y_t = f\left(W_y h_t\right) \tag{2}$$

$$h_t = \sigma(W_h h_{t-1} + W_x x_t) \tag{3}$$

---

[7] http://www.thithtoolwin.com/.
[8] https://my.wikipedia.org/.

Bidirectional LSTMs: It is an extension of the LSTM model that can learn from the past and future information for a specific task.

CNN-LSTM: (Hassan A, 2018) proposed a joint CNN and LSTM framework to produce the feature map by CNN and to capture long term dependencies by LSTM.

5.3 Experimental Result and Discussion

The experiment is accomplished on Google Cloud Laboratory[9] that does not require to configure the Jupyter notebook by using, Keras[10], a model-level library. The performance of the CNN-Bi-LSTM model is compared with comparison models described in Sect. 5.2 as listed in Table 4. The highest performance sores for each dataset are highlighted in bold. According to the experiments, the proposed model improves accuracy in four out of five datasets. The CNN model performs equally with the proposed model in two datasets. The CNN-LSTM combined model performs better in two out of five datasets. We also measure the training time of each model. According to the measurement results, the CNN model requires the minimum training time because we used only one convolution layer. Although CNN-Bi-LSTM model performed better in three datasets than the remaining models, it requires more time for training than CNN-LSTM, RNN and CNN models. Average training time of each model are listed in Table 5.

Table 4. Comparison of average testing accuracy

Datasets	Bi-LSTM	CNN	CNN-Bi-LSTM	CNN-LSTM	RNN
7Day Daily	98.07	98.13	98.11	**98.14**	97.74
DVB	91.92	90.24	**92.14**	91.58	82.97
The Voice	95.39	**95.67**	**95.67**	95.06	94.98
Thit Htoo Lwin	96.12	**96.49**	**96.49**	96.29	94.95
Myanmar Wiki	94.01	93.99	93.80	**94.21**	86.76

Table 5. Comparison of average training time

Datasets	Bi-LSTM	CNN	CNN-Bi-LSTM	CNN-LSTM	RNN
7Day Daily	18 min 6 s	1 min 47 s	17 min 6 s	9 min 9 s	5 min 22 s
DVB	19 min	1 min 20 s	13 min 22 s	6 min 50 s	3 min 23 s
The Voice	13 min 19 s	1 min 11 s	11 min 6 s	6 min 23 s	3 min 14 s
Thit Htoo Lwin	19 min 55 s	1 min 34 s	18 min 1 s	9 min 53 s	5 min 9 s
Myanmar Wiki	18 min 55 s	1 min 31 s	16 min 56 s	9 min 43 s	5 min 11 s

---

[9] https://colab.research.google.com.
[10] https://keras.io/.

6 Conclusion

This paper presents a joint CNN-Bi-LSTM model that take advantages of CNN to extract feature and Bi-LSTM to capture long term context information from both past and future information. A series of the experiment is performed by comparing the pro-posed model with CNN, Bi-LSTM, RNN, CNN-LSTM models in term of accuracy on five Myanmar articles datasets. According to the experiment, the proposed system per-forms better in three out of five datasets. The CNN model requires minimum training time than the remaining models and CNN-Bi-LSTM model takes more time than CNN, RNN, and CNN-LSTM models.

Acknowledgement. We deeply thank the anonymous reviewers for sharing their precious time to check our manuscript. We greatly thank the researchers who released pre-trained vectors publicly and these resources helpful for low resources languages. We greatly thank the friends who assist to collect and annotate Myanmar text datasets.

References

1. Aye, Y.M., Aung, S.S.: Enhanced sentiment classification for informal Myanmar text of restaurant reviews. In: 16th International Conference on Software Engineering Research, Management and Applications (SERA), pp. 31–36. IEEE (2018). https://doi.org/10.1109/SERA.2018.8477231
2. Conneau, A., Schwenk, H., Barrault, L., Lecun, Y.: Very deep convolutional networks for text classification. In: The European Chapter of the Association for Computational Linguistics, EACL 2017 (2017). https://doi.org/10.18653/v1/e17-1104
3. Grave, E., Bojanowski, P., Gupta, P., Joulin, A., Mikolov, T.: Learning word vectors for 157 languages. In: Proceedings of the Eleventh International Conference on Language Resources and Evaluation, LREC-2018 (2018)
4. Hassan, A., Mahmood, A.: Convolutional recurrent deep learning model for sentence classification. IEEE Access **6**, 13949–13957 (2018). https://doi.org/10.1109/ACCESS.2018.2814818
5. Heinzerling, B., Michael, S.: BPEmb: tokenization-free pre-trained sub-word embeddings in 275 languages. In: Proceedings of the Eleventh International Conference on Language Resources and Evaluation, LREC-2018, pp. 31–36 (2018). https://doi.org/10.11588/data/V9CXPR
6. Joulin, A., Grave, E., Bojanowski, P., Mikolov, T.: Bag of tricks for efficient text classification. In: Proceedings of the 15th Conference of the European Chapter of the Association for Computational Linguistics: Volume 2, Short Papers, pp. 427–431 (2017). https://doi.org/10.18653/v1/e17-2068
7. Khine, A.H., Nwet, K.T., Soe, K.M.: Automatic Myanmar news classification. In: 15th International Conference on Computer Applications, ICCA 2017, pp. 401–408 (2017)
8. Kim, Y.: Convolutional neural networks for sentence classification. In: Proceedings of the 2014 Conference on Empirical Methods in Natural Language Processing (EMNLP), pp. 1746–1751 (2014). https://doi.org/10.3115/v1/D14-1181
9. Kyaw, T.N., Nyo, N.N.: Myanmar spam filtering based on Naïve Bayesian learning algorithm (MSFNBLA). In: 14th International Conference on Computer Applications, ICCA 2016 (2016)

10. Lai, S., Liheng, X., Kang, L., Jun, Z.: Recurrent convolutional neural networks for text classification. In: The Twenty-Ninth AAAI Conference on Artificial Intelligence (2015)
11. Mon, A.N., Pa, W.P., Thu, Y.K.: Exploring the effect of tones for Myanmar language speech recognition using convolutional neural network (CNN). In: Hasida, K., Pa, W.P. (eds.) PACLING 2017. CCIS, vol. 781, pp. 314–326. Springer, Singapore (2018). https://doi.org/10.1007/978-981-10-8438-6_25
12. Phyu, S.P., Nwet, K.T.: Article classification in Myanmar language. In: The Proceeding of 2019 International Conference on Advanced Information Technologies (ICAIT), pp. 188–193. IEEE (2019). https://doi.org/10.1109/AITC.2019.8920927
13. Song, X., Petrak, J., Roberts, A.: A deep neural network sentence level classification method with context information. In: Proceedings of the 2018 Conference on Empirical Methods in Natural Language Processing, pp. 900–904 (2018). https://doi.org/10.18653/v1/D18-1107
14. Yu, T., Nwet, K.T.: Annotation and sentiment analysis for Myanmar news. In: 16th International Conferences on Computer Applications, ICCA 2018
15. Zhang, X., Zhao, J., LeCun, Y.: Character-level convolutional networks for text classification. In: Advances in Neural Information Processing Systems, pp. 649–657 (2015)

Open Information Extraction as Additional Source for Kazakh Ontology Generation

Nina Khairova[1(✉)] iD, Svitlana Petrasova[1] iD, Orken Mamyrbayev[2] iD,
and Kuralay Mukhsina[3] iD

[1] National Technical University "Kharkiv Polytechnic Institute",
Kyrpychova Street, Kharkiv 61002, Ukraine
nina_khajrova@yahoo.com, svetapetrasova@gmail.com
[2] Institute of Information and Computational Technologies, 125, Pushkin Street,
050010 Almaty, Republic of Kazakhstan
morkenj@mail.ru
[3] Al-Farabi Kazakh National University, 71 Al-Farabi Avenue, Almaty, Republic of Kazakhstan
kuka_ai@mail.ru

Abstract. Nowadays, structured information that obtains from unstructured texts and Web context can be applied as an additional source of knowledge to create ontologies. In order to extract information from a text and represent it in the RDF-triplets format, we suggest using the Open Information Extraction model. Then we consider the adaptation of the model to fact extraction from unstructured texts in the Kazakh language. In our approach, we identify lexical units that name the participants of the action (the Subject and Object) and semantic relations between them based on words characteristics in a sentence. The model provides semantic functions of the action participants via logical-linguistic equations that express the relations of the grammatical and semantic characteristics of the words in a Kazakh sentence. Using the tag names and some syntactic characteristics of words in the Kazakh sentences as the values of the predicate variables in corresponding equations allows us to extract Subjects, Objects and Predicates of facts from texts of Web content. The experimental research dataset includes texts extracted from Kazakh bilingual news websites. The experiment shows that we can achieve the precision of facts extraction over 71% for Kazakh corpus.

Keywords: Open Information Extraction · RDF-triplets · Unstructured text · Logical-linguistic equations · Kazakh bilingual news websites

1 Introduction

Nowadays, the problem of information and fact extraction remains unsolved. Existing models and algorithms for fact extraction depend on the degree of a document structuring. In this way, we can divide text documents into: (1) well-structured texts, which often content tabular data; (2) semi-structured text documents described a specific domain, and (3) unstructured text document of any domain [1].

Generally, there are robust algorithms [2, 3] for fact extraction from well-structured text documents. At the same time, despite the constant growth of interest in researches

© Springer Nature Switzerland AG 2020
N. T. Nguyen et al. (Eds.): ACIIDS 2020, LNAI 12033, pp. 86–96, 2020.
https://doi.org/10.1007/978-3-030-41964-6_8

of information extraction from Web content, there is no general and well-grounded app-roach for structured information extraction from unstructured texts [4, 5]. This growing interest is primarily caused by the huge volumes of unstructured text information avail-able in corporate and Internet networks (according to some sources, there are more than 85% of such texts). Additionally, increasing interest in researches of fact identification and extraction from unstructured texts is largely due to the expansion of areas of their use.

For instants, fact extraction from unstructured texts can be a serious additional source for ontologies generation based on Web content knowledge. Recent approaches of Open Information Extraction (Open IE) extract a fact as a triplet of Subject-Predicate-Object, where the Object and Subject are usually represented by nouns or noun phrases, while the Predicate is mostly expressed by a verb. This view of fact corresponds to an RDF graph (Fig. 1).

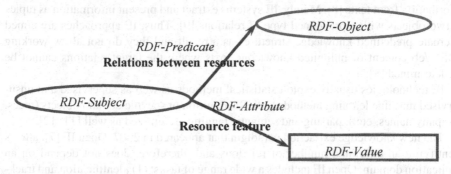

Fig. 1. The RDF diagram of a fact triplet, which corresponds to the concept of a fact in Open IE approach

Yet, the current approaches to structured information extraction from unstructured texts are either based on a limited number of predefined facts (IE) or use existing NLP tools for each specific language (Open IE). Solving both IE and Open IE tasks, a large labeled corpora of a particular language are required.

In our approach, we propose a logical-linguistic model for fact expression in a sen-tence of the natural language. This model implements the general approach of Open IE, namely, extraction of the unlimited domain-independent number of facts from texts.

This study focuses on the adaptation of our fact extraction model for the texts of the Kazakh language, which is the language with limited linguistic resources and, obviously, demands additional sources for Kazakh ontologies generation.

In order to estimate the effectiveness of the model, we utilize relatively small experimental corpus included texts extracted from Kazakh bilingual news websites.

The remainder of the paper is organized as follows. Section 2 gives an overview of the related works, corresponding with IE and Open IE challenges. Section 3 describes the usage of our model within the general approach of Open IE and its implementation for the Kazakh language. Section 4 introduces the working corpus and describes its usage in our experiment. In the last Sect. 5, the scientific and practical contributions of the research, its limitations and future work are discussed.

2 Related Work

The problem of information extraction from unstructured texts can be divided into two basic approaches: Information Extraction (IE) and Open IE. Both of these technologies allow considering large volumes of texts that contain relatively small amount of factual information.

Herewith IE can be thought as a special kind of Information Retrieval (IR), when the query is formulated in advance. However, IE creates a data structure, describing facts, from a set of processed documents, whereas the result of IR is a set of links to documents that match the query.

First IE systems were mostly domain-oriented and based on the knowledge, generated in advance. The example of such an approach is one of the first IE systems. Working with texts on Latin American terrorism, it exploited pre-developed morphological and semantic patterns [6, 7]. Modern IE systems also use a predefined set of rules to extract information from texts [8]. Mainly, IE systems extract and present information as tuples of two objects with a predefined type of relations [9]. Thus, IE approaches are aimed to create predefined knowledge structures as a result and they do not allow working with Web content of unlimited knowledge texts where the target relations cannot be predetermined [8].

IE technologies usually exploit statistical methods as well as supervised and unsupervised machine learning methods [10]. Recognition of specific domain objects (faces, company names, etc.), parsing and semantic tagging are utilized as well [11, 12].

The new knowledge extraction paradigm that appeared in 2007, Open IE [7], allows identifying an unlimited number of relations and, therefore, does not depend on an application domain. Open IE includes a wide range of tasks: (1) identification and tracking of entities, (2) identification of their relations and characteristics, (3) detection and characterization of events.

The most of Open IE applications use NLP tools such as POS-tagging and Dependency parsing [13, 14], employing lexical restrictions [15] or semantic annotations [16] to minimize the large number of possible specific relations [17].

The reasons for ineffectiveness of statistical methods in solving Open IE problems are as follows. Due to the fact that statistical methods consider the document as an unordered "bag-of-words" in IR and text classification or clustering tasks [18], some knowledge, related to grammar and semantics, is lost. The second reason is the obvious need to extract facts not from the whole text but from sentences. This approach is associated with the fact presentation as a triplet: Subject–Relation-Object. In this paradigm, knowledge of a certain domain is a collection of information about the objects or subjects of this domain, their essential properties and relations presented in separate sentences. The third reason for the low effectiveness of using statistical methods is the synonymy and ambiguity of language units, which leads to the frequent occurrence of hidden facts in the text.

Today, the problem of fact extraction is studied for all languages; it has a high level of implementation not only for English texts but also for many others. For example, an experiment was conducted in [19] for assessing the adequacy of measuring the factual density of 50 randomly selected Spanish documents in the CommonCrawl corpus. In a recent study [20], densities of simple and complex facts were considered as characteristics of measuring the quality of Russian Wikipedia articles. In [20], the first Open IE system was introduced to extract fact triplets from Chinese texts.

Despite the available research results, however, there are no multilingual standard Open IE methods and approaches [19], in particular, for languages with limited linguistic resources such as the Kazakh language.

3   Using the Model Within General Approach of Open Information Extraction

3.1   Mathematical Means of the Model

The Open IE approach extracts triplets of Subject - Predicate - Object without defining specific relation types in advance. Since this kind of facts is usually expressed by various unregulated constructions of the natural language, we identify lexical units that name the participants of the action (the Subject and Object), and semantic relations between them in the sentence.

We distinguish four semantic types of facts extracted from the text. Each of them is expressed by different structures of natural languages [21] (Fig. 2).

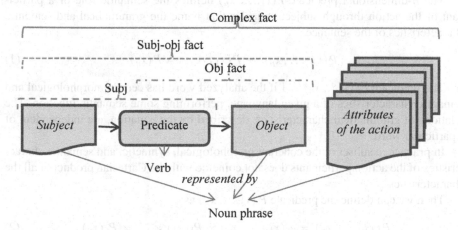

Fig. 2. The scheme of the formalization of four semantic types of facts in unstructured text.

The first semantic type of facts, called *subj-fact*, is expressed by the smallest grammatical sentence. It includes only a noun phrase that defines the Subject as the initiator of the action, and the Predicate that defines the action. The second type of facts, called *obj-fact*, is also identified in the smallest grammatical sentence. It includes the Predicate and a noun phrase defining the participant of the action, i.e. the Object of the action. The third semantic type of facts, called *subj-obj*, is expressed by a sentence including two noun groups that name both the Subject and Object of the action, and the Predicate. The last type of facts is a *complex* fact extracted from a sentence that includes more than two noun groups. It names the Subject, Object and action Predicate as well as some attributes of the action (time, place, mode of action, etc.).

To set semantic relations in a fact, we suggest applying semantic functions expressed as the ratio of morphological and semantic categories of sentence participants by means of the algebra of finite predicates. Its formulas consist of symbols, predicate variables, signs of disjunction, conjunction, negation, logical constants 0 and 1. The predicate is basic in this algebra for recognizing the subject a by the predicate variable x_i: $x_i^a = 1$ if $x_i = a$, and $x_i^a = 0$ otherwise, where $i = \{1, 2,..., n\}$, n is the number of variables [22].

Identifying an object, the predicate is introduced on a given finite universe U of elements, $a \in U$. In set theory, the universe means the concept of universal set, which contains all the entities. The universe U of the complex Kazakh language system, considered in our model, includes predefined sets of words, morphemes, collocation characteristics, and many similar discrete and finite linguistic objects. In particular, the universe U includes a finite, discrete, deterministic set of grammatical and lexical characteristics of words of the Kazakh sentence, influencing their semantic roles $M = \{m_1, ..., m_n\}$, where n is the number of these characteristics. The relations between these characteristics can be represented as a Cartesian product.

Let us introduce the predicate system S on the set M so that any predicate $P(x_i) \in S$ equals 1 on the set of sentence words with grammatical and semantic information corresponding to a certain semantic role and equals 0 otherwise.

The n-dimensional predicate $P(x_1,..., x_n)$ defines the semantic role of a participant of the action through subject variables that name the grammatical and semantic characteristics of the sentence:

$$P(x_1, \ldots, x_n) \to P(x_1) \wedge \ldots \wedge P(x_n) \tag{1}$$

The predicate $P(x_1,..., x_n) = 1$ if the analyzed word has certain morphological and semantic characteristics of a given language, performing some semantic function. The relations of grammatical characteristics, described by the equation, are independent of a particular word.

In practice, a subset of the coherent morphological, syntactic, and semantic characteristics of the action participants does not coincide with the Cartesian product of all the characteristics.

Then we can define the predicate $P(x_1,..., x_n)$ as:

$$P(x_1, \ldots, x_n) = \gamma_k(x_1, \ldots, x_n) \times P_1(x_1) \times \ldots \times P_n(x_n), \tag{2}$$

where $k \in [1, h]$, h is the number of participants and attributes of the action. The predicate $\gamma_k(x_1,..., x_n) = 1$ if the conjunction of the grammatical characteristics of the sentence words shows a certain semantic role of the participant (the Subject, Object) or the attribute of the action, and $\gamma_k(x_1,..., x_n) = 0$ otherwise. Thus, if the relations between the grammatical characteristics of the Kazakh sentence words do not express any fact element, they are removed from the formula (2) by the predicate $\gamma_k(x_1,..., x_n)$.

The semantic functions of the participants and attributes of the action are expressed by the relations of the grammatical and semantic characteristics of the words in the sentence of particular natural language. However, due to the existing difference in morphology and syntax, there are features of the model implementation for each specific language.

We analyzed the implementation of our logical-linguistic OIE model for English [23] and Russian [20] languages. In this study, we consider its adaptation for the texts of the Kazakh language.

3.2 Implementation of the Model for the Kazakh Language

Adapting the developed OIE model for fact extraction from Kazakh texts, we introduce the irreducible set M of ten predicate variables that define the grammatical and semantic features of sentence words. They affect the semantic role of action participants. Most of these features are expressed by affixes in the language structure.

The Kazakh language model is represented by a large number of predicate variables due to the agglutinativeness of the language. This means that each word-forming morpheme has its own specific morphological or semantic meaning (for example, person, case, number). Using a large number of predicate variables in the model is also based on the need to distinguish not only action participants but also different types of actions (Action or Predicate) in the Kazakh language.

Table 1 shows the predicate variables and their values ranges defined in the model.

Using the set of predicative variables $\{x, f, z, a, n, c, y, d, m, b\}$ introduced for the Kazakh language, we can transform Eq. (2) to the following form:

$$P = \gamma_k \times P_x(x) \times P_y(y) \times P_z(z) \times P_f(f) \times P_m(m) \times P_n(n) \times P_a(a) \times P_b(b)$$
$$\times P_c(c) \times P_d(d)$$

$$(3)$$

Then, we define the predicate of the action initiator or the Subject of the fact as γ_{1K}:

$$\gamma_{1K} = (x^1 \vee x^2 \vee x^3) z^{Nom} (c^{tar} \vee c^{ter} \vee c^{dar} \vee c^{der} \vee c^{lar} \vee c^{ler} \vee c^0) \quad (4)$$

The semantic role of the Object of the fact in the Kazakh phrase, i.e. the person or object of the action is defined as γ_{2K}:

$$\gamma_{2K} = (x^0 \vee x^2 \vee x^3)(z^{Gen} \vee z^{Acc})(y^{NoV} \vee y^{NoN} \vee y^{NCom} \vee y^{NDer} \vee y^0)$$
$$\wedge (c^{tar} \vee c^{ter} \vee c^{dar} \vee c^{der} \vee c^{lar} \vee c^{ler} \vee c^0) a^{NSim}$$

$$(5)$$

Forming the logical-linguistic equation of the Predicate of the action in the Kazakh phrase is based on the definition of the fact. According to the definition, a fact is a real, concrete single event that happened or will happen. Thus, we consider only the indicative mood of verbs and do not take into consideration the imperative, optative, conditional moods that exist in the Kazakh language.

The predicate γ_{VK} defines a combination of semantic and grammatical features of the central part of a fact triplet, namely an action or a fact Predicate:

$$\gamma_{VK} = (x^{-1} \vee x^{-2} \vee x^{-3})((f^{tur} \vee f^{otur} \vee f^{jatyr} \vee f^{jur})_m^{Pr F} z^{Vad}$$
$$\vee (y^{Oad} \vee y^{FuCo})_m^{Pr Fl} \vee y^{FuCo}(_m^{Pr Fl} \vee m^{Pr Flfedi}) \vee y^y(f^{edi} \vee f^{eken})$$
$$\vee (y^{Vad} m^{Pr Fl}(p^{mic} \vee p^0)) \vee m^{PoFl}((y^{Vart} \vee y^{Vpa} \vee y^{Vpas})$$
$$\vee f^{edi}(n^{joq} \vee n^{emes} \vee n^{me} \vee n^0) \vee (y^{Part} \vee y^{Vad} \vee f^{otur} \vee f^{tur} \vee f^{jatyr}$$
$$\vee f^{jur} \vee f^{ParP} \vee f^{UnFu})))$$

$$(6)$$

Table 1. The predicate variables and their values ranges defined in the Open IE model for the Kazakh language

Variables	Features	Values
x	The location of the analyzed word in a phrase	Shows a word position in a sentence, "minus" means the start of the count from the end of the sentence; 0 shows any other position of the word except the first three and the last three words in the sentence
f	The feature of an auxiliary verb in the phrase	*aux* shows the existence of any of 35 auxiliary verbs of the Kazakh language in the analyzed phrase
z	The grammatical case of the Kazakh noun	*Nom* – nominative, *Gen* – genitive, *Dat* – dative, *Acc* – accusative, *Ela* – local, *Ins* – instrumental, *Abl* – ablative
a	The types of the Kazakh nouns declensions	*NSim* is a simple declension of nouns, *NPos* is a possessive declension of nouns
n	The feature of the negative sentence	*me* and *emes* are signs of a negative sentence, represented by two different lists of words or particles
c	The feature of plural suffixes	*tar, ter, dar, der, lar, ler* show the presence of a plural suffix with the same name in the analyzed word
y	The derivational suffixes for verbs, nouns, participles, adverbials	*UnFu, FuCo* are features of a suffix of uncertain future tense and future conjecture tense in the analyzed word; *Psuf* and *Usuf* are features of one of 189 productive or one of 65 unproductive suffixes from specific lists in the analyzed verb; *NoN, NoV, Ncom, Nder* are features of the noun generation (*NoN* – from a noun, *NoV* – from a verb, *Nder* is a feature of some expression); *Part, ParP* are features of the participle generation by means of two different lists of suffixes; *VaP, Oad, Vad* are features of the verbal participle generation by means of three different lists of suffixes; *Vpas* is a feature of one of 20 verb suffixes in the analyzed word; y is a sign of the existence of suffix of the infinitive verb form; 0 is a sign of a verb stem
d	The subjunctive action of the analyzed verb	*shi* shows a suffix of the subjunctive in the analyzed verb and 0 shows lack of such suffixes
m	A personal predicative or possessive flexion of the analyzed verb and verbal forms	*PrFl/PoF* show a personal predicative/possessive flexion of analyzed participles, verbal adverbs, main and auxiliary verbs
b	The supplementary semantics of the analyzed action	*mic* denotes the guessed action, *se* denotes the conditional mood and 0 denotes the lack of some supplementary semantics of the analyzed verb

Figure 3 shows an example of the model implementation for the Kazakh sentences. In the Kazakh phrase "Operatorlar úide myltyq tapty", according to formula (6), the verb "tapty" represents an action (past perfect tense). According to Eq. (5), the noun "Operatorlar" is identified as the subject of the action or the subject of the fact. The predicate γ_{2K} (5) identifies the noun "mylty", as an object of the fact.

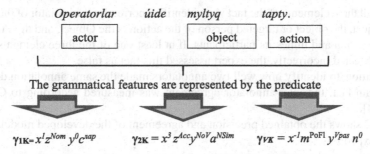

$$\gamma_{1K} = x^l z^{Nom} y^0 c^{nap} \qquad \gamma_{2K} = x^3 z^{Acc} y^{NoV} a^{NSim} \qquad \gamma_{VK} = x^{-l} m^{PoFl} y^{Vpas} n^0$$

Fig. 3. An example of the fact identification in the Kazakh phrase. The predicate γ_{1K} defines the grammatical features of the Subject action, the predicate γ_{2K} defines the Object and γ_{VK} is the Predicate of the fact.

4   Source Data and Experimental Results

The experimental research dataset includes a pilot parallel corpus of Russian-Kazakh texts extracted from Kazakh bilingual news websites inform.kz, azattyq.org, patrul.kz, zakon.kz, caravan.kz, lenta.kz, nur.kz by the parser, based on the Python BeautifulSoup library. Information collection time: from June 2018 to June 2019. The choice of sites is grounded on: (1) the reliability of the sites; (2) the ability to select specialized criminal texts; (3) the ability to switch between Kazakh and Russian languages. The volume of the corpus is about 500 thousand words. Since this study considers the implementation of the logical-linguistic model for the Kazakh texts, in the experiment we used only the Kazakh part of the corpus, which includes about 225 thousand words.

In order to get the values of the subject variables of formulas (4)–(6), tokenization and POS-tagging of texts were carried out. Tokenization was conducted by the tokenize module of the NLTK Python library. For POS-tagging of Kazakh texts, we developed a tagger based on the RegexpTagger class of the NLTK Python package. Figure 4 shows a fragment of a regular expression that allows identifying some forms of nouns in Kazakh sentences.

> patterns=[(r'.*бен$','NN'), ('r.* пенен$','NN'), ('r.* басшылык$','NN'), (r'.* ipкону$','NN'), (r'.* тармен$','NN'), (r'.* герлермен$','NN'), (r'.* здар$','NN')]

Fig. 4. A fragment of a regular expression that allows identifying some noun forms in Kazakh sentences.

Using the tag names and some syntactic characteristics of words in the Kazakh sentences as the values of the predicate variables in corresponding equations allows us to extract Subjects, Objects and Predicates of facts from texts of Web content.

Since the training corpus is created using the model, only the precision indicator was used to evaluate the effectiveness of the model. To assess the results of the model, approximately a thousand facts were randomly selected from the list of facts that were automatically extracted from the corpus. The expert evaluated the extracted fact as true if the fact triplet was identified correctly and false otherwise. A fact is considered to be correct if all three elements of the fact were identified correctly: the initiator of the action is the Subject, the object or targeted person of the action is the Object, and the Predicate names the action and unites its participants. If at least one of the three elements of the fact was detected incorrectly, the expert assessed this fact as false.

In addition, to identify how well two annotators made the same annotation decision for a certain fact, the inter-annotator agreement was measured according to Cohen's Kappa [24].

Table 2 shows the obtained precision and agreement of the developed model for the Kazakh text corpus.

Table 2. Evaluation of the experimental results.

Language	Size, words	Precision	Agreement
Kazakh	225 000	71.0%	0.72

5 Conclusions and Future Works

The main result of this research is the adaptation of the developed logical-linguistic model for fact triples extraction from unstructured Kazakh texts. This model, created within Open IE approach, allows extracting the unlimited domain-independent number of facts from sentences of the Kazakh Web content.

Representing the structured information extracted in our model as the Subject-Predicate-Object fact allows exploiting it to form automatic RDF triplets, i.e. automatic ontology generation. In this case, in the ontology RDF graph, the word, whose semantics is described by Eq. (6), will form the RDF-Predicate, the noun corresponding to Eq. (4) will form the RDF-Subject, and the noun described by Eq. (5) will represent the RDF-Object of the triplet.

Extracted from Kazakhstan news websites, the constructed text corpus and conducted experiment show that the precision of the model is more than 71%, with the agreement coefficient of about 72%. The precision is thought to be slightly increased by improving the results of POS-tagging of texts.

In future studies, we intend to formulate and experimentally verify the logical-linguistic equations that identify the attributes of the fact in the Kazakh sentence, such as time, place of the action, etc. This problem is a more complex challenge than the fact core formation (the Subject-Predicate-Object triplet). The reason is the lack of strict determinacy of grammatical features expressing semantics of the fact attributes. We can assume that the solution to this problem will require the integrated use of regular expressions and logical-linguistic equations.

In addition, to increase the experimental reliability, our further work will extend the domain of the texts studied and compare the results of the model implementation for Russian and Kazakh texts of the constructed parallel corpus.

References

1. Sint, R., Schaffert, S., Stroka, S., Ferstl, R.: Combining unstructured, fully structured and semi-structured information in semantic wikis. In: Proceedings of the 4th Semantic Wiki WorkShop (SemWiki) at the 6th European Semantic Web Conference, ESWC (2009)
2. Crestan, E., Pantel, P.: Web-scale knowledge extraction from semi-structured tables. In: WWW 2010 Proceedings of the 19th International Conference on World Wide Web, pp. 1081–1082 (2010)
3. Wong, Y.W., Widdows, D., Lokovic, T., Nigam, K.: Scalable attribute-value extraction from semi-structured text. In: 2009 IEEE International Conference on Data Mining Workshops, pp. 302–307 (2009)
4. Phillips, W., Riloff, E.: Exploiting strong syntactic heuristics and co-training to learn semantic lexicons. In: Proceedings of the conference on Empirical Methods in Natural Language Processing (EMNLP) (2002)
5. Jones, R., Ghani, R., Mitchell, T., Riloff, E.: Active learning with multiple view feature sets. In: ECML 2003 Workshop on Adaptive Text Extraction and Mining (2003)
6. ARPA. Proceedings of the 3rd Message Undestanding Conference (1991)
7. Etzioni, O., Banko, M., Soderland, S., Weld, D.: Open information extraction from the web. Commun. ACM **51**(12), 68–74 (2008)
8. Fader, A., Soderland, S., Etzioni, O.: Identifying relations for open information extraction. In: Proceedings of the Conference on Empirical Methods in Natural Language Processing, Edinburgh, Scotland, UK, pp. 1535–1545 (2011)
9. Duc-Thuan, V., Ebrahim, B.: Open information extraction. In: Encyclopedia with Semantic Computing and Robotic intelligence, vol. 1, no. 1 (2016)
10. Shinzato, K., Sekine, S.: Unsupervised extraction of attributes and their values from product description. In: Sixth International Joint Conference on Natural Language Processing, IJCNLP 2013, pp. 1339–1347 (2013)
11. Liu, L., Ren, X., Zhu, Q., et al.: Heterogeneous supervision for relation extraction: a representation learning approach. In: Proceedings of the 2017 Conference on Empirical Methods in Natural Language Processing, pp. 46–56 (2017)
12. Wang, X., Zhang, Y., Chen, Y.: A survey of truth discovery in information extraction (2018)
13. Gamallo, P., Garcia, M., Fernandez-Lanza, S.: Dependency-based open information extraction. In: Proceedings of the Joint Workshop on Unsupervised and Semi-Supervised Learning in NLP, pp. 10–18 (2012)
14. Akbik, A., Loser, A.: KrakeN: N-ary facts in open information extraction. In: Proceedings of the Joint Workshop on Automatic Knowledge Base Construction and Web-scale Knowledge Extraction, pp. 52–56 (2012)
15. Fader, A., Soderland, S., Etzioni, O.: Identifying relations for open information extraction. In: Proceedings of the Conference on Empirical Methods in Natural Language Processing, pp. 1535–1545 (2011)
16. Angeli, G., Premkumar, M.J., Manning, C.D.: Leveraging linguistic structure for open domain information extraction. In: Proceedings of the 53rd Annual Meeting of the Association for Computational Linguistics, pp. 344–354 (2015)
17. Gashteovsk, K., Gemulla, R., Del Corro, L.: MinIE: minimizing facts in open information extraction. In: Proceedings of the Conference on Empirical Methods in Natural Language Processing, pp. 2630–2640 (2017)
18. Mooney, R.J., Bunescu, R.: Mining knowledge from text using information extraction. ACM SIGKDD Explor. Newslett. **7**(1), 3–10 (2005). Natural language processing and text mining
19. Gamallo, P., Garcia, M.: Multilingual open information extraction. In: Portuguese Conference on Artificial Intelligence, pp. 711–722 (2015)

20. Khairova, N., Lewoniewski, W., Węcel, K.: Estimating the quality of articles in russian wikipedia using the logical-linguistic model of fact extraction. In: Abramowicz, W. (ed.) BIS 2017. LNBIP, vol. 288, pp. 28–40. Springer, Cham (2017). https://doi.org/10.1007/978-3-319-59336-4_3

21. Khairova, N., Lewoniewski, W., Węcel, K., Orken, M., Kuralai, M.: Comparative analysis of the informativeness and encyclopedic style of the popular web information sources. In: Abramowicz, W., Paschke, A. (eds.) BIS 2018. LNBIP, vol. 320, pp. 333–344. Springer, Cham (2018). https://doi.org/10.1007/978-3-319-93931-5_24

22. Khudhair, A.T.: The intelligence theory mathematical apparatus formal BASE. Adv. Inf. Syst. 1(1), 38–43 (2017)

23. Khairova, N.F., Petrasova, S., Gautam, A.P.S.: The logical-linguistic model of fact extraction from English texts. In: Dregvaite, G., Damasevicius, R. (eds.) ICIST 2016. CCIS, vol. 639, pp. 625–635. Springer, Cham (2016). https://doi.org/10.1007/978-3-319-46254-7_51

24. Regneri, M., Wang, R.: Using discourse information for paraphrase extraction. In: Proceedings of the 2012 Joint Conference on Empirical Methods in Natural Language Processing and Computational Natural Language Learning, pp. 916–927 (2012)

Comparing High Dimensional Word Embeddings Trained on Medical Text to Bag-of-Words for Predicting Medical Codes

Vithya Yogarajan[1]([⊠]) [ID], Henry Gouk[2][ID], Tony Smith[1][ID], Michael Mayo[1], and Bernhard Pfahringer[1][ID]

[1] Department of Computer Science, University of Waikato, Hamilton, New Zealand
vy1@students.waikato.ac.nz
[2] School of Informatics, University of Edinburgh, Edinburgh, Scotland

Abstract. Word embeddings are a useful tool for extracting knowledge from the free-form text contained in electronic health records, but it has become commonplace to train such word embeddings on data that do not accurately reflect how language is used in a healthcare context. We use prediction of medical codes as an example application to compare the accuracy of word embeddings trained on health corpora to those trained on more general collections of text. It is shown that both an increase in embedding dimensionality and an increase in the volume of health-related training data improves prediction accuracy. We also present a comparison to the traditional bag-of-words feature representation, demonstrating that in many cases, this conceptually simple method for representing text results in superior accuracy to that of word embeddings.

Keywords: Word embeddings · Binary classification · Machine learning for health

1 Introduction

Recent years have seen significant growth in the use of machine learning techniques to better understand health care and improve quality of service—primarily due to the increase in the availability of large quantities of electronic health records (EHRs). Secondary analysis of EHRs has the potential to improve a variety of healthcare aspects, including patient care, medical outcomes, surgical outcomes, risk management, clinical decision support and medical diagnoses. However, the free-form text content of EHRs poses many challenges not typically addressed by conventional natural language processing (NLP). Due to the complexity and variations presented in the data, and the legal and ethical aspects associated with the use of this data, the analysis of EHRs has not seen benefits as significant as those enjoyed by more common application domains.

© Springer Nature Switzerland AG 2020
N. T. Nguyen et al. (Eds.): ACIIDS 2020, LNAI 12033, pp. 97–108, 2020.
https://doi.org/10.1007/978-3-030-41964-6_9

Word embeddings are often used for solving problems that involve extracting high-level knowledge from free-form text data. These word embeddings are typically trained on corpora composed of general language, such as archives of English Wikipedia, that are unlikely to be representative of the way language is used in EHRs and other related healthcare data sources. Using embeddings trained on general text for tasks that involve specialised language results in a domain shift, which will typically cause suboptimal performance [21]. Ideally, if we want to classify documents derived from EHRs then we should train the embeddings on a large collection of free-form text extracted from EHRs. For various legal and ethical reasons, this is not possible: the collections of health records available for research purposes are not large enough to train high quality word embeddings.

The contributions of this work are two-fold: (i) we demonstrate that training word embeddings on health-related corpora provides an increase in accuracy compared to embeddings trained on general text—particularly when the dimesionality of the embeddings is increased; (ii) it is shown that the bag of words representation is often as effective, if not more so, than dense word embeddings when applied to medical code prediction.

2 Related Work

Many NLP tasks, health-related or otherwise, use word embeddings to represent text data, due to their ability to encode semantic similarity between words. Word embeddings represent a single word or sub-word as a vector based on the context in which it appears. Examples of use of word embeddings for health applications include: learning medical concepts such as diagnosis codes, medication codes, procedure codes [6], early detection of heart failure [5], and medical event detection [13]. Many previous techniques have used the word2vec [19,20] or GloVe [22] packages for training embeddings. One issue with the methods employed by word2vec and GloVe is that they cannot produce embeddings for words that were not seen during training. In contrast, fastText [2,16,17] makes use of character level n-grams, which enables one to generate accurate embeddings for words that do not appear in the training vocabulary. The use of character-level n-grams is of particular importance in the medical domain, where a significant number of compound words are used [4,25,28].

Generally applications of NLP in health use general text to train word embeddings. In cases where the health-related text is used to train word embeddings, most published models only use between 200 and 400 dimensions [1,18]. Recent studies show that the use of large corpora from more than one source can improve the performance of embeddings [3,27]. Chen et al. [4] and Zhang et al. [28] provide embeddings on health-related texts, with word embeddings of 700- and 200-dimensional embeddings respectively. Zhang et al. [28] make use of the sub-word information during the training of word embeddings. We also make use of sub-word information during the training of word embeddings; however, in contrast to Zhang et al. [28], we present high dimensional word embeddings. Also, our word embeddings make use of large corpora of health-related text from multiple sources. We present comparisons of F-measures using these recently published word embeddings for the prediction of medical codes to our word embeddings.

Purushotham *et al.* [23] use Medical Information Mart for Intensive Care (MIMIC) III to present benchmark models on clinical prediction tasks such as mortality prediction, forecasting length of stay, and ICD-9 code group prediction. MIMIC III is one of the largest publicly available medical databases, containing both structured data and free-form text records [7,9,15]. We make use of the free-form text hospital discharge reports contained in MIMIC III, along with the corresponding ICD-9 diagnosis codes.

3　Representing Text

The bag of words (BOW) approach is a simple method for representing text that does not consider the order that words occur in a document. A document is represented as a sparse vector where each element stores either the number of occurrences of a word, or a binary value indicating that the word is present in the document. BOW is considered to be a relatively simple yet effective method [8,17].

Embedding words in vector spaces that encode semantics has become popular in recent years. In general, for many NLP tasks, continuous word representations trained on large unlabelled datasets have been shown to improve performance relative to other representations [2,16,17]. Figure 1 provides a pictorial example of how these vector spaces may be organised. The use of word embeddings is motivated by the distributional hypothesis [12], which states that there is a higher chance that words with similar meaning will occur in similar contexts. By examining a large corpus, it is possible to learn embeddings that capture the semantic similarity between words, as inferred by the contexts they are seen in. Word embeddings provide a means for effective representation learning without the complexity of deep neural networks, and can be trained efficiently on large datasets [19].

FastText [2] is one popular system for learning word embeddings. It supports both the skip-gram with negative sampling (SGNS) and continuous bag of words (CBOW) methods for training word embeddings. In contrast to word2vec, where distinct word embeddings are learnt directly from words, fastText represents each word as a bag of character *n*-grams, and word embeddings are obtained by summing these character *n*-gram representations. More information on fastText is provided by Bojanowski *et al.* [2]. For example, the tri-grams for the word "apple" are "app", "ppl", and "ple". The resulting word embedding vector for "apple" will be the sum of the vectors of each of these three tri-grams. This modelling choice enables fastText to produce vectors even for novel words that were not present in the training data, as long as at least some of the n-grams have been seen before. It has been shown that fastText can achieve accuracies similar to deep learning classifiers, while being a lot more efficient to train [17].

The classification problems encountered in natural language processing typically involve predicting labels for entire documents, rather than individual words.

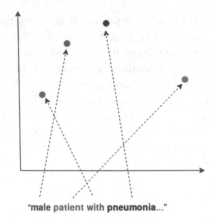

Fig. 1. Visual representation of word embeddings, where each word is mapped to a vector. For simplicity only a 2-D representation is used for embeddings.

As such, one must define a representation for documents that can be easily constructed using the embeddings learned for words. In this work, we obtain document embeddings by computing the vector sum of the embeddings for each word in the document. This vector sum is then normalised to have length one, to ensure that documents of different lengths have representations of similar magnitudes.

4   Data

MIMIC III is used in this study both for classification experiments and for training word embeddings. The most recent version of this dataset, MIMIC III, is one of the most comprehensive publicly available medical databases [7,9,15]. It contains de-identified health records of 49,785 adult patient admissions (age >15) and 7,870 neonatal admissions to critical care units. The data was collected at the Beth Israel Deaconess Medical Center between 2001 and 2012. It includes information such as demographics, laboratory test results, procedures, medications, and physician notes. For this research, we are interested in the discharge summaries of patients admitted to the hospital.

The TREC precision medicine/clinical decision support track 2017 (TREC 2017) [24] provides a considerable corpus of health-related free-form text. This includes 26.8 million published abstracts of medical literature listed on PubMed Central, 241,006 clinical trials documents, and 70,025 abstracts from recent proceedings focused on cancer therapy from AACR (American Association for Cancer Research) and ASCO (American Society of Clinical Oncology). The dataset from the TREC 2017 competition is used here for training word embeddings.

Medical codes, such as the International Classification of Diseases (ICD-9) codes, are widely used to describe diagnoses of patients [14]. Most hospitals

Table 1. Percentage of occurrence of ICD-9 code groupings in unique hospital admissions in MIMIC III. The total number of hospital admissions with a recorded discharge summary is 52,710. E and V codes are referring to external causes of injury and supplemental classification.

ICD-9	%	ICD-9	%	ICD-9	%
Circulatory (circ)	78.40	Digestive (diges)	38.80	Muscular (musc)	17.99
E and V (e+v)	69.09	Blood (bld)	33.56	Prenatal (pren)	17.07
Endocrine (endo)	66.51	Symptoms (symp)	31.36	Neoplasms (neop)	16.37
Respiratory (resp)	46.63	Mental (ment)	29.66	Skin (skin)	12.02
Injury (inj)	41.42	Nervous (nerv)	29.10	Congenital (cong)	5.41
Genitourinary (gen)	40.29	Infectious (inf)	26.96	Pregnancy (preg)	0.31

manually assign the correct codes to patient records based on doctors' clinical diagnosis notes. Hence, the use of machine learning techniques to predict ICD codes from free-form medical text and thus automating the medical coding process has become an important research avenue.

MIMIC III contains ICD-9 annotations to indicate the diagnoses and diseases of admitted patients. There are 6,984 distinct ICD-9 codes reported in MIMIC III, among the more than 50,000 patient admission records found in this database. These can be grouped into 18 categories, as shown in Table 1 along with the frequencies of these groups. Records typically have more than one code assigned. This work focuses on the application of labelling discharge summaries, as these are the most readily available free-form text records in the MIMIC III dataset.

5 Experiments

We consider the 18 categories of medical codes, presented in Table 1, as 18 separate binary classification problems. That is, each group of ICD-9 codes from the MIMIC III discharge summaries is predicted in isolation. FastText is used for training word embeddings and representing documents, and the Waikato Environment for Knowledge Analysis (WEKA) [11,26] framework is used to train classifiers on these documents. This section discusses the experimental setup in more detail.

5.1 Data Pre-processing

In order to maximise the use of free-form medical text "as is," we minimise preprocessing. One of the significant issues of data mining medical text in free-form is the use of acronyms and abbreviations. Simple changes such as converting uppercase letters to lowercase, or omitting full stops can result in a completely different meaning. For example, "Ab" is used to refer to an antibody, while "AB"

Table 2. Word embeddings trained by us (top), from previous work (middle), or concatenations thereof (bottom). Dimension details are presented, as are training times and word embeddings model sizes.

Models	Dimensions	Source data	Train time	Model size
M300	300	MIMIC	1 h	5G
T300	300	TREC	7 h	13G
TM300	300	TREC+MIMIC	9 h	15G
T600	600	TREC	13 h	23G
TM600	600	TREC+MIMIC	16 h	30G
T900	900	TREC	19 h	35G
TM900	900	TREC+MIMIC	23 h	54G
W300 [10]	300	Wiki	–	7G
BWV200 [4, 28]	200	PubMed[a]+MIMIC	–	26G
BSV700 [4, 28]	700	PubMed+MIMIC	–	21G
T300+M300	600	TREC+MIMIC	8 h	18G
W300+T300+M300	900	Wiki+TREC+MIMIC	8+ h	25G
T900+W300	1,200	Wiki+TREC	19+ h	42G
TM900+W300	1,200	Wiki+TREC+MIMIC	23+ h	61G

[a]https://www.ncbi.nlm.nih.gov/pubmed/

is used to refer to abortion. As word embeddings are case sensitive, we keep the text as is for both training and experiments to capture maximum meaning. We do not perform downcasing, nor do we remove special characters or full stops. We only remove extra spaces and unwanted newline characters, as fastText uses new line characters to separate examples.

5.2   Training Word Embeddings

Our embeddings are trained to the exact same specifications as the Wikipedia and common crawl fastText models in [10]. We make use of both MIMIC III and TREC 2017 datasets to train our word embeddings. The sizes of these datasets are 4 GB and 24 GB, respectively. The word embeddings are trained using the CBOW method, character n-grams of length 5, a window of size 5, ten negative samples per positive sample, and with various settings for the number of dimensions. The learning rate used for training these models is 0.05, with the exception of M300 (see Table 2), where the learning rate is 0.03. We also include two very recently published (2019) medical text trained word embeddings of dimension size 200 and 700 [4, 28] for comparison.

Table 2 presents details of the embedding trained by us, previously published word embeddings, and the concatenated word embeddings. Concatenated embeddings are word embeddings formed by concatenating multiple word embeddings. For example, in the T300+M300, the first 300 elements are the word vectors obtained using the TREC dataset, and the second 300 elements are taken from the embeddings trained on MIMIC III. The table includes details

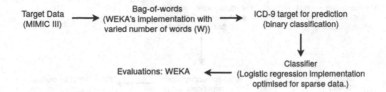

Fig. 2. Flow chart of using bag-of-words for prediction.

on dimensions, input data, training time[1] and the size of the model. Both the size of the input data and the number of dimensions influence the training times and model sizes.

5.3 Experimental Process and Classification

Figures 2 and 3 present flowcharts of using BOW and word embeddings for predicting ICD-9 groups from MIMIC III discharge summaries. We use a total of 52,710 discharge summaries, with text length ranging from a few sentences to close to twenty pages. WEKA's implementation of BOW is used with a varied number of words. We use ten-fold cross-validation and classifiers as implemented in WEKA.

We use logistic regression with ridge value of 1 for word embeddings experiments. We experimented with the use of random forests, with various parameter choices, as well as other ridge values for logistic regression. However, we found logistic regression was performing well and was providing consistent F-measures across a range of different ridge values. The purpose of this research is not to achieve the highest possible F-measures, but to show that the high dimensional word embeddings trained on medical text do provide advantages in health applications relative to low dimensional embeddings trained on general text. Hence, we only present results for logistic regression.

For BOW due to the sparsity of the data for large dictionary sizes (such as 100,000 or 600,000 words) we use an implementation of logistic regression optimised for sparse data.

6 Results

This section presents an overview of our experimental results. Table 3 provides a comparison of F-measure for predicting ICD-9 groups from free-form MIMIC III discharge summaries for 300-dimensional and 600-dimensional embeddings. For 300-dimensional embeddings, W300 are word embeddings that are trained by fastText on Wikipedia and other common crawl text. W300 embeddings are readily available for use in any application. Except for the circulatory label, which is the most frequent one (78.4%), word embeddings specially trained on medical

---

[1] Training was run on a 4 core Intel i7–6700K CPU @ 4.00 GHz with 64 GB of RAM.

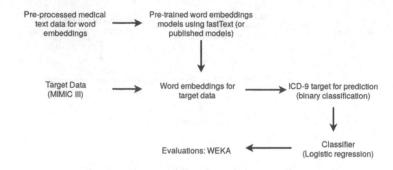

Fig. 3. Flow chart of using word embeddings for prediction.

corpora have better F-measures. Overall T300 provides better F-measures than other 300-dimensional word embeddings for most ICD-9 groups. When compared to the recently published BWV200, we found that our 300-dimensional word embeddings performed better for all categories, and on par for E and V.

For 600-dimensional word embeddings, Table 3 presents comparisons across embeddings obtained in a single training phase (T600 and TM600), and word embeddings obtained via concatenation (T300+M300). We compare our 600 dimension word embeddings to the published 700 dimensional word embeddings (BSV700). F-measures of our 600 dimensional word embeddings are on par with or better than those of the recently published high dimensional word embeddings.

Table 4 presents a comparison for predicting ICD-9 code from free-form discharge summaries in MIMIC III with various dimensions of word embeddings, different number of words for BOW, and between word embeddings and BOW. For word embeddings, the best F-measures for 600-, 900- and 1200- dimensional embeddings are presented for each ICD-9 group. We also indicate which model produced the best 900-dimensional and 1,200-dimensional embeddings (see Table 2 for details of word embeddings, input data and model dimensionality). Generally, the higher the dimensionality, the better the F-measures are for predicting the ICD-9 groups.

For BOW, F-measures of dictionary sizes 1,000, 10,000, 100,000 and 600,000 are presented for all 18 ICD-9 groups. BOW with 600,000 number of words is the largest possible number of features. There is an increase in F-measure as the size of the dictionary increases. A dictionary size of 600,000 results in the best F-measures for all ICD-9 category except pregnancy.

In comparison to word embeddings, F-measures of BOW is consistently better for all ICD-9 groups that occur for less than 42% of the examples. In terms of BOW performance, pregnancy is the most interesting ICD-9 group. Its frequency is only 0.31% in the MIMIC III dataset. Text in this group is very specific, including uniquely identifying words such as delivery, labour and birth, and is probably one possible explanation to the success of BOW for predicting the pregnancy label.

Table 3. A comparison of F-measures for predicting ICD-9 groups using 200, 300, 600 and 700-dimensional word embeddings are presented. BWV200 and BSV700 are both published word embeddings and are compared with 300-dimensional word embeddings and 600-dimensional word embeddings, respectively. We use bold to indicate the best F-measures among low dimensional groups of word embeddings (200–300) and the higher dimensional word embeddings (600–700). The best F-measure across all presented word embeddings is underlined for each category.

ICD-9	BWV200	W300	M300	T300	TM300	T600	TM600	T3+M3	BSV700
circ	0.931	**0.932**	**0.932**	**0.932**	0.931	**<u>0.935</u>**	0.934	0.924	0.931
e+v	**0.829**	0.828	**0.829**	**0.829**	0.828	**<u>0.832</u>**	**0.832**	**0.832**	0.831
endo	0.847	0.845	0.846	**0.849**	0.846	**<u>0.851</u>**	0.849	0.850	0.847
resp	0.774	0.774	0.774	**0.778**	0.772	**<u>0.789</u>**	0.788	0.787	0.776
inj	0.660	0.649	**0.663**	0.662	0.660	0.675	0.676	0.677	**<u>0.682</u>**
gen	0.721	0.716	0.724	**0.731**	0.725	**<u>0.740</u>**	**0.740**	**0.740**	0.732
diges	0.679	0.692	0.693	**0.696**	0.692	**<u>0.712</u>**	0.705	0.710	0.696
bld	0.557	0.566	**0.573**	0.570	0.570	0.593	0.589	**<u>0.594</u>**	0.586
symp	0.475	0.486	0.482	**0.487**	0.483	0.504	0.502	0.500	**<u>0.505</u>**
ment	0.533	0.530	0.530	**0.542**	0.539	**<u>0.577</u>**	0.576	**0.577**	0.559
nerv	0.530	0.534	0.527	**0.543**	0.531	**<u>0.571</u>**	0.558	0.564	0.553
inf	0.634	0.634	0.641	**0.647**	**0.647**	0.663	0.659	**<u>0.664</u>**	0.648
musc	0.254	0.274	0.258	**0.294**	0.267	**<u>0.338</u>**	0.314	0.319	0.315
pren	0.589	0.590	0.588	**0.594**	0.587	0.601	0.597	0.598	**<u>0.603</u>**
neop	0.693	0.688	0.702	**0.705**	0.690	0.728	0.721	**<u>0.732</u>**	0.727
skin	0.343	0.335	0.344	**0.346**	0.344	0.389	0.384	0.386	**<u>0.397</u>**
cong	0.365	0.371	0.369	**0.391**	0.350	**<u>0.438</u>**	0.406	0.435	0.424
preg	0.525	0.502	0.543	**0.565**	0.512	0.579	0.566	**<u>0.599</u>**	0.586

7 Discussion

In this paper, we investigate how the source domain used for training word embeddings impacts the performance of medical text classification. We also demonstrate the effect that embedding dimensionality plays in determining the accuracy of the resulting classifiers. The prediction of ICD-9 codes from discharge summaries of MIMIC III is used as an example health application to show that high dimensions, especially trained on health-related corpora, have better F-measures compared to word embeddings with lower dimensions or are trained on general text such as Wikipedia. We also compare our word embeddings with recently published word embeddings and show that our embeddings perform better for most ICD-9 groups and are very similar for others. Reasons for such differences include pre-processing of input data, parameter selection, and source of the datasets used for training the embeddings. We also present comparisons with BOW, where we observe that F-measures obtained using BOW

Table 4. A comparisons of F-measure for ICD-9 groups between word embeddings with varied dimensions (left) and BOW with varied number of words (right). For word embeddings the best F-measure across 900-dimensional and 1200-dimensional word embeddings for each category are presented. Corresponding best 900- and 1,200-dimensional models are also listed. For details of the models see Table 2. We use bold to indicate the best F-measures among varied dimensional word embeddings and varied number of words for BOW. The best F-measure across all is underlined for each category.

ICD-9	Word embeddings					BOW			
	600	900	Best 900-dim model	1,200	Best 1,200-dim model	1,000	10,000	100,000	600,000
circ	0.935	**0.937**	T900	0.936	T9W3	0.930	0.920	0.931	**0.932**
e+v	0.832	**0.833**	W3T3M3	**0.833**	T9W3	0.801	0.788	0.808	**0.812**
endo	0.851	0.853	T900	0.854	T9W3	0.814	0.825	0.840	**0.845**
resp	0.789	0.792	W3T3M3	0.794	TM9W3	0.763	0.775	0.788	**0.792**
inj	0.677	0.684	WTM+T900	**0.689**	TM9W3	0.642	0.675	0.693	0.697
gen	0.740	0.748	TM9	**0.751**	TM9W3	0.733	0.751	0.769	0.775
diges	0.712	0.724	T900	**0.730**	T9W3	0.701	0.728	0.748	0.752
bld	0.594	0.601	TM9	**0.607**	T9W3	0.558	0.595	0.608	0.614
symp	0.504	0.514	W3T3M3	**0.517**	T9W3	0.476	0.507	0.524	0.528
ment	0.577	0.592	TM9	**0.606**	TM9W3	0.567	0.616	0.635	0.639
nerv	0.571	0.577	T900	**0.586**	T9W3	0.491	0.594	0.628	0.629
inf	0.664	0.671	T900	**0.677**	T9W3	0.606	0.667	0.693	0.698
musc	0.338	0.354	T900	**0.372**	T9W3	0.344	0.476	0.488	0.489
pren	0.601	0.607	W3T3M3	**0.608**	Both	0.574	0.557	0.620	0.623
neop	0.732	0.741	W3T3M3	**0.746**	T9W3	0.665	0.713	0.766	0.773
skin	0.389	0.418	T900	**0.435**	T9W3	0.438	0.483	0.526	0.530
cong	0.438	**0.465**	T900	0.463	T9W3	0.348	0.485	0.519	0.529
preg	0.599	0.593	W3T3M3	**0.605**	TM9W3	0.737	0.705	0.709	0.726

are consistently better that word embeddings for all ICD-9 groups that occur for less than 42% of the examples. In general, word embeddings are favoured over BOW as word embeddings are known to capture the meaning of text content, and can better utilise a range of classifiers compared to BOW. However, the results also indicate that for some categories, such as pregnancy, where the data is rather imbalanced, and very specific vocabulary is used, BOW may be the better option.

We also present the sizes and training times required for training word embeddings. Model sizes and training times are both influenced by the input data size and the number of dimensions generated, and can become quite large. The main reason for the large model sizes is the use of hash tables for storing character n-gram information. FastText does provide ways to reduce the final word embeddings model sizes, however such compression necessarily also impacts on accuracy.

For this research, we considered ICD-9 groupings as a set of binary classification problems. Alternatively they could also be represented as a single multi-label classification problem. Each unique patient admitted to the hospital can have more than one diagnosis and be categorised into different groups or have more than one diagnosis from the same ICD-9 group. Our main aim for this research was to investigate high dimensional word embeddings trained in the medical text and hence treating ICD-9 grouping as a binary classification problem was sufficient. However, to optimise the accuracy of predicting ICD-9 code from the free-form medical text, it will be essential to also investigate it as a hierarchical multi-label classification problem in future work.

References

1. Beam, A.L., et al.: Clinical concept embeddings learned from massive sources of multimodal medical data. arXiv preprint arXiv:1804.01486 (2018)
2. Bojanowski, P., Grave, E., Joulin, A., Mikolov, T.: Enriching word vectors with subword information. arXiv preprint arXiv:1607.04606 (2016)
3. Cao, Y., Huang, L., Ji, H., Chen, X., Li, J.: Bridge text and knowledge by learning multi-prototype entity mention embedding. In: Proceedings of the 55th Annual Meeting of the Association for Computational Linguistics (Volume 1: Long Papers), pp. 1623–1633 (2017)
4. Chen, Q., Peng, Y., Lu, Z.: BioSentVec: creating sentence embeddings for biomedical texts. In: 7th IEEE International Conference on Healthcare Informatics (2019)
5. Choi, E., Schuetz, A., Stewart, W.F., Sun, J.: Using recurrent neural network models for early detection of heart failure onset. J. Am. Med. Inform. Assoc. JAMIA **24**(2), 361–370 (2017). https://doi.org/10.1093/jamia/ocw112
6. Choi, Y., Chiu, C.Y.I., Sontag, D.: Learning low-dimensional representations of medical concepts. AMIA Summits on Transl. Sci. Proc. 41–50 (2016)
7. MIT Critical Data: Secondary Analysis of Electronic Health Records. Springer, Cham (2016). https://doi.org/10.1007/978-3-319-43742-2_30
8. Goldberg, Y.: Neural network methods for natural language processing: Synth. Lect. Hum. Lang. Technol. **10**(1), 1–309 (2017)
9. Goldberger, A.L., et al.: PhysioBank, PhysioToolkit, and PhysioNet: components of a new research resource for complex physiologic signals. Circulation **101**(23), e215–e220 (2000)
10. Grave, E., Bojanowski, P., Gupta, P., Joulin, A., Mikolov, T.: Learning word vectors for 157 languages. In: Proceedings of the International Conference on Language Resources and Evaluation (LREC 2018) (2018)
11. Hall, M., Frank, E., Holmes, G., Pfahringer, B., Reutemann, P., Witten, I.: The WEKA data mining software: an update. ACM SIGKDD Explor. Newsl. **11**(1), 10–18 (2009)
12. Harris, Z.S.: Distributional structure. Word **10**(2–3), 146–162 (1954). https://doi.org/10.1080/00437956.1954.11659520
13. Jagannatha, A.N., Yu, H.: Bidirectional RNN for medical event detection in electronic health records. In: North American Chapter Meeting, pp. 473–482. Association for Computational Linguistics (2016)
14. Jensen, P.B., Jensen, L.J., Brunak, S.: Mining electronic health records: towards better research applications and clinical care. Nat. Rev. Genet. **13**(6), 395 (2012)

15. Johnson, A.E., et al.: MIMIC-III, a freely accessible critical care database. Sci. Data **3**, 160035 (2016)
16. Joulin, A., Grave, E., Bojanowski, P., Douze, M., Jégou, H., Mikolov, T.: Fasttext.zip: compressing text classification models. arXiv preprint arXiv:1612.03651 (2016)
17. Joulin, A., Grave, E., Bojanowski, P., Mikolov, T.: Bag of tricks for efficient text classification. arXiv preprint arXiv:1607.01759 (2016)
18. Mencía, E.L., De Melo, G., Nam, J.: Medical concept embeddings via labeled background corpora. In: Proceedings of the Tenth International Conference on Language Resources and Evaluation (LREC 2016), pp. 4629–4636 (2016)
19. Mikolov, T., Chen, K., Corrado, G., Dean, J.: Efficient estimation of word representations in vector space. arXiv preprint arXiv:1301.3781 (2013)
20. Mikolov, T., Sutskever, I., Chen, K., Corrado, G.S., Dean, J.: Distributed representations of words and phrases and their compositionality. In: Advances in Neural Information Processing Systems, pp. 3111–3119 (2013)
21. Pakhomov, S.V., Finley, G., McEwan, R., Wang, Y., Melton, G.B.: Corpus domain effects on distributional semantic modeling of medical terms. Bioinformatics **32**(23), 3635–3644 (2016)
22. Pennington, J., Socher, R., Manning, C.: Glove: global vectors for word representation. In: Proceedings of the 2014 Conference on Empirical Methods in Natural Language Processing (EMNLP), pp. 1532–1543 (2014)
23. Purushotham, S., Meng, C., Che, Z., Liu, Y.: Benchmark of deep learning models on large healthcare mimic datasets. arXiv preprint arXiv:1710.08531 (2017)
24. Roberts, K., et al.: Overview of the TREC 2017 precision medicine track. NIST Special Publication, pp. 500–324 (2017)
25. Shi, H., Xie, P., Hu, Z., Zhang, M., Xing, E.P.: Towards automated ICD coding using deep learning. arXiv preprint arXiv:1711.04075 (2017)
26. Witten, I., Frank, E., Hall, M., Pal, C.: Data Mining: Practical Machine Learning Tools and Techniques, 4th edn. Morgan Kaufmann Publishers Inc., San Francisco (2016)
27. Yamada, I., Shindo, H., Takeda, H., Takefuji, Y.: Joint learning of the embedding of words and entities for named entity disambiguation. arXiv preprint arXiv:1601.01343 (2016)
28. Zhang, Y., Chen, Q., Yang, Z., Lin, H., Lu, Z.: BioWordVec, improving biomedical word embeddings with subword information and MeSH. Sci. Data **6**(1), 52 (2019)

Dyslexic Frequency Signatures in Relaxation and Letter Writing

N. B. Mohamad[1], Khuan Y. Lee[1,2]([✉]) [iD], W. Mansor[1,2], Z. Mahmoodin[3],
C. W. N. F. Che Wan Fadzal[1], and S. Amirin[4]

[1] Faculty of Electrical Engineering, Universiti Teknologi MARA, Shah Alam, Malaysia
leeyootkhuan@uitm.edu.my
[2] Computational Intelligence EK, Pharmaceutical and Lifesciences Communities of Research,
Universiti Teknologi MARA, Shah Alam, Malaysia
[3] Medical Engineering Technology Department, Universiti Kuala Lumpur British Malaysian
Institute, Kuala Lumpur, Malaysia
[4] Dyslexia Association Malaysia, Kuala Lumpur, Malaysia

Abstract. Dyslexia is a neurological disorder with impact to a child's confidence
in life. It can be reduced if detected at an early stage. This difficulty with learning
due to impairment to the left hemisphere of the brain, associated with language
processing, is often misunderstood as being lazy. Electroencephalography (EEG)
is one of technologies popularly employed to study dyslexia. Most previous works
on dyslexia are either based on subjective or psychometric methods and focused
on reading and spelling. Our research here aims to study the brain activity of
the dyslexic children during relaxation and letter writing, by comparing the EEG
frequency content between normal and dyslexic children. The uniqueness and
credibility of our experiment are tasks being adopted from the diagnostic manual
of the Dyslexia Association of Malaysia (DAM). Also, the electrode placement
has been optimized to four, i.e. C3, C4, P3 and P4 along the neurological pathway
for writing activity. It is found that the frequency ranges of EEG recorded during
relaxation is 8–13 Hz, in the alpha-subband and that during non-segmented writing
is 13–28 Hz, in beta-subband. In relaxed state, the EEG amplitude indicates that
the normal HighIQ group is more relaxed than the dyslexic Capable group. Higher
activity is found in the frequency pattern of EEG during the writing tasks than
during relaxation for both normal HighIQ/dyslexic Capable group. The dyslexic
Capable group is observed to exert more stress to do the task. The two peaks in the
alpha and beta-subband in the Fast Fourier Transform (FFT) envelope of EEG from
all groups can be explained by pausing to think in between writing. Based on these
peaks, the normal group shows use of high neural resources at the beginning of the
task while the dyslexic group shows that at the beginning and end of the task. It
is also observed that the EEG bandwidth is wider for the normal HighIQ/dyslexic
Capable group than the normal AverageIQ/dyslexic Average group.

Keywords: Dyslexia · Writing · Electroencephalogram (EEG) · Fast Fourier
Transform (FFT)

N. T. Nguyen et al. (Eds.): ACIIDS 2020, LNAI 12033, pp. 109–119, 2020.
https://doi.org/10.1007/978-3-030-41964-6_10

1 Introduction

Dyslexia is of neurological origin and manifests itself as a difficulty in learning, in particular oral reading and written writing, according to the International Dyslexia Association [1]. Dyslexia is usually accompanied by Dysgraphia, an inability to capture the written language, thus making writing difficult. Dysgraphia generally demonstrates as poor handwriting and problems with spelling [2]. In United States, about 5–17.5% of school-age children were found to be dyslexic and 80% of these children have specific learning disabilities [3, 4]. In Malaysia, the Education Ministry reported 5–10% of school-going children (or 314,000) were dyslexic and the number have been on the rise [5]. Besides, the Dyslexia Association of Malaysia (DAM) unraveled a finding from their records of diagnosis that, one out of every 20 students at least was found suffering from dyslexia, indicating severity of the problem [5]. Dyslexia is alleged to be the main reason for failure in school, as the children find motor coordination, concentration and personal organization challenging. This discourages them from reading and writing, which imposes enormous pressure on them [6] and hence leads to more complex psychopathological disorders [7]. This potentially becomes a crucial obstacle to the development of an educated, competent and competitive society towards employment chances and a loss to the nation, because intelligence of dyslexics is reckoned in the upper stratum.

There is a strong consensus that the difficulty in dyslexia shows a deficit in language processing area [8, 9] located in the left hemisphere of human brain, which plays a pivotal role in processing reading and writing activities, from fMRI images [10, 11]. Adults with dyslexia also shows failure of left hemisphere posterior brain systems during reading, using functional brain imaging [12]. This coincides with neuroscience theory that the normal neurological pathway during reading or writing activates the left hemisphere of human brain: it initiates from the primary visual cortex, then continues to the Angular Gyrus's area, Wernicke's area, Broca's area and terminates at the motor cortex [13]. Current methods to diagnose dyslexia are mostly subjective and psychometric based, such as Combined Raven's Test [14], the Twenty Statement Test (TST) adopted by the city of Bytom [15], Specific Learning Difficulties conducted by the British Dyslexia Association [16], or the Alphabetic-Phonetic-Structural-Linguistic (APSL) program used in UK dyslexia centers [17].

Electroencephalogram (EEG) is a measurement of the electrical activity of the brain by placing conductive metal electrodes on the human scalp [18]. Previous works have reported on the use of EEG to explain the characteristics of dyslexic children in reading. Converging evidence reported [19] the dyslexic group being less accurate at non-word reading in comparison to control group, based on Duncan's Multiple Range test. Researchers in [8] studied brain activation patterns in dyslexic and non-impaired children during pseudo-word and real-word reading tasks. It re-ported that reading skill was positively correlated with the magnitude of activation in the left occipito-temporal region. Sklar et al. [20] discovered from reading tasks that dyslexic group presented auto-spectral between 16–32 Hz while the normal group tended to have greater energy. There was also attempt to extract EEG spectral features to identify the trait of dyslexic children [20, 21]. In spite of these, the writing deficit in dyslexic children has not been explored.

Our previous works have explored a range of techniques to extract features during relaxation and writing, such as Fast Fourier Transform (FFT) [22–24], Short Time Fourier Transform (STFT) [25] and also Wavelet Packet Analysis [26]. Here, our work aims to study the brain state of the dyslexic children in writing, by comparing the frequency content of EEG between the normal and dyslexic children. The tasks used in this experiment are adopted from the diagnostic manual of the DAM. Section 2 described the data collection, data acquisition and theoretical background. Section 3 compares and discusses FFT features extracted from dyslexic and control group.

2 Methodology

2.1 Data Collection

Data were collected from five normal and five dyslexic children after informed con-sent and ethic approval were obtained from their parents. Subjects were recruited between 7–11 years old. All of them are right-handed and have healthy medical history. The normal subjects were grouped into the HighIQ and AverageIQ group, according to their scholastic grade. Data from dyslexic subjects were collected from the Centers of DAM. They were grouped according to class division by the DAM, into Average and Capable group. Children with history of neurological problems or under medication, were excluded from the study.

2.2 Data Acquisition

Subjects were invited to sit on a chair and relax before being instructed to perform six tasks. This paper only focuses on two of the tasks: relaxation and writing letters: s, w, j, t, m, q, and e. These letters are reckoned as challenging to the dyslexics.

Task to relax the subjects involved closing their eyes while listening to relaxing music for 120 s after taking their seat. Subjects were restrained from movement during data collection. In the writing task, the subjects were asked to look at the letters (s, w, j, t, m, q and e) displayed on the screen of the computer and then write the letters on a piece of paper. This took up 40 s per letter, with five seconds between repetition (eight times of writing letter). Children were first introduced to the equipment and practiced before performing the actual tasks. This is to abide by the compliance with EEG procedure that focuses on decreasing anticipatory anxiety and desensitizing the children to the components of procedure.

The EEG was recorded using Ag/AgCl electrodes placed on the scalp inserted with gGAMMA gel to improve the electrode skin impedance, as in Fig. 1. The electrodes were placed according to the International 10–20 Electrode System. The reference electrode was located on the right ear, while the ground electrode on the forehead (FPz). Electrode C3 (central left), C4 (central right), P3 (parietal left) and P4 (parietal right) were applied. They were chosen as they were situated on the neurological pathway for reading/writing and engaged during writing. C3 and C4 are associated with sensory and motor, while P3 and P4 with word interpretation that records perception and differentiation activities [4, 18]. The EEG signals were sampled and filtered using gMOBIlab hardware equipped with low-noise bio-signal amplifiers and 16-bit ADC with 256 Hz sampling frequency.

Fig. 1. EEG recording session for a dyslexic child performing the letter writing task.

2.3 Fast Fourier Transform (FFT)

Fast Fourier Transform is widely used in the analysis of EEG signals from subjects with brain related disorders, for example to analyze the EEG background activity in autism disorders [27]. Here, the EEG signals, recorded in time domain, were converted into frequency domain, to analyze its frequency content during writing, using FFT. Raw data were visually inspected for artifacts from electrocardiograms (ECG) and eye blinks, which were filtered. A bandpass FIR filter with frequency range between 8–30 Hz was used.

FFT is an algorithm which speeds up the calculation of discrete Fourier transform (DFT) (1).

$$X(k) = \sum_{n=0}^{N-1} x[n] . W_N^{nk} \tag{1}$$

X[k] is evaluated from $0 \le k \le N - 1$, which is a periodic function with a limited number of distinct values.

Cooley and Tukey introduced the concept of decimation for computing FFT algorithm in 1956 [15]. This concept divided the FFT algorithms into four types: Divide and Conquer, Radix-2 FFT algorithm, Radix-4 FFT algorithm and Split Radix-8 FFT algorithm. Equations (2) and (3) shows Radix-2 FFT algorithm that is derived from Divide and Conquer algorithm for our research. The original equation of DFT is replaced with n = (2r) for even term and n = (2r + 1) for odd term to produce (2), where X(k) is evaluated from $0 \le k \le N - 1$ and G(k) and H(k) is even and odd terms.

$$X(k) = \sum_{r=0}^{(N/2)-1} x(2r) . W_{N/2}^{rk} + W_N^k \sum_{r=0}^{(N/2)-1} x(2r + 1) . W_{N/2}^{rk}$$

$$= G(k) + W_N^k H(k) \tag{2}$$

$$W_N^{2rk} = e^{-j2\pi/N.(2rk)} = e^{-j2\pi/(N/2).(rk)} = W_{N/2}^{rk}$$

To obtain the Radix-2 FFT algorithm as in (3), N is set to 2, so that X(k) is evaluated from $0 \le k \le 1$, then the Divide and Conquer algorithm in (2) is replaced with k = 0 and k = 1.

$$X(0) = G(0) + W_N^k H(0) = x(0) + x(1), \quad k = 0$$

$$X(1) = G(1) + W_N^k \, H(1) = x(0) - x(1), \quad k = 1 \tag{3}$$

Radix-2 FFT algorithm can also be represented using a butterfly signal flow for 2-point DFT as shown in Fig. 2, where x(0) and x(1) are signals in time domain and X(0) and X(1) are signals in frequency domain.

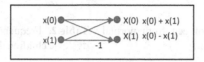

Fig. 2. A butterfly signal flow for 2-point DFT

3   Results and Discussions

The frequency pattern of EEG from electrodes C3, C4, P3, and P4 for normal and dyslexic children during relaxation is shown in Fig. 3a, b, c, and d respectively. The frequency range during relaxation is observed to lie within the alpha-subband, i.e. 8–13 Hz [18]. Four conditions of EEG amplitude are observed at channel C3, C4, P3 and P4. Firstly, the EEG of normal HighIQ group shows higher amplitude in comparison to normal AverageIQ group (see Fig. 3a and b). Secondly, the EEG of dyslexic Capable group shows higher amplitude in comparison to Average group (see Fig. 3c and d). From evaluation of EEG between normal HighIQ group and dyslexic Capable group, the former presents higher amplitude (see Fig. 3a and c). Relaxing performance in the normal AverageIQ group depicts higher amplitude of EEG than dyslexic Average group (see Fig. 3b and d). Results above demonstrate the normal groups are able to relax better

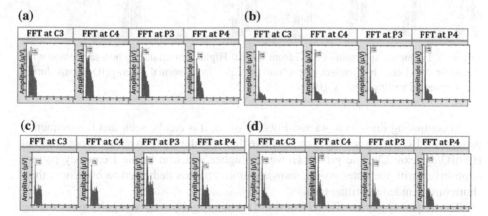

Fig. 3. a. Frequency spectrum of EEG from normal HighIQ group during relaxation (Left). b. Frequency spectrum of EEG from normal AverageIQ group during relaxation (Right). c. Frequency spectrum of EEG from dyslexic Capable group during relaxation (Left). d. Frequency spectrum of EEG from dyslexic Average group during relaxation (Right).

than the dyslexic groups. Based on social science study, children with brain related disorders do have problem to relax [20, 28]. However, this study has not gone deeper by comparing the sub-categories, like ours.

Tables 1 and 2 tabulate frequencies obtained during relaxation from normal and dyslexic subjects respectively. Results show these frequencies stay between 8–13 Hz, within the alpha subband known for relaxation [20, 28].

Table 1. Frequency range of eeg from normal children during relaxation (left).

Electrode	Normal	
	HighIQ	AverageIQ
C3	10–13 Hz	9–11 Hz
C4	9–10 Hz	9–11 Hz
P3	9–10 Hz	8–9 Hz
P4	8–10 Hz	8–9 Hz

Table 2. Frequency range of eeg from dyslexic children during relaxation (right).

Electrode	Dyslexic	
	Capable group	Average group
C3	9–10 Hz	8–9 Hz
C4	9–10 Hz	8–10 Hz
P3	9–10 Hz	8–9 Hz
P4	10–11 Hz	8–10 Hz

Figures 4a, b, 5a and b show the non-segmented (continuous 40 s) frequency pattern of EEG from normal and dyslexic children during the letter writing task. Seven letters are designated for the letter writing task: s, w, j, t, m, q, and e. Due to space constraint, only the results of writing letter 's' is shown here.

(a)   **(b)**

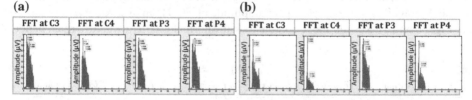

Fig. 4. a. Frequency spectrum of EEG from normal HighIQ group during non-segmented writing letter 's' (Left); b. Frequency spectrum of EEG from normal AverageIQ group during non-segmented writing letter 's' (Right)

In comparing Figs. 3a to 4a and Figs. 3c to 5a, it is can be seen that the frequency pattern of EEG during writing is more active than that during relaxation for both normal HighIQ/dyslexic Capable group. However, higher variation in the frequency pattern is observed with the latter group, showing more effort is dedicated to overcome their shortcoming in letter writing task.

The frequency pattern of EEG from both normal HighIQ and AverageIQ group during letter writing task drew on two peaks, one in alpha- and beta-subband each (see Fig. 4a and b). The peak in the alpha-subband is attributed to moments at which the subjects pause during writing. The amplitude is highest at the beginning of writing task, showing higher neural activity.

(a)   **(b)**

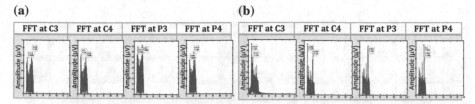

Fig. 5. a. Frequency spectrum of EEG from dyslexic Capable group during non-segmented writing letter 's' (Left); b. Frequency spectrum of EEG from dyslexic Average group during non-segmented writing letter 's' (Right)

Table 3. Frequency range (Hz) of eeg from normal HighIQ and AverageIQ group during non-segmented writing letter 's'(left).

Electrode	Normal	
	HighIQ	AverageIQ
C3	13–20 Hz	13–26 Hz
C4	13 – 18 Hz	13–21 Hz
P3	14–19 Hz	13–27 Hz
P4	13–19 Hz	13 – 22 Hz

Table 4. Frequency range (Hz) of eeg from capable reader and average reader during non-segmented writing letters (right).

Electrode	Dyslexic	
	Capable group	Average group
C3	13–27 Hz	13–28 Hz
C4	13–25 Hz	25–26 Hz
P3	13–27 Hz	19–26 Hz
P4	13–27 Hz	19–26 Hz

From Table 3, it can be observed that the frequency range for normal HighIQ group lies between 13–20 Hz, while that for normal AverageIQ group is 13–27 Hz, both lie within the beta-subband for writing activity [20, 28].

With reference to Fig. 5a and b, similar to the above finding, the frequency pattern of EEG from dyslexic during letter writing task also contains two peaks in the alpha- and beta-subband. Unlike the normal group, one of the peaks is high at the beginning while the other at the end of the letter writing task.

From Table 4, it can be observed that the EEG frequency range for the dyslexic Capable group is 13–27 Hz meanwhile the frequency range for the dyslexic Average group is 13–28 Hz, both lie within the beta-subband for writing activity [20, 28].

In addition, results on segmented EEG signals during writing letters task from normal and dyslexic subjects are shown in Fig. 6a, b, c, and d correspondingly. The EEG signals are segmented for every 5 s from a length of 40 s. Each subject is asked to repeat writing letter for eight times to maintain optimum compliance with the EEG procedure. Different EEG spectral patterns are obtained at electrode C3, C4, P3 and P4.

For normal HighIQ (see Fig, 6a), frequency spectra present only one high peak at the beginning of task. This is consistent throughout the 40 s data collection. This evidence implies that normal HighIQ group are able to sustain the same writing pattern at every repetition during data collection.

Frequency spectra of EEG from normal AverageIQ in Fig. 6b display many peaks, showing high neural activation. This illustrates that this group makes constant efforts

(a)

(b)

(c)

(d)

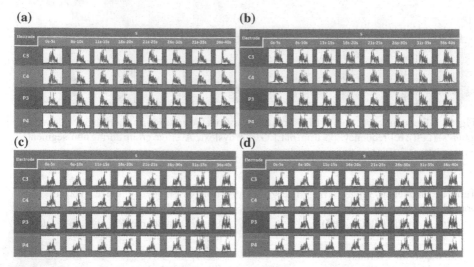

Fig. 6. a. Frequency spectrum of EEG from normal HighIQ group during segmented writing letter 's' (Top Left). b. Frequency spectrum of EEG from normal AverageIQ group during segmented writing letter 's' (Top Right). c. Frequency spectrum of EEG from dyslexic Capable group during segmented writing letter 's' (Bottom Left). d. Frequency spectrum of EEG from dyslexic Average group during segmented writing letter 's' (Bottom Right)

throughout the 40 s data collection. Results in Table 5 also supports point mentioned, which 16–23 Hz frequency obtained, while frequencies for versus group lies between 16–18 Hz.

Results on dyslexic Capable group during segmented letter writing task drew on two peaks for the first 30 s, one each in alpha- and beta-subband (see Fig. 6c). The last 10 s depicts maximum neural activity of dyslexic, which may be their final effort to fulfill the task. Inconsistent frequency spectral result demonstrates subjects start to loss focus of writing task.

With reference to Fig. 6d, the segmented frequency spectra look similar to those for the dyslexic Capable group, with one peak each in the alpha- and beta-subband.

Table 5. Frequency range (Hz) of eeg from normal and dyslexic children during segmented writing letter 's'

Electrode	Normal		Dyslexic	
	HighIQ	AverageIQ	Capable	Average
C3	17–18 Hz	17–23 Hz	15–24 Hz	13–24 Hz
C4	16–18 Hz	16–19 Hz	15–24 Hz	24–25 Hz
P3	16–17 Hz	16–20 Hz	14–24 Hz	19–25 Hz
P4	16–18 Hz	17–18 Hz	14–25 Hz	24–25 Hz

In contrast to the dyslexic Capable group with frequency between 14–24 Hz, the dyslexia Average group display a wider range of 13–25 Hz, as in Table 5.

4 Conclusions

The analysis of EEG signals from four electrodes (C3, C4, P3 and P4) along the neural pathway for writing (segmented and non-segmented), for normal and dyslexic children during relaxation and letter writing task, has been compared. The frequency range of EEG recorded during relaxation is 8–13 Hz, while that during writing is 13–28 Hz, within the alpha- and beta-subband, accorded for the different neural activity. During relaxation, it is found that the EEG of normal HighIQ group shows higher amplitude than the dyslexic Capable group, showing that the normal HighIQ group is able to relax better than the dyslexic Capable group. From results of non-segmented writing the letter 's', it can be observed that the frequency pattern of EEG is more active than that during relaxation, for both normal HighIQ/dyslexic Capable group. However, higher activity in the beta-subband is observed in the dyslexic Capable group, showing more effort is dedicated to overcome their deficiency in letter writing. The FFT envelope of EEG from both normal HighIQ/dyslexic Capable and AverageIQ/dyslexic Average group drew on two peaks in the alpha- and beta-subband, due to pause in between writing. For the normal group, the amplitude is highest at the beginning of writing task, showing use of high neural resources. For the dyslexic group, the frequency pattern displays one of the peaks at the beginning while the other at the end of the letter writing task. It is also observed that the bandwidth of the EEG from the normal HighIQ/dyslexic Capable group is wider than that of the normal AverageIQ/dyslexic Average group. Segmented frequency spectrum showing results supporting evidence on non-segmented frequency spectrum, in which the dyslexic group exhibited higher frequency in comparison to the normal group.

Acknowledgment. The author would like to thank the Research Management Institute, Universiti Teknologi MARA, Malaysia, for the Research Entity Initiative Grant (600-IRMI/REI 5/3 (02 1/2018); the Faculty of Electrical Engineering, Universiti Teknologi MARA, Malaysia, for the support and assistance given to the authors in carrying out this research; Dyslexia Association Centre Malaysia (DAM) for their assistance and permission in providing subjects and advice, without which our research would be impossible.

References

1. Berninger, V.W., Nielsen, K.H., Abbott, R.D., Wijsman, E., Raskind, W.: Writing problems in developmental dyslexia: under-recognized and under-treated. J. Sch. Psychol. **48**, 1–21 (2008). https://doi.org/10.1016/j.jsp.2006.11.008
2. Molfese, V., Molfese, D., Molnar, A., Beswick, J.: Developmental dyslexia and dysgraphia. In: Encyclopedia of Language & Linguistics, pp. 485–491. Elsevier Ltd. (2006). https://doi.org/10.1016/B0-08-044854-2/04166-3
3. Shaywitz, S.E.: Dyslexia. N. Engl. J. Med. **338**, 307–312 (1998). https://doi.org/10.1056/NEJM199801293380507

4. Cortiella, C.: The State of Learning Disabilities. National Center for Learning Disabilities, New York, NY, pp. 1–37 (2011). https://www.americaspromise.org/sites/default/files/d8/The%20State%20of%20Learning%20Disabilities.pdf
5. Suet, L.K.: Living with Dyslexia in Malaysia (2007). http://dyslexiainmalaysia.wordpress.com/2007/07/13/star-article-7-july-2005./
6. Kooijman Valesca, J.C., Johnson, E.K., Hagoort, P., Cutler Anne, H.: Predictive brain signals of linguistic development. Front. Psychol. **4**, 4–25 (2013). https://doi.org/10.3389/fpsyg.2013.00025
7. Gaggi, O., Galiazzo, G., Palazzi, C., Facoetti, A., Franceschini, S.: A serious game for predicting the risk of developmental dyslexia in pre-readers children. In: 21st International Conference on Computer Communications and Networks (ICCCN 2012), pp. 1–5 (2012)
8. Shaywitz, B.A., et al.: Disruption of posterior brain systems for reading in children with developmental dyslexia. Biol. Psychiatry **52**, 101–110 (2002). https://doi.org/10.1016/s0006-3223(02)01365-3
9. Clark, A.D.: Diseases and Disorders Dyslexia. Lucent Books. pp. 1–107 (2005). ISBN 1590180402
10. Fadzal, C.W.N.F.C., Mansor, W., Khuan, L.Y.: Review of brain computer interface application in diagnosing dyslexia. In: IEEE Control and System Graduate Research Colloquium (ICSGRC 2011), pp. 124–128 (2011). https://doi.org/10.1109/icsgrc.2011.5991843
11. Menon, V., Desmond, J.: Left superior parietal cortex involvement in writing, integrating fMRI with lesion evidence. Cogn. Brain. Res. **12**, 337–340 (2001). https://doi.org/10.1016/s0926-6410(01)00063-5
12. Simos, P.G., Breier, J.I., Fletcher, J.M., Bergman, E., Papanicolaou, A.C.: Cerebral mechanisms involved in word reading in dyslexic children: a magnetic source imaging approach. Cereb. Cortex **10**, 809–816 (2000). https://doi.org/10.1093/cercor/10.8.809
13. Marshall, A.: The Dyslexic Reader. Davis Dyslexia Association International, vol. 33, pp. 1–24 (2003). https://www.scribd.com/document/2938104/The-Dyslexic-Reader-2003-Issue-33
14. Sun, Z., et al.: Prevalence and associated risk factors of dyslexic children in a middle-sized city of china: a cross-sectional study. PLoS ONE **8**, e56688 (2013). https://doi.org/10.1371/journal.pone.0056688
15. Marszalek, W.A.: New Educ. Rev. **17**, 264–275 (2012). ISBN 17326729. Polish
16. Benjamin, S.: Screening and Assessments (2017). http://www.bdadyslexia.org.uk/about-dyslexia/schools-colleges-and-universities/sreening-and-assessments.html
17. Umar, R.S.B., Mokhtar, F.: D-Mic: a mobile learning application for dyslexic children. Malays. J. Mob. Learn. **1**, 15 (2010). https://scholar.google.com/scholar?cluster=13554038619762685869&hl=en&oi=scholarr
18. Teplan, M.: Fundmentals of EEG measurement. Measur. Sci. Rev. **2**, 1–11 (2002). https://www.semanticscholar.org/paper/FUNDAMENTALS-OF-EEG-MEASUREMENT-Teplan/2eda7ede68dac52e32065c72fbdd6ae67da36e10
19. Hazan, V., Adlard, A.: Speech perception in children with specific reading difficulties (dyslexia). In: Proceeding Fourth International Conference on Spoken Language (ICSLP 1996), vol. 1, pp. 165–168 (1996). https://doi.org/10.1080/713755750
20. Sklar, B., Hanley, J., Simmons, W.W.: A computer analysis of eeg spectral signatures from normal and dyslexic children. IEEE Trans. Biomed. Eng. **BME-20**, 20–26 (1973). https://doi.org/10.1109/TBME.1973.324247
21. Rippon, G., Brunswick, N.: Trait and state EEG indices of information processing in developmental dyslexia. Int. J. Psychophysiol. **36**, 251–265 (2000). https://doi.org/10.1016/S0167-8760(00)00075-1

22. Che Wan Fadzal, C.W.N.F., Mansor, W., Lee, K.Y., Mohamad, S., Mohamad, N., Amirin, S.: Comparison between characteristics of EEG signal generated from dyslexic and normal children. In: 2012 IEEE EMBS Conference on Biomedical Engineering and Sciences (IECBES 2012), pp. 943–946 (2012). https://doi.org/10.1109/iecbes.2012.6498210
23. Che Wan Fadzal, C.W.N.F., Khuan, L.Y., Mansor, W.: Frequency content analysis of brainwave C3 and P3 for dyslexia related writing disorder. In: 2012 IEEE-EMBS International Conference on Biomedical and Health Informatics (BHI 2012), pp. 309–312 (2012). https://doi.org/10.1109/bhi.2012.6211574
24. Che Wan Fadzal, C.W.N.F., Mansor, W., Lee, K.Y., Mohamad, S., Amirin, S.: Frequency analysis of EEG signal generated from dyslexic children. In: 2012 IEEE Symposium on Computer Applications and Industrial Electronics (ISCAIE 2012), pp. 202–204 (2012). https://doi.org/10.1109/iscaie.2012.6482096
25. Che Wan Fadzal, C.W.N.F., Mansor, W., Khuan, L.Y., Zabidi, A.: Short-time Fourier Transform analysis of EEG signal from writing. In: 2012 IEEE 8th International Colloquium on Signal Processing and its Applications (CSPA 2012), pp. 525–527 (2012). https://doi.org/10.1109/cspa.2012.6194785
26. Fuad, N., Mansor, W., Lee, K.Y., Mohamad, N.B.: Wavelet packet analysis of EEG signals from children during writing. In: IEEE 9th International Colloquium on Signal Processing and its Applications (CSPA 2013), pp. 359–361 (2013). https://doi.org/10.1109/isci.2013.6612408
27. Sheikhani, A., Behnam, H., Mohammadi, M.R., Noroozian, M., Pari, G.: Analysis of quantitative Electroencephalogram background activity in Autism disease patients with Lempel-Ziv complexity and Short Time Fourier Transform measure. In: 4th IEEE/EMBS International Summer School and Symposium on Medical Devices and Biosensors (ISSS-MDBS 2007), pp. 111–114 (2007). https://doi.org/10.1109/issmdbs.2007.4338305
28. Karim, I., Abdul, W., Kamaruddin, N.: Classification of dyslexic and normal children during resting condition using KDE and MLP. In: 5th International Conference on Information and Communication Technology for the Muslim World (ICT4M 2013), pp. 1–5 (2013). https://doi.org/10.1109/ict4m.2013.6518886

Decision Support and Control Systems

A Trading Framework Based on Fuzzy Moore Machines

Iván Calvo, Mercedes G. Merayo[iD], and Manuel Núñez[✉][iD]

Departamento Sistemas Informáticos y Computación,
Universidad Complutense de Madrid, Madrid, Spain
{ivcalvo,mgmerayo,manuelnu}@ucm.es

Abstract. The everlasting competition between investing strategies has seen a remarkable impulse after automated trading algorithms took their place. Any failure in this kind of algorithms may end up implying huge monetary losses. Because of that, these systems may represent an important application of formal methods. Furthermore, considering the inherent uncertainty of stock markets and the usual imprecision in the definition of many investment strategies, any attempt to model these software systems is very challenging. In this paper we propose a complete framework, built upon the formalism of fuzzy automata, that can be used to define and evaluate a variety of automatic trading strategies based on the observation of candlestick patterns.

Keywords: Fuzzy automata · Automated trading · Candlestick patterns

1 Introduction

The use of models helps to develop more reliable systems. This claim is well-known in most engineering disciplines [25]. It is interesting to realize that in Computer Science, despite the complexity of the current systems, it is not assumed that a *blueprint* should be used to guide the development process [22]. There is plenty of work in the academia introducing different approaches and techniques to support the formal development of computer systems and validate the correctness of the developed systems [11]. We think that the lack of good tools to support the theoretical approaches is an important deterrent to achieve a widespread use of formal methods in industry, although there are some remarkable experiences [3,23,30]. Therefore, our first aim when introducing a new formalism should be to provide tools to support its use.

There are many situations where the development of a system inherently needs to consider imprecise information. For example, this is the case if we have

This work has been supported by the Spanish MINECO-FEDER (grant numbers DArDOS, TIN2015-65845-C3-1-R and FAME, RTI2018-093608-B-C31) and the Region of Madrid (grant number FORTE-CM, S2018/TCS-4314) co-funded by EIE Funds of the European Union.

N. T. Nguyen et al. (Eds.): ACIIDS 2020, LNAI 12033, pp. 123–134, 2020.
https://doi.org/10.1007/978-3-030-41964-6_11

to represent a simple system just tossing a coin. In order to formally design and analyze this kind of systems, *fuzzy logic* [34, 35] is very useful because, in addition to having an underlying mathematical theory, it has been shown that it can appropriately model systems where we need to use a certain degree of *imprecision*. We are interested in equipping a classical formalism with fuzzy logic capabilities so that we can model imprecision. Fortunately, there are many variants *fuzzifying* variants of finite automata [1, 13, 28, 29, 33]. Our previous work [6, 8, 9], which we use as initial step of the formalism presented in this paper, was built on top of these approaches and introduced a fuzzy version of classical Mealy machines. We successfully used our formalism to analyze heart data, extracted from electrocardiograms, in order to detect abnormal behaviors.

The next step in our research is to evaluate the versatility of our framework. In previous work we focused on the behaviour of medical systems. In this paper we consider a completely different field of application: we study how our framework can be used to specify a simple trading system and evaluate its suitability. To be more precise, since we do not need a formalism as expressive as the one that we used before, we will consider a fuzzy variant of Moore machines, a formalism slightly simpler than Mealy machines. We will present a simple trading model where we analyse the values of a certain stock, given as *candlesticks*, in order to detect a certain pattern. Candlesticks include four different pieces of information concerning how a certain stock was trading during a day. The *body* of the candlestick denotes the gap between the open and closing prices of the analysed stock. A green/red color denotes that the first price was lower/higher than the second one. In addition, the *shadow* of the candlestick shows the high and low prices for the session. Figure 1 shows the candlesticks corresponding to twenty sessions of the Apple stock.

Currently, there exists a rich literature explaining in detail trading strategies based on the observation of candlesticks patterns [5, 31]. In this paper, our main goal is not to specify a complex strategy but to show how our framework can encompass in a simple way a whole family of strategies, so that the user can fine-tune certain parameters according to the specific stock of interest. Specifically, we will define a framework where investment decisions will be based on the detection of *hammers*. A hammer is observed when a stock trades during a session significantly lower than its opening, but within the session is able to close near opening price. In order to consider that a candlestick has a hammer shape, it is usually assumed that the lower shadow must be at least twice the size of its body. The strategy looking for hammers considers that they mark a possible bottom in the value of a stock. More precisely, this strategy considers that the price of the stock might rise after we observe a hammer preceded by a sequence, at least three, of sessions where the value of the stock has declined.

In this paper we present a new formalism, a natural evolution and simplification of previous work, to represent fuzzy systems and apply it to specify a parametric strategy to trade a stock taking into account the occurrence of hammers in candlesticks series. Our model is parametric because it allows to trade different percentages of the total bankroll in a single operation. In the simplified version, $n = 1$, the investor buys/sell the complete bankroll in each single operation. For a general value of $n \in \mathbb{N}$, at a certain point of time the investor has, for a certain $0 \leq m \leq n$, $\frac{m}{n}$ of the bankroll invested in a stock and $\frac{n-m}{n}$ in liquidity. A single operation will buy/sell stock for $\frac{1}{n}$ of the bankroll. The framework is fully supported by a tool that allows users to process candlesticks information so that different strategies can be backtested.

The rest of the paper is structured as follows. In Sect. 2 we review basic concepts from fuzzy theory and introduce a variant of fuzzy automata, based on Moore machines, that is particularly well suited for the definition of trading strategies. In Sect. 3 we show how our formalism can represent candlesticks and patterns associated with different types of hammers. In Sect. 4 we present our strategy. Finally, in Sect. 5 we present our conclusions and sketch some lines for future work.

2 Fuzzy Moore Machines

In this section we review some concepts associated with the definition of fuzzy automata and present our new formalism. The interested reader is referred to our previous work [8,9] for more details. *Fuzzy relations* play the role of applying *soft* constraints. This is made by setting some δ, which expresses the level of tolerance that can be afforded when applying the given constraint. This means that, given some *crisp* relation, such as $x \leq y$, the fuzzy relation will *partially* hold even when $x > y$, provided that the difference between x and y is still lower than δ. More precisely, in this specific example, we define a function $\mathbb{R} \times \mathbb{R} \longrightarrow [0,1]$ to be the fuzzy version of the $x \leq y$ relation:

$$\overline{x \leq y}^{\delta} \equiv \begin{cases} 1 & \text{if } x < y \\ \frac{\delta + y - x}{\delta} & \text{if } y \leq x \leq y + \delta \\ 0 & \text{if } y + \delta < x \end{cases}$$

Similarly, we can define the fuzzy analogous of the $x \geq y$ and $x = y$ relations:

$$\overline{x \geq y}^{\delta} \equiv \begin{cases} 1 & \text{if } x > y \\ \frac{\delta + x - y}{\delta} & \text{if } y \geq x \geq y - \delta \\ 0 & \text{if } y - \delta > x \end{cases}$$

$$\overline{x = y}^{\delta} \equiv \begin{cases} 0 & \text{if } x \leq y - \delta \\ \frac{x - y + \delta}{\delta} & \text{if } y - \delta < x \leq y \\ \frac{-x + y + \delta}{\delta} & \text{if } y < x \leq y + \delta \\ 0 & \text{if } y + \delta < x \end{cases}$$

In addition to fuzzy versions of the usual *arithmetic relations*, we need a fuzzy version of *Boolean relations* such that *partially true* statements can be joined together, forming logical conjunctions. The fuzzy analogous of the *crisp* Boolean conjunction is the concept of *triangular norms*, also known as *t*-norms.

Definition 1. *A t-norm is a function*

$$\triangle : [0,1] \times [0,1] \longrightarrow [0,1]$$

satisfying the following properties:

- *Commutativity: $x \triangle y = y \triangle x$ for all $x, y \in [0,1]$.*
- *Associativity: $(x \triangle y) \triangle z = x \triangle (y \triangle z)$ for all $x, y, z \in [0,1]$.*
- *Identity: $1 \triangle a = a$ for all $a \in [0,1]$.*
- *Monotonicity: if $x_1 \leq y_1$ and $x_2 \leq y_2$ then we have $x_1 \triangle x_2 \leq y_1 \triangle y_2$ for all $x_1, x_2, y_1, y_2 \in [0,1]$.*

In this paper we will consider only the Hamacher t-norm, defined as $(x, y) \mapsto \frac{xy}{x+y-xy}$. This t-norm happens to be strictly monotonic, which means that the $\leq$ signs are actually strict $<$ signs in the monotonicity condition.

The goal of introducing these notions is to finally have a *fuzzy* way to impose constraints on the behaviour of the system under certain inputs. These *fuzzy constraints* are defined as follows:

Definition 2. *A fuzzy constraint is a formula consisting of fuzzy relations, possibly combined with t-norms, which may contain free variables and constant values within the $[0,1]$ interval.*

Next, we will discuss the way in which we decided to *fuzzify* the usual notions brought from automata theory. In order to fix the alphabet, a distinction between *input* and *output* actions was used in our previous work. This distinction is useful to specify the way in which certain system outputs correspond to the given inputs. The way in which outputs are *raised* upon traversing some particular transition between states bears a slight resemblance with classical Mealy machines. It is at this point where we diverge from previous work: while we still want the model to serve as a specification of the correspondence between *inputs* and *outputs*, we will make no distinction between the symbols of the alphabet. Instead of that, each *state* of the automaton will be associated with a particular output value. This value will be raised every time that this state is traversed. With this distinction, the proposed formalism can be seen as a *fuzzified* version of Moore machines. Therefore, we will define our alphabet to be the set of *events* in our model.

Definition 3. *An event is a collection of variables, which are instantiated by the environment according to some predefined set of observations and expressions, together with a fuzzy constraint whose free variables are those variables. We identify each event with a name preceded by the # symbol.*

Let us illustrate this concept with an example that will be later used in the proposed model.

Example 1. We will denote by #opportunityCost the event consisting in an empty set of variables together with the constant constraint whose truth value is always 0.7.

Now, we have the elements to define our version of *fuzzy Moore machine*.

Definition 4. *A fuzzy Moore machine is a tuple*

$$(S, \mathcal{E}, \triangle, f, s_0, T)$$

whose components denote:

- *S is a finite set of states.*
- *$\mathcal{E}$ is a finite set of events.*
- *$\triangle$ is a strictly monotonic t-norm.*
- *f is a function taking a state from S and returning its associated output value.*
- *s_0 is the initial state.*
- *$T \subseteq S \times \mathcal{E} \times S$ is the set of transitions.*

The semantics of these machines is quite intuitive. The action of the environment must be formalized as a sequence of functions $\mathcal{O}_1, \ldots, \mathcal{O}_n$ mapping each expression to its instantiation value at the n-th time step. Having defined this, the satisfaction degree of an event #ev at a time step n, denoted by $\mu_n(\#\text{ev})$, will be defined in the natural way (the explicit formal definition can be found in our previous work [8]).

Finally, let us define some notation to establish the correspondence between observed events and produced outputs.

Definition 5. *Let $F = (S, \mathcal{E}, \triangle, f, s_0, T)$ be a fuzzy Moore machine, $s_1, \cdots, s_n$ be a sequence of states in S and $\#ev_1, \cdots, \#ev_n$ be a sequence of events. If for all $1 \leq k \leq n$ we have that $\epsilon_k := \mu_k(\#ev_k) > 0$ and $(s_{k-1}, \#ev_k, s_k) \in T$ then we write*

$$\#ev_1, \cdots, \#ev_n \Rightarrow_\epsilon f(s_1), \cdots, f(s_n)$$

where the value of ϵ is given by $\epsilon_1 \triangle \cdots \triangle \epsilon_n$.

3   Modeling Candlestick Patterns

In this section we will show how fuzzy constraints can be used to give a formal definition of various candlesticks patterns. Since the formalism is open to different ways in which the environment may instantiate the parameters of each individual observation, in this section we will review the basic elements present in our trading environment and discuss the precise definition of the observations that will be later used to define our model in Sect. 4. The definition of the observations is expressed in a formal language based on arithmetic expressions. We think that this formal language is expressive enough to define almost any possible trading strategy.

First, we will assume that we can obtain the *open, close, high* and *low* value of the stock under consideration on any day prior to the current, n-th, day. Note that this information is indeed freely available from multiple sources (including the Nasdaq website and The Wall Street Journal).

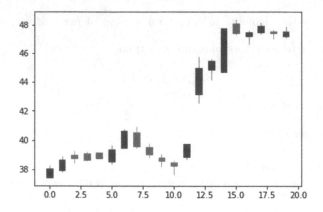

Fig. 1. Candlesticks chart from Apple.

Definition 6. *Let k be a natural number and let e be an expression of the form $(open|close|high|low) \sim k$. We denote by $\mathcal{O}_n(e)$ the corresponding value at the day $n - k$.*

We can extend the $\mathcal{O}$ notation to any arithmetical expression in a compositional way. Let e, e_1, e_2 be expressions of the previous form, $\neg$ be an unary arithmetic operator and $\bigcirc$ be a binary operator. Then we have

$$\mathcal{O}_n(\neg e) = \neg \mathcal{O}_n(e) \qquad \text{and} \qquad \mathcal{O}_n(e_1 \bigcirc e_2) = \mathcal{O}_n(e_1) \bigcirc \mathcal{O}_n(e_2)$$

Similarly, we will also consider real valued constant expressions. Finally, We define an observation to be any composition of these expressions.   □

The expressive power of these formulas can be shown with a couple of illustrative examples.

Example 2. The difference between the value of the asset at the market close of the last two days can be expressed with the following formula:

$$(\text{close} \sim 1) - (\text{close} \sim 2)$$

A moving average of the midpoint of the range of prices of each time interval can be expressed as follows:

$$\frac{\frac{low\sim1+high\sim1}{2} + \frac{low\sim2+high\sim2}{2} + \frac{low\sim3+high\sim3}{2}}{3}$$

In order to define our strategy, we need to provide a formal definition of the *shape* of a hammer pattern and the one corresponding to a doji pattern. Since both patterns are based on the relative size of the components of the candlestick, we will define them by using fuzzy constraints defined over relative size observations.

Table 1. Relopen and relclose corresponding to Fig. 1 (top) and satisfaction degree of hammer and doji patterns in candlesticks corresponding to Fig. 1 (bottom)

	1	2	3	4	5	6	7	8	9	10	11	12	13	14	15	16	17	18	19	20
relopen	0.01	0.11	0.67	0.89	1.00	0.13	0.02	0.73	0.77	0.70	0.91	0.13	0.20	0.49	0.00	0.76	0.56	0.16	0.80	0.17
relclose	0.70	0.83	0.37	0.15	0.09	0.76	0.91	0.11	0.24	0.45	0.69	0.99	0.76	0.92	1.00	0.11	0.86	0.75	0.60	0.62

	1	2	3	4	5	6	7	8	9	10	11	12	13	14	15	16	17	18	19	20
hammer	0.00	0.01	0.09	0.00	0.00	0.04	0.00	0.00	0.00	0.21	0.55	0.04	0.13	0.52	0.00	0.00	0.56	0.08	0.43	0.08
doji	0.01	0.00	0.56	0.00	0.00	0.09	0.00	0.11	0.23	0.65	0.68	0.00	0.20	0.38	0.00	0.07	0.56	0.15	0.71	0.35

Definition 7. *We define the **relopen** observation by the formula:*

$$\frac{open \sim 1 - low \sim 1}{high \sim 1 - low \sim 1}$$

*We define the **relclose** observation by the formula:*

$$\frac{close \sim 1 - low \sim 1}{high \sim 1 - low \sim 1}$$

In Table 1 (top) we see the **relopen** and **relclose** values corresponding to the candlesticks shown in Fig. 1.

Once these relative size observations have been defined, we can give a *fuzzy* definition of *hammer* and *doji* patterns.

Definition 8. *The grade of confidence associated with the event of observing a **hammer** is given by the fuzzy constraint*

$$\overline{relopen \geq 0.8}^{0.7} \triangle \overline{relclose = 1}^{0.7}$$

Similarly, the grade of confidence associated with the event of observing a doji is given by the fuzzy constraint

$$\overline{relopen = relclose}^{0.7}$$

In Table 1 (bottom) we show the grade of confidence of each candlestick shown in Fig. 1 being a *hammer* and a *doji*.

4   Strategy Definition and Evaluation

In this section we present a basic trading strategy, using our formalism, and show how a candlestick pattern recognition approach can be defined. In Fig. 2 we show the outline of our model, where some parameters of the corresponding automaton have been removed for the sake of clarity. First, we would like to mention that this model has only three states. Note that the numbers inside the states do not represent the name of the state but its associated output value. This means that a model might have different states associated to the same output value, which can be useful when modelling systems in which there are situations that are fundamentally different but may imply the same outcome. The three possible outcomes of our model, that is, the values of the set $\{0, \frac{1}{2}, 1\}$, encode the proportion of money invested in the stock under consideration. For example, an outcome of 1 at a given time indicates that we can invest all our available funds in the corresponding stock. On the contrary, an outcome of 0 indicates that we have to sell all our investment in the considered stock. The implications of a $\frac{1}{2}$ outcome are a little bit more subtle. Since the value of the stocks is asked to be equal to that of the available cash, a raise in the price of the asset can imply a sell order, while a fall in its price should imply a buy order. The initial state of our model is the one associated with a 0 outcome, meaning that the strategy starts with the account owning no stocks and, therefore, full liquidity. In this simple model, we assume that we only sell stocks that have been previously bought, ruling out the possibility of *short selling*. In order to illustrate our model, in this paper we have considered as stock of interest Apple Inc. (denoted by *AAPL*).

As mentioned in Sect. 3, the key element of our strategy is the observation of *hammer* patterns. Since an observation of a *hammer* after a downtrend may predict the reversal of the trend, the strategy will be based on waiting until one of these patterns appears. If a *hammer* appears, then there are two situations: the trend before it may be falling or not. In the first case, we want to buy as many stocks as possible, hence a transition moving to the 1 state is triggered. In the second case, we have to balance whether the risk of buying after no downtrend is too high or not. If it is not, then only a half of the available cash will be spent, so that a transition moving to the $\frac{1}{2}$ state will be triggered.

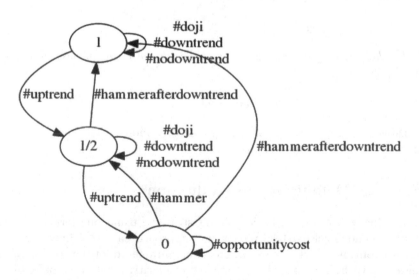

Fig. 2. Topology of our model.

The way in which our formalism manages these decisions is, as explained in Sect. 2, to assign a confidence value to each possible sequence of outputs. After each time step, the current state associated with the highest grade of confidence is the one producing the output. In this case, that output is the target amount of capital that should be invested in the stock.

The described strategy has been successfully implemented and its python source code is available on github (https://github.com/FINDOSKDI/trading). The source code is an adaptation of the one from our previous tool, AUNTY [7], which was used to represent and execute models of our other version of this formalism, based on fuzzy Mealy machines. The code of the tool, without its graphical user interface, was inserted into a *jupyter notebook*, as shown in Fig. 4, and adapted in order for it to be able to connect with *OHLC* datasets contained in *pandas DataFrames*.

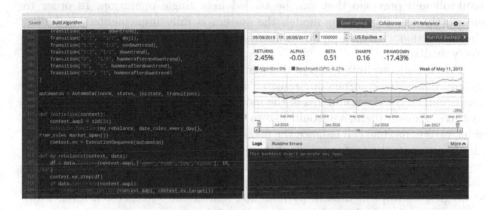

Fig. 3. Formalism integration with quantopian.

```python
In [1]: def fuzzyleq(y,delta):
            return lambda x: 1 if x <= y else \
                0 if x > y + delta else \
                1 - (x - y) / delta

        def fuzzyor(lam1, lam2):
            return lambda x: max([lam1(x), lam2(x)])

        def fuzzyand(lam1, lam2):
            return lambda x: min([lam1(x), lam2(x)])

        def fuzzynot(lam1):
            return lambda x: 1 - lam1(x)
        def fuzzyeq(x,y,delta):
            return min([fuzzyleq(y,delta)(x), fuzzyleq(x,delta)(y)])
        def fuzzygeq(z,delta):
            return fuzzynot(fuzzyleq(z-delta,delta))

        def fuzzyinterval(y,z,delta): # z < x < y
            return fuzzyand(fuzzyleq(y,delta),fuzzygeq(z,delta))

        def Hama(a,b):
            if a == 0 and b == 0:
                return 0
            return (a*b)/(a+b-a*b)
```

Fig. 4. Fragment of code from our framework, being integrated in a jupyter notebook.

It is worth to mention that our framework is integrated with quantopian, as shown in Fig. 3. After each trading interval, the quantopian algorithm API provides a DataFrame with historical data corresponding to the previous intervals. This DataFrame is then fed to the automaton, which establishes the most suitable execution corresponding to that DataFrame and all previous ones. Then, the final state of that execution raises its corresponding output value, which signals the desired trading order to the quantopian algorithm API.

5   Conclusions and Future Work

In this paper we have used a variant of fuzzy automata to represent a simple trading system where decisions about trading are *fuzzy*. The trading strategy is based on the observation of *hammers*, an indicator of a change of trend. Our strategy is parameterised so that the user can choose the portion of the available bankroll and portfolio that can be traded in a single operation. In order to ensure the usability of our model, the system is fully implemented, including the connection with quantopian to incorporate real data about candlesticks, and freely available.

This is only our initial step in this line of work and, therefore, there are many potential ways to continue our research in this field. A first group of improvements can target the usefulness of the model by incorporating more complicated features. Among them, we would like to introduce ways to calibrate beta, consider variations than can balance the investment between different stocks, strategies to optimize the conformation of the portfolio and introduce hard constraints regarding stop losses. In order to implement these improvements, we will consider as initial steps previous work at the implementation level, where different strategies are implemented in python [20], and at the specification level, using other appropriate high level formalisms [2, 4, 24].

A second group of improvements consider a more focused, and formal, analysis of the information. Specifically, in this paper we have presented a formalism to represent trading systems but we have not considered approaches to formally analyse these systems and to decide whether a certain *real* system (in this case, a certain stock) follows the behaviour specified by the system. First, we would like to take into account time and probabilities in a more formal approach. In particular, we would like to establish whether the quotes of a certain stock *timely/probabilistically conforms* to a trading strategy [19, 32]. In this line of work, we would like to use formal testing approaches to analyse the relation between a stock and a trading strategy. Interestingly enough, we cannot use classical testing approaches where the tester interacts with the system because they would be unrealistic: except in very restricted markets, a typical trader cannot strongly influence a stock. Therefore, we need to use a *passive* approach where the tester observes the system without interacting with it [17, 26, 27]. Similarly, we need to use approaches to test in the distributed architecture [16, 18] because real-time data can be collected from sources distributed along different physical locations. We would also like to use recent work on testing from FSMs using Information Theory [21] to evaluate the likelyhood of fault masking in the studied strategies. Finally, we would like to consider recent work on mutation testing [10,12,14,15] to assess the validity of a given strategy when compared to likely worse ones given by *mutants*.

References

1. Andrés, C., Llana, L., Núñez, M.: Self-adaptive fuzzy-timed systems. In: 13th IEEE Congress on Evolutionary Computation, CEC 2011, pp. 115–122. IEEE Computer Society (2011)
2. Bernal, A., Cambronero, M.E., Núñez, A., Cañizares, P.C., Valero, V.: Improving cloud architectures using UML profiles and M2T transformation techniques. J. Supercomput. **75**(12), 8012–8058 (2019)
3. Bonfanti, S., Gargantini, A., Mashkoor, A.: A systematic literature review of the use of formal methods in medical software systems. J. Softw. Evol. Process **30**(5), e1943 (2018)
4. Boubeta-Puig, J., Díaz, G., Macià, H., Valero, V., Ortiz, G.: MEdit4CEP-CPN: an approach for complex event processing modeling by prioritized colored petri nets. Inf. Syst. **81**, 267–289 (2019)
5. Bulkowski, T.N.: Encyclopedia of Candlestick Charts. Wiley, New York (2008)
6. Calvo, I., Merayo, M.G., Núñez, M.: An improved and tool-supported fuzzy automata framework to analyze heart data. In: Nguyen, N.T., Hoang, D.H., Hong, T.-P., Pham, H., Trawiński, B. (eds.) ACIIDS 2018. LNCS (LNAI), vol. 10751, pp. 694–704. Springer, Cham (2018). https://doi.org/10.1007/978-3-319-75417-8_65
7. Calvo, I., Merayo, M.G., Núñez, M.: AUNTY: a tool to automatically analyze data using fuzzy automata. In: 3rd International Conference on Computational Intelligence and Applications, ICCIA 2018, pp. 102–106. IEEE Computer Society (2018)
8. Calvo, I., Merayo, M.G., Núñez, M.: A methodology to analyze heart data using fuzzy automata. J. Intell. Fuzzy Syst. **37**, 7389–7399 (2019)
9. Calvo, I., Merayo, M.G., Núñez, M., Palomo-Lozano, F.: Conformance relations for fuzzy automata. In: Rojas, I., Joya, G., Catala, A. (eds.) IWANN 2019. LNCS, vol. 11506, pp. 753–765. Springer, Cham (2019). https://doi.org/10.1007/978-3-030-20521-8_62
10. Cañizares, P.C., Núñez, A., Merayo, M.G.: Mutomvo: mutation testing framework for simulated cloud and HPC environments. J. Syst. Softw. **143**, 187–207 (2018)
11. Cavalli, A.R., Higashino, T., Núñez, M.: A survey on formal active and passive testing with applications to the cloud. Ann. Telecommun. **70**(3–4), 85–93 (2015)
12. Delgado-Pérez, P., Rose, L.M., Medina-Bulo, I.: Coverage-based quality metric of mutation operators for test suite improvement. Software Qual. J. **27**(2), 823–859 (2019)
13. Doostfatemeh, M., Kremer, S.C.: New directions in fuzzy automata. Int. J. Approximate Reasoning **38**(2), 175–214 (2005)
14. Gómez-Abajo, P., Guerra, E., de Lara, J., Merayo, M.G.: A tool for domain-independent model mutation. Sci. Comput. Program. **163**, 85–92 (2018)
15. Gutiérrez-Madroñal, L., García-Domínguez, A., Medina-Bulo, I.: Evolutionary mutation testing for IoT with recorded and generated events. Softw. Pract. Exp. **49**(4), 640–672 (2019)
16. Hierons, R.M., Merayo, M.G., Núñez, M.: Controllable test cases for the distributed test architecture. In: Cha, S.S., Choi, J.-Y., Kim, M., Lee, I., Viswanathan, M. (eds.) ATVA 2008. LNCS, vol. 5311, pp. 201–215. Springer, Heidelberg (2008). https://doi.org/10.1007/978-3-540-88387-6_16

17. Hierons, R.M., Merayo, M.G., Núñez, M.: An extended framework for passive asynchronous testing. J. Log. Algebraic Methods Program. **86**(1), 408–424 (2017)
18. Hierons, R.M., Merayo, M.G., Núñez, M.: Bounded reordering in the distributed test architecture. IEEE Trans. Reliab. **67**(2), 522–537 (2018)
19. Hierons, R.M., Núñez, M.: Implementation relations and probabilistic schedulers in the distributed test architecture. J. Syst. Softw. **132**, 319–335 (2017)
20. Hilpisch, Y.: Python for Finance: Analyze Big Financial Data. O'Reilly Media, Sebastopol (2014)
21. Ibias, A., Hierons, R.M., Núñez, M.: Using squeeziness to test component-based systems defined as finite state machines. Inf. Softw. Technol. **112**, 132–147 (2019)
22. Lamport, L.: Who builds a house without drawing blueprints? Commun. ACM **58**(4), 38–41 (2015)
23. Lecomte, T., Deharbe, D., Prun, E., Mottin, E.: Applying a formal method in industry: a 25-year trajectory. In: Cavalheiro, S., Fiadeiro, J. (eds.) SBMF 2017. LNCS, vol. 10623, pp. 70–87. Springer, Cham (2017). https://doi.org/10.1007/978-3-319-70848-5_6
24. López, N., Núñez, M., Rodríguez, I., Rubio, F.: A formal framework for e-barter based on microeconomic theory and process algebras. In: Unger, H., Böhme, T., Mikler, A. (eds.) IICS 2002. LNCS, vol. 2346, pp. 217–228. Springer, Heidelberg (2002). https://doi.org/10.1007/3-540-48080-3_19
25. Magnani, L., Bertolotti, T. (eds.): Handbook of Model-Based Science. Springer, Cham (2017). https://doi.org/10.1007/978-3-319-30526-4
26. Merayo, M.G., Hierons, R.M., Núñez, M.: Passive testing with asynchronous communications and timestamps. Distrib. Comput. **31**(5), 327–342 (2018)
27. Merayo, M.G., Hierons, R.M., Núñez, M.: A tool supported methodology to passively test asynchronous systems with multiple users. Inf. Softw. Technol. **104**, 162–178 (2018)
28. Mordeson, J.N., Malik, D.S.: Fuzzy Automata and Languages: Theory and Applications. Chapman & Hall/CRC, Boca Raton (2002)
29. Mraz, M., Lapanja, I., Zimic, N., Virant, J.: Fuzzy numbers as inputs to fuzzy automata. In: 18th International Conference of the North American Fuzzy Information Processing Society, NAFIPS 1999, pp. 453–456. IEEE Computer Society (1999)
30. Newcombe, C., Rath, T., Zhang, F., Munteanu, B., Brooker, M., Deardeuff, M.: How Amazon web services uses formal methods. Commun. ACM **58**(4), 66–73 (2015)
31. Nison, S.: Japanese Candlestick Charting Techniques: A Contemporary Guide to the Ancient Investment Techniques of the Far East, 2nd edn. Prentice Hall, Upper Saddle River (2001)
32. Núñez, M., Rodríguez, I.: Conformance testing relations for timed systems. In: Grieskamp, W., Weise, C. (eds.) FATES 2005. LNCS, vol. 3997, pp. 103–117. Springer, Heidelberg (2006). https://doi.org/10.1007/11759744_8
33. Wee, W.G., Fu, K.S.: A formulation of fuzzy automata and its application as a model of learning systems. IEEE Trans. Syst. Sci. Cybern. **5**(3), 215–223 (1969)
34. Zadeh, L.A.: Fuzzy sets. Inf. Control **8**(3), 338–353 (1965)
35. Zadeh, L.A.: Fuzzy Sets, Fuzzy Logic, and Fuzzy Systems. Advances in Fuzzy Systems - Applications and Theory, vol. 6. World Scientific Press, Hackensack (1996)

Fuzzyfication of Repeatable Trust Game

Anna Motylska-Kuźma[✉] ⓘ and Jacek Mercik ⓘ

WSB University in Wroclaw, Wroclaw, Poland
{anna.motylska-kuzma,jacek.mercik}@wsb.wroclaw.pl

Abstract. The purpose of the work is to check what consequences in the interpretation of the results of experiments under the so-called trust games has a fuzzy approach. In previous considerations it was assumed that players' behaviors are implementations of specific probability distributions. For many reasons, the probabilistic approach can be considered inappropriate and too simplistic when explaining the choice of strategy. The study reinterpreted the obtained results of experiments assuming that the described behavior model is a model based on the fuzzy set theory.

Keywords: Trust game · Fuzzy approach · Cooperation

1 Introduction

The classical trust game has a two players P1 and P2. Player P1 is given some amount of money (a), which he can transfer in some part or whole to Player P2. The value of the money he transfers is multiplied by multiplier (m). Player 2 then decides how much money to transfer back to Player 1.

In one-shot game under the assumptions of "economic rationality", i.e. each player maximizes his/her expected reward, Player 2 should not return any money to Player 1 and hence Player 1 should not transfer any money. However, results from experiments indicate that humans do not behave in this way. Their actual behavior results from other factors, e.g. trust and level of aversion of in equality.

In existing trust games, it is assumed that the decisions taken by their participants are subject to certain probability distributions. For example, Markowska - Przybyła and Ramsey [20] claim that if Player P1 is egalitarian enough and will highly assess the probability of the player sharing the win, then he is able to give P2 the entire amount received. If he is not sure of the reciprocity of the Player P2, then he will share only part of his amount. However, in a situation where the Player P1 does not show an adequate level of egalitarianism, then he will keep the entire amount received for himself. These and other results of theoretical considerations are not confirmed in the case of real decisions made by participants. The observed actual results of the conducted experiments (see [8, 21, 22]) show that the Player P1 gives on average about 50% of the received amount, while the Player P2 only 30%. Therefore, other factors must influence the decisions the participants make. These include, among others examining the impact of individual player reputation on decisions [14], the impact of intention and time pressure on the decision to share the prize [10], the relationship between risk aversion or being

© Springer Nature Switzerland AG 2020
N. T. Nguyen et al. (Eds.): ACIIDS 2020, LNAI 12033, pp. 135–144, 2020.
https://doi.org/10.1007/978-3-030-41964-6_12

cheated and the decision to give away some of the profit [4], etc. A slightly different approach was used in the work of Motylska–Kuzma, Mercik and Buczek [23], in which the examined factor influencing the distribution of amounts transferred was the time of making decisions. Noting the lack of unambiguous answers and confronting it with the fact that the results of the above analyses were obtained by statistical analysis of the aggregate results of individual games, one should consider whether the perspective of a single player would not be more appropriate here. Quantitative assessment of his moves may not meet the basic assumption in statistical analysis, which is the assumption of homogeneity and representativeness of the tested sample. It can be assumed that during the games individual players should be treated individually and not as representatives of a wider population. This assumption meets the fuzzy approach [9], where the behavior of individual players is assessed linguistically. Therefore, the purpose of this study is to transform data describing the decisions of individual players into fuzzy data and then to verify whether such an approach (closer to the natural description of individual players) does not change the results obtained previously.

2   Repeatable Trust Game

In the classical trust game two players (an investor - P1 and a trustee - P2) decide to transfer or keep the amount received. P1 receives a certain base amount (a_j, e.g. 100) which he can keep or share with P2 (decision p1). The amount (p1), then, is multiplied by a specified multiplier, known for players (m_j, e.g. 3) and transferred to the account of the second player. In the next step, P2 retains or transfers some or all of the amount received to the first player (decision p2). Depending on value of the basic amount and the multiplier, cooperation between players will bring them specific benefits or losses, as shown in Fig. 1.

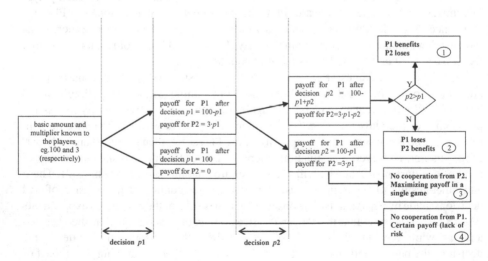

Fig. 1. Decision chart of a single game assuming that the multiplier is greater than 1

In a single game there are 4 result situations (marked with numbers in circles – Fig. 1). Looking at the P1 Player's preferences and assuming his rationality in making decisions, we can see that the most preferred situation will be 1, then 4, 2 and finally 3. For P2, this ranking will look as follows: 2, 1, 3, 4. With P2 having no impact on situation 4. If we expand our considerations to more games in the same pairs, then Player P2 will benefit from cooperation with Player P1 in the next game only if the result of the previous game is the base amount (a_j) in the next game. In our experiment, this is not the case. Thus, the P2 Player never benefits from cooperation with the P1 Player. His willingness to give away some of his profit may be dictated only by altruism or the awareness that players may switch roles in the next round and failure to cooperate may favor the P1 Player's "retaliation" strategy (after changing roles). Due to this fact, it is worth focusing mainly on the P1 Player and his willingness to cooperate.

Because the assumption of full rationality of participants does not translate into observations from practice, and is even considered unrealistic to achieve (see [3, 11] or [24]), therefore study of the sources of this irrationality and its consequences seems fully justified. What's more, in the econometric approach, in the value we assign to the convergence factor, i.e. fitting the model to the learning data ($\varphi^2 = 1 - R^2$, where R^2 is the coefficient of determination [27]: the smaller the value φ^2 the better the fit of the model to the data), we are explicitly talking about the existence of factors not included in the model. Because in models based on real data almost always $\varphi^2 > 0$, which means that there are factors affecting the explained variable not included in the model and their impact strength and meaning are not subject to statistical classification. It seems that the fuzzy approach allows us to take into account all potential factors of a given model [9].

3 Fuzzy Numbers and Their Arithmetic

Let $\mathbb{R}$ be $(-\infty; \infty)$, i.e. the set of all real numbers.

The fuzzy number, denote by $\tilde{X}$ is a fuzzy subset of $\mathbb{R}$ with member function $u_{\tilde{X}} : \mathbb{R} \to [0, 1]$ satisfying the following conditions:

- There exists at least one number $a_0 \in \mathbb{R}$ such that $u_{\tilde{X}}(a_0) = 1$
- $u_{\tilde{X}}(x)$ is nondecreasing on $(-\infty, a_0)$ and nonincreasing on (a_0, ∞)
- $u_{\tilde{X}}(x)$ is upper semi-continuos, i.e. $\lim_{x \to x_0^+} u_{\tilde{X}}(x) = u_{\tilde{X}}(x_0)$ if $x_0 < a_0$ and $\lim_{x \to x_0^-} u_{\tilde{X}}(x) = u_{\tilde{X}}(x_0)$ if $x_0 > a_0$
- $\int_{-\infty}^{\infty} u_{\tilde{X}}(x) dx < \infty$

Following Zadeh [31] and Dubois and Prade [9], a triangular fuzzy number is an $A = (m, \alpha, \beta)$, where m is the most probable value and the α and β – left and right spread, respectively.

The sum of two triangular fuzzy numbers $\tilde{X} = (m_X, \alpha_X, \beta_X)$ and $\tilde{Y} = (m_Y, \alpha_Y, \beta_Y)$ is a triangular fuzzy number $\tilde{X} + \tilde{Y} = (m_X + m_Y, \alpha_X + \alpha_Y, \beta_X + \beta_Y)$.

The multiplication of two triangular fuzzy numbers $\tilde{X} = (m_X, \alpha_X, \beta_X)$ and $\tilde{Y} = (m_Y, \alpha_Y, \beta_Y)$ is a triangular fuzzy number $\tilde{X} \otimes \tilde{Y} = (m_X \cdot m_Y, \alpha_X \cdot \alpha_Y, \beta_X \cdot \beta_Y)$, if the mx, my, αx, αy, βx, βy are the positive numbers.

Let Rj and Wj, where j = 1, 2, . . ., n, be the fuzzy rating and fuzzy weighting given to factor j, respectively. Following the procedure of Lin and Chen [16] consolidation of the fuzzy numbers is calculated as the fuzzy attractiveness ratio (FAR) as:

$$FAR = \frac{\sum_{j=1}^{n}(W_j \otimes Rj)}{\sum_{j=1}^{n} W_j} \tag{1}$$

In our case, the fuzzy numbers include the factors, thus the FAR will be calculated not as a weight average but the simple average of all membership of the fuzzy numbers.

Once the FAR has been calculated, this value can be approximated by a similar close linguistic term (LT) from the fuzzy values predefined scale. Several methods for approximating the FAR with an appropriate corresponding linguistic term have been proposed. The Euclidean distance will be used since it is the most intuitive from the human perception of approximation and the most commonly used method. The distance between FAR and each fuzzy number member of LT can be calculated as follows:

$$D(FAR, LTi) = \left\{ \sum_{x=1}^{t} (f\text{FAR}(x) - f\text{LT}i(x))^2 \right\}^{1/2} \tag{2}$$

4 Experiment Design

To implement the experiment we have created the appropriate software to support and automate the games. At the beginning, the participants of the game made registration and each participant had a unique login and password to enable bidding and viewing the results of previous games. They had also the unique number, thus the players were not anonymous for us. Then, a random selection of player pairs and setting of game parameters was made, i.e. basic amount (a_j) and multiplier (m_j). The course of the game was as follows (see [22])

- Player 1 from the pair i-th was informed by the system about the conditions of the game and he could transfer the amount p_1 to the Player 2 from the same pair. To make a decision the P1 had one week, which means that he could seriously think over his response. Not all randomly drawn participants of the pairs cooperated and performed the first moves appropriate for the Player 1. Such players were gradually eliminated from the system, which meant that the number of players taking part in the competition finally decreased. As part of the whole experiment, new players were not added after the experiment began,
- Player 2 from i-th pair was informed by the system about the decision p1 and the final amount, which he obtained (p1 multiplied by multiplier mj). He had also one week to make a decision and transfer the selected amount (p_2) to Player 1 from the same pair. It happened that the Player 2 did not cooperate. Then only the value p_1 was used in analyzes. An inactive P2 was not eliminated from the set of players and could in the future take part in games as both the first and the second player, after the end of a given game individual results were saved on the account of each participant of the game. Each of them could at any time view their results and analyze their previous behavior.

Using the statistical approach [22], we obtained from the experiment the following features:

- the Player P1's decision does not depend on the multiplier value
- the Player P1's decision does not depend on the basic amount
- the Player P2's decision does not depend on the multiplier value
- the Player P2's decision does not depend on the basic amount
- the correlation between the decision made before and after changing the role is relatively high
- Player P2 is not guided by the values received from Player P1, although he knows how the amount transferred to him is generated
- it cannot be excluded that Player P1 is not guided by efficiency in his decisions

5   Fuzzyfication of Repeatable Trust Game

The fuzzy approach to games as such can be found in many works. Beginning with fuzzy coalition introduced by Aubin [2], fuzzy games studied among others by Butnariu [5, 6], Butnariu and Kroupa [7], Tsurumi et al. [25] or Li and Zhang [15]. Yu and Zhang [28, 29] defined fuzzy core for any kinds of games with fuzzy coalition. Mareš [17, 18] with Vlach [19] focused on the uncertainty of the value of some characteristic factors associated with cooperative games. They treated fuzzy values mainly as payoffs for individual players. In another work, Gladysz et al. [12, 13] used a fuzzy approach to solve selected Shapley value problems in group decision making.

One of the conclusions from the introduction of a fuzzy approach to games is the observation that the presumed assumption of players' knowledge of the amount of expected withdrawal before making any decisions is simply unrealistic. There are many indefinite and uncertain factors that affect the decision making process, therefore players only have fragmentary and imprecise information about actual payoffs. Thus, according to Yu et al. [30] not only the decision to cooperate, but also its result is blurred. By imposing on the above observations an experiment with the repeatable trust game and assuming that the cooperation is the result of trust, which can be expressed in everyday language (very small, small, medium, large, very large), using fuzziness it is possible to check if Player P1 is willing cooperation with P2 Player. For this purpose, we adopt the following assumptions regarding linguistic assessment (after [1]) of the level of willingness to cooperate, i.e. the likelihood of specific situations:

- very low (VL): (0.05, 0.01, 0.10),
- low (L): (0.10, 0.05, 0.25),
- medium (M): (0.30, 0.25, 0.50),
- high (H): (0.60, 0.50, 0.75),
- very high (VH): (0.80, 0.75, 1.0).

The level of willingness is here very similar to the level of trust, thus in setting the numbers we were guided by the results of the research made by the World Bank [26] about the trust. Note that the left spread for "very low" and the right spread for "very high" are not symmetrical, because the ratings relate to probability and thus must be cut accordingly. In other cases, we assume that players are symmetrical in their assessments, i.e. they have no tendency to underestimate or overestimate their scores.

In the next step, it is necessary to transform the decisions $p1$ and $p2$ into fuzzy numbers, treating them as expressed by the players level of trust in relation to the partner in a given game.

Thus, we have a base amount $a_j > 0$, which is fixed for the game j and multiplier $m_j > 0$, which is also fixed for the game j ($j \in \langle 1, 4 \rangle$). In the game j Player P1 make the decision $p1$. We know about it that:

- $0 \leq p1 \leq 1 - p1$ is a percent of basic amount a_j
- $p1$ is the realization of a random variable X_j^1 describing the behavior of the first players in j-th game.

We change this value into triangular fuzzy number in such way:

(a) the real $p1$ is the most probable value (x)
(b) the left and the right spread (l_x and r_x) are the lowest and highest value (accordingly) of decision made by particular Player 1 in experiments with given Player 2 for a given m_j and a_j

In j-th game, Player P2 bets on the value $p2$. We know about it, that:

- $0 \leq p2 \leq 1 - p2$ is a percent of the multiplication of p1 and multiplier (m_j)
- $p2$ is the implementation of a random variable X_j^2 describing the behavior of the second players in j-th game.

We change this value into triangular fuzzy number in the same way as p1.

Aggregating decisions p1 i p2 for all games using FAR we get:

FAR(p1) = (0.463, 0.426, 0.500)
FAR(p2) = (0.331, 0.295, 0.367)

Then, we approximate FAR to the adopted assessment scale, in accordance with Eq. (2).

D(FAR(p1),VL) = 0.7097	D(FAR(p2),VL) = 0.4811
D(FAR(p1),L) = 0.5793	D(FAR(p2),L) = 0.3565
D(FAR(p1),M) = 0.2399	**D(FAR(p2),M) = 0.1438**
D(FAR(p1),H) = 0.2945	D(FAR(p2),H) = 0.5109
D(FAR(p1),VH) = 0.6845	D(FAR(p2),VH) = 0.9098

If the smallest distance FAR from LT determines the right level of willingness of a given player to cooperate, for Player P1 we get that he is willing to cooperate with P2 at an average level. For the Player P2 we obtain the same. Comparing these results with previous calculations using a traditional statistical approach, we can see that the conclusions are slightly different, i.e. from classical statistical approach Player P1 shows a greater level of confidence in Player P2, and thus the desire to cooperate. If we use the fuzzy approach, the willingness to cooperate is the same regardless the player. The P2 Player, despite the above-mentioned rational groundlessness of undertaking such cooperation, shows an equal level of trust. These observations may be due to the fact that the P2 Player does not know if in the next game the roles will not be swapped and is afraid of "retaliation" on the part of the partner.

By grouping the decisions of the P1 Player according to the experience criterion, i.e. those in which P1 has no previous experience in competitions with the Player P2 and those in which the Player P1 already has previous experience, and by using the above FAR calculation procedure, it is possible to verify whether the willingness to cooperate changes as experience increases. From our data the trust level of Player P1 if he has no previous experience is rather high $(D(FAR(p1),H) = 0.2581)$, while the trust level of the same player after few games with Player P2 decreases to medium level $(D(FAR(p1),M) = 0.1192)$. From comparing this observations with the average level of decision p2, we can conclude that P1 is somehow guided by efficiency in his decisions. This is similar to the conclusions which we have from the classic statistical approach.

Due to the fact that in the conducted games we manipulated the multiplier and the base amount, it is worth checking whether the willingness to cooperate in the pairs of the Player P1 and P2 is the same if we change the multiplier m_j, i.e. when it is definitely closer to 1 or even below one. Theoretically, looking at the structure of player preferences, it should be the opposite, i.e. both Player P1 and Player P2 should show a lack of willingness to cooperate. Looking at the Player P1 and his preferences, if the multiplier $m_j \leq 1$, then in situation 1 Player P1 loses in cooperation, because it is impossible that p2 > p1. Situations 2 and 3, as in the case of the multiplier $m_j > 1$, lead to losses. Therefore, the only acceptable situation is situation number 4. To sum up, the ranking of the situation in such a game would look like this: 4, 2, 3, 1 (where situation number 1 is not possible) (see Fig. 1).

By dividing p1 decisions into 3 groups depending on the height of the multiplier (m_j) and by aggregating values in individual groups using FAR we get:

$FAR(p1\_m=0.8) = (0.429, 0.381, 0.476)$
$FAR(p1\_m=2.75) = (0.489, 0.461, 0.518)$
$FAR(p1\_m=3) = (0.480, 0.446, 0.512)$

Then, we approximate FAR to the adopted assessment scale, in accordance with Eq. (2) and obtain for $m_j = 2.75$ that willingness to cooperate of the P1 Player is rather high $(D(FAR(p1\_m=2.75),H) = 0.2601)$, while at $m_j = 0.8$ and $m_j = 3$ this desire decreases to average $((D(FAR(p1\_m=0.8),M) = 0.1854; (D(FAR(p1\_m=3),M) = 0.2667)$. Using the traditional statistical approach, the hypothesis about the impact of the multiplier on player's decisions was rejected. When changing the approach to fuzzy, it can be seen that the decisions of Player P1 are somehow different to the height of the multiplier but there is no logic of this difference. Thus, we can conclude that the willingness to cooperate with Player P2 does not depend on the multiplier.

In the next step we check the decisions p2 and the influence of the multiplier on its level. By dividing p2 decisions into 3 groups depending on the height of the multiplier (mj) and by aggregating values in individual groups using FAR we get:

$FAR(p2\_m=0.8) = (0.317, 0.254, 0.379)$
$FAR(p2\_m=2.75) = (0.388, 0.337, 0.439)$
$FAR(p2\_m=3) = (0.305, 0.292, 0.317)$

From the approximation FAR to the adopted assessment scale, we obtain the same level of willingness to cooperate of the P2 Player independently from the value of multiplier. This willingness is rather medium ($D(FAR(p2\_m=0.8),M) = 0.1219$), $D(FAR(p2\_m=2.75),M) = 0.1376$; $D(FAR(p2\_m=3),M) = 0.1875$). Using the traditional statistical approach, the hypothesis about the impact of the multiplier on player's decisions was rejected. When changing the approach to fuzzy, we can also conclude that the Player's P2 willingness to the cooperate with Player P1 does not depend on the multiplier.

6 Conclusions

Summing up the results obtained, we can see that the following conclusions did not change their nature with the change from a probabilistic approach to a fuzzy one:

– the willingness to cooperate is the same regardless the player,
– Player P1 is somehow guided by efficiency in his decisions,
– Player's P1 willingness to the cooperate with Player P2 does not depend on the multiplier,
– Player's P2 willingness to the cooperate with Player P1 does not depend on the multiplier.

They are probably not dependent on the way players are modeled and are universal. However, we still do not know how stable (i.e. independent of the model used to describe players' behavior: probabilistic or fuzzy model) are the results related to the following issues:

– independence of the $p2$ value from the initial values,
– the impact of other factors on the observed variability of amount of the Player P2 as well
– influence of $p1$ value on P2 Player behavior.

These results require further experimentation.

References

1. Al-Mutairi, M.S., Hipel, K.W., Kamel, M.S.: Trust and cooperation from a fuzzy perspective. Math. Comput. Simul. **76**(5–6), 430–446 (2008)

2. Aubin, J.P.: Mathematical Methods of Game and Economic Theory (revised edition). North-Holland, Amsterdam (1982)
3. Binmore, K.: Does Game Theory Work?. The MIT Press, Cambridge (2007)
4. Bohnet, I., Zeckhauser, R.: Trust, risk and betrayal. J. Econ. Behav. Organ. **55**(4), 467–484 (2004)
5. Butnariu, D.: Fuzzy games: a description of the concept. Fuzzy Set Syst. **1**, 181–192 (1978)
6. Butnariu, D.: Stability and Shapley value for n-persons fuzzy game. Fuzzy Set Syst. **4**, 63–72 (1980)
7. Butnariu, D., Kroupa, T.: Shapley mapping and the cumulative value for n-person games with fuzz coalition. Eur. J. Oper. Res. **186**, 288–299 (2008)
8. Camerer, C.F.: Behavioral Game Theory. Experiments in Strategic Interaction. Princeton University Press, Princeton (2003)
9. Dubois, D., Prade, H.: Possibility theory, probability theory and multiple-valued logics: a clarification. Ann. Math. Artif. Intell. **32**, 35–66 (2001)
10. Gazdag, B.A., Haude, M., Hoegl, M., Muethel, M.: I do not want to trust You, but I do: on the relationship between trust intent, trusting behavior, and time pressure. J. Bus. Psychol. **34**(5), 731–743 (2019)
11. Gintis, H., Bowles, S., Boyd, R.T., Fehr, E.: Moral Sentiments and Material Interests: The Foundations of Cooperation in Economic Life. The MIT Press, Cambridge (2005)
12. Gładysz, B., Mercik, J., Stach, I.: Fuzzy Shapley value-based solution for communication network. In: Nguyen, N.T., Chbeir, R., Exposito, E., Aniorté, P., Trawiński, B. (eds.) ICCCI 2019. LNCS (LNAI), vol. 11683, pp. 535–544. Springer, Cham (2019). https://doi.org/10.1007/978-3-030-28377-3_44
13. Gładysz, B., Mercik, J., Ramsey, D.: A fuzzy approach to some Shapley value problems in group decision making. In: Algaba, E., Fragnelli, V., Sánchez-Soriano, J. (eds.) Handbook of the Shapley Value. CRC Press, Taylor & Francis Group (2019)
14. King-Casas, B., Tomlin, D., Anen, C., Camerer, C.F., Quartz, S.R., Montague, P.R.: Getting to know you: reputation and trust in a two-person economic exchange. Science **308**(5718), 78–83 (2005)
15. Li, S., Zhang, Q.: A simplified expression of the Shapley functions for fuzzy game. Eur. J. Oper. Res. **196**, 234–245 (2009)
16. Lin, C., Chen, Y.: Bid/No-bid decision-making—a fuzzy linguistic approach. Int. J. Project Manag. **22**, 585–593 (2004)
17. Mareš, M.: Fuzzy coalition structures. Fuzzy Set Syst. **114**, 23–33 (2000)
18. Mareš, M.: Fuzzy Cooperative Games: Cooperation with Vague Expectations. Physica-Verlag, New York (2001)
19. Mareš, M., Vlach, M.: Linear coalition games and their fuzzy extensions. Int. J. Uncertainty Fuzziness Knowl. Based Syst. **9**, 341–354 (2001)
20. Markowska–Przybyła, U., Ramsey, D.: A game theoretical study of generalised trust and reciprocation in Poland. I. theory and experimental design. Oper. Res. Decis. (3) (2014)
21. Markowska–Przybyła, U., Ramsey, D.: A game theoretical study of generalized trust and reciprocation in Poland. II. A description of the study group. Oper. Res. Decis. (2) (2015)
22. Motylska-Kuzma, A., Mercik, J., Sus, A.: Repeatable trust game – preliminary experimental results. In: Nguyen, N.T., Gaol, F.L., Hong, T.-P., Trawiński, B. (eds.) ACIIDS 2019. LNCS (LNAI), vol. 11431, pp. 488–498. Springer, Cham (2019). https://doi.org/10.1007/978-3-030-14799-0_42
23. Motylska-Kuźma, A., Mercik, J., Buczek, A.: Repeated trust game – statistical results concerning time of reaction. In: Nguyen, N.T., Kowalczyk, R., Mercik, J., Motylska-Kuźma, A. (eds.) Transactions on Computational Collective Intelligence XXXIV. LNCS, vol. 11890, pp. 74–89. Springer, Heidelberg (2019). https://doi.org/10.1007/978-3-662-60555-4_6

24. Tommassini, M.: Games, evolution, and society. Rendiconti Del Seminario Matematico **66**(3), 229–258 (2008)
25. Tsurumi, M., Tanino, T., Inuiguchi, M.: A Shapley function on a class of cooperative fuzzy games. Eur. J. Oper. Res. **129**, 596–618 (2001)
26. World Value Survey (2014). www.worldvaluesurvey.org. Accessed 5 Sept 2019
27. Wright, S.: Correlation and causation. J. Agric. Res. **20**(7), 557–585 (1921)
28. Yu, X., Zhang, Q.: The fuzzy core in games with fuzzy coalitions. J. Comput. Appl. Math. **230**, 173–186 (2009)
29. Yu, X., Zhang, Q.: An extension of fuzzy cooperative games. Fuzzy Set Syst. **161**, 1614–1634 (2010)
30. Yu, X., Zhang, Q., Zhou, Z.: Cooperative game with fuzzy coalition and payoff value in the generalized integral form. J. Intell. Fuzzy Syst. **33**, 3641–3651 (2017)
31. Zadeh, L.A.: Fuzzy sets. Inf. Control **8**, 338–353 (1965)

Online Auction and Optimal Stopping Game with Imperfect Observation

Vladimir Mazalov[1,2] and Anna Ivashko[1(✉)]

[1] Institute of Applied Mathematical Research of the Karelian Research Centre of the Russian Academy of Sciences, Pushkinskaya Str., 11, Petrozavodsk 185910, Russia
{vmazalov,aivashko}@krc.karelia.ru
[2] School of Mathematics and Statistics, Qingdao University,
Institute of Applied Mathematics of Shandong, Qingdao 266071, China

Abstract. The paper examines a multi-stage game-theoretic model of an auction where the participants (players) set minimum threshold price levels above which they are ready to sell. Price offerings are a sequence of independent and identically distributed random variables. A two-person game in which each player is interested in selling at a price higher than the competitor's is considered. Optimal threshold pricing strategies and expected payoffs of the players are determined. Numerical modeling results are presented.

Keywords: Optimal stopping · Imperfect observation · Zero-sum game · Auction · Game with priority

1 Introduction

One of the perspectives of interest for researchers in studying various socio-economic situations is the analysis of the participants' behavior in decision-making. Decision-making situations arise in sequential choice problems, such as buying/selling goods or services, job search, choice of spouse or business partner. In the process of choice one needs to decide on the time moment at which to stop the search. A useful tool when studying such problems with several participants is game-theoretical models with optimal stopping. These models take into account important aspects of the choosing process and define the optimal stopping time.

Different schemes based on competitive prices are used with resources buying/selling or renting. The examples are auctions and tenders, competition for computational resources and storage. An important task is to model the competitive behavior of participants in various types of auctions. Online auctions are the best source for examples of modern markets. In an auction, it is crucial for a participant to determine his/her optimal strategy in order to increase the odds to win. Various auction types are used. In some online auction setups a bidder

---

Supported by "The Double-Hundred Talent Plan" of the Shandong Province, China (grant no. WST2017009).

does not receive information about the exact price offered by a buyer. An example, from the seller's perspective, is the Priceline auction. Each seller can name a minimum threshold price level at which they are ready to sell. The deal in the auction is made if the price offered by a bidder is above the named threshold value. An essential task here is to investigate the behavior of the participants of such an auction.

We suggest a game-theoretic model of an auction where sellers (players) announce the minimum threshold price at which they are ready to sell, while bidders sequentially offer prices modeled by a sequence of independent and identically distributed random variables. The deal is made and the seller sells the item to the buyer if the price offered is above the threshold value named by the player. We apply the minmax criterion, in which we are interested in the strategies that maximize the chances of selling the item at a price higher than the competitor's.

We consider the following multi-stage zero-sum optimal stopping game with two players. Two Players, 1 and 2, observe sequentially a known number N of independent and identically distributed random variables from a known continuous distribution with the aim to choose a higher value than the one chosen by the opponent. Each of the players can choose one observation at most. The random variables cannot be perfectly observed. At each stage a random variable is sampled the player is informed only whether it is higher or lower than some level specified by him. If the observed value is higher than this level, a player accepts this observation. Otherwise, he/she rejects it. If they both want to accept the same observation the priority is given to a specified player, say Player 1. If one player chooses the observation he/she stops, while the other continues sampling. If both players reject the current observation, they both move to the next stage. If the process has been going on until the last stage, one of the players must make the choice. The class of optimal strategies and the suitable gain function for the problem are constructed.

The remainder of this paper is structured as follows. Section 2 offers a review of related works. The optimal stopping game with imperfect observation is described in Sect. 3. Section 4 deals with a case of two observations. Next, in Sects. 5 and 6, we describe the behavior of the players in the case one of them has stopped. We present the optimal thresholds and the corresponding gain functions. Section 7 is devoted to the optimal strategies and gains in the case neither of the players has stopped yet. The numerical results are reported.

2   Related Works

Optimal stopping game-theoretic models are often used to study the behavior of participants in socio-economic systems. The optimal stopping problems where the participants have to take the decision about choosing an item that fulfills certain criteria are also known as best-choice problems. Such problems often arise in economics, finance, sociology and politics.

The players' optimal strategies in best-choice problems are often derived using dynamic programming, a method that solves a complicated problem by breaking it down into simpler sub-problems recursively defined. This method has various applications in studying different game settings, such as quitting games (Solan and Vieille [1]), house-selling problem (Sofronov [2]), job-search problem (Immorlica et al. [3]), mate choice problem (Alpern et al. [4]), and it is successfully used in economic contexts for best-choice games and auctions (Whitmeyer [5], Riedel [6], Harrell et al. [7]).

One distinctive class of problems is associated with TV shows. It should interest economists who study human decision-making (Seregina et al. [8], Mazalov and Ivashko [9], Tenorio and Cason [10], Bennett and Hickman [11]). E.g., each player in the game "The Price is Right" must decide whether or not to spin the wheel again, just as job searchers must decide whether or not to search again, and managers must decide whether or not to continue investing in projects.

The mathematical best-choice models where players do not know the exact value of the observed random variables are called problems with imperfect observations. Various problems with imperfect observations have been considered in the literature. Porosinski [12] and Sakaguchi [13] described various one-player problems with imperfect observations in their papers. Papers by Enns and Ferenstein [14], Neumann et al. [15], and by Mazalov et al. [16] investigated the problem with imperfect observations and a priority. Porosinski and Szajowski [17] described a game with a random priority.

Seregina et al. [8], Mazalov and Ivashko [9], Ivashko et al. [18] consider models related to various types of auctions and tenders. In papers by Seregina et al. [8], Mazalov and Ivashko [9] auctions with different amounts of available information about the behavior of the players are studied and applied to the TV show "The Price is Right". The paper Ivashko et al. [18] applies optimal stopping methods to the problem of choosing the optimal offer in an online auction for cloud computing resources.

In this paper, we suggest a new game-theoretic model with optimal stopping and imperfect observations, where one of the players has priority, and where the minmax criterion for optimal decision-making is applied.

3   Optimal Stopping Game with Imperfect Observation

Let $X_1, X_2, \ldots, X_N$, $N \in \mathbf{N}$, be a sequence of independent and identically distributed random variables with a common and known distribution defined on a probability space $(\Omega, \mathcal{F}, P)$. The sequence of random variables is sequentially sampled by two players (Player 1 and Player 2). However, the observations are imperfect and the exact values are unknown. The players' strategies at each stage n are the thresholds x_n and y_n. If the current observation X_n exceeds the threshold named by the player, he/she accepts the observation. Player 1 has priority, i.e. if X_n exceeds the thresholds of both players, the observation goes to Player 1. If one of the players stops, the other player continues choosing. His aim is to choose an observation that is greater than the earlier-stopping player has had. The problem is a generalization of the single player optimal stopping problem with imperfect observation (Porosinski [12], Sakaguchi [13]).

The strategies of Player 1 and Player 2 in this game are represented, respectively, by the collections $(x_1, x_2, ..., x_N)$ and $(y_1, y_2, ..., y_N)$ of thresholds. We take a zero-sum game where the gain function is determined as follows. The player obtaining the observation with the highest value gets $+1$, and -1 otherwise.

Let $\mathbf{T}$ be the set of stopping times with respect to $\{\mathcal{F}_n\}_{n=1}^N$, where $\mathcal{F}_n = \sigma\{X_1, X_2, ..., X_N\}$ is a σ-field of information at the moment n, $n = 1, 2, ..., N$.

Since the observations are imperfect, we adopt $\mathbf{S} = \{\tau \in \mathbf{T} : \tau = \min\{1 \leq n \leq N : X_n \geq x_n\}\}$ as the class of adequate strategies for a player. We assume that the players have sets of strategies $\mathbf{S}$ and the above described priority of Player 1 will be involved in the gain function.

Here we can assume without loss of generality that the observed random variables have a uniform distribution and the set of strategies $\mathbf{S}$ is equivalent to $[0,1]^n$. This is possible because know that any continuous distribution can be reduced to uniform on $[0,1]$ by appropriate scaling.

It is obvious that an optimal x_N must be equal to 0, i.e. if the players do not stop before N, then Player 1 takes X_N and wins with probability 1. Assume for convenience that $y_N = 0$. Consequently, if $N = 1$ then $x_N = 0$, $y_N = 0$.

4   Case $N = 2$

Let $N = 2$ and Players 1 and 2 choose the strategies x and y, respectively, with $y < x$. Because $x_N = y_N = 0$, the payoff of Player 1 in the zero-sum game is equal to

$$
\begin{aligned}
H^{(2)}(x, y) = \int\limits_0^1 &\Big[I_{\{0 \leq t \leq y\}} + (P\{X_N > t\} - P\{X_N < t\}) I_{\{y < t \leq x\}} \\
&+ (P\{X_N < t\} - P\{X_N > t\}) I_{\{x < t \leq 1\}} \Big] dt \\
= y + &\int\limits_y^x (1 - 2t) dt + \int\limits_x^1 (2t - 1) dt = 2x - 2x^2 + y^2,
\end{aligned}
\tag{1}
$$

where the random variable $I_A(\omega) = \begin{cases} 1, \text{if, } \omega \in A; \\ 0, \text{otherwise.} \end{cases}$

Here, the first term under the integral corresponds to the payoff of Player 1 if both players have rejected the observation and moved on to the next stage of the game. The second term corresponds to the payoff of Player 1 if he rejects the observation at this stage, while Player 2 accepts it. The third term is the payoff of Player 1 if he accepts the current observation.

In convex zero-sum games, the maxmin value of the game equals the minmax value. Hence, equilibrium is found by solving the optimization problem

$$
H^{(2)} = \max_x \min_y H^{(2)}(x, y).
$$

Evidently the payoff function (1) has the equilibrium $x_1 = 1/2, y_1 = 0$ and value of the game is equal to

$$H^{(2)} = \max_x \min_y \{2x - 2x^2 + y^2\} = 0.5.$$

We will use this method for the general case. Be warned, however, that in the case $n > 2$ the optimal behavior of a player before and after one of them stops would be different. First, we will analyze this problem when one of the players has made the decision and stops at a stage $n < N$.

5   Optimal Behavior of Player 2 in the Case Player 1 Stopped First

Suppose that Player 1 stops at a stage $n < N$ (it means that $X_n \geq x_n$) and let us find the optimal behavior of Player 2 that minimizes the payoff of Player 1. The strategy of Player 2 is determined by the sequence of levels $y_{n+1}, ..., y_N$.

Player 1 gets
+1 when $\{X_{n+1} < y_{n+1}, ..., X_{k-1} < y_{k-1}, X_k \geq y_k, X_n > X_k\}$; $k = n+1, ..., N$,
−1 when $\{X_{n+1} < y_{n+1}, ..., X_{k-1} < y_{k-1}, X_k \geq y_k, X_n < X_k\}$, $k = n+1, ..., N$,
or 0 otherwise.

Denote by $H_1^{(n)}(x; y_{n+1}, ..., y_N)$ the expected payoff of Player 1 given that $X_n \geq x$ in the case Player 1 stopped first.

Thus, $H_1^{(n)}(x; y_{n+1}, ..., y_N)$ is equal to

$$H_1^{(n)}(x; y_{n+1}, ..., y_N)$$
$$= \int_x^1 [\sum_{k=n+1}^N (P\{X_{n+1} < y_{n+1}, ..., X_{k-1} < y_{k-1}, X_k \geq y_k, X_k < t\}$$
$$- P\{X_{n+1} < y_{n+1}, ..., X_{k-1} < y_{k-1}, X_k \geq y_k, X_k > t\})] dt$$
$$= (1 - y_{n+1})(x - y_{n+1}) + \sum_{k=n+2}^{N-1} \prod_{i=n+1}^{k-1} y_i (1 - y_k)(x - y_k) + \prod_{i=n+1}^{N-1} y_i (1 - x^2)$$
$$= x - x^2 \prod_{i=n+1}^{N-1} y_i - \sum_{k=n+1}^{N-1} \prod_{i=n+1}^{k} y_i (1 - y_k).$$

$$(2)$$

The aim of Player 2 is to minimize (2). Calculating the derivatives of (2) in every argument and equaling it to zero we find that the optimal behavior of Player 2 is determined by the vector $(y_{n+1}, ..., y_N)$ in the form

$$y_N = 0, \quad y_{N-1} = \frac{1 + x^2}{2}, \quad y_i = \frac{1 + y_{i+1}^2}{2}, i = n + 1, ..., N - 2. \qquad (3)$$

Substituting optimal values from (3) into (2), we find the payoff of Player 1 for the case when he stops at stage n and Player 2 plays optimally. From (2)–(3) we get that

$$H_1^{(n)}(x) = \min H_1^{(n)}(x; y_{n+1}, ..., y_N)$$
$$= x - \prod_{i=n+1}^{N-1} y_i(x^2 + 1 - y_{N-1}) - \sum_{k=n+1}^{N-2} \prod_{i=n+1}^{k} y_i(1 - y_k). \tag{4}$$

Using (3) we can rewrite (4) in the form

$$H_1^{(n)}(x) = x - y_{N-1}^2 \prod_{i=n+1}^{N-2} y_i - \sum_{k=n+1}^{N-2} \prod_{i=n+1}^{k} y_i(1 - y_k). \tag{5}$$

Arguing the same way as above we get

$$H_1^{(n)}(x) = x - y_{n+1}^2. \tag{6}$$

Hence, the next result is valid.

Theorem 1. *The optimal behavior of Player 2 in the case Player 1 stops first at stage n depends on the levels satisfying the relations in (3) and the optimal payoff is equal to (6).*

The next assertion follows from (3).

Corollary 1. *The optimal levels of Player 2 satisfy the relations*

$$x < y_{N-1} < y_{N-2} < ... < y_{n+1} < 1. \tag{7}$$

6   The Optimal Behavior of Player 1 in the Case Player 2 Stopped First

Suppose now that Player 2 was the first to stop at stage n. This is possible only if $y_n = y < x = x_n$ and $y \le X_n < x$. Obviously, the optimal levels must be lower or equal to the value x. Let us calculate the expected payoff of Player 1 in this case if he uses the vector of levels $(x_{n+1}, ..., x_N)$. In this case Player 1 gets
   +1 when $\{X_{n+1} < x_{n+1}, ..., X_{k-1} < x_{k-1}, X_k \ge x_k, X_k > X_n\}$, $k = n + 1, ..., N$,
   −1 when $\{X_{n+1} < x_{n+1}, ..., X_{k-1} < x_{k-1}, X_k \ge x_k, X_k < X_n\}$, $k = n + 1, ..., N$ or
   0 otherwise.
   Denote by $H_2^{(n)}(x; x_{n+1}, ..., x_N)$ the expected payoff of Player 1 given that $y \le X_n < x$ in the case Player 2 stopped first. Hence, $H_2^{(n)}(x; x_{n+1}, ..., x_N)$ is equal to

$$H_2^{(n)}(x; x_{n+1}, ..., x_N) = \int\limits_y^x [P\{X_{n+1} \ge x_{n+1}, X_{n+1} \ge t\}$$

$$-P\{X_{n+1} \ge x_{n+1}, X_{n+1} < t\}$$

$$+ \sum_{k=n+2}^{N-1} (P\{X_{n+1} < x_{n+1}, ..., X_{k-1} < x_{k-1}, X_k \ge x_k, X_k \ge t\}$$

$$-P\{X_{n+1} < x_{n+1}, ..., X_{k-1} < x_{k-1}, X_k \ge x_k, X_k < t\})$$

$$+P\{X_{n+1} < x_{n+1}, ..., X_{N-1} < x_{N-1}, X_N \ge t\} \tag{8}$$

$$-P\{X_{n+1} < x_{n+1}, ..., X_{N-1} < x_{N-1}, X_N < t\}]dt$$

$$= (x - y) - (x - x_{n+1})^2 - \sum_{k=n+2}^{N-1} \prod_{i=n+1}^{k-1} x_i (x - x_k)^2 + \prod_{i=n+1}^{N-1} x_i (y^2 - x^2)$$

$$= (x - y) - (x^2 - y^2) \prod_{i=n+1}^{N-1} x_i - (x - x_{n+1})^2 - \sum_{k=n+2}^{N-1} \prod_{i=n+1}^{k-1} x_i (x - x_k)^2.$$

Let us now find the optimal strategy of Player 1 that maximizes (8). To this end we can solve the system of equations

$$\frac{\partial H_2^{(n)}}{\partial x_i} = 0, i = N - 1, ..., n + 1.$$

We have

$$\frac{\partial H_2^{(n)}}{\partial x_{N-1}} = \prod_{i=n+1}^{N-2} x_i \left[-x^2 + y^2 + 2(x - x_{N-1}) \right] = 0$$

hence,

$$x_{N-1} = x - \frac{x^2 - y^2}{2}. \tag{9}$$

The equation $\frac{\partial H_2^{(n)}}{\partial x_{N-2}} = 0$ is equivalent to

$$x_{N-1} \left[-x^2 + y^2 + 2x - x_{N-1} \right] - x^2 + 2(x - x_{N-2}) = 0. \tag{10}$$

Substituting (9) into (10), we get

$$x_{N-1}^2 - x^2 + 2(x - x_{N-2}) = 0,$$

or

$$x_{N-2} = x - \frac{x^2 - x_{N-1}^2}{2}.$$

Arguing the same way as above we will obtain the following proposition.

Theorem 2. *The optimal behavior of Player 1 in the case Player 2 stopped first at stage n is defined by the levels satisfying the relations*

$$x_{N-1} = x - \frac{x^2 - y^2}{2}, ..., x_i = x - \frac{x^2 - x_{i+1}^2}{2}, i = n+1, ..., N-2. \tag{11}$$

and the optimal payoff is equal to

$$H_2^{(n)}(x, y) = (x - y) - (x^2 - x_{n+1}^2). \tag{12}$$

Proof. The relations (11) can be easily obtained by induction as above. Let us prove (12).

Present (8) in the form

$$H_2^{(n)}(x, y) = (x-y) - (x-x_{n+1})^2 - \left[(x^2 - y^2) \prod_{i=n+1}^{N-1} x_i + \sum_{k=n+2}^{N-1} \prod_{i=n+1}^{k-1} x_i (x - x_k)^2 \right]. \tag{13}$$

Denote

$$J_{N-1} = \sum_{k=n+2}^{N-1} \prod_{i=n+1}^{k-1} x_i (x - x_k)^2.$$

We can now rewrite the expression in square brackets in (13) in the form

$$\begin{aligned}
(x^2 &- y^2) \prod_{i=n+1}^{N-1} x_i + J_{N-1} \\
&= (x^2 - y^2) \prod_{i=n+1}^{N-1} x_i + \prod_{i=n+1}^{N-2} x_i (x - x_{N-1})^2 + J_{N-2} \\
&= \prod_{i=n+1}^{N-1} x_i (x^2 - y^2 + x_{N-1} - 2x) + x^2 \prod_{i=n+1}^{N-2} x_i + J_{N-2} \\
&= (x^2 - x_{N-1}^2) \prod_{i=n+1}^{N-2} x_i + J_{N-2} \\
&\cdots\cdots\cdots\cdots\cdots\cdots\cdots\cdots\cdots\cdots\cdots \\
&= (x^2 - x_{n+3}^2) \prod_{i=n+1}^{n+2} x_i + J_{n+2} \\
&= (x^2 - x_{n+2}^2) x_{n+1}.
\end{aligned} \tag{14}$$

Substituting (14) into (13) we get

$$H_2^{(n)}(x, y) = (x-y) - (x-x_{n+1})^2 - (x^2 - x_{n+2}^2)x_{n+1} = (x-y) - (x^2 - x_{n+1}^2). \tag{15}$$

The theorem is proved.

7   Construction of Minimax Strategies

Thus, we have found the optimal behavior of a player in the case his opponent stopped earlier than him. Optimal strategies are determined by relations

(3) and (11). Let us now find the optimal stopping rules for the case where neither player has stopped yet. We shall use the dynamic programming method.

Suppose that up to stage n neither player has stopped yet, and they are using the strategies $x_n = x$ and $y_n = y$ respectively ($y < x$). We have a game on the unit square $[0,1]^2$ and the value of this game $H^{(n)}$ satisfies the Bellman equation of the form

$$
\begin{aligned}
H^{(n)} &= \sup_x \inf_y \left\{ \int_0^y H^{(n+1)} dt + H_2^{(n)}(x,y) + H_1^{(n)}(x) \right\} \\
&= \sup_x \inf_y \left\{ H^{(n+1)} y + (x-y) - (x^2 - x_{n+1}^2) + (x - y_{n+1}^2) \right\}, n = 1, ..., N-2,
\end{aligned}
$$
(16)

with x_{n+1}, y_{n+1} satisfying the relations (3), (11), and the final condition $H^{(N-1)} = 0.5$ (see Sect. 4).

Equation (16) is the basis for constructing the optimal strategies of the players.

It is more convenient to rewrite Eqs. (16) in the form where the parameter $t = N - n + 1$ — number of stages until the end.

$$
H^{(2)} = 0.5, \quad H^{(t)} = \sup_x \inf_y \left\{ H^{(t-1)} y + (x-y) - (x^2 - x_{t-1}^2) + x - y_{t-1}^2 \right\}, \quad (17)
$$

with $\{x_t, y_t\}$ satisfying the relations

$$
x_2 = x - \frac{x^2 - y^2}{2}, \quad x_t = x - \frac{x^2 - x_{t-1}^2}{2}, t = 3, ..., N-1, \quad (18)
$$

$$
y_2 = \frac{1 + x^2}{2}, \quad y_t = \frac{1 + y_{t-1}^2}{2}, t = 3, ..., N-1. \quad (19)
$$

We can rewrite (18)–(19) in the equivalent form

$$
x_1 = y, \quad x_t = x - \frac{x^2 - x_{t-1}^2}{2}, t = 2, ..., N-1, \quad (20)
$$

$$
y_1 = x, \quad y_t = \frac{1 + y_{t-1}^2}{2}, t = 2, ..., N-1. \quad (21)
$$

Example 1. Consider the process of selling a good or service in an auction, where $N = 3$. Participants of the auction (Player 1 and Player 2) define their minimum price values at which they are ready to sell at each stage. At the first stage, for $t = 3$ we obtain from (17)–(19)

$$
H^{(3)} = \sup_x \inf_y \left\{ \frac{y}{2} + (x-y) - (x^2 - (x - \frac{x^2 - y^2}{2})^2 + x - (1 + x^2)/2 \right\}.
$$

Computing shows that the optimal strategies here are $x_3 = 0.679, y_3 = 0.453$, and the value of the game is $H^{(3)} = 0.440$ (see Table 1).

Thus, the minimum price levels of Player 1 and Player 2 at the first stage are $x_3 = 0.679$ and $y_3 = 0.453$, respectively. If the offer from bidder X_1 at this stage is above 0.679, then the deal is made with Player 1. Player 2, remaining in the game, proceeds to the second stage, and continues observing the bidders' price offerings. The minimum price value at which Player 2 would be ready to sell the item at stage two is determined by Eq. (3).

If the price offer from bidder X_1 at the first stage is within the interval of 0.453 to 0.679, the deal is made with Player 2. In that case, Player 1 remains in the game alone, and determines the minimum price level at which he/she is ready to sell using the formula (11).

If, on the other hand, bidder X_1 at the first stage offers a price below 0.453, no deal is made with either player. In that case, both players proceed to the next stage, and the minimum price values at which they would be ready to sell the item at this stage are $x_2 = 0.5$ and $y_2 = 0$, respectively. This procedure is then repeated.

At the last stage, both players want to sell the item, wherefore $x_3 = 0$ and $y_3 = 0$. In this case, since Player 1 has the priority of choice, he is the one to make the deal at the last stage.

Numerical analysis of relations (17)–(19) gives the following results presented in Table 1.

Table 1. Minimax strategies of Players and the value of the game for some values of N.

N	1	2	3	4	5	6	7	8	9	10	20	30	40
x_N	0	0.5	0.679	0.768	0.818	0.851	0.874	0.890	0.903	0.913	0.957	0.972	0.979
y_N	0	0	0.453	0.612	0.700	0.755	0.793	0.821	0.842	0.859	0.932	0.955	0.966
$H^{(N)}$	1	0.5	0.440	0.415	0.401	0.393	0.387	0.382	0.379	0.376	0.365	0.361	0.359

As indicated in the Table, the value of the game is $H^{(N)} > 0$, which means that the first player benefits the most from the game. This is corroborated by the priority of choice he/she has. This also explains the fact that the first player's threshold x_N is not lower than the respective threshold y_N for the second player. Where $N = 1$, the value of the game is 1, meaning that Player 1 wins if the sampling process has only one stage. Note also that the thresholds of both players increase along with N. As the number of stages N decreases, players tend to become more cautious about the threshold values they announce.

8   Conclusion

We study a game-theoretic model of an auction where two competing sellers wish to sell an item at a price higher than the opponent's. Bidders stage in one after another, offering their prices for the item. The sellers have no prior knowledge of the bidder's offers, which represent a sequence of random variables. This situation is modeled by a zero-sum game, in which sellers (players) name the minimum threshold values of the price at which they would be ready to sell. The deal is made if the price offered by a bidder is above the named threshold value. One of the players is supposed to have priority is striking the deal. The paper describes the optimal behavior of the participants of such an auction. Players' optimal strategies were obtained in the form of multiple threshold strategies, and players' optimal payoffs were determined for different situations in the game. The optimal strategies of both players were numerically simulated for games with different numbers of stages. The results were interpreted in relation to the behavior of the players in the auction. According to our results, this game is the most beneficial for the first player, who enjoys decision-making priority over the opponent.

The results can later be expanded to the case with several players, and other problem setups can be analyzed, such as problems with random priority, or models with other gain functions. Another possible path for further research is to apply the results to determine the optimal behavior of participants in real-life auctions and online auctions for goods or computing resources.

References

1. Solan, E., Vieille, N.: Quitting games. Math. Oper. Res. **26**(2), 265–285 (2001). https://doi.org/10.1287/moor.26.2.265.10549
2. Sofronov, G.: An optimal sequential procedure for a multiple selling problem with independent observations. Eur. J. Oper. Res. **225**(2), 332–336 (2013). https://doi.org/10.1016/j.ejor.2012.09.042
3. Immorlica, N., Kleinberg, R., Mahdian, M.: Secretary problems with competing employers. In: Spirakis, P., Mavronicolas, M., Kontogiannis, S. (eds.) WINE 2006. LNCS, vol. 4286, pp. 389–400. Springer, Heidelberg (2006). https://doi.org/10.1007/11944874_35
4. Alpern, S., Katrantzi, I., Ramsey, D.: Partnership formation with age-dependent preferences. Eur. J. Oper. Res. **225**(1), 91–99 (2013). https://doi.org/10.1016/j.ejor.2012.09.012
5. Whitmeyer, M.: A competitive optimal stopping game. B.E. J. Theor. Econ. **18**(1), 1–15 (2018). https://doi.org/10.1515/bejte-2016-0128
6. Riedel, F.: Optimal stopping with multiple priors. Econometrica **77**(3), 857–908 (2009). https://doi.org/10.3982/ECTA7594
7. Harrell, G., Harrison, J., Mao, G., Wang, J.: Online auction and secretary problem. In: International Conference on Scientific Computing, pp. 241–244 (2015)
8. Seregina, T., Ivashko, A., Mazalov, V.: Optimal stopping strategies in the game "the price is right". Proc. Steklov Inst. Math. **307**(Suppl. 1), 1–15 (2019)
9. Mazalov, V., Ivashko, A.: Equilibrium in n-person game of showcase-showdown. Probab. Eng. Inform. Sci. **24**, 397–403 (2010). https://doi.org/10.1017/S0269964810000045
10. Tenorio, R., Cason, T.N.: To spin or not to spin? Natural and laboratory experiments from "the price is right". Econ. J. **112**(476), 170–195 (2002). https://doi.org/10.1111/1468-0297.0j678

11. Bennett, R.W., Hickman, K.A.: Rationality and the "price is right". J. Econ. Behav. Organ. **21**(1), 99–105 (1993). https://doi.org/10.1016/0167-2681(93)90042-N
12. Porosiński, Z.: Full-information best choice problems with imperfect observation and a random number of observations. Zastos. Matem. **21**, 179–192 (1991)
13. Sakaguchi, M.: Best choice problems with full information and imperfect observation. Math. Japonica **29**, 241–250 (1984)
14. Enns, E., Ferenstein, E.: On a multi-person time-sequential game with priorities. Sequential Anal. **6**, 239–256 (1987). https://doi.org/10.1080/07474948708836129
15. Neumann, P., Porosinski, Z., Szajowski, K.: On two person full-information best choice problem with imperfect information. In: Petrosjan, L.A., Mazalov, V.V. (eds.) Game Theory and Application II, pp. 47–55. Nova Science Publishers, New York (1996)
16. Mazalov, V., Neumann, P., Falko, I.: Optimal stopping game with imperfect information. Far-Eastern Math. Rep. **6**, 74–86 (1998)
17. Porosiński, Z., Szajowski, K.: Random priority two-person full-information best choice problem with imperfect observation. Applicationes Mathematicae **27**(3), 251–263 (2000)
18. Ivashko, E., Tchernykh, A., Ivashko, A., Safonov, G.: Cost-efficient strategy in clouds with spot price uncertainty. Matematicheskaya Teoriya Igr i Ee Prilozheniya **11**(3), 5–30 (2019)

Metaheuristics for Discovering Favourable Continuous Intravenous Insulin Rate Protocols from Historical Patient Data

Hongyu Wang[1], Lynne Chepulis[2], Ryan G. Paul[2,3], and Michael Mayo[1(✉)]

[1] Department of Computer Science, University of Waikato, Hamilton, New Zealand
michael.mayo@waikato.ac.nz
[2] Waikato Medical Research Center, University of Waikato, Hamilton, New Zealand
[3] Waikato Regional Diabetes Service, Waikato District Health Board,
Hamilton, New Zealand

Abstract. Metaheuristic search algorithms such as particle swarm optimisation algorithm and covariance matrix adaptation evolution strategy are used to discover improved strategies for setting intravenous insulin rates of hospital in-patients with diabetes. We describe an approach combining and extending two existing methods recently reported in the literature: the Glucose Regulation for Intensive Care Patients (GRIP) method, and a favourability metric used for comparing competing strategies using historical medical records. We demonstrate with a dataset of blood glucose level/insulin infusion rate time series records from sixteen patients that new and significantly better insulin infusion strategies than GRIP can be discovered from this data.

1 Introduction

At our local hospital, admitted patients previously diagnosed with type 1 or type 2 diabetes are placed on a continuous intravenous (IV) insulin treatment regime regardless of admission reason. The aim of the IV treatment is to maintain the patient's glycemic control (i.e. blood glucose levels) in a normal range which is typically 6–10 mmol/L. To achieve this, the patient's blood glucose levels are closely monitored by medical staff and the continuous IV infusion rate is regularly adjusted (nominally once every hour) either upwards, to decrease the probability of *hyperglycemia*, defined as glucose levels being too high, or downwards, to limit the chance of *hypoglycemia*, caused by glucose levels being too low. The current practice at the hospital is to have medical staff follow a "protocol" involving written instructions and a lookup table so that the next suitable infusion rate can be determined. The protocol also identifies the conditions under which a doctor needs to be called to take over immediate care of the patient (e.g. in the event that the patient experiences severe hypoglycemia).

The research question that we set out to answer in this paper is how to devise new strategies for calculating the optimal insulin infusion rate for these patients in a mathematically consistent way. Our ultimate aim is to come up with new

© Springer Nature Switzerland AG 2020
N. T. Nguyen et al. (Eds.): ACIIDS 2020, LNAI 12033, pp. 157–169, 2020.
https://doi.org/10.1007/978-3-030-41964-6_14

computational protocols which maximise the patient's time in the normal range and work better than the hospital's current practice.

There are several possible means of approaching this problem. In this paper, we use metaheuristic search algorithms in conjunction with historical insulin infusion rate/blood glucose level time series data recorded by the hospital. The aim of the metaheuristics is to discover a strategy that *would have* performed better than the actual (current) strategy that was used for the specific patients in the time series data.

We use as one of our baselines an existing continuous IV insulin treatment protocol known as Glucose Regulation for Intensive Care Patients (GRIP) [7], and we combine this with a recently published method for evaluating such strategies against historical medical data [8]. GRIP utilises a simple linear formula for calculating infusion rate adjustments and is designed to be applicable at any point in time when an insulin infusion adjustment needs to be made, so long as blood glucose levels and past insulin infusion rate data are available for the previous four hours.

The novel contribution of this paper is to show that metaheuristic algorithms (specifically particle swarm optimisation [1] and covariance matrix adaptation evolution strategy [5]) can be used to optimise the parameters of GRIP, thus improving the strategy with respect to the historical data. We can then expand and generalise the GRIP formula to produce a more flexible rate adjustment rule and optimise that as well. The result is a set of new strategies that significantly outperform GRIP in our experiments based on the historical data.

2   Background

In this section the GRIP protocol [7] is described in detail. We also outline the "favourability" method from the literature used to assess insulin rate adjustment protocols [8]. Additionally, Table 1 defines key symbols used in the algorithms and equations and should be referred to throughout this section.

2.1   Glucose Regulation for Intensive Care Patients

GRIP [7] is an algorithm-based protocol for calculating suitable insulin infusion rate adjustments for patients receiving insulin intravenously. Although originally designed with an ICU setting in mind, we utilise GRIP in a more general non-ICU setting in which any in-patient with diabetes must be treated.

In GRIP, the recommended change in insulin infusion rate for continuous IV is determined by the mean insulin infusion rate of the last four hours ($\overline{I_{-4h}}$), the current blood glucose level measured in mmol/L (G_0), the target blood glucose level also in mmol/L (G_{target}, currently set to 6.5 mmol/L), and the change in blood glucose levels in the last four hours ($\Delta_{4h}G$). GRIP's output is the insulin infusion rate change (ΔI) measured in units of insulin per hour (U/hr), which is a standard rate measurement for insulin infusion. This is primarily a

Table 1. Definitions of commonly used terms and symbols in this paper. Note that while time index t is shown explicitly, most of these terms can also be indexed by patient.

Term	Definition
ΔI, $\overline{I_{-4h}}$, G_0, G_{target}, $\Delta_{4h}G$	Terms used in GRIP, see Sect. 2.1 for definition
$\beta = [\beta_0, \ldots \beta_{n-1}]$	Numeric vector defining the strategy being optimised ($n = 4$ or $n = 8$ depending on the strategy)
t	Time
bg_t	Historical blood glucose level at time t
I_t	Historical insulin infusion rate at time t
ΔI_t^{β}	Infusion rate change suggested by the strategy being tested at time t
I_t^{β}	Recommended insulin infusion rate change at time t
δ	Threshold for determining whether two insulin rate adjustments are equal, set to 0.1 (10%) in this paper
r_{hypo}	Threshold for hypoglycemia, set to 6.0 mmol/L
r_{hyper}	Threshold for hyperglycemia, set to 8.0 or 10.0 mmol/L
$\text{fav}_t(\beta)$	Favourability at time t, defined by Algorithm 1
λ	Penalty size term for limiting magnitude of ΔI_t^{β}
$\text{obj}(\beta)$	The objective function to maximise during optimisation

recommendation for medical staff which may be overruled at any time according to medical judgement. The GRIP formula itself is shown in Eq. 1.

$$\Delta I = (1 + 0.25 \times \overline{I_{-4h}}) \times (0.2(G_0 - G_{target}) + 0.3 \times \Delta_{4h}G) \qquad (1)$$

GRIP is essentially a linear combination of the four input variables. Further constraints are usually added to the output of the GRIP formula, for example rounding large positive values of ΔI down to 1.5 U/hr [7]. In this paper we treat these constraints as optional "post processing" of the recommendations that are specific to the hospital itself. For optimisation purposes we therefore ignore these constraints and focus solely on the formula given in Eq. 1.

2.2 Favourability-Based Evaluation of Insulin Infusion Protocols

Wong et al. [8] proposed a methodology for comparing computer-based protocols for continuous IV insulin infusion rate adjustments. The basic idea is to compute the *relative favourability* of one algorithm (e.g. GRIP) compared to another algorithm (e.g. the method currently in use at the local hospital) by examining and comparing the recommendations produced by both algorithms over a range of historical scenarios.

To perform this comparison, we first of all need historical data from a number of different patients over the course of their treatment – twelve or more hours worth of data per patient is ideal. Each patient's dataset must consist of a pair of time series, one measuring the patient's blood glucose levels at specific time points, and the other measuring the continuous insulin IV rate at the same time points. Figure 1 gives an example data from one such patient. The time points are separated by approximately one hour in our data.

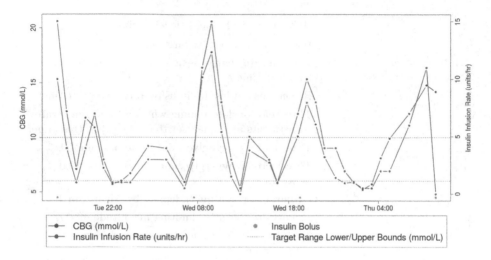

Fig. 1. Example visualisation of an insulin infusion session for a patient. Red horizontal lines indicate bounds for the normal range.

Once the historical data is ready, we take our algorithm to be evaluated and do the following: for each patient and each point in time we calculate the recommendation that the algorithm *would have made* had it been applied at that point. Since the hospital uses a different protocol to the algorithm being evaluated, the recommendation from our algorithm and the historical record (i.e. what actually happened) will most likely differ. This is the key point.

In order to assess which of the methods (GRIP or hospital method) produced the better recommendation at a time t, the current blood glucose level in the time series is compared to the next one (i.e. at $t + 60\,\text{min}$). Essentially, if the future blood glucose level rose above the normal range in the time series, then the algorithm which would have recommended the higher infusion rate is preferred. In other words, if the algorithm being evaluated would have suggested a higher rate adjustment than what actually happened in the historical record then that algorithm is more favourable because clearly the amount of insulin administered was insufficient to reduce the blood glucose level. Conversely, if the future blood glucose level fell below normal range in the time series, then the algorithm which would have recommended the lower infusion rate is preferred since this would have reduced hypoglycemia risk. If both algorithms recommend approximately the same rate change (which in our experiments is $\delta = 0.1$ or 10%) then they can be considered equivalent.

By considering all historical data points and all patients in this way, the overall performance of one algorithm vs. the method used to generate the historical data can be estimated. This sum of individual outcomes is what we refer to here as the "favourability" of the algorithm.

Algorithm 1 formalises the process for clarity. The algorithm depicts the steps taken to calculate the relative favourability of a strategy denoted by a vector β at a single point in time t, and is essentially the same approach as described by Wong et al.'s [8] but presented as pseudocode. The algorithm should be applied to every consecutive point in time for every patient in the historical data and the outputs summed.

Function fav$_t(\beta)$:
  $f \leftarrow$ case_low$(I_t, I_t^\beta, bg_{t+1})$ +
   case_high$(I_t, I_t^\beta, bg_{t+1})$ +
   case_target$(I_t, I_t^\beta, bg_{t+1})$
  return f;

Function case_low$(I_t, I_t^\beta, bg_{t+1})$:
  if $bg_{t+1} < r_{hypo}$ **then**
    if $I_t^\beta \leq I_t \cdot (1 - \delta)$ **then**
      return 1
    end
  end
  return 0

Function case_high$(I_t, I_t^\beta, bg_{t+1})$:
  if $bg_{t+1} > r_{hyper}$ **then**
    if $I_t^\beta \geq I_t \cdot (1 + \delta)$ **then**
      return 1
    end
  end
  return 0

Function case_target$(I_t, I_t^\beta, bg_{t+1})$:
  if $bg_{t+1} \geq r_{hypo}$ & $bg_{t+1} \leq r_{hyper}$
  then
    if $I_t^\beta \leq I_t \cdot (1 - \delta)$ **then**
      return 1
    end
  end
  return 0

Algorithm 1: Function definition for computing relative favourability of a strategy β at time t for a specific patient. Constants are defined in Table 1.

Note that in Wong et al.'s published version of Algorithm 1, the cases where the next blood glucose level are on target and below target are essentially the same. This implies that only the upper bound of the target range (r_{hyper}) is significant in the favourability calculation, which is a weakness of the approach. For this preliminary investigation we keep strictly to the framework as published with the exception of setting r_{hyper} to 8.0 mmol/L which gave better results.

2.3   Meta-heuristics for Continuous Optimisation

For optimising the GRIP variants that we propose in Sect. 4, we use the Distributed Evolutionary Algorithms in Python (DEAP) framework [4]. Our experiments involved two algorithms, namely a specific type of particle swarm optimisation (PSO) called Constriction Coefficient Particle Swarm Optimisation [1] and the covariance matrix adaption evolution strategy (CMA-ES) [5]. The former optimiser, PSO, maximises an objective function by manipulating a population of "particles" where each particle is a potential solution with a position and a velocity. Particles remember their "local" best solution so far and additionally the "global" best solution is also recorded. CMA-ES, on the other hand, optimises the objective by maintaining and updating a covariance matrix and two evolution paths, and accordingly it evolves a solution population according to the covariance matrix.

3   Dataset

The dataset used in this project was collected from our local hospital (Waikato District Health Board, New Zealand). Time series datasets from sixteen patients were collected, each consisting of multiple blood glucose level readings and insulin infusion rate adjustments. The median number of insulin infusion rate adjustments was 23, with the maximum number of adjustments being 88 and the minimum being 7. The average amount of time that a patient was in a normal glycemic state was 49.6%, suggesting that there is room for improvement.

The data contains as its attributes for each time point t: the date/time of the insulin infusion rate adjustment; the recorded blood glucose level at each adjustment; the insulin infusion rate of each adjustment; and the date/time of each insulin bolus (tablet or injection). Although we currently have data on insulin boluses we have not used them in the analysis presented here because GRIP essentially ignores this data. Figure 1 depicts the time series data for one patient.

4   Method

The novel methodology proposed in this paper is to generalise the GRIP formula so that it can be optimised. The objective for optimisation is the relative favourability of the current protocol vs. the historical data, as described in Sect. 2.2.

For experimental comparison purposes, we include two baseline strategies, A and B. Strategy A, shown by Eq. 2, proposes no change to the current insulin rate. In other words, the next insulin rate used is the same as the previous one. We expect this strategy to perform poorly, but include it as a "sanity check" for the other strategies. The second baseline method, Strategy B, is the standard GRIP formula given by Eq. 1.

$$\Delta I = 0 \tag{2}$$

Next, we define two generalisations of GRIP with tunable parameters. The first of these, Strategy C, replaces the constants in the original GRIP formula with variables. Since there are four constants in the GRIP formula, there are correspondingly four variables in Strategy C. This strategy is defined by Eq. 3.

$$\Delta I^\beta = (\beta_0 + \beta_1 \times \overline{I_{-4h}}) \times (\beta_2(G_0 - G_{target}) + \beta_3 \times \Delta_{4h}G) \tag{3}$$

Since Strategy C is relatively low dimensional with only four β values to tune, we next propose Strategy D which is higher dimensional and therefore potentially more flexible and able to represent a larger space of potential protocols. Strategy D is defined by Eq. 4.

$$
\begin{aligned}
\Delta I^\beta =&\beta_0 + \beta_1 \times \overline{I_{-4h}} + \beta_2 \times (G_0 - G_{target}) + \beta_3 \times \Delta_{4h}G \\
&+ \beta_4 \times \Delta_{4h}G \times (G_0 - G_{target}) \\
&+ \beta_5 \times \Delta_{4h}G \times \overline{I_{-4h}} \\
&+ \beta_6 \times \overline{I_{-4h}} \times (G_0 - G_{target}) \\
&+ \beta_7 \times \Delta_{4h}G \times (G_0 - G_{target}) \times \overline{I_{-4h}}
\end{aligned}
\tag{4}
$$

Basically, Strategy D represents a mathematical expansion of Eq. 3. If Eq. 3 were expanded, then it would consist of a linear sum of terms, where each term is either an individual input (e.g. $\overline{I_{-4h}}$) or an interaction term (e.g. $\Delta_{4h}G \times \overline{I_{-4h}}$). To capture the error between the target blood glucose level G_{target} and the current blood glucose level G_0, we keep $(G_0 - G_{target})$ together as a single term. This makes sense since G_{target} is a constant during the course of a single patient's treatment. This leaves three individual terms ($\overline{I_{-4h}}$, $\Delta_{4h}G$ and $(G_0 - G_{target})$) to appear in the expansion of Eq. 3 either by themselves or in interaction with one or more of the other terms, leading to a total of eight different terms. Thus, Strategy D represents an eight dimensional continuous optimisation problem while Strategy C is a four dimensional problem.

Finally, we come to the definition of the objective function that the meta-heuristic algorithms will be used to maximise. For any strategy β, the objective is defined as follows: firstly, we sum over all patients and compute the relative favourability of the strategy against the historical data for each time point; secondly, we sum over these individual favourabilities. The gives a total favourability score.

However, optimising for favourability alone does not produce ideal strategies because we also want to penalise large swings in the insulin infusion rate. Ideally, insulin rate adjustments should be modest. To illustrate why this is a problem, suppose that two different strategies recommend rate adjustments of $\Delta I = 5.0$ and $\Delta I = 2.5$ respectively, and according to Algorithm 1 both adjustments are favourable with respect to the historical data. In this case, the smaller value of ΔI is to be preferred because the higher value increases the risk of hypoglycemia.

The two strategies that are equivalent in terms of favourability can therefore be distinguished by adding a penalty term to the objective function, which is defined simply as the sum of the squared insulin rate adjustments multiplied by a constant λ. When $\lambda = 0$, the objective value of a strategy is exactly the strategy's favourability; however for values of $\lambda > 0$, the objective value is penalised according to the magnitudes of the ΔI recommendations made.

The overall final value of the objective function for both Strategies C and D is given by Eq. 5. Note that in the equation we add an additional index for the patient since we are summing over all patients.

$$\text{obj}(\beta) = \sum_{p \in patients} \left(\sum_t \text{fav}_{t,p}(\beta) - \lambda \sum_t ((\Delta I_{t,p}^\beta)^2) \right) \tag{5}$$

5   Results

Experimental results for Strategies C and D are given in Tables 2, 3, 4 and 5. For these strategies, we varied both the λ values and the population sizes used by PSO and CMA-ES. For both PSO and CMA-ES, the number of iterations was fixed at 100. The tables show the best and median favourability scores as well as the final penalty terms, for ten repetitions of each meta-heuristic/strategy/λ value combination.

An inspection of the tables reveals that in terms of favourability, Strategy D clearly outperforms Strategy C. The best favourability for Strategy C observed in our experiments was 280 (out of a maximum of 430), compared to 309 for Strategy D. Focusing on the favourability of Strategy D (Table 4), we can see that the CMA-ES algorithm clearly and significantly outperforms PSO in most cases except those where λ is relatively large.

Table 2. Best favourability results for strategy C optimisation

Best fav (Median fav)		λ					
		10^{-6}	10^{-5}	10^{-4}	10^{-3}	10^{-2}	10^{-1}
PSO	80	**279** (278)	**277** (276)	**271** (271)	**262** (259)	**229** (223)	**184** (175.5)
	160	**279** (278)	**277** (275.5)	**272** (271)	**264** (262.5)	**231** (223.5)	**185** (177)
	320	**279** (279)	**277** (276.5)	**274** (271)	**264** (261.5)	**233** (226.5)	**184** (179)
CMA-ES	80	**280** (278.5)	**277** (276.5)	**271** (271)	**260** (260)	**228** (227.5)	**180** (180)
	160	**279** (279)	**277** (277)	**271** (271)	**260** (260)	**233** (228)	**180** (180)
	320	**280** (279)	**277** (277)	**271** (271)	**260** (260)	**231** (226)	**180** (180)

On the grounds that PSO may need further iterations to achieve comparable results with CMA-ES, we re-ran the PSO/Strategy D experiments with 1,000 iterations instead of 100. However, the results did not reach the favourability levels obtained by CMA-ES and therefore we can conclude that CMA-ES is the superior optimisation technique for this problem.

Table 3. Best penalty sum results for strategy C optimisation

Min sum (Median sum)		λ					
		10^{-6}	10^{-5}	10^{-4}	10^{-3}	10^{-2}	10^{-1}
PSO	80	7.4e5	1.2e5	3.3e4	9.3e3	1.7e3	1.8e2
		(8.3e5)	(2.6e5)	(6.4e4)	(1.2e4)	(2.1e3)	(2.6e2)
	160	3.0e5	1.1e5	3.2e4	1.1e4	1.7e3	2.1e2
		(7.9e5)	(1.7e5)	(6.4e4)	(1.5e4)	(2.0e3)	(2.7e2)
	320	3.6e5	1.2e5	4.3e4	9.6e3	1.7e3	2.3e2
		(7.8e5)	(2.6e5)	(6.0e4)	(1.4e4)	(2.3e3)	(2.8e2)
CMA-ES	80	7.0e5	1.2e5	5.6e4	1.1e4	2.0e3	2.5e2
		(1.2e6)	(2.4e5)	(5.6e4)	(1.1e4)	(2.3e3)	(2.5e2)
	160	7.0e5	1.9e5	5.6e4	1.1e4	1.9e3	2.5e2
		(7.3e5)	(2.5e5)	(5.6e4)	(1.1e4)	(2.3e3)	(2.5e2)
	320	7.0e5	1.1e5	5.6e4	1.1e4	1.9e3	2.5e2
		(7.6e5)	(2.5e5)	(5.6e4)	(1.1e4)	(2.1e3)	(2.5e2)

Table 4. Best favourability results for strategy D optimisation

Best fav (Median fav)		λ					
		10^{-6}	10^{-5}	10^{-4}	10^{-3}	10^{-2}	10^{-1}
PSO	80	303	298	290	275	253	219
		(300)	(294.5)	(282.5)	(258)	(209)	(156.5)
	160	302	300	300	286	264	192
		(299.5)	(297)	(286.5)	(268)	(203)	(156.5)
	320	305	301	293	277	263	191
		(302)	(297)	(290.5)	(264)	(229)	(162.5)
CMA-ES	80	309	305	304	290	248	202
		(309)	(305)	(304)	(287.5)	(243)	(194)
	160	309	306	305	290	248	200
		(309)	(305)	(304)	(290)	(248)	(194)
	320	309	306	304	290	248	202
		(309)	(305)	(301)	(290)	(248)	(200.5)

Examining next the penalty sums for Strategies C and D shown in Tables 3 and 5 we can see that there is significant variation. Low values of λ such as 10^{-6} result in very high penalty sums such as 7.0×10^5 in the case of CMA-ES and Strategy C. This implies that the ΔI recommendations are very high. On the other hand, much greater λ values produce much smaller penalty sums (e.g. 2.5×10^2 in the case of some of the CMA-ES runs) as expected, but the corresponding favourabilities of those strategies are lower. Therefore, selection of a single optimal strategy requires trading off the total penalty sum against the total favourability. Tables 4 and 5 show that for CMA-ES, favourability decreases

Table 5. Best penalty sum results for strategy D optimisation

Min sum (Median sum)		λ					
		10^{-6}	10^{-5}	10^{-4}	10^{-3}	10^{-2}	10^{-1}
PSO	80	**3.5e5**	**1.9e5**	**4.5e4**	**6.3e3**	**4.6e3**	**1.2e3**
		(9.3e5)	(3.1e5)	(9.9e4)	(3.2e4)	(8.9e3)	(4.3e3)
	160	**3.4e5**	**2.0e5**	**4.2e4**	**1.7e4**	**1.7e3**	**1.0e3**
		(6.4e5)	(3.8e5)	(8.7e4)	(2.4e4)	(6.6e3)	(4.0e3)
	320	**3.4e5**	**1.7e5**	**3.2e4**	**1.2e4**	**3.3e3**	**8.8e2**
		(1.0e6)	(2.5e5)	(1.0e5)	(2.2e4)	(7.5e3)	(4.4e3)
CMA-ES	80	**1.3e6**	**2.3e5**	**7.0e4**	**9.3e3**	**1.7e3**	**1.9e2**
		(1.5e6)	(2.7e5)	(7.7e4)	(1.2e4)	(2.1e3)	(2.1e2)
	160	**1.5e6**	**1.2e5**	**6.5e4**	**1.2e4**	**2.0e3**	**1.8e2**
		(1.5e6)	(2.4e5)	(7.6e4)	(1.2e4)	(2.4e3)	(2.1e2)
	320	**1.5e6**	**1.2e5**	**6.5e4**	**1.2e4**	**2.4e3**	**1.8e2**
		(1.5e6)	(2.3e5)	(7.1e4)	(1.2e4)	(2.4e3)	(2.2e2)

Table 6. Sample recommendations from the dataset. Note that the optimisation was performed with the target set to 6.5 mmol/L.

Case	$\overline{I_{-4h}}$	G_0	$\Delta_{4h}G$	Rec. ΔI^β
1	3.7	11.8	−8.8	4.1
2	4.3	6.3	−2.1	−2.8
3	8.2	15.0	−6.2	6.7
4	5.3	13.7	0.2	7.0
5	3.0	9.9	2.8	1.9
6	1.6	12.4	3.0	5.3
7	8.5	14.7	0.9	8.1
8	2.0	11.8	1.4	4.5
9	3.5	4.2	−4.4	−6.2

relatively slowly as λ increases from 10^{-6} to 10^{-3}. When $\lambda = 10^{-2}$, however, there is a significant drop in favourability. Therefore a value of $\lambda = 10^{-3}$ seems reasonable.

We also note that for the best performing algorithm CMA-ES, the population size does not appear to have a significant impact. In Table 4, there is no significant difference between CMA-ES' best performances with population sizes of 80 and 320 where $\lambda = 10^{-3}$. Therefore we can conclude that CMA-ES is relatively robust to the population size parameter.

In contrast to these positive results, our fixed baselines Strategy A (no rate change) and Strategy B (the original GRIP formula) achieved favourabilities of 103 and 207 respectively, both of which are significantly worse than the best results from Strategies C and D.

As a final analysis, we selected a single optimal strategy produced by one of our runs (Strategy D, CMA-ES, population size 320, $\lambda = 10^{-3}$) and further examined it. Table 6 shows nine randomly selected examples from the data for illustration purposes.

From this table, we can see that most of the recommendations seem reasonable. For example, in Case 1, where blood glucose levels are initially well above the 6.5 mmol/L target but decreasing over the past four hour period, moderate increases in insulin rate are recommended. Case 2, on the other hand, represents a patient who is already below target but has been receiving some insulin over the past four hours. In this case, moderate negative adjustments are recommended. Case 4 shows a patient with high initial glucose levels that have not changed much in the past four hours, and therefore significant increments in insulin pump rate are recommended. Finally, case 9 shows a patient at significant risk of severe hypoglycemia following a significant drops in glucose level. In this case, a large negative adjustment is recommended which effectively sets the insulin pump rate to zero (since negative pump rates do not make sense). The remaining cases show similar sensible recommendations being made.

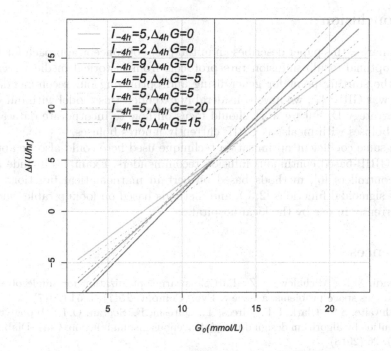

Fig. 2. Visualisation of one optimised version of strategy D

Figure 2 is a visualisation of the same strategy used to produce the recommendations in Table 6. To produce the visualisation, we fixed values of $\overline{I_{-4h}}$ and $\Delta_{4h}G$ to different constants and plotted the resulting linear relationships between G_0 and ΔI. Note that the model itself is not linear in practice, since $\overline{I_{-4h}}$ and $\Delta_{4h}G$ would be constantly changing over time for a particular patient. The true model is a curved surface in four dimensions, but nonetheless, this type of visualisation enables us to see some of the model's properties.

The $\overline{I_{-4h}}$ and $\Delta_{4h}G$ inputs both appear to influence the slope of the $G_0/\Delta I$ relationship. In particular, negative values of $\Delta_{4h}G$ (indicating a decrease in blood glucose levels over the past four hours) and large values of $\overline{I_{-4h}}$ (indicating high insulin infusion rates over the past four hours) result in a gentler slope and therefore smaller rate increments when an increase is recommended. The figure shows that when $\overline{I_{-4h}}$ is set to 5.0, the protocol recommends no change when G_0 is roughly 8.0, regardless of $\Delta_{4h}G$. This indicates that the protocol considers the scenario where the blood glucose level is around 8.0 mmol/L and the average infusion rate has been around 5.0 U/hr to be a relatively stable state, and recommends keeping the same insulin infusion rate until there are further glucose changes. For the plot, we selected $\Delta_{4h}G$ values of 5.0 and -5.0, since they are "typical" values for this variable in the data, while -20.0 and 15.0 are rare extreme values.

6 Conclusion

To summarise, this paper describes an initial proof of concept approach for identifying optimal insulin infusion rate protocols from historical medical records. While the starting point for generalising and optimising our recommendation models was GRIP [7], we have considerably extended this model with our optimised Strategy D. Future work should focus on how to incorporate data about insulin boluses with meals, as GRIP currently ignores boluses.

The same coefficient optimisation technique used here could also be applied to non-GRIP-based continuous infusion recommenders. Examples include PID-based controllers [6], methods based on certain mathematical functions (e.g. roughly sigmoidal functions [2,3]), and methods based on lookup tables such as that currently in use by the local hospital.

References

1. Bonyadi, M.R., Michalewicz, Z.: Particle swarm optimization for single objective continuous space problems: a review. Evol. Comput. **25**(1), 1–54 (2017)
2. Braithwaite, S.S., Clark, L.P., Idrees, T., Qureshi, F., Soetan, O.T.: Hypoglycemia prevention by algorithm design during intravenous insulin infusion. Curr. Diab. Rep. **18**(5), 26 (2018)
3. Devi, R., Zohra, T., Howard, B.S., Braithwaite, S.S.: Target attainment through algorithm design during intravenous insulin infusion. Diab. Technol. Ther. **16**(4), 208–218 (2014)
4. Fortin, F.A., De Rainville, F.M., Gardner, M.A., Parizeau, M., Gagné, C.: DEAP: evolutionary algorithms made easy. J. Mach. Learn. Res. **13**, 2171–2175 (2012)
5. Hansen, N.: The CMA evolution strategy: a tutorial. CoRR abs/1604.00772 (2016). http://arxiv.org/abs/1604.00772

6. Steil, G.M., Deiss, D., Shih, J., Buckingham, B., Weinzimer, S., Agus, M.S.: Intensive care unit insulin delivery algorithms: why so many? How to choose? J. Diab. Sci. Technol. **3**(1), 125–140 (2009)
7. Vogelzang, M., Zijlstra, F., Nijsten, M.W.: Design and implementation of GRIP: a computerized glucose control system at a surgical intensive care unit. BMC Med. Inform. Decis. Mak. **5**(1), 38 (2005)
8. Wong, A.F., Pielmeier, U., Haug, P.J., Andreassen, S., Morris, A.H.: An in silico method to identify computer-based protocols worthy of clinical study: an insulin infusion protocol use case. J. Am. Med. Inform. Assoc. **23**(2), 283–288 (2016)

Continuous-Time Approach to Discrete-Time PID Control for UPS Inverters - A Case Study

Marian Błachuta[1]([✉]) [iD], Zbigniew Rymarski[2] [iD], Robert Bieda[1] [iD],
Krzysztof Bernacki[2] [iD], and Rafal Grygiel[1] [iD]

[1] Chair of Automation and Robotics, Silesian University of Technology,
16 Akademicka St., Gliwice 44-101, Poland
blachuta@polsl.pl
[2] Chair of Electronics, Electrical Engineering and Microelectronics, Silesian
University of Technology, 16 Akademicka St., Gliwice 44-101, Poland

Abstract. A quasi-continuous-time approach to the design of a digital PID control for DC/AC inverters with LC filter that use semiconductor switches producing PWM output voltage is presented. It is shown that an appropriately chosen continuous-time model of the digital controller with the PWM converter behaves like the actual discrete-time system, which allows for a simple controller design. SIMULINK models are used to validate this approach for a nonlinear rectifier load. The outcomes are compared with a real inverter experiment.

Keywords: Inverter · PWM · THD · Modeling · Simulation · PID control · Time scale separation

1 Introduction

DC/AC inverters are the most important parts of Uninterruptible Power Supply systems (UPS), which backup power to the load when the line voltage fails. They convert DC voltage supplied from the battery into the standard AC line voltage. The output voltage control is achieved by means of the power converters that include semiconductor (mainly MOSFET or IGBT) switches and passive components (coils and capacitors) storing and delivering the electrical energy in the subsequent parts of switching periods. There are three algorithms of switching the four transistors in the H-bridge inverter, called Pulse Width Modulation (PWM) schemes [4]. The first scheme is the most useful for the instantaneous control systems (there is no problem when voltage is crossing zero), the third scheme results in the lowest power losses in the switching transistors and the best EMC of the inverter [9]. When only the output voltage is measured the

Supported by the Polish National Centre for Research and Development, grant TANGO3/427467/NCBR/2019, and Polish Ministry of Science and Higher Education, grant no. 02/010/BK-19/0143.

© Springer Nature Switzerland AG 2020
N. T. Nguyen et al. (Eds.): ACIIDS 2020, LNAI 12033, pp. 170–181, 2020.
https://doi.org/10.1007/978-3-030-41964-6_15

inverter control is SISO type [7]. While [1] advocates the One Sample Ahead Preview (deadbeat) controller, the most common is PID like control [13]. In the advanced control systems not only the inverter output voltage is measured but as well the inductor current and the output current and all of them are the input variables of the MISO output voltage control. The modern solution of the MISO control is the Passivity Based Control [5,6,8,10,11] that keeps system passive – the derivative of its internal energy (stored in inductors and capacitors) described by Hamiltonian should be negative. The disadvantage of the MISO control systems is the requirement of the adjective currents sensors and additional measuring channels as well as the more extensive data processing that requires more powerful microprocessor.

The target of the authors is to verify a quasi-continuous-time design technique to design high-performance single-loop discrete-time PID control.

The paper, extending the results of [2], is focused on performance measured as the value of the Total Harmonic Distortion (THD) of the output voltage under standard nonlinear rectifier load.

2   Test Bed

In order to compare our results with the technical reality we refer to the experimental setup whose simplified schematic diagram is depicted in Fig. 1. The control system is based on the STM 32F407VGT6 microcontroller working at frequency 168 MHz. The PWM modulator input frequency equals to 84 MHz. For the PWM frequency $f_C = 25600$ Hz there are 512 switching cycles in the fundamental period 20 ms of the 50 Hz AC output. The maximum number in the PWM comparator equals to $84000000/25600 = 3281$. As a result, the resolution of the modulator equals to $1/3280$. At the beginning of the PWM pulse a program interrupt is generated that triggers output voltage sampling, A/D conversion, and computation of the digital control signal stored in registers of both PWM comparators to determine the width of pulses in the next switching period. This program organization results in a one-step delay in the controller.

A galvanic isolation is provided in the path of the output voltage measurement system and in the path of discrete control PWM output. Unfortunately, the analog isolation amplifier using signal switching at the frequency of 500 kHz produces a significant high-frequency output ripple with peak-to-peak values of 20 mV. Therefore, an additional low pass filter is put in the measurement path. However, since its dynamics are very fast, they are neglected when modeling the entire control loop.

3   Modeling

As far as modeling is concerned, we neglect signal quantization of both the ADC and PWM modulator. Moreover, although the paper is focused on the performance under rectifier load, the theoretical analysis and controller design is performed for purely resistive load assuming two different resistance values.

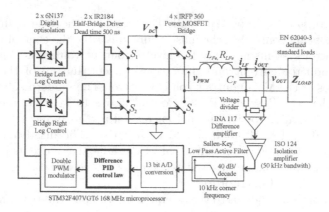

Fig. 1. Experimental setup. L_{Fe}=1 mH, R_{Fe}=1 Ω, C_F = 50 μF

This leads to two transfer functions of the LC(R) filter and allows using the linear control theory. Our approach assumes successive approximation of the PWM controlled system by a PAM controlled one eventually approximated by a continuous time system whose parameters depend on the sampling period.

Unfortunately, both the rectifier load and jump-wise switching of the load resistance make the system nonlinear and can not be modeled using linear blocs. Therefore, at the simulation stage, a part of the system consisting of the LC filter and nonlinear load are modeled as an electric circuit, while the rest consists of standard SIMULINK blocs. The interface between them is realized using `Measurement` and `Controlled Voltage Source` blocks.

3.1   Model of the PWM Modulated System

The SIMULINK model of the inverter with the discrete-time PID controller $H_c(z)$ and a load is depicted in Fig. 2. The controller operates on integer numbers. The controller features a one step delay. The value of the measurement path gain $k_D = [V_{out}]/V_{out}$), where $[V_{out}]$ is the integer output from the ADC, was

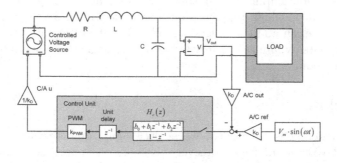

Fig. 2. Model of a discrete-time PWM control system

determined experimentally as $k_D = 110.8[1/V]$. Since the maximum value of the ADC output is 4096, the maximum measurable value of the output voltage is 36.96 [V]. It is also assumed that the nominal value of the supply voltage is $V_{DC} = 40[V]$. This voltage expressed in the measurement channel units equals to $[V_{DC}] = k_D V_{DC} = 4433.053$. Since for f = 25600 [Hz] the pulse has full width we define $k_{PWM} = [V_{DC}]/1640 = 2.7030$ as the PWM gain.

3.2 Model of the PAM Modulated System

At high sampling rates, i.e. sampling periods very short compared with the plant dynamics, the particular form of the control signal between two sampling instants is irrelevant, and what counts is its area. This allows approximation of the PWM control system by a Pulse Amplitude Modulated (PAM) one, as depicted in Fig. 3, with pulses produced by the Zero Order Hold (ZOH). It forms a linear sampled-data system that admits the classical $\mathcal{Z}$ transform approach, and is an intermediate form between discrete-time PWM modulated model and a quasi-continuous-time (QCT) one.

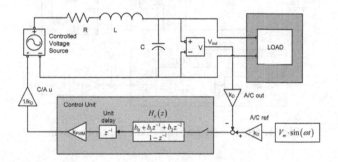

Fig. 3. PAM model of the PWM control system

3.3 Quasi-Continuous-Time Model

The final stage of our approach relies on the approximation of a PAM system by its QCT counterpart

$$\tilde{K}(s) = (1 - s\frac{h}{2})\frac{1 - s\frac{h}{2}}{1 + s\frac{h}{2}}K(s), \tag{1}$$

where the first factor stands for the Zero Order Hold and the second for the one step delay in the controller.

Consider a discrete-time PID controller with the transfer function $H_c(z)$

$$H_c(z) = \frac{b_0 z^2 + b_1 z + b_2}{z(z - 1)} = \frac{b_0 + b_1 z^{-1} + b_2 z^{-2}}{1 - z^{-1}} \tag{2}$$

Its QCT counterpart bases on the substitution

$$z = \frac{1 + s\frac{h}{2}}{1 - s\frac{h}{2}} \tag{3}$$

This gives

$$\tilde{C}(s) = \tilde{k}\frac{(s + \tilde{c}_1)(s + \tilde{c}_2)}{s(s + \frac{2}{h})}. \tag{4}$$

Given $\tilde{k}$, $\tilde{c}_1$ and $\tilde{c}_2$ the coefficients of (2) can be expressed as follows

$$b_0 = \frac{\tilde{k}}{8}(2 - \tilde{c}_1 h)(2 - \tilde{c}_2 h), \; b_1 = -\frac{\tilde{k}}{4}(4 - \tilde{c}_1\tilde{c}_2 h^2), \; b_2 = \frac{\tilde{k}}{8}(2 + \tilde{c}_1 h)(2 + \tilde{c}_2 h). \tag{5}$$

The resulting simulation model is depicted in Fig. 4.

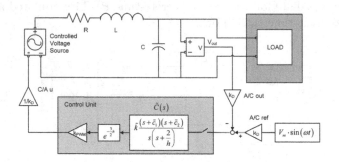

Fig. 4. QCT model of the PAM control system

Assuming a constant resistive load R_1, the LC filter has the transfer function:

$$K(s) = \frac{1}{LCs^2 + (\frac{L}{R_1} + RC)s + 1 + \frac{R}{R_1}}, \tag{6}$$

Denote $K_{500}(s)$ the transfer function for $R_1 = 500\;\Omega$ and $K_{45}(s)$ for $R_1 = 45\;\Omega$ Then there is

$$K_{500}(s) = \frac{1.961 \cdot 10^7}{s^2 + 1039\,s + 1.965 \cdot 10^7} \tag{7}$$

$$K_{45}(s) = \frac{1.961 \cdot 10^7}{s^2 + 1431\,s + 2 \cdot 10^7} \tag{8}$$

The poles of $K_{500}(s)$ and $K_{45}(s)$ are shown in the format $s_{1,2} = \sigma(1 \pm j\theta)$ in Table 1. Their impulse responses are plotted in Fig. 5. Based on the approach described in [3] we assume a PID controller with slightly less oscillatory zeros

Table 1. Plant poles for various values of R_1

(a) $K_{500}(s)$	(b) $K_{45}(s)$	σ_b/σ_a	θ_b/θ_a
$-520(1 \pm j8.5)$	$-716(1 \pm j6)$	1.4	0.7

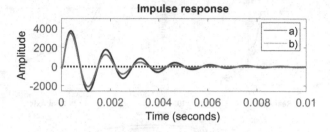

Fig. 5. Impulse responses of $K(s)$, (a) $R_1 = 500\ \Omega$, (b) $R_1 = 45\ \Omega$

and real parts twice as far from the origin as the actual poles of the LCR plant in the following form

$$\tilde{C}(s) = 34.4914 \frac{(s + 1453(1 + j3.2))(s + 1453(1 - j3.2))}{s(s + 51200)} \tag{9}$$

We also assume the QCT transfer function of the system to be controlled in the form presented in (1). The controller gain is chosen so that the gain margin equals to 2. Appropriate closed-loop root loci are displayed in Fig. 6 for both $K_{550}(s)$ and $K_{45}(s)$. Note the double pole at $-2/h$ and double zero at $2/h$. The roots collected in Table 2 may be split into two parts: the slow and the fast ones. The slow roots are in the vicinity of the controller parameters $-c_1$ and $-c_2$, and their values are practically independent of the load R_1. Fast roots have about 3 times greater damping factor and are almost independent of R_1. There is also a very fast real root at about -10000, whose dynamics is completely negligible compared to remaining roots. Notice that the harmful effect of sampling and time necessary for control signal computation exhibited as a double positive zero at $s = 2/h$ limits increase of the controller gain because of the danger of the stability loss. This limits the control performance. Therefore, it can be concluded that increase of the sampling frequency can bring better disturbance attenuation.

Table 2. Closed loop poles

Roots	(a) $R_1 = 500\,\Omega$	(b) $R_1 = 45\,\Omega$	σ^b/σ^a	θ^b/θ^a
v. fast	-106229	-106343	1	–
Fast	$-3666\ (1 \pm j4.41)$	$-3852\ (1 \pm j4.27)$	1.05	0.97
Slow	$-1543\ (1 \pm j3.28)$	$-1498\ (1 \pm j3.32)$	0.97	1.01

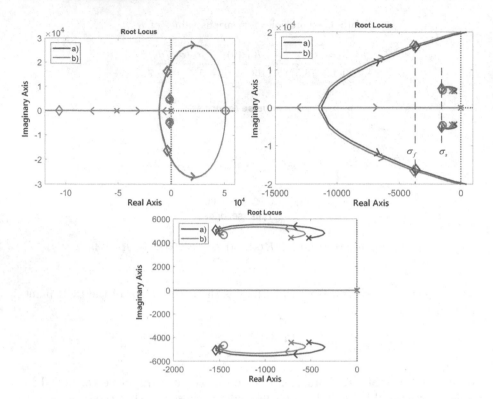

Fig. 6. Subsequent zooms of the root loci for (a) $R_1 = 500\ \Omega$ and (b) $R_1 = 45\ \Omega$. Circles denote zeros, crosses poles and diamonds closed-loop poles for the controller gain 34.4914. Notice the time-scale separation between the slow (σ_s) and fast (σ_f) roots, and closeness of the closed-loop poles and controller zeros in spite of R_1

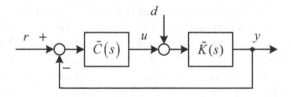

Fig. 7. System with the fictitious disturbance $d(t)$

Consider the block diagram of the system with a fictitious disturbance depicted in Fig. 7. Notice that if the control signal were $u(t) = -d(t) = -1(t)$ then the disturbance would be completely compensated and the control error would equal to 0. Therefore the faster control signal reaches -1 the smaller is the control error. This requires the occurrence of a time-scale separation between the control and the output signals as shown in Fig. 8 for system as in Fig. 7. This feature is universal for systems with second order plants and PID controllers, see [3].

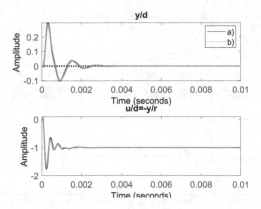

Fig. 8. Response to $d(t) = 1(t)$. Notice the time-scale separation among $y(t)$ and $u(t)$, and insensitivity of both variables to R_1

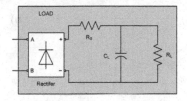

Fig. 9. Nonlinear rectifier load, $R_s = 1\,\Omega$, $R_L = 47\,\Omega$, $C_L = 430\,\mu F$

It is believed that excellent properties of this control system will assure good results also for the nonlinear load of Fig. 9. This claim is checked experimentally.

4 Simulations and Experiments

In order to validate the method for the rectifier load presented in Fig. 9, simulations were executed using SIMULINK schemes displayed in Figs. 4, 3 and 2. The simulation results are then compared with experiments performed on real inverter of Fig. 1.

In the case of QCT approach simulated using the SIMULINK model of Fig. 4, in order to avoid non-proper subsystems the first two factors of the plant model in (1) are replaced by their time-delay approximation

$$(1 - s\frac{h}{2})\frac{1 - s\frac{h}{2}}{1 + s\frac{h}{2}} \simeq e^{-\frac{1}{2}sh}e^{-sh} = e^{-\frac{3}{2}sh}. \tag{10}$$

The discrete transfer function $H_c(z)$ of the PID controller based on (9) is

$$H_c(z) = \frac{18.3810z^2 - 34.1789z + 16.4228}{z(z-1)}. \tag{11}$$

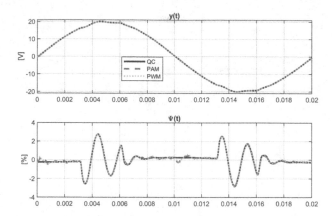

Fig. 10. Comparison of simulation results for quasi-continuous-time QCT, PAM and PWM models. For further explanation see the comments above.

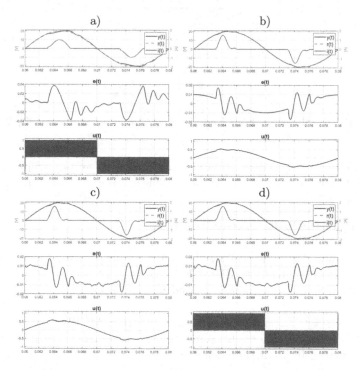

Fig. 11. Results of simulations: (a) open loop vs closed loop (b) QCT (c) PAM (d) PWM

The results displayed in Fig. 11 are collected in Fig. 10. It is worth noting that there is practically no difference between two discrete-time systems with PWM and PAM, and their quasi-continuous-time approximation. The only difference is a slightly different behavior around reference signal zero crossing, which is due

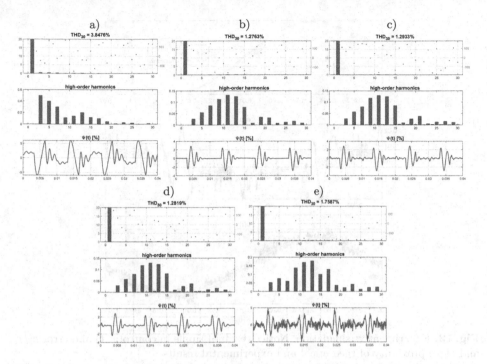

Fig. 12. Harmonics distribution and THD. (a) open loop: $THD_{30} = 3.85$ vs closed loop: (b) QCT $THD_{30} = 1.28$ (c) PAM: $THD_{30} = 1.29$ (d) simulation PWM: $THD_{30} = 1.28$ (e) experiment PWM: $THD_{30} = 1.75$

to the lack of synchronization of zero crossing and sampling instants. Generally speaking, the results confirm high value of the applied quasi-continuous time approach.

Denote the Fourier series of the inverter output $y(t)$ as

$$y(t) = \sum_{i=0}^{\infty} A_i \cos(i\omega t + \varphi_i). \tag{12}$$

Then the Total Harmonic Distortion (THD_H) coefficient

$$THD_H = \frac{\sqrt{\sum_{i=2}^{H} A_i^2}}{A_1}, \tag{13}$$

that aggregates the quality of control is characterized in the next series of plots. Additionally, the distortion function $\psi(t)$ determined as

$$\psi(t) = \frac{y(t) - A_1 \cos(\omega t + \varphi_1)}{A_1} 100\%, \tag{14}$$

is displayed in Fig. 12.

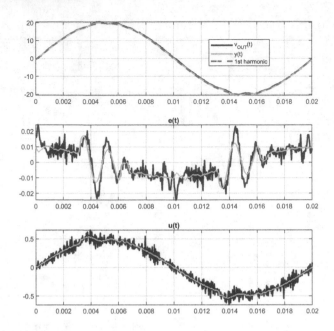

Fig. 13. Experiment vs simulation. Notice residual ripple in control $u(t)$ and error $e(t)$ and close proximity of theoretical and experimental results

Comparison of an experimental result obtained on real inverter with PWM simulation is presented in Fig. 13.

5   Summary

Quasi-continuous-time method supported by root locus proved to be efficient tools for synthesis of a discrete-time PID controller for the UPS inverter. The main point is that the common effect of processing the discrete-time data and PWM voltage modulation can be modeled by rational transfer functions in the Laplace variable 's'. This allows use of the time-scale separation technique [3], [12] for implementation of a digital PID controller.

Performed real inverter experiments match with SIMULINK simulation results and confirm that a well tuned simple single-loop PID controller suffices to attain high control performance.

References

1. Ben-Brahim, L., Yokoyama, T., Kawamura, A.: Digital control for UPS inverters. In: The Fifth International Conference on Power Electronics and Drive Systems, PEDS 2003, vol. 2, 17–20 November 2003, pp. 1252–1257 (2003)
2. Blachuta, M., Rymarski, Z., Bieda, R., Bernacki, K., Grygiel, R.: Design, modeling and simulation of PID control for DC/AC inverters. In: International Conference MMAR 2019, Miedzyzdroje, Poland (2019)

3. Blachuta, M., Bieda, R., Grygiel, R.: PID regulatory control design for a double tank system based on time-scale separation. In: Nguyen, N.T., Gaol, F.L., Hong, T.-P., Trawiński, B. (eds.) ACIIDS 2019. LNCS (LNAI), vol. 11431, pp. 420–430. Springer, Cham (2019). https://doi.org/10.1007/978-3-030-14799-0_36
4. Van der Broeck, H.W., Miller, M.: Harmonics in DC to AC converters of single phase uninterruptible power supplies. In: 17th International Telecommunications Energy Conference, INTELEC 1995, pp. 653–658 (1995)
5. Gui, Y., Wei, B., Li, M., Guerrero, J.M., Vasquez, J.C.: Passivity-based coordinated control for islanded AC microgrid. Appl. Energy **229**(C), 551–561 (2018). https://doi.org/10.1016/j.apenergy.2018.07.115
6. Komurcugil, H.: Improved passivity-based control method and its robustness analysis for single-phase uninterruptible power supply inverters. IET Power Electron. **8**(8), 1558–1570 (2015)
7. Luo, F.L., Ye, H., Rashid, M.: Digital Power Electronics and Applications. Elsevier Academic Press, Amsterdam (2006)
8. Meshram, R.V., et al.: Port-controlled phasor Hamiltonian modeling and IDA-PBC control of solid-state transformer. IEEE Trans. Control Syst. Technol. **27**, 161–174 (2017). https://doi.org/10.1109/TCST.2017.2761866
9. Rymarski, Z.: Measuring the real parameters of single-phase voltage source inverters for UPS systems. Int. J. Electron. **104**(6), 1020–1033 (2017). https://doi.org/10.1080/00207217.2017.1279232
10. Rymarski, Z., Bernacki, K., Dyga, L., Davari, P.: Passivity-based control design methodology for UPS systems. MDPI-Energies **12**(22), 1–19 (2019). https://doi.org/10.3390/en12224301
11. Serra, F.M., De Angelo, C.H., Forchetti, D.G.: IDA-PBC control of a DC-AC converter for sinusoidal three-phase voltage generation. Int. J. Electron. **104**(1), 93–110 (2017)
12. Yurkevich, V.D.: Design of Nonlinear Control Systems with the Highest Derivative in Feedback. World Scientific, Singapore (2004)
13. Zhou, D., Song, Y., Blaabjerg, F.: Modeling and control of three-phase AC/DC converter including phase-locked loop. In: Blaabjerg, F. (ed.) Control of Power Electronic Converters and Systems, vol. I, pp. 117–152. Elsevier, Academic Press (2018). Chapter 5

On Stabilizability of Discrete Time Systems with Delay in Control

Artur Babiarz[1]($\boxtimes$)(iD), Adam Czornik[1](iD), and Jerzy Klamka[2](iD)

[1] Silesian University of Technology, Akademicka 2A, 44-100 Gliwice, Poland
{artur.babiarz,adam.czornik}@polsl.pl
[2] Institute of Theoretical and Applied Informatics, Polish Academy of Sciences,
Bałtycka 5, 44-100 Gliwice, Poland
jerzy.klamka@iitis.pl

Abstract. The paper deals with discrete time-invariant systems with a delay in the control variable. The relations between different types of controllability and stabilizability are presented and discussed. The results are related to asymptotic null controllability, bounded feedback stabilizability and small feedback stabilizability for linear discrete-time systems with delay in control. The main tool employed is the technique of reducing the delayed equation to a delay-free equation. Thanks to this idea the criteria for bounded feedback stabilizability and small feedback stabilizability for the delayed systems are expressed in the appropriate properties of delay-free systems. Main results are analogical of this one proved in [18] for discrete time-invariant delay-free systems and to those from [14] for continuous-time systems. One of the additional result of this paper provides a criterion for controllability of discrete time system with delay in control. An important contribution of this paper is the indication of further generalizations of the obtained results.

Keywords: Stabilizability · Controllability · Delayed system · Discrete system

1 Introduction

One of the fundamental problems in control theory is designing of an efficient feedback which guarantees control strategy capable to stabilize the possibly unstable system and guarantee a certain level of performance. This problem is well-known and it was extensively investigated in the literature [2, 4, 13, 14] for the delay-free systems.

The research presented here was done by authors as parts of the projects funded by the National Science Centre in Poland granted according to decision UMO-2017/27/B/ST6/00145 (JK), DEC-2017/25/B/ST7/02888 (AB) and Polish Ministry for Science and Higher Education funding for statutory activities 02/990/BK_19/0121 (AC).

N. T. Nguyen et al. (Eds.): ACIIDS 2020, LNAI 12033, pp. 182–190, 2020.
https://doi.org/10.1007/978-3-030-41964-6_16

However, delay system constitute, nowadays, an important class of mathematical models of real phenomena. Many applications of delayed systems in engineering, mechanics and economics are presented in [5]. Delay is very often encountered in different technical systems, such as electric, pneumatic and hydraulic networks, chemical processes, long transmission lines, etc. [11]. Delays are inherent in many physical and engineering systems. In particular, pure delays are often used to ideally represent the effects of transmission and transportation. This is because these systems have only limited time to receive information and react accordingly. Such a system cannot be described by purely differential equations, but has to be treated with differential-difference equations or the so-called differential equations with difference variables. The basic theory concerning the stability of systems described by equations of this type was developed by Pontryagin in 1942. Also, important work has been written by Bellman and Cooke in 1963, [3]. The presence of time delays in a feedback control system leads to a closed-loop characteristic equation which involves the exponential type transcendental terms. The exponential transcendentally brings infinitely many isolated roots, and hence it makes the stability analysis of time-delay systems a challenging task. It is well recognized that there is no simple and universally applicable practical algebraic criterion, like the Routh-Hurwitz criterion for stability of delay-free systems, for assessing the stability of linear time-invariant time-delayed systems. On the other side, the existence of pure time delay, regardless if it is present in the control or/and state, may cause an undesirable system transient response, or generally, even an instability. Numerous reports have been published on this matter, with a particular emphasis on the application of Lya punovs second method, or on using the idea of matrix measure [8]. The analysis of time-delay systems can be classified such that the stability or stabilization criteria involve the delay element or not. In other words, delay independent criteria guarantee global asymptotic stability for any time-delay that may change from zero to infinity.

This arguments motivate us to investigate certain properties of discrete linear equation with delay in control. We study asymptotic null controllability, bounded feedback stabilizability and small feedback stabilizability. The main idea we use is to convert the delay equation into a delay-free equation.

We call a system asymptotically null controllable with bounded control if there exists a control, which steers the solution asymptotically to origin. A natural question is whether the control can be realised in a feedback form or in a form of feedback with small feedback gain. This questions are the main topic of this paper. References closely related to these questions are [10, 16]. In these papers results on semi-global stabilizability and global stabilizability for some special systems are presented for delay-free systems. In the context of stabilization of discrete-time systems is worth to notice the paper [15], where the authors consider the problem of stabilization of a linear time-invariant system given by transfer function, by a first-order feedback controller. The paper contains a description of stabilizing controller in the controller parameter space. The solution is describe by Chebyshev representations of the characteristic equation in the unit circle. Moreover, it is shown that the set can be computed explicitly. Also stabilization of discrete-time invariant system with delay is investigated in [19]. The main results of this paper are about observer based output feedback stabilization. Based on predictor feedback theory a design method of the controller is proposed. Two classes of controllers, namely, the memory observer and memory less observer are considered. The problem of stabilization of time-invariant discrete linear systems with time-varying delay is considered in [17], where the authors use the H_∞ approach to obtain conditions for the output feedback stabilization. The paper [9] focuses on the stabilization problem of discrete linear time-invariant systems with multiply delays in the control variable and multiplicative noise. The authors first transform the original system into a delay-free system and next use the linear quadratic technique. The main result of this paper states that the system can be stabilized in the mean-square sense if and only if the set of solutions of the Riccati difference equations is convergent.

In the present paper we investigate a discrete time-invariant linear system with delay in the control variable and we describe relations between controllability and stabilizability. Following the idea from [9] we use the technique of converting the original delay system to a delay-free one. In the best knowledge of the authors such questions for discrete-time systems with delay have not been investigated in the literature. Similar questions for continuous-time systems with delay have been investigated in [1, 7, 12].

2   Main Results

In this paper we consider the following equations

$$x(k+1) = Ax(k) + B_0 u(k) + B_1 u(k-1) \tag{1}$$

$$y(k+1) = \hat{A}y(k) + \hat{B}v(k) \tag{2}$$

$k = 0, 1, ...$, where the states $x(k)$, $y(k) \in \mathbb{R}^n$, controls $u(k) \in \mathbb{R}^m$, $v(k) \in \mathbb{R}^p$ and A, $\hat{A}$, B_0, B_1, $\hat{B}$ are given matrices of appropriate sizes. In case of Eq. (1) we put $u(-1) = 0$. For initial condition $x(0) = x_0 \in \mathbb{R}^n$, $y(0) = y_0 \in \mathbb{R}^n$ and fixed controls u, v the appropriate solutions of (1) and (2) are denoted by $x(\cdot, x_0, u)$ and $y(\cdot, y_0, v)$. If the control u in (1) has the form $u(k) = L_1(x(k))$ or $u(k) = L_2(x(k), x(k-1))$ for certain functions $L_1 : \mathbb{R}^n \to \mathbb{R}^m$, $L_2 : \mathbb{R}^{2n} \to \mathbb{R}^m$ then we say that u is a feedback control and then we call L_1 and L_2 a feedback. Similarly a control $v(k) = \hat{L}_1(y(k))$ or $\hat{L}_2(y(k), y(k-1))$ in (2), where $\hat{L}_1 : \mathbb{R}^n \to \mathbb{R}^p$ and $\hat{L}_2 : \mathbb{R}^{2n} \to \mathbb{R}^p$ are called feedback controls for (2) and then $\hat{L}_1$ and $\hat{L}_2$ are called feedbacks. If we apply a feedback $u(k) = L(x(k))$ to system (1) then we obtain system

$$x(k+1) = Ax(k) + B_0 L(x(k)) + B_1 L(x(k-1)) \tag{3}$$

which is called closed-loop system. In the same way we define the closed-loop system for system (2). If the closed-loop system is stable then the feedback is called a stabilizing feedback.

The main idea of this paper is to connect solutions of (1) and (2) according to the following theorem.

Theorem 1. *Sequence $x(\cdot, x_0, u)$ is the solution of (1) if and only if $x(\cdot, x_0, u)$ is the solution of (2) with $\hat{A} = A$, $\hat{B} = \begin{bmatrix} B_0 & B_1 \end{bmatrix}$, $y(0) = x_0$, $p = 2m$ and*

$$v(k) = \begin{bmatrix} u(k) \\ u(k-1) \end{bmatrix} \quad k = 0, 1, \dots \ .$$

Proof. Suppose that $x(\cdot, x_0, u)$ is the solution of (1) and let $y(\cdot, y_0, v)$ be the solution of (2) with $\hat{A} = A$, $\hat{B} = \begin{bmatrix} B_0 & B_1 \end{bmatrix}$, $y(0) = x_0$, $p = 2m$ and v is defined as in the theorem. For $k = 0$ we have $x(0, x_0, u) = x_0 = y(0, y_0, v)$. Suppose that $x(l, x_0, u) = y(l, y_0, v)$ for certain $l \in \mathbb{N}$, then

$$x(l+1, x_0, u) = Ax(l, x_0, u) + B_0 u(l) + B_1 u(l-1)$$

$$= Ay(l, x_0, v) + \begin{bmatrix} B_0 & B_1 \end{bmatrix} \begin{bmatrix} u(l) \\ u(l-1) \end{bmatrix}$$

$$= \hat{A}y(l, x_0, v) + \hat{B}v(l) = y(l+1, x_0, v).$$

Suppose now that $y(\cdot, y_0, v)$ is the solution of (2) with $\hat{A} = A$, $\hat{B} = \begin{bmatrix} B_0 & B_1 \end{bmatrix}$, $y(0) = x_0$, $p = 2m$ and v is defined as in the theorem. For $k = 0$ we have $x(0, x_0, u) = x_0 = y(0, y_0, v)$. Suppose that $x(l, x_0, u) = y(l, y_0, v)$ for certain $l \in \mathbb{N}$, then

$$y(l+1, x_0, v) = \hat{A}y(l, y_0, v) + \hat{B}v(k)$$

$$= Ax(l, x_0, u) + \begin{bmatrix} B_0 & B_1 \end{bmatrix} \begin{bmatrix} u(l) \\ u(l-1) \end{bmatrix}$$

$$= Ax(l) + B_0 u(l) + B_1 u(l-1) = x(l+1, x_0, u).$$

The proof is completed.

Definition 1. *[18] We say that system (1) is*

- *asymptotically null controllable with bounded controls (ANCBC) if there is a bounded subset $\mathcal{U}$ of $\mathbb{R}^m$ which contains zero in its interior such that, for each initial state $x_0 \in \mathbb{R}^n$ there exists a sequence $u(\cdot) = u(0), u(1), \dots$ with all values $u(t) \in \mathcal{U}$, which steers the solution $x(t)$ asymptotically to the origin i.e.*

$$\lim_{k \to \infty} x(k, x_0, u) = 0.$$

- *bounded feedback stabilizable (BFS) if there exists a bounded locally Lipschitz feedback $L : \mathbb{R}^n \to \mathbb{R}^m$ that the closed-loop system is asymptotically stable.*
- *small feedback stabilizable (SFS) if for every $\varepsilon > 0$ there exists a stabilizing feedback $L : \mathbb{R}^n \to \mathbb{R}^m$ such that $\|L(x)\| \leq \varepsilon$ for all $x \in \mathbb{R}^n$.*

In an analogical way we define ANCBC, BFS and SFS for system (2). The next theorem proved in [18] shows that this properties are equivalent for system (2).

Theorem 2. *The following conditions are equivalent*

- (2) *is SFS,*
- (2) *is BFS,*
- (2) *is ANCBC.*

The next three theorems describe relations between ANCBC, BFS and SFS for system (1) and similar properties of system (2).

Theorem 3. *Suppose that system* (1) *is SFS then for* (2) *with* $\hat{A} = A$, $\hat{B} = \begin{bmatrix} B_0 \ B_1 \end{bmatrix}$ *for every* $\varepsilon > 0$ *there exists a stabilizing feedback* $\overline{L} : \mathbb{R}^{2n} \rightarrow \mathbb{R}^m$, $u(k) = \overline{L}\left(x(k), x(k-1)\right)$ *such that* $\|L(x)\| \leq \varepsilon$ *for all* $x \in \mathbb{R}^{2n}$.

Proof: Let us fix $\varepsilon > 0$. Suppose that system (1) is SFS and let $\overline{L} : \mathbb{R}^n \rightarrow \mathbb{R}^m$ be the stabilizing feedback satisfying $\|\overline{L}(x)\| \leq \frac{\varepsilon}{\sqrt{2}}$ for all $x \in \mathbb{R}^n$. Let us define a feedbeck $L : \mathbb{R}^{2n} \rightarrow \mathbb{R}^{2m}$ in (2) as follows

$$L(x, y) = \begin{bmatrix} \overline{L}(x) \ \overline{L}(y) \end{bmatrix}, \ x, \ y \in \mathbb{R}^n.$$

For the control $v(k) = L(y(k), y(k-1))$ we know by Theorem 1 that the solutions of (1) and (2) coincide for the same initial condition. Moreover

$$\|L(x, y)\| = \left\| \begin{bmatrix} \overline{L}(x) \ \overline{L}(y) \end{bmatrix} \right\| \leq \sqrt{2} \max \left\{ \|\overline{L}(x)\|, \|\overline{L}(y)\| \right\} \leq \varepsilon.$$

The proof is completed.

Theorem 4. *Suppose that system* (1) *is BFS then for* (2) *with* $\hat{A} = A$, $\hat{B} = \begin{bmatrix} B_0 \ B_1 \end{bmatrix}$ *there exists a bounded locally Lipschitz feedback* $L : \mathbb{R}^{2n} \rightarrow \mathbb{R}^m$ *such that the closed-loop system is asymptotically stable.*

Proof: The proof is analogical as the proof of Theorem 1 additionally we have to use the fact that if $\overline{L} : \mathbb{R}^n \rightarrow \mathbb{R}^m$ is bounded and locally Lipschitz, then $L : \mathbb{R}^{2n} \rightarrow \mathbb{R}^m$, $L(x, y) = \begin{bmatrix} \overline{L}(x) \ \overline{L}(y) \end{bmatrix}$, $x, y \in \mathbb{R}^n$ is also bounded and locally Lipschitz.

Theorem 5. *If system* (1) *is ANCBC then* (2) *with* $\hat{A} = A$, $\hat{B} = \begin{bmatrix} B_0 \ B_1 \end{bmatrix}$ *is ANCBC.*

Proof: Suppose that (1) is ANCBC. Let $\mathcal{U}$ be the set from the definition of ANCBC and let us define the subset $\hat{\mathcal{U}} \subset \mathbb{R}^{2m}$ by

$$\hat{\mathcal{U}} = \left\{ \begin{bmatrix} u_1 \\ u_2 \end{bmatrix} : u_1, \ u_2 \in \mathcal{U} \right\}.$$

From the properties of the set $\mathcal{U}$ it is clear that the set $\hat{\mathcal{U}}$ contains zero and it is bounded. We will show that for each $y_0 \in \mathbb{R}^n$ there exist a control sequence $(v(k))_{k \in \mathbb{N}}$, $v(k) \in \hat{\mathcal{U}}$ such that

$$\lim_{k \to \infty} y(k, y_0, v) = 0. \tag{4}$$

Let us fix $y_0 \in \mathbb{R}^n$ and consider a sequence $(u(k))_{k \in \mathbb{N}}$, $u(k) \in \hat{\mathcal{U}}$ for all $k \in \mathbb{N}$, that stabilizes system (1) for initial condition y_0. Let us define a sequence $(v(k))_{k \in \mathbb{N}}$ as follows

$$v(0) = \begin{bmatrix} u(0) \\ 0 \end{bmatrix} \tag{5}$$

and

$$v(k) = \begin{bmatrix} u(k) \\ u(k-1) \end{bmatrix}. \tag{6}$$

From Theorem 1 we know that $x(k, y_0, u) = y(k, y_0, v)$ and therefore (4) holds. The proof is completed.

Finally we will present a result about relations between controllability of systems (1) and (2). We start with the definitions of this concept.

Consider certain subset $\mathcal{G}$ of the set of all sequences which elements are in $\mathbb{R}^m$.

Definition 2. *We say that system* (1) *is $\mathcal{G}$-controllable in time N, $N \in \mathbb{N}$ if for all $x_0, x_1 \in \mathbb{R}^n$ there exists a control $u \in \mathcal{G}$ such that*

$$x(N, x_0, u) = x_1.$$

When $\mathcal{G}$ is the set of all sequences of vectors from $\mathbb{R}^m$, then we will say that (1) *is controllable in time N*

Analogically we define controllability of (2). To formulate the next theorem which presents relation between controllability of systems (1) and (2) let us introduce certain special set $\overline{\mathcal{G}}$ of controls in $\mathbb{R}^{2m}$ consisting of all sequences $(u(k))_{k \in \mathbb{N}}$,

$$u(k) = [u_1(k), ..., u_{2m}(k)]^T$$

satisfying

$$[u_1(k), ..., u_m(k)]^T = [u_{m+1}(k+1), ..., u_{2m}(k+1)]^T.$$

Theorem 6. *System* (1) *is controllable in time N if and only if system* (2) *with $\hat{A} = A$, $\hat{B} = \begin{bmatrix} B_0 & B_1 \end{bmatrix}$ is $\overline{\mathcal{G}}$-controllable in time N.*

Proof. Suppose that system (1) is controllable in time N. Let us fix $x_0, x_1 \in \mathbb{R}^n$ and let $u = (u(k))_{k \in \mathbb{N}}$ be a control such that

$$x(N, x_0, u) = x_1.$$

If we apply control

$$v(k) = \begin{bmatrix} u(k) \\ u(k-1) \end{bmatrix}, \quad k = 0, 1, ...$$

in system (2) with initial condition x_0, then according to Theorem 1, solutions of (1) and (2) coincides. In particular

$$x(N, x_0, u) = y(N, x_0, v)$$

and therefore
$$y(N, x_0, v) = x_1.$$

It is also clear that $v \in \overline{\mathcal{G}}$ what implies that (2) with $\hat{A} = A$, $\hat{B} = \begin{bmatrix} B_0 & B_1 \end{bmatrix}$ is $\overline{\mathcal{G}}$-controllable in time N.

Suppose now that system (2) is $\overline{\mathcal{G}}$-controllable in time N. Let us fix $x_0, x_1 \in \mathbb{R}^n$ and let $v = (v(k))_{k \in \mathbb{N}}$ be a control from $\overline{\mathcal{G}}$ such that

$$y(N, x_0, v) = x_1.$$

Denote

$$v(k) = \begin{bmatrix} u(k) \\ u(k-1) \end{bmatrix}, \ k = 0, 1, \dots$$

where $u(k) \in \mathbb{R}^m$, $k = 0, 1, \dots$. Notice that this notation is correct since $v \in \overline{\mathcal{G}}$. According to Theorem 1, solutions of (1) and (2) coincides. In particular

$$x(N, x_0, u) = y(N, x_0, v)$$

and therefore

$$x(N, x_0, v) = x_1.$$

The proof is completed.

We will illustrate the last theorem on an example.

Example 1. Consider the following system (1)

$$x(k+1) = \begin{bmatrix} 1 & 2 \\ 3 & 4 \end{bmatrix} x(k) + \begin{bmatrix} 2 \\ -1 \end{bmatrix} u(k) + \begin{bmatrix} -1 \\ 1 \end{bmatrix} u(k-1). \tag{7}$$

We are going to show that the system is controllable in time 2. According to the Theorem 6 we have to prove that the following delay-free system

$$y(k+1) = \begin{bmatrix} 1 & 2 \\ 3 & 4 \end{bmatrix} y(k) + \begin{bmatrix} 2 & -1 \\ -1 & 1 \end{bmatrix} v(k) \tag{8}$$

is $\mathcal{G}$-controllable in time 2. Since

$$v(k) = \begin{bmatrix} u(k) \\ u(k) \end{bmatrix} = \begin{bmatrix} 1 \\ 1 \end{bmatrix} u(k)$$

and

$$\begin{bmatrix} 2 & -1 \\ -1 & 1 \end{bmatrix} \begin{bmatrix} 1 \\ 1 \end{bmatrix} = \begin{bmatrix} 1 \\ 0 \end{bmatrix}$$

then $\mathcal{G}$-controllability in time 2 of system (8) is equivalent to controllability in time 2 of the following system

$$z(k+1) = Az(k) + B\nu(k),$$

where:

$$A = \begin{bmatrix} 1 & 2 \\ 3 & 4 \end{bmatrix}$$

and

$$B = \begin{bmatrix} 1 \\ 0 \end{bmatrix}.$$

Finally, controllability in time 2 of the last system follows from the classical Kalman controllability condition (see [6]) since

$$[B \quad AB] = \begin{bmatrix} 1 & 1 \\ 0 & 3 \end{bmatrix}.$$

3 Conclusions

In this paper we study problems of ANCBC, BFS, SFS and controllability of system (1). The origin of this paper is in the results of [14,18] where the authors obtained a complete picture of the relation between global stabilization and controllability of discrete and continuous delay-free time-invariant systems. Here, we were able only to provide necessary conditions for ANCBC, BFS and SFS of delayed systems in terms of analogical properties of delay-free systems. In that purpose we adapted the idea, from [6], of converting the original systems to an appropriate delay-free system. As it was shown in [14] and [18] for delay-free systems the concept of ANCBC, BFS and SFS are equivalent. The problem of equivalence of ANCBC, BFS and SFS for system (1) is an open problem. Another important direction of further research should be finding of generalizations of our results to time-varying systems as well as to systems with time-varying delay. The goal of this research should be a theory analogical to this which is known for delay-free system (see [14,18]).

References

1. Artstein, Z.: Linear systems with delayed controls: a reduction. IEEE Trans. Autom. Control **27**(4), 869–879 (1982)
2. Barabanov, N.E.: On quadratic stabilizability of linear dynamical systems. Siberian Math. J. **37**(1), 1–16 (1996)
3. Bellman, R., Cooke, K.L.: Differential Difference Equations. New York (1963)

4. Bernussou, J., Peres, P., Geromel, J.: Stabilizability of uncertain dynamical systems: the continuous and the discrete case. IFAC Proc. Volumes **23**(8, Part 3), 159–164 (1990). 11th IFAC World Congress on Automatic Control, Tallinn, 1990 - Volume 3, Tallinn, Finland
5. Górecki, H.: Analysis and Synthesis of Time Delay Systems. Wiley, New York (1989)
6. Klamka, J.: Controllability of Dynamical Systems. Kluwer Academic Publishers, Dordrecht (1991)
7. Kwon, W., Pearson, A.: Feedback stabilization of linear systems with delayed control. IEEE Trans. Autom. Control **25**(2), 266–269 (1980)
8. Lee, T., Dianat, S.: Stability of time-delay systems. IEEE Trans. Autom. Control **26**(4), 951–953 (1981)
9. Li, L., Zhang, H.: Stabilization of discrete-time systems with multiplicative noise and multiple delays in the control variable. SIAM J. Control Optim. **54**(2), 894–917 (2016)
10. Lin, Z., Saberi, A.: Semi-global exponential stabilization of linear discrete-time systems subject to input saturation via linear feedbacks. Syst. Control Lett. **24**(2), 125–132 (1995)
11. Malek-Zavarei, M., Jamshidi, M.: Time-Delay Systems: Analysis, Optimization and Applications. Elsevier Science Inc., New York (1987)
12. Olbrot, A.: On controllability of linear systems with time delays in control. IEEE Trans. Autom. Control **17**(5), 664–666 (1972)
13. Silva, J., Silva, L., Scola, I.R., Leite, V.: Robust local stabilization of discrete-time systems with time-varying state delay and saturating actuators. Math. Probl. Eng. **2018**(3), 1–9 (2018). Article ID 5013056
14. Sussmann, H.J., Sontag, E.D., Yang, Y.: A general result on the stabilization of linear systems using bounded controls. IEEE Trans. Autom. Control **39**(12), 2411–2425 (1994)
15. Tantaris, R.N., Keel, L.H., Bhattacharyya, S.P.: Stabilization of discrete-time systems by first-order controllers. IEEE Trans. Autom. Control **48**(5), 858–860 (2003)
16. Tsirukis, A.G., Morari, M.: Controller design with actuator constraints. In: Proceedings of the 31st IEEE Conference on Decision and Control, vol. 3, pp. 2623–2628 (1992)
17. Wu, M., He, Y., She, J.H.: Stability and stabilization of discrete-time systems with time-varying delay. In: Stability Analysis and Robust Control of Time-Delay Systems, pp. 147–161. Springer, Heidelberg (2010). https://doi.org/10.1007/978-3-642-03037-6_7
18. Yang, Y., Sontag, E.D., Sussmann, H.J.: Global stabilization of linear discrete-time systems with bounded feedback. Syst. Control Lett. **30**(5), 273–281 (1997)
19. Zhou, B.: Stabilization of discrete-time systems with input and output delays. Truncated Predictor Feedback for Time-Delay Systems, pp. 237–272. Springer, Heidelberg (2014). https://doi.org/10.1007/978-3-642-54206-0_9

A Simple yet Efficient MCSes Enumeration with SAT Oracles

Miyuki Koshimura[1]([⊠])[iD] and Ken Satoh[2]

[1] Kyushu University, Fukuoka, Japan
koshi@inf.kyushu-u.ac.jp
[2] National Institute of Informatics, Tokyo, Japan
ksatoh@nii.ac.jp

Abstract. The enumeration of the maximal satisfiable subsets (MSSes) or the minimal correction subsets (MCSes) of conjunctive normal form (CNF) formulas is a cornerstone task in various AI domains. This paper presents an algorithm that enumerates all MCSes with SAT oracles. Our algorithm is simple because it follows a plain algorithm without any techniques that decrease the number of calls to a SAT oracle. The experimental results show that our proposed method is more efficient than state-of-the-art MCS enumerators on average to deal with Partial-MaxSAT instances.

Keywords: Minimal correction subset · Enumeration · SAT oracle

1 Introduction

A set of constraints that cannot be simultaneously satisfied is over-constrained. A minimal correction subset (MCS) of an over-constrained system is a minimal[1] set of constraints whose removal restores the system's consistency [7]. The remaining set of constraints is satisfied, and thus it becomes a maximal satisfiable subset (MSS) of the system. In terms of an unsatisfiable conjunctive normal form (CNF) propositional formula, an MCS is a minimal set of clauses such that, once removed, the rest of the formula is satisfiable.

Enumerating the MCSes or MSSes of an unsatisfiable CNF propositional formula is a cornerstone task in various AI domains. On the theoretical side, they are computational bases of circumscription; Computing all the MSSes of the negation of the minimized propositions given a propositional theory provides a circumscriptive formula with the disjunctions of all the conjunctions of the negations of propositions in each MSS [1]. This method can be used to compute a cautious formula in "closed world assumption" such that the formula is satisfied in all the models. On the application side, in constraint processing, soft constraints are used to represent user preferences over solutions, and we need to compute their MSSes [2]. In model-based diagnosis [19], if we add an auxiliary proposition in the condition of each component that expresses normality, given a diagnosis, we compute a set of normal components by computing the

---

[1] We always consider set-inclusion minimality in this paper unless stated otherwise.

© Springer Nature Switzerland AG 2020
N. T. Nguyen et al. (Eds.): ACIIDS 2020, LNAI 12033, pp. 191–201, 2020.
https://doi.org/10.1007/978-3-030-41964-6_17

MSSes for these auxiliary propositions, and hence we can detect the abnormal components of the system [14].

In this paper, we deal with the enumeration of all MCSes with SAT oracles, which are used by many approaches in the literature [10,13,16,17]. Each approach is regarded as a variant of a MaxSAT in the sense that a MaxSAT solution corresponds to a minimum-sized MCS where MaxSAT finds a model that satisfies as many clauses as possible.

Clearly, in these approaches, the time-consuming part lies in the multiple calls to a SAT oracle. To reduce the number of calls, several techniques have been proposed: backbone literals [13], core caching [17], premise caching [18], model rotation [5], and so on.

We propose an algorithm that enumerates MCSes. It is simple because it follows a plain algorithm and does not use any techniques to decrease the number of calls to a SAT oracle. Our algorithm relies on making sequences of calls to a SAT oracle. Each sequence ends with a SAT oracle whose result is unsatisfiable, and an MCS is obtained at that time. In each SAT oracle, an MCS candidate is obtained and its size is decreased by *at least* one. This differs from previous works that follow such similar (basic) algorithms as BLS in [17]. Such similar works decreased the size by *at most* one. Thus, we expect that our algorithm will need fewer SAT oracles than the previous ones.

Experimental results show that on average, our proposed algorithm is more efficient than the two state-of-the-art MCS enumerators, Enum-ELS-RMR-Cache [10] and LBX-Cache [18], for dealing with Partial-MaxSAT instances and less efficient for plain MaxSAT instances. When the number of instances for which all the MCSes have been enumerated, the proposed method is superior to the enumerators for both Partial-MaxSAT and plain MaxSAT instances. The proposed method is also superior in memory efficiency.

The remainder of this paper is organized as follows. The technical background and the well-known MSS and MCS concepts are briefly reviewed in the preliminaries. In Sect. 3, we discuss related work. Section 4 presents a MCSes enumeration algorithm. Then, we present our experimental study. We conclude in Sect. 6 with future work.

2   Preliminaries

In this paper, a problem is given by a propositional formula in a conjunctive normal form (CNF). A CNF formula is a conjunction ($\wedge$) of clauses. A clause is a disjunction ($\vee$) of literals. Clause α is called a unit if its length is one, i.e., $|\alpha| = 1$. A literal is either a variable or its negation ($\neg$). We regard a clause as the set of literals in it. Clause α subsumes clause β iff $\alpha \subseteq \beta$. We also regard a CNF formula as a set of clauses that constitute the formula.

The set of variables in a set of clauses Σ is denoted by $vars(\Sigma)$. Assignment μ of the variables $vars(\Sigma)$ is mapping $vars(\Sigma) \mapsto \{0, 1\}$. Assignment μ is extended to literals, clauses, and sets of clauses as follows: $\mu(\neg x) = 1$ iff $\mu(x) = 0$, and $\mu(\neg x) = 0$ iff $\mu(x) = 1$, for all $x \in vars(\Sigma)$. For clause α, $\mu(\alpha) = 1$ iff $\mu(l) = 1$ for some $l \in \alpha$, and otherwise $\mu(\alpha) = 0$. For a set of clauses Σ, $\mu(\Sigma) = 1$ iff $\mu(\alpha) = 1$ for all $\alpha \in \Sigma$, and otherwise $\mu(\Sigma) = 0$. Clause α is satisfied by μ iff $\mu(\alpha) = 1$, and otherwise it is falsified. A set of clauses Σ is satisfied by μ iff $\mu(\Sigma) = 1$, otherwise, Σ is falsified. Σ is satisfiable iff there exists at least one assignment μ that satisfies Σ.

This paper deals with two types of clauses: hard and soft. Hard clauses must be satisfied and soft clauses should be satisfied as much as possible. The core, MUS, MSS, and MCS cross-related concepts are defined as follows. Let Σ be a set of clauses that consists of set Σ_1 of hard clauses and set Σ_2 of soft clauses, that is, $\Sigma = \Sigma_1 \cup \Sigma_2$. We assume that Σ_1 is always satisfiable throughout our paper.

Definition 1 (Core). Set of clauses Σ' is a *core* of Σ iff $\Sigma_1 \subseteq \Sigma' \subseteq \Sigma$ and Σ' is unsatisfiable.

Definition 2 (MUS). A *minimal unsatisfiable subset* (MUS) Σ' of Σ is a core of Σ such that $\forall \alpha \in \Sigma' (\Sigma' \setminus \{\alpha\}$ is satisfiable)

Definition 3 (MSS). A *maximal satisfiable subset* (MSS) Φ of Σ is a satisfiable subset of Σ containing Σ_1 ($\Sigma_1 \subseteq \Phi \subseteq \Sigma$) such that $\forall \alpha \in \Sigma \setminus \Phi$, $\Phi \cup \{\alpha\}$ is unsatisfiable.

An MSS is an extension of Σ_1 to the limit with the elements of Σ_2 while keeping its satisfiability.

Definition 4 (MCS). A *minimal correction subset* (MCS) Ψ of Σ is set $\Psi \subseteq \Sigma$ whose complement in Σ, that is, $\Sigma \setminus \Psi$ is an MSS of Σ.

An MCS is a subset of Σ_2 because the corresponding MSS includes Σ_1. An MCS is a minimal subset of Σ_2 such that Σ becomes satisfiable as soon as all the elements in the MCS are removed from Σ.

Example 1 ([10]). Let Σ be an unsatisfiable CNF that is formed by a set of clauses $\{\alpha_1, \alpha_2, \alpha_3, \alpha_4, \alpha_5, \alpha_6\}$, where $\alpha_1 = a \lor b$, $\alpha_2 = \neg a \lor b$, $\alpha_3 = a \lor \neg b$, $\alpha_4 = \neg a \lor \neg b$, $\alpha_5 = \neg b$, $\alpha_6 = b$. The MCSes of Σ are $\{\alpha_1, \alpha_6\}$, $\{\alpha_2, \alpha_6\}$, $\{\alpha_3, \alpha_5\}$, and $\{\alpha_4, \alpha_5\}$. The MUSes of Σ are $\{\alpha_1, \alpha_2, \alpha_3, \alpha_4\}$, $\{\alpha_1, \alpha_2, \alpha_5\}$, $\{\alpha_3, \alpha_4, \alpha_6\}$, and $\{\alpha_5, \alpha_6\}$.

MCS enumerators often exploit so-called *clause selectors* as follows. For each soft clause $\alpha \in \Sigma_2$, a fresh variable s_α is introduced and the clause is augmented by the negation of the variable. This yields a new set of clauses $\Sigma_2^S = \{\alpha \lor \neg s_\alpha \mid \alpha \in \Sigma_2\}$. A SAT oracle determines the satisfiability of $\Sigma_1 \cup \Sigma_2^S$ instead of Σ.

Under assumption $s_\alpha = 1$, the satisfiability of $\alpha \lor \neg s_\alpha$ is consistent with that of α. Then $\alpha \lor \neg s_\alpha$ is activated (resp., deactivated) when the variable s_α is set to 1 (resp. 0). Thus, extracting an MSS corresponds to an increase in the activated soft clauses to a limit while keeping the satisfiability of $\Sigma_1 \cup \Sigma_2^S$. In other words, extracting an MCS is equivalent to a decrease in the deactivated soft clauses to the limit. Based on this perspective, the algorithm in Sect. 4 enumerates the MCSes.

3 Related Work

Our work is fueled by the minimal model (MM) generation [11] based on the following observation. An MCS corresponds to an MM with respect to a set of selectors. In [11], Koshimura et al. proposed an algorithm that enumerates all the MMes and implemented it with a MiniSAT 2.1 SAT solver [8]. Unfortunately, they only evaluated it for computing a single MM for job-shop scheduling problems and did not evaluate it for enumerating MMes.

From the viewpoint of MCSes enumeration, our algorithm closely resembles the clause D based (CLD) algorithm [13]. A clause D corresponds to a blocking clause in our algorithm. The difference between them is that the D clause consists of the literals in the soft clauses of an MCS candidate, but the blocking clause consists of the selectors of the soft clauses. Thus, the D clause's length is the number of distinct literals in the soft clauses, and the length of the blocking clause is the number of the soft clauses. Each blocking clause is added to the set of clauses from which the MCSes were enumerated, which affects subsequent enumerations. Each D clause is used only for computing the next D clause or an MCS. The CLD algorithm originally used a disjoint set of unsatisfiable cores, and backbone literals. In this respect, our algorithm is a plain CLD algorithm. According to a recent study in [18], CLD only achieved good performance when the number of MCSes enumerated by both the CLD and a state-of-the-art MCS enumerator is small.

Caching (or memorization), which is a well-known general concept, has been successfully applied for speeding up procedures for several problems. To decrease the number of calls to a SAT oracle, Previti et al. [17] proposed caching unsatisfiable cores (core caching) that were met during a search within MCSes enumeration with a SAT oracle. Their implementation is called mcscache-els. Before summoning a SAT oracle to determine a formula's satisfiability, mcscache-els checks the consistency of the cache with a set of unit clauses, each of which is a selector extracted from the formula. If the result is inconsistent, the SAT oracle is avoided. Since each core only contains selectors, the consistency checking is cheaper than the SAT oracle. Experimental results showed that mcscache-els clearly outperformed the CLD when the number of enumerated MCSs increases. They also proposed another caching called premise set caching (PS caching) [18]. Their implementation LBX-Cache noticeably outperforms mcscache-els on a great majority of benchmarks.

Grégoire et al. [10] proposed a new technique that boosts MCSes enumeration which is based on the form of so-called model rotation paradigm [5]. Their implementation is a state-of-the-art MCS enumerator called Enum-ELS-RMR-Cache, which combines several techniques other than model rotation, including backbone literals and caching. Their experimental results showed that Enum-ELS-RMR-Cache outperformed mcscache-els for almost all the benchmarks. In Sect. 5, we compare our algorithm, LBX-Cache and Enum-ELS-RMR-Cache.

Another approach is called core-guided MCSes enumeration [15]. In it, selectors are gradually introduced to soft clauses during a search within MCSes enumeration while the above methods introduce them before search. This is based on the well-known core-guided MaxSAT algorithm [9]. In the core-guided approach, an unsatisfiable core is obtained on the termination of a SAT oracle. For each soft clause in the core, its selector is introduced and added to the clause. Then, the next call to a SAT oracle is invoked. We obtain an MCS when a SAT oracle returns satisfiable. This approach enumerates MCSes in increasing order of size. Norodytska et al. [16] proposed an efficient MCS and an MUS enumerator based on core-guided MCS enumeration with several new techniques.

Finally, we refer to a quite different approach [3] to enhance MCS enumeration. This refines a SAT solver itself as suitable for an intensive assumption-based incremental SAT solving task, such as computing MUSes. Our algorithm in Sect. 4 introduces as many selectors as the number of soft clauses. A subset of these selectors is passed to the SAT solver as an assumption. Thus, our algorithm will benefit from the refinement. We used this refinement, which was introduced into a Glucose SAT solver [4] as an incremental solving mode, in our experimental study (Sect. 5).

4 MCSes Enumeration

Algorithm 1. Enum-MCS (Enumerate All MCSes)

Input: $\Sigma \, (\leftarrow \Sigma_1 \cup \Sigma_2)$ // Σ_1: set of hard clauses, Σ_2: set of soft clauses
Output: all MCSes of Σ
1: $\Sigma_2^S \leftarrow \{\alpha \vee \neg s_\alpha \mid \alpha \in \Sigma_2\}$; // with s_α fresh variables
2: $S \leftarrow \{s_\alpha \mid \alpha \in \Sigma_2\}$; // set of selectors
3: $A \leftarrow \emptyset$; // set of selectors for MSS candidate
4: $B \leftarrow \emptyset$; // set of selectors for MCS candidate
5: **while true do**
6:     $(st, \mu) \leftarrow SAT(\Sigma_1 \cup \Sigma_2^S, A)$; // SAT oracle
7:     **if** $st = TRUE$ **then**
8:         $A \leftarrow \{s \mid \mu(s) = 1, s \in S\}$;
9:         $B \leftarrow \{s \mid \mu(s) = 0, s \in S\}$;
10:        $\Sigma_1 \leftarrow \Sigma_1 \cup (\bigvee_{s \in B} s)$; // blocking clauses
11:     **else if** $A = \emptyset$ **then**
12:        **return**
13:     **else**
14:        $output(\{\alpha \mid s_\alpha \in B\})$; // an MCS is found
15:        $A \leftarrow \emptyset$;
16:     **end if**
17: **end while**

Algorithm 1 outlines the algorithm to enumerate all MCSes. $SAT(\Sigma_1 \cup \Sigma_2^S, A)$ is a SAT oracle that determines the satisfiability of $\Sigma_1 \cup \Sigma_2^S$ under assumption $s = 1$ for all $s \in A$ (line 6). When its result is satisfiable, we set st to $TRUE$ and obtain an assignment through μ. Otherwise, st is set to $FALSE$.

A and B in the algorithm are sets of selectors. We refer to the set of soft clauses that correspond to A, i.e., $\{\alpha \mid \mu(s_\alpha) = 1\}$ as A_{MSS}, while that corresponding to B, i.e., $\{\alpha \mid \mu(s_\alpha) = 0\}$ as B_{MCS}. A_{MSS} is a MSS candidate, and B_{MCS} is a MCS candidate. The algorithm explores an MCS with assignment μ as a starting point. It repeats lines 6 to 10 as long as SAT returns $TRUE$. A gradually increases, A_{MSS} approaches an MSS, B gradually decreases, B_{MCS} approaches an MCS. Note that A and B are partitions of S. Each time one assignment is obtained, an additional clause is created (line 10). Its role is to block a B' larger than B from being obtained.

If SAT returns $FALSE$ and $A \neq \emptyset$, current B is minimal because no solution smaller than B. Thus, we conclude that corresponding B_{MCS} is an MCS (line 14). $A \leftarrow \emptyset$ resets the MSS candidate and restarts the search (line 15). If SAT returns $FALSE$ and $A = \emptyset$, there are no more MCSes, and the algorithm terminates (lines 11, 12).

4.1 Substitute the Selector of a Unit Soft Clause with the Clause

We can substitute the selector of a unit soft clause with the clause itself [6]. This decreases the number of selectors and may improve the efficiency. To achieve this, we need to change the following two lines in Algorithm 1:

$$1: \Sigma_2^S \leftarrow \{\alpha \vee \neg s_\alpha \mid \alpha \in \Sigma_2, |\alpha| > 1\}$$
$$\cup \{\alpha \mid \alpha \in \Sigma_2, |\alpha| = 1\};$$
$$2: S \leftarrow \{s_\alpha \mid \alpha \in \Sigma_2, |\alpha| > 1\}$$
$$\cup \{\alpha \mid \alpha \in \Sigma_2, |\alpha| = 1\};$$

5 Experimental Study

We implemented our algorithm in C++ and used Glucose 3.0 [4], which is based on Minisat [8], as a backend SAT solver. We selected 1090 benchmarks from a previous work [10]; 493 are plain MaxSAT (MS) ones that only consist of soft clauses, and the remaining 597 are Partial MaxSAT (PMS) ones that consist of both hard and soft clauses.

As mentioned in Sect. 4, blocking clause B in Algorithm 1 gradually decreases during the repetition of lines 6 to 10. This implies that the current blocking clause always subsumes its previous (old) ones. We can ignore the subsumed clauses. In the current implementation, the old blocking clause is removed from Σ_1. Thus, only the last blocking clause in the repetition (which corresponds to an MCS) remains in Σ_1.

All experiments were conducted on Intel Xeon E3-1246v6 (3.70 GH) with 32 GB memory on Linux Ubuntu. Time-outs were set to 1800 s for each algorithm run on an instance; memory-outs were set to 8 GB for each such run. These limitations are identical as those in the previous work [10].

Table 1. Average numbers of variables and clauses

	# Variables	# Clauses		
		Hard	Soft	(Unit)
MS	156,844	0	496,827	(3,030)
PMS	16,032	100,134	10,759	(10,758)

Table 1 shows the average numbers of variables and clauses of the MS and PMS instances. The numbers in parentheses are the average numbers of the unit soft clauses. Almost all of the soft clauses of the instances in PMS are units.

Table 2. Total numbers of MCSes enumerated, SAT oracles, and their ratio

	LBX-Cache	Enum-ELS-RMR-Cache	Proposed algorithm			
			B	B+Inc	US	US+Inc
MS	198,228	**349,156**	59,711	85,297	60,173	80,735
	88,156	401,598	568,399	865,993	567,967	818,751
	(0.44)	(1.15)	(9.52)	(10.15)	(9.44)	(10.14)
PMS	375,480	494,309	295,870	280,603	**637,801**	627,593
	1,234,544	1,023,506	1,187,627	935,170	1,536,802	1,507,640
	(3.29)	(2.07)	(4.01)	(3.33)	(2.41)	(2.40)

top: #MCSes enumerated ($\times 10^3$),   middle: #SAT oracles ($\times 10^3$),   bottom: middle/top

We compared our proposed algorithm with two state-of-the-art MCS enumerators, LBX-Cache[2]. [18] and Enum-ELS-RMR-Cache[3] [10]. Table 2 shows the total numbers of enumerated MCSes and called SAT oracles, and their ratio in parentheses. The ratio represents the average number of SAT oracles that are required to extract one MCS. We ran four versions of the proposed algorithm. The B version is a straight implementation of Algorithm 1, and the US version uses the unit clauses themselves as substitutes for their selector. +Inc indicates that the incremental solving mode of Glucose is turned on and enhances the performance of the SAT oracles that are invoked with many selectors as assumptions.

As shown in Table 2, the number of MCSes increased from 295,870 thousand to 637,801 thousand for the PMS instances by the US version. Recall that almost all the soft clauses in PMS are units. The US version does affect such an instance without really affecting the MS instances where almost none of the soft clauses are units. +Inc shows a good effect on the MS instances for which many selectors are introduced. The number of MCSes increased from 59,711 thousand to 85,297 thousand for the B version. As a result, the performance of the B+Inc version is the best for MS, and the US version is the best for PMS among the four versions.

We compared the number of MCSes enumerated by the proposed algorithm with those by LBX-Cache and Enum-ELS-RMR-Cache. The number by the B+Inc version is half of that by LBX-Cache and a quarter of that by Enum-ELS-RMR-Cache for MS, and by the US version the amounts are 1.7 and 1.3 times as many as those by LBX-Cache and Enum-ELS-RMR-Cache for PMS. Thus, the US version outperforms these state-of-the-art MCS enumerators for PMS, although all four versions are inferior to them for MS. This inferiority was caused by the number of SAT oracles required to extract one MCS. LBX-Cache and Enum-ELS-RMR-Cache needed only 0.44 and 1.15 oracles for the MS instances, but the four versions needed about 10 oracles on average. If the number of SAT oracles is reduced to the same as that of the US version for PMS, the performance of the proposed algorithm can be improved

---

[2] It is available from https://www.cs.helsinki.fi/group/coreo/lbx-cache/.

[3] It is available from http://www.cril.fr/enumcs/. We also obtained the 1090 benchmarks from the same site.

and be compatible with these two enumerators. Note that since the proposed algorithm needs at least two SAT oracles to extract one MCS, ratios 2.41 and 2.40 in Table 2 seem relatively good.

Table 3. Numbers of instances completed and stopped by memory-out

	LBX-Cache	Enum-ELS-RMR-Cache	Proposed algorithm			
			B	B+Inc	US	US+Inc
MS	45	52	57	**58**	56	**58**
	94	93	**2**	16	**2**	13
PMS	188	191	182	186	**193**	**193**
	0	72	**0**	19	22	28

top: #completed,   bottom: #memory-out

Table 3 shows the number of instances for which enumeration was completed and those that stopped due to memory limitations. Obviously, our proposed algorithm is superior to Enum-ELS-RMR-Cache in terms of memory efficiency. 93 and 72 runs of Enum-ELS-RMR-Cache were stopped for the MS and PMS instances. At most 16 and 28 runs of the four versions were stopped for them. LBX-Cache consumes more memory for MS and less for PMS than the others. Its memory consumption may heavily depend on the number of soft clauses in the benchmark instance since the average number of soft clauses in MS is about 46 times greater than that in PMS.

Based on the comparison between B and B+Inc or US and US+Inc, +Inc consumes more memory because the SAT solver with +Inc retains longer learned clauses than without +Inc.

Figures 1, 2, and 3 show the corresponding log scale scatter plots[4]. Point (a, b) corresponds to an instance for which one on the horizontal axis enumerated a MCSes and another on the vertical axis enumerated b MCSes. In all the plots, the more points below the $x = y$ line, the better the one on the horizontal axis performed compared to one on the vertical axis. The left subfigures show the MS results and the right ones show the PMS results.

Figure 1 compares the proposed algorithm and LBX-Cache. The number of points below the $x = y$ line exceeds that above the line for both MS and PMS. This implies that the proposed algorithm is superior to LBX-Cache in view of the number of instances the enumerate more MCSes. This superiority does not conflict with the results in Table 2 where the total numbers of the enumerated MCSes are evaluated. There are not a few instances of MS for which LBX-Cache enumerates more than 10^6 MCSes, while the proposed algorithm does so for fewer than 10^6. These instances increase the total number of enumerated MCSes. Figure 2 compares the proposed algorithm and Enum-ELS-RMR-Cache. Our proposed algorithm is superior for PMS;

---

[4] We replaced 0 with 0.5 during the plotting because the logarithm is defined only for positive numbers. Thus, point $(0.5, 0.5)$ represents an instance for which both methods failed to find an MCS.

(MS: 493 instances)   (PMS: 597 instances)

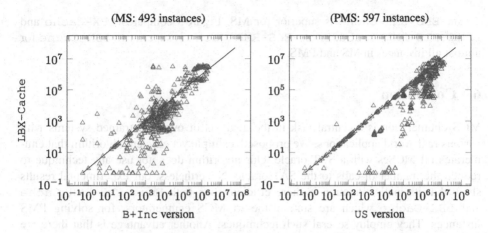

Fig. 1. Proposed algorithm vs. LBX-Cache

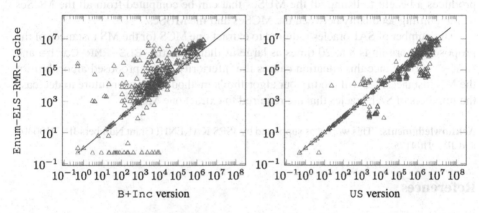

Fig. 2. Proposed algorithm vs. Enum-ELS-RMR-Cache

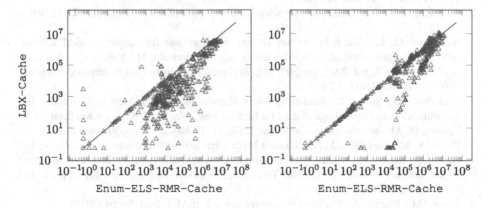

Fig. 3. Enum-ELS-RMR-Cache vs. LBX-Cache

Enum-ELS-RMR-Cache is superior for MS. Figure 3 compares LBX-Cache and Enum-ELS-RMR-Cache. Enum-ELS-RMR-Cache is superior to LBX-Cache for almost all instances in MS and PMS.

6 Conclusion

MCS enumeration is a central task in the analysis of over-constrained systems with various real-world applications. We proposed a simple yet efficient algorithm that enumerates all MCSes with a SAT oracle. Our algorithm does not use any technique to reduce the number of calls to the SAT oracle. Nevertheless, our experimental results show that its implementation is more efficient than both Enum-ELS-RMR-Cache and LBX-Cache, which are state-of-the-art MCS enumerators, for solving PMS instances. They employ several such techniques. Another advantage is that there are more instances for which all MCSes have been enumerated for both MS and PMS. This produces a benefit to listing all the MUSes that can be computed from all the MCSes using a hitting set duality between the MCSes and the MUSes.

The number of SAT oracles required to extract one MCS for the MS instances of the proposed algorithm is 8 to 20 times as large as that of Enum-ELS-RMR-Cache and LBX-Cache. Since this situation causes the inferiority of the proposed algorithm for the MS instances, we will modify our algorithm's method in the near future to decrease the numbers of SAT oracles that are required to extract one MCS.

Acknowledgments. This work was supported by JSPS KAKENHI Grant Numbers JP17K00307 and JP19H04175.

References

1. Alviano, M.: Model enumeration in propositional circumscription via unsatisfiable core analysis. TPLP **17**(5–6), 708–725 (2017)
2. Argelich, J.: Max-sat formalisms with hard and soft constraints. AI Commun. **24**(1), 101–103 (2011)
3. Audemard, G., Lagniez, J.-M., Simon, L.: Improving glucose for incremental SAT solving with assumptions: application to MUS extraction. In: Järvisalo, M., Van Gelder, A. (eds.) SAT 2013. LNCS, vol. 7962, pp. 309–317. Springer, Heidelberg (2013). https://doi.org/10. 1007/978-3-642-39071-5_23
4. Audemard, G., Simon, L.: Predicting learnt clauses quality in modern SAT solvers. In: Boutilier, C. (ed.) Proceedings of the 21st International Joint Conference on Artificial Intelligence, IJCAI 2009, Pasadena, California, USA, 11–17 July 2009, pp. 399–404 (2009)
5. Belov, A., Marques-Silva, J.: Accelerating MUS extraction with recursive model rotation. In: Bjesse, P., Slobodová, A. (eds.) International Conference on Formal Methods in Computer-Aided Design, FMCAD 2011, Austin, TX, USA, 30 October–02 November 2011, pp. 37–40. FMCAD Inc. (2011)
6. Berre, D.L., Parrain, A.: The sat4j library, release 2.2. JSAT **7**(2–3), 59–64 (2010)
7. Birnbaum, E., Lozinskii, E.L.: Consistent subsets of inconsistent systems: structure and behaviour. J. Exp. Theor. Artif. Intell. **15**(1), 25–46 (2003)

8. Eén, N., Sörensson, N.: An extensible SAT-solver. In: Giunchiglia, E., Tacchella, A. (eds.) SAT 2003. LNCS, vol. 2919, pp. 502–518. Springer, Heidelberg (2004). https://doi.org/10. 1007/978-3-540-24605-3_37
9. Fu, Z., Malik, S.: On solving the partial MAX-SAT problem. In: Biere, A., Gomes, C.P. (eds.) SAT 2006. LNCS, vol. 4121, pp. 252–265. Springer, Heidelberg (2006). https://doi. org/10.1007/11814948_25
10. Grégoire, É., Izza, Y., Lagniez, J.: Boosting MCSes enumeration. In: Lang, J. (ed.) [12], pp. 1309–1315
11. Koshimura, M., Nabeshima, H., Fujita, H., Hasegawa, R.: Minimal model generation with respect to an atom set. In: Peltier, N., Sofronie-Stokkermans, V. (eds.) Proceedings of the 7th International Workshop on First-Order Theorem Proving, FTP 2009, Oslo, Norway, 6–7 July 2009. CEUR Workshop Proceedings, vol. 556 (2009). CEUR-WS.org
12. Lang, J. (ed.): Proceedings of the Twenty-Seventh International Joint Conference on Artificial Intelligence, IJCAI 2018, 13–19 July 2018, Stockholm, Sweden (2018). ijcai.org
13. Marques-Silva, J., Heras, F., Janota, M., Previti, A., Belov, A.: On computing minimal correction subsets. In: Rossi, F. (ed.) Proceedings of the 23rd International Joint Conference on Artificial Intelligence, IJCAI 2013, Beijing, China, 3–9 August 2013, pp. 615–622. IJCAI/AAAI (2013)
14. Marques-Silva, J., Janota, M., Ignatiev, A., Morgado, A.: Efficient model based diagnosis with maximum satisfiability. In: Yang, Q., Wooldridge, M.J. (eds.) Proceedings of the Twenty-Fourth International Joint Conference on Artificial Intelligence, IJCAI 2015, Buenos Aires, Argentina, 25–31 July 2015, pp. 1966–1972. AAAI Press (2015)
15. Morgado, A., Liffiton, M., Marques-Silva, J.: MaxSAT-based MCS enumeration. In: Biere, A., Nahir, A., Vos, T. (eds.) HVC 2012. LNCS, vol. 7857, pp. 86–101. Springer, Heidelberg (2013). https://doi.org/10.1007/978-3-642-39611-3_13
16. Narodytska, N., Bjørner, N., Marinescu, M.V., Sagiv, M.: Core-guided minimal correction set and core enumeration. In: Lang, J. (ed.) [12], pp. 1353–1361
17. Previti, A., Mencía, C., Järvisalo, M., Marques-Silva, J.: Improving MCS enumeration via caching. In: Gaspers, S., Walsh, T. (eds.) SAT 2017. LNCS, vol. 10491, pp. 184–194. Springer, Cham (2017). https://doi.org/10.1007/978-3-319-66263-3_12
18. Previti, A., Mencía, C., Järvisalo, M., Marques-Silva, J.: Premise set caching for enumerating minimal correction subsets. In: McIlraith, S.A., Weinberger, K.Q. (eds.) Proceedings of the Thirty-Second AAAI Conference on Artificial Intelligence, (AAAI-18), the 30th Innovative Applications of Artificial Intelligence (IAAI-18), and the 8th AAAI Symposium on Educational Advances in Artificial Intelligence (EAAI-18), New Orleans, Louisiana, USA, 2–7 February 2018, pp. 6633–6640. AAAI Press (2018)
19. Reiter, R.: A theory of diagnosis from first principles. Artif. Intell. **32**(1), 57–95 (1987)

Alpha-N: Shortest Path Finder Automated Delivery Robot with Obstacle Detection and Avoiding System

Asif Ahmed Neloy$^{(\boxtimes)}$ ⓘ, Rafia Alif Bindu ⓘ, Sazid Alam ⓘ, Ridwanul Haque ⓘ, Md. Saif Ahammod Khan ⓘ, Nasim Mahmud Mishu ⓘ, and Shahnewaz Siddique

Department of Electrical and Computer Engineering, North South University, Plot-15, Block-B, Bashundhara Residential Area, Dhaka, Bangladesh
{asif.neloy,rafia.bindu,sazid.alam,ridwanul.haque,saif.ahammod, nasim.mishu,shahnewaz.siddique}@northsouth.edu

Abstract. *Alpha-N* - A self-powered, wheel-driven *Automated Delivery Robot (ADR)* is presented in this paper. The ADR is capable of navigating autonomously by detecting and avoiding objects or obstacles in its path. It uses a vector map of the path and calculates the shortest path by **Grid Count Method (GCM)** of Dijkstra's Algorithm. *Landmark determination with Radio Frequency Identification (RFID)* tags are placed in the path for identification and verification of source and destination, and also for the re-calibration of the current position. On the other hand, an *Object Detection Module (ODM)* is built by Faster R-CNN with VGGNet-16 architecture for supporting path planning by detecting and recognizing obstacles. The *Path Planning System (PPS)* is combined with the output of the GCM, the *RFID Reading System (RRS)* and also by the binary results of ODM. This PPS requires a minimum speed of 200 RPM and 75 s duration for the robot to successfully relocate its position by reading an RFID tag. In the result analysis phase, the ODM exhibits an accuracy of *83.75%,* RRS shows *92.3%* accuracy and the PPS maintains an accuracy of *85.3%.* Stacking all these 3 modules, the ADR is built, tested and validated which shows significant improvement in terms of performance and usability comparing with other service robots.

Keywords: Mobile robot · Obstacle avoiding system · RFID · Automated Delivery Robot · Dijkstra's Algorithm · Grid Count Method · Faster R-CNN · VGGNet-16

1 Introduction

Over the past few years, there has been an increasing number of robotic researches in Bangladesh. Especially a number of studies are conducted for Service Robots, Military Robots and also for the Rescue Robots. But it is rare to see robots in Bangladesh that are autonomous and capable of making complex decisions. An autonomous intelligent robot that can go from one place to another within a constructed map while avoiding an obstacle, is quite formidable to build. It comes with many challenges for solving this

© Springer Nature Switzerland AG 2020
N. T. Nguyen et al. (Eds.): ACIIDS 2020, LNAI 12033, pp. 202–213, 2020.
https://doi.org/10.1007/978-3-030-41964-6_18

complex research topic. Moreover, the potential use of this kind of robot is in the food or parcel delivery systems which can be introduced in many rides sharing companies like Pathao Foods, Uber Eats, Food Panda in Bangladesh and also in developing countries [1]. Considering this huge opportunity and vacancy, this proposed ADR is studied.

This research particularly aims in the interaction between autonomous navigations using object detection. In such a problematic scenario, the robot has to detect and recognize objects as well as estimate their position by avoiding them. Although Object Detection and Recognizing are largely used in recent studies, most of them typically assume that the object is either already segmented from the background or that it occupies a large portion of the image. In ADR, to locate an object in any environment is very important because neither of the above predictions is correct since the distance to the object and also its size in the image can vary significantly. Therefore, the robot has to detect objects while moving with various frame rates and with the same trained object with different angles and sizes. This paper especially denotes to this problem and methods.

To module this complication, the proposed methodology uses Faster R-CNN having VGGNet-16 architecture for object detection that is especially suitable for detecting objects in a natural environment with complex systems [2], as it is able to cope with problems such as complex model, varying pixels and object perceptions. The construction of this type of system hasn't studied yet. Faster R-CNN uses selective search to find out the region proposals within an object. So, Faster R-CNN is much faster than its predecessors [2]. On the other hand, VGGNet-16 is a 16-layer model which shows the classification/localization accuracy within very little complex modeling. Therefore, it can be improved by increasing the depth of CNN in spite of using small receptive fields in the layers. For Alpha-N, a combination of both early learners and lower complexity with High Frame per Seconds (FPS) is considered for choosing the architecture. Faster R-CNN and VGG Net with 16 layers provide good results for this purpose which is illustrated in Sect. 5.

Due to the complex setup of the outdoor or even in the indoor environment, the path planning needs to recalibrate once every one cycle of planning is successfully completed. A possible solution for this recirculation is proposed where modeling approaches have been using as a cognitive map. Also, simultaneous localization and mapping (SLAM) and host localizations are proposed in existing studies [3, 4]. But semantic and geometrical aspects are important for PPS with influential output from ODM. The ADR uses the vector map of the path between source and destination and calculates the shortest path using Dijkstra's Shortest Path Algorithm (GCM).

RFID Reader/Writer RC522 SPI S50 is considered for the operation of RRS. The output of the ODM and ODM controls the speed of the motor to scan and read the RFID tag. So, altogether Alpha-N combines the idea of an ADR having different object detection, avoiding, pathfinding algorithms to provide a solid fundamental improvement in the field of Industrial Robotics for Bangladesh.

2 Related Work

This paper extending the Previous work done by Neloy et al. [5]. Previous work illustrates an Automated delivery robot made with RFID Scanner and Ultrasonic Sensor that can find its path from one point to another while avoiding obstacles, without any lines to follow. Although there exists a large number of works related to mobile robots with an object detection system or autonomous movement [6, 7]. But there are still no fully operational systems that can operate robustly and long-term in indoor and outdoor environments combining obstacles detection and avoiding system. The current trend in the development of robots is divided onto single parts where a particular problem is studying. But the approach presented in this paper occupies both object detection along with autonomous navigation hand to hand.

There are some examples of such systems [8, 9] where the robot is able to acquire and facilitate autonomous movement. Different to our methods, the works presented here [10], is a method of simultaneous localization and mapping (SLAM) to integrate both object detection and autonomous floor mapping. Along with SLAM, a Panoramic scan that automatically searches and generates the most precise and fast route which enables the robot to realize autonomous path planning path and so reaches the destination with the shortest route around obstacles [11]. A navigation system without a Grid map is also studied by [12]. These Autonomous robots are capable of move freely in a given circumstance, but most of them are not integrated with the idea of locating the distance between source and destination [13]. Also, previous work proposed by Mac et al. [14] isn't capable of adopting the idea of autonomous navigation by detecting obstacles from streaming or live video.

Most of the work on Dijkstra's algorithm resolves the pathfinding in a contained track. Also, the pathfinding problem exists in an environment where obstacles and constraints are situated statically. Kümmerle et al. presented a navigation system for mobile robots that are designed to operate in crowded city environments and pedestrian zones [3]. They generated the local occupancy map by applying Dijkstra's algorithm to compute the distance to the nodes of the SLAM graph. A novel hierarchical global path planning approach for mobile robots in a cluttered environment is proposed by Morales et al. [4]. The authors' proposed Dijkstra's algorithm to find a collision-free path used as input reference for the next level. This study is very similar to ours but no module is offered to verify object location. Another study by Nazarahari et al. proposed a hybrid approach for path planning of multiple mobile robots in continuous environments [15]. The proposed path length, smoothness, and safety are combined to form a multi-objective path planning problem. This study is also incomputable with autonomous movement in large complex systems.

The ADR we are presenting in this work needs to rank or navigate on the path based on the robust outliers of object recognition and avoid dynamic objects. No study is conducted where the robot both uses path planning consists of the shortest pathfinding and detecting, avoiding objects in that particular path. This paper bridges the gap between autonomous navigation accessing the shortest path passively observing the obstacles, viewing conditions and verifying by RFID tags in any dynamic environments.

Object recognition and detection have been extensively improved over a few years. Most of the approach is taken in deep learning techniques, among them, Faster R-CNN, Faster R-CNN, VGG Net are mostly studied which are region proposal based popular methods [2, 16]. An object detection algorithm based on its color for the control of a mobile robotic platform with Android in a mobile robot is already studied by

Gonzalez et al. [17]. Vision techniques are introduced to produce this work. Holz et al. propose a system for depalletizing and a complete pipeline for detecting and localizing objects as well as verifying that the acquired object does not deviate from the known object model [18]. They demonstrated the depalletizing of automotive and other pre-fabricated parts with both high reliability and efficiency. All of the states of art models symbolize the local method of object detection and recognizing. To do so, the object must appear large enough in the camera image or local features cannot be extracted from it. So, these models aren't capable to work on moving robots or may work with very little efficiency and accuracy. Moreover, the existing detector frameworks aren't built to generate any feedback to support path planning or localization. One of the contributions in this paper is the dynamic system that makes uses both approaches in a combined framework of path planning and detecting objects that let the methods complement each other.

The general idea of the overall system proposed in this paper is to learn action policies for the robot to locate a user-specified target destination from an initial point and reach the destination while avoiding obstacles by following an adequate algorithm. Also, it can identify the source and destination by RFID scanning. After delivering the items, the robot will backtrack its path to return on its source. Also, this extending study is improving both navigation, path analysis and can add a major contribution in both the hybrid navigation system constructed by Dijkstra's algorithm and avoiding obstacles by detecting objects. The outcome from both object detection and navigation provides precise performance for both indoor and outdoor environments.

3 Shortest Path Calculation by Vector Mapping and Dijkstra's Algorithm

The floor on which the Alpha-N will work on is first modeled into a graph. Each room is a node. Dijkstra's shortest path algorithm is modified to find the shortest distance between two nodes. First, Alpha-N needs a distance between source and destination. Suppose the Source is and the destination is . The Alpha-N will calculate the shortest distance by Vector Mapping using the following algorithm [19]-

Step 1. The pathfinder introduces an Auxiliary Vector D, each of the component $D[i]$, the shortest path is discovered in the current starting point U, to each end V_i, then the $D[i]$ is the weight of the arc, otherwise $D[i] \rightarrow \infty$ and the length is:

$$D[i] = M_i^{in}\{D[i]|U_i \in V\} \tag{1}$$

This is the shortest path from U to V and stats as - $P(U_i, V)$.

Step 2. If the adjacency matrix arc with weights represent the directed graph and arc of the graph doesn't exist, then the arc [i][j] will be ∞. The initial value from map U to map V remains as V_i and possible shortest path:

$$D[i] = arcs[Locate\ Vex(G, v)[i]]vi \in V \tag{2}$$

Step 3. If the shortest path changes from the current V_i to V_K, then the path will be updated as:

$$D[j] + arcs[j][k] < D[k] \tag{3}$$

Repeat until it reaches the initial state and updates the equation as:

$$D[k] = D[j] + arcs[j][k] \qquad (4)$$

Using this updated $D[k]$ value, Dijkstra's Algorithm gets the weighted value of each node from source to destination. So, the value is denoting path cost for each decision. Shortest Path Grid Movement through Vector Map (Source to Destination and Destination to Source) is based on the 2-D grip. It defines the row and column between source and distance. So, this method calculates row to row and column to column distance to measure the actual path (Fig. 1).

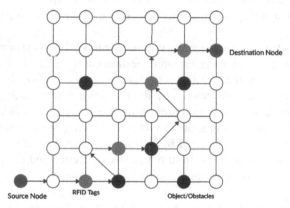

Fig. 1. A sample path following implementation using vector mapping and Grid Count from Eqs. 1 to 4.

4   Principles of the RFID Reading System (RRS)

The RFID tags can store information which supplies to any reader that is within a proximity range which can be up to approximately 15 m. A special type of RFID Tag is used in this research which contains 6-digit. Based on the pattern of the Digit, the robot can make decisions. A brief classification of the tag digits is discussed in Table 1.

Table 1. RFID Tag values

Position ($xx - yy - zz$)	Values		Decision
xx	xx values always 00		Initial or source node
yy	00	01	The current node is situated just after the source node,
	01	10	Middle node
	10	11	The destination node is after this current node
zz	zz values always 11		Destination node

Based on the values, if the tag value is *00-01-11*, Alpha-N is in the source node, if *00-10-11* then Alpha-N is reached the destination and if 01-10-11 Alpha-N is in the path from source to destination.

5 Construction of ODM

The standard process of object detection and recognition for this research is consist of three steps:

- Detect the candidate regions of the object from the live streaming.
- Predict the class label of each region using bounding box regression and provide the final output.
- Denote each RPI to the next output for the path planning system.

However, there is two problem statement required for this task. The active perceiver setting for the models needs to detect specific objects for avoiding them. Following this observation, the authors adopt a deep neural network with leaser latency that simply takes account of the problem statements [20].

Convolutional Networks and Region Proposal Network (RPN). Based   on   the CAFFE framework [21], the output size of each convolutional and pooling layer can be calculated precisely by the following Eqs. 5 and 6

$$output_{size} = \left\lfloor \frac{input_{size} + 2 \times pad - \left[dilation \times \left(kernel(k)_{size} - 1\right) + 1\right]}{stride} \right\rfloor + 1$$

(5)

$$output_{pool} = \left\lceil \frac{input_{size} + 2 \times pad - kernel(k)_{size}}{stride} \right\rceil + 1 \qquad (6)$$

where $\lfloor \rfloor$ and $\lceil \rceil$ are denoted as the floor and ceil function, respectively. For generating region proposals, each kernel location, $k = 9$, anchor boxes are used with 3 scales of 128, 256 and 512 values having 3 aspect ratios of $1:1, 1:2, 2:1$. Also, the area for each pooling area $(middle) = \frac{h}{H} \times \frac{w}{W}$ is obtained from the proposal [2]. Equation 5 and 6) provide the output area for each pooling 2×3 or 3×3 after rounding from the input Region of interest pooling (ROI) of 8×8. So, the output area for each pooling is 3×3 from the model.

Classes and Bounding Boxes Prediction. The final network predicts object class (classification) and Bounding boxes (Regression) from the model. This final output is also optimized by Stochastic Gradient Descent (SGD) which minimizes the convolution layers, RPN weights, and fully connected NN weights [2]. The depth of the feature map is 32 (9 anchors $\times$ 4 positions). Smooth-L1 loss function on the position (x, y) having $top - left$ of the box (Eqs. 7 and 8) is proposed in this paper based on Eqs. 5 and 6)

$$L_{loc}\left(t^u, v\right) = \sum_{i \in \{x,y,w,h\}} smooth_{L_1}\left(t_i^u - v_i\right) \tag{7}$$

$$\text{in which, } smooth_{L_1}(x) = \begin{cases} 0.5x^2, \ if |x| < 1 \\ |x| - 0.5 \, otherwise, \end{cases} \tag{8}$$

The Classification and regression loss is the combination of the overall loss of RPN. The relation between this loss function and model accuracy is showed in Fig. 3.

5.1 Sample Dataset and Basic Setup

This paper uses the dataset of NYU V2 [22] for training and SUN RGB-D [23] for testing the model. The datasets are collected using an RGB-D camera in several indoor environments. A basic preprocessing and sample experimental setup are followed by previous work of van Beers et al. [24]. A total number of 1400 of unique examples with 85 Classes are trained for the model. Figure 2 shows sample outputs from the model after the training phase. The output is primarily displayed or tested trough Pi camera.

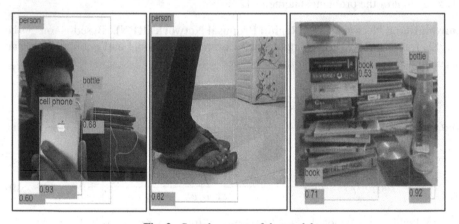

Fig. 2. Sample output of the model.

5.2 Training Parameter Setting and Accuracy

The pre-trained VGG16 model on ImageNet and Caffe Framework are applied during the training phase to initialize the parameters which are the forward channel of the network. The coefficient of the $smooth_{L_1} the$ loss function is set to 1, the learning ratio is set to 0.0075 after several experiments. Figure 3 shows the accuracy observation through training and testing curve. Overall accuracy is compared with two closely related papers. This score is recorded after test samples are fed to the model and the number of mistakes (zero-one loss) the model makes is accuracy. The problem of overfitting is minimized for this model by finding the optimum iterations with maximum accuracy and stops the epoch where when the accuracy is not improving, given a threshold. Previous work from Cho et al. [25] started this whole process.

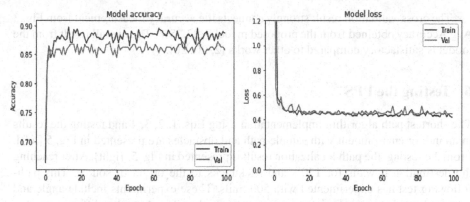

Fig. 3. Model accuracy [left] and model loss [right].

Figure 3 exhibits early learner trends in the accuracy curve. The model could be trained more as the trend for accuracy on both curves is still rising for the last few epochs. Also, the model has not yet over-learned the training examples of boxplot, showing comparable skill on curve rising. Moreover, from the loss function graph, a relatively good comparable performance can be seen. the training might stop according to $smooth_{L_1}$ if it finds an overfitting score. Finally, the validation of the model is inspected by *Precision* and *Recall Curve* with metric proposed by Rezatofighi et al. [26]. Table 2 and Fig. 4. show all the results derived from the validation process.

Summary Statistics	
Number of Cases	88
Number Correct	67
Accuracy	83.75
Sensitivity	87%
Specificity	82%
Pos Cases Missed	5
Neg Cases Missed	8
Fitted ROC Area	0.814
Empiric ROC Area	0.865

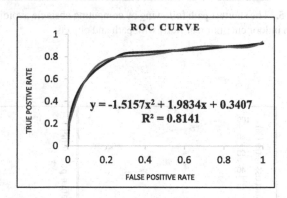

$$y = -1.5157x^2 + 1.9834x + 0.3407$$
$$R^2 = 0.8141$$

Fig. 4. ROC curve observation with statistics.

Table 2. 2-Fold cross-validation results of Faster R-CNN

n-Fold	Recall	Precision	Avg recall	Avg precision	FP	FN
1st fold	83.96	83.96	83.94	83.95	2110.2	2111.5
2nd fold	83.92	83.93				

The cross-validation result strongly connects the accuracy measurement from Fig. 3 Avg Accuracy obtained from the proposed model is 83.75%. The overall result from the model is satisfactory compared to other works [2, 20, 26].

6 Testing the PPS

The shortest path algorithm implementation using Eqs. 1, 2, 3, 4 and testing the results in an indoor environment with sample-path and obstacles are presented in Fig. 5 [left]. From the testing, the path localization result is presented in Fig. 5 [right]. After reaching the destination it waits for 120 s and backtracks to the origin or source. This path-following test has experimented with 500 trails. These experiments include angle and distance error in terms of relative path and tag number (Fig. 6) with a different number of grids (Fig. 7). The results exhibit the overall accuracy and performance of the PPS.

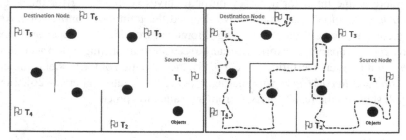

Fig. 5. The shortest path following by computing obstacle avoiding algorithm [right] is simulated in an indoor environment [left] with path and objects.

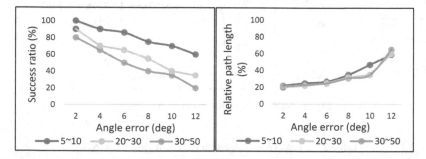

Fig. 6. 3 experiments constructed with 500 trials having a different number of RFID Tag numbers (5–50). Success rate demonstrates in terms of angle [left] and path distance [right]. The figure shows the best performance (92.3%) with a 20–30 tag number and a relative path distance of 60 m.

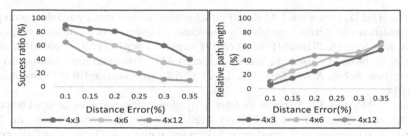

Fig. 7. 3 experiments constructed with 500 trials by varying the *grid number* (4–12). Success rate demon-states in terms of traveled Distance. The figure shows the best performance (85.3%) with 4–3 *grids* and a relative distance of 40 m.

7 Conclusion and Future Work

This paper presents Alpha-N, an ADR with Shortest path Finder using Grid Count Algorithm, object or obstacle detecting module using Faster R-CNN with VGGNet-16, obstacle avoiding algorithm build with Improved Dynamic Window Approach (IDWA) and an Artificial Potential Field (APF) along with object detection. The primary objective of this research is to present a robot, which contributes to the research and industrial aspects of the robotics domain in Bangladesh and also in developing countries. It is shown that the overall experimental results of the system show satisfactory results and outcomes. Also, the results and working principle can be improved by examining the different environments and deploy on a large scale of entities.

In the future, all the frameworks will be tested in a hazardous environment with different scenarios to improve the sustainability of the system. More improvement of the design, framework, and modules will be studied as well.

References

1. Islam, M.Z.: Meals by mail getting popular. https://www.thedailystar.net/business-/food-panda-bd-pathao-online-food-delivery-services-getting-popular-1714402
2. Ren, S., He, K., Girshick, R., Sun, J.: Faster R-CNN: towards real-time object detection with region proposal networks. IEEE Trans. Pattern Anal. Mach. Intell. **39**, 1137–1149 (2017). https://doi.org/10.1109/tpami.-2016.2577031
3. Pillai, S., Leonard, J.: Monocular SLAM supported object recognition. https://arxiv.org/abs/1506.01732. https://doi.org/10.15607/rss.2015.xi.034
4. Morales, Y., Carballo, A., Takeuchi, E., Aburadani, A., Tsubouchi, T.: Autonomous robot navigation in outdoor cluttered pedestrian walkways. J. Field Robot. **26**, 609–635 (2009). https://doi.org/10.1002/rob.20301
5. Neloy, A.A., Arman, A., Islam, M.S., Motahar, T.: Automated mobile robot with RFID scanner and self obstacle avoiding system. Int. J. Pure Appl. Math. **118**(18), 3139–3150 (2018)
6. Wang, L., et al.: Multi-channel convolutional neural network based 3D object detection for indoor robot environmental perception. Sensors **19**(4), 893 (2019). https://doi.org/10.3390/s19040893
7. Choi, J.-H., Choi, B.-J.: Indoor moving and implementation of a mobile robot using hall sensor and Dijkstra algorithm. IEMEK J. Embed. Syst. Appl. **14**(3), 151–156 (2019). https://doi.org/10.14372/-IEMEK.2019.14.3.151

8. Santos, D.H.D., Goncalves, L.M.G.: A gain-scheduling control strategy and short-term path optimization with genetic algorithm for autonomous navigation of a sailboat robot. Int. J. Adv. Rob. Syst. **16**, 172988141882183 (2019). https://doi.org/10.1177/1729881418821830

9. Oishi, S., Inoue, Y., Miura, J., Tanaka, S.: SeqSLAM: view-based robot localization and navigation. Robot. Auton. Syst. **112**, 13–21 (2019). https://doi.org/10.1016/j.robot.2018.10.014

10. Kato, Y., Morioka, K.: Autonomous robot navigation system without grid maps based on double deep Q-network and RTK-GNSS localization in outdoor environments. In: 2019 IEEE/SICE International Symposium on System Integration (SII) (2019). https://doi.org/10.1109/SII.2019.8700426

11. Wang, L.: Automatic control of mobile robot based on autonomous navigation algorithm. Artif. Life Robot. (2019). https://doi.org/10.1007/s10015-019-00542-0

12. Chou, J.-S., Cheng, M.-Y., Hsieh, Y.-M., Yang, I.-T., Hsu, H.-T.: Optimal path planning in real time for dynamic building fire rescue operations using wireless sensors and visual guidance. Autom. Constr. **99**, 1–17 (2019). https://doi.org/10.1016/j.autcon.2018.11.020

13. Murtra, A.C., Tur, J.M.M., Sanfeliu, A.: Action evaluation for mobile robot global localization in cooperative environments. Robot. Auton. Syst. **56**, 807–818 (2008). https://doi.org/10.1016/j.robot.2008.06.009

14. Mac, T.T., Copot, C., Tran, D.T., Keyser, R.D.: Heuristic approaches in robot path planning: a survey. Robot. Auton. Syst. **86**, 13–28 (2016). https://doi.org/10.1016/j.robot.2016.08.001

15. Nazarahari, M., Khanmirza, E., Doostie, S.: Multi-objective multi-robot path planning in continuous environment using an enhanced genetic algorithm. Expert Syst. Appl. **115**, 106–120 (2019). https://doi.org/10.1016/j.eswa.2018.08.008

16. Girshick, R., Donahue, J., Darrell, T., Malik, J.: Rich feature hierarchies for accurate object detection and semantic segmentation. In: IEEE Conference on Computer Vision and Pattern Recognition (CVPR) (2014). https://doi.org/10.1109/CVPR.2014.81

17. Gonzalez, E.M.A., Hermosilla, D.M.: Object detection algorithm for a mobile robot using Android. ITEGAM- J. Eng. Technol. Ind. Appl. (ITEGAM-JETIA) **4** (2018). https://doi.org/10.5935/2447-0228.201828

18. Holz, D., Topalidou-Kyniazopoulou, A., Rovida, F., Pedersen, M.R., Kruger, V., Behnke, S.: A skill-based system for object perception and manipulation for automating kitting tasks. In: 2015 IEEE 20th Conference on Emerging Technologies & Factory Automation (ETFA) (2015). https://doi.org/10.1109/ETFA.2015.7301453

19. Wang, H., Yu, Y., Yuan, Q.: Application of Dijkstra algorithm in robot path-planning. In: 2011 Second International Conference on Mechanic Automation and Control Engineering (2011). https://doi.org/10.1109/mace.2011.5987118

20. Redmon, J., Divvala, S.K., Girshick, R.B., Farhadi, A.: You only look once: unified, real-time object detection. CoRR, abs/1506.02640 (2015)

21. Komar, M., Yakobchuk, P., Golovko, V., Dorosh, V., Sachenko, A.: Deep neural network for image recognition based on the Caffe framework. In: 2018 IEEE Second International Conference on Data Stream Mining & Processing (DSMP) (2018). https://doi.org/10.1109/dsmp.2018.8478621

22. Silberman, N., Hoiem, D., Kohli, P., Fergus, R.: Indoor segmentation and support inference from RGBD images. In: Fitzgibbon, A., Lazebnik, S., Perona, P., Sato, Y., Schmid, C. (eds.) ECCV 2012. LNCS, vol. 7576, pp. 746–760. Springer, Heidelberg (2012). https://doi.org/10.1007/978-3-642-33715-4_54

23. Song, S., Lichtenberg, S.P., Xiao, J.: SUN RGB-D: a RGB-D scene understanding benchmark suite. In: 2015 IEEE Conference on Computer Vision and Pattern Recognition (CVPR) (2015). https://doi.org/10.1109/cvpr.2015.7298655

24. Van Beers, F., Lindström, A., Okafor, E., Wiering, M.: Deep neural networks with inter-section over union loss for binary image segmentation. In: Proceedings of the 8th International Conference on Pattern Recognition Applications and Methods (2019). https://doi.org/10.5220/0007347504380445
25. Cho, S., Baek, N., Kim, M., Koo, J., Kim, J., Park, K.: Face detection in nighttime images using visible-light camera sensors with two-step faster region-based convolutional neural network. Sensors **18**, 2995 (2018). https://doi.org/10.3390/s18092995
26. Rezatofighi, H., Tsoi, N., Gwak, J., Sadeghian, A., Reid, I., Savarese, S.: Generalized inter-section over union: a metric and a loss for bounding box regression. In: Proceedings of the IEEE Conference on Computer Vision and Pattern Recognition, pp. 658–666 (2019)

Image Representation for Cognitive Systems Using SOEKS and DDNA: A Case Study for PPE Compliance

Caterine Silva de Oliveira[1](✉), Cesar Sanin[1](✉), and Edward Szczerbicki[2](✉)

[1] The University of Newcastle, Newcastle, NSW, Australia
caterine.silvadeoliveira@uon.edu.au,
cesar.maldonadosanin@newcastle.edu.au
[2] Gdansk University of Technology, Gdansk, Poland
edward.szczerbicki@newcastle.edu.au

Abstract. Cognitive Vision Systems have gained significant interest from academia and industry during the past few decade, and one of the main reasons behind this is the potential of such technologies to revolutionize human life as they intend to work under complex visual scenes, adapting to a comprehensive range of unforeseen changes, and exhibiting prospective behavior. The combination of these properties aims to mimic the human capabilities and create more intelligent and efficient environments. Nevertheless, preserving the environment such as humans do still remains a challenge in cognitive systems applications due to the complexity of such process. Experts believe the starting point towards real cognitive vision systems is to establish a representation which could integrate image/video modularization and virtualization, together with information from other sources (wearable sensors, machine signals, context, etc.) and capture its knowledge. In this paper we show through a case study how Decisional DNA (DDNA), a multi-domain knowledge structure that has the Set of Experience Knowledge Structure (SOEKS) as its basis can be utilized as a comprehensive embedded knowledge representation in a Cognitive Vision System for Hazard Control (CVP-HC). The proposed application aims to ensure that workers remain safe and compliant with Health and Safety policy for use of Personal Protective Equipment (PPE) and serves as a showcase to demonstrate the representation of visual and non-visual content together as an experiential knowledge in one single structure.

Keywords: Cognitive vision systems · Knowledge representation · SOEKS · DDNA · PPE compliance · Hazard control

1 Introduction

Cognitive Vision Systems have gained considerable interest from academia and industry during the past few decade, and one of the main reasons behind this is the potential of such technologies to revolutionize human life as they intend to work under complex visual scenes, adapting to a comprehensive range of unforeseen changes, and exhibiting

© Springer Nature Switzerland AG 2020
N. T. Nguyen et al. (Eds.): ACIIDS 2020, LNAI 12033, pp. 214–225, 2020.
https://doi.org/10.1007/978-3-030-41964-6_19

prospective behavior [1]. The combination of these properties aims to mimic the human capabilities and create more intelligent and efficient environments [2].

Nonetheless, preserving the environment such as humans do still remains a challenge in cognitive systems applications due to the complexity of such process. It involves understanding the context and gathering visual and other sensorial information available and translating it into knowledge to be useful. Moreover, past experiences also plays an important role when it comes to perception [3] and must also be considered as an important element in this process. Smart cognitive systems that have been proposed so far oversight the potential of using these experiences to enrich the application with smartness while, at the same time, creating decisional fingerprints. This would allow the system knowledge growth through daily operation autonomously, just like human experience do in real life [4].

Experts believe the starting point towards real cognitive vision systems is to establish a representation which could integrate image/video modularization and virtualization, together with information from other sources (wearable sensors, machine signals, context, etc.) and capture its knowledge. In this context, Decisional DNA (DDNA), a multi-domain knowledge structure based on experience, has been extended to the visual domain to be used as a comprehensive embedded knowledge representation for Cognitive Systems [5]. DDNA has the Set of Experience Knowledge Structure (SOEKS) [6] as its basis and allow the creation of a multi-modal space composed of information from different sources, such as contextual, visual, auditory etc., in a form of a structure and explicit experiential knowledge [7].

The applicability of such representation have been tested over a Cognitive Vision Platform for Hazard Control (CVP-HC). The CVP-HC is scalable yet adaptable platform capable of working in a variety of video analysis scenarios whilst meeting specific safety requirements of industries [8]. This platform aims to assist the safety management process in industrial environments, and the special case of PPE compliance is presented in this paper.

This paper is organized as follow: In Sect. 2, some fundamental concepts are presented, including the evolution of systems towards augmented cognitive technologies and the challenge of representation and management of knowledge in these systems. The proposed representation based on SOEKS and DDNA is also explained. In Sect. 3 a case study for the case of PPE compliance is presented, including its applicability, design and experimental results achieved so far. Finally, in Sect. 5 conclusions and future work is presented.

2 Fundamental Concepts

In order to offer a more complete view, we briefly introduce concepts that have driven the proposed research as well as the technologies involved.

2.1 From Computer Vision to Cognitive Vision Systems

The use of computer vision techniques can support automatic detection and tracking of objects and people with reasonable accuracy [9–13]. Visual sensing facilities, such as video cameras can gather a large amount of data, such as video sequences or digitized visual information that, with support of machine learning technologies and powerful machines, can operate in real time [14]. For those reasons, computer vision systems have been a research focus for a long time in surveillance systems, human detection, and tracking.

However, computer vision systems have their own inherent limits, especially those whose task is to work in unidentified environments and deal with unknown scenarios and specifications. Besides the significant improvements in computer vision technologies, they are still challenged by issues such as occlusion or position accuracy; and background changes result in the necessity of adapting the algorithms for different conditions, clients and situations. To date, the creation of a general-purpose vision system with the robustness and resilience comparable to human vision still remains a challenge [13].

In this context, methods incorporating prior knowledge and context information have gained interest. The understanding about scene composition in an image (which set of objects are present) can improve recognition performance about the scene where they are inserted [15]. For instance, the presence of multiple cutlery items in an image can aid the recognition of a kitchen image. This relationship is held both ways, as contextual knowledge can also offer insights about the function of an object in a scene, reducing the impacts of sensor noise or occlusions [16]. These technologies are known as knowledge-based systems. For instance, an automatic semantic and flexible annotation service able to work in a variety of video analysis with little modification to the code using Set of Experience Knowledge Structure (SOEKS) was proposed in work by Zambrano et al. [17]. This system is a pathway towards cognitive vision and it is composed, basically, by the combinations of detection algorithms and an experience based approximation.

The design of a general-purpose vision system with the robustness and resilience of the human vision is still a challenge. One of the latest trends in computer vision re-search to mimic the human-like capabilities is the joining of cognition and computer vision into cognitive computer vision. Cognitive Systems have been defined as "a system that can modify its behavior on the basis of experience" [18]. Although, most experts tend to agree that such systems only exists in theory, that is, systems that can independently process, reason and create in the same capacity as the human brain has not yet been implemented successfully [19].

In this scenario, the concept of Augmented Intelligence, also known as Cognitive Augmentation or Intelligence Amplification (IA) comes into play [20]. For any specific application humans being and machines have both their own strengths and weaknesses. Machines are very efficient in numerical computation, information retrieval, statistical reasoning, with almost unlimited storage. Machines can capture many categories of information from the environment through various sensors, such as range sensors, visual sensors, vibration sensors, acoustic sensors, and location sensors [21]. On the other hand, humans have their own cognitive capabilities which includes consciousness, problem-solving, learning, planning, reasoning, creativity, and perception. These cognitive functions allows humans to learn from last experiences and use this experiential knowledge to adapt to new situations and to handle abstract ideas to change their environment. Therefore, the combination of both human experiential knowledge and information collected by a system can be used to enhance smartness of systems and for improved decision making [22]. Figure 1 shows the steps towards Augmented Cognitive Vision and a synthesis of components involved in each stage.

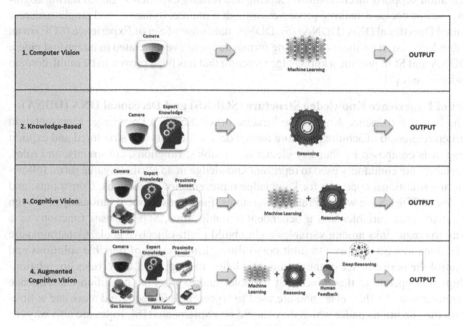

Fig. 1. Steps towards Augmented Cognitive Vision.

2.2 Knowledge Representation for Cognitive Systems

The implementation of cognitive vision systems require the design of functionalities for knowledge engineering (acquisition and formalism), recognition and categorization, reasoning about events for decision making, and goal specification, all of which are concerned with the semantics of the relationship between the visual agents and their environments i.e. context [2]. These functionalities direct cognitive vision systems towards purposeful behavior, adaptability, anticipation, such as human beings.

In this context, knowledge and leaning are central to cognitive vision. To be readily articulated, codified, accessed and shared, knowledge must be represented in an explicit and structured way [23]. In addition, the choice of a suitable representation greatly facilitates obtaining methods that efficiently learn the relevant information available. Therefore, an appropriate knowledge representation is crucial for the success in designing of cognitive systems.

Nevertheless, most approaches that have been proposed on past years, even though they present some principles for intelligent cognitive vision, they fail in providing a unique standard that could integrate image/video modularization, its virtualization, and capture its knowledge [6]. To address these issues an experience-based technology that allows a standardization of image/video and the entities within together with any other information as a multi-source knowledge representation (required for the further development of cognitive vision) without limiting their operations to a specific domain and/or following a vendor's specification has been proposed [24]. This representation supports mechanisms for storing and reusing experience gained during cognitive vision decision-making processes through a unique, dynamic, and single structure called Decisional DNA (DDNA) [5]. DDNA makes use of Set of Experience (SOE) in an extended version for the use of storing formal decision events related to image and video. DDNA and SOE provide a knowledge structure that has been proven to be multi-domain independent [7].

Set of Experience Knowledge Structure (SOEKS) and Decisional DNA (DDNA).
The Set of Experience Knowledge Structure (SOEKS) is a knowledge representation structure created to acquire and store formal decision events in a structured and explicit way. It is composed by four key elements: variables, functions, constraints, and rules. Variables are commonly used to represent knowledge in an attribute-value form, following the traditional approach for knowledge representation. Functions, Constraints, and Rules of SOEKS are ways of relating variables. Functions define relationships between a set of input variables and a dependent variable; thus, SOEKS uses functions as a way to create links among variables and to build multi-objective goals. Constraints are functions that act as a way to limit possibilities, limit the set of possible solutions and control the performance of the system in relation to its goals. Lastly, rules are relationships that operate in the universe of variables and express the condition-consequence connection as "if-then-else" and are used to represent inferences and associate actions with the conditions under which they should be implemented [6]. Rules are also ways of inputting expert knowledge into the system. The Decisional DNA consists is a structure capable of capturing decisional fingerprints of an individual or organization and has the SOEKS as its basis. Multiple Sets of Experience can be collected, classified, organized and then grouped into decisional chromosomes, which accumulate decisional strategies for a specific area of an organization. The set of chromosomes comprise, finally, what is called the Decisional DNA (DDNA) of the organization [5].

3   Case Study: PPE Safety Compliance

Hazards are present in all workplaces and can result in serious injuries, short and long-term illnesses, or death [25]. Reports HSE UK report has shown that over 80% of reported workplace injuries are sustained due to a person not wearing correct protective clothing [26]. In this context, the verification of PPE compliance becomes essential in the management of safety to ensure the occupational health of workers. Technologies to support its practical and automated implementation have emerged as a need, but the current technologies available still face considerable limitations [9, 13, 15, 27].

The combination of vision and sensor data together with the resulting necessity for explicit and formal representations builds a central element of an autonomous system for detection and tracking of laborers in workplaces environments. To be able to perform in a variety of plants and scenes, making sure employees remain safe and compliant with Health & Safety policy without the necessity of recoding the application for each specific case scenario, the system must be adaptable and perceive the environment as automatically as possible and change its behavior accordingly. However, computer vision systems have their own inherent limits, especially those whose task is to work in unidentified environments and deal with unknown scenarios and specifications [28].

The gaps of current systems may be filled by connecting the probabilistic area of detection of events with the logical area of formal reasoning in a Cognitive Vision Platform for Hazard Control (CVP-HC) [28]. This platform verifies the PPE compliance in variety of video analysis scenarios whilst meeting specific safety requirements of industries [24]. The proposed system is based on the Set of Experience Knowledge Structure (SOEKS or SOE in short) and Decisional DNA (DDNA).

3.1 Applications

Automated verification of PPE compliance can be useful in a variety of industries (e.g. Oil & Gas, Manufacturing & Production, Construction, Engineering, Pharmaceuticals, etc.) and applied in a range use case scenarios to ensure employees remain safe [29]. Below we exemplify two main applications that the proposed solution can address.

Access Control. With cameras positioned above an entrance/exit of a site or facility, the system is able to visually verify that laborers are wearing the protective equipment according to the safety requirements of that industry/area before allowing entry. In case of any equipment being missed at the point of entry, then the system will not permit a gate to open and will advise which items must be worn in order to enable access. Once all the mandatory equipment are detected the access is granted. The visual information from the cameras can be combined with other sensor data to give extra information about crucial required equipment (e.g. oxygen mask when oxygen level is critically low).

Continuous Monitoring. Another solution can address the continuous monitoring of works by the use of cameras and other sensor data covering the site or facility to ensure that employees remain wearing the required PPE in a given context. If laborers remove a required equipment then the system will recognize this in real-time and carry out an action based on a set of given preferences or recommendations. For instance, an alert can be sent directly to the employee or manager for correction on site; the event can be logged for future reports and analysis, etc. If sensors detect any abnormality, which changes the status of the required equipment, workers can also be advised of that for a quick action.

3.2 Representation of Variables, Constrains, Functions and Rules

For the case study in analysis, a set of variables, functions, constrains and rules are represented as a Set of Experience Knowledge Structure (SOEKS). SOEKS allows the representation, use, storing and retrieval of visual and non-visual knowledge content together in one single standardized structure [24].

Variables. The variables in our system are composed by each image/frame being analyzed, body parts of workers, and annotations of each Personal Protective Equipment

(PPE). In addition, we include, as part of the set of variables, the calculation of area of intercept A_I between the bounding boxes containing a body part and a corresponding PPE, as well as the area of each PPE in the scene, which is defined respectively by:

$$A_I = \begin{aligned} & max(0, min(ppe_{xmax}, bp_{xmax}) - max(ppe_{xmin}, bp_{xmin}))* \\ & max(0, min(ppe_{ymax}, bp_{ymax}) - max(ppe_{ymin}, bp_{xmin})) \end{aligned} \tag{1}$$

$$A_{ppe} = ((ppe_{xmax} - ppe_{xmin}) * (ppe_{ymax} - ppe_{ymin})) \tag{2}$$

Finally, the last two variables considered are: the dependent variable resulting from the creation of the overlap function $O_{I,ppe}$ (Eq. 3), and the safety status of scene, to be defined by the set of rules. Both variables will explained in the following subsections.

Functions. As defined before, function establishes relationships among input and dependent variables as a way to find more elements of decision-making that reduce the possibility of duality, while facilitating knowledge elicitation [6]. In our application, for each body part of a person detected there may be a range of compatible surrounding PPEs that can be associated with it, including ones belonging to other people in the scene. For instance, let's imagine a scene where two people are being detected, one is wearing a respirator and another one is not (the second's person respirator is placed next to them, on the floor). In this case we have four interceptions being computed and inputted into the system, producing different states in relation to the safety status of the scene. In this case, it is necessary to reduce the possibilities of duality in finding an optimal unique set of variables that identifies a unique state while reducing ambiguity [6]. Therefore, we calculate the overlap between the areas of intercept A_I and PPEs A_{ppe} as a function, which objective is to maximize the area of overlap, associating the PPE to the closest conforming body part.

The maximum overlap $O_{I,ppe}$ between intercept and corresponding PPE goes from 0 (disjoint) to 1 (complete overlap) is calculated as following:

$$O_{I,ppe} = \left\{ [max] \frac{A_I}{A_{ppe}} \right\} \tag{3}$$

Table 1 shows values of maximum $O_{I,helmet}$ for a sequence of frames and the status of *wearing/not wearing* associated with them.

Constraints. In our analysis, we only consider the XY plane, i.e. no depth information is taking into consideration. When not taking the Z plane, protective equipment on the background may be wrongly associated with the body parts even being meters distant on the depth plane and vice versa. To minimize the set of possible misleading associations of body parts and PPEs that are distant from each other on the Z plane, we create a set of constraints. These constraints restrict the possible size of the PPE that can be associated with each body part being detected.

Rules. To ensure flexibility and as well as to attend each specific requirements of different industries and scenarios, a set of rules is created. These rules are also a way of allowing expert knowledge to be included in the system reasoning as they can be easily changed and adjusted to attend specific requisites and situations. For this analysis in specific, the following set of rules are considered:

Table 1. Examples of [max] $O_{I,\text{helmet}}$ and respective wearing/not wearing status.

Frame					
$[max]O_{I,\text{helmet}}$	0.49	0.44	0.40	0.00	0.00
Wearing helmet?	YES	YES	YES	NO	NO

Rule 1:
IF $O_{I,respirator}$ > threshold
  THEN safety_status = SAFE
ELSE safety_status = UNSAFE

Rule 2:
IF $O_{I,helmet}$ > threshold
  THEN safety_status = SAFE
ELSE safety_status = UNSAFE

Rule 3:
IF $O_{I,respirator}$ > threshold &
$O_{I,helmet}$ > threshold
  THEN safety_status = SAFE
ELSE safety_status = UNSAFE

Rule 4:
IF $O_{I,hivis}$ > threshold & $O_{I,boot}$
> threshold
  THEN safety_status = SAFE
ELSE safety_status = UNSAFE

Rule 5:
IF $O_{I,respirator}$ > threshold &
$O_{I,helmet}$ > threshold & $O_{I,googles}$
> threshold
  THEN safety_status = SAFE
ELSE safety_status = UNSAFE

Rule 6:
IF $O_{I,harness}$ > threshold &
$O_{I,helmet}$ > threshold & $O_{I,glove}$
> threshold
  THEN safety_status = SAFE
ELSE safety_status = UNSAFE

The *threshold* that defines wearing/not wearing is set to 0.4 for all overlaps in this analysis but can be modified to better suit each application's requirement.

A summary of all variables, functions, constraints and rules considered in this analysis is presented in Table 2.

3.3 Experimental Results

The system has been tested over collection of frames (representing different industrial settings) of successful detections of body parts and PPEs. Only successful detections of PPEs are considered, as the goal at this stage is to evaluate the reasoning only. These images have been tested for two different set of rules, totalizing 150 observations.

Table 3 shows examples of the outputs representing the safety status of the frame in analysis for the given rule. Body parts are represented on blue rectangles and PPEs as green rectangles on the input frames.

Table 2. Set of variables, functions, constraints and rules considered in analysis.

Elements	Term
Variable	Image
	Body Parts: head, forearm, legs, torso etc.
	PPEs: boot, earmuff, respirator, etc.
	Area of intercept A_I between body part and PPE
	Area of PPE A_{ppe}
	$O_{I,ppe}$
	safety_satatus of the scene
Function	Maximum overlap $O_{I,ppe}$
Constraint	Size of PPEs relative to body part
Rule	Set of Rules (1, 2, 3, 4, 5 and 6)

Table 3. Output of system for each given set of rules.

	Rule 1	Rule 2	Rule 3	Rule 4	Rule 5	Rule 6
Frame						
Required equipment	Respirator	Helmet	Respirator and Helmet	High Visibility Clothes and Boots	Respirator, Helmet and Googles	Harness, Helmet and Gloves
Output	SAFE	UNSAFE	UNSAFE	SAFE	UNSAFE	SAFE

The outputs were manually verified to check the suitability of such approach. It has been measured the number of True Positive (TP), which is the number of frames tagged correctly as UNSAFE; True Negative (TN), the number of frames marked appropriately as SAFE; False Positive (FP), which is amount of frames that should have been identified as SAFE by the system but wrongly outputted the status as UNSAFE; and finally False Negative (FN), the number of frames the system tagged as UNSAFE mistakenly.

The sensitivity and specificity rates also known by True Positive Rate (TPr) and True Negative Rate (TNr) respectively, have also been calculated [30]. Table 4 shows the results for evaluation of performance.

Given a set of successful detections, the methodology works effectively in recognising the safety status of the scene. For real time applications, the wrong status of safety may happen due to wrong status of each variable inputted into the system reasoning

Table 4. Evaluation of performance.

Parameter	TP	TN	FP	FN	TP$_r$	TN$_r$	Accuracy
Value	97	50	2	1	98.98%	96.15%	98.00%

(e.g. wrong detections of body parts and PPEs) or mistakes in the interpretation of these variables during the reasoning process. One of the advantages of explicit representation of knowledge is the possibility to evaluate the causes of unreasonable outputs by checking the status of each variable involved. This way, if the issues are found to be related to the status of the variables, calibration the classifiers can be done as well as adjust on the data gathering process that could lead to such mislead. In addition, if the status of variables are found to be accurate, correction to reasoning can be made by adding a new set of constrains, functions or rules that adjust the output to the correct value for future observations.

4 Conclusions

In this paper we have shown through a case study how Decisional DNA (DDNA), a multi-domain knowledge structure that has the Set of Experience Knowledge Structure (SOEKS) as its basis can be utilized as a comprehensive embedded knowledge representation in a Cognitive Vision System for Hazard Control (CVP-HC). The proposed application aims to ensure that workers remain safe and compliant with Health and Safety policy for use of Personal Protective Equipment (PPE) and serves as a showcase to demonstrate the representation of visual and non-visual content together as an experiential knowledge in one single structure. At this point the implementation is working in offline mode, i.e. the application has been tested over images coming from a database. For next steps, more complex scenarios will be explored for the creation of more complexes set of rules and analyses of the results presented for online operation of the system in which the input images and context variables are gathered from video cameras and sensors in real time.

References

1. Sanin, C., Haoxi, Z., Shafiq, I., Waris, M.M., de Oliveira, C.S., Szczerbicki, E.: Experience based knowledge representation for Internet of Things and Cyber Physical Systems with case studies. Future Gener. Comput. Syst. **92**, 604–616 (2019)
2. Vernon, D.: The space of cognitive vision. In: Christensen, H.I., Nagel, H.-H. (eds.) Cognitive Vision Systems. LNCS, vol. 3948, pp. 7–24. Springer, Heidelberg (2006). https://doi.org/10.1007/11414353_2
3. Gregory, R.L.: Eye and Brain: The Psychology of Seeing. McGraw-Hill, Blacklick (1973)
4. De Oliveira, C.S., Sanin, C., Szczerbicki, E.: Visual content representation and retrieval for Cognitive Cyber Physical Systems. In: 23rd International Conference on Knowledge-Based and Intelligent Information & Engineering Systems (2019)
5. Sanin, C., et al.: Decisional DNA: a multi-technology shareable knowledge structure for decisional experience. Neurocomputing **88**, 42–53 (2012)

6. Sanin, C., Szczerbicki, E.: Experience-based knowledge representation SOEKS. Cybern. Syst. **40**(2), 99–122 (2009)
7. Sanin, C., Szczerbicki, E.: Decisional DNA and the smart knowledge management system: a process of transforming information into knowledge. In: Gunasekaran, A. (ed.) Techniques and Tool for the Design and Implementation of Enterprise Information Systems, pp. 149–175. IGI Global, New York (2008)
8. de Oliveira, C.S., Sanin, C., Szczerbicki, E.: Cognition and decisional experience to support safety management in workplaces. In: Świątek, J., Borzemski, L., Wilimowska, Z. (eds.) ISAT 2018. AISC, vol. 853, pp. 266–275. Springer, Cham (2019). https://doi.org/10.1007/978-3-319-99996-8_24
9. Han, S., Lee, S.: A vision-based motion capture and recognition framework for behavior-based safety management. Autom. Constr. **35**, 131–141 (2013)
10. Ciresan, D.C., Meier, U., Masci, J., Maria Gambardella, L., Schmidhuber, J.: Flexible, high performance convolutional neural networks for image classification. In: IJCAI Proceedings-International Joint Conference on Artificial Intelligence, vol. 22, no. 1, p. 1237, July 2011
11. Little, S., et al.: An information retrieval approach to identifying infrequent events in surveillance video. In: Proceedings of the 3rd ACM Conference on International Conference on Multimedia Retrieval, pp. 223–230. ACM, April 2013
12. Krizhevsky, A., Sutskever, I., Hinton, G.E.: ImageNet classification with deep convolutional neural networks. In: Advances in Neural Information Processing Systems, pp. 1097–1105 (2012)
13. Mosberger, R., Andreasson, H., Lilienthal, A.J.: Multi-human tracking using high-visibility clothing for industrial safety. In: 2013 IEEE/RSJ International Conference on Intelligent Robots and Systems (IROS), pp. 638–644. IEEE, November 2013
14. Chen, L., Hoey, J., Nugent, C.D., Cook, D.J., Yu, Z.: Sensor-based activity recognition. IEEE Trans. Syst. Man Cybern. Part C (Appl. Rev.) **42**(6), 790–808 (2012)
15. Zambrano, A., Toro, C., Nieto, M., Sotaquirá, R., Sanín, C., Szczerbicki, E.: Video semantic analysis framework based on run-time production rules – towards cognitive vision. J. Univ. Comput. Sci. **21**(6), 856–870 (2015)
16. Aditya, S., Yang, Y., Baral, C., Aloimonos, Y., Fermuller, C.: Image Understanding using vision and reasoning through Scene Description Graph. Comput. Vis. Image Underst. **173**, 33–45 (2017)
17. de Oliveira, C.S., Sanin, C., Szczerbicki, E.: Flexible knowledge–vision–integration platform for personal protective equipment detection and classification using hierarchical convolutional neural networks and active leaning. Cybern. Syst. **49**(5–6), 355–367 (2018)
18. Hollnagel, E., Woods, D.D.: Joint Cognitive Systems: Foundations of Cognitive Systems Engineering. CRC Press, Boca Raton (2005)
19. Cole, G.S.: Tort liability for artificial intelligence and expert systems. Computer/LJ **10**, 127 (1990)
20. Ashby, W.R.: An Introduction to Cybernetics. Chapman & Hall Ltd., London (1961)
21. Yu, Y., et al.: Intelligence-augmented rat cyborgs in maze solving. PLoS ONE **11**(2), e0147754 (2016)
22. Pathak, N.: The future of AI. In: Artificial Intelligence for NET: Speech, Language, and Search, pp. 247–259. Apress, Berkeley (2017)
23. Brézillon, P., & Pomerol, J. C.: Contextual knowledge and proceduralized context. In Proceedings of the AAAI-99 Workshop on Modeling Context in AI Applications, Orlando, Florida, USA, July. AAAI Technical Report (1999)
24. de Oliveira, C.S., Sanin, C., Szczerbicki, E.: Towards Knowledge Formalization and Sharing in a Cognitive Vision Platform for Hazard Control (CVP-HC). In: Nguyen, N.T., Gaol, F.L., Hong, T.-P., Trawiński, B. (eds.) ACIIDS 2019. LNCS (LNAI), vol. 11431, pp. 53–61. Springer, Cham (2019). https://doi.org/10.1007/978-3-030-14799-0_5

25. Safe Work Australia: Australian Work Health and Safety Strategy 2012–2022. Creative Commons (2012)
26. Health and Safety Executive (HSE): Measuring the Effectiveness of HSES Field Activities. HSE Occasional Paper Series, OPll, Health and Safety Commission. HSE, London (2018)
27. DeJoy, D.M.: Behavior change versus culture change: divergent approaches to man-aging workplace safety. Saf. Sci. **43**(2), 105–129 (2005)
28. de Oliveira, C.S., Sanin, C., Szczerbicki, E.: Contextual knowledge to enhance workplace hazard recognition and interpretation in a cognitive vision platform. Procedia Comput. Sci. **126**, 1837–1846 (2018)
29. Au, K.W., et al.: U.S. Patent No. 9,695,981. U.S. Patent and Trademark Office, Washington, DC (2017)
30. Cortes, C., Mohri, M.: AUC optimization vs. error rate minimization. In: Advances in Neural Information Processing Systems, pp. 313–320 (2004)

Harmony Search Algorithm with Dynamic Adjustment of PAR Values for Asymmetric Traveling Salesman Problem

Krzysztof Szwarc$^{(\boxtimes)}$ (iD) and Urszula Boryczka (iD)

Institute of Computer Science, University of Silesia in Katowice,
ul. Bedzinska 39, 41-200 Sosnowiec, Poland
{krzysztof.szwarc,urszula.boryczka}@us.edu.pl
http://ii.us.edu.pl/

Abstract. This paper describes an improvement to the Harmony Search algorithm, which has been adjusted to effectively solve a problem with indisputable practical significance, i.e. Asymmetric Traveling Salesman Problem (ATSP). We modify the technique structure, enabling the value of PAR parameter to be changed dynamically, which has an impact on the frequency of greedy movements during the construction of another harmony. The article demonstrates the effectiveness of the described approach and presents a comparative study of three sets of characteristic PAR values used during the method execution. The research was conducted on a 'test bed' consisting of nineteen instances of the ATSP.

Keywords: Dynamic adjustable parameters · Harmony Search · Asymmetric Traveling Salesman Problem

1 Introduction

The object of research in this paper is an interesting metaheuristic, which draws inspiration from the process of musical improvisation—Harmony Search (HS). In recent years the technique has found numerous applications in solving utilitarian issues, such as weapon-target assignment problem [3], evaluation and prediction of performance of diamond wire saw [11], task prioritisation in job shop [5] and flow shop [16], project scheduling [7] and sudoku solving [14]. However, in this manuscript, we focus on improving the special version of HS that was adapted to solve instances of the Asymmetric Traveling Salesman Problem (ATSP). We pay special attention to the ATSP because it belongs to the class of $\mathcal{NP}$-hard problems and is characterized by practical significance (e.g. modelling the process of mobile collection of e-waste [12] and municipal waste [15]).

The effectiveness of the HS adjusted to solve the ATSP instances (proposed in the paper [1]), prompted work on additional improvements of the technique, eliminating the imperfections of the developed approach. The noticeable decrease

© Springer Nature Switzerland AG 2020
N. T. Nguyen et al. (Eds.): ACIIDS 2020, LNAI 12033, pp. 226–238, 2020.
https://doi.org/10.1007/978-3-030-41964-6_20

in the efficiency of the method for tasks described by the occurrence of over three hundred vertices and the relatively good performance achieved for some of these tasks by algorithms based on the greedy selection of subsequent locations imply that the process of searching for solutions in HS should more intensively use knowledge about the problem. The examined technique contains a mechanism that supports the construction of a solution that is based on selection of the least distant node, and it is activated throughout the period of algorithm execution with constant PAR probability (which, however, turned out to be too small for bigger instances of the problem—A constant increase of the value may, however, cause the HS to get stuck in a local optimum, especially when solving small problems). In order to improve the analysed method, we decided to allow the dynamic modification of the PAR value, which in consequence enabled efficient balancing between exploration and exploitation of the solution space.

The purpose of this article is to increase the effectiveness of the algorithm proposed in a previous paper [1] by developing an effective approach to dynamically changing the value of PAR, taking into consideration the occurrence of specific mechanisms used in the analysed technique.

The paper is structured as follows: the introduction to the subject, is followed by the formulation of ATSP, a description of the classical HS algorithm, the characterisation of HS with the dynamically changing value of PAR adjusted for solving ATSP instances, the research methodology, a discussion of the results obtained, and finally the conclusions and a discussion of planned further work.

2   Formulation of the Asymmetric Traveling Salesman Problem

The definition of the Traveling Salesman Problem presented previously in [19] was adapted for the purpose of the conducted research. In accordance with this definition, a search for the shortest (minimal length) oriented cycle containing all n cities in the directed graph $D = (N, A)$ (where the arc weights are given by c_{ij} $(i, j \in \{1, 2, \ldots, n\})$) is assumed. The asymmetric variant of the problem enables the occurrence of $c_{ij} \neq c_{ji}$ irregularities.

The presence of the edge connecting the i and j nodes in the constructed solution is represented by the decision variable x_{ij}, which adapts the following values:

$$x_{ij} = \begin{cases} 1 \text{ if edge } (i,j) \text{ is part of the route,} \\ 0 \text{ otherwise.} \end{cases} \tag{1}$$

The objective function which assumes minimisation of the total of edge weights forming the commercial agent's route was formulated as follows:

$$\sum_{i=1}^{n} \sum_{j=1}^{n} c_{ij} x_{ij} \to min. \tag{2}$$

In order to ensure that all cities were visited by the salesman, the following limiting conditions were added only once:

$$\sum_{i=1}^{n} x_{ij} = 1, \quad j = 1, \ldots, n, \quad \sum_{j=1}^{n} x_{ij} = 1, \quad i = 1, \ldots, n. \tag{3}$$

Elimination of the possibility to create solutions representing separate cycles instead of one combined cycle was carried out by means of additional restrictions referred to as MTZ:

$$1 \leq u_i \leq n - 1, \quad u_i - u_j + (n-1)x_{ij} \leq n - 2, \quad i, j = 2, \ldots, n. \tag{4}$$

It uses u_i variables, in order to designate the order in which vertex i is visited.

3   Classical Formulation of Harmony Search

HS is a very interesting metaheuristic which, despite a relatively short period of existence (it was first described in 2000 in [6]), has caused a lot of controversy. The method is considered by some researchers (e.g. [17,18]) to be only a special version of the Evolution Strategies (ES), and is accused of not being innovative.

HS assumes a similarity between the process of searching for the global optimum by algorithmic methods and jazz improvisation. Its mode of action is described by the harmony memory (HM). Each of the HMS elements in HM (called harmony) possess a given number of pitches, which correspond to the value of decision variables of the relevant result. Each harmony is regarded as a complete solution of the problem with the value of the objective function determined from its components.

At the initial stage of method execution, the HM is randomly generated, and particular HM elements are sorted on the basis of their objective function values, such that the element in the first position is described with the best result. The performance of the described instructions initiates the iterative process of creating new solutions.

The knowledge stored in the HM structure is used during the development of another result, based on the analogy to the process of improvising new harmony in music. The solution is based on an iterative selection of the following pitch, using two parameters: $HMCR$ (harmony memory consideration rate) and PAR (pitch adjustment rate). According to the $HMCR$ probability, the pitch i is selected on the basis of the values in position i, in particular, the HM components (otherwise the available value is generated randomly). When creating a solution based on the elements of HM, the pitch may be modified with the given PAR probability (the value is changed on the basis of bw, which depends on the problem representation).

After completing the development of a new solution, the value of its objective function is compared with the relevant parameter describing the ultimate component belonging to the HM. If a better result is created, it will replace the worst result located in the HM structure. The operation is followed by the sorting of the elements of HM (in order to put the new harmony in the right place of HM). The process of generating further solutions is carried out for IT iterations, which is followed by returning the best result (located in the first position of HM).

The significant similarity between ES and HS was considered by many researchers (e.g. [8,17,18]); however, a significant difference in the functioning of exploitation and exploration strategies appearing in the above-mentioned techniques was emphasized in the paper [8]. HS has three operators with the frequency of launching controlled by $HMCR$ and PAR values, whereas the $(\mu+1)$ strategy (which most closely resembles HS) has only two obligatory operations-crossover and mutation.

4 Harmony Search Algorithm with Dynamic Adjustment of PAR Values for Asymmetric Traveling Salesman Problem

This section consists of two parts. The first part contains a description of the HS structure adjusted to effectively solve the ATSP, whereas the second part contains proposals for modifications enabling the application of dynamic values of PAR in the examined technique.

4.1 The Adaptation of the Harmony Search for Asymmetric Traveling Salesman Problem

This paper concerns modification of the algorithm proposed in the paper [1], which assumes that pitches are represented by integers corresponding to the numbers of particular cities located on the salesman's route. The sequence of a commercial agent's travel corresponds to the sequence of locations in a particular harmony.

The process of creating a new harmony assumes including the sequence of appearance of nodes which is done through a selection of the following pitch value on the basis of the created list of available vertices, appearing in the stored solutions directly after the last location that belongs to the developed result. Using the established structure, the city is selected by means of the roulette wheel method (the probability of node approval depends on the value of the objective function of the entire solution, analogous to the approach presented in the publication [9]), or any unvisited node is drawn (when the created list of cities is empty). The selection (made among the available nodes) of the city situated nearest to the last visited location was adapted as a modification of the pitch (related to PAR parameter) in the created solution, thereby supplementing the metaheuristic with the greedy approach.

The mechanism for resetting the elements of HM was introduced so that the algorithm does not get stuck in a local optimum. It is activated after R iterations from the time that the result in HM is replaced (in other words, the memory is reset when no new solution of sufficiently short length has been found after

R iterations). The process assumes preserving the best result and drawing all other solutions.

4.2 Harmony Search with Dynamic Adjustment of Values for the Asymmetric Traveling Salesman Problem

A dynamic approach to the change of HS parameter values was examined previously in the paper [4], where its effectiveness was demonstrated in comparison with using static $HMCR$ and PAR values. The value of particular parameters in iteration i was determined on the basis of the following formula:

$$P = P_{min} + \frac{(P_{max} - P_{min})}{maxIterations} \cdot i, \tag{5}$$

where: P_{min} is the minimum permissible value of P (corresponding to $HMCR$ or PAR respectively), P_{max} constitutes the maximum permissible value of P, and $maxIterations$ represents the stop condition expressed by the maximum number of iterations.

In other work [10] dynamic modification of the values of bw and PAR was performed. The change in bw assumed value reduction along with performance of subsequent iterations and was based on the formula (6). The PAR update was based on the formula (7).

$$bw = bw_{max} - (bw_{max} - bw_{min}) \cdot \frac{i}{maxIterations}, \tag{6}$$

where bw_{min} is the minimum permissible value of bw, and bw_{max} is its maximum value.

$$PAR = (PAR_{max} - PAR_{min}) \cdot \frac{arctan(i)}{\frac{\pi}{2}} + PAR_{min}, \tag{7}$$

where PAR_{min} is the minimum permissible value of PAR, and PAR_{max} is its maximum value.

Based on a literature analysis, the decision was made to direct HS at an early stage of execution by reducing the value of PAR parameter (responsible for the application of greedy approach) along with the performance of subsequent algorithm iterations, which in consequence increased exploration at the later stages of execution. Because analysed approach enables the HM elements to be reset, we assumed that making the PAR values conditional on the number of performed iterations is ineffective (after drawing new solutions, it is recommended to move again to the exploitation stage), which prevented the application of formulas (5), (6) and (7). The following formula was used to dynamically modify PAR (the approach was marked as DHS):

$$PAR = \begin{cases} PAR - c & \text{if } PAR \geq PAR_{prog}, \\ PAR & \text{otherwise.} \end{cases} \tag{8}$$

where c refers to the value adjustment step, and PAR_{prog} is the lowest value of PAR from which c is deducted.

When the HM elements were reset, it was assumed that the initial value of PAR (PAR_b) parameter would be restored. Additionally, the risk of moving to the exploitation stage too quickly may be reduced by updating the values only if the created solution is worse than the last harmony stored in HM (we used this approach during research). The proposed approach to designing DHS is presented in Fig. 1.

5   Methodology of Research

The effectiveness of the proposed approach was examined on the basis of nineteen tasks representing the ATSP instances (their characteristics are presented in Table 1). Each model test was solved 30 times.

Table 1. Characteristics of 'test bed' (based on [13])

No.	Name	Number of vertices	Optimum
1	br17	17	39
2	ftv33	34	1286
3	ftv35	36	1473
4	ftv38	39	1530
5	p43	43	5620
6	ftv44	45	1613
7	ftv47	48	1776
8	ry48p	48	14422
9	ft53	53	6905
10	ftv55	56	1608
11	ftv64	65	1839
12	ft70	70	38673
13	ftv70	71	1950
14	kro124p	100	36230
15	ftv170	171	2755
16	rbg323	323	1326
17	rbg358	358	1163
18	rbg403	403	2465
19	rbg443	443	2720

The average error, which constitutes the measurement of algorithm quality, was expressed by means of the following formula:

$$Average\ error = \frac{average\ objective\ function\ value - optimum}{optimum} \cdot 100\%. \qquad (9)$$

Pseudocode of Harmony Search with Dynamic Adjustment of PAR value for ATSP

1: $iterations = 0$
2: $iterationsFromTheLastReplacement = 0$
3: **for** $i = 0; i < HMS; i + +$ **do**
4:     $HM[i]$=stochastically generate feasible solution
5: **end for**
6: Sort HM
7: **while** $iterations < IT$ **do**
8:     $H = \emptyset$
9:     $H[0]$=first city
10:     **for** $i = 1; i < n; i + +$ **do**   ▷ n - number of cities
11:         Choose random $r \in (0, 1)$
12:         **if** $r < HMCR$ **then**
13:             $list$=create list containing vertices occurring after $H[i-1]$ in HM
14:             **if** $list.length > 0$ **then**
15:                 $H[i]$=choose element $\in list$ according to the roulette wheel
16:             **else**
17:                 $H[i]$=choose randomly available city $\notin H$
18:             **end if**
19:             Choose random $k \in (0, 1)$
20:             **if** $k < PAR$ **then**
21:                 $H[i]$=find nearest and available city from $H[i-1]$
22:             **end if**
23:         **else**
24:             $H[i]$=choose randomly available city $\notin H$
25:         **end if**
26:     **end for**
27:     **if** $f(H)$ is better than $f(HM[HMS - 1])$ **then**
28:         $HM[HMS - 1] = H$
29:         Sort HM
30:         $iterationsFromTheLastReplacement = 0$
31:     **else**
32:         $iterationsFromTheLastReplacement + +$
33:         Update PAR based on formula (8)
34:     **end if**
35:     **if** $iterationsFromTheLastReplacement = R$ **then**
36:         **for** $i = 1; i < HMS; i + +$ **do**
37:             $HM[i]$=stochastically generate feasible solution
38:         **end for**
39:         Sort HM
40:         $iterationsFromTheLastReplacement = 0$
41:         $PAR = PAR_b$
42:     **end if**
43:     $iterations + +$
44: **end while**
45: return $HM[0]$

Fig. 1. Pseudocode of HS with dynamic adjustment of PAR value for ATSP

Moreover, the variations in the distribution of objective function values were determined using the coefficient of variation V, specified with the following formula:

$$V = \frac{sample\ standard\ deviation}{sample\ mean\ value} \cdot 100\%. \tag{10}$$

The following parameter values were assumed based on the previous work [1]: $R = 1000$, $HMS = 5$, $HMCR = 0.98$ and $PAR = 0.25$. Additionally, relying on our previous experience [2], we used $IT = 1000000$. Based on the conducted research, the following values for the DHS parameters were used: $c = 0.002$, PAR_b was set to 0.5 or 0.75, whereas PAR_{prog} was set to 0.25 or 0.002. Three DHS variants, selected on the basis of experimental research, were analysed:

1. DHS1 ($PAR_b = 0.5$ and $PAR_{prog} = 0.25$).
2. DHS2 ($PAR_b = 0.75$ and $PAR_{prog} = 0.25$).
3. DHS3 ($PAR_b = 0.75$ and $PAR_{prog} = 0.002$).

The algorithms were implemented using $C\#$, and the research was conducted on a Lenovo Y520 laptop whose parameters are presented in Table 2.

Table 2. Parameters of the laptop used for conducting research

No.	Parameter	Value
1	Processor	Intel Core i7-7700HQ
2	RAM	32 GB (SO-DIMM DDR4, 2400 MHz)
3	HDD	1000 GB SATA 7200 RPM, 240 GB SSD M.2 PCIe
4	OS	Windows 10 Home 64-bit

6   Results

The average error obtained with the DHS variants and HS with a constant PAR value (equal to 0.25; other parameter values were identical as in each DHS variant) are presented in Table 3. We found that HS only obtained better results than any DHS variant for five tasks. Regardless of the parameter values used, better results (from the perspective of the summary average error) than those determined by HS were obtained by enabling dynamic change of the PAR values. Particular attention should be drawn to the observation that DHS1 obtained worse results than HS for six tasks, DHS2 for seven and DHS3 for eight. The lowest summary value of the average error is characteristic for DHS2; however, the difference between DHS2 and DHS3 is relatively small (0.15%), indicating the effectiveness of both sets of values.

The detailed results obtained by particular DHS variants are presented in Tables 4, 5 and 6 (for DHS1, DHS2 and DHS3 variants, respectively). The highest average value of the coefficient V is characteristic for variant DHS1 (1.34%),

Table 3. Summary of average errors obtained by particular DHS and HS variants

Test name	Average error			
	HS	DHS1	DHS2	DHS3
br17	**0**	**0**	**0**	**0**
ftv33	3.63	3.11	**1.3**	3.4
ftv35	**1.35**	1.38	1.49	1.49
ftv38	**1.44**	2.01	2.73	2.79
p43	**0.05**	**0.05**	**0.05**	**0.05**
ftv44	1.76	1.81	0.9	**0.62**
ftv47	**1.95**	2.06	2.14	2.15
ry48p	**0.89**	1.03	0.97	1.4
ft53	10.06	8.56	7.88	**6.25**
ftv55	2.92	**2.38**	3.01	3.97
ftv64	3.09	2.42	3.03	**2.34**
ft70	4.35	4.28	4.65	**3.84**
ftv70	**4.89**	4.97	5.73	6.25
kro124p	8.88	**8.31**	8.53	9.44
ftv170	19.15	17.58	17.49	**17.47**
rbg323	53.84	39.33	**29**	29.28
rbg358	77.01	54.99	39.64	**39.31**
rbg403	29.27	28.85	**27.93**	28.54
rbg443	30.45	30.38	**30.26**	31.01
Average	13.42	11.24	**9.83**	9.98

Table 4. Detailed results for DHS1

Test name	Objective function value				
	Avg.	Min.	Max.	Sample std. dev.	V
br17	39	39	39	0	0
ftv33	1326.03	1286	1388	32.46	2.45
ftv35	1493.27	1473	1499	6.83	0.46
ftv38	1560.7	1540	1603	15.25	0.98
p43	5622.83	5620	5623	0.59	0.01
ftv44	1642.13	1613	1708	27.47	1.67
ftv47	1812.67	1777	1853	23.9	1.32
ry48p	14569.87	14507	14767	65.92	0.45
ft53	7495.8	7143	7779	207.89	2.77

(continued)

Table 4. (*continued*)

Test name	Objective function value				
	Avg.	Min.	Max.	Sample std. dev.	V
ftv55	1646.3	1608	1731	32.93	2
ftv64	1883.5	1856	1943	25.44	1.35
ft70	40328.93	39457	40619	243.24	0.6
ftv70	2046.93	1968	2096	37.5	1.83
kro124p	39240.7	37927	40028	562.19	1.43
ftv170	3239.4	3095	3539	95.84	2.96
rbg323	1847.57	1793	1910	29.27	1.58
rbg358	1802.53	1746	1875	30.46	1.69
rbg403	3176.23	3112	3237	25.67	0.81
rbg443	3546.4	3460	3615	35.45	1

Table 5. Detailed results for DHS2

Test name	Objective function value				
	Avg.	Min.	Max.	Sample std. dev.	V
br17	39	39	39	0	0
ftv33	1302.67	1286	1388	29.41	2.26
ftv35	1495	1490	1499	4.28	0.29
ftv38	1571.8	1547	1603	18.14	1.15
p43	5623	5620	5627	0.95	0.02
ftv44	1627.53	1613	1674	12.88	0.79
ftv47	1814	1776	1853	20.75	1.14
ry48p	14561.2	14507	14670	40.92	0.28
ft53	7449.27	7188	7815	144.57	1.94
ftv55	1656.37	1608	1699	22.23	1.34
ftv64	1894.77	1856	1963	31.31	1.65
ft70	40472.43	40123	40748	147.69	0.36
ftv70	2061.77	2018	2096	25.01	1.21
kro124p	39321.7	38275	40578	471.68	1.2
ftv170	3236.93	3134	3371	60.46	1.87
rbg323	1710.5	1675	1768	18.16	1.06
rbg358	1624	1554	1667	27	1.66
rbg403	3153.5	3086	3238	31.78	1.01
rbg443	3543.03	3507	3597	27.65	0.78

Table 6. Detailed results for DHS3

Test name	Objective function value				
	Avg.	Min.	Max.	Sample std. dev.	V
br17	39	39	39	0	0
ftv33	1329.67	1286	1390	44.48	3.35
ftv35	1495	1473	1514	7.32	0.49
ftv38	1572.7	1549	1607	13.37	0.85
p43	5622.97	5622	5624	0.32	0.01
ftv44	1623.07	1613	1674	10.64	0.66
ftv47	1814.2	1776	1878	28.19	1.55
ry48p	14623.6	14507	15026	100.49	0.69
ft53	7336.87	7170	7537	103.77	1.41
ftv55	1671.9	1635	1732	26.36	1.58
ftv64	1881.97	1848	1958	31.74	1.69
ft70	40157.13	39809	40712	192.97	0.48
ftv70	2071.83	2016	2113	24.95	1.2
kro124p	39649.5	38651	40591	508.46	1.28
ftv170	3236.3	3112	3472	80.98	2.5
rbg323	1714.3	1682	1762	19.15	1.12
rbg358	1620.23	1574	1656	21.15	1.31
rbg403	3168.43	3126	3249	27.68	0.87
rbg443	3563.5	3496	3601	23.78	0.67

whereas the lowest value is characteristic for DHS2 (1.05%), which implies that the results determined with DHS2 are easier to predict.

The percentage surpluses of the objective function values determined by particular DHS models in comparison with the optimum were subject to the Wilcoxon Signed-Rank Test, using R and the wilcox.test function (the following parameter values were applied: paired = TRUE, correct = F, alternative = 'less', exact = F). The following method variants (marked as $M1$ and $M2$) were subject to the process, and 0.05 was used for the level of test significance (the obtained p-values described with the lower result are distinguished by using bold, and they indicate that we accept the alternative hypothesis according to which $M1$ obtained lower results than $M2$). The results of the performed test are presented in Table 7. HS obtained worse results than DHS, while DHS1 and DHS3 obtained worse results than DHS2.

Table 7. Results of Wilcoxon Signed-Rank Test for the conducted research

$M1\backslash M2$	HS	DHS1	DHS2	DHS3
HS	N/A	1	1	0.999999
DHS1	**4.66E-26**	N/A	0.999944	0.893981
DHS2	**1.36E-16**	**5.64E-05**	N/A	**0.001472**
DHS3	**1.37E-06**	0.106019	0.998528	N/A

7 Conclusions

The paper demonstrates the effectiveness of the proposed approach to dynamically changing PAR, enabling efficient balancing of exploration and exploitation of the HS solution space, adjusted for solving ATSP instances (in comparison with HS with a constant PAR value, the summary average error decreased from 13.42 to 9.83%). We recommend the following parameter values for DHS: $PAR_b = 0.75$, $PAR_{prog} = 0.25$ and $c = 0.002$. The parameters gave both the lowest summary average error and the lowest value of V.

Further work related to DHS should include precise adjustment of parameter values to the analysed 'test bed' and adjustment of the technique for solving other combinatorial optimization problems.

References

1. Boryczka, U., Szwarc, K.: The adaptation of the harmony search algorithm to the ATSP. In: Nguyen, N.T., Hoang, D.H., Hong, T.-P., Pham, H., Trawiński, B. (eds.) ACIIDS 2018. LNCS (LNAI), vol. 10751, pp. 341–351. Springer, Cham (2018). https://doi.org/10.1007/978-3-319-75417-8_32
2. Boryczka, U., Szwarc, K.: The adaptation of the harmony search algorithm to the atsp with the evaluation of the influence of the pitch adjustment place on the quality of results. J. Inf. Telecommun. **3**(1), 2–18 (2019)
3. Chang, Y., Li, Z., Kou, Y., Sun, Q., Yang, H., Zhao, Z.: A new approach to weapon-target assignment in cooperative air combat. Math. Probl. Eng. **2017** (2017). https://doi.org/10.1155/2017/2936279. 17 pages
4. Daham, B.F., Mohammed, M.N., Mohammed, K.S.: Parameter controlled harmony search algorithm for solving the Four-Color Mapping Problem. Int. J. Comput. Inf. Technol. **3**(6), 1398–1402 (2014)
5. Gaham, M., Bouzouia, B., Achour, N.: An effective operations permutation-based discrete harmony search approach for the flexible job shop scheduling problem with makespan criterion. Appl. Intell. **48**(6), 1423–1441 (2018)
6. Geem, Z.W.: Optimal design of water distribution networks using harmony search. Ph.D. thesis, Korea University (2000)
7. Geem, Z.W.: Multiobjective optimization of time-cost trade-off using harmony search. J. Constr. Eng. Manag. **136**(6), 711–716 (2010)
8. Kim, J.H.: Harmony search algorithm: a unique music-inspired algorithm. Procedia Eng. **154**, 1401–1405 (2016). 12th International Conference on Hydroinformatics (HIC 2016) - Smart Water for the Future

9. Komaki, M., Sheikh, S., Teymourian, E.: A hybrid harmony search algorithm to minimize total weighted tardiness in the permutation flow shop. In: 2014 IEEE Symposium on Computational Intelligence in Production and Logistics Systems (CIPLS), pp. 1–8 (2014)
10. Li, X., Qin, K., Zeng, B., Gao, L., Wang, L.: A dynamic parameter controlled harmony search algorithm for assembly sequence planning. Int. J. Adv. Manuf. Technol. **92**(9), 3399–3411 (2017)
11. Mikaeil, R., Ozcelik, Y., Ataei, M., Shaffiee Haghshenas, S.: Application of harmony search algorithm to evaluate performance of diamond wire saw. J. Mining Environ. **10**(1), 27–36 (2019)
12. Nowakowski, P., Szwarc, K., Boryczka, U.: Vehicle route planning in e-waste mobile collection on demand supported by artificial intelligence algorithms. Transp. Res. Part D Transp. Environ. **63**, 1–22 (2018)
13. Osaba, E., Diaz, F., Onieva, E., Carballedo, R., Perallos, A.: A population meta-heuristic with adaptive crossover probability and multi-crossover mechanism for solving combinatorial optimization problems. IJAI **12**, 1–23 (2014)
14. Rojas-Morales, N., Rojas, M.C.R.: Improving harmony search algorithms by using tonal variation: the case of Sudoku and MKP. Connection Sci. **30**(3), 245–271 (2018)
15. Syberfeldt, A., Rogström, J., Geertsen, A.: Simulation-based optimization of a real-world travelling salesman problem using an evolutionary algorithm with a repair function. Int. J. Artif. Intell. Expert Syst. (IJAE) **6**(3), 27–39 (2015)
16. Wang, L., Pan, Q.K., Tasgetiren, M.F.: Minimizing the total flow time in a flow shop with blocking by using hybrid harmony search algorithms. Expert Syst. Appl. **37**(12), 7929–7936 (2010)
17. Weyland, D.: A rigorous analysis of the harmony search algorithm: how the research community can be misled by a "novel" methodology. Int. J. Appl. Metaheuristic Comput. **1**(2), 50–60 (2010)
18. Weyland, D.: A critical analysis of the harmony search algorithm—how not to solve sudoku. Oper. Res. Perspect. **2**, 97–105 (2015)
19. Öncan, T., Altınel, I.K., Laporte, G.: A comparative analysis of several asymmetric traveling salesman problem formulations. Comput. Oper. Res. **36**(3), 637–654 (2009)

Graph-Based Optimization of Public Lighting Retrofit

Adam Sędziwy⬤ and Leszek Kotulski(✉)⬤

AGH University of Science and Technology, Kraków, Poland
{sedziwy,kotulski}@agh.edu.pl

Abstract. Modernization of public lighting (also referred to as a retrofit) such as replacing high-pressure sodium lamps with LED ones, burdens the budgets of municipalities. For that reason such investments are usually made in phases, spanned over a period of several years, which scopes are determined by financial resources. It should be remarked that selection of lamps for modernization is based on various criteria such as lamp aging, power efficiency of an installation but also some high-level objectives such as a payback period of an investment. In this work we propose the scalable computational, graph-based approach which enables optimizing lamp modernization schedule, towards reduced payback time. The presented results are based on analysis of real-life data of nearly 10,000 streetlights.

Keywords: Graphs · Graph methods · LED lighting · Roadway lighting · Lighting retrofit

1 Introduction

Public lighting installations are subject to continuous modernization works such as replacement of an aging equipment: poles, arms, power lines etc. The common case is replacing the high-intensity discharge (HID) lamps with LED-based (light emitting diode) ones. In the latter case, the additional profit, besides equipment renewal, is obtaining substantial energy savings which imply reduced operational costs and CO_2 emission. Although modernization (also referred to as a *retrofit*) of the public lighting infrastructures can yield spectacular energy savings which can reach 80%, municipalities cannot perform, in the real-life cases, a single-phase lamp change (replacing all HIDs to LEDs, at once), due to the costs which usually exceed municipal budget resources. Instead, some scheduling strategy is made, e.g., basing on an equipment status or lamp locations. Keeping in mind that investment costs and expected energy savings belong to key criteria during modernization of public lighting, one has to prepare optimized lamp setups and top rated replacement schedule. In a result of completing the former task one gets planned lamp powers, necessary for achieving the second goal. Determining proper lamp configurations, however, requires performing high complexity computations which cannot be made in an acceptable time. To address this issue

© Springer Nature Switzerland AG 2020
N. T. Nguyen et al. (Eds.): ACIIDS 2020, LNAI 12033, pp. 239–248, 2020.
https://doi.org/10.1007/978-3-030-41964-6_21

and reduce the calculation time we propose a graph-based approach. Moreover, we propose a retrofit planning method which allows reducing an investment's payback period.

It relies on selection of luminaires which contribute the most to achieving the highest return of investment (ROI) factor. The optimization considered here combines either criteria: energy consumption and business perspective. Increasing ROI is achieved just by reducing the power usage and related costs. The resources obtained thanks to a decreased payback period, allow launching modernization of a next group of street lights.

The tests carried out for the real-life data show that an appropriate selection of luminaires to be replaced, can decrease an OPEX by up to several dozen percents. To enable finding the most profitable modernization schedule, conforming with good practices, we will use a graph model to represent both roadway network and lighting infrastructure.

The structure of the work is following. In the next section a brief overview of power reduction methods will be presented. Section 3 contains theoretical approach to optimized retrofit scheduling. In this case, all practical requirements such as simultaneous lamp replacement for particular streets, are neglected. In Sect. 4 the *layout graph* model is introduced and a retrofit schedule generation is presented for real-life data. The practical constraints are also taken into account. The last section contains the summary and propositions of further research areas.

2   Background of Lighting Installation Optimization

As said in the previous section, the optimization of lighting installations was a subject of multiple works. Their common denominator was reducing the power usage which straightforwardly implies the lower energy-related costs and greenhouse gases emission [7–10]. An additional and mandatory requirement for the resultant lamp configurations was that the lighting system performance fulfills relevant standards, e.g., [1–3]. Although the principal objective of an optimization is usually reducing the power usage, also the other criteria (i.e., objective functions) could be considered instead, for example, investment costs or payback period. The problem solution outline, however, stayed unchanged.

Energy efficiency optimization of lighting infrastructure is a task characterized by the high computation time, particularly for large sized installations containing tens of thousands of lamps. It is mainly due to the potentially high search space size. To reduce computation time and thus make calculations doable in an acceptable time, the graph formalism was introduced [11–13].

The graph models applied for both representing urban spaces and lighting infrastructure, and for control purposes, were presented in multiple works [11, 14, 16]. Those models used graphs and hypergraphs which enabled describing a broad set of entities, physical and abstract ones, ranging from buildings, lighting equipment, areas (streets, squares, junctions) to control states [15].

The analysis presented in next sections relies on the real-life installation, consisting of 9,803 and illuminating 1,786 areas. The basic parameters required for a business analysis are: (i) the energy price: € 0.10/kWh, (ii) the annual operational time of a lighting installation: 4,200 h for the considered city, (iii) an investment time span: 10 years. It is assumed that due to financial limitations, a municipality replaces approximately 1,000 lamps per year. There are two prices of new LED fixtures: € 234, for common fixture types (9,518 pcs. in the considered case) and € 316, for high wattage types (285 pcs. in the considered case).

3   Computing Modernization Effectiveness - Theoretical Approach

In this section we compare the best and the worst (in terms of power savings) retrofit schedule. It is assumed here that for each lamp both fixture powers are known: before and after a retrofit. Thanks to this the power change ΔP can be computed for each fixture. Having the above we sort all fixtures descending by ΔP. Replacement process is carried out according to the following scheme. The sorted list of luminaires is divided into the sets s_i ($i = 1, 2, \ldots, 10$), each one containing 1,000 items (except $|s_{10}| = 803$):

$$s_1 = \{f_1, f_2, \ldots, f_{1000}\}, s_2 = \{f_{1001}, f_{1002}, \ldots, f_{2000}\}, \ldots,$$

where f_i denotes an i-th fixture on the previously sorted list. Table 1 presents the comparison of particular sets: ΔP_i values, annual energy reductions, resultant cost savings but also the cumulated prices of new fixtures replacing the old, HID ones. In the first year luminaires from s_1 are replaced, in the second year - lamps from s_2 and so forth, until the tenth year, when lamps from the last group, s_{10}, are retrofitted.

Table 1. Installation power before and after modernization, broken down into lamp groups (optimized lamp grouping)

Set	Installed power [W]		ΔP [%]	ΔP [W]	Annual savings [kWh]	Annual savings [€]	Tot. fixture cost [€]
	Before	After					
s_1	214,449	25,297	88	189,152	794,438	81,862	237,553
s_2	176,712	31,359	82	145,353	610,482	62,907	234,848
s_3	169,900	46,899	72	123,001	516,604	53,233	234,192
s_4	127,627	24,285	81	103,342	43,4037	44,725	246,077
s_5	120,490	21,866	82	98,624	414,222	42,683	240,667
s_6	91,061	12,927	86	78,134	328,163	33,815	234,274
s_7	83,320	10,012	88	73,308	307,893	31,727	234,192
s_8	83,202	12,125	85	71,077	298,523	30,761	234,192
s_9	83,350	14,892	82	68,458	287,525	29,628	234,192
s_{10}	68,940	21,023	70	47,917	201,252	20,738	188,958

Table 2. Optimized replacement schedule

Year	Investment cost [€]	Annual savings [€], broken down into lamp sets										Tot. annual savings [€]
		s_1	s_2	s_3	s_4	s_5	s_6	s_7	s_8	s_9	s_{10}	
1	237,553	81,862	-	-	-	-	-	-	-	-	-	81,862
2	234,848	81,862	62,907	-	-	-	-	-	-	-	-	144,769
3	234,192	81,862	62,907	53,233	-	-	-	-	-	-	-	198,002
4	246,077	81,862	62,907	53,233	44,725	-	-	-	-	-	-	242,728
5	240,667	81,862	62,907	53,233	44,725	42,683	-	-	-	-	-	285,411
6	234,274	81,862	62,907	53,233	44,725	42,683	33,815	-	-	-	-	319,226
7	234,192	81,862	62,907	53,233	44,725	42,683	33,815	31,727	-	-	-	350,953
8	234,192	81,862	62,907	53,233	44,725	42,683	33,815	31,727	30,761	-	-	381,714
9	234,192	81,862	62,907	53,233	44,725	42,683	33,815	31,727	30,761	29,628	-	411,342
10	188,958	81,862	62,907	53,233	44,725	42,683	33,815	31,727	30,761	29,628	20,738	432,080
Total	**€ 2,319,145**											**€ 2,848,087**

Having input data shown in Table 1 we can see the financial simulation of the replacement schedule for particular lamp groups, s_i (Table 2). The second column contains the investment cost broken into consecutive years (as also shown in Table 1) and its cumulated value. In the last column, the aggregated (by lamp group) annual money savings are presented for particular years. Their total value is also shown.

In the second scenario, the "pessimistic" replacement schedule is considered. In this case sets $\{s_i\}$ are defined as:

$$s_1 = \{f_{9803}, f_{9802}, \ldots, f_{8804}\}, s_2 = \{f_{8804}, f_{8803}, \ldots, f_{7805}\}, \ldots$$

where f_i denotes an i-th fixture on the list of fixtures, sorted descending by ΔP.

Taking into account corresponding input data (Table 3), we calculate scheduled savings in consecutive years (Table 4).

Table 3. Comparison of installation's powers before and after modernization, broken down into lamp groups ("pessimistic" scenario)

Set	Installed power [W]		ΔP [%]	ΔP [W]	Annual savings [kWh]	Annual savings [€]	Tot. fixture cost [€]
	Before	After					
s_1	85,387	24,196	72	61,191	257,002	26,483	235,094
s_2	83,254	14,234	83	69,020	289,884	29,871	234,192
s_3	83,458	11,986	86	71,472	300,182	30,932	234,192
s_4	83,096	9,327	89	73,769	309,831	31,926	234,192
s_5	98,126	16,093	84	82,033	344,538	35,503	235,340
s_6	124,272	24,216	81	100,056	420,235	43,303	245,422
s_7	128,969	24,096	81	104,873	440,466	45,388	240,258
s_8	176,534	47,603	73	128,931	541,509	55,800	234,684
s_9	176,178	27,626	84	148,552	623,917	64,291	234,356
s_{10}	179,777	21,307	88	158,470	665,574	68,584	191,417

Table 4. Pessymistic replacement schedule

Year	Investment cost [€]	Annual savings [€], broken down into lamp sets										Tot. annual savings [€]
		s_1	s_2	s_3	s_4	s_5	s_6	s_7	s_8	s_9	s_{10}	
1	235,094	26,483	-	-	-	-	-	-	-	-	-	26,483
2	234,192	26,483	29,871	-	-	-	-	-	-	-	-	56,354
3	234,192	26,483	29,871	30,932	-	-	-	-	-	-	-	87,286
4	234,192	26,483	29,871	30,932	31,926	-	-	-	-	-	-	119,212
5	235,340	26,483	29,871	30,932	31,926	35,503	-	-	-	-	-	154,715
6	245,422	26,483	29,871	30,932	31,926	35,503	43,303	-	-	-	-	198,018
7	240,258	26,483	29,871	30,932	31,926	35,503	43,303	45,388	-	-	-	243,405
8	234,684	26,483	29,871	30,932	31,926	35,503	43,303	45,388	55,800	-	-	299,205
9	234,356	26,483	29,871	30,932	31,926	35,503	43,303	45,388	55,800	64,291	-	363,496
10	191,415	26,483	29,871	30,932	31,926	35,503	43,303	45,388	55,800	64,291	68,584	432,080
Total	€ 2,319,145											€ 1,980,253

Comparing Tables 2 and 4, the following conclusions can be drawn.

1. The optimized schedule yields the total energy savings being 44% higher (thus the OPEX is reduced by 44%), compared to the "pessimistic" schedule.
2. The investment payback period is 9 years for the optimized case and 11 years for the "pessimistic" one. It means, for the latter case, that after closing the 10-years investment time span, only 84% of incurred costs is covered by obtained energy savings.
3. Although the CAPEX is the same for bot cases, for the optimized scenario, however, the cumulated investment costs (€ 1,898,290) are covered by cumulated energy savings (€ 2,004,562) after the seventh year. Subsequent retrofit phases, carried out after this period, yield a net profit.

4   Graph Model for Practical Approach to Optimizing Modernization Profitability

The analysis presented in the previous section was made under the assumption that the planned LED fixture powers are known which is usually not true. In these circumstances, selecting appropriate lamp groups, $\{s_i\}$, is unfeasible due to its complexity. Namely, the number of variants to be verified can be computed straightforward as:

$$N = \binom{9,803}{1,000} \times \binom{8,803}{1,000} \times \cdots \times \binom{1,803}{1,000}.$$

To resolve this issue some premises for lamp aggregation have to be used. Such premises can rely on lamp locations with respect to streets, identifying parent power cabinets and/or power line circuits which lamps are plugged to, lighting classes of the nearest streets and so on. To use all those data we need a graph structure which is capable of storing them and which supports further processing.

4.1   Practical Constraints and the Graph Model

As shown above, determining appropriate lamp groups for retrofit scheduling, leads to a combinatorial explosion. To avoid this, one has to use a suitable formal representation which helps establish logical links, based on an environment context, between elements of a lighting system, and thus substantially reduces time required for determining $\{s_i\}$.

Besides the aggregation premises mentioned above, two additional constraints present in real-life scenarios can be helpful:

Constraint 1: Luminaires located alongside a given street, belong to a single s_i.
Constraint 2: Lamps belonging to a given s_i are neighboring rather than "isolated" among the ones from outside of s_i. Frequently, satisfying the former constraint implies fulfilling this one.

To manage the above calculations, with various types of entities, relationships and, last but not least, a high complexity, we need to define a handy graph model, supporting efficient computer methods of grouping luminaires, planned for retrofitting.

Definition 1. *Layout graph is a tuple of the form* $G = (V, E, \mathrm{lab}, \mathrm{att}, \mathcal{L}, \mathcal{A})$, *where:*

1. $V = A \cup L \cup C$ *is a nonempty set of vertices representing areas, lamps and power cabinets respectively. Nonempty sets* A, L, C *are pairwise disjoint;*
2. $E = E_A \cup E_L \cup E_C$ *is a nonempty set of edges representing structural relations:* $E_A = \{\{u, v\}, u, v \in A\}, E_L = \{\{u, v\}, u \in A, v \in L\}, E_C = \{\{u, v\}, u \in C, v \in L\}$. *The* E_A *set represents relations of neighboring of areas;* E_L *represents relations of being illuminated by;* E_C *represents relations of being powered by (or alternatively, being connected to). Note that cases of edges* $\{u, v\}$, *where* $u, v \in L$ *or* $u, v \in C$, *are excluded by the definition;*
3. $\mathrm{lab} : V \cup E \rightarrow \mathcal{L}$ *is a labeling function for edges and nodes,* $\mathcal{L}$ *is a set of labels. The* lab *function assigns to a vertex/edge a pair, consisting of an unique identifier and a label (which doesn't need to be unique);*
4. $\mathrm{att} : V \cup E \rightarrow 2^{\mathcal{A}}$ *is an attributing function for edges and nodes,* $\mathcal{A}$ *is a set of attributes. The* att *function assigns a set of attributes to a vertex/edge, which carry all relevant information about an entity: geometric characteristics, physical properties, quantitative information about spatial relations etc.*

Remark. Unlike in the other works (see [14]), for more readability, we define both lab and att functions jointly, for nodes and edges.

Figure 1a presents the sample real-life scheme of a lighting installation. It has a hierarchical structure with power cabinets at the top level (we do not consider power lines above this level). To each cabinet several circuits are attached. The fixtures are located at the bottom level. Each fixture is identified unambiguously (in the scale of entire city) by a triple consisting of cabinet identifier, circuit number and lamp number in a circuit. As we focus on fixture replacement, a distinction between a lamp and a fixture has to be made. Intuitively, a lamp is associated with a pole and it can contain several fixtures.

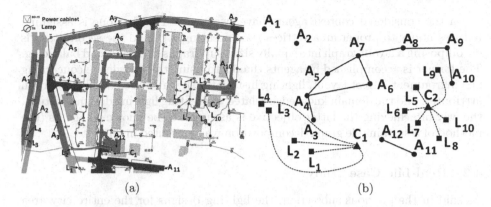

(a)   (b)

Fig. 1. The sample lighting system scheme (a) and its initial layout graph with E_A and E_C (dashed lines) edges (b). A_i's denote illuminated areas, C_i's denote power cabinets, L_i's denote lamps (for readability only a few lamps were denoted). Gray shaded areas are streets and diagonal line-filled ones on left image represent buildings

In the context of lighting design and design optimization, the following attributes of nodes are important. For area nodes we consider the elements of $att(v \in A)$: spatial properties such as GIS (geographic information system) coordinates, sizes, number of lanes, for roads and streets, surface reflective properties etc. Among interesting attributes of lamp nodes, $att(v \in L)$, are: GIS coordinates, pole height, arm length, fixture model, degree of wear of lamp parts, nominal lamp power, energy billing data of corresponding control cabinet, control schedule etc.

Building a layout graph is made in two phases. First, the basic information about area structure is created, i.e., nodes A and edges E_A are set. Analogously, a lighting installation structure is constructed: C and L sets together with E_C. Up to this moment (see Fig. 1b) the layout graph construction process is straightforward. More complicated actions are undertaken in its second phase, when relations among lighting points and areas (E_L) have to be identified.

In the second phase of a layout graph construction, each fixture node, $u \in L$, is ascribed to at least one vertex $v \in A$. This assignment is based on composite premises, containing such criteria as distances from a given fixture to the nearest areas, luminaire type (e.g., street, decorative etc.), area lighting class [2], parent power cabinet and so forth.

As recreating a full layout graph (tens of thousands of lamp nodes, thousands of area nodes, hundreds of power cabinet nodes) may lead to the significant processing workload, it is necessary to use a computational method reducing which makes computations doable in a reasonable time. The proven approach is distributing a graph model (see e.g., [4,12]). The particularly useful formalism, due to quadratic complexity of its basic operations [5], is the representation referred to as Replicated Complementary Graphs (RCG). Details of this formalism are beyond the work scope, but they may be found in [4]. The another important property of RCG model, except a good complexity of operations performed on its graphs, is possibility of using it in conjunction with multi-agent systems.

In the considered context agents have two tasks. The first one is assigning fixtures nodes to proper area vertices (i.e., creating E_L) and the second task is decomposing a layout graph into equally sized subgraphs to enable load balancing. The former is accomplished by agents thanks to using the knowledge concerning an environment (street layout, lighting installation), which is embedded in graph attributes, and the domain knowledge related to creating lighting designs. In turn, the way of achieving the latter objective is assumed to be known to agents (the method of a sustainable graph decomposition was presented in [6]).

4.2   Real-Life Case Study

As said in the previous subsection, the lighting designs for the entire city area have to be made, when planned LED powers are not known. Creating a lighting design was discussed and analyzed in previous works, [12,14] and will not be discussed here.

When setting lamp groups s_i for a real-life case, the condition $|s_i| = 1,000$ is weakened by adding the 2% size margin, to enable satisfying Constraint 1. Thus, in the performed tests it was set $|s_i| \leq 1,020$.

The selection process is made as follows.

1. Considered problem requires aggregating 1,786 illuminated areas into 499 "super-areas". This aggregations is made on the basis of the area adjacency criterion. Each "super-area" has a set of lamps ascribed, say σ_i.
2. The $\{\sigma_i\}$ series is sorted descending by the cumulated power reduction, related to the HID to LED replacement.
3. We define: $s_j = \bigcup\limits_{m=1}^{i(j)} \sigma_m$ for $j = 1, 2, \ldots, 10$, where

$$|s_j| \leq 1,020 \text{ and } \sum_{m=1}^{i(j)+k} |\sigma_m| > 1,020 \text{ for } k > 0.$$

4. As lamp dimmings vary, depending on neighboring lighting points (there may exist simultaneously both HID and LED light sources in a neighborhood of s_j), power savings are recalculated for s_j.

Table 5 presents investment costs per s_i and obtained annual savings during the retrofit duration period.

As seen, the resultant savings are not so high as for the theoretical approach, but they still remain significant, compared to the "pessimistic" scenario, namely 32% higher. Moreover they enable covering the whole fixture-related outlay, before closing an investment time span.

Having the above one can not only setup a lighting standard-compliant installation but also compute potential gains, in terms of energy, money, CO_2 emission and other, business parameters, achieved thanks to a planned retrofit.

Table 5. Retrofit schedule for the real-life case. Numbers of lamps for particular sets are placed in brackets

Year	Investment cost [€]	Annual savings [€], broken down into lamp sets										Tot. annual savings [€]
		s_1 (1,004)	s_2 (979)	s_3 (938)	s_4 (1,002)	s_5 (1,020)	s_6 (986)	s_7 (1,012)	s_8 (1,009)	s_9 (1,013)	s_{10} (840)	
1	249,309	57,083	-	-	-	-	-	-	-	-	-	57,083
2	229,930	57,083	51,084	-	-	-	-	-	-	-	-	108,166
3	227,541	57,083	51,084	50,154	-	-	-	-	-	-	-	158,321
4	234,660	57,083	51,084	50,154	45,797	-	-	-	-	-	-	204,118
5	238,876	57,083	51,084	50,154	45,797	45,350	-	-	-	-	-	249,467
6	231,241	57,083	51,084	50,154	45,797	45,350	41,524	-	-	-	-	290,991
7	237,002	57,083	51,084	50,154	45,797	45,350	41,524	40,705	-	-	-	331,696
8	236,300	57,083	51,084	50,154	45,797	45,350	41,524	40,705	37,942	-	-	369,637
9	237,237	57,083	51,084	50,154	45,797	45,350	41,524	40,705	37,942	35,080	-	404,717
10	197,049	57,083	51,084	50,154	45,797	45,350	41,524	40,705	37,942	35,080	27,363	432,080
Total	€ 2,319,145											€ 2,606,275

5 Conclusions

In this work we focused on the problem of optimization of a retrofit schedule, planned for a 10-years time span. We showed that the difference between power savings which can be achieved after 10 years, in theoretical, optimistic and pessimistic scenario, is 44%. In real-life cases, however, one is constrained among other, by common-sense constraints such as necessity of replacing in a single retrofit phase all lamps located alongside a given street. To determine an appropriate retrofit order and to avoid combinatorial explosion one has to aggregate lamps planned to retrofit basing on additional, environment-based system properties. The natural and most suitable formal representation which can be used for this purpose are graphs. The calculations which were made in compliance with the discussed constrained allowed to achieve 32% improvement, compared to theoretical "pessimistic" level.

For the sake of simplicity, in this work we made an implicit assumption concerning constant lamp efficiencies and prices, during the entire 10-years period. In a real life case, however, the replacement planning process should be repeated year by year, for all remaining lamps, planned for retrofitting.

It should be noted that there exist areas for further research. The first one is analysis of retrofits which, besides the lamp replacement, introduce control capabilities to a lighting installation. The second question, potentially interesting for municipalities, is selecting a fixture vendor which guarantees minimized payback period.

References

1. Australian/New Zealand Standard: AS/NZS 1158.1.1:2005 Lighting for roads and public spaces Vehicular traffic (Category V) lighting - Performance and design requirements. SAI Global Limited (2005)

2. European Committee For Standarization: Road Lighting. Performance requirements, EN 13201-2:2015 (2015)
3. Illuminating Engineering Society of North America (IESNA): American National Standard Practice For Roadway Lighting, RP-8-14. IESNA, New York (2014)
4. Kotulski, L.: On the control complementary graph replication. In: Mazurkiewicz, J., et al. (eds.) Models and methodology of system dependability, Monographs of System Dependability, vol. 1. Oficyna Wydawnicza Politechniki Wrocławskiej, Wrocław (2010)
5. Kotulski, L., Sędziwy, A.: Parallel graph transformations supported by replicated complementary graphs. In: Dobnikar, A., Lotrič, U., Šter, B. (eds.) ICANNGA 2011. LNCS, vol. 6594, pp. 254–264. Springer, Heidelberg (2011). https://doi.org/10.1007/978-3-642-20267-4_27
6. Kotulski, L., Sedziwy, A.: Agent framework for decomposing a graph into the equally sized subgraphs. In: FCS (2008)
7. Peña-García, A., Gil-Martín, L., Hernández-Montes, E.: Use of sunlight in road tunnels: an approach to the improvement of light-pipes' efficacy through heliostats. Tunn. Undergr. Space Technol. **60**, 135–140 (2016). https://doi.org/10.1016/j.tust.2016.08.008.   http://www.sciencedirect.com/science/article/pii/S0886779815302121
8. Pena-García, A., Gómez-Lorente, D., Espín, A., Rabaza, O.: New rules of thumb maximizing energy efficiency in street lighting with discharge lamps: the general equations for lighting design. Eng. Optim. **48**(6), 1080–1089 (2016). https://doi.org/10.1080/0305215X.2015.1085715.   http://dx.doi.org/10.1080/0305215X.2015.1085715
9. Rabaza, O., Pena-García, A., Pérez-Ocón, F., Gómez-Lorente, D.: A simple method for designing efficient public lighting, based on new parameter relationships. Expert Syst. Appl. **40**(18), 7305–7315 (2013). http://goo.gl/10UQGm
10. Salata, F., et al.: Energy optimization of road tunnel lighting systems. Sustainability **7**(7), 9664 (2015). https://doi.org/10.3390/su7079664
11. Sędziwy, A.: Effective graph representation for agent-based distributed computing. In: Jezic, G., Kusek, M., Nguyen, N.-T., Howlett, R.J., Jain, L.C. (eds.) KES-AMSTA 2012. LNCS (LNAI), vol. 7327, pp. 638–647. Springer, Heidelberg (2012). https://doi.org/10.1007/978-3-642-30947-2_69
12. Sędziwy, A.: Sustainable street lighting design supported by hypergraph-based computational model. Sustainability **8**(1), 13 (2016)
13. Sędziwy, A., Kotulski, L.: Graph-based optimization of energy efficiency of street lighting. In: Rutkowski, L., Korytkowski, M., Scherer, R., Tadeusiewicz, R., Zadeh, L.A., Zurada, J.M. (eds.) ICAISC 2015. LNCS (LNAI), vol. 9120, pp. 515–526. Springer, Cham (2015). https://doi.org/10.1007/978-3-319-19369-4_46
14. Sedziwy, A., Kotulski, L.: Towards highly energy-efficient roadway lighting. Energies **9**(4), 263 (2016). https://doi.org/10.3390/en9040263. https://doi.org/10.3390%2Fen9040263
15. Wojnicki, I., Ernst, S., Kotulski, L., Sędziwy, A.: Advanced street lighting control. Expert Syst. Appl. **41**(4, Part 1), 999–1005 (2014). https://doi.org/10.1016/j.eswa.2013.07.044. http://www.sciencedirect.com/science/article/pii/S0957417413005319
16. Wojnicki, I., Kotulski, L.: Street lighting control, energy consumption optimization. In: Rutkowski, L., Korytkowski, M., Scherer, R., Tadeusiewicz, R., Zadeh, L.A., Zurada, J.M. (eds.) ICAISC 2017. LNCS (LNAI), vol. 10246, pp. 357–364. Springer, Cham (2017). https://doi.org/10.1007/978-3-319-59060-8_32. http://bit.ly/2QnbhZV

Solving the Unrelated Parallel Machine Scheduling Problem with Setups Using Late Acceptance Hill Climbing

Mourad Terzi[1], Taha Arbaoui[1(✉)], Farouk Yalaoui[1], and Karima Benatchba[2]

[1] Logistique et Optimisation des Systèmes Industriels (LOSI-ICD),
Université de Technologie de Troyes,
12 rue Marie Curie, CS 42060, 10004 Troyes, France
{mourad.terzi,taha.arbaoui,farouk.yalaoui}@utt.fr
[2] Laboratoire des Méthodes de Conception de Systèmes (LMCS), Ecole Nationale
Supérieure d'Informatique, Algiers, Algeria
k_benatchba@esi.dz

Abstract. We propose a Late Acceptance Hill-Climbing (LAHC) approach to solve the unrelated parallel machine scheduling problem with sequence and machine-dependent setup times. LAHC is an iterative list-based single-parameter metaheuristic that exploits information from one iteration to another to decide whether the new candidate solution is accepted. A dynamic job insertion heuristic is used to generate initial solutions. Three local search operators (job swap between different machines, job swap within the same machine and job insertion from one machine to another) are used to improve solutions. A Variable Neighborhood Descent (VND) method is proposed to improve the candidate solution and accelerate the convergence of the LAHC. To the best of our knowledge, this is the first application of LAHC to parallel machine scheduling problems. We evaluate and compare the proposed algorithm against the best methods from the literature. Having a single parameter which makes it simpler than all existing approaches, the proposed method outperforms existing methods on most of the tested benchmark instances.

1 Introduction

In the last two decades, several heuristics and metaheuristics were proposed to solve parallel machine scheduling problems. The unrelated parallel machine scheduling problem with sequence and machine-dependent setup times $(Rm|s_{ijk}|C_{max})$ [9] is a generalization of the identical and uniform parallel machine scheduling problem. $Rm|s_{ijk}|C_{max}$ consists of a set N of n jobs $N = \{1, 2, ..., n\}$ and a set M of m machines $M = \{1, 2, ..., m\}$. Each job is a single task that has to be processed on exactly one machine. In the case of unrelated parallel machines, a job j can have a different processing time from one machine to another. Setup times depend on both the machine and the pair of jobs, i.e. the setup time between jobs j and k on machine i is different from

© Springer Nature Switzerland AG 2020
N. T. Nguyen et al. (Eds.): ACIIDS 2020, LNAI 12033, pp. 249–258, 2020.
https://doi.org/10.1007/978-3-030-41964-6_22

the setup time between jobs k and j on the same machine. Similarly, the setup time between jobs j and k on machine i is different from the setup time between job j and k on another machine l [16]. The considered objective function is the minimization of the makespan. Motivated by its effectiveness and its simplicity, the Late Acceptance Hill-Climbing (LAHC) method is used to solve the $R_m|s_{ijk}|C_{max}$. To the best of our knowledge, this work is the first application of LAHC on parallel machine scheduling problems. The approach, based on three operators to generate the new candidate solution and a Variable Neighborhood Descent [14] to accelerate its convergence speed, is presented and evaluated.

The remainder of this paper is organized as follows. In Sect. 2, a brief literature review on the unrelated parallel machine scheduling problem with setups is given. Section 3 details the proposed Late Acceptance Hill-Climbing approach. Section 4 is devoted to the computational experiments, where we compare the proposed approach with existing approaches and show that it outperforms the best methods of the literature. Finally, the conclusion is to be found in Sect. 5.

2   Literature Review

Several exact and heuristic methods have been proposed to address the unrelated parallel machine scheduling problems with setup times. Helal et al. [7] presented a tabu search metaheuristic to solve the $R_m|s_{ijk}|C_{max}$. In Rabadi et al. [12], the authors presented a metaheuristic called Meta-Raps. Results showed that Meta-Raps found the optimal solution for the small problems and outperformed the solutions obtained by the then-existing heuristics for a larger problems. Ying et al. [17] proposed a simulated annealing for parallel machine scheduling problems with setup times. The authors showed that their method outperforms Meta-Raps. Arnaout et al. [3] proposed a two-stage Ant Colony Optimization (ACO1) for the same problem. Their study was limited to specific problem instances, in which setup times and processing times were generated from the same uniform distribution and the ratio N/M was large. The proposed ACO1 reached better results compared to the tabu search [7] and Meta-RaPS [12]. In [2], the same authors presented a second Ant Colony Optimization (ACO2), which is an extension of ACO1. ACO2 used a pheromone re-initialization technique to avoid a local optima convergence. ACO2 was tested on a large set of instances in which there are different setup times and processing times dominance criteria. Moreover, there is no restriction for the ratio N/M. ACO2 was compared to the tabu search [7], Meta-RaPS [12], simulated annealing [17] and ACO1. Experiments proved the superiority of the enhanced ACO2.

Vallada and Ruiz [16] introduced a Mixed Integer Programming (MIP) and a Genetic Algorithm (GA) to solve $R_m|s_{ijk}|C_{max}$. GA includes two local search methods. The first was applied according to a probability p_{ls} in the generation phase and after the mutation. The second was used to enhanced the crossover operator. Computational experiments showed that proposed GA obtained better results compared to the best known methods. An Immune-inspired algorithm for the $R_m|s_{ijk}|C_{max}$ was proposed by Diana et al. [5]. The algorithm was compared to GA [16], ACO2 [2] and simulated annealing [17]. Experiments performed showed that the Immune-inspired algorithm reached better results than all the other methods. The main drawback of all these approaches is the number of parameters to be tuned to reach their best performance. The approach's performance is therefore dependent on the parameter tuning and changes from one instance to another since the same parameter configuration, once chosen, is applied to all instances.

Compared to heuristic approaches, fewer exact methods have been proposed, mainly due to the difficulty encountered to solve large-scale instances. Martello et al. [10] presented a Branch and Bound (B&B) for makespan minimization in parallel machine environment, the lower bound is based on a Lagrangian relaxation. Shim and Kim [13] proposed a Branch-and-Bound method for the identical parallel machine scheduling problem with total tardiness minimization. Na et al. [11] provided a B&B to minimize the total weighted tardiness. The proposed B&B included an efficient function for the lower bound and a heuristic for upper bound. Recently, Tran et al. [15] proposed a logic-based benders decomposition and a branch-and-check method to solve $R_m|s_{ijk}|C_{max}$. Their approaches were able to solve instances of 100 jobs to optimality. Fanjul-Peyro et al. [6] introduced a mathematical programming approach that is able to solve large instances in reasonable time.

For more details and works on unrelated parallel machine scheduling problems, we refer the reader to the recent survey on scheduling problems with setups [1].

3 Overview of the Proposed Approach

As described in [4], LAHC is an iterative metaheuristic which uses the history of the search process by saving the values of the current solution's costs in a candidate list called Fa of length LFa. LAHC has been shown effective compared to other metaheuristics, thanks to a simple and efficient acceptance criterion which allows it to escape from local optima. LAHC accepts the new candidate solution s_i^* at an iteration i if its quality is better than the current solution s_i or if it is better than the s', where s' is the solution of the iteration $i - LFa$ (i.e. $C(s_i^*) \leqslant C(s_i)$ or $C(s_i^*) \leqslant C(s_{i \bmod LFa})$). The length of the candidate list LFa is the single input parameter of the approach, giving it the advantage of simple tuning. The pseudocode of LAHC is shown in Algorithm 1 [4].

3.1 Solution Encoding and Construction Phase

A parallel machine schedule can be represented using a matrix of M rows. Each row corresponds to a given machine and contains a list of jobs assigned to this machine. The index of a job j in a row m determines its processing order in a machine m. The solution's cost is $C_{max} = max(C_1, .., C_i, .., C_M)$ where C_i is the completion time of machine i, calculated using row i.

Algorithm 1. Late Acceptance Hill Climbing Approach

Produce an initial solution s;
Calculate the initial cost $C(s)$;
Set maximum number of iterations $maxIter$;
Specify LFa ;
for $k \leftarrow 0$ **to** $LFa - 1$ **do**
$\quad \mid \quad f_k = C(s)$;
end
$i = 0$;
repeat
$\quad \mid \quad$ Construct a solution s_i^*;
$\quad \mid \quad$ Calculate $C(s_i^*)$;
$\quad \mid \quad$ v = i mod LFa;
$\quad \mid \quad$ **if** $C(s_i^*) < C(s_i)$ *or* $C(s_i^*) < C(s_v)$ **then**
$\quad \mid \quad \quad \mid \quad s_i = s_i^*$
$\quad \mid \quad$ **end**
$\quad \mid \quad$ **if** $C(s_i^*) < C(s_v)$ **then**
$\quad \mid \quad \quad \mid \quad s_v = s_i^*$
$\quad \mid \quad$ **end**
$\quad \mid \quad$ i++;
until $i > maxIter$;

As mentioned above, LAHC starts its execution by generating an initial candidate solution. In order to have a good quality initial solution, we used the greedy heuristic "Dynamic Machine insertion" (DMI) based on the heuristic proposed in [8]. DMI showed its performance in [5] when it was used to generate the initial immunological memory. DMI generates a solution by assigning, at each iteration, a single job to a given machine in the partial solution S_p. DMI starts by creating a list N of the jobs which are yet to be assigned to S_p. At each iteration, DMI computes for each job j of N its best pair of (machine, position) if we add it to the partial solution S_p. The q best jobs are assigned to the Restricted Candidate List RCL. Next, DMI randomly selects a job j from RCL and inserts it in S_p and updates N. DMI stops its execution when there is no more jobs in the list N.

3.2 Neighbourhood and Local Search Operators

Each LAHC iteration begins by generating a new candidate solution in the neighborhood of the current one. To achieve this, we start by applying the following operators:

1. **Internal swap**: Interchanges the positions of two jobs j and k assigned to the same machine i.
2. **External swap**: Interchanges two jobs j and k assigned to two different machines i and i'. Job j of machine i is inserted in position of job k in machine i' and vice versa.

3. **External insertion**: Inserts a job j of machine i in another machine i'.

In each operator, the choice of jobs and machines is randomly performed.

A Variable Neighborhood Descent (VND) is applied to the new candidate solution. The proposed VND, as shown in Algorithm 2, is based on the three operators described above to explore the neighborhood of the current candidate solution. The VND method includes the following five operators:

1. **Internal swap exploration**: Applies the *Internal swap* movement between the jobs of the machine m, where $C_m = C_{max}$. **Internal swap exploration** keeps the movement that improves best of the current solution.
2. **External swap exploration**: Applies *External swap* movement between the jobs of the machine m that has $C_m = C_{max}$ and jobs of a machine l that has $C_l \neq C_{max}$. The best movement is returned.
3. **External insertion exploration**: Inserts each job of the machine m, where $C_m = C_{max}$, in each position of a machine l that has $C_l \neq C_{max}$ randomly selected. **External insertion exploration** keeps the movement that improves best the current solution.
4. **Balancing**: Inserts a maximum number of jobs of the machine m, where $C_m = C_{max}$, in the machine l that has the lower completion time.

Algorithm 2. Proposed Variable Neighborhood Descent

$c = 0$;
while $c < 5$ do
    if $c == 0$ then
        | σ = Internal_swap neighborhood_exploration (s_i)
    end
    if $c == 1$ then
        | σ = Inter_machine_insertion (s_i)
    end
    if $c == 2$ then
        | σ = Internal_insertion_neighborhood_exploration (s_i)
    end
    if $c == 3$ then
        | σ = External_swap_neighborhood_exploration (s_i)
    end
    if $c == 4$ then
        | σ = Balancing(s_i)
    end
    if $C(\sigma) < C(s_i)$ then
        $S = \sigma$;
        $c = 0$;
    else
        $c = c + 1$;
    end
end
Return S;

---

5. **Inter machine insertion**: In contrary to the first four methods which are based on machine m that has $C_m = C_{max}$, this operator improves the input solution by the insertion of each job of each machines in all positions of all the others machines and keeps the insertion which gives the best makespan.

4   Experiments

Our proposed method was implemented using Python3. All the experiments were performed on Intel(R) Xeon(R) Gold 5120, 2.2 GHz. 92 GB of RAM. Parallel programming was not used.

Our experiments were based on the instances given in [16]. They consist of 1640 instances divided into two subsets, Small and Large. The small set consists of 640 instances that have the number of jobs in $\{6, 8, 10, 12\}$ and the number of machines in $\{2, 3, 4, 5\}$. The large set consists of 1000 instances that have the number of jobs in $\{50, 100, 150, 200, 25\}$ and the number of machines in $\{10, 15, 20, 25, 30\}$. The values of the setup times were uniformly generated from the intervals $\{1 - 9\}$, $\{1 - 49\}$, $\{1 - 99\}$ and $\{1 - 124\}$.

Table 1. Mean RPD values in the first experiments on the small instances.

Class	T(s)	LAHC	Immune
6×2	0.3	0%	0%
6×3	0.45	0%	0%
6×4	0.6	**−0.8%**	0%
6×5	0.75	2.6%	**0%**
8×2	0.4	**−0.1%**	0%
8×3	0.6	**−0.5%**	0%
8×4	0.8	0%	0%
8×5	1	0.1%	**0%**
10×2	0.5	0%	0%
10×3	0.75	**−0.3%**	0%
10×4	1	**−0.9%**	0%
10×5	1.25	**−0.3%**	0%
12×2	0.6	**−0.8%**	0%
12×3	0.9	**−0.5%**	0%
12×4	1.2	**−0.7%**	0%
12×5	1.5	**−0.2%**	**−0.16%**

To the best of our knowledge, the Immune-inspired algorithm proposed in [5] represents the best performing approach on these instances. Therefore, we use it as basis of comparison for our proposed Late Acceptance Hill-Climbing

Table 2. Mean RPD values in the first experiments on the large instances.

Class	T(s)	LAHC	Immune
50 × 10	13	**−2.9%**	−2.2%
50 × 15	19	**−8%**	−6.5%
50 × 20	26	**−10.3%**	−10.1%
50 × 25	31	−12.6%	**−12.7%**
50 × 30	38	−10.8%	**−11%**
100 × 10	25	**−1.3%**	0.3%
100 × 15	38	**−7.7%**	−5.3%
100 × 20	50	**−12.7%**	−8.7%
100 × 25	63	**−16.7%**	−12.9%
100 × 30	75	**−19.3%**	−15.3%
150 × 10	38	**−1.3%**	0.4%
150 × 15	56	**−6.8%**	−3.6%
150 × 20	75	**−12.1%**	−8.1%
150 × 25	94	**−17.5%**	−12.8%
150 × 30	113	**−20.5%**	−13.9%
200 × 10	50	**−1.8% %**	0.2%
200 × 15	75	**−6.7%**	−4%
200 × 20	100	**−12.1%**	−8%
200 × 25	125	**−17.1%**	−12.2%
200 × 30	150	**−20.6%**	−14.5%
250 × 10	63	**−1.8% %**	0.4%
250 × 15	94	**−7.3%**	−4.5%
250 × 20	125	**−11.8%**	−8.2%
250 × 25	156	**−17%**	−12.2%
250 × 30	188	**−20.8%**	−15%

approach. To make a consistent and fair comparison, we implemented the best version (CL_4) of the Immune-inspired algorithm [5].

The comparison was based on Relative Percentage Deviation (RPD), which is the difference between the makespan found by the algorithm $Method_{sol}$ and the best solution from the literature $Best_{lit}$ (Eq. (1)). As in [5], we have executed each method 5 times on each instance. For each execution, RPD and the mean RPD of each class of instances are calculated. A class consists of a set of the instances that have the same number of jobs and machines.

$$RPD = \frac{Method_{sol} - Best_{lit}}{Best_{lit}} \tag{1}$$

To take advantage as much as possible of the five VND operators, we ran LAHC several times using different orders of the five operators inside VND on he large set. We practically observed that the order used in Algorithm 2 is the best order.

Our experiments were conducted in two phases. At the first, we ran the proposed LAHC and the implemented CL_4 by fixing the stopping criterion (time limit) as proposed in [16] (Eq. (2)). The parameter calibration of the implemented immune-inspired algorithm was the same as those used in [5]. For the LAHC, the LFa is set to 1000. The mean of the relative percentage deviation of our first experiments are shown in Tables 1 and 2. In both tables, the first column provides the considered class. The second one represents the average computing time for all the instances of the class, which is common to both approaches. The two last columns are devoted to the mean RPD values of LAHC and the immune-inspired algorithm.

$$T_{ime} = n * (m/2) * t \ (ms) \tag{2}$$

Analyzing the RPD's mean values in Table 1, which is devoted to the results of small instances, we can conclude that the proposed LAHC reaches better solutions than the immune-inspired algorithm on a large number of classes. Looking at Table 2 which presents the results obtained for large instances, the same conclusion can be made. Except instances of the classes 50×25 and 50×30 in which the implemented CL_4 reaches better solutions, the proposed LAHC is able to reach the best results on the remaining classes. This superiority explains the efficiency of using three different movements and a sophisticated VND based on five neighborhood operators.

In order to compare the proposed LAHC and the immune-inspired algorithm thoroughly, we conducted another experiment on the large instances where the stopping criterion (time limit) was fixed to ten times the time considered in [16]. The objective behind the choice of such a criterion is to give a sufficient time to the methods to reach their best solutions. To be able to reach such a time limit, the LFa was fixed to 10000. The mean values of the relative percentage deviation of these second experiments are shown in Table 3.

One can notice that providing both methods with more time allowed them to reach a better performance. This shows the importance of the choice of the stopping criterion. Moreover, results also show that the immune-inspired algorithm reaches better mean RPD values compared to LAHC on more classes, such as 150×10. However, LAHC remains very competitive and provides the best results for more than half of the classes.

From the first and the second experiments, one can conclude that LAHC reaches competitive solutions and is able to outperform the immune-inspired algorithm when the time limit is limited.

Table 3. Mean RPD values in the second experiment on the large instances.

Class	T(s)	LAHC	Immune
50 × 10	130	**−4.3%**	**−4.3%**
50 × 15	190	**−8.9%**	−8.3%
50 × 20	260	−11.4%	**−11.7%**
50 × 25	310	−14.2%	**−15%**
50 × 30	380	−13.6%	**−14.3%**
100 × 10	250	−3%	**−4.9%**
100 × 15	380	−9.5%	**−9.9%**
100 × 20	500	**−14.3%**	−13.6%
100 × 25	630	**−18.8%**	−17.7%
100 × 30	750	**−21%**	−19.2%
150 × 10	380	−2.8%	**−4.8%**
150 × 15	560	−7.9%	**−8.9%**
150 × 20	750	**−13.6%**	−12.8%
150 × 25	940	**−18.8%**	−17%
150 × 30	1130	**−21.8%**	−18.9%
200 × 10	500	−3.2%	**−4.7%**
200 × 15	750	−8.4%	**−9.2%**
200 × 20	1000	**−13.2%**	−13.1%
200 × 25	1250	**−18.3%**	−17%
200 × 30	1500	**−21.7%**	−18.7%
250 × 10	630	−3.2%	**−4.3%**
250 × 15	940	−8.7%	**−9.5%**
250 × 20	1250	−12.9%	**−13.1%**
250 × 25	1560	**−17.9%**	−16.4%
250 × 30	1880	**−21.9%**	−18.6%

5   Conclusion

In this paper, we proposed an efficient Late-acceptance Hill Climbing metaheuristic to solve the unrelated parallel machines scheduling problem with setup times. LAHC was based on an effective five-operator variable neighborhood descent. To demonstrate the efficiency of LAHC, we compared it to the approach provided in [5], considered as one of the best methods in the literature, on the literature benchmarks. The results of the proposed LAHC showed the importance of including advanced local search methods to improve solutions' quality. Moreover, they showed that LAHC is able to outperform the best methods of the literature, particularly when methods are given short times.

Acknowledgments. This research has been funded by a grant from Aube French Department and Troyes Champagne Metropole (TCM).

References

1. Allahverdi, A.: The third comprehensive survey on scheduling problems with setup times/costs. Eur. J. Oper. Res. **246**(2), 345–378 (2015)
2. Arnaout, J.-P., Musa, R., Rabadi, G.: A two-stage ant colony optimization algorithm to minimize the makespan on unrelated parallel machines-Part II: enhancements and experimentations. J. Intell. Manuf. **25**(1), 43–53 (2014)
3. Arnaout, J.-P., Rabadi, G., Musa, R.: A two-stage ant colony optimization algorithm to minimize the makespan on unrelated parallel machines with sequence-dependent setup times. J. Intell. Manuf. **21**(6), 693–701 (2010)
4. Burke, E.K., Bykov, Y.: The late acceptance hill-climbing heuristic. Computing Science and Mathematics, University of Stirling, Technical report No. CSM-192 (2012)
5. Diana, R.O.M., de França Filho, M.F., de Souza, S.R., de Almeida Vitor, J.F.: An immune-inspired algorithm for an unrelated parallel machines' scheduling problem with sequence and machine dependent setup-times for makespan minimisation. Neurocomputing **163**, 94–105 (2015)
6. Fanjul-Peyro, L., Ruiz, R., Perea, F.: Reformulations and an exact algorithm for unrelated parallel machine scheduling problems with setup times. Comput. Oper. Res. **101**, 173–182 (2019)
7. Helal, M., Rabadi, G., Al-Salem, A.: A tabu search algorithm to minimize the makespan for the unrelated parallel machines scheduling problem with setup times. Int. J. Oper. Res. **3**(3), 182–192 (2006)
8. Kurz, M.E., Askin, R.G.: Heuristic scheduling of parallel machines with sequence-dependent set-up times. Int. J. Prod. Res. **39**(16), 3747–3769 (2001)
9. Lee, Y.H., Pinedo, M.: Scheduling jobs on parallel machines with sequence-dependent setup times. Eur. J. Oper. Res. **100**(3), 464–474 (1997)
10. Martello, S., Soumis, F., Toth, P.: Exact and approximation algorithms for makespan minimization on unrelated parallel machines. Discrete Appl. Math. **75**(2), 169–188 (1997)
11. Na, D.-G., Kim, D.-W., Jang, W., Chen, F.F.: Scheduling unrelated parallel machines to minimize total weighted tardiness. In: 2006 IEEE International Conference on Service Operations and Logistics, and Informatics, pp. 758–763. IEEE (2006)
12. Rabadi, G., Moraga, R.J., Al-Salem, A.: Heuristics for the unrelated parallel machine scheduling problem with setup times. J. Intell. Manuf. **17**(1), 85–97 (2006)
13. Shim, S.-O., Kim, Y.-D.: A branch and bound algorithm for an identical parallel machine scheduling problem with a job splitting property. Comput. Oper. Res. **35**(3), 863–875 (2008)
14. Talbi, E.-G.: Metaheuristics: From Design to Implementation, vol. 74. Wiley, Hoboken (2009)
15. Tran, T.T., Araujo, A., Beck, J.C.: Decomposition methods for the parallel machine scheduling problem with setups. INFORMS J. Comput. **28**(1), 83–95 (2016)
16. Vallada, E., Ruiz, R.: A genetic algorithm for the unrelated parallel machine scheduling problem with sequence dependent setup times. Eur. J. Oper. Res. **211**(3), 612–622 (2011)
17. Ying, K.-C., Lee, Z.-J., Lin, S.-W.: Makespan minimization for scheduling unrelated parallel machines with setup times. J. Intell. Manuf. **23**(5), 1795–1803 (2012)

Computer Vision Techniques

Study on Digital Image Evolution of Artwork by Using Bio-Inspired Approaches

Julia Garbaruk[1]([✉]) [iD], Doina Logofătu[1]([✉]) [iD], Costin Bădică[2]([✉]) [iD], and Florin Leon[3]([✉]) [iD]

[1] Department of Computer Science and Engineering, Frankfurt University of Applied Sciences, 60318 Frankfurt am, Germany
logofatu@fb2.fra-uas.de
[2] Department of Computer Sciences and Information Technology, University of Craiova, 200285 Craiova, Romania
[3] Faculty of Automatic Control and Computer Engineering, Gheorghe Asachi Technical University of Iaşi, 700050 Iaşi, Romania

Abstract. Whether for optimizing the speed of microprocessors or for sequence analysis in molecular biology—evolutionary algorithms are used in astoundingly many fields. Also the art was influenced by evolutionary algorithms—with principles of natural evolution works of art can be created or imitated, whereby initially generated art is put through an iterated process of selection and modification. This paper covers an application in which given images are emulated evolutionary using a finite number of semi-transparent overlapping polygons, which also became known under the name "Evolution of Mona Lisa". In this context, different approaches to solve the problem are tested and presented here. In particular, we want to investigate whether Hill Climbing Algorithm in combination with Delaunay Triangulation and Canny Edge Detector that extracts the initial population directly from the original image performs better than the conventional Hill Climbing and Genetic Algorithm, where the initial population is generated randomly.

Keywords: Evolution of Mona Lisa · Evolving images · Evolutionary algorithms

1 Introduction

The idea of constructing images using evolving polygons was first introduced by Johansson in 2008 [1]. Using an evolutionary approach, more precisely an evolutionary strategy, he managed to create a copy of the world famous painting Mona Lisa from only 50 polygons.

The color and the coordinates of the initial polygons were chosen randomly and the transparency and overlapping aspects allowed for overlapping pixels which result in different color, following the rules of color mixing. The population of the original implementation consisted of only two individuals—a parent

© Springer Nature Switzerland AG 2020
N. T. Nguyen et al. (Eds.): ACIIDS 2020, LNAI 12033, pp. 261–270, 2020.
https://doi.org/10.1007/978-3-030-41964-6_23

set of polygons and a child set of polygons created by mutating the color and coordinates of the parent set of polygons. From these, the set was selected, which came closest to Mona Lisa.

Through steady evolution of the polygons, Johansson reached a very good approximation of Mona Lisa after ca. 1,000,000 generations.

The problem is quite extraordinary, as it represents an interface between two very different and seemingly incompatible disciplines: computer science and art (Fig. 1).

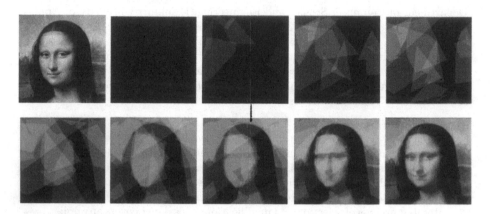

Fig. 1. Gradual transformation of the polygons into the Mona Lisa painting (Johansson's experiment).

Also from the point of view of computer science, it is a very interesting problem, since it combines various aspects of computer science, such as evolutionary algorithms and image processing. Another interesting feature is that it reveals the limits of human possibilities: Especially at the beginning of polygon evolution, when the overall picture still has no clearly recognizable shapes, it is very difficult for us to register small improvements and to judge which image most closely resembles Mona Lisa. Even later potential improvements often are noticeable only if there are a certain number of generations between two pictures.

2   Previous Work

As already mentioned, the Roger Johansson blog laid the foundation for this problem in 2008. "Evolution of Mona Lisa" was received with great interest by the IT community and was hotly debated. Soon other programmers tried to reprogram it. The most outstanding is probably the variant from alteredqualia.com [3] with its implementation in JavaScript, which is accessible via the webbrowser for everyone and has some interesting features. For example, you can set the mutation rate, the number of polygons and the vertices there and then see in real time how the image is evolving, what fitness the current image has and how many mutations have already been made (Fig. 2).

Fig. 2. Two pictures, which were created after 301 generations (l.) and 348 generations (r.)—only the computer can determine with certainty, which of them looks more like Mona Lisa.

After initial popularity in the IT community, the interest in this issue later seemed to have dropped. A formal definition of the problem followed only ten years later, in 2018, by Thuan et al. [2]. In their work, they also identified some issues in Johansson's original solution and the solutions of other programmers, who tried to solve this problem, namely the inefficiency and the slow convergence, which means that the program needs more and more generations to carry out small improvements as time progresses. In their view, the reason for this is, among other things, that the creation of diversity in the population of candidate solutions was always given a higher priority than the natural selection process, in which the overall quality of the population is increased by removing bad candidates. The latter, however, is more effective, according to the authors. At the same time they also present their own approach to "Evolution of Mona Lisa", namely a natural selection strategy inspired by Simulated Annealing. Complementing a classical Genetic Algorithm, this approach provides a mechanism that prevents Genetic Algorithm from getting stuck in a local optimum. To achieve this, they introduce an additional probability of preferring a worse solution over a better one. This way they give a solution the chance to move away from a local optimum in order to find a better optimum. This probability of choosing the worse solution falls during the evolutionary process, that is, as the number of generations increases. Numerous tests have shown that this approach delivers better results than the already existing.

3   Proposed Approaches

In this work we tried to solve the problem using a combination of Hill Climbing Algorithm with Delaunay Triangulation and Canny Edge Detector. By using Canny Edge Detector and Delaunay Triangulation, we can generate a simplified pattern of output before mutating it using the Hill Climbing Algorithm. This is contrasted with the conventional Hill Climbing Algorithm as well as a classical Genetic Algorithm.

In the following all approaches will be presented shortly.

3.1 Hill Climbing Algorithm

Hill Climbing is a simple, heuristic method to solve optimization problems. In this technique, the solution we start with is improved step by step until some condition is maximized. In each step, a local change is carried out and only adopted if the resulting solution candidate is better suited. The Hill Climbing Algorithm can be thought of as a simple evolutionary algorithm with only one individual—in this case a set of polygons—and one mutation operation to mutate the color or the coordinates of the polygons.

3.2 Canny Edge Detector

The Canny edge detector is an algorithm for edge detection that is widely used in digital image processing. It delivers an image which ideally contains only the edges of the input image [4,6]. The algorithm is involving five steps: Smoothing of input image by Gaussian filter, finding gradients of the image, non-maximum supression, double tresholding and then edge tracking by hysteresis.

3.3 Delaunay Triangulation

Delaunay triangulation is a common technique for creating a triangle mesh from a set of points. With this method, points in $\mathbb{R}^2$ are meshed into triangles so that within the circle on which the three triangle points lie (perimeter of the triangle), no other points are contained [4].

3.4 Genetic Algorithm

Genetic algorithms (GAs) are stochastic, metaheuristic optimization methods whose operation is inspired by the evolution of natural creatures. These algorithms are suitable for searching very large solution spaces. Like in natural genetics, in GAs a set of possible solutions (population) undergoes recombination and mutation, producing new children. Each individual is assigned a fitness value based on its objective function value. The individuals with the higher fitness have a higher probability to mate and yield more "fitter" individuals, which leads in the long run to an improvement of the objective function values (in our case color and position of the polygons).

3.5 Hill Climbing Algorithm with Canny Edge Detection and Delaunay Triangulation

In this variant, the output image is segmented into triangles before the actual mutation of the attributes takes place. First, we obtain image edges by a Canny edge detection procedure. Next, a Delaunay triangulation of the edge point set is performed so that no triangle edge intersects an edge between neighboring edge pixels. The triangles thus created are then colored with the average color of the pixels enclosed by the triangle. In this way, the program starts with a picture (initial population), which is already very close to the original, as you can see in Fig. 3.

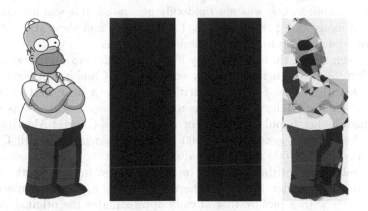

Fig. 3. Result of evolving after a few mutations (from left to right): Original image, Genetic Algorithm, Hill Climbing Algorithm and a combination of Hill Climbing Algorithm, Canny Edge Detection and Delaunay Triangulation.

3.6   Fitness Function

The fitness function we used compares the original image ($p1$) with the polygon based image ($p2$) pixel by pixel. The returned value represents the total difference of the colors as a floating-point number and serves as an indicator of whether the evolving picture is converging toward the original. The smaller the value, the more similar the compared pixels and thus the images.

$$f = \frac{\sum_{i=0}^{W} \left(\sum_{j=0}^{H} \left(\sum_{k=0}^{2} (p1_{i,j,k} - p2_{i,j,k})^2 \right) \right)}{W \cdot H} \tag{1}$$

4   Experimental Results

For our first test, we used the image of the popular cartoon character Homer Simpson shown in Fig. 3. The size of this image is 306×842 pixels. We tested it in three different algorithms—Hill Climbing Algorithm, Genetic Algorithm and a combination of Hill Climbing Algorithm and Canny Edge Detector and Delaunay Triangulation over a period of 50,000 generations. We chose this number because our experiments have shown that after 50,000 generations, there are usually no major improvements in terms of fitness.

In Hill Climbing Algorithm we had only one randomly generated individual, where polygons were mutated in their color and coordinates.

In the Genetic Algorithm, on the other hand, our population consisted of 50 different individuals, where polygons were not only mutated but also recombined through a crossover operation.

The combined algorithm again consisted of only one individual, which, unlike the other two approaches, was not randomly generated, but was obtained from the original image using triangulation. The polygons and their attributes were then mutated according to the Hill Climbing principle.

It has been shown that there is only a minimal difference in fitness achieved during/after 50,000 generations between the Hill Climbing Algorithm and Genetic Algorithm. The Genetic Algorithm performs a little better, but it is also very computationally intensive, as more operations need to be performed due to much higher population size. For example, the Genetic Algorithm took 9.8 h in this test for generating 50,000 generations, while the Hill Climbing Algorithm reached 50,000 generations after only 1.4 h.

The combined algorithm performs in this case best in all aspects: The calculations took just under 40 min. Furthermore, thanks to the triangulation, the program starts with a picture that already approximates the original very well and therefore has a very low fitness, i.e a very good fitness. However, the result does not get much better than the result of triangulation image as you can see in Fig. 4.

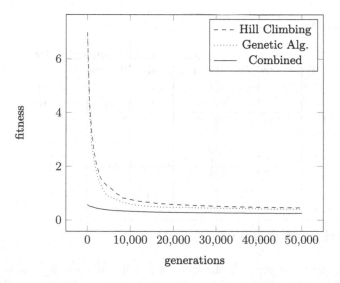

Fig. 4. Evolving of Homer Simpson picture from Fig. 3 (306 × 842 pixels): a fitness - generation number relationship for the three presented approaches.

Since the results can be influenced by the size and some other properties of the image such as texture, amount of detail etc., we tested the three approaches on some other randomly selected images with the size of 100 × 100 (images 1 to 5), 150 × 150 (images 6 to 10) and 200 × 200 (images 11 to 15) pixels to test our hypothesis that the combined approach performs best in all aspects.

Table 1. Fitness of 10 Images after 50,000 Generations in percent.

No.	HCA	GA	HCA + DT + CED
1 (flower)	0.2358	0.2036	0.1773
2 (little boy)	0.1567	0.1355	0.1325
3 (racoon)	0.1067	0.0922	0.0916
4 (a Disney princess)	0.1057	0.0966	0.0592
5 ("Sunflowers" by Van Gogh)	0.1743	0.1655	0.1689
6 (strawberry)	0.2362	0.1908	0.1824
7 (abstract)	0.4068	0.2863	0.2337
8 (woman)	0.2741	0.1975	0.1491
9 (parrot)	0.0584	0.0523	0.0316
10 (ship)	0.2778	0.2091	0.2037
11 (mountains)	0.1182	0.1013	0.0904
12 (treasure)	0.4371	0.3910	0.3836
13 (sheep)	0.0777	0.0648	0.0585
14 (balloons)	0.2186	0.1898	0.1131
15 (man)	0.1398	0.1037	0.0952

The test conditions such as the number of generations, number of individuals etc. remained the same as in the first test (Table 2).

With the help of the obtained numbers our assumption that the combined approach provides the best results in terms of fitness could be confirmed.

Although the differences to other two algorithms do not seem to be that great when looking at the numbers, there are significant differences when looking at images produced: The combined algorithm produces more accurate images with much clearer edges and contrasts than the other two algorithms as can be seen in Fig. 5.

In terms of time, the pure Hill Climbing Algorithm seems to perform best, contrary to the initial assumption. However, it must be said that the time required depends not least on the processor speed and some other factors, which is why the presented numbers can only show an approximate ratio between the different approaches.

The time-waster of the combined approach is probably the fact that this approach, unlike the other two, works with a dynamically determined number of polygons, which is usually very high—depending on the size of the picture several hundred to over 1000. As a consequence, this increases the computation time.

Table 2. Required time for 50,000 Generations in minutes (images from Table 1).

No.	HCA	GA	HCA + DT + CED
1 (flower)	5.5	43.3	16.63
2 (little boy)	5.5	48.68	16.41
3 (racoon)	15.16	135.53	44.45
4 (a Disney princess)	18.1	145.45	39.11
5 ("Sunflowers" by Van Gogh)	17.15	127.98	87.03
6 (strawberry)	26.6	205.80	61.76
7 (abstract)	22.68	171.47	45.07
8 (woman)	17.61	156.63	60.38
9 (parrot)	17.04	146.5	63.12
10 (ship)	21.58	181.77	79.05
11 (mountains)	39.9	193.33	85.66
12 (treasure)	33.82	299.02	124.1
13 (sheep)	32.01	200.8	57.36
14 (balloons)	41.43	274.55	74.27
15 (man)	31.1	236.71	77.46

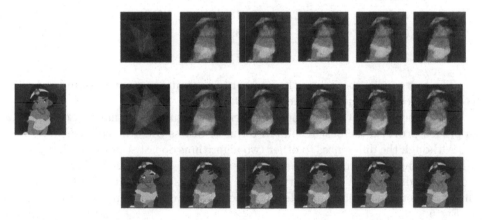

Fig. 5. Comparison of the results of evolving during the first 50,000 generations in ten thousand steps (image 4): 1st row—Hill Climbing Algorithm, 2nd row—Genetic Algorithm, 3rd row—Combination of Hill Climbing Algorithm, Delaunay Triangulation and Canny Edge Detector.

5   Conclusion and Future Work

The aim of this work was to introduce a new approach to solving the "Evolution of Mona Lisa" problem and to compare it with two other known solution strategies—Hill Climbing Algorithm and Genetic Algorithm.

It could be shown that the approach presented by us provides better fitness for a given number of generations and thus better approximates the original image. For this the three algorithms were tested on different pictures of different size.

As a next step, it would be exciting to compare our results with the results of L.G. Thuan and his colleagues, who presented the Simulated Annealing approach to this problem (see the "Previous Work" section). For this, however, first of all a common basis must be created, for example one would have to agree on a uniform fitness function etc., in order to make the results comparable.

It would also be possible to combine our approaches. This should not be a problem, because they turn on two different screws: While our approach describes how to generate a suitable initial population, the work of L.G. Thuan and his colleagues deals with the selection process of an individual for the next generation after a successful mutation.

The efficiency of the program could also be improved. A big time waster is determining the fitness of a picture. With a 200×200 pixel image with the 3 color values per pixel and the transparency value, there are already 160,000 values that are taken into account each time the fitness is determined. For example, to speed up this process, one could scale down the image before calculating the fitness.

All in all, we find that this topic is quite exciting and deserves to be further researched. In any case, it is also very well suited for academic purposes, for example to get acquainted with the functionality of evolutionary algorithms. Practical applications would also be conceivable, for example in image compression.

References

1. Genetic Programming: Evolution of Mona Lisa. https://rogerjohansson.blog/2008/12/07/genetic-programming-evolution-of-mona-lisa. Accessed 27 Sept 2019
2. Lam, G.T., Balabanov, K., Logofătu, D., Badica, C.: Novel nature-inspired selection strategies for digital image evolution of artwork. In: Nguyen, N.T., Pimenidis, E., Khan, Z., Trawiński, B. (eds.) ICCCI 2018. LNCS (LNAI), vol. 11056, pp. 499–508. Springer, Cham (2018). https://doi.org/10.1007/978-3-319-98446-9_47
3. Evolving Mona Lisa Vizualization. https://alteredqualia.com/visualization/evolve/. Accessed 27 Nov 2019
4. Hole, K.R., Gulhane, V.S., Shellokar, N.D.: Application of genetic algorithm for image enhancement and segmentation. Int. J. Adv. Res. Comput. Eng. Technol. (IJARCET) **2**(4), 1342 (2013)
5. Russell, S.J., Norvig, P.: Artifcial Intelligence: A Modern Approach. Prentice Hall, Upper Saddle River (2004)
6. Canny, J.F.: A variational approach to edge detection. In: AAAI, vol. 1983 (1983)

7. Ho, S.Y., Chen, Y.C.: An efficient evolutionary algorithm for accurate polygonal approximation. Pattern Recogn. **34**, 2305–2317 (2001)
8. Gerkey, B.P., Thrun, S., Gordon, G.: Parallel stochastic hillclimbing with small teams. In: Parker, L.E., Schneider, F.E., Schultz, A.C. (eds.) Multi-Robot Systems: From Swarms to Intelligent Automata, vol. 3, pp. 65–77. Springer, Dordrecht (2005). https://doi.org/10.1007/1-4020-3389-3_6

3D Model-Based 6D Object Pose Tracking on RGB Images

Mateusz Majcher and Bogdan Kwolek[✉]

AGH University of Science and Technology, 30 Mickiewicza, 30-059 Krakow, Poland
majcher@agh.edu.pl
http://home.agh.edu.pl/~bkw/contact.html

Abstract. In this paper, we present a 3D-model based algorithm for 6D object pose estimation and tracking on segmented RGB images. The object of interest is segmented by U-Net neural network trained on a set of manually delineated images. A Particle Swarm Optimization is used to estimate the 6D object pose by projecting the 3D object model and then matching the projected image with the image acquired by the camera. The tracking of 6D object pose is formulated as a dynamic optimization problem. In order to keep necessary human intervention minimal, we use an automated turntable setup to prepare a 3D object model and to determine the ground-truth poses. We compare the experimental results obtained by our algorithm with results achieved by PWP3D algorithm.

Keywords: Tracking 6D pose of object · Image segmentation · Optimization

1 Introduction

Estimating the 6-DoF pose (3D rotations + 3D translations) of an object with respect to the camera is extremely challenging task. It is very important problem in robotic applications since a robotic arm needs to know the location and orientation to detect and to move objects in its vicinity. 3D information about object position and orientation permits task planning, obstacle avoidance and object grasping. Having on regard its significance a variety of research efforts have been devoted to tackling the 6D pose estimation problem from computer vision community [1], robotics community [2] and augmented reality [3]. In virtual reality applications a precise object pose is required for interaction with an object as well as for determining the initial pose for tracking.

In conventional approaches, the object pose is recovered on the basis of template matching or local-features, where image attributes are matched against templates or features from the 3D model of the object. The pose estimation is then done through a selection the best matching viewpoint onto the object or on the basis of 2D-3D correspondences between such local features and a Perspective-n-Point (PnP) algorithm [4]. The correspondence-based approaches require rich texture features. They estimate the pose by solving the PnP with

© Springer Nature Switzerland AG 2020
N. T. Nguyen et al. (Eds.): ACIIDS 2020, LNAI 12033, pp. 271–282, 2020.
https://doi.org/10.1007/978-3-030-41964-6_24

the recovered to 2D-3D correspondences, often in a RANSAC [4] framework for outlier rejection. Hinterstoisser et al. [5,6] introduced holistic template-based methods that can cope with texture-less objects in 6D pose recovering in cluttered scenes. Another more recent approach to pose estimation on RGB images is discussed in [7]. It is based on learning so-called object coordinates. Although this approach outperforms strategies based on template matching, its runtime performance is far away from real-time.

Deep convolutional neural network-based approaches have shown promising results on many computer vision tasks over the last few years [8], particularly in object detection as well as classification and segmentation tasks. Wohlhart and Lepetit [9] proposed a CNN not for estimation of the object pose directly but to determine a feature descriptor such that the Euclidean distance between the descriptors corresponding to two poses are bigger when the poses are dissimilar and smaller if the poses are similar. A first attempt to use a Convolutional Neural Network (CNN) for direct regression of 6DoF object poses was PoseCNN [10]. PoseCNN decouples the pose estimation into estimating the translation and rotation in end-to-end framework. It uses features discovered by convolutional layers of the VGG16 and then three different branches. Two convolutional branches perform a semantic segmentation and 2D center voting for handling occlusions. The third branch comprises a RoI pooling and a fully-connected architecture, which regresses each RoI to a quaternion describing the rotation. In general, two main CNN-based trends to estimation of 6D pose of objects have emerged: either regressing the 6D object pose from the image directly [10] or predicting 2D key-point locations in the image [11], from which the object pose can be determined by the PnP algorithm.

Most of the RGB image-based approaches to 6D pose recovery has focused on accuracies of the algorithms as well as processing times. The majority of present techniques to 6D object pose estimation ignore temporal information and provide only a single hypothesis for object pose [12]. In [13] both the robot pose and object pose were determined using planar primitives to achieve object grasping by a humanoid robot. In order to give robots greater spatial perception so they can manipulate objects and navigate through space more accurately a Rao-Blackwellized Particle Filter (PoseRBPF) for object pose estimation has been proposed in [12]. There are several publicly available datasets for benchmarking the performance of algorithms for 6D object pose estimation, including OccludedLinemod [1], YCB-Video [10]. However, most of the current datasets focuses on a small set of objects. In general, a trend towards building a large-scale datasets covering all kinds of objects, and relevant to ImageNet in the image classification domain can be observed. Having on regard a need to prepare a model of particular object quickly in order to perform object grasping, with fast adoption to novel set of objects, in this paper we discuss a 3D model-based approach. We investigate the problem of 6-DOF object pose estimation and tracking from RGB images, where the object of interest is rigid and a 3D model of the object is known. The object is segmented from the background using an U-Net convolutional neural network. The network is trained using a small set of object images.

A Particle Swarm Optimization (PSO) [14,15] is then used to estimate the 6D object pose by projecting the 3D object model and then matching the projected image with the image acquired by the camera. The tracking of 6D object pose is formulated as a dynamic optimization problem. In order to keep necessary human intervention minimal, we use an automated turntable setup to prepare the 3D object model as well as to determine the ground-truth poses.

2   Neural Network for Object Segmentation

The objects were segmented using an U-Net neural network [16] in which we can distinguish a down-sampling (encoding) path and an up-sampling (decoding) path, see Fig. 1. The down-sampling path has five convolutional blocks. Each block comprises convolutional layers with 3×3 filters and stride equal to 1. Down-sampling is performed by max pooling with stride 2×2 that is applied on the end of each block except the last one. In the up-sampling path, each block begins with a deconvolutional layer with 3×3 filter and 2×2 stride, which doubles the dimension of feature maps in both directions and decreases the number of feature maps by two. In each up-sampling block, two convolutional layers reduce the number of feature maps, which arise as a result of concatenation of deconvolutional feature maps and the feature maps from corresponding block in the encoding path. Finally, a 1×1 convolutional layer is used to diminish feature number to two. The neural network was trained on RGB images of size 288×512. In order to reduce training time, as well as to prevent overfitting and increase performance of the U-Net we added Batch Normalization (BN) [17] after each Conv2D. The neural network has been trained using data augmentation.

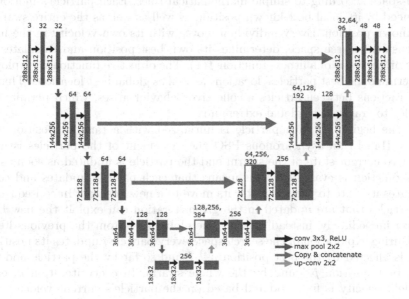

Fig. 1. Architecture of U-Net used for object segmentation.

The pixel-wise cross-entropy has been used as the loss function for object segmentation:

$$\mathcal{L}_{CE} = -\frac{1}{N} \sum_{i=1}^{N} [y_i \log(\hat{y}_i) + (1 - y_i) \log(1 - \hat{y}_i)] \tag{1}$$

where N stands for the number of training samples, y is true value and $\hat{y}$ denotes predicted value. The motivation for using the cross entropy loss function is that it has nice differentiable properties, and it can be easy optimized using back-propagation algorithm.

3 6D Object Pose Tracking

At the beginning of this Section we discuss 6D object pose tracking using particle swarm optimization. Afterwards, we present the fitness function.

3.1 6D Object Pose Tracking Using Particle Swarm Optimization

Particle Swarm Optimization (PSO) [14,15] is a global optimization meta-heuristic and stochastic method, which is based on swarm intelligence. It is derivative–free, population–based computational method, which demonstrated a high potential in optimization of unfriendly non–convex functions. It optimizes a problem by iteratively trying to improve a candidate solution with respect to a given measure of quality. The optimal solution is seek by a population of particles exploring candidate solutions, through moving particles around in the search-space according to simple mathematical rules. Each particle's motion is influenced by its local best known position as well as well as the entire swarm's best known position. Every individual moves with its own velocity in the mul-tidimensional search space, determines its own best position and calculates its fitness on the basis of a fitness function $f(x)$. The objective function is employed to determine the best particles' locations as well as global best locations. Thanks to interactions between articles a collective behavior arises, which permits the particles to gravitate to global extremum.

At the beginning, each particle is initialized with a random position and velocity [14]. In an asynchronous PSO the movement of the particles is with respect to current state of the swarm and the particle is updated as soon as its fitness function is evaluated. This means that each particle updates and com-municates its state to particles after its move to a new position. In consequence, the particles that are updated in the given iteration can exploit the new best position immediately, instead of using the global best from the previous itera-tion. During exploration of the search space every particle i updates its position, which is affected by the best position $p^{(i)}$ found so far by the particle and the global best position $\hat{p}$ found by the whole swarm. In every iteration k, each particle's velocity is first updated based on the particle's current velocity, the particle's local information and global information discovered so far by the entire population. Then, the particle updates its position using the updated velocity. In the ordinary PSO, the position and velocity of the particle are determined in the following manner:

$$v_j^{(i)}(k+1) = wv_j^{(i)}(k) + c_1r_{1,j}^{(i)}(p_j^{(i)}(k) - x_j^{(i)}(k)) + c_2r_{2,j}^{(i)}(\hat{p}_j - x_j^{(i)}(k)) \qquad (2)$$

$$x_j^{(i)}(k+1) = x_j^{(i)}(k) + v_j^{(i)}(k+1) \qquad (3)$$

where w is a positive inertia weight, $v_j^{(i)}$ is the velocity of particle i in dimension j, $r_{1,j}^{(i)}$ and $r_{2,j}^{(i)}$ are uniquely generated random numbers with the uniform distribution in the interval $[0.0, 1.0]$, c_1, c_2 are positive, cognitive and social constants, respectively, $p^{(i)}$ is the best position found so far by the particle i, whereas $\hat{p}$ stands for the best position that was found by any member of the whole population.

Equation (2), which updates the particle velocity has three main components. The first component, which is frequently referred to as inertia weight determines the contribution of the particle's previous velocity to its velocity it the current time step, i.e. moving in the same direction. It permits balancing between exploration and exploitation. The second component is referred as cognitive and it pulls the particle towards the best position $p^{(i)}$ that was found formerly. The last component, called social, attracts the particle towards the best position $\hat{p}$ found by any particle of the population. After determining $x_j(k+1)$ using (3), the best position $p^{(i)}(k+1)$ of particle i is calculated as follows:

$$p^{(i)}(k+1) = \begin{cases} p^{(i)}(k) & \text{if} \quad f(x^{(i)}(k+1)) \geq f(p^{(i)}(k)) \\ x^{(i)}(k+1) & \text{if} \quad f(x^{(i)}(k+1)) < f(p^{(i)}(k)) \end{cases} \qquad (4)$$

A topology with the global best has been selected due to its faster convergence in comparison to neighborhood best one. After updating the local best position of particle i using (4) the global best position $\hat{p}$ is updated and then further used to calculate updates until the iteration does not finish.

The algorithm presented above is usually used for solving static optimization problems. The motion tracking can be attained by dynamic optimization and incorporating the temporal continuity information into the ordinary PSO. Consequently, it can be achieved by a sequence of static PSO-based optimizations, followed by re-diversification of the particles to cover the potential poses that can arise in the next time step. The re-diversification of the particle i can be obtained on the basis of normal distribution concentrated around the best particle location $\hat{p}$ in time $t-1$, which can be expressed as: $x^{(i)} \leftarrow \mathcal{N}(\hat{p}, \Sigma)$, where $x^{(i)}$ stands for particle's location in time t, Σ denotes the covariance matrix of the Gaussian distribution, whose diagonal elements are proportional to the expected movement.

3.2 Cost Function

PSO has already been applied in several model-based applications, among others object detection [18] and 3D pose refinement using rendering and texture-based matching [19]. The most computationally demanding operation in 3D model-based 6D object pose tracking is calculation of the objective function. A considerable acceleration of the calculation of the fitness function can be achieved on modern GPU devices [20, 21]. In this work, we pay more attention to object segmentation as well as tracking accuracy of 6D object pose, and thus we focus on CPU implementation of the fitness function in order to simplify design as well as evaluation of the algorithm.

In PSO-based approach each particle represents a hypothesis about possible 6D object pose. In the evaluation of the particle's fitness score the projected 3D model is matched with the current image observations. The fitness score depends on the amount of overlap between the segmented object in the current image and the projected and rasterized 3D model in the hypothesized pose. The amount of overlap is calculated through checking the overlap degree from the object shape to the rasterized model as well as from the rasterized model to the object shape. The larger the overlap is, the larger is the fitness value. The objective function reflects also the normalized distance between the model's projected edges and the closest edges in the image. It is calculated on the basis of the edge distance map [20].

The fitness score is calculated on the basis of following expression:

$$\left(0.5 * \frac{P_{outside}}{P_{model}} + 0.5 * \frac{P_{empty}}{P_{seg}}\right)^{w_1} * \left(\frac{K}{P_K}\right)^{w_2} \tag{5}$$

where $P_{outside}$ stands for number of pixels projected from model that are outside of segmented object on the image, see also see Fig. 2 (left), P_{model} denotes number of pixels in the projected model, P_{empty} is number of pixels of segmented object on the image that are not covered by projected model, P_{seg} stands for number of pixels from segmented object, K is sum of L2 distance transform values from projected model's edges to object edges on the image, see Fig. 2 (right), P_K denotes number of edge pixels in the projected object outline, and $w_1 = 0.4$, $w_2 = 0.6$ are exponents that were determined experimentally. The rasterization of the 3D model is performed using OpenGL.

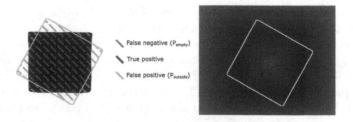

Fig. 2. Fitness function.

4   Experimental Results and Discussion

The experimental evaluation has been conducted on three objects: a box, a bottle and a duck. The objects are texture-less or almost texture-less. 3D models of the objects were prepared using the Kinect 2.5D RGB-D camera [22] and SfM techniques. The camera has been calibrated using the OpenCV library [23]. The ground truths of the object poses have been determined using measurements provided by a turnable device. Each object has been observed from three different camera views, see Fig. 3. Nineteen images for each camera view were registered. The objects were rotated in range $0° \dots 180°$. During object rotation, every ten degrees an image has been acquired with corresponding rotation angle. Each experiment has been repeated three times, and the estimated poses were averaged for each considered angle.

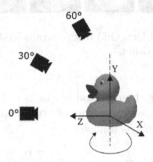

Fig. 3. Experiments setup.

In order to prepare the segmentation model the objects were observed by the camera from different views. For each object more than one hundred manually delineated objects were prepared and then used to train models for object segmentation. The models trained in such a way were then used to segment the considered objects from the background. Afterwards, a dataset consisting of RGB images with the corresponding ground truth data has been prepared for the evaluation of the algorithm for 6D object pose estimation and tracking.

4.1   Object Segmentation

A single U-Net neural network discussed in Sect. 2 has been trained to segment the considered object using the set of manually segmented images. Figure 4 depicts some example RGB images with corresponding images segmented by the U-Net. The depicted images are from the test subset and were not included in the training subset. Table 1 presents the Dice scores achieved on the test subset of the dataset. The Dice similarity coefficient for two sets A and B can be expressed in the following manner:

$$dice(A, B) = \frac{2 * |intersection(A, B)|}{(|A| + |B|)} \quad (6)$$

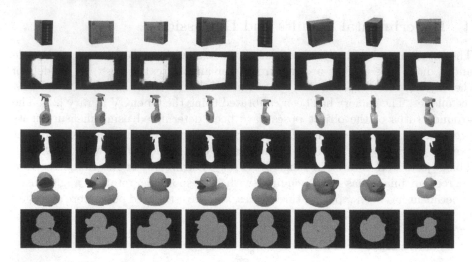

Fig. 4. Segmentation of the object. Odd rows contain RGB images whereas even rows contain images segmented by U-net.

where $|A|$ and $|B|$ stand for the cardinal of the set A and B, respectively. The Dice score can also be expressed in terms of true positives (TP), false positives (FP) and false negatives (FN) as follows:

$$dice(A, B) = \frac{2 * TP}{2 * TP + FP + FN} \tag{7}$$

The test subset contains thirty images for each object. As we can observe in Table 1, better segmentation results were achieved for U-Net trained separately for each of the considered objects. Having on regard a better usefulness of a single U-Net for segmentation of all three objects, in the following subsections we present the experimental results achieved on the basis of the single U-Net.

Table 1. Dice scores on the test sub-dataset.

	Box	Bottle	Duck
U-Net for each object	0.985	0.964	0.974
Single U-Net for all objects	0.946	0.972	0.978

4.2   Evaluation Metric for 6D Pose Estimation

We evaluated the quality of 6-DoF object pose estimation using ADD score [2]. ADD is defined as an average Euclidean distance between model vertices transformed using the estimated pose and the ground truth pose. This means that it expresses the average distance between the 3D points transformed using the estimated pose and those obtained with the ground-truth one. It is defined as follows:

$$ADD = \text{avg}_{x \in M}\|(Rx + t) - (\hat{R}x + \hat{t})\|_2 \tag{8}$$

where M is a set of 3D object model points, t and R are the translation and rotation of a ground truth transformation, respectively, whereas $\hat{t}$ and $\hat{R}$ correspond to those of the estimated transformation.

4.3 Experimental Evaluation

Table 2 presents experimental results that were obtained by our algorithm in estimation of 6D pose of box, bottle and duck objects in single RGB images. The 6D object pose estimate is considered valid if the ADD is smaller than ten percent of object's diameter. For the duck the threshold is equal to 1.0 cm, but because of somewhat larger errors we report scores for threshold set to 1.5 cm. Each result is an average of five independent runs of the PSO algorithm with unlike initialization. The average has been calculated for each object. As we can observe a little bit worse results were obtained for the duck object. One of the reasons is that this object is a little bit smaller than remaining objects. The experiments were conducted on RGB images of size 288×512. The camera has been calibrated using OpenCV [23]. The presented results were achieved by PSO consisting of fifty particles and executing twenty iterations. The initial particle positions were sampled from the following initial searching space: y-axis $-10° \ldots 190°$, x and z axes $-30° \ldots 30°$. As we can observe, despite initial poses quite different from true poses the algorithm is capable of finding satisfactory results.

Table 2. Scores [%] achieved by our algorithm for box, bottle and duck objects.

Angle [deg]	0	10	20	30	40	50	60	70	80	90	Avg.
Box (ADD < 3.5 cm)	80	80	100	100	80	100	100	100	80	20	84
Bottle (ADD < 2.6 cm)	100	80	100	100	100	100	100	80	40	20	82
Duck (ADD < 1.5 cm)	100	80	80	100	100	80	80	60	80	100	86

Afterwards, we conducted experiments consisting in tracking the 6D object pose in sequences of RGB images. The pose tracking of the box object has been done by our algorithm and by the PWP3D algorithm [24], which has been developed for real-time segmentation and tracking of 3D objects. The poses in current frames were estimated on the basis of estimates from the previous frame. The 3D box pose has been estimated in three different experiments with unlike initializations in the first frame and then averaged. Table 3 presents the experimental results. It presents also the averages from three independent runs for each considered object rotation. In a first scenario the object pose has been estimated for object rotations about ten degrees, whereas in the second one it has been estimated every twenty degrees. As we can observe, the ADD scores attained by the PWP3D algorithm are far worse in comparison to ADD scores achieved by our algorithm. In particular, the PWP3D can not cope with bigger inter-frame rotations or rapid motions.

Table 3. Scores [%] achieved by PWP3D algorithm and our algorithm on the box object.

Angle [deg]	0	10	20	30	40	50	60	70	80	90	Average
PWP3D, exp. 1	67	67	67	100	100	67	67	100	33	67	73.3
PWP3D, exp. 2	67	–	67	–	33	–	0	–	0	–	33.0
Our alg., exp. 3	67	78	89	89	100	56	89	67	22	44	70.1
Our alg., exp. 4	89	–	100	–	78	–	78	–	44	–	78.0

Finally, we conducted experiments consisting in tracking the 6D poses of objects observed from different views. The experiments were performed on sequence of images acquired in advance and stored in mp4 files. For every camera view, the 6D pose of each object has been tracked on ten and nineteen images. The objects were rotated about the vertical axis with poses changed about ten degrees. As we can notice, the poses of symmetrical objects were estimated with satisfactory accuracies with rotations about vertical symmetry axes in range $0° \ldots 180°$. The discussed experiment demonstrated a potential of the algorithm and its usefulness for robotic applications consisting in object grasping by a robot (Table 4).

Table 4. Scores [%] achieved in 6D object tracking for box [ADD < 3.5 cm], bottle [ADD < 2.6 cm] and duck [ADD < 1.5 cm].

Tracking score [%]	Box	Bottle	Duck
Angle $0 \ldots 90°$	0.733	0.867	0.856
Angle $0 \ldots 180°$	0.807	0.877	0.719

The complete system for 6D pose estimation has been implemented in C/C++ and python. The system runs on an ordinary PC with GPU. The images for training and evaluating the segmentation algorithm as well as extracted objects with corresponding ground-truth for evaluating the 6D pose estimation are freely available at: http://home.agh.edu.pl/~majcher/src/aciids.

5   Conclusions

We have presented a 3D model based algorithm for 6D object pose estimation on RGB images. The object has been segmented using U-Net neural network.

The 6D object pose estimation has been performed by PSO algorithm. We have presented the segmentation results obtained by the U-Net. We presented results achieved by our algorithm as well as the PWP3D algorithm. The results achieved by our algorithm are far better than results achieved by the PWP3D algorithm. In future work we are going to perform tracking of the 6D object pose using particle filter combined with particle swarm optimization. The initialization of the tracking will be done on the basis of pose regression neural networks.

Acknowledgment. This work was supported by Polish National Science Center (NCN) under a research grant 2017/27/B/ST6/01743.

References

1. Brachmann, E., Krull, A., Michel, F., Gumhold, S., Shotton, J., Rother, C.: Learning 6D object pose estimation using 3D object coordinates. In: Fleet, D., Pajdla, T., Schiele, B., Tuytelaars, T. (eds.) ECCV 2014. LNCS, vol. 8690, pp. 536–551. Springer, Cham (2014). https://doi.org/10.1007/978-3-319-10605-2_35
2. Hinterstoisser, S., et al.: Model based training, detection and pose estimation of texture-less 3D objects in heavily cluttered scenes. In: Lee, K.M., Matsushita, Y., Rehg, J.M., Hu, Z. (eds.) ACCV 2012. LNCS, vol. 7724, pp. 548–562. Springer, Heidelberg (2013). https://doi.org/10.1007/978-3-642-37331-2_42
3. Marchand, E., Uchiyama, H., Spindler, F.: Pose estimation for augmented reality: a hands-on survey. IEEE Trans. Vis. Comput. Graph. **22**(12), 2633–2651 (2016)
4. Fischler, M.A., Bolles, R.C.: Random sample consensus: a paradigm for model fitting with applications to image analysis and automated cartography. Commun. ACM **24**(6), 381–395 (1981)
5. Hinterstoisser, S., et al.: Multimodal templates for real-time detection of texture-less objects in heavily cluttered scenes. In: International Conference on Computer Vision, pp. 858–865 (2011)
6. Model-Based Training, Detection and Pose Estimation of Texture-less 3D Objects in Heavily Cluttered Scenes. In: Proceedings of the Asian Conference on Computer Vision (ACCV) (2012)
7. Brachmann, E., Michel, F., Krull, A., Yang, M., Gumhold, S., Rother, C.: Uncertainty-driven 6D pose estimation of objects and scenes from a single RGB image. In: IEEE Conference on Computer Vision and Pattern Recognition (CVPR), pp. 3364–3372 (2016)
8. Pouyanfar, S., et al.: A survey on deep learning: algorithms, techniques, and applications. ACM Comput. Surv. **51**(5), 1–36 (2018)
9. Wohlhart, P., Lepetit, V.: Learning descriptors for object recognition and 3D pose estimation. In: Conference on Computer Vision and Pattern Recognition, pp. 1–10 (2015)
10. Xiang, Y., Schmidt, T., Narayanan, V., Fox, D.: PoseCNN: a convolutional neural network for 6D object pose estimation in cluttered scenes. In: Robotics: Science and Systems XIV (RSS) (2018)
11. Rad, M., Lepetit, V.: BB8: a scalable, accurate, robust to partial occlusion method for predicting the 3D poses of challenging objects without using depth. In: IEEE International Conference on Computer Vision, ICCV, pp. 3848–3856 (2017)

12. Deng, X., Mousavian, A., Xiang, Y., Xia, F., Bretl, T., Fox, D.: PoseRBPF: a rao-blackwellized particle filter for 6D object pose tracking. In: Robotics: Science and Systems (RSS) (2019)
13. Gritsenko, P., Gritsenko, I., Seidakhmet, A., Kwolek, B.: Plane-based humanoid robot navigation and object model construction for grasping. In: Leal-Taixé, L., Roth, S. (eds.) ECCV 2018. LNCS, vol. 11129, pp. 649–664. Springer, Cham (2019). https://doi.org/10.1007/978-3-030-11009-3_40
14. Kennedy, J., Eberhart, R.: Particle Swarm Optimization. In: Proceedings of IEEE International Conference on Neural Networks, pp. 1942–1948. IEEE Press, Piscataway (1995)
15. Sengupta, S., Basak, S., Peters, R.A.: Particle swarm optimization: a survey of historical and recent developments with hybridization perspectives. Mach. Learn. Knowl. Extr. $\mathbf{1}$(1), 157–191 (2019)
16. Ronneberger, O., Fischer, P., Brox, T.: U-Net: convolutional networks for biomedical image segmentation. In: Navab, N., Hornegger, J., Wells, W.M., Frangi, A.F. (eds.) MICCAI 2015. LNCS, vol. 9351, pp. 234–241. Springer, Cham (2015). https://doi.org/10.1007/978-3-319-24574-4_28
17. Ioffe, S., Szegedy, C.: Batch normalization: Accelerating deep network training by reducing internal covariate shift. In: ICML, vol. 37, pp. 448–456 (2015)
18. Ugolotti, R., Nashed, Y.S.G., Mesejo, P., Ivekovič, V., Mussi, L., Cagnoni, S.: Particle swarm optimization and differential evolution for model-based object detection. Appl. Soft Comput. $\mathbf{13}$(6), 3092–3105 (2013)
19. Zabulis, X., Lourakis, M.I.A., Stefanou, S.S.: 3D pose refinement using rendering and texture-based matching. In: Chmielewski, L.J., Kozera, R., Shin, B.-S., Wojciechowski, K. (eds.) ICCVG 2014. LNCS, vol. 8671, pp. 672–679. Springer, Cham (2014). https://doi.org/10.1007/978-3-319-11331-9_80
20. Rymut, B., Kwolek, B., Krzeszowski, T.: GPU-accelerated human motion tracking using particle filter combined with PSO. In: Blanc-Talon, J., Kasinski, A., Philips, W., Popescu, D., Scheunders, P. (eds.) ACIVS 2013. LNCS, vol. 8192, pp. 426–437. Springer, Cham (2013). https://doi.org/10.1007/978-3-319-02895-8_38
21. Tan, Y.: GPU-based parallel implementation of swarm intelligence algorithms. Morgan Kaufmann, San Francisco (2016)
22. Izadi, S., et al.: KinectFusion: real-time 3D reconstruction and interaction using a moving depth camera. In: Proceedings of the 24th Annual ACM Symposium on User Interface Software and Technology, New York, NY, USA, pp. 559–568. ACM (2011)
23. Bradski, G., Kaehler, A.: Learning OpenCV: Computer Vision in C++ with the OpenCV Library, 2nd edn. O'Reilly Media Inc, Sebastopol (2013)
24. Prisacariu, V.A., Reid, I.D.: PWP3D: real-time segmentation and tracking of 3D objects. Int. J. Comput. Vision $\mathbf{98}$(3), 335–354 (2012)

Angiodysplasia Segmentation on Capsule Endoscopy Images Using AlbuNet with Squeeze-and-Excitation Blocks

Sirichart Gobpradit$^{(\boxtimes)}$ and Peerapon Vateekul$^{(\boxtimes)}$

Department of Computer Engineering, Faculty of Engineering, Chulalongkorn University,
Chulalongkorn University Big Data Analytics and IoT Center (CUBIC),
Bangkok 10330, Thailand
6170970021@student.chula.ac.th, peerapon.v@chula.ac.th

Abstract. Angiodysplasia is a small primary lesion in the gut, which may cause gastrointestinal bleeding. Wireless capsule endoscopy is one of the best tools to capture images of these lesions. Since it generates thousands of images, it is crucial to segment angiodysplasia automatically. Recently, AlbuNet, a deep learning network, has shown a promising result and considered as the state-of-the-art technique. In this paper, we aim to enhance AlbuNet from two angles. First, squeeze-and-excitation is similar to the concept of attention on different channels, so it can combine variants of extracted features. Second, a pre-processing step to enhance an image's quality is proposed by applying a computer vision technique called "contrast limit adaptive histogram equalization (CLAHE)". The experiment was conducted on two benchmarks: MICCAI 2017 and 2018 datasets and evaluated in terms of Dice coefficient and Jaccard index scores. The results showed that our model outperformed a baseline technique, AlbuNet, on both datasets.

Keywords: Angiodysplasia · Deep learning · Medical image semantic segmentation · Convolutional neural network

1 Introduction

The most common vascular lesion in the gastrointestinal tract is angiodysplasia. This condition could potentially be asymptomatic, but it may cause gastrointestinal (GI) bleeding [1]. As in Fig. 1 [2], wireless capsule endoscopy (WCE) is a small pill-like tool used for capturing images along digestive tracts and send to experts for diagnosis. WCE may produce over 50,000 frames for each patient [3], depending on the frames per second of the device [4]. The physician needs approximately 1.5 h to review this endoscopy.

Image segmentation is the task for identifying the location of target objects in a given image. The image can be divided into various parts called segments. By using the segmented image, we can make use of the essential segments for image processing in order to find the object we are interested in. Image segmentation creates a mask of pixel-wise for each object in the image, allowing us to see the shape of the object.

© Springer Nature Switzerland AG 2020
N. T. Nguyen et al. (Eds.): ACIIDS 2020, LNAI 12033, pp. 283–293, 2020.
https://doi.org/10.1007/978-3-030-41964-6_25

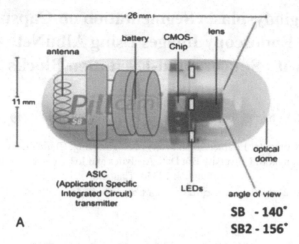

Fig. 1. Schematic structure of the capsule [2]

There are many algorithms for image segmentation, such as region-based segmentation, edge detection segmentation, clustering-based segmentation, and deep learning-based segmentation.

Traditionally, there is a computer vision-based segmentation method developed for angiodysplasia captured by capsule endoscopy [5], which is region-based segmentation for finding regions of interest and classifying them using various features in colon images.

Formerly deep learning-based image segmentation models have been implemented using convolutional neural networks (CNN), which outperforms traditional computer vision-based methods in various medical-image tasks, including for the colon. It has shown promising results in many computer vision tasks, for example, U-Net [6] is a deep learning-based model for cell image segmentation, Attention U-Net [7] is a deep learning-based model for pancreas segmentation, and Y-Net [8] is a deep learning-based model for polyps image segmentation.

Recently a state-of-the-art model deep learning-based image segmentation model for the angiodysplasia task has been proposed, which is called "AlbuNet" [9]. We utilize this model as a baseline model to compare our proposed method because it outperforms other methods in angiodysplasia image segmentation.

In this paper, we proposed a deep learning-based solution in combination with image pre-processing for angiodysplasia image segmentation from captured frames from wireless capsule endoscopy. The datasets we used to measure the performance of our model are MICCAI 2017 and MICCAI 2018 Endoscopic Vision Sub-Challenge datasets [10, 11]. Based on our knowledge, this paper is the first that uses the MICCAI 2018 as a dataset for measuring performance. We will present it as a baseline dataset for other researchers.

2 Related Works

In this section, we provide details of the traditional computer vision method and semantic segmentation networks in the colon as a timeline.

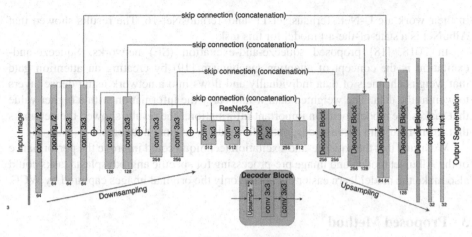

Fig. 2. AlbuNet model from [9]

Before we start talking about the networks in the colon tasks, we will introduce the traditional deep learning-based model used as a baseline model in many medical academic papers. Historically, in the ImageNet challenge in 2012 [12], deep learning models became more widespread and popular. In 2015, [6] proposed the U-Net model, a deep learning model for biomedical image segmentation. This model contains a multi-channel feature map using convolutional layers. There are two parts in this model that scale-down and then scale-up feature maps, called "encoder" and "decoder", respectively. The model has a link between each encoder and decoder to allow information to flow using a concatenate operation. In their work, they used the model to segment cells in the image from the International Symposium on Biomedical Imaging (ISBI) dataset, and it outperformed other models in the paper.

In 2017, [5] presented a system for the automatic detection of angiodysplasia lesions from images retrieved by wireless capsule endoscopy. They used image pre-processing, selection of a potential region of interest, feature extraction and selection, and classification of the potential region of interest with a boosted decision tree classification method to detect angiodysplasia lesions. The pre-processing image methods that were used in this work are histogram equalization (HE), contrast limit adaptive histogram equalization (CLAHE) [13], and RGB decorrelation stretch.

In 2018, [14] proposed the TernausNet-11 model, a modification of the U-Net model, which used pre-trained VGG [15] weights from other tasks to initial the weights of the new model to improve the performance of the model [16]. The pre-trained weights used in their paper came from the ImageNet and Carvana datasets. They compared three models: no pre-trained weights, ImageNet pre-trained weights, and Carvana pre-trained weights. The results showed that pre-trained weights from either ImageNet or Carvana gave better results than the no pre-trained weighted model.

In 2018, [9] proposed a deep learning model for localizing angiodysplasia lesions. They used a modification of the U-Net model with pre-trained ResNet-34 [17] encoder called "AlbuNet" on the MICCAI 2017 dataset. They compared their model as in Fig. 2 to other models to show that their model outperformed other models. Other models used in their work are U-Net, TernausNet-11, and TernausNet-16. The results showed that AlbuNet is a state-of-the-art model for this task.

In 2018, [18] proposed squeeze-and-excitation (SE) networks. Squeeze-and-excitation is the concept of attention mechanism [19] by creating an attention gate that weighs channels of data individually and flows into a network using dense layers to weight the data. The weighted data would carry the information with a higher value that makes the model focus on important information. The results showed the networks that, combined with SE methods, outperformed the original networks.

We found that the squeeze-and-excitation technique could improve the performance of the AlbuNet model, and image pre-processing for specific angiodysplasia tasks could also make the model learn easier than using only the original images captured by WCE.

3   Proposed Method

In this section, we aim to leverage techniques from various papers to improve the performance of the model for angiodysplasia.

We adopt the model framework from [9], extracted from their repository, to construct our baseline. We aim to improve image pre-processing and the model for angiodysplasia image segmentation. We describe our method by separated it into two parts, as follows—Sect. 3.1. Image pre-processing shows the computer vision techniques applied to images; and Sect. 3.2. Model shows the combination of model building.

3.1   Image Pre-processing

In the project, we apply multiple augmentation methods to images. The methods which are in our scope are random flipping, random rotating, and random hue. Also, the essential image pre-processing technique is to crop the image at the center of it to 512×512 pixels. Next, another image pre-processing we do is adding CLAHE from [5] to images before the training. CLAHE is an advancement in adaptive histogram equalization (AHE) in which the contrast amplification is limited. CLAHE divided the image into blocks with the specific dimension input value. It computes the contrast transform function for each block individually. The contrast amplification is limited by specific clipping input values. Figure 3 demonstrates the CLAHE applied to images.

3.2   Model

In the project, the model, AlbuNet, is a modification of U-Net, which involved pre-trained ResNet-34 as encoders. The model contains unequal channels from encoder concatenate with the decoder in some layers. We added a squeeze-and-excitation block from [18] between each layer of encoders, decoders, and links between encoders and decoders. Squeeze-and-excitation (SE) block is an attention gate building from network layers, which is an average pooling layer, a linear or a dense layer with a ReLU activation function, and a linear or a dense layer with a Sigmoid activation function tiled together respectively. SE block structure is shown in Fig. 4. This helps to add weights to channels

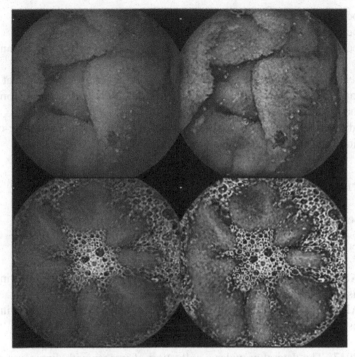

Fig. 3. Examples of original images (left) and applied CLAHE images (right) from MICCAI 2017 [10] (top) and MICCAI 2018 [11] (bottom)

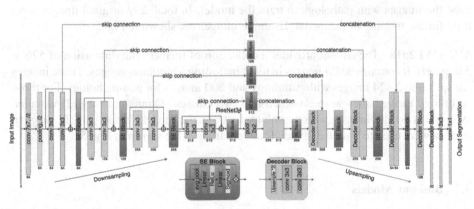

Fig. 4. AlbuNet model with squeeze-and-excitation blocks [9], SE blocks are red block colors (Color figure online)

and eventually enable the model to focus important information flow from each layer in the network. The model is shown in Fig. 4.

We add the SE block into each layer and skip connections as a result of our experiments. We test three models with differently placed SE blocks as follows:

(1) SE blocks between encoder layers
(2) SE blocks between encoder layers and between decoder layers
(3) SE blocks between encoder layers, between decoder layers and skip connections

By increasing the number of SE blocks in each experiment, we found that the third model performed the best, so we chose the third model as the new model to compare the baseline model in this paper.

By combining the methods described above, the model can archive state-of-the-art for the angiodysplasia image segmentation task.

4   Experiment Design

4.1 Datasets

In this paper, we used two datasets for angiodysplasia as follows:

1. MICCAI 2017 Endoscopic Vision Challenge: Sub-Challenge Gastrointestinal Image Analysis (GIANA): Angiodysplasia localization task
2. MICCAI 2018 Endoscopic Vision Challenge: Sub-Challenge Gastrointestinal Image Analysis: Small Bowel Lesion Localization task

MICCAI 2017. The dataset provides a collection of images with dimension of 576 × 576 pixels. It contains 1,198 images in total, including annotation images. Those images are split into 598 images with pathology and 600 images for non-pathology. We only used the images with pathology to train the model. In total, 299 original images were used for the training of the model. Example images are shown in Fig. 5.

MICCAI 2018. The dataset provides a collection of images with dimension of 576 × 576 pixels. It contains 3,024 images in total, including annotation images. Those images are split into 2,424 images with pathology and 600 images for non-pathology. For those pathology images, they were classified into two classes (1) inflammatory (2) vascular. We only used the images with pathology to train the model. In total, 1,212 original images were used to train the model. Example images are shown in Fig. 5.

4.2 Baseline Models

To evaluate the performance of our methods, we compared our proposed methods with results from [9], which are results from the AlbuNet model.

For hyperparameters tuning, we adopted the hyperparameters settings created by [9]. We performed 5-fold cross-validation as a procedure to estimate the performance of the models. We trained each model using Adam optimizer with a learning rate of 0.0001 for the first ten epochs and 0.00001 for other epochs.

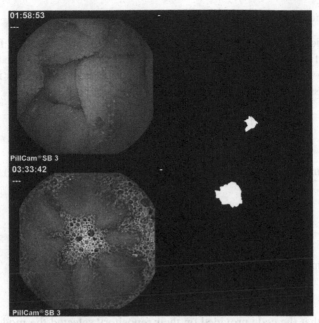

Fig. 5. Examples of original images with lesion and annotation images from MICCAI 2017 [10] (top) and MICCAI 2018 [11] (bottom) datasets

4.3 Evaluation Metrics

We used the Dice coefficient (Dice) and Jaccard Index or Intersect over Union (IoU) as metrics, similar to [9]. The Dice equations are shown in Eqs. (1) and (2). The IoU equations are shown in Eqs. (3) and (4). Since the WCE technique does not require a real-time process, we will focus on the accuracy of the model instead of an inference time.

By given two sets, A and B, Dice equation is defined as follows:

$$Dice = \frac{2|A \cap B|}{|A| + |B|} \qquad (1)$$

when applied to binary data, Dice equation is defined as follows:

$$Dice = \frac{2TP}{2TP + FP + FN} \qquad (2)$$

where TP is true positive, FP is false positive, and FN is false negative.

By given two sets, A and B, IoU equation is defined as follows:

$$IoU = \frac{|A \cap B|}{|A \cup B|} \qquad (3)$$

When applied to binary data, IoU equation is defined as follows:

$$IoU = \frac{TP}{TP + FP + FN} \qquad (4)$$

where TP is true positive, FP is false positive, and FN is false negative.

4.4 Loss Function

For loss function, we also used the same loss function in [9]. The loss function is a combination of binary cross entropy and Jaccard function from [20]. The binary cross entropy equation is shown in Eq. (5). The loss function equation is shown in Eq. (7).

Binary cross entropy function equation is defined as follows:

$$H = -\frac{1}{n} \sum_{i=1}^{n} [y_i \log(\hat{y}_i) + (1 - y_i)\log(1 - \hat{y}_i))] \tag{5}$$

Adapted function for a discrete object, pixels in an image, from Eq. (3) is defined as follows:

$$J = \frac{1}{n} \sum_{i=1}^{n} \left(\frac{y_i \hat{y}_i}{y_i + \hat{y}_i - y_i \hat{y}_i} \right) \tag{6}$$

The loss function equation is defined as follows:

$$L = H - \log(J) \tag{7}$$

We found that the code provided on their repository selected the model at the last epoch of the training process for evaluation of the metric scores. This procedure of evaluation should be changed to the model that provides the best or the least loss validation score for each fold. So, we added this procedure to keep the best validation loss model for each fold and used the saved models as the final models to evaluate the metrics instead. The reason we chose the best validation loss model over the last trained models as they proposed because the last trained models tend to overfit the data more than the best validation loss models.

5   Experiment Results

The results of metrics, Dice, and IoU scores are shown in Table 1. For the MICCAI 2017 dataset and Table 2. for the MICCAI 2018 dataset. The compared prediction images are shown in Fig. 6.

Table 1. MICCAI 2017 segmentation results, results are in %.

Model	Dice	IoU
AlbuNet (baseline)	85.07	75.63
AlbuNet + SE	86.09	76.60
AlbuNet + SE + CLAHE	**86.17**	**76.64**

The results showed the performance improvement of the model using our proposed methods because we focused on various perspectives for building the model and finding

Table 2. MICCAI 2018 segmentation results, results are in %.

Model	Dice	IoU
AlbuNet (baseline)	62.99	51.91
AlbuNet + SE	65.62	54.61
AlbuNet + SE + CLAHE	**66.20**	**55.11**

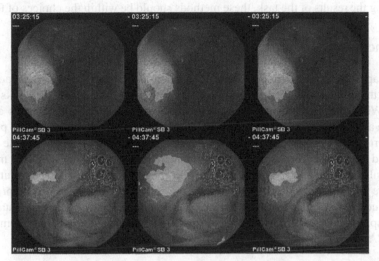

Fig. 6. Examples of ground truth images with a lesion (left) compared to prediction result images from AlbuNet (middle) and Ours (right).

solutions for each part. Our process started from the image pre-processing, namely CLAHE, to enhance images before the training. Secondly, we added SE blocks to weight data channels to help model focus on relevant data flow in the network.

For image pre-processing, we used the CLAHE function from the OpenCV library. It required two parameters contrast limit and block size. The contrast limit is a parameter to avoid noise amplified, and the block size is the size of the divided block. By combining these parameters, the results could be changed and be applied to other image segmentation tasks. Before applying CLAHE to images, we converted images from RGB color space to YUV color space, then applied CLAHE before converting them back to the original RGB color space.

For adding SE blocks, we experimented with a combination of SE blocks explained in Sect. 3.2. These blocks are hyperparameters to be tuned when building the model. By adding a combination of SE, blocks could also change the results of the model because the SE block is an attention gate that amplified the data passing through the gate, so if we add these gate on the right places and with the right amounts, it could improve performance of the model. On the other hand, if we miss the place and amount, it could also reduce the performance of the model too.

The reason for different scores from MICCAI 2017 and 2018, in our opinion, is because of the variety of two of the image classes in the MICCAI 2018 dataset, which is more than MICCAI 2017 that has only one image class dataset. Even with human vision, the locations of the lesion are difficult to be determined in some pathology images.

The MICAI 2018 dataset is suitable for challenging the models for improvement in the angiodysplasia task.

In conclusion, these methods we proposed required experiment tuning such as the hyperparameters for specific models and datasets that may be time-consuming, but for improving the results of the task, these methods should be within the choices of testing.

6 Conclusion

In this paper, we proposed methods to improve model performance. We proposed to enhance them by two methods: using image pre-processing and add SE blocks to the model. With these performance improvements, we obtained a model that is able to diagnose angiodysplasia with a higher rate of accuracy. This will benefit many patients and experts in the field. Even though the proposed methods required time to tune for the optimized parameters, for the perspective of enhancing the performance of the models, they should be considered as collections of the choices for hyperparameters tuning in the future experiments as well. The code implemented in this paper will be provided upon request. Finally, the MICCAI 2018 dataset is a unique and challenging dataset for angiodysplasia image segmentation tasks due to the increasing complexity comparing to MICCAI 2017 dataset.

References

1. Hussein Al-Hamid, M.F., Department of Gastroenterology, Providence Hospital: Angiodysplasia of the Colon. https://emedicine.medscape.com/article/170719-overview (2019)
2. Kim, S.H., Cha, Y.S., Lee, Y., Kim, H., Yoon, I.N.: Successful treatment of central retinal artery occlusion using hyperbaric oxygen therapy. Clin. Exp. Emerg. Med. 5(4), 278–281 (2018). https://doi.org/10.15441/ceem.17.271
3. Medical Advisory, S.: Wireless capsule endoscopy: an evidence-based analysis. Ont. Health Technol. Assess. Ser. 3(2), 1–35 (2003)
4. Fernandez-Urien, I., et al.: Capsule endoscopy capture rate: has 4 frames-per-second any impact over 2 frames-per-second? World J. Gastroenterol. 20(39), 14472–14478 (2014). https://doi.org/10.3748/wjg.v20.i39.14472
5. Noya, F., Álvarez-González, M.A., Benítez, R.: Automated angiodysplasia detection from wireless capsule endoscopy. In: 2017 39th Annual International Conference of the IEEE Engineering in Medicine and Biology Society (EMBC), 11–15 July 2017, pp. 3158–3161 (2017)
6. Ronneberger, O., Fischer, P., Brox, T.: U-Net: convolutional networks for biomedical image segmentation. In: Navab, N., Hornegger, J., Wells, William M., Frangi, Alejandro F. (eds.) MICCAI 2015. LNCS, vol. 9351, pp. 234–241. Springer, Cham (2015). https://doi.org/10.1007/978-3-319-24574-4_28
7. Oktay, O., et al.: Attention U-Net: learning where to look for the pancreas (2018)
8. Mohammed, A.K., Yayilgan, S.Y., Farup, I., Pedersen, M., Hovde, Ø.: Y-Net: a deep convolutional neural network for polyp detection. CoRR abs/1806.01907 (2018)

9. Shvets, A.A., Iglovikov, V.I., Rakhlin, A., Kalinin, A.A.: Angiodysplasia detection and localization using deep convolutional neural networks. In: 2018 17th IEEE International Conference on Machine Learning and Applications (ICMLA), 17–20 December 2018, pp. 612-617 (2018)
10. MICCAI 2017 Endoscopic Vision Challenge: Angiodysplasia detection and localization (2017). https://endovissub2017-giana.grand-challenge.org/Angiodysplasia-ETISDB/
11. MICCAI 2018 Endoscopic Vision Challenge: Angiodysplasia detection and localization (2018). https://giana.grand-challenge.org/WCE/
12. Krizhevsky, A., Sutskever, I., Hinton, G.: ImageNet classification with deep convolutional neural networks. In: Neural Information Processing Systems, vol. 25 (2012). https://doi.org/10.1145/3065386
13. Zuiderveld, K.: Contrast limited adaptive histogram equalization. In: Paul, S.H. (ed.) Graphics Gems IV, pp. 474–485. Academic Press Professional, Inc., San Diego (1994)
14. Iglovikov, V., Shvets, A.: TernausNet: U-Net with VGG11 encoder pre-trained on ImageNet for image segmentation (2018)
15. Simonyan, K., Zisserman, A.: Very deep convolutional networks for large-scale image recognition. arXiv: 1409.1556 (2014)
16. Yosinski, J., Clune, J., Bengio, Y., Lipson, H.: How transferable are features in deep neural networks? In: Proceedings of the 27th International Conference on Neural Information Processing Systems, vol. 2, Montreal, Canada
17. He, K., Zhang, X., Ren, S., Sun, J.: Deep residual learning for image recognition. In: 2016 IEEE Conference on Computer Vision and Pattern Recognition (CVPR), 27–30 June 2016, pp. 770-778 (2016)
18. Hu, J., Shen, L., Sun, G.: Squeeze-and-excitation networks. In: 2018 IEEE/CVF Conference on Computer Vision and Pattern Recognition, 18–23 June 2018, pp. 7132-7141 (2018)
19. Xu, K., et al.: Show, attend and tell: neural image caption generation with visual attention. In: Proceedings of the 32nd International Conference on Machine Learning, vol. 37, Lille, France (2015)
20. Iglovikov, V., Mushinskiy, S., Osin, V.: Satellite imagery feature detection using deep convolutional neural network: a kaggle competition (2017)
21. Dray, X., et al.: CAD-CAP: une base de données française à vocation internationale, pour le développement et la validation d'outils de diagnostic assisté par ordinateur en vidéocapsule endoscopique du grêle. Journées Francophones d'Hépatogastroentérologie et d'Oncologie Digestive. Paris, 22–25 mars 2018
22. Leenhardt, R., et al.: CAD-CAP: a 25000 Images Database Serving the Development of Artificial Intelligence for Capsule Endoscopy, Endoscopy international Open (2020, in press)

GOHAG: GANs Orchestration for Human Actions Generation

Aziz Siyaev$^{(\boxtimes)}$ (iD) and Geun-Sik Jo (iD)

Inha University, Incheon, Republic of Korea
azizsiyaev.ai@gmail.com, gsjo@inha.ac.kr

Abstract. Generative Adversarial Networks (GANs) made a huge contribution to the development of content creation technologies. Important place in this advancement takes video generation due to the need for human animation applications, automatic trailer or movie generation. Therefore, taking advantage of various GANs, we proposed own method for human movement video generation GOHAG: GANs Orchestration for Human Actions Generations. GOHAG is an orchestra of three GANs, where Poses generation GAN (PGAN) creates a sequence of poses, Poses Optimization GAN (POGAN) optimizes them, and Frames generation GAN (FGAN) attaches texture for the sequence, creating a video. The proposed method generates a smooth and plausible video of high-quality and showed potentials among modern techniques.

Keywords: Generative Adversarial Network · Poses Generation · Video generation

1 Introduction

Generative Adversarial Networks (GANs) [1] shown rapid development in content creation tasks like images and video generation. Among them, special attention takes video generation due to the development of various applications for humans and objects animation. Video generation task requires special approach, unlike images generation. Even though a video is a set of frame-images combined together, it has various factors to consider for accomplishing generation tasks such as scene dynamics, relationship with previous and over-all movements, and consistency of the content. This field of study requires deep research and analysis, however, notable results have been proposed already.

Various research works [2–5] attempted to provide methods for video generation. [2] used latent variables generation and their transformation into a video. [3, 4] tried to generate video's direct pixels. [5] generated a sequence of frames from random vectors. However, research works [6, 7] focused on the generation of the human actions demonstrated that the dynamics of the video should be controlled and managed with a human pose. [6, 7] both applied similar pipeline of video frames generation: first, produce a sequence of poses, and next, use the poses for scene creation. Both works proposed their own methods though. [6] introduced a GAN that takes the initial pose of movement as input and produces a sequence of poses to be feed to the next GAN for frames creation.

© Springer Nature Switzerland AG 2020
N. T. Nguyen et al. (Eds.): ACIIDS 2020, LNAI 12033, pp. 294–305, 2020.
https://doi.org/10.1007/978-3-030-41964-6_26

On the contrary, [7] for the poses generation stage applied two GANs that work together for the human skeleton movement generation. One GAN maps noise to a pose, at the same time another one generates noise and reuses the first one. However, we proposed yet another method for human action generation.

In this research work, we proposed a method GOHAG: GANs orchestration for Human Actions Generation. Using special data representation for movement, we applied the collaboration of three GANs for human actions video generation task. The first network - PGAN (Pose generation GAN) produces poses representation, the second one - POGAN (Poses Optimization GAN) optimizes the results of PGAN, and the third one - FGAN (Frame generation GAN) converts skeleton picture to the image frame. In comparison with existing works, GOHAG provides a new perspective for the poses generation by considering this task as the image synthesis, where the generated image encapsulates information about a particular movement. Applying GOHAG orchestration of GANs for human action generation helps to create a human movement without providing initial pose as it was done in [6]. What is more, instead of learning random vectors that represent a particular action, like in [7], GOHAG learns to generate a pattern of a movement.

In the Experiments section, we made the evaluation of GOHAG on the public dataset and, using different metrics, and compared its performance with [7]. We demonstrated that the developed framework is capable of producing promising results in comparison with state-of-the-art video generator.

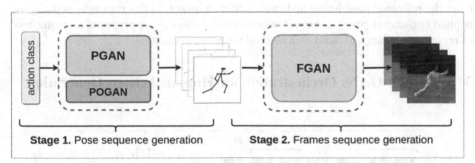

Stage 1. Pose sequence generation   **Stage 2.** Frames sequence generation

Fig. 1. GOHAG: two-staged development method overview. Stage 1 takes user-defined action class and generates human poses, and next, Stage 2 produces the textured frames

2 Related Works

The introduction of GANs [1] triggered the development of various research works in a video generation task, and there have been many approaches to do that. [2] proposed a GAN that can learn semantic representations of a video and produce their own. The network produces a set of latent variables that later are mapped to the frames in videos, which later combined into a single video. [3] proposed an unsupervised method of video generation using captions. Taking advantage of short-term and long-term temporal and spatial context, researchers at [3] could generate video preserving consistency between frames. [5] used an approach of decomposition of content and motion in their video generation GAN. Using a sequence of random vectors that include motion and content part, the proposed framework in [5] performs the vector-to-frames mapping. According to a particular action, the motion part changes, whereas the content part remains fixed. Mentioned works showed promising results, however, the two-stage human video generation proposed by [6] and [7] showed that a pose-controllable scene generation, where the final frames created by texturing skeleton pose, produces stable and more plausible results.

[6] applied GAN for poses generation, which takes an initial pose and action class as input and synthesizes pose sequence as an encoder-decoder manner. Pose sequence encoded using several convolutional layers, embedded with action class label, and after several residual blocks, results passed through decoder part with a series of fractionally-striped spatial convolutional layers. In addition, for stable and better learning, their poses generator network applied an integrated LSTM module. Frames generation GAN is done by conditioning the network with human pose and a corresponding input image.

[7] applied double-GAN architecture for poses generation stage. The first GAN takes an action class, concatenates it with random noise and produces a single pose. At the same time, the second GAN responsible for the "proper" latent random vector generation, which is going to be used by the first GAN. Therefore, the second GAN reuses their first GAN multiple times to create a sequence of poses. For the texturing part, this work applied U-Net like network with convolutional autoencoders with skip connections that takes the reference pose image and outputs human image. In this research, as [6, 7] we applied two-staged pose-guided video generation, however, we proposed a unique way of representing data and build own method for poses generation.

3   GOHAG: GANs Orchestration for Human Actions Generation

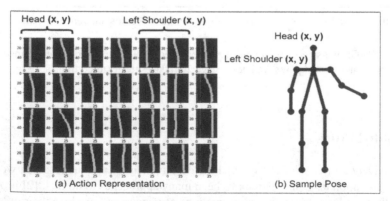

Fig. 2. Human action representation. Action Representation (a) shows the pattern of change of each pose point of the sample pose (b) for particular human movement

GANs Orchestration for Human Actions Generation (GOHAG) method consists of two main parts. First, we represent human actions in a special way. Instead of focusing on the generation of exact skeleton points in the space, we learn a pattern of a particular human action based on the skeleton movement. Second, we proposed the two-staged GANs orchestration for human action generation (see Fig. 1). Stage 1 takes user-defined action class as input and provides to proposed PGAN (Poses generation GAN) that works together with POGAN (Poses Optimization GAN) for pose sequence generation. PGAN generates pose pattern, whereas POGAN optimizes them to reduce noise. Stage 2 of GOHAG concentrates on texture generation. FGAN (Frames generator GAN) takes the sequence of generated skeleton poses and produces frames by overlaying texture with pose.

3.1 Action Representation

Proper data representation plays a crucial role in particular tasks. In the case of human action generation, trying to generate a sequence of poses just by learning skeleton points' position in the space is not efficient. Instead, we proposed a special representation of human movements, which constructs a pattern of changes of skeleton points within a particular action.

Figure 2 shows a sample of action representation. Assuming that pose skeleton (b) in Fig. 2 takes an action (i.e. walking), on the left, we illustrated its corresponding action representation (a). The sample pose (b) consists of 14 points in 2D or 28 points in 1D space, and Action Representation (a) demonstrates how each point of skeleton pose changes within a number of frames. In Fig. 2 we illustrated a human stretching action example within 60 frames and encapsulate each point movement in $S \times N$ picture, where N is the number of frames and S is a scaling factor for transformation of the original pose point. In this work, we applied S as 40. Analyzing points of the head movement, you can see that x point stays the same, however, y changes its position. Therefore, every human action will be converted to the proposed representation and used in the Poses Generation stage.

3.2 Poses Generation

Poses Generation stage comprises of two GANs that work together. The PGAN takes a user-defined action class, embeds it with noise and generates pose sequence representation. Next, POGAN performs an optimization process on the generated sequence to remove noise and enhance the produced action patterns from PGAN.

PGAN. Poses generator GAN is the main part of GOHAG, which is responsible for smooth poses generation according to the given action class. The generator of PGAN tries to reproduce action in the form of action representation (see Fig. 2) and fool the discriminator network. The discriminator at the same time learns to identify whether the generated pose sequence is real or fake.

Figure 3 describes the architecture of PGAN. The proposed PGAN is Conditional GAN [8] that has a deep architecture. Both generator and discriminator networks are

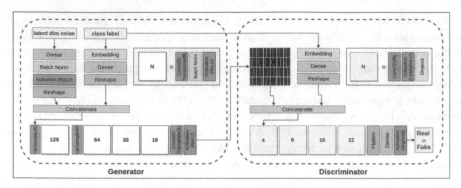

Fig. 3. Architecture of PGAN

conditioned with a user-defined class label and trained together on binary cross-entropy loss.

The generator gets a class label and normally distributed latent dimension noise (in our work latent dimension is 100) and after performing Embedding, Dense layers for the class label, Dense, Batch Normalization, ReLU for latent noise, Reshapes them in the same dimension for Concatenation. Next, the output passed through the series of blocks that contain Conv2DTranspose, Batch Normalization, and ReLU (number of filters for convolutions defined on the block, kernel size is 5, striding is 2). The output of the network and class label are then passed to the discriminator network.

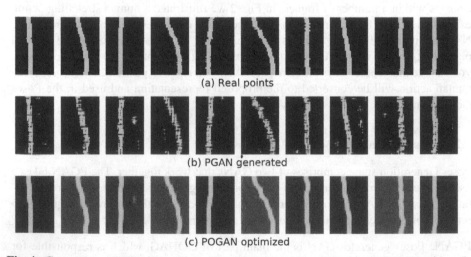

Fig. 4. Comparison of POGAN optimized pose sequence (c) generated by PGAN (b) with real sequence (a)

The discriminator network works with produced pose sequences of the generator and the initial action class label. It performs the same steps with the action label as the generator and concatenates with produced output. Performing the series of convolution block layers which consist of Conv2D, LeakyReLU, Dropout (LeakyReLU is 0.2, Dropout is 0.5), and increasing the number of filters for convolutions (described in Fig. 3) lead the discriminator to produce a single statement (fake or real). The architecture of PGAN was developed empirically. We noticed that more filters in convolutional layers will not always lead to better convergence, and it's crucial to find a balance between generator and discriminator networks.

POGAN. Poses Optimization GAN plays a considerable role in smooth and plausible human action generation. The main objective of POGAN is to optimize the output produced by PGAN. Figure 4(b) shows a sample result of PGAN. Comparing the real points' pattern and PGAN-generated, PGAN produces acceptable output, however, there is a noise that affects the smoothness of the final pose sequence. This demonstrates the need for the optimization step.

POGAN is a conditional CycleGAN [9]. CycleGANs showed great potential and abilities in different use cases, therefore, we applied the architecture of [9] for POGAN with the adopted dimensions for feeding our data. The main idea behind optimization - considering this step as the style changing task, where we convert noised input into a smooth poses sequence. For training POGAN we used paired mapping of different noised level input and expected styled output. As a result of the optimization step, the noised output of PGAN described in Fig. 4(b) is converted to smooth and cleaned pose sequence representation illustrated in (c). The resultant output of PGAN and POGAN in the Pose Generation stage produces the sequential skeleton poses for the next stage of human action generation.

3.3 Frames Generation

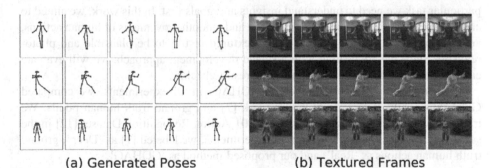

(a) Generated Poses   (b) Textured Frames

Fig. 5. Sample results from the Poses Generation stage (a) and the according textured poses from the Frames Generation stage (b)

In the Frames Generation stage of GOHAG, we convert generated poses into the sequence of frames. Third GAN - FGAN overlays the texture on each skeleton for a given set of poses.

For the development of FGAN, we used NVIDIA's pix2pixHD [10], which is intended to make photorealistic image-to-image translation. [10] applied high-resolution image synthesis with Conditional GANs [8]. We used the pix2pixHD network as a baseline for building FGAN. Modifying input layers of [10], we proposed Frames generator GAN that takes pose skeleton image as input and generates scene frame. Since the output of the first stage of GOHAG is a poses sequence in our own data representation (see Fig. 2(a)), we convert them into a set of skeleton pose images (see Fig. 5(a)). Feeding a skeleton pose image to FGAN will return the textured frame on that pose. Training of FGAN requires two sets of images: skeleton poses images and according set of textured frames. We noted that the performance of FGAN is dependable from previous networks. The stability of results produced by FGAN increases with the variety of examples. We illustrated the sample results of FGAN in Fig. 5(b) based on generated poses of PGAN&POGAN in (a).

Comparing recent works on poses generation [6, 7], GOHAG has its own specialties. [6] has an LSTM module inside of their network which needs to get the initial pose for the generation of the sequence, however, GOHAG can produce a movement just using a latent noise. Researchers at [7] applied two GANs for the poses generation stage. One GAN manipulates random noise by concatenating a class of action and mapping it into a pose. Second GAN creates noise for sequence. In the case of GOHAG, we try to map a random noise concatenated with a class of movement to the action representation image (Fig. 2). GOHAG considers poses generation as image generation, which has patterns that describe a particular movement. Next, we performed a quantitative and qualitative analysis of the GOHAG and described it in the experiments section.

4   Experiments

Evaluation of the GANs is not straightforward. [11] states that for the evaluation of each particular task we need to understand insights and goals first. In this work, we aimed to create a method that can generate sequential and smooth movements of human actions, and, at the same time, we want the final textured results to be plausible and photorealistic. Since we are dealing with a staged development approach, we will evaluate each step separately, and also perform total result-check.

To demonstrate the performance of GOHAG, in all experiments, we compared GOHAG's results with state-of-the-art work [7] and ground truth human points. We trained models on chosen classes of UCF 101 Action Recognition Dataset [12] in the same way as it was done in [7]. For all experiments, we labeled [7] as 'Deep', ground truth human actions as 'Real', and our proposed method as 'GOHAG'.

4.1   Dataset

Evaluation of GOHAG is done with the public dataset. We used UCF101 Action Recognition Dataset [12] in order to train and compare the proposed method with another work. We selected three action classes (jumping jack, jumping rope, taichi) each containing 4–5 short videos where one object performing an action. In order to prepare the dataset for GOHAG, we framed all videos and estimated poses with state-of-the-art human pose estimator OpenPose [13] and extracted 28 human skeleton points. We prepared two sets of the dataset for each stage of GOHAG. First, the dataset of the sequence of poses and corresponding action labels for Stage 1. Second, the dataset of pose skeletons and its textured frames for Stage 2. We trained GOHAG and all experiments on a PC with NVidia RTX 2080 Ti GPU, Intel Core i3-8100 CPU with 4 cores, 8 GB RAM. Stage 1 training took about 2 h with 1000 epochs, whereas Stage 2 – about 5 h with 50 epochs. Next, we performed the evaluation process.

4.2 Evaluation of Poses Generation

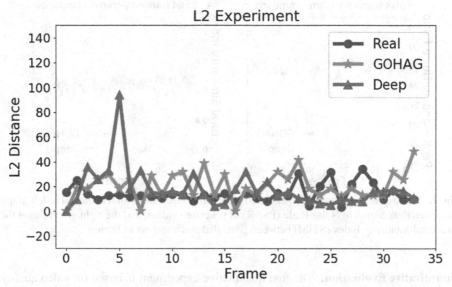

Fig. 6. L2 distances of consecutive poses. High points on the graphs show rapid change between consecutive poses, and low points represent slow or absence of motion

The output of the Poses generation stage is a sequence of poses. Since we are aiming to have smooth and sequential human action generation, we want GOHAG to produce stable pace results. To check this property, we proposed an L2 distances experiment (see Fig. 6) that measured the distance between each consecutive pose within 35 frames. High L2 distance represents rapid change between poses; low distance - slow or no motion. We compared the plots of the proposed model (GOHAG) with [7] (Deep) and ground truth movement (Real). The experiment shows that both methods produce poses' pace within a range of Real graph change. The average L2 distances of Real is 16.5, Deep is 20.94 and GOHAG is 14.57. At the same time, the standard deviations of L2 distances of Real, Deep and GOHAG are 9.81, 11.11 and 7.11 respectively. This demonstrates that GOHAG is able to generate smooth movements that describe original action (see Fig. 5(a)) with a closer pace to original than Deep since GOHAG's results are closer to Real.

4.3 Evaluation of Frames Generation

Experiments on Frames Generation stage is an evaluation of the results from the full pipeline because the performance of FGAN depends on the output from PGAN&POGAN. Therefore, we generated sample poses from the Poses generation stage and produced textured frames with the help of FGAN.

Evaluation of Frames Generation stage consists of two parts: a quantitative and qualitative analysis. Quantitative evaluation represents the experiments where we compare generated videos with real ones in terms of their quality, diversity, and distribution. Qualitative evaluation checks the plausibility and realism of generated frames.

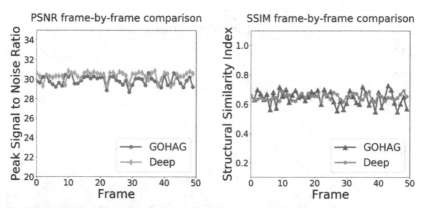

Fig. 7. Video quality evaluation of generated videos in comparison with real ones. The left graph illustrates Peak Signal to Noise Ratio (PSNR) of generated videos, and the right graph shows the Structural Similarity Index (SSIM) between generated and a real set of frames

Quantitative Evaluation. The first quantitative experiment is based on video quality evaluation. Using Peak Signal to Noise Ratio (PSNR) [14] and Structural Similarity Index (SSIM) [14] metrics, we compared generated video with real video in a frame-by-frame manner with-in 50 frames (see Fig. 7). PSNR metric shows how well a generated image was reconstructed from the original frame. PSNR metric usually used as the simulation of human perception in judging a reconstruction quality of images after compression. Generally, PSNR is the ratio of signal to noise, where a signal is the original data, and the noise is the error of compression. Therefore, the greater the value of PSNR is better. In our case, we used this metric to compare generated and real frames. Results (Fig. 7 (left graph)) illustrate that, in the PSNR metric, Deep performs negligibly better than GOHAG. The average PSNR value for GOHAG is 29.95, Deep is 30.35, and their standard deviations are 0.46 and 0.44 for GOHAG and Deep, accordingly. At the same time, we performed the video quality evaluation experiment with the SSIM metric (Fig. 7 (right graph)). SSIM is a perception-based model intended to show how similar two images are. PSNR estimates absolute errors, however, SSIM deals with an image compression or degradation as a change in its structural information (spatially close inter-dependent pixels). We used SSIM to find similarity between generated and real images (Fig. 7 (right graph)). The graph demonstrates that GOHAG's plot has higher peaks than

Deep. GOHAG achieved an average of 0.651 similarity, where Deep is 0.648. Their standard deviations are 0.04 and 0.02 for GOHAG and Deep respectively. The results in the video quality experiment demonstrate that both methods (Deep and GOHAG) produce nearly equal performance and high-quality generation output.

The second experiment for quantitative evaluation is based on Inception Score (IS) [15] and Fréchet Inception Distance (FID) [16]. Inception Score is a metric proposed in [15] for evaluating the performance of GANs. IS illustrates how diverse the generated images are, and how distinctly they look like from each other. IS takes the name from the Inception classifier, where it feeds the images to the Inception network and returns the probability distribution of a label for the image. According to the distribution and probabilities computes Inception Score. In addition, we applied another metric FID [16] that calculates the distance between two distributions (real and generated images). For IS and FID experiments, we used all images generated by GOHAG, Deep [7] and Real images. We summarized the results in Table 1.

Table 1. Comparison table Deep, Real and GOHAG results based on IS and FID

Example images	Inception Score (**IS**)	Fréchet Inception Distance (**FID**)
Real	1.41	0
Deep	1.12	106
GOHAG	**1.35**	**98**

The results of Table 1 show low IS because we trained only with 3 action classes. However, these experiments created equal conditions for comparison between GOHAG and Deep generated videos. In both metrics (IS and FID) GOHAG outperforms Deep. GOHAG's IS closer to the Real and at the same time, its FID has a smaller value. This shows that the generated distribution of GOHAG images is closer to the ground truth images.

Qualitative Evaluation. Qualitative evaluation of GOHAG and Deep is based on the users' responses. For this experiment, we used the psychophysics approach described in [17], which is intended to evaluate the perceptual similarity and variability of the images. In [17], researchers provided users real and synthetic sets of images and asked them to find a real one. The size of the images in the sets was changed from small to large, which led to an increase in the percentage of a correctly identified real set. Following the methodology in [17], we conducted an online survey, where the users asked to find a set of real images. For each question, we provided the sets of generated (by GOHAG or Deep) and real images with the same image size in a set. Users did not know about the comparison experiment of GOHAG and Deep, so they just had to distinguish a real set of images. The main objective of the evaluation is to see for which images (GOHAG or Deep) users can hardly distinguish between generated and real ones. The online survey was conducted in a social network, where we received responses from 31 people. Figure 8 summarizes a user evaluation. As shown in a graph, with increasing

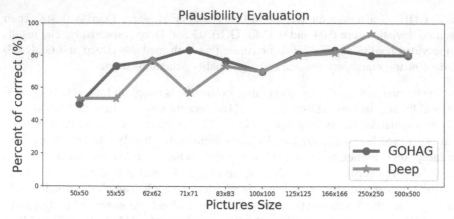

Fig. 8. User evaluation summary

the size of the picture in the set, users started correctly identify generated images. With 50×50 images it is almost random responses (about 50%) since it is hard to distinguish due to very small images. However, for larger size images like 500×500 about 80% could correctly identify real ones. Analyzing the graphs in Fig. 8, the lower graph is better since we want users' responses to be incorrect (confuse generated images as real). Deep performs better in sets with smaller size image, however, starting from 100×100 GOHAG shows similar or better results. This result shows that while comparing generated pictures in the high dimensions, GOHAG produces higher quality and more realistic images rather than Deep.

5 Conclusion

In this research work, we proposed the method for human action video generation GOHAG. GOHAG is the two-staged action generation method that orchestrating GANs for producing pose sequences and texturing them to generate a video. GOHAG consists of three GANs: Poses generation GAN (PGAN) that produces poses, in the form of proposed movement representation, from a user-defined action class; Poses Optimization GAN that optimizes the result of PGAN, and Frames generation GAN (FGAN) that takes the sequence of poses from the previous stage and convert them into textured frame sequence. We evaluated produced results quantitatively and qualitatively on different experiments. Results showed that GOHAG outperforms state-of-the-art work with stable and smooth poses generation. The proposed method GOHAG generates high-quality image frames and achieves the production of plausible and realistic human actions.

Acknowledgements. This research was supported by the MSIT (Ministry of Science and ICT), Korea, under the ITRC (Information Technology Research Center) support program (IITP-2017-0-01642) supervised by the IITP (Institute for Information & communications Technology Promotion).

References

1. Goodfellow, I., et al.: Generative adversarial nets. In: NIPS (2014)
2. Saito, M., Matsumoto, E., Saito, S.: Temporal generative adversarial nets with singular value clipping. In: ICCV (2017)
3. Marwah, T., Mittal, G., Balasubramanian, V.N.: Attentive semantic video generation using captions. In: 2017 IEEE International Conference on Computer Vision (ICCV), pp. 1435–1443. IEEE (2017)
4. Vondrick, C., Pirsiavash, H., Torralba, A.: Generating videos with scene dynamics. In: Lee, D.D., Sugiyama, M., Luxburg, U.V., Guyon, I., Garnett, R. (eds.) Advances in Neural Information Processing Systems 29, pp. 613–621. Curran Associates, Inc. (2016). http://papers.nips.cc/paper/6194-generating-videos-with-scene-dynamics.pdf
5. Tulyakov, S., Liu, M.Y., Yang, X., Kautz, J.: MoCoGAN: decomposing motion and content for video generation. arXiv preprint arXiv:1707.04993 (2017)
6. Yang, C., Wang, Z., Zhu, X., Huang, C., Shi, J., Lin, D.: Pose guided human video generation. In: Ferrari, V., Hebert, M., Sminchisescu, C., Weiss, Y. (eds.) ECCV 2018. LNCS, vol. 11214, pp. 204–219. Springer, Cham (2018). https://doi.org/10.1007/978-3-030-01249-6_13
7. Cai, H., Bai, C., Tai, Y.-W., Tang, C.-K.: Deep video generation, prediction and completion of human action sequences. In: Ferrari, V., Hebert, M., Sminchisescu, C., Weiss, Y. (eds.) ECCV 2018. LNCS, vol. 11206, pp. 374–390. Springer, Cham (2018). https://doi.org/10.1007/978-3-030-01216-8_23
8. Mirza, M., Osindero, S.: Conditional generative adversarial nets. arXiv preprint arXiv:1411.1784 (2014)
9. Zhu, J., Park, T., Isola, P., Efros, A.A.: Unpaired image-to-image translation using cycle-consistent adversarial networks. In: ICCV (2017)
10. Wang, T., Liu, M., Zhu, J., Tao, A., Kautz, J., Catanzaro, B.: High-resolution image synthesis and semantic manipulation with conditional GANs. In: CVPR (2018)
11. Theis, L., Oord, A.V.D., Bethge, M.: A note on the evaluation of generative models. In: ICLR (2016)
12. Soomro, K., Zamir, A.R., Shah, M.: UCF101: a dataset of 101 human actions classes from videos in the wild. arXiv preprint arXiv:1212.0402 (2012)
13. Cao, Z., Hidalgo, G., Šimon, T., Wei, S., Sheikh, Y.: Realtime multi-person 2D pose estimation using part affinity fields. In: CVPR (2017)
14. Horé, A., Ziou, D.: Image quality metrics: PSNR vs. SSIM. In: ICPR (2010)
15. Salimans, T., Goodfellow, I.J., Zaremba, W., Cheung, V., Radford, A., Chen, X.: Improved techniques for training GANs. arXiv preprint arXiv:1606.03498 (2016)
16. Heusel, M., Ramsauer, H., Unterthiner, T., Nessler, B., Hochreiter, S.: GANs trained by a two time-scale update rule converge to a local nash equilibrium. In: NIPS (2017)
17. Gerhard, H.E., Wichmann, F.A., Bethge, M.: How sensitive is the human visual system to the local statistics of natural images? PLoS Comput. Biol. 9(1), e1002873 (2013)

Machine Learning and Data Mining

Finger-Vein Classification Using Granular Support Vector Machine

Ali Selamat[1(✉)], Roliana Ibrahim[2], Sani Suleiman Isah[2], and Ondrej Krejcar[3]

[1] Malaysia Japan International Institute of Technology (MJIIT), Universiti Teknologi Malaysia, Jalan Sultan Yahya Petra, Kuala Lumpur, Malaysia
aselamat@utm.my

[2] School of Computing, Faculty of Engineering, UTM and Media and Games Center of Excellence (MagicX), Universiti Teknologi Malaysia, Johor Bahru, Malaysia

[3] Faculty of Informatics and Management, University of Hradec Kralove, Rokitanskeho 62, 500 03 Hradec Kralove, Czech Republic

Abstract. The protection of control and intelligent systems across networks and interconnected components is a significant concern. Biometric systems are smart systems that ensure the safety and protection of the information stored across these systems. A breach of security in a biometric system is a breach in the overall security of data and privacy. Therefore, the advancement in improving the safety of biometric systems forms part of ensuring a robust security system. In this paper, we aimed at strengthening the finger vein classification that is acknowledged to be a fraud-proof unimodal biometric trait. Despite several attempts to enhance finger-vein recognition by researchers, the classification accuracy and performance is still a significant concern in this research. This is due to high dimensionality and invariability associated with finger-vein image features as well as the inability of small training samples to give high accuracy for the finger-vein classifications. We aim to fill this gap by representing the finger vein features in the form of information granules using an interval-based hyperbox granular approach and then apply a dimensionality reduction on these features using principal component analysis (PCA). We further apply a granular classification using an improved granular support vector machine (GSVM) technique based on weighted linear loss function to avoid overfitting and yield better generalization performance and enhance classification accuracy. We named our approach PCA-GSVM. The experimental results show that the classification of finger-vein granular features provides better results when compared with some state-of-the-art biometric techniques used in multimodal biometric systems.

Keywords: Granular computing · Cybersecurity · Biometric · Finger-vein · Support vector machines

1 Introduction

Biometrics forms part of an agenda for guaranteed security of information and user privacy [1]. For decades, biometric systems use human physiological traits such as fingerprints, voice, face, and iris for identification and verification purposes [2]. Biometric

© Springer Nature Switzerland AG 2020
N. T. Nguyen et al. (Eds.): ACIIDS 2020, LNAI 12033, pp. 309–320, 2020.
https://doi.org/10.1007/978-3-030-41964-6_27

systems have a significant advantage in security-related applications such as; access control, forensic analysis, border security, fraud identity, and detection and prevention of terrorists [3]. Several biometric authentication systems are used based on human physiological and behavioral characteristics such as fingerprints recognition, face recognition, voice recognition, iris recognition, hand geometry, and finger-vein recognition [4]. The finger-vein authentication system has been identified as the most secured, with higher accuracy, low cost, and above all, long term stability [3].

However, despite these high potentials of finger-vein recognition, it also presents drawbacks and limitations at every stage of the recognition process. For example, in the image acquisition stage, the quality of infrared images, which as a result of closeness with the camera, cause optical blurring, limited texture information, and position guidance of the finger all affect recognition performance [5]. Also, accompanying the report based on the results of shadow produced by various thicknesses of the finger muscles, tissues and bones surrounding the vein that has been captured in the infrared images of a finger vein, affects recognition performance [6]. Besides that, finger vein identification systems arc vulnerable to spoofing attacks [3]. Moreover, all the issues highlighted are in the form of classification problem which arises as a result of the need for a robust matching process with high recognition performance and high accuracy.

On the other hand, Granular Computing (GrC), as advanced by the works of [7], is widely used in pattern recognition to solve classification problems. When a problem is complex, and it involves high dimensional data, it is best addressed by a granular computing approach [8]. The problem, such as classification of biometric data is a complex problem due to the high dimensionality of images involved as well as uncertainty in the feature representations. Moreover, a biometric application forms part of a cybersecurity system [9]. In our previous paper [10], we highlighted how granular computing could be of significant importance to cybersecurity problems. Therefore, we argue that for improved efficiency in biometric recognition systems, GrC is the best approach. Finger vein image is considered as high dimensional due to the influence of shadow produced by various thicknesses of the finger muscles, tissues, and bones surrounding the vein and blurry boundaries with different illumination. These will result in degraded performance and lead to less accurate results.

In this paper, we proposed a novel finger-vein classification technique based on granular computing. We considered finger-vein authentication as a pattern classification problem due to persistence in degraded performance as a result of the feature complexity of finger-vein images. On that basis, we propose a novel approach by representing the finger vein features in the form of information granules using an interval-based hyperbox granular approach and then apply a dimensionality reduction and feature extraction on these features using principal component analysis (PCA). We further apply a granular classification using an improved granular support vector machine (GSVM) technique based on weighted linear loss function to avoid overfitting and yield better generalization performance and enhance classification accuracy. We named our approach PCA-GSVM. A granular support vector machine (GSVM) [11] can tackle the low-efficiency learning problem in SVM and as well be able to have an improved generalization performance [11]. Tang et al. [12], first proposes granular support vector machines by combining statistical theory and granular computing theory, which involve splitting the feature space into subspaces and build the SVM for each of the subspaces thereby achieving better generalizations ability. Many works use GSVM to take advantage of tackling imbalanced data, high dimensionality, and generalization ability [13–16]. However, to the best of our knowledge, no one has applied GSVM in finger vein classification even though been used widely in other classification tasks.

The general approach to finger-vein identification is a four-process stage; image acquisition, preprocessing, feature extraction, and matching [3]. We intend to alter the sequence by merging to a two-stage process called; preparation stage and pattern classi-fication stage, as depicted in Fig. 1. In this work, a region of interest (ROI) localization and image enhancement is performed during preprocessing. We then identify the homo-geneous region in the feature space using interval-valued class membership grade [17] based on the hyperbox granular approach. Then we characterize the classification pro-cess by determining the bounds for all the created granular prototypes using an improved GSVM based on the weighted linear loss function. And lastly, we aggregate those with similar class membership and apply global matching for high recognition performance.

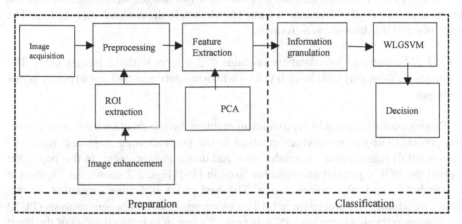

Fig. 1. nl Proposed approach for finger-vein pattern classification using our approach

This paper, to the best of our knowledge, is the first to propose the use of a single modal biometric trait – finger-vein – in a granular computing perspective to enhance the recognition process. The paper is organized as follows: in Sect. 2, we introduce the proposed granular support vector machine approach to finger-vein classification; in Sect. 3, we present the experimental studies and results; and finally, the conclusion and some recommended feature work in Sect. 4.

2 Proposed Granular Approach for Finger-Vein Pattern Classification

This paper proposes a granular computing approach to finger-vein biometric authentica-tion from the classification point of view. The process of classification provides a clear distinction between the regions of high homogeneity otherwise refer to as confidence region and the region of uncertainty. The proposed approach is mainly focused on the classification process, which follows after the feature extraction process of the finger-vein recognition system. The process is divided into two broad stages; preparation stage (comprising of image acquisition, preprocessing, and feature extraction) and classification stage (which involves granulation, use of GSVM classifier, and classification results) as depicted in Fig. 1. These stages are briefly described in the following subsections. However, we o emphasis on the granular support vector machine classification stage of the process. The following paragraphs present the proposed approach in detail.

2.1 Preparation Stage

In the preparation stage, which comprising of image acquisition, preprocessing, and feature extraction is first preceded with the choice of the relevant database. In this study, the image datasets are from the FV-USM database [18], which was collected and structured accordingly. All the images are converted JPG files and as well scaled to 320×240 pixels each for uniformity. A total of 4,748 images were collectively in the database and we used 900 instances from 150 classes in the evaluation procedure. The description of the database is as follows:

- *FV-USM database:* The database contains 492 fingers with 12 images each. The images are 8-bit grayscale level JPG files with a resolution of 640×480 pixels in raw images.

Preprocessing is next, which provides an enabling environment for the feature extraction process. The major activities involved in the preprocessing stage are; region of interest (ROI) segmentation, normalization, and image enhancement. In this paper, we applied the ROI segmentation algorithm used in [19]. Figure 2 shows the finger-vein images before and after preprocessing. The next phase is feature extraction. In this phase, we adopted the procedure in [20] on using principal component analysis (PCA) as a feature extraction technique. PCA is very efficient in reducing the size of the input and as well for feature extraction by creating a feature space. PCA is also known to have low computational time that is very much relevant in biometric systems. However, PCA encounters a problem when presented with nonlinear relationships and become relatively inefficient [21]. The PCA uses Kerhunen-Loeve transform in identifying feature vectors and classify the images according to the Euclidean distance between feature vectors. Therefore, to obtain the feature vectors, the following procedure is followed:

(i)   Let n rows and n columns belong to M finger-vein images. Let $⌉_1, ⌉_2, \ldots, ⌉_M$ be the images with column vectors having $(n^2 x 1)$ dimension. And to calculate the mean finger-vein image μ we use:

$$\mu = \frac{1}{M} \sum_{i=1}^{M} ⌉_i \tag{1}$$

(ii) Using the mean acquired from the finger-vein image, we calculate the finger-vein image distance ϕ_1 called column vector.

$$\phi_i =]_i - \mu \qquad (2)$$

(iii) All column vectors are further assembled in matrix form $k = [\phi_1, \phi_2, \ldots, \phi_M]$ with $(n^2 x M)$ with a covariance matrix C calculated as:

$$C = K \cdot K^2 \qquad (3)$$

Furthermore, to avoid high computational complexity that arises as a result of calculating each eigenvalue and eigenvectors of n^2 we form a matrix of C in $(the M x M)$ dimension [22].

$$C = K^2 \cdot K \qquad (4)$$

(iv) To form the eigen-finger-vein space [23], we calculate M eigenvalues (λ_i) and M eigenvectors (V_j) of C. Let $E = [V_1, V_2, \ldots, V_M]$ represent a combined matrix of eigenvectors C with $M x M$ dimension, and the eigen-finger-vein space can be calculated by $D = [D_1, D_2, \ldots, D_M]^T$ given as:

$$D = V \cdot K^T \qquad (5)$$

Where D is the row is vectors and stands for eigen-finger-vein of finger-vein images in the training set.

(v) We calculate the feature vectors of W using:

$$D = D \cdot K \qquad (6)$$

For $W = [w_1, w_2, \ldots, w_M]$ with dimension $M x M$ corresponding to each finger-vein image in the training set. In this study, we perform a comparison of test images using the eigen-finger-vein and the feature vectors acquired with the finger-veins in training set by the following steps:

(a) Consider $]_T$ to be the test image with $n^2 x 1$ dimension and to calculate the distance of the test image from the mean finger-vein image we use:

$$\phi_T =]_T - \mu \qquad (7)$$

(b) We follow with the projection of the test image in the eigen-finger-vein space to obtain its feature vectors W_T given as:

$$W_T = D \cdot \phi_T \qquad (8)$$

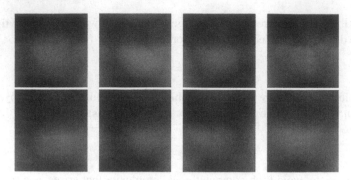

Fig. 2. Extracted Finger vein

(c) At this point, determining the similarity of W_T to each W_i in matrix W is very crucial to find the resembled test image in the training set.

However, current approaches use support vector machines for this test, but the procedure is flawed with computationally complex quadratic programming problem $0(n^3)$ solving. Therefore, this study used granular support vector machines (GSVM), as presented in Sect. 2.2.

2.2 Information Granulation Using Interval-Based Hyperbox Granular Approach

In this section, one of the most recent methods of information granulation based on the principle of justifiable granularity (PJG) [24] is exploited. The PJG is a formalism that advocates the creation of an information granule that is rich in experimental data and meaning for knowledge derivation and discovery. However, creating an information granule using this formalism is confronted with conflicting requirements of balance between coverage and specificity of an information granule Ω. Such a condition requires trade-off by applying a multiplicative index as:

$$M = cov(\Omega) x spec(\Omega) \tag{9}$$

$$Where \, cov(\Omega) = \sum_{k:x \in \Omega=[a,b]} wk \tag{10}$$

$$and \, spec(\Omega) = f(|b-a|) \tag{11}$$

We consider M as an objective function of the interval between upper and lower boundaries a and b, respectively, for a weighted data Z as given in Eq. (4).

$$M(b) = \left(\sum_{k:m<xk\leq b} Wk\right) x f(|b-m|) \tag{12}$$

where m is a numerical representation of the data Z and f is a non-increasing function given as $f(u) = e^{-\alpha u}, \infty \geq 0$

$$M(a) = \left(\sum_{k:a\leq xk<m} Wk\right) x e^{-\alpha|m-a|} \tag{13}$$

This process requires the optimization of Eqs. (12) and (13) in which ∞ is consistently used as a threshold or granularity (size) to either determine the impact of specificity and coverage. It should be noted that different values of ∞ is put in the range of [0, 1], which means if $\infty = 0$, the $f(u) = 1$ that specifies the created interval information granule (Ω) is not bounded by either experimental data and specificity.

To maximize the objective function (b) with respect to the granularity ∞ [0, 1] given as:

$$\alpha_{max}(b) = \max\{\alpha_{1max}(b), \alpha(b), \alpha(b), \ldots \alpha(p-1)\alpha_{2max}(b)\} \qquad (14)$$

Equation (6) is synonymous to maximizing the lower boundary (a) to determine the maximum granularity value of:

$$\alpha(a) - \alpha_{max}(a) \qquad (15)$$

In this study, a family of internal information granules is generated around the numerical representation of the weighted data in a range of 0 to 1. It is based on the acquired information granule that we build the support vectors, which is explained in the next Sect. 2.3.

2.3 Granular Support Vector Machine Classification

Support vector machines (SVM) have received considerable attention in tackling classification problems [25]. Using SVM involves the application of hyperplane, where the distinction between correct and non-correct samples is made. An essential advantage of using SVM is its ability to handle disperse data and a considerable capacity to handle overfitting. However, SVM is computationally complex in solving a quadratic programming problem $0(n^3)$ (where n is the size of the entire training sample), and its biasness leads to misclassification in high dimensional feature space. Therefore the research on SVM has promoted the advancement of granular support vector machine (GSVM), which combines statistical theory and granular computing [26] to enhance generalization and facilitate segregation of nonlinear inseparable problems into a series of linear separable issues. GSVM was first proposed by [26], which constructs a sequence of information granules and then builds support vector machines in each information granule to maximize the margins and have better generalization capabilities. The use of GSVM as a learning model will facilitate the reduction of the problem complexity and as well as increase classification efficiency. It is noteworthy to say that the creation of an optimal information granule remains a crucial problem, and to tackle complex classification problems in areas of high-level security such as biometric recognition requires the creation of an optimal information granule as a foundation for successful classification.

We argue that a successful classification using granular support vector machines requires a well-designed optimal information granule in addition to the original power of GSVM to tackle information loss that is very evident in biometric data. The design of GSVM follows a top-down approach where the hyperplane is divided according to the statistical margin maximization principle. The quadratic problem involved can be solved using the Wolfe dual formulation [27].

$$Maximize\; w_d = \sum_i \alpha - 1/2 \sum_{i,j} \alpha_i \alpha_j, q_i q_j \, p(x_i, x_j) \qquad (16)$$

Subject to

$$0 \leq \alpha_i \leq M \tag{17}$$

$$\sum_i \alpha_i q_i = 0 \tag{18}$$

To maximize the margin width, then:

$$M = \sum_i^S \alpha_i q_i x_i \tag{19}$$

$$margin\ width = \frac{2}{m} \tag{20}$$

Where S represents the number of support vectors.

A straight forward approach to split the hyperplane based on the granular concept is by building support vector machines as clusters in the feature space tagging them as cluster1 and cluster2. Therefore, an aggregative margin width (AMW) would be defined as:

$$Aggregative\ margin\ width = \frac{1}{m_1 + m_2} \tag{21}$$

Where m_1, m_2 are weights calculated by (19) for the cluster1 and cluster2 support vector machines. An evident drawback in this approach is finding the optimal splitting hyperplane and determining the most substantial aggregative margin width value. In [26], a grid search heuristic optimization algorithm is used in optimizing the hyperplane. However, grid search suffers some drawbacks as a result of an increase in dimensionality, which hinders a guaranteed perfect solution.

Therefore, in search of optimal hyperplane:

$$p_1, p_2, \ldots, p_m \left(x^{k\_min}, x^{k\_max} \right) \tag{22}$$

Where $\left(x^{k\_min}, x^{k\_max} \right)$ are the minimum and maximum value for the training dataset, respectively.

3 Experimental Results

In this study, the finger-vein image database was acquired from the University Sains Malaysia USM-FV database [18] upon request. The database contains a total of 4,748 images. The images are 8-bit grayscale level JPG files with a resolution of 320×240 pixels in raw images. A detailed guide on how to use the database was presented in [18]. The experiments for the PCA feature vector extraction were coded in eclipse java JDK 8, while GSVM classification is achieved by MATLAB R2016b on a PC with Intel® Core™ i7-3770 CPU @ 3.40 GHz installed with 16 GB RAM.

In this study, we used images from the extracted vein folder of the repository, and we further create a sub-folder containing 900 finger images from 150 different fingers, which we consequently used the PCA feature extraction method on these images. During the GSVM classification with Matlab, we used 10-fold cross-validation and an RBF kernel with a gamma value of 0.1. Table 1 shows the recognition rates. We repeat the experimentation by changing the kernel function for the proposed GSVM, and we consider radial basis function (RBF) kernel, linear polynomial kernel (LinPol) kernel and quadratic polynomial (QuadP) kernel for GSVM.

Table 1. Classification methods, recognition rates

Technique	Kernel function	Recognition rate
PCA-GSVM	Radial basis function	92.1%
PCA-GSVM	Linear polynomial	92.6%
PCA-GSVM	Quadratic polynomial	94.6%

Furthermore, as is evident from Table 1, the peak level achieved is (94.6%) in the recognition rate using PCA-GSVM (QuadP) method. And also PCA-GSVM (RBF) recorded the least rate with (92.1%). Considering previous works of [28, 29] on finger-vein databases presented in Table 2, a relatively higher recognition rate than [28] by 8% increase in recognition rate. However, [29], which uses a deep convolution neural network, achieves a higher recognition rate of 96.8%. But the increase in [29] is as a result of their database and as well a small amount of training data involved. We, however, identify the performance in biometric recognition is database dependent, which indicates, different databases will generate different results even with the best available approach.

Table 2. Classification methods, recognition rates, and database

Authors	Feature extraction method	Classification method	Recognition rate	Database
Das et al. (2019) [28]	CNN	PCA	72.97%	FV-USM
Wang et al. (2018) [29]	DCNN	PCA + SVM	96.8%	Own database
This study	PCA	GSVM	94.6%	FV-USM

4   Concluding Remarks and Future Work

Biometric systems have gained considerable acceptance as a high level secured form of authentication in the management of unauthorized access to information systems.

In this paper, a robust unimodal finger-vein recognition system is presented using a granular computing approach. To the best of our knowledge, this is the first attempt to use a granular computing approach to finger-vein recognition as a unimodal form of biometrics. The GSVM is used on a binary classification problem after a PCA-based feature extraction method and has shown significant improvement when compared to the previous application on biometric recognition. However, in this paper, spoof attacks are not considered, which contributes to the degradation of performance in biometric recognition. We also state that the performance in biometric identification is database dependent, which indicates, the different databases will generate different results even with the best available approach in the future we would integrate the proposed method with an anti-spoofing mechanism to provide a robust recognition system. And also, a multimodal approach is envisioned using a similar approach in the future, where finger-vein and finger knuckle-print would be considered to create a robust biometric fusion scheme.

Acknowledgments. This research has been funded by Universiti Teknologi Malaysia (UTM) under Research University Grant Vot-20H04, Malaysia Research University Network (MRUN) Vot 4L876 and the Fundamental Research Grant Scheme (FRGS) Vot 5F073 supported under Ministry of Education Malaysia. The work is partially supported by the SPEV project, University of Hradec Kralove, FIM, Czech Republic (ID: 21xx-2020). We are also grateful for the support of Ph.D. student Sebastien Mambou in consultations regarding application aspects.

References

1. Onuiri, E.E., Idowu, S.A., Komolafe, O.: Electronic health record systems, and cyber- security challenges. In: International Conference African Development Issues, pp. 98–105 (2015)
2. Syazana-Itqan, K., Syafeeza, A.R., Saad, N.M., Hamid, N.A., Saad, W.H.B.M.: A review of finger-vein biometrics identification approaches. Indian J. Sci. Technol. **9** (2016). https://doi.org/10.17485/ijst/2016/v9i32/99276
3. Shaheed, K., Liu, H., Yang, G., Qureshi, I., Gou, J., Yin, Y.: A systematic review of finger vein recognition techniques. Information **9**, 213 (2018). https://doi.org/10.3390/info9090213
4. Yang, J, Wei, J., Shi, Y.: Accurate ROI localization and hierarchical hyper-sphere model for finger-vein recognition. Neurocomputing, 1–11. https://doi.org/10.1016/j.neucom.2018.02.098
5. Hong, H.G., Lee, M.B., Park, K.R.: Convolutional neural network-based finger-vein recognition using NIR image sensors. Sensors (Switzerland) **17** (2017). https://doi.org/10.3390/s17061297
6. Liu, Z., Yin, Y., Wang, H., Song, S., Li, Q.: Finger vein recognition with manifold learning. J. Netw. Comput. Appl. **33**, 275–282 (2010). https://doi.org/10.1016/j.jnca.2009.12.006
7. Zadeh, L.A.: Toward a theory of fuzzy information granulation and its centrality in human reasoning and fuzzy logic. Fuzzy Sets Syst. **90**, 111–127 (1997). https://doi.org/10.1016/S0165-0114(97)00077-8
8. Liu, S., Pedrycz, W., Gacek, A., Dai, Y.: Development of information granules of higher type and their applications to granular models of time series. Eng. Appl. Artif. Intell. **71**, 60–72 (2018). https://doi.org/10.1016/j.engappai.2018.02.012
9. Gavrilova, M.L.: Biometric-based authentication for cyberworld security: challenges and opportunities. Can. Def. Foreign Aff. Inst., 1–9 (2014). https://doi.org/10.1103/PhysRevE.64.041902

10. Isah, S.S., Selamat, A., Ibrahim, R., Anuar, S.: Granular computing approach to cybersecurity problem, In: Fujita, E.H.-V. H. (ed.) New Trends in Intelligent Software Methodologies, Tools and Techniques, vol. 303. IOS Press Ebooks, Granada Spain, pp. 215–225 (2018). https://doi.org/10.3233/978-1-61499-900-3-215

11. Guo, H., Wang, W.: Granular support vector machine: a review. Artif. Intell. Rev. **51**(1), 19–32 (2017). https://doi.org/10.1007/s10462-017-9555-5

12. Tang, Y., Jin, B., Sun, Y., Zhang, Y.-Q.: Granular support vector machines for medical binary classification problems. In: IEEE Symposium Computational Intelligence Bioinformatics Computational Biology, pp. 73–78 (2005). https://doi.org/10.1109/cibcb.2004.1393935

13. Tang, Y., Jin, B., Zhang, Y.-Q., Fang, H., Wang, B.: Granular support vector machines using linear decision hyperplanes for fast medical binary classification. In: 14th IEEE International Conference Fuzzy Systems 2005. FUZZ 2005, pp. 138–142 (2005). https://doi.org/10.1109/FUZZY.2005.1452382

14. Tang, Y., Jin, B., Zhang, Y.Q.: Granular support vector machines with association rules mining for protein homology prediction. Artif. Intell. Med. **35**, 121–134 (2005). https://doi.org/10.1016/j.artmed.2005.02.003

15. Tang, Y., Zhang, Y.Q.: Granular support vector machines with data cleaning for fast and accurate biomedical binary classification. In: 2005 IEEE International Conference Granular Computing, pp. 262–265 (2005). https://doi.org/10.1109/GRC.2005.1547281

16. Huang, H., Ding, S., Jin, F., Yu, J.: A novel granular support vector machine based on mixed Kernel function. Int. J. Digit. Content Technol. Appl. **6**, 484–492 (2012). https://doi.org/10.4156/jdcta.vol6.issue20.52

17. Pedrycz, W.: Granular Computing: Analysis and Design of Intelligent Systems. CRC Press, 2016. https://books.google.com.my/books?id=yV7NBQAAQBAJ

18. Asaari, M.S.M., Suandi, S.A., Rosdi, B.A.: Fusion of band limited phase only correlation and width centroid contour distance for finger-based biometrics. Expert Syst. Appl. **41**, 3367–3382 (2014). https://doi.org/10.1016/j.eswa.2013.11.033

19. Gupta, P., Gupta, P.: An accurate finger vein based verification system. Digit. Sig. Process. A Rev. J. **38**, 43–52 (2015). https://doi.org/10.1016/j.dsp.2014.12.003

20. Gumus, E., Kilic, N., Sertbas, A., Ucan, O.N.: Evaluation of face recognition techniques using PCA, wavelets and SVM. Expert Syst. Appl. **37**, 6404–6408 (2010). https://doi.org/10.1016/j.eswa.2010.02.079

21. Kurşun, O., Favorov, O.V.: SINBAD automation of scientific discovery: from factor analysis to theory synthesis. Nat. Comput. **3**, 207–233 (2004). https://doi.org/10.1023/B:NACO.0000027756.50327.26

22. Turk, M., Pentland, A.: Eigenfaces for recognition. J. Cogn. Neurosci. **3**, 71–86 (1991)

23. Khan, M., Subramanian, R., Khan, N.: Low dimensional representation of dorsal hand vein features using principal component analysis (PCA). World Acad. Sci. **3**, 1001–1007 (2009)

24. Pedrycz, W., Homenda, W.: Building the fundamentals of granular computing: a principle of justifiable granularity. Appl. Soft Comput. J. **13**, 4209–4218 (2013). https://doi.org/10.1016/j.asoc.2013.06.017

25. Roy, A., Singha, J., Devi, S.S., Laskar, R.H.: Impulse noise removal using SVM classification based fuzzy filter from grayscale images. Sig. Process. **128**, 262–273 (2016). https://doi.org/10.1016/j.sigpro.2016.04.007

26. Tang, Y., Jin, B., Sun, Y., Zhang, Y.-Q.: Granular support vector machines for medical binary classification problems. In: 2014 Symposium on Computational Intelligence in Bioinformatics Computational Biology, CIBCB 2004, pp. 73–78 (2004). https://doi.org/10.1109/CIBCB.2004.1393935

27. Aguilar, J.F.: Adapted fusion schemes for multimodal biometric, n.d

28. Das, R., Piciucco, E., Maiorana, E., Campisi, P.: Convolutional neural network for finger-vein-based biometric identification. IEEE Trans. Inf. Forensics Secur. **14**, 360–373 (2018). https://doi.org/10.1109/TIFS.2018.2850320
29. Wang, J., Yang, K., Pan, Z., Wang, G., Li, M., Li, Y.: Minutiae-based weighting aggregation of deep convolutional features for vein recognition. IEEE Access **6**, 61640–61650 (2018). https://doi.org/10.1109/ACCESS.2018.2876396

Mining Non-redundant Periodic Frequent Patterns

Michael Kofi Afriyie[1], Vincent Mwintieru Nofong[1(✉)], John Wondoh[2],
and Hamidu Abdel-Fatao[1]

[1] University of Mines and Technology, P. O. Box 237, Tarkwa, Ghana
vnofong@umat.edu.gh
[2] University of South of Australia, Adelaide, Australia

Abstract. Discovering periodic frequent patterns has been useful in
various decision making. Traditional algorithms, however, often report
a large number of such patterns, most of which are often redundant
since their periodic occurrences can be inferred from other periodic frequent patterns. Employing such redundant periodic frequent patterns
in decision making would often be detrimental if not trivial. To address
this challenge and report only non-redundant periodic frequent patterns,
this paper employs the concept of deduction rules in mining the set
of non-redundant periodic frequent patterns. A Non-redundant Periodic
Frequent Pattern Miner (NPFPM) is subsequently proposed to achieve
this purpose. Experimental analysis on benchmark datasets show that
NPFPM is efficient and can effectively prune the set of redundant periodic frequent patterns.

Keywords: Frequent patterns · Periodic frequent patterns ·
Non-redundance

1 Introduction

Mining frequent patterns has been widely researched on over the past years. Several techniques and approaches have been proposed purposely for discovering
interesting categories of frequent patterns for various applications. Typical of such
approaches and techniques and can be found in works such as [1, 3, 15, 20, 22–24].

Despite the usefulness of frequent patterns in identifying patterns that occur
frequently in databases, they always fail to reveal the shapes of patterns' occurrences in databases. The shapes of patterns' occurrences in databases are often
needed in some vital decision making. In crime data analysis for instance, though
frequent pattens mined from such data will reveal the frequently occurring crimes
with time, they will not be able to reveal the occurrence shapes of crimes.
Revealing the occurrence shapes of crimes could be useful in decision making
towards curbing future crimes. The usefulness of the occurrence shapes of patterns in decision making brought about research on discovering periodic frequent
patterns.

© Springer Nature Switzerland AG 2020
N. T. Nguyen et al. (Eds.): ACIIDS 2020, LNAI 12033, pp. 321–331, 2020.
https://doi.org/10.1007/978-3-030-41964-6_28

Discovering periodic frequent patterns (PFPs) in transactional databases has been widely researched on. Over the past years, several approaches and techniques have been developed for discovering different categories of periodic frequent patterns in works such as [2,5,7,12,18,19]. Notwithstanding the numerous existing techniques available for mining PFPs, one particular challenge which still exist is reducing the number of redundant PFPs often reported. This is a challenge as existing approaches mostly discover and report a huge number of periodic frequent patterns - most of which are often redundant since their periodicities can be inferred from the periodicities of their proper subsets. Reporting and employing these redundant periodic frequent patterns in decision making would not only consume memory but could be detrimental if they are false positive periodic frequent pattens.

This paper addresses this challenge of reducing the number of redundant periodic frequent patterns by employing deduction rules in mining our defined non-redundant periodic frequent patterns. Our defined non-redundant periodic frequent patterns are devoid of redundant information and their periodicities cannot be inferred from other periodic frequent patterns.

The main contributions of this paper in the discovery of PFPs are:

- It introduces the concept of non-redundant PFPs (as the set of periodic frequent patterns devoid of redundant information) which achieves a size reduction in the number of reported PFPs.
- It proposes and develops NPFPM, an efficient algorithm for discovering the set of non-redundant periodic frequent patterns.

The rest of the paper is presented as follows. The related works are presented in Sect. 2 while Sect. 3 introduces the non-redundant PFPs. Section 4 presents our experimental analysis while Sect. 5 outlines our conclusions.

2   Related Work

The associated notations for periodic frequent pattern mining is as follows.

Let $I = \langle i_1, i_2, ..., i_n \rangle$ be a set of literals, called items. Then, a transaction is a nonempty set of items. A pattern S is a set of items satisfying some conditions of measures like frequency. A pattern is of length-k if it has k items, for example, $S = \{a, b, d, e, f\}$ is a length-5 pattern.

Given a transactional database of k transactions, $\mathbf{D} = <n_1, n_2, n_3, ..., n_k>$, where each n_m in $\mathbf{D}$ is identified by m called transaction identifier (TID), the *cover* of a pattern S in $\mathbf{D}$, $cov_\mathbf{D}(S)$, is the set of TIDs of transactions that contain S. That is,

$$cov_\mathbf{D}(S) = \{m : n_m \in \mathbf{D} \wedge S \subseteq n_m\} \qquad (1)$$

where $|cov_{\mathbf{D}}(S)|$ is often referred to as the *support count* of $S \in \mathbf{D}$. The *support* of a pattern $S \in \mathbf{D}$, $sup_{\mathbf{D}}(S)$, is defined as,

$$sup_{\mathbf{D}}(S) = \frac{|cov_{\mathbf{D}}(S)|}{|\mathbf{D}|} \tag{2}$$

Given a user desired minimum support (ε), a pattern $S \in \mathbf{D}$ is said to be frequent if $sup_{\mathbf{D}}(S) \geq \varepsilon$.

For any given pattern S in a transactional database $\mathbf{D}$ with $cov_{\mathbf{D}}(S)$ as its coverset, the notation $e.cov_{\mathbf{D}}(S)$ is used to indicate the extension of $cov_{\mathbf{D}}(S)$ by inserting a starting time 0 and the last time m to $cov_{\mathbf{D}}(S)$. That is,

$$e.cov_{\mathbf{D}}(S) = \{0 \cup cov_{\mathbf{D}}(S) \cup m\} \tag{3}$$

where $m = |\mathbf{D}|$. The last time, m will be duplicated if it is already in $cov_{\mathbf{D}}(S)$. For instance, given $|\mathbf{D}| = 6$ and $cov_{\mathbf{D}}(S) = \{1, 3, 4, 5, 6\}$, then, $e.cov_{\mathbf{D}}(S) = \{0\} \cup \{1, 3, 4, 5, 6\} \cup \{6\} = \{0, 1, 3, 4, 5, 6, 6\}$.

Let $(m_j, m_{j+1}) \in e.cov_{\mathbf{D}}(S)$ be two consecutive transaction IDs (occurrence times) of S in $\mathbf{D}$, then $p_j^S = m_{j+1} - m_j$ is the jth period of S in $\mathbf{D}$. The set of all periods of S, that is, P^S, obtained from its extended cover is denoted as:

$$P^S = \{p_1^S, p_2^S, \cdots, p_{r-1}^S, p_r^S\} \tag{4}$$

where $r = |e.cov_{\mathbf{D}}(S)| - 1$.
For example, given $e.cov_{\mathbf{D}}(S) = \{0, 1, 3, 4, 5, 6, 6\}$, then $p_1^S = (1-0) = 1$, $p_2^S = (3-1) = 2$, $p_3^S = (4-3) = 1$, $p_4^S = (5-4) = 1$, $p_5^S = (6-5) = 1$, $p_6^S = (6-6) = 0$, giving $P^S = \{1, 2, 1, 1, 1, 0\}$. Thus, for any pattern S, it can be derived that:

$$|P^S| = |cov_D(S)| + 1 \tag{5}$$

To mine periodic frequent patterns, Tanbeer et al. in [19] defined a periodicity measure on patterns as follows.

Definition 1. *[19] Given a database $\mathbf{D}$, a pattern S and its set of periods P^S in $\mathbf{D}$, the periodicity of S, $Per(S)$, is defined as, $Per(S) = \max\{p | p \in P^S\}$.*

With Definition 1, Tanbeer et al. [19] defined a PFP as a frequent pattern whose periodicity is not greater than the user defined maximum periodicity threshold, *maxPer*.

Though discovering periodic frequent patterns using the maximal period proposed in [19] has been used in works such as [5,6,8,11,18], authors in [16] argued that discovering PFPs based on the proposition in [19] is inappropriate since it is based on the maximum time-interval for which a pattern does not occur in a database as the patterns periodicity. Subsequently, authors in [16] introduced their defined periodicity of a pattern under the name *regularity* as follows.

Definition 2. *[16] Given a database $\mathbf{D}$, a pattern S and its set of periods P^S in $\mathbf{D}$, the regularity of S, $Reg(S)$, is defined as $Reg(S) = var(P^S)$, where $var(P^S)$ is the variance of P^S.*

The regular frequent patterns introduced in Definition 2 by Rashid et al. [16] have been used in works such as [9,17] for discovering periodic (regular) frequent patterns.

Nofong in [12] however argued that periodic frequent pattern mining algorithms based on the propositions in both [19] and [16] will always mine and return PFPs with totally distinct periods. To enable mine and return periodic frequent patterns having similar periodicities, Nofong [12] defined a PFP as follows.

Definition 3. *[12] Given a database* **D***, minimum support threshold* ε*, periodicity threshold* p*, difference factor* p_1*, a pattern* S *and* P^S*, S is a periodic frequent pattern if* $sup_{\mathbf{D}}(S) \geq \varepsilon$*,* $(p - p_1) \leq Prd(S) - std(P^S)$ *and* $Prd(S) + std(P^S) \leq (p + p_1)$*.*

where, $Prd(S)$ (the mean of P^S, that is, $\bar{x}(P^S)$) is the periodicity of S and $std(P^S)$ the standard deviation in P^S.

With Definition 3, Nofong [12] further incorporated the productiveness measure [21] in defining the productive periodic frequent patterns.

Recently, Fournier-Viger et al. in [2] introduced PFPM, a periodic frequent pattern miner with novel pruning techniques. Unlike other existing periodic frequent pattern mining algorithms proposed in [12,16,19], PFPM introduced the *minimum*, *maximum* and *average* periodicity measures for mining user desired periodic frequent patterns.

As mentioned previously, the propositions in [2,12,16,19] and works relying on these propositions ([4–6,8,9,13,14,17,18]) have challenges of; reporting redundant PFPs and difficulty in early termination during the PFP mining process. To the best of our knowledge, no work exists which addresses the issue of eliminating the set of redundant PFPs and reporting only the set nonredundant periodic frequent patterns. This paper thus proposes and defines the non-redundant periodic frequent patterns towards ensuring only periodic frequent patterns without redundant information are reported.

3   Non-redundant Periodic Frequent Patterns

For any given dataset, we adopt Definition 3, the definition of a periodic frequent pattern proposed in [12].

Though Definition 3 will detect and report the set of PFPs with similar periods, some PFPs may be periodic due to their proper subsets being periodic. Such PFPs which would be containing redundant information, may be trivial if not detrimental[1] in decision making. To be able to detect the set of non-redundant PFPs, we employ the concept of frequent generators and define a non-redundant PFP as follows.

Definition 4. *Given a periodic frequent pattern set,* $Per_{\mathbf{D}} = \{S_1, S_2, \dots S_j\}$*, a periodic frequent pattern,* S_n*, is a non-redundant periodic frequent pattern if* $\nexists S_u \in Per_{\mathbf{D}}$ *such that,* $S_u \subset S_n$ *and,* $sup_{\mathbf{D}}(S_n) = sup_{\mathbf{D}}(S_u)$*.*

---

[1] They would be detrimental in decision making if they are false positively periodic.

Though these are the set of generator PFPs, we term them non-redundant as their periodicities cannot be derived or obtained from their subset PFPs.

The advantages of employing the concept of frequent generators in mining our defined non-redundant PFPs include:

- Finding an early termination mechanism in the periodic frequent pattern discovery process - based on the anti-monotonic property of frequent generators;
- Reporting periodic frequent patterns devoid of redundant information and whose periodicities cannot be derived or inferred from other PFPs;
- Reducing the number of "likely false positive" periodic frequent patterns being reported, and;
- Reporting periodic frequent patterns that are more preferable in model selection [10].

3.1 Pruning Redundant Periodic Frequent Patterns

Given S as a periodic frequent pattern, we use the Redundance-Test function below to test if S is a redundant or a non-redundant periodic pattern as follows:

Function Redundance-Test

  Input: Set of periodic frequent patterns in D, Per_D and a PFP, $S \in D$
  Output: S.class $\in$ [Non-redundant, Redundant]
1 Create S.class $= null$
2 Let $\Gamma = Per_D$
3 **if** S *is a length-1 pattern* **then**
4   **if** $sup_D(S) = sup(\emptyset)$ **then**
5   S.class = Redundant
6   **else**
7   S.class = Non-redundant
8 **else**
9   **if** $\exists S_l \in Per_D | S_l \subset S \land sup_D(S_l) = sup_D(S)$ **then**
10   S.class = Redundant
11   **else**
12   S.class = Non-redundant
13 **return** $S.class$

As shown in Line 1 of the Redundance-Test function, the class of S is created as null, and Γ in Line 2 assigned as the set of all PFPs. If S is a length-1 periodic frequent pattern, S is classified as a redundant PFP in Line 5 if $sup_D(S) = sup(\emptyset)$ (where $sup(\emptyset) = 1.0$), else, S is classified as a non-redundant PFP in Line 7.

If S is not a length-1 PFP, S is classified as a redundant PFP in Line 10, if its support can be derived from any of its subset periodic frequent patterns. That is, S is classified redundant if there exists S_l in Γ such that, $S_l \subset S$ and $sup_D(S_l) = sup_D(S)$. If the conditions in Line 9 are not met, S is classified as non-redundant in Line 12 since its periodicity cannot be derived from its subsets that are also periodic. In Line 13, the Redundance-Test function returns the classification of S as either a redundant or a non-redundant PFP.

3.2   Mining the Non-redundant Periodic Frequent Patterns

To be able to mine the set of non-redundant PFPs defined in Definition 4, we adopt and modify the PPFP algorithm proposed in [12]. The Redundance-Test function previously discussed is incorporated into PPFP as NPFPM. Similar to the PPFP algorithm, NPFPM employs two steps in discovering the non-redundant PFPs:

1. Identifying the set of frequent length-1 items from the given database, and
2. Discovering the set of all non-redundant PFPs from frequent length-1 items.

These two steps are shown in Algorithms 1 and 2 respectively.

Algorithm 1: NPFPM(D, ε, p, p_1)

Input: Dataset D, min. support ε, periodicity, p and difference factor, p_1
Output: Non-redundant PFP set Per_D

1 Create HashMap h_n /* to store all length-1 items in D */
2 Create set L
3 **for** *each* transaction $T \in D$ **do**
4   **for** *each* length-1 item $a_y \in T$ **do**
5   **if** $a_y \notin h_n$ **then**
6   Create $cov_D(a_y) = \{$ TID of $a_y\}$ /* TID = Transaction ID */
7   Add $(a_y, cov_D(a_y))$ to h_n
8   **else**
9   Let $(a_y, cov_D(a_y)) = h_n(a_y)$
10   Udate $cov_D(a_y)$ as $cov_D(a_y) = cov_D(a_y) \cup$ TID of a_y
11   Update h_n with $(a_y, cov_D(a_y))$
12 **for** *each* item $a_y \in h_n$ **do**
13   Let $(a_y, cov_D(a_y)) = h_n(a_y)$
14   **if** $sup_D(a_y) \geq \varepsilon$ **then**
15   Add $(a_y, cov_D(a_y))$ to L
16 Sort L in descending order of items
17 MinePFPs(L, ε, p, p_1)
18 **return** Per_D

The operations and functions of these two algorithms do not differ much from those used in [12]. The major difference is found in Lines 13 and 19 of Algorithm 2 where the Redundance-Test function is used in NPFPM instead of the productiveness test in PPFP. Since the processes involved in Algorithms 1 and 2 are similar to those in [12] and to avoid repetition, we refer readers to [12] for more details on the operations of the two algorithms.

Algorithm 2: MinePFPs(L, ε, p, p_1)

Input: Set L, periodicity, p, difference factor, p_1, and minimum support ε

Output: Non-redundant PFP set Per_D

1 Create Per_D

2 Create set TempL $= \emptyset$

3 Let $P_{a_n}[0, b]$ be the the length-b prefix of a_n

4 **if** $|L| = 0$ **then**

5 | **return** Per_D

6 **else**

7 | **while** $|L| > 0$ **do**

8 | **for** $k = 0$ *to* $|L|$-*1* **do**

9 | Let $(a_k, cov_D(a_k)) = L[k]$

10 | **if** $|a_k| = 1$ **then**

11 | Obtain P^{a_k} from $e.cov_D(a_k)$

12 | Evaluate $Prd(a_k)$ and $std(P^{a_k})$ from P^{a_k}

13 | **if** a_k *is periodic and non-redundant* **then**

14 | Add a_k to Per_D

15 | **for** $l = (k + 1)$ *to* $|L|$-*1* **do**

16 | Let $(a_l, cov_D(a_l)) = L[l]$

17 | **if** $P_{a_k}[0, |a_k|$-$1] = P_{a_l}[0, |a_l|$-$1]$ **then**

18 | Create $S = (a_k \cup a_l, cov_D(a_k) \cap cov_D(a_l))$

19 | **if** $sup_D(S) \geq \varepsilon$ *and* S *is non-redundant* **then**

20 | Add S to TempL

21 | Get P^S from $e.cov_D(S)$

22 | Evaluate $Prd(S)$ and $std(P^S)$ from P^S

23 | **if** S *is periodic* **then**

24 | Add S to Per_D

25 | $L =$ TempL

26 | TempL.clear()

27 **return** Per_D

---

4 Experimental Analysis

For our experimental analysis, we used the following implementations:

- NPFPM: This is an implementation of the proposed non-redundant periodic frequent pattern mining algorithm based on the approach in [12]. For any given dataset and user thresholds, NPFPM mines and returns all non-redundant periodic frequent patterns.
- PFP*: This is an implementation for detecting all periodic frequent patterns with similar periodicities based on Definition 3 without incorporating the productiveness or non-redundance measure. For any given dataset and user thresholds, PFP* mines and returns all periodic frequent patterns.
- PPFP: This is the implementation of the approach proposed in [12]. For any given dataset and user thresholds, PPFP discovers and reports the set of all productive PFPs.

Experimental analysis were conducted with regards to (i) runtime performance and (ii) reported patterns.

The following datasets were used for the experimental analysis:

- Kosarak10K: partly dense with 10,000 transactions
- Kosark45K: partly dense with 45,000 transactions
- Tafeng Nov. 2000: very sparse with 31,807 transactions for the month of November 2000.

4.1   Runtime Performance: PFP Discovery

Figures 1a, b and c show the execution time of the three implementations on the Kosarak10K, Kosarak45k and Tafeng datasets respectively.

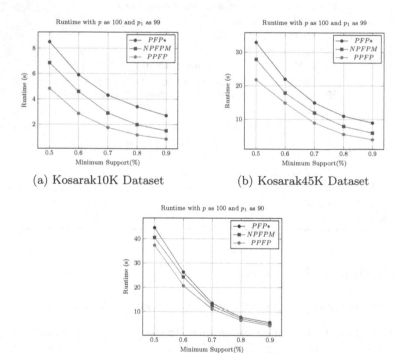

(a) Kosarak10K Dataset   (b) Kosarak45K Dataset

(c) Tafeng Nov 2000 Dataset

Fig. 1. Periodic Frequent Pattern Discovery: Runtime

As can be seen in Figs. 1a, b and c, NPFPM (which mines the set of only non-redundant periodic frequent patterns) is more efficient in mining PFPs compared to PFP*. PPFP however is slightly more efficient than NPFPM because the non-redundance test used in NPFPM is computationally more expensive than productiveness test employed in PPFP. It is worth noting that the runtimes of all three algorithms do not differ so much in the Tafeng dataset (as seen in Fig. 1c) because it is relatively very sparse and all algorithms are reporting same number of PFPs (see Table 3).

4.2 Reported Patterns: PFP Discovery

Tables 1, 2 and 3 show the number of reported periodic frequent patterns for NPFPM, PPFP and PFP* in the three datasets described above.

We observed that in very sparse datasets (Tafeng dataset - see Table 3), all compared approaches report same number of PFPs. In the Kosarak10K and Kosarak45K datasets, NPFPM reports a smaller number of periodic frequent patterns compared to PFP*. PPFP however reports a smaller number of periodic frequent patterns than NPFPM as the productiveness measure is more restrictive than the non-redundance measure.

As can be observed in Tables 1 and 2, with the non-redundance measure, NPFPM is able to prune the set of PFPs with redundant information and as such, reports a much smaller set of periodic frequent patterns compared to PFP*.

Table 1. Kosarak10K dataset

ε	NPFPM	PPFP	PFP*
	$p=100$	$p=100$	$p=100$
	$p_1=99$	$p_1=99$	$p_1=99$
0.8%	114	104	210
0.7%	114	104	210

Table 2. Kosarak45K dataset

ε	NPFPM	PPFP	PFP*
	$p=100$	$p=10$	$p=100$
	$p_1=99$	$p_1=99$	$p_1=99$
0.8%	173	141	188
0.7%	173	141	188

Table 3. Tafeng Nov 2000 dataset

ε	NPFPM	PPFP	PFP*
	$p=130$	$p=130$	$p=130$
	$p_1=120$	$p_1=120$	$p_1=120$
0.8%	20	20	20
0.7%	26	26	26

5   Conclusion

Non-redundant periodic frequent patterns are the set of periodic frequent patterns whose periodic occurrence cannot be inferred from their subset PFPs. In this work, we employed deduction rules in identifying our defined set of non-redundant periodic frequent patterns. We subsequently propose and develop a Non-redundant Periodic Frequent Pattern Miner (NPFPM) for mining the set of non-redundant PFPs. Our experimental results on benchmark datasets show that NPFPM is efficient and reports a smaller set of non-redundant periodic frequent patterns compared to the set of all periodic frequent patterns. In our future works, we will investigate on measures that can be employed in memory efficient mining of interesting periodic frequent patterns.

References

1. Agrawal, R., Imieliński, T., Swami, A.: Mining association rules between sets of items in large databases. SIGMOD Rec. **22**(2), 207–216 (1993). ACM
2. Fournier-Viger, P., et al.: PFPM: discovering periodic frequent patterns with novel periodicity measures. In: Proceedings of the 2nd Czech-China Scientific Conference 2017. InTech
3. Han, J., Pei, J., Yin, Y.: Mining frequent patterns without candidate generation. ACM SIGMOD Rec. **29**(2), 1–12 (2000). ACM
4. Ismail, W.N., Hassan, M.M., Alsalamah, H.A., Fortino, G.: Mining productive-periodic frequent patterns in tele-health systems. J. Netw. Comput. Appl. **115**, 33–47 (2018)
5. Uday Kiran, R., Krishna Reddy, P.: Towards efficient mining of periodic-frequent patterns in transactional databases. In: Bringas, P.G., Hameurlain, A., Quirchmayr, G. (eds.) DEXA 2010. LNCS, vol. 6262, pp. 194–208. Springer, Heidelberg (2010). https://doi.org/10.1007/978-3-642-15251-1_16
6. Kiran, R.U., Kitsuregawa, M.: Novel techniques to reduce search space in periodic-frequent pattern mining. In: Bhowmick, S.S., Dyreson, C.E., Jensen, C.S., Lee, M.L., Muliantara, A., Thalheim, B. (eds.) DASFAA 2014. LNCS, vol. 8422, pp. 377–391. Springer, Cham (2014). https://doi.org/10.1007/978-3-319-05813-9_25
7. Kiran, R.U., Kitsuregawa, M.: Discovering Quasi-periodic-frequent patterns in transactional databases. In: Bhatnagar, V., Srinivasa, S. (eds.) BDA 2013. LNCS, vol. 8302, pp. 97–115. Springer, Cham (2013). https://doi.org/10.1007/978-3-319-03689-2_7
8. Kiran, R.U., Reddy, P.K.: An alternative interestingness measure for mining periodic-frequent patterns. In: Yu, J.X., Kim, M.H., Unland, R. (eds.) DASFAA 2011. LNCS, vol. 6587, pp. 183–192. Springer, Heidelberg (2011). https://doi.org/10.1007/978-3-642-20149-3_15
9. Kumar, V., Valli Kumari, V.: Incremental mining for regular frequent patterns in vertical format. Int. J. Eng. Tech. **5**(2), 1506–1511 (2013)
10. Li, J., Li, H., Wong, L., Pei, J., Dong, G.: Minimum description length principle: generators are preferable to closed patterns. In: Proceedings of the 21st National Conference on Artificial Intelligence, pp. 409–414 (2006)
11. Lin, J.C.W., Zhang, J., Fournier-Viger, P., Hong, T.P., Zhang, J.: A two-phase approach to mine short-period high-utility itemsets in transactional databases. Adv. Eng. Inf. **33**, 29–43 (2017)

12. Nofong, V.M.: Discovering productive periodic frequent patterns in transactional databases. Ann. Data Sci. **3**(3), 235–249 (2016)
13. Nofong, V.M.: Fast and memory efficient mining of periodic frequent patterns. In: Sieminski, A., Kozierkiewicz, A., Nunez, M., Ha, Q.T. (eds.) Modern Approaches for Intelligent Information and Database Systems, SCI, vol. 769, pp. 223–232. Springer, Cham (2018)
14. Nofong, V.M., Wondoh, J.: Towards fast and memory efficient discovery of periodic frequent patterns. J. Inf. Telecommun. **3**(4), 480–493 (2019)
15. Pei, J., Han, J., Lu, H., Nishio, S., Tang, S., Yang, D.: H-mine: hyper-structure mining of frequent patterns in large databases. In: Proceedings IEEE International Conference on Data Mining, pp. 441–448, IEEE (2001)
16. Rashid, M.M., Karim, M.R., Jeong, B.-S., Choi, H.-J.: Efficient mining regularly frequent patterns in transactional databases. In: Lee, S., Peng, Z., Zhou, X., Moon, Y.-S., Unland, R., Yoo, J. (eds.) DASFAA 2012. LNCS, vol. 7238, pp. 258–271. Springer, Heidelberg (2012). https://doi.org/10.1007/978-3-642-29038-1_20
17. Rashid, M.M., Gondal, I., Kamruzzaman, J.: Regularly frequent patterns mining from sensor data stream. In: Lee, M., Hirose, A., Hou, Z.-G., Kil, R.M. (eds.) ICONIP 2013. LNCS, vol. 8227, pp. 417–424. Springer, Heidelberg (2013). https://doi.org/10.1007/978-3-642-42042-9_52
18. Surana, A., Kiran, R.U., Reddy, P.K.: An efficient approach to mine periodic-frequent patterns in transactional databases. In: Cao, L., Huang, J.Z., Bailey, J., Koh, Y.S., Luo, J. (eds.) PAKDD 2011. LNCS (LNAI), vol. 7104, pp. 254–266. Springer, Heidelberg (2012). https://doi.org/10.1007/978-3-642-28320-8_22
19. Tanbeer, S.K., Ahmed, C.F., Jeong, B.-S., Lee, Y.-K.: Discovering periodic-frequent patterns in transactional databases. In: Theeramunkong, T., Kijsirikul, B., Cercone, N., Ho, T.-B. (eds.) PAKDD 2009. LNCS (LNAI), vol. 5476, pp. 242–253. Springer, Heidelberg (2009). https://doi.org/10.1007/978-3-642-01307-2_24
20. Tseng, F.C.: Mining frequent itemsets in large databases: the hierarchical partitioning approach. Expert Syst. Appl. **40**(5), 1654–1661 (2013)
21. Webb, G.I.: Self-sufficient itemsets: an approach to screening potentially interesting associations between Items. ACM Trans. Knowl. Discov. Data **4**(1), 3:1–3:20 (2010)
22. Zaki, M.J.: Scalable algorithms for association mining. IEEE Trans. Knowl. Data Eng. **12**(3), 372–390 (2000)
23. Zaki, M.J., Parthasarathy, S., Ogihara, M., Li, W.: Parallel algorithms for discovery of association rules. Data Min. Knowl. Disc. **1**(4), 343–373 (1997)
24. Zaki, M.J., Gouda, K.: Fast vertical mining using diffsets. In: Proceedings of the 9th ACM SIGKDD International Conference on Knowledge Discovery and Data Mining, pp. 326–335 (2003)

On Robustness of Adaptive Random Forest Classifier on Biomedical Data Stream

Hayder K. Fatlawi[1,2] and Attila Kiss[1]

[1] Faculty of Informatics, Department of Information Systems, Eötvös Loránd
University, Budapest, Hungary
hayder.fatlawi@uokufa.edu.iq, kiss@inf.elte.hu
[2] Center of Information Technology Research and Development,
University of Kufa, Najaf, Iraq

Abstract. Data Stream represents a significant challenge for data analysis and data mining techniques because those techniques are developed based on training batch data. Classification technique that deals with data stream should have the ability for adapting its model for the new samples and forget the old ones. In this paper, we present an intensive comparison for the performance of six of popular classification techniques and focusing on the power of Adaptive Random Forest. The comparison was made based on four real medical datasets and for more reliable results, 40 other datasets were made by adding white noise to the original datasets. The experimental results showed the dominant of Adaptive Random Forest over five other techniques with high robustness against the change in data and noise.

Keywords: Classification · Biomedical · Data stream · Ensemble modeling

1 Introduction

A series of researches and projects in medical science and Information Technology are starting a relationship between the healthcare industry and the IT industry that will rapidly lead to a better and interactive relation among patients, their doctors and health institutions. Data mining has a significant role in medical data processing and analysis that mostly aims to predict possibility of diseases or diagnosing them. Typical data mining techniques depend on the assumption that the data is a random and the samples generate from a stationary distribution, while this assumption is violated in most of available datasets [1].

Digital Universe Study [2], reported that there was over 2.8ZB of data were generated and processed in 2012, with expected growth to 15 times by 2020. Stream data represents a major source for a huge amount of data, especially

A. Kiss was also with J. Selye University, Komarno, Slovakia.

© Springer Nature Switzerland AG 2020
N. T. Nguyen et al. (Eds.): ACIIDS 2020, LNAI 12033, pp. 332–344, 2020.
https://doi.org/10.1007/978-3-030-41964-6_29

in medical related data in which the wearable devices, diagnosing tools, mobile phones and monitoring devices can provide continuous samples of data. Mining this size of data could be an infeasible process because the limitation of computation resources, also the mining model should be adaptive to learn from a new cases and ignoring old ones using a mechanism that called concept drift [3,4]. In this work, six data mining techniques are used to work with the data stream.

1.1 Related Works

In [5] a boosting framework was developed which can be utilized to create online boosting algorithms. Logistic Regression, Least Squares Regression, and Multiple Instance Learning were derived and tested with a wide range of experiments that showed empirical evidence to get similar performance as the standard batch boosting algorithms.

ADWIN Bagging [6] improved using Hoeffding AdaptiveTrees which can learn from data streams adaptively over time. An error change detector for classifiers was added for speeding up the time of adapting of Adaptive-Size Hoeffding Tree (ASHT) Bagging. The improvements was tested by making an evaluation on synthetic and real-world datasets contain ten million examples.

The feasibility of many classification techniques for analyzing biosignals like EEG and ECG as an infinite data streams was studied in [7]. An evaluation is made between traditional and stream-based classification that showed a decline of accuracy traditional technique after new data arrives. The impact of data mining techniques with biomedical data streams was investigated by [8]. A simulation is used for comparing between two types biomedical data (case-based and stream-based) and the results showed that case based had better accuracy but slower in execution time.

In [9] a clinical-support-system that depends on data stream mining techniques was described by proposing a new system called VFDT which has the ability to analysis this kind of data and produces real time prediction. It improved the capability of typical decision tree to save the relationship between the history records and leaf nodes. [10] improved decision tree by an incremental VFDT for medical data. It has a major difference that for splitting operation there was no dependency on the number of reading samples. The result of their work showed better accuracy resulted from their method but more execution time in compare with VFDT.

Random Forest ensemble learning algorithm was improved in [11] for dealing with imbalanced and binary classification tasks. The empirical error is used as the measurement to obtain class weights of the classifier for medical data and the results of the proposed method had a high accuracy classifying according to F1 and Recall. In [12] medical data streams are processed by a framework that is dealing with the cumulative frequency queries over to support the online medical decision. The proposed framework includes two parts: data summarization and dynamic maintenance, and results demonstrate the efficiency of the proposed approach.

2 Basic Concepts in Stream Data Mining

2.1 Data Stream Constraints

Unlike with batch data, stream data faces many constraints as follow; (1) infinite arrival of data samples make storing them impossible, (2) the fast arrival of data samples require process each sample in real time, (3) the possibility of changing items' distribution over time in which the old data would be useless for the current status. Thereby, the perfect classification model should produce maximum accuracy in fastest time and minimum computational resources [5].

2.2 Concept Drift

Concept drift refers to that the data is being gathered may change from time to time, every time according to some minimum persistence. Changes may occur during time in which the old training examples become irrelevant to the current state and the learning system should forget such kind of information. There are two important issues related with the change: causes of change, and the rate of change [3].

2.3 ADWIN (ADaptive sliding WINdow)

It is an estimation technique that aim for detecting the change in a data stream based on sliding window with adaptive size. It has a qualified and significant method to tracking the average of bits in the stream. In this technique, the length of windows is not updated as long as the average value inside the window doesn't change [3].

2.4 Hoeffding Tree

Hoeffding Tree or Very Fast Decision Tree (VFDT) is a variation from typical decision tree designed for stream data. The learning of this techniques depends on replacing leaves of the tree with decision nodes. Each terminal node (leaf) in the tree stores enough information statistics about features values which is used by heuristic function to perform splitting test. After reaching an new data instance, it will transfer starting from the root until reaching to a specific leaf node. In this point, the statistics information will be evaluated and a new decision node may be created based on this evaluation [3]. It is very popular to utilize VFDT as a base learner for ensemble classification model, thereby, the ensemble techniques in this work used VFDT as well.

2.5 Ensemble Modeling

Ensemble modeling aims to build a strong accumulative classifier from many weak classifiers.

Oza Bagging is an ensemble learning method that is an improvement of the popular Bagging ensemble method for effectively handle data streams. In traditional Bagging technique, there are K classifiers training on K different datasets, created by drawing L samples from the L-sized training set with replacement. In the online bagging, the strategy is to simulate this task by training every arriving data sample K times [13].

Adaptive Random Forest is a variation from typical random forest algorithm for data stream mining tasks. The main idea is to utilize Hoeffding trees, which have the ability of adapting with distribution changes, as the base classifier for the bagging ensemble method [5]. For detecting the change in a data stream, ADWIN is used in this techniques. It depends on Online Bagging as a resampling method and a drift monitor for change detecting per each tree [5]. Boosting is another mechanism for ensemble modeling and AdaBoost is the most popular boosting technique.

Oza Boost technique extended typical AdaBoost to the online setting. The main idea of this algorithm was to model arrival of data examples as sampling (with replacement) from a Poisson process, where more difficult examples are given a higher mean.

2.6   K-Nearest Neighbor

K Nearest Neighbor (k-NN) algorithm can be coped with heterogeneous concept drift. It distinguishes between current concept and former concept, and preserves both of them. As a result of that, it has advantage in comparison with other methods which discard the previous knowledge leading to more mistakes in case of reoccurring drift [14].

2.7   Naive Bayes

A naive Bayes classifier is a simple probabilistic classifier that is built by applying Bayes' theorem with naive independence assumption. Despite this strong assumption, it is very effective in many real applications. This classifier relies on the estimation of the conditional probabilities. This estimation can be provided on a data stream using a "supervised quantiles summary" [15].

3   Methodology

The main aim of this work is to investigate the most reliable and accurate classification techniques for data stream mining tasks. Thereby, there are two main stage; the first one is to apply some of preprocessing procedures to prepare the medical data for mining process. The second stage is to build the six popular classification techniques and compare the performance of them based on streaming real batch datasets. Figure 1 illustrates all the steps of our work.

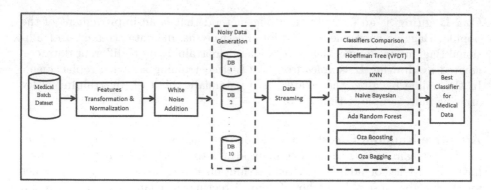

Fig. 1. Comparison of classification techniques based on medical data stream

3.1 Stage One: Data Preprocessing

The First stage concerns with preparing and generating a new dataset from the original one by using features transformation, Normalization, and the white noise concept. The aim of this stage is that even a large dataset cannot be sufficient for finding the best techniques and for that we need to create many datasets for learning the classification models. This stage can be described as following steps:

Categorical to Numerical Should Transformation. To add the noise values to the data, features should be in a numeric form. So, in this step every Categorical (Textual values) was converted to numerical values. For binary features values like (Yes, No), the simplest coding is used and the values became (1,0). For multiple values (more than two values), frequency of each distinct value (1.N) for each feature was calculated. After that coding was used in which the most frequent distinct value had value (N) and less frequent had (1) value.

Data Normalization. Range of features values can be different such as age has values between 1–150 while the yearly income can be between (1-10000000). For that, we need to apply normalization to prevent any dominant for one of the features during statistical calculation that performed during classifier building. the new range for all feature values were between −1 and 1.

White Noise Generation. For each data set, noise value was added in which the mean of those values for each feature will be 0. The standard deviation (STD) represents intensity of the noise and the gradual increase of it will provide many new datasets. The range (0.01–0.1) will be used for STD values to generate 10 datasets from the original one.

Data Streaming. According to evaluate the performance of the classifiers, the batch datasets will be converted to a stream. The frequency of samples (stream size) was configured based on the total number of data instances in the datasets in which high frequency was used with the large number of instances.

3.2   Stage Two: Building Classification Techniques

Techniques which used in this work can be classified into two categories; (1) Single model classifier, (2) Ensemble model classifier. The first group includes: Hoeffding Tree, Naive Bayesian, and K-NN classifiers. On the other side, the second group includes; Adaptive Random Forest, Oza Boosting, and Oza Bagging classifiers. Even with the high computational requirements, ensemble mechanism tends to have more reliable performance in comparative with single classifiers. The base classifier for the bagging and boosting ensemble classifiers was be Hoeffding Tree. The stream size is configured according to the number of data instances in the dataset, in which, if the instances increase then stream size also increase.

Hoeffding Tree Classifier Building as presented by [16] includes two type of nodes: internal and terminal nodes. Each terminal node (leaf) in the tree stores enough information statistics about features values. This information is used by heuristic function to perform splitting test. After reaching new data instance, it will transfer starting from the first node (root) until reaching to a specific leaf node. In this point, the statistics information is evaluated and a new decision node may be created based on this evaluation. The statistical information about evaluation of splitting in this node is updated by decision nodes.

Ensemble Model Building using online Bagging of [13] K base classifiers is created, any new data sample could be chosen according to a Poisson(1) distribution. The classification decision of the ensemble bagging model is based on voting of all K base classifiers with equal weight for all of them. It gives every new data example an initial weight w=1, then it is passed to the first weak learner. If this data example is misclassified, it's weight is increased before passing it to the next weak learner. The base learner in our comparison was Hoeffding Tree classifier and the size of ensemble that used was ten learners. Adaptive Random forest was built depending on [6] which utilized Online Bagging's resampling method but the difference was in adaptive method. Hidden trees was built in the moment that a warning was detected. Replacing of trees was made only if there was a detection of a drift in the stream.

Online Boosting building according to [13] includes K base classifiers and the training data weighted according to the ability of classify them correctly. In the current classifier, if the samples set of training data X is misclassified, the weight of it's samples was changed to have the half of the total weight in the next classifier. The remaining samples which had been classified correctly got the half remaining weight.

Naive Bayesian Building. A graphical model method is used as presented by [15] to extend Naive Bayesian with data stream using a stochastic estimation. The method is incremental and produces a Weighted Naive Bayes Classifier for data stream. It has weights for both variable and class. In the graphical model, the first layer represents a linear layer for a weighted sum of each class. Stochastic gradient descent is used for optimization of the weights.

K-NN Building. In our comparison, K-NN Classifier is used as developed by [14] in which the accuracy of prediction is maximized. Specific models from current state of the concepts is combined with others from the past.

4   Implementation and Experimental Results

4.1   Medical Datasets Description

According to validate the proposed model, four real medical datasets are used. The first one was EEG Eye State dataset that contains 15 features and 14980 data instances with binary class values. The second dataset was Thyroid Disease dataset that had 21 features and 7200 instances with three class values. Skin Segmentation dataset was the third that had 4 features and 245057 instances with binary class values. The fourth was hypothyroid dataset and it had 26 features, 3163 instances, and binary class values. The first three dataset are available in UCI data repository [17] while the last one available in kaggle [18].

4.2   Data Analysis Platform

In this work, three major tools were utilized to perform the comparison; Waikato Environment for Knowledge Analysis (Weka), Massive Online Analysis (MOA), and Sklearn. Weka Platform is an open source software for data analysis tasks including Classification, Clustering and Association Rules. It is developed by University of Waikato using java programming language. It was utilized in this work for preprocessing operations (Transformation and Normalization). MOA Platform is an improvement for Weka platform for mining data stream. It provides many of popular mining techniques, stream generator, and concept drift detection techniques, in our comparison, it performed the data streaming and implementation of classification techniques. Sklearn is a python free library for machine learning tasks. It contains many of classification techniques such as random forest and boosting. Sklearn was used in this work for adding white noise values to data.

4.3   Case Study One

In this case, six of popular classification techniques are applied with the original four datasets. For this task, Massive Online Analysis platform is used to convert batch datasets to a data stream, then to train the classifier based on that stream. Different configurations are used for datasets according to its number of instances, for EEG Eye State and Skin Segmentation datasets were configured as a stream with size 100 sample/sec, while hypothyroid and Thyroid Disease datasets were configured to have 10 sample/sec stream size. Four measurements are used for evaluating the performance of the classification techniques; mean of correctly classified instances, mean of F1 score, mean of precision, and mean of Recall. The Tables 1, 2, 3, 4 and Fig. 2 clarify the performance of each classification technique with the original four datasets.

Table 1. Comparison of the performance of each classification technique with EEG Eye State dataset

Technique	Correctly classified	F1 Score	Precision	Recall
ADARandomForest	98.7515	96.8030	92.5949	96.8572
K-NN	95.4494	90.9678	86.5407	90.6723
OzaBagging	94.0752	89.1728	87.0007	89.1419
Hoeffman Tree	75.9037	71.2037	66.6565	72.4120
OzaBoost	72.7193	76.2884	77.0834	77.2055
Naive based	50.0529	57.0754	61.5685	54.7204

Table 2. Comparison of the performance of each classification technique with Skin Segmentation dataset

Technique	Correctly classified	F1 Score	Precision	Recall
ADARandomForest	99.998	99.9959	99.9758	99.9956
K-NN	99.9776	99.9773	98.6185	99.9704
OzaBagging	99.9727	99.96	99.9044	99.9581
Hoeffman Tree	99.9494	99.9357	99.5617	99.9299
Oza Boost	99.9575	99.9392	99.933	99.9343
Naive Bayesian	95.2974	97.1962	67.9254	95.3057

The results above shown that Adaptive Random Forest had the best performance according to the four measurements. According of the mean of correctly classified samples Adaptive Random Forest presented 98.75 rate with EEG state Eye data stream and 99.998 rate with skin segmentation data stream.

Another important observation from the results that is stability of Adaptive Random Forest during training process. The minimum accuracy during 150 training processes on 150 samples sets was 95.4 while the accuracy of K-NN, Heoffman Tree, OzaBagging, OzaBoosting, Naive Bayesian) were (88.6, 46.7, 75.3, 0,0) respectively. Fig. 3 illustrates this observation.

Table 3. Comparison of the performance of each classification technique with hypothyroid dataset

Technique	Correctly classified	F1 Score	Precision	Recall
ADARandomForest	99.68151	99.3759	99.2346	99.5769
K-NN	95.9275	95.1928	91.7478	96.1143
OzaBagging	98.2211	97.7739	65.5441	98.1558
Hoeffman Tree	99.0534	98.517	96.376	99.0345
Oza Boost	98.2211	97.783	65.5441	98.1558
Naive Bayesian	97.9982	97.6545	65.5118	98.1172

Table 4. Comparison of the performance of each classification technique with Thyroid dataset

Technique	Correctly classified	Kappa	Kappa Temporal	Kappa M
ADARandomForest	94.9981	94.9981	60.8699	94.9981
K-NN	93.6299	93.6299	52.1417	93.6299
OzaBagging	93.8357	93.8357	52.8027	93.8357
Heoffman Tree	93.6537	93.6537	51.7060	93.6537
Oza Boost	96.0968	84.9073	87.2015	78.6781
Naive Based	93.0586	93.05867	46.7857	93.0586

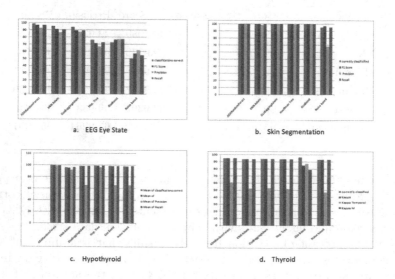

a. EEG Eye State

b. Skin Segmentation

c. Hypothyroid

d. Thyroid

Fig. 2. Performance of six techniques with original datasets

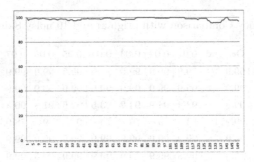

Fig. 3. Performance of Adaptive Random Forest during EEG Eye State training

4.4 Case Study Two

This case study is related with evaluating the performance of the classifiers against gradually change in data by using white noise. For this task, Weka platform is used to perform two of preprocessing steps; features transformation and normalization. Then MOA is used to convert batch datasets to a data stream, then to train the classifier based on that stream. Similarly to case study one, EEG Eye State and Skin Segmentation datasets were configured as a stream with size 100 sample/sec, while Hypothyroid and Thyroid Disease datasets were configured to have 10 sample/sec stream size. Also the same Four measurements are used for evaluating the performance of the classification techniques. The Tables 5, 6, 7, 8, 9 and Fig. 4 clarify the performance of each classification technique with the original four datasets.

From the results above, we can conclude the following: (1) mostly, Adaptive Random Forest had the best accuracy among the six classifiers; (2) it had a significant stability and robustness against the white noise;(3) K-NN classifier also had a good accuracy but with less stability and slow performance (4) Naive Bayesian had the worst accuracy and stability among the classifiers but it has fast performance.

Table 5. Accuracy Comparison with original and 10 noisy EEG Eye State dataset

Tech/STD	Original	0.01	0.02	0.03	0.04	0.05	0.06	0.07	0.08	0.09	0.1
AdaRandomForest	98.7	98.1	98	97.9	97.7	98	98	97.8	97.8	97.8	97.8
K-NN	95.4	84.8	84	84	85.2	85.2	85.2	86.3	85.3	85.5	85.6
OzaBagging	94	84.6	84.9	84.9	84.5	85	83.8	84.9	84.9	84.9	85
Hoeffman Tree	75.9	60.6	60.4	59.2	59.8	60.3	63.2	58.9	61.9	60.7	60.8
Ozaboosting	72.7	81.8	77.7	79.9	82.3	80.3	82.8	80.5	78.9	78.2	78
Naive Bayesian	50	49.3	50.4	51.4	51.8	52.3	52.8	53	53.3	53.7	53.8

Table 6. Accuracy Comparison with original and 10 noisy Skin Seg. dataset

Technique/STD	Original	0.01	0.02	0.03	0.04	0.05	0.06	0.07	0.08	0.09	0.1
Ada RandomForest	99.9	99.9	99.9	99.9	99.9	99.9	99.9	99.9	99.9	99.9	99.9
K-NN	99.9	99.9	99.9	99.9	99.9	99.9	99.9	99.9	99.9	99.9	99.9
Naive Bayesian	99.9	93.4	94.9	94.2	93.5	92.6	91.8	90.8	89.9	89.2	88.3
Ozaboosting	99.9	53.6	79	79	79	0	79	78.9	79	79	79
Ozabagging	99.9	99.9	99.9	99.9	99.9	99.9	99.9	99.9	99.9	99.9	99.9
Hoeffman Tree	95.2	99.8	99.9	99.9	99.9	99.9	90.5	99.9	99.9	99.9	99.9

Table 7. Accuracy Comparison with original and 10 noisy hypothyroid dataset

Tech/STD	Original	0.01	0.02	0.03	0.04	0.05	0.06	0.07	0.08	0.09	0.1
AdaRandomForest	99.6	99.4	99.5	99.5	99.6	99.4	99.4	99.4	99.4	99.3	99.3
K-NN	95.9	95.8	95.9	95.9	96.2	95.5	95.2	95.5	94.9	94.8	94.8
OzaBagging	98.2	94.4	93.4	94.2	92.7	94.8	94.2	94.9	94.5	92.8	92.8
Hoeffman Tree	99	91.9	92	91.8	92.1	92.2	92.1	92.1	92	92	92
Ozaboosting	99.5	98.1	98.2	98	98.1	98.6	97.9	97.9	97.6	97.6	97.6
Naive Bayesian	97.9	46.7	51.4	51.1	52.9	52.8	51.9	55.5	53.5	53	53

Table 8. Accuracy Comparison with original and 10 noisy Thyroid dataset

Tech/STD	Original	0.01	0.02	0.03	0.04	0.05	0.06	0.07	0.08	0.09	0.1
AdaRandomForest	94.9	93.9	93.4	93.4	93.3	92.7	92.7	92.6	92.6	92.6	92.6
K-NN	93.6	93.2	93.2	93	93	93	92.8	92.8	92.8	92.7	92.8
OzaBagging	93.8	92.6	92.6	92.6	92.6	92.6	92.6	92.6	92.6	92.6	92.6
Hoeffman Tree	93.6	92.6	92.6	92.6	92.6	92.6	92.6	92.6	92.6	92.6	92.6
Ozaboosting	96	90.5	91.5	91.5	91.4	91.4	91	91.4	91.4	92.6	91.4
Naive Bayesian	93	36.1	41.5	45	48.2	49.5	55.5	61.7	64.4	69	70.9

Table 9. Comparison of classification execution time (seconds) with four datasets

Technique/Dataset	Thyroid	EEG Eye State	Skin Segmentation	Hypothyroid
Ozabag	0.77	1.91	16.83	0.75
Ozaboos	1.16	3.98	15.33	0.55
Hoeffding tree	0.08	0.34	0.84	0.06
K-NN	3.59	10.3	140	3.03
Naive Bayesian	0.05	0.12	0.58	0.05
AdaRandomForest	1.42	4.05	29.08	0.44

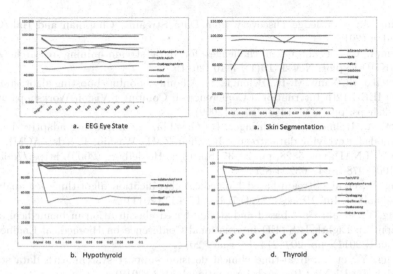

Fig. 4. Accuracy Comparison with original and 10 noisy Four datasets

5   Conclusion

The effective contribution of data analysis systems to disease prediction and decision support in health institutions has led to the continuous development of these technologies. As data stream is an important source of these analyzes, the development of technologies to deal with data stream rather than batch datasets takes a reasonable of the current research interest in data science. This paper presents an attempt to investigate the power of popular data classification techniques especially Random Forest, including the strength in handling a large volume of data. It also identifies the best and most stable technology to resist the gradual change in data stream. The results showed a significant advantage and stability of Adaptive Random Forest technique over other techniques. 0.98 of the experiments showed the goodness of this technique With great stability during data stream training.

Acknowledgment. The project was supported by the European Union, co-financed by the European Social Fund (EFOP-3.6.3-VEKOP-16-2017-00002).

References

1. Hulten, G., Spencer, L., Domingos, P.: Mining time-changing data streams. In: Proceedings of the Seventh ACM SIGKDD International Conference on Knowledge Discovery and Data Mining, pp. 97–106. ACM (2001)
2. Gantz, J., Reinsel, D.: The digital universe in 2020: big data, bigger digital shadows, and biggest growth in the far east. IDC iView: IDC Analyze the future **2007**(2012), 1–16 (2012)

3. Gama, J.: Knowledge Discovery from Data Streams. Chapman and Hall/CRC, London (2010)
4. Krempl, G., et al.: Open challenges for data stream mining research. ACM SIGKDD Explor. Newsl. **16**(1), 1–10 (2014)
5. Babenko, B., Yang, M.-H., Belongie, S.: A family of online boosting algorithms. In: 2009 IEEE 12th International Conference on Computer Vision Workshops, ICCV Workshops, pp. 1346–1353. IEEE (2009)
6. Bifet, A., Holmes, G., Pfahringer, B., Gavaldà, R.: Improving adaptive bagging methods for evolving data streams. In: Zhou, Z.-H., Washio, T. (eds.) ACML 2009. LNCS (LNAI), vol. 5828, pp. 23–37. Springer, Heidelberg (2009). https://doi.org/10.1007/978-3-642-05224-8_4
7. Fong, S., et al.: Stream-based biomedical classification algorithms for analyzing biosignals. J. Inf. Process. Syst. **7**(4), 717–732 (2011)
8. Hang, Y., et al.: Case-based and stream-based classification in biomedical application. In: Eighth IASTED International Conference on Biomedical Engineering (Biomed 2011), pp. 207–214. February 2011
9. Zhang, Y., et al.: Real-time clinical decision support system with data stream mining. In: BioMed Research International 2012 (2012)
10. Cazzolato, M.T., Ribeiro, M.X.: A statistical decision tree algorithm for medical data stream mining. In: Proceedings of the 26th IEEE International Symposium on Computer-Based Medical Systems, pp. 389–392. IEEE (2013)
11. Zhu, M., et al.: Class weights random forest algorithm for processing class imbalanced medical data. IEEE Access **6**, 4641–4652 (2018)
12. Al-Shammari, A., Zhou, R., Liu, C., Naseriparsa, M., Vo, B.Q.: A framework for processing cumulative frequency queries over medical data streams. In: Hacid, H., Cellary, W., Wang, H., Paik, H.-Y., Zhou, R. (eds.) WISE 2018. LNCS, vol. 11234, pp. 121–131. Springer, Cham (2018). https://doi.org/10.1007/978-3-030-02925-8_9
13. Oza, N.C.: Online bagging and boosting. In: 2005 IEEE International Conference on Systems, Man and Cybernetics. vol. 3, pp. 2340–2345, IEEE (2005)
14. Losing, V., Hammer, B., Wersing, H.: KNN classifier with self adjusting memory for heterogeneous concept drift. In: 2016 IEEE 16th International Conference on Data Mining (ICDM), pp. 291–300. IEEE (2016)
15. Salperwyck, C., Lemaire, V., Hue, C.: Incremental weighted naive bays classifiers for data stream. In: Lausen, B., Krolak-Schwerdt, S., Böhmer, M. (eds.) Data Science, Learning by Latent Structures, and Knowledge Discovery. SCDAKO, pp. 179–190. Springer, Heidelberg (2015). https://doi.org/10.1007/978-3-662-44983-7_16
16. Domingos, P., Hulten, G.: Mining high-speed data streams. In: Kdd. vol. 2, p. 4 (2000)
17. Irvine UC: Machine Learning Repository. July 2019. url: https://archive.ics.uci.edu/ml/index.php
18. kaggle Rebosotiry: Public Datasets. July 2019. url: https://www.kaggle.com/datasets

Sentence Writing Test for Parkinson Disease Modeling: Comparing Predictive Ability of Classifiers

Aleksei Netšunajev[1], Sven Nõmm[2(✉)], Aaro Toomela[3], Kadri Medijainen[4], and Pille Taba[5]

[1] Tallinn University of Technology, Akadeemia tee 3, 12618 Tallinn, Estonia
aleksei.netsunaev@taltech.ee
[2] Department of Software Science, Tallinn University of Technology,
Akadeemia tee 15a, 12618 Tallinn, Estonia
sven.nomm@taltech.ee
[3] School of Natural Sciences and Health, Tallinn University,
Narva mnt. 25, 10120 Tallinn, Estonia
aaro.toomela@tlu.ee
[4] Institute of Sport Sciences Physiotherapy,
University of Tartu, Puusepa 8, 51014 Tartu, Estonia
kadri.medijainen@ut.ee
[5] Department of Neurology and Neurosurgery,
University of Tartu, Puusepa 8, 51014 Tartu, Estonia
pille.taba@kliinikum.ee

Abstract. The present paper is devoted to the modeling of the sentence writing test to support diagnostics of Parkinson's disease. Combination of the digitalized fine motor tests and machine learning based analysis frequently lead the results of very high accuracy. Nevertheless, in many cases, such results do not allow proper interpretation and are not fully understood by a human practitioner. One of the distinctive properties of the proposed approach is that the set of features consists of parameters that may be easily interpreted. Features that represent size, kinematics, duration and fluency of writing are calculated for each individual letter. Furthermore, proposed approach is language agnostic and may be used for any language based either on Latin or Cyrillic alphabets. Finally, the feature set describing the test results contains the parameters showing the amount and smoothness of the fine motions which in turn allows to precisely pin down rigidity and unpurposeful motions.

1 Introduction

Parkinson disease (PD) is known to be a wide-spread neurodegenerative disorder. While the cure for it is not available at the moment, timely diagnose is very important as it allows to relive the patient from many symptoms affecting the quality of everyday life. Progressing PD usually affects amount and smoothness of the motions. Patients exhibit rigidity, tremor and slowness of the motor movements both on the gross and fine motor levels, meaning that handwriting and

© Springer Nature Switzerland AG 2020
N. T. Nguyen et al. (Eds.): ACIIDS 2020, LNAI 12033, pp. 345–357, 2020.
https://doi.org/10.1007/978-3-030-41964-6_30

drawing are affected at the first stage [6]. For that reason a handwriting tests is one of the diagnostic tools available for doctors [9].

Several tests have been digitalized during the recent years, for example, Luria's alternating series test [8] and sentence writing tests [13]. Some of the tests were analyzed quantitatively, see for instance [8,14]. Various studies show that it is possible to set up a classifier that discriminates reasonably well between healthy control (HC) and PD patients based on fine motor drawing tests [7]. Handwriting of people affected by PD has been studied substantially. Several handwriting impairments are known to characterize the patients. One of them is *micrographia* that is reductions in writing size. A reduction in letter size is fairly simple to detect with conventional paper-and-pen tools. Even though [19] report micrographia to be found in nearly half of the PD cohort, its exact prevalence is not clear and may vary substantially. [17] argue that PD patients reduce the size of their handwriting strokes when concurrent processing load increases. Micrographia may result in consistent reduction in the size of letters [5] as well as disability to keep the fixed size of letters for consecutive characters.

With the adoption of tablets with stylus new predictors describing the writing style of a person arise. For example, iPad brings in altitude and azimuth angles of the Apple Pencil (see [7]), that may give additional information on drawing ability. Thus researchers started questioning whether writing size is the most important predictor for PD. Over time the focus has shifted from the analysis of only letter size to the analysis of a set of kinematic features of handwriting. Velocity and acceleration are the main kinematic properties studied. [5] document that kinematic features differentiate better between control participants and PD patients than the traditional measure of static writing size. Furthermore, overview made by [16] convince that studies based on kinematic analysis of handwriting have revealed that patients with PD may have abnormalities in velocity, fluency, and acceleration in addition to micrographia. In the present paper specific type of the handwriting test is analyzed. The test is based on written full sentence. Traditionally the test is conducted using pen and paper with the practitioner evaluating the outcome subjectively. However, there are numerous possibilities to conduct the test using a digital pen. Current paper uses the samples collected with the help of an iPad.

Even though several studies analyzed digital data of fine motor movement tests, data originating from handwriting tests where the full sentence is written are not fully explored in the literature. For example, [13] consider only a very limited amount of features and [3] is mainly interested in how new pressure measure improves the predictive ability. While it may be tempting to limit with words to extract features it is possible to extract individual letters and construct features based on the letters. This will provide more information for the model and may improve its predictive ability.

The paper extends the literature in three directions. First, an algorithm for letter extraction for the samples of the handwritten sentence is developed. We use our own custom letter extraction system as standard approaches may not be fruitful due to specific handwriting of PD patients. Second, in the feature engineering phase interpretation to most important features is given. Finally, classification analysis with classical machine learning algorithms is performed.

We leave deep learning approach aside as it would require an additional step of inference to produce features that may be interpreted. With our approach we operate on features with interpretation straight away and find that kinematic ones have the highest explanatory power in this task. The features that describe fluency of writing follow. Random forest with a prediction accuracy of 88.5% is the model that outperform the other selected classifiers based on K-fold cross-validation. The rest of the paper is organized as follows. Section 2 describes the data used for the current analysis. Section 3 outlines the algorithm used for letter extraction. Feature engineering and training of classifier are discussed in Sect. 4. Final Sect. 5 concludes.

2   Data Acquisition

Digital version of the sentence writing test requires one to write given sentence on the screen of tablet PC using the stylus pen. The sentence is chosen from the literature usually taught during the first years of school education. This guarantees that all the subjects with the same native language know the sentence very well. In the frameworks of the present research subjects who's native language is Estonian were tested. The sentence reads 'Kui Arno isaga koolimajja jõudsid, olid tunnid juba alanud.' which means "When Arno with his father arrived to the school lessons has already started". The data is collected using an iPad application that records the state of Apple Pencil at certain discretization level. The observations include floating point numbers that describe the way a person is performing the test. Acquired data is stored in the form of numeric array where the rows correspond to different time instances and columns to the data attributes. Figure 1 shows an example of the full sentence. The dataset consists of 11 PD patients and 8 healthy controls of approximately the same age mean value 68 years old.

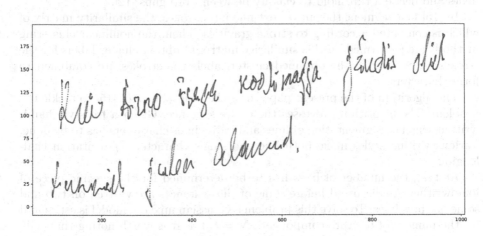

Fig. 1. Full sentence written in a tilted way

The most important feature of the data is that the points may be chronologically ordered and thus it is possible to calculate time changes for a sequence of points. By the same token, sorting the points based on time gives the sequence of points in the order the person has written them. This is a very important feature of the data as it will be used to construct the letters extraction algorithm.

While the data is collected for individuals who have PD it is quite likely that the recorded tests will be of different quality. The algorithm for letters extraction should take these peculiarities into account. It may happen that the person is unable to maintain a straight line and the lines become tilted. The line may form either a positive or negative angle with the imaginary x-axis. Figure 1 presents an example of the sentence written in a tilted way. It is worth noting that the sentence in Fig. 1 is written on two lines. In principle, this is not always the case. In the analyzed sample the sentence is written on three lines for most of the cases. No cases with of a sentence fitting on a single line were observed.

3   Individual Letter Recognition

Unsupervised clustering is a common approach for this task. Several known problems, such as non-linear separation boundaries between characters or overlapping of strokes may result in poor separation of characters. These are the reasons for the standard distance based clustering methods to fail. Density-based methods such as DBSCAN are not working reasonably well due to a non-elliptic distribution of points that comprise a character. Probability-based mixture model for character extraction may fail in the situations where characters are hardly separable which may often be the consequence of PD. [2] conclude that standard clustering algorithms may be outperformed by specialized one in word segmentation. Most of the proposed techniques for word extraction in the literature consider a spatial measure of the gap between successively connected components and define a threshold to classify between word gaps [12].

In [15] first segment the entire text line into strokes, the similarity matrix of which is computed according to stroke gravities. Then, the nonlinear clustering methods are performed on this similarity matrix to obtain cluster labels for the strokes. According to the obtained cluster labels the strokes are combined to form characters.

The algorithm of the present paper uses the general idea of [15] to tackle the problem. The algorithm processes the entire sentence. Given a freestyle handwritten sentence segmentation of lines and individuals characters has to be done. Various writing styles make both line and single character segmentation challenging.

At first, the number of lines has to be determined for the specific piece of handwriting. As discussed before some of the sentences are written on two and some on three lines. To solve this problem a Gaussian mixture model is estimated for the number of mixture components $N = 2, 3, 4$. It is worth noting that only y coordinates are used as observed variables in the mixture models. The idea of taking only a single coordinate is based on the histogram of these coordinates.

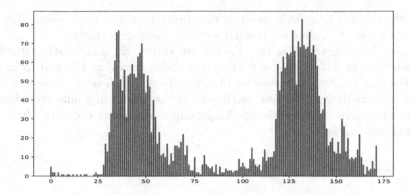

Fig. 2. Histogram of the y-coordinates of a sentence

Figure 2 shows an example of the histogram for the y coordinates of a sentence. Two humps of points that may be modeled as Gaussians are clearly visible. The decision over N is made as an $\arg\max_N$ MeanSilhoutteScore(N).

Processing of individual lines goes next. The separation of a line into smaller parts is based on the calculated time changes Δt of points for the whole line. An example of the series is shown on Fig. 3. It becomes obvious that there are bigger chunks of points that are separated with longer time intervals. These larger chunks are groups of letters, sometimes full words, sometimes parts of words. This depends very much on the way the person writes. Separating the points that belong to a common chunk applying a threshold on Δt makes up words for most lines. It is very important to pick the right separation threshold. With various candidates examined, the final value is chosen to be 20% of the maximum change.

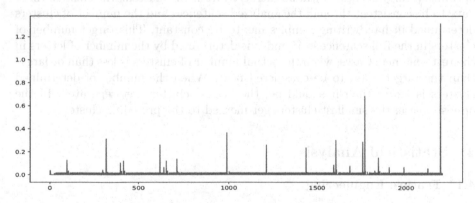

Fig. 3. An example of the time change

Further, the chunks of points extracted in the previous step are processed. These are separated further based on the distance of points. The idea is to understand where the person who is writing made a gap between points. This is very likely to be the gap between letters. Note that small clusters such as dots over letters i or j arise as a result of this procedure. For that reason clusters having less than 85 points are merged with the preceding cluster.

It may be observed that the clusters are rather big and mostly represent the sequences of letters that are written without removing the pen from the iPad. For that reason, the separation of large clusters into several smaller ones is performed next. It is done based on the size of the cluster of points, from bigger clusters more letters are extracted. An example of clusters obtained after this step is shown in Fig. 4.

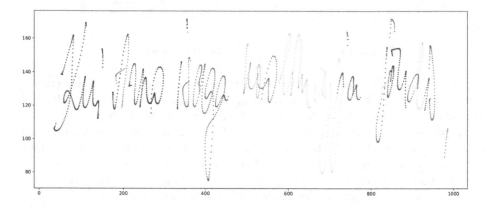

Fig. 4. From larger clusters to smaller clusters

As the final step, the number of clusters determined in the sentence should be optimized. As the features will be extracted from the clusters, the clusters those have to be consistent through the analyzed sentences and the number of clusters determined in handwriting samples has to be constant. This target number of clusters in the full sentence is 48 and it is determined by the number of letters in the test sentence. Cases, when the actual number of clusters is less than or larger than the target, have to be considered next. When the number of determined clusters is larger than it should be, the largest clusters get separated. In the opposite case, the smallest clusters get merged to the preceding clusters.

4   Statistical Analysis

4.1   Feature Engineering

In this subsection, the exhaustive list of features calculated for each sentence is described. Certain features are constructed for the sentence as a whole whereas the majority of features are derived from the individual letters. It should be clear that the features calculated in the paper were already used in the literature to analyze one or another aspect of the handwriting. The novelty of the paper is in the feature extraction from small clusters approximating letters extracted by the algorithm developed in the previous section. The overall feature set may be divided into three subsets. The first subset consists of features that are commonly used to conduct the test in its classic form by means of the paper and pencil. This set describes the ability of the tested individual to keep the same size of the letters during writing. For example the angle between the lines bounding the text from below and above. The angle between the line which bounds written sentence from below and a horizontal edge of the screen describes the ability of the patient to write in a straight line. Together with the measure of time these features constitute the first subset.

As the sentence may be consisting of two or three lines these calculations were performed for the first two lines only. The regression lines are useful for calculating the angles that are formed by the imaginary x-axis and the regression lines. Figure 5 visualizes the estimated regressions for a tilted sentence. In total six values of angles in degrees are included in the set of features. These features are related to micrographia and their meaning is well known to the practitioners.

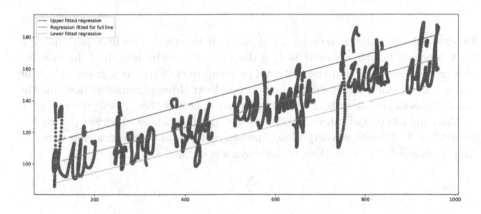

Fig. 5. Regression fitted for a line written in tilted way

The second subset is proposed by [3] and it includes different average parameters describing fine motor motions of the writing and drawing process. Finally, the third set of features constitute integral like parameters which accumulate absolute values of the velocities, accelerations, and jerks along the tangent vector to the writing trajectory. Example of tangent vectors for a letter is shown in Fig. 6. These features are complemented by the ratios allowing to relate their values to the particular drawing or writing task. This feature subset referred as *Motion mass* parameters was initially proposed by [10] to model the changes in gross motor motions, later in [8] and [7] the set was adopted and extended for the case of fine motor movements. For the sake of self-sufficiency we formally define *motion mass* parameters and explain their meaning. Motion mass parameters are usually associated with a certain part of the test. This part of the test may be defined either by time interval or may be a meaningful part of the test,

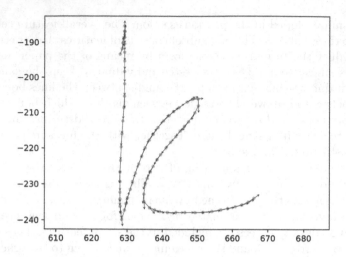

Fig. 6. An example of tangent vectors for the letter k

for example drawing or writing an element of the test. The first parameter is the trajectory length, denoted as L, it describes the entire length of the drawing observed during a certain time interval or being part of the test. It describes the amount of motion performed by the pen tip. Next three parameters describe the smoothness of the motion. The following parameter is the *velocity mass*. Let T be the time interval of interest and t are the time instances observed during the interval of T. Denote velocity along the tangent vector to the drawing curve at time instance $t \in T$ as v_t then *velocity mass* is defined as:

$$V = \sum_{t \in T} |v_t| \tag{1}$$

In the same manner *acceleration-* and *jerk-* (change of acceleration) masses are defined

$$A = \sum_{t \in T} |a_t|; \quad J = \sum_{t \in T} |j_t| \tag{2}$$

where a_t is the acceleration at time t and j_t is the jerk mass along the tangent vector.

A motion may be not-smooth in different ways this justifies the necessity to have various parameters describing smoothness of the motion. The original definition of the *motion mass* parameters also included ratios of the trajectory length to the Euclidean distance between the first and last points of the motions and ratio of the acceleration mass to the same distance. These measures are less relevant to the current work and therefore omitted. Instead of this, the logic of motion mass is applied to the pressure that is observed when the pen tip of the stylus touches the screen. The following list summarizes the features that are calculated for individual letters.

- Size features: trajectory is the sum of Euclidean distance between all points that comprise the letter; slope mass is the sum of $\Delta y/\Delta x$, slope mean, mean of first difference of slopes;
- Kinematic features: velocity mass is the sum of velocities for the letter, acceleration mass is the sum of accelerations for the letter, pressure mass is the sum of pressure values applied to the pen, velocity mean, acceleration mean, pressure mean, pressure extrema, velocity extrema, acceleration extrema;
- Duration features: duration is the time interval between the first and last point of the letter
- Fluency features: jerk mass is sum of jerk value for the letter, jerk mean, jerk extrema.

This leads to a total of 48 letters × 17 letter features + 6 sentence based angles of writing = 822 features that are derived from the sentence and letters. Obviously, for the majority of classical machine learning techniques, a such number of features will inevitably cause the curse of dimensionality. Also interpreting decisions made on the basis of a large number of features may be difficult. Which leads the necessity of a proper feature selection.

4.2   Feature Selection

Feature selection is performed by Fisher scoring ([1]). Features with the score above a threshold of 0.5 are included in predictive models. Table 1 shows the Fisher score for the sentence-specific features. It becomes clear that none of those are included in the predictive models as the threshold is not exceeded. It is worth noting that [7] report the angles of drawing to be of very limited importance with a much lower Fisher score than the one in Table 1.

Table 1. Fisher's score of the sentence specific features

Line	Feature	Fisher's scr.
1	∠ of upper reg. line	0.195
	∠ of middle reg. line	0.116
	∠ of lower reg. line	0.284
2	∠ of upper reg. line	0.205
	∠ of middle reg. line	0.100
	∠ of lower reg. line	0.195

Table 2. Ranking of letter specific features based on Fisher score

Feature	Nr. above threshold
Velocity mean	42
Acceleration mean	37
Duration	4
Jerk mass	4
Jerk mean	3
Velocity Extrema	2

Table 2 shows the number of individual letters based features to exceed the threshold of the Fisher score. The features that exceed the threshold most frequently are kinematic features associated with the first moment of velocity and acceleration. [3,13,18] and [7] report kinematic features to be important with velocity and acceleration being the most important features in the last two papers. We find that duration and fluency features follow kinematic ones with a substantially lower number of occurrences. In contrast to [3, 11] and [14] we do not find pressure related measures to be informative which is consistent with [7].

4.3   Classification Analysis

The classification model in a general form may be written as

$$y_i = f(X_i) + u_i \tag{3}$$

where y_i is the binary variable indicating whether the person has PD or not, X_i is a vector of features, $f(\cdot)$ is a (nonlinear) function to be estimated and u_i is the residual. The most parsimonious model to estimate $f(\cdot)$ may be k-nearest neighbors or logistic regression. Further decision trees, random forests and support vector machines may be used for classification. More complicated models such as adaptive boosting may be useful as well. See [1] and [4] for an overview of models and estimation techniques.

In the present paper the following classifiers are trained to obtain the one that performs best for the task in hand:

- Logistic regression
- K-nearest neighbors classifier (KNN)
- Decision tree
- Random forest classifier
- Support vector machine with linear basis kernel SVM_L and with radial basis kernel denoted by SVM_{RBF}
- Adaptive boosting

In the present paper we use training for similarities in features. While training for differences in features may show a better result in some circumstances, the task under investigation excludes such an approach. The subjects are asked to write the same sentence and there are no orthographic mistakes in the analyzed sentences.

For validation purposes, the K-fold cross-validation technique is used. The original dataset is divided into mutually exclusive K subsets that are called folds. Each fold is then used as a validation set with other $K - 1$ folds being the training sets. The metric to discriminate between the models is the prediction accuracy obtained on the training set. The accuracy of each model is the mean accuracy of the model over all folds. Given that the data is scarce and there are just 19 observations the validation is performed for $K = 3$.

The accuracy of models is shown in Table 3. The model with the highest accuracy is the random forest consisting of trees with a maximum depth of 5. In the related literature [7] use a similar set of models. Also in that study random forest is found to be the best model in terms of accuracy. To some extent, it is not surprising that the ensemble model outperforms the individual models as it averages out the variance of the prediction. [3] limit their attention to KNN, adaptive boosting and SVM and report SVM to outperform the other models.

Table 3. Accuracy of models obtained by K-fold validation with $K = 3$

Model	Accuracy
Random forest	0.885
Logit	0.838
SVM_L	0.771
Decision tree	0.714
KNN	0.742
Adaptive boosting	0.695
SVM_{RBF}	0.580

Even though the dataset may seem small, we believe that the results constitute an important building block the area of research. While the model that performs best may change with the addition of new data, the results of feature engineering remain intact.

5   Conclusions

In the paper, we propose a novel approach to model sentence writing test that is used for Parkinson's disease diagnostics. We analyze digitally conducted tests that were done by PD patients and healthy controls. The objective is to come up with a set of features that have high predictive power and may be interpreted and find a classifier that maximizes prediction accuracy. The novelty of the paper is in producing features on the basis of letters that are extracted from the full sentence. This part is quite sophisticated as it requires several steps. First, the number of lines that the sentence is written on is determined. Bigger chunks of letters are extracted next on the basis of the time that had passed during writing the letters. Those sets of letters are separated into smaller parts taking into account distance between points. Finally, the number of extracted parts is optimized to meet the target number of letters. It is worth noting that the approach is language agnostic and is based on the properties of the data that are recorded while writing. Features that represent size, kinematics, duration and fluency of writing are calculated for each individual letter. The feature selection is performed based on Fisher score and the kinematic features are found to be of primary importance. This is consistent with the general paradigm shift discussed in [16] and findings in [7]. Interestingly the fact that a person has written the sentence in a tilted way does not seem to be important in PD prediction. With the set of features that have high predictive power in hand, we perform a horse race of seven classifiers. We find the random forest to perform the best with the accuracy of 88.5% on K-fold cross-validation.

References

1. Aggarwal, C.C.: Data Mining. Springer, Cham (2015). https://doi.org/10.1007/978-3-319-14142-8
2. Al-Dmour, A., Fraij, F.: Segmenting arabic handwritten documents into text lines and words. Int. J. Adv. Comput. Technol. **6**(3), 109–119 (2004)
3. Drotar, P., Mekyska, J., Rektorova, I., Masarova, L., Smékal, Z., Faundez-Zanuy, M.: Evaluation of handwriting kinematics and pressure for differential diagnosis of parkinson's disease. Artif. Intell. Med. **67**, 39–46 (2016). https://doi.org/10.1016/j.artmed.2016.01.004
4. Hastie, T., Tibshirani, R., Friedman, J.: The Elements of Statistical Learning. SSS. Springer, New York (2009). https://doi.org/10.1007/978-0-387-84858-7
5. Letanneux, A., Danna, J., Velay, J.L., Viallet, F., Pinto, S.: From micrographia to Parkinson's disease dysgraphia. Mov. Disord. **29**(12), 1467–1475 (2014). https://doi.org/10.1002/mds.25990
6. Moustafa, A.A., Chakravarthy, S., Phillips, J.R., Gupta, A., Kcri, S., Polner, B., Frank, M.J., Jahanshahi, M.: Motor symptoms in parkinson's disease: a unified framework. Neurosci. Biobehav. Rev. **68**, 727–740 (2016). https://doi.org/10.1016/j.neubiorev.2016.07.010
7. Nõmm, S., Bardõš, K., Toomela, A., Medijainen, K., Taba, P.: Detailed analysis of the luria's alternating seriestests for parkinson's disease diagnostics. In: 2018 17th IEEE International Conference on Machine Learning and Applications (ICMLA), pp. 1347–1352, December 2018. https://doi.org/10.1109/ICMLA.2018.00219
8. Nõmm, S., Toomela, A., Kozhenkina, J., Toomsoo, T.: Quantitative analysis in the digital luria's alternating series tests. In: 2016 14th International Conference on Control, Automation, Robotics and Vision (ICARCV), pp. 1–6, November 2016. https://doi.org/10.1109/ICARCV.2016.7838746
9. Nackaerts, E., et al.: Validity and reliability of a new tool to evaluate handwriting difficulties in Parkinson's disease. Plos One **12**(3), 1–14 (2017). https://doi.org/10.1371/journal.pone.0173157
10. Nõmm, S., Toomela, A.: An alternative approach to measure quantity and smoothness of the human limb motions. Est. J. Eng. **19**(4), 298–308 (2013)
11. Rosenblum, S., Samuel, M., Zlotnik, S., Erikh, I., Schlesinger, I.: Handwriting as an objective tool for parkinson's disease diagnosis. J. Neurol. **260**, 2357–2361 (2013). https://doi.org/10.1007/s00415-013-6996-x
12. Seni, G., Cohen, E.: External word segmentation of off-line handwritten text lines. Pattern Recogn. **27**(1), 41–52 (1994). https://doi.org/10.1016/0031-3203(94)90016-7
13. Smits, E., et al.: Standardized handwriting to assess bradykinesia, micrographia and tremor in parkinson's disease. PLoS ONE **9** (2014). https://doi.org/10.1371/journal.pone.0097614
14. Stepień, P., Kawa, J., Wieczorek, D., Dabrowska, M., Sławek, J., Sitek, E.J.: Computer aided feature extraction in the paper version of luria's alternating series test in progressive supranuclear palsy. In: Pietka, E., Badura, P., Kawa, J., Wieclawek, W. (eds.) ITIB 2018. AISC, vol. 762, pp. 561–570. Springer, Cham (2019). https://doi.org/10.1007/978-3-319-91211-0_49
15. Tan, J., Lai, J.H., Wang, C.D., Wang, W.X., Zuo, X.X.: A new handwritten character segmentation method based on nonlinear clustering. Neurocomputing **89**, 213–219 (2012). https://doi.org/10.1016/j.neucom.2012.02.026

16. Thomas, M., Lenka, A., Kumar Pal, P.: Handwriting analysis in Parkinson's disease: current status and future directions. Mov. Disord. Clin. Pract. 4(6), 806–818 (2017). https://doi.org/10.1002/mdc3.12552
17. Van Gemmert, A., Hans-Leo, T., George, S.: Parkinsonian patients reduce their stroke size with increased processing demands. Brain Cogn. 47(3), 504–512 (2001). https://doi.org/10.1006/brcg.2001.1328
18. Lange, K.W., et al.: Brain dopamine and kinematics of graphomotor functions. Human Mov. Sci. 25, 492–509 (2006). https://doi.org/10.1016/j.humov.2006.05.006
19. Shukla, A.W., Ounpraseuth, S., Okun, M., Gray, V., Schwankhaus, J.: Micrographia and related deficits in parkinson's disease: a cross-sectional study. BMJ Open 2(3), e000628 (2012). https://doi.org/10.1136/bmjopen-2011-000628

Confidence in Prediction: An Approach for Dynamic Weighted Ensemble

Duc Thuan Do[1], Tien Thanh Nguyen[2(✉)] , The Trung Nguyen[3],
Anh Vu Luong[4], Alan Wee-Chung Liew[4], and John McCall[2]

[1] School of Applied Mathematics and Informatics,
Hanoi University of Science and Technology, Hanoi, Vietnam
[2] School of Computing Science and Digital Media,
Robert Gordon University, Aberdeen, UK
t.nguyen11@rgu.ac.uk
[3] School of Information and Communication Technology,
Hanoi University of Science and Technology, Hanoi, Vietnam
[4] School of Information and Communication Technology,
Griffith University, Gold Coast, Australia

Abstract. Combining classifiers in an ensemble is beneficial in achieving better prediction than using a single classifier. Furthermore, each classifier can be associated with a weight in the aggregation to boost the performance of the ensemble system. In this work, we propose a novel dynamic weighted ensemble method. Based on the observation that each classifier provides a different level of confidence in its prediction, we propose to encode the level of confidence of a classifier by associating with each classifier a credibility threshold, computed from the entire training set by minimizing the entropy loss function with the mini-batch gradient descent method. On each test sample, we measure the confidence of each classifier's output and then compare it to the credibility threshold to determine whether a classifier should be attended in the aggregation. If the condition is satisfied, the confidence level and credibility threshold are used to compute the weight of contribution of the classifier in the aggregation. By this way, we are not only considering the presence but also the contribution of each classifier based on the confidence in its prediction on each test sample. The experiments conducted on a number of datasets show that the proposed method is better than some benchmark algorithms including a non-weighted ensemble method, two dynamic ensemble selection methods, and two Boosting methods.

Keywords: Supervised learning · Classification · Ensemble method · Ensemble learning · Multiple classifier system · Weighted ensemble

1 Introduction

In recent years, learning with an ensemble of classifiers (EoC) has enjoyed increased attention in the machine learning community due to its advantage in achieving better prediction than using a single classifier [11]. In ensemble

© Springer Nature Switzerland AG 2020
N. T. Nguyen et al. (Eds.): ACIIDS 2020, LNAI 12033, pp. 358–370, 2020.
https://doi.org/10.1007/978-3-030-41964-6_31

method, the diverse classifiers are obtained by learning different algorithms on a training set (heterogeneous ensemble) or learning one algorithm on many different training sets (homogeneous ensemble) [12]. Each learning algorithm learns a classifier with the aim of describing the relationship between the features and the class label of the training observations. The generated classifier returns the output in the form of crisp labels (0–1 class memberships) or posterior probabilities (fuzzy class memberships) [8]. A combiner is then used to aggregate the outputs of all classifiers to obtain the final decision.

In the combining method, simple averaging can be conducted on the classifiers' output by assigning equal weights for all individual classifiers. In ensemble systems where individual classifiers exhibit nonidentical strength, unequal weights in the aggregation may achieve a better performance than simple averaging [11]. In this work, we focus on weighted ensemble in which the prediction of each classifier is associated with a weight when combining for final decision. In fact, weighted ensemble is a special case of ensemble pruning (which is also known as selective ensemble or ensemble selection) where the weights on some classifiers are set to zero. In ensemble pruning, the EoC can be obtained via static or dynamic approach. In detail, the static approach learns one optimal EoC on the training data and uses it to assign a label for all test samples. This, therefore, limits the flexibility of the selection procedure. Meanwhile, the dynamic approach selects a classifier or an EoC with the most competencies in a defined region associated with each test sample. Although this approach provides more flexibility than the static approach, the performance of the dynamic approach is dependent on the performance of the techniques that define the region of competence (RoC) [3]. A new weighted ensemble method which benefits from the advantage of both static and dynamic selective method i.e. learning the optimal condition on the training data to select classifiers and then determining particular weights for each test sample would be beneficial.

Our idea for weighted ensemble is based on the observation that classifiers in an ensemble are generated from different methodologies, and therefore have different confidence level in their predictions. On a particular dataset, some classifiers can provide very high confidence in the classification while others can have difficulty in decision when assigning a label for a test sample based on their outputs. We come up with an idea to encode the level of confidence of a classifier by associating with each classifier a credibility threshold, computed from the entire training set by minimizing the entropy loss function with the gradient descent method. We also measure the confidence level of each classifier's prediction on each test sample i.e. how confident it makes the decision. The confidence in the prediction is then compared to the credibility threshold to determine whether the output of the classifier should be included in the aggregation. In a procedure to assign a label for a test sample, when a classifier is attended, its confidence level and the credibility threshold will be used to compute the weight to show its contribution to the aggregation. By this way, the proposed method integrates both static and dynamic approach of ensemble selection through finding the credibility threshold like in the static methods and assigning particular weights for classifiers on each test sample like in dynamic methods.

The contribution of this work are: (i) We propose a measure to qualify the confidence in the output of each classifier. (ii) We propose a dynamic weighted ensemble method to select classifiers based on the confidence of its prediction. (iii) We formulate the optimization problem on the convex entropy loss function to search for the credibility threshold (iv) Experiments on a number of datasets demonstrate that the proposed method is better than several well-known benchmark algorithms.

2   Background and Related Work

In ensemble system, the outputs of classifiers are combined to obtain the final discriminative decision. Traditionally, simple combining methods like Sum and Vote are frequently applied to the outputs of base classifiers to predict class labels [13]. In fact, the simple combining method is the special case of the weighted combining method where the classifiers are treated equally in the aggregation, i.e. all classifiers make the equal contributions in the final collaborated decision. In weighted combining methods, each classifier can put different weight on the prediction result and the combining algorithm works by taking M weighted linear combinations of posterior probabilities for the M classes. Several approaches have been proposed to find the weights. In [15], Ting et al. proposed MLR method which depends on solving M Linear Regression models corresponding to the M classes based on meta-data and the training data labels in crisp form to find these combining weights. Yijing et al. [18] proposed the new weighted combining rules in which the weight of each classifier is computed based on its performance on the training data measured by Area under the ROC Curve (AUC). Wu [17] proposed a new ensemble learning paradigm that takes into account information about the performance ordering of the base classifiers reported in previous literature. By measuring the similarity between two learning tasks, the performance ranking of the trained classifiers of a given learning task can be inferred so as to obtain the optimal combining weights of the trained classifiers. Nguyen et al. [10] weighed the base classifiers generated on projected data of training observations by the linear regression model.

Boosting is also a family of weighed ensemble methods. The idea of this approach is to learn weak classifiers with respect to a distribution to form a strong classifier. When weak classifiers are combined, they will have weights which usually are related to the weak classifiers' accuracy. Some well-known examples of the boosting approach are AdaBoost [5] where the weak classifier is tweaked to handle previously misclassified samples, LPBoost [4] where the margin between training samples of different classes is maximized via linear programming, and RUSBoost [14] where imbalanced datasets are handled by learning from skewed training data.

Ensemble pruning is a special case of weighted ensemble in which the weights of some classifiers are set to zero. The purpose of ensemble pruning is to search for a suitable subset of classifiers that is better than using the whole ensemble. In this technique, a single classifier or an EoC can be obtained via static or dynamic approach. The static approach selects only one subset of classifiers during the training phase and uses it to predict for all unseen samples. In the past years, many statistic ensemble selection methods have been proposed to search for the optimal or sub-optimal subset of ensembles, and they can be grouped into three categories: ordering-based methods [9], clustering-based methods [1], and optimization-based methods via mathematical programming [19], probabilistic pruning [2], or heuristic search [6]. Zhang et al. [19] formulated the ensemble pruning problem as a quadratic integer programming problem and used semi-definite programming to acquire an approximate solution. Although this method outperforms the other heuristics in the author's evaluation, fixing the number of selected base classifiers is a hindrance to efficient performance. Chen et al. [2] propose a probabilistic pruning method which includes a sparsity-inducing prior distribution introduced over the combination weights. The maximum a posteriori estimation of the weights is then acquired by the Expectation Propagation algorithm.

On the other hand, in the dynamic approach, a classifier or an EoC is selected to classify each test sample based on the competence level of the classifiers computed according to some criteria on a local region of the feature space [3]. Here the region to compute competence can be defined by kNN methods [3,7] and potential functions [16]. Comparison experiments indicated that a simple dynamic selection method like KNORA Union can be competitive or sometimes outperforms more complex methods.

3   Proposed Method

3.1   Problem Formulation

Given the training set $\mathcal{D}$ with N data points and K learning algorithms $\mathcal{K} = [\mathcal{K}_k]$. The base classifier h_k is generated by training $\mathcal{K}_k$ on $\mathcal{D}$. Denote $\mathbf{P} = [p_{k,j}]$, $p_{k,j} = P(y_j = 1|\mathbf{x})$ as the prediction of h_k for a sample $\mathbf{x}$ to class label $y_j = 1$. For example, prediction vector $(0.3, 0.6, 0.1)$ of a classifier for a sample in a 3-class classification problem means that probabilities this sample belongs to class y_1, y_2, and y_3 are 0.3, 0.6, and 0.1, respectively. To measure the confidence on the prediction of h_k on $\mathbf{x}$, we define $\mathbf{e} = [e_k]$ as the difference between the maximum value among the predictions and the average of the other values. In fact, the class label is assigned based on the maximum value of the posterior probability. By defining e_k, we aim to measure the convincing decision in this decision strategy. The higher e_k results in the bigger gap between the maximum value of the posterior probability and the average of the others, making it the more convincing decision.

$$\begin{cases} e_k = p_{k,s} - \frac{1}{M-1}\sum_{j=1, j\neq s}^{M} p_{k,j} \\ s = \operatorname{argmax}_{j=1,\dots,M} p_{k,j} \end{cases} \tag{1}$$

Proposition 1: e_k is bounded in $[0,1]$

As mentioned above, we measure the confidence level in the prediction of each classifier and then compare with the credibility threshold associated with this classifier. If the confidence level is higher than the threshold, the classifier will be attended to the aggregation. By this way, we define a '*relu-like*' function a_k for activation calculation concerning the confidence threshold $\beta = [\beta_k]$:

$$a_k = \max(0, e_k - \beta_k) \quad k = 1, \ldots, K \tag{2}$$

The combining vector $\boldsymbol{pc} = [pc_j]$ from all K classification probability vectors for the class y_j $(j = 1, \ldots, M)$ is:

$$pc_j = \sum_{k=1}^{K} a_k p_{k,j} \tag{3}$$

Clearly, when $e_k > \beta_k$, the function $a_k > 0$ that means classifier h_k is activated in the combination in (3) and its contribution to the combination is the value of the activated function a_k. The proposed method is a dynamic weighted ensemble since the weight of each classifier in combining vector is different on each test sample via the different confidence in its prediction. We use softmax function to transform the combination vector to the ensemble classification probability $\boldsymbol{pe} = [pe_j]$ as:

$$pe_j = \frac{e^{pc_j}}{\sum_{j=1}^{M} e^{pc_j}} \tag{4}$$

In this work, the credibility threshold is found by minimizing the convex entropy loss function. This loss function on a data point $(\mathbf{x}, \mathbf{y})$ in which $\mathbf{y}$ is one-hot vector of class label of $\mathbf{x}$ is given by:

$$\mathcal{L}(\boldsymbol{\beta}) = -\sum_{j=1}^{M} y_j \log pe_j \tag{5}$$

3.2 Optimization

We use the gradient descent approach to solve the optimization problem for the function in (5). First, the entropy loss function is transformed to:

$$
\begin{aligned}
\mathcal{L}(\boldsymbol{\beta}) &= -\sum_{j=1}^{M} y_j \log pe_j = -\sum_{j=1}^{M} y_j \log \frac{e^{pc_j}}{\sum_{j=1}^{M} e^{pc_j}} \\
&= -\sum_{j=1}^{M} \left(y_j pc_j - y_j \log \sum_{j'=1}^{M} e^{pc_{j'}} \right) \\
&= -\sum_{j=1}^{M} y_j pc_j + \log \sum_{j'=1}^{M} e^{pc_{j'}} \left(\text{Due to} \sum_{j=1}^{M} y_j = 1 \right)
\end{aligned}
$$

We compute the gradient of the lost function at each data point $(\mathbf{x}, \mathbf{y})$. The gradient of $\mathcal{L}$ along $\boldsymbol{\beta}$ is $\frac{\partial \mathcal{L}}{\partial \boldsymbol{\beta}} = \left[\frac{\partial \mathcal{L}}{\partial \beta_k} \right]$ in which each subgradient is computed by:

$$\frac{\partial \mathcal{L}}{\partial \beta_k} = -\sum_{j=1}^{M} y_j \frac{\partial pc_j}{\partial \beta_k} + \frac{\sum_{j=1}^{M} \frac{\partial e^{pc_j}}{\partial \beta_k}}{\sum_{j=1}^{M} e^{pc_j}} \tag{6}$$

In detail:

$$\frac{\partial pc_j}{\partial \beta_k} = \frac{\partial \left(\sum_{k=1}^{K} a_k p_{k,j} \right)}{\partial \beta_k}$$

$$= \frac{\partial (a_k p_{k,j})}{\partial \beta_k} \left(\text{Due to } \frac{\partial (a_{k'} p_{k',j})}{\partial \beta_k} = 0 \ \forall k' \neq k \right)$$

$$= \frac{\partial (\max (0, e_k - \beta_k) p_{k,j})}{\partial \beta_k}$$

$$= \begin{cases} 0, & \text{if } e_k \leq \beta_k \\ -p_{k,j}, & \text{if } otherwise \end{cases} := g_{k,j} \tag{7}$$

$$\frac{\partial e^{pc_j}}{\partial \beta_k} = e^{pc_j} \frac{\partial pc_j}{\partial \beta_k} = e^{pc_j} g_{k,j} \tag{8}$$

Replace (6) by results in (7) and (8), we have:

$$\frac{\partial \mathcal{L}}{\partial \beta_k} = -\sum_{j=1}^{M} y_j g_{k,j} + \frac{\sum_{j=1}^{M} e^{pc_j} g_{k,j}}{\sum_{j=1}^{M} e^{pc_j}}$$

$$= -\sum_{j=1}^{M} y_j g_{k,j} + \sum_{j=1}^{M} pe_j g_{k,j}$$

$$= -\langle \mathbf{y}, \mathbf{g}_k \rangle + \langle \mathbf{pe}, \mathbf{g}_k \rangle = \langle \mathbf{pe} - \mathbf{y}, \mathbf{g}_k \rangle \tag{9}$$

where $\mathbf{g}_k = [g_{k,j}]$

To calculate the gradient of $\mathcal{L}$ along β for a mini-batch of n data points, we take the average of gradients of these points as:

$$\frac{\partial \mathcal{L}}{\partial \beta_k} := \frac{1}{n} \sum_{i=1}^{n} \langle \mathbf{pe}^{(i)} - \mathbf{y}^{(i)}, \mathbf{g}_k^{(i)} \rangle \tag{10}$$

Now, we update β_k according to gradient descent method with learning rate η_k at k^{th} iteration:

$$\beta_{k+1} = \beta_k - \eta_k \frac{\partial \mathcal{L}}{\partial \beta_k} \tag{11}$$

In this work, we applied the proposed weighted ensemble method to the heterogeneous ensemble systems where several different learning algorithms learn on one training set to obtain the base classifiers. As these classifiers perform differently on each dataset because of the differences in learning strategies, it is expected to obtain better results than simple aggregation [11]. First, we learn base classifiers $\mathcal{H} = [h_k]$ on $\mathcal{D}$ using the given learning algorithms. The meta-data $\mathbf{P}$ of $\mathcal{D}$ is generated via the T-fold cross validation procedure. Specifically, $\mathcal{D}$ is divided into T disjoint parts. The meta-data of observations in one part is then created by the classifiers generated by training the K learning algorithms on the complement. All meta-data sets from each part are concatenated to generate the meta-data $\mathbf{P}$.

Having $\mathbf{P}$ in hand, we compute the e_k from $\mathbf{P}$ for each $\mathbf{x} \in \mathcal{D}$ by using Eq. (1). The threshold β_k is initialized to 0. The algorithm loops via a number of single passes through the full training set (*epochs*). In each epoch, we create the random permutation for N observations and then shuffle on $\mathbf{P}$ and associated class labels $\mathbf{y}$ using this permutation. Based on the batch size n, one permutation is divided in to $\left(nb = \frac{N}{n}\right)$ mini-batch. On each mini-batch, using the value of shuffled $\mathbf{P}$ and $\mathbf{y}$, we compute the weights $\boldsymbol{a} = [a_k]$ of the base classifiers. The posterior probabilities $pb_{k,j}$ of observations in the mini-batch is combined via the weights a_k to obtain the pc_j (Eq. (3)). The gradient of the lost function is updated based on condition in Eq. (7). Each β_k finally is updated by using Eq. (11).

The classification process works in a straightforward way. Each testing sample first is predicted by the base classifiers in $\mathcal{H}$ to obtain the posterior probabilities. We then compute the value of $[e_k]$ on the prediction results and compute the weight vector $[a_k]$ by Eq. (2). The base classifier with $e_k \leq \beta_k$ will not be attended in the final ensemble for the testing sample, and therefore contribute nothing to the final aggregation. The other base classifiers will join to the final ensemble with their contribution $(e_k - \beta_k)$. The class label is assigned by returning the class label associated with the maximum of the combination vector (3).

4   Experimental Studies

4.1   Experimental Setup

Eighteen real world and synthetic datasets are used in the experiment. For the six real-world datasets (Chess-krvk, DownJones-1985–2003, Electricity, Letter, Penbased Skin_NonSkin) we collected a number of datasets from the UCI[1] and OPENML[2] data sources. For the synthetic datasets, we used MOA library[3] to generate the data. The detailed information of the datasets is summarized in Table 1.

We applied the proposed method to a heterogeneous ensemble system, generated by using 3 learning algorithms named Linear Discriminant Analysis (denoted by LDA), Naïve Bayes, and kNearest Neighbors (where the value of k was set to 5, denoted as kNN5). The proposed method in this case was denoted by Proposed Method3. For the mini-batch approach in the mini-batch gradient descent method, we initialized learning rate $\eta = 0.001$, the credibility threshold $\boldsymbol{\beta} = [\beta_0] = \mathbf{0}$.

We performed extensive comparative studies using a number of existing algorithms as benchmarks: two homogeneous ensemble method named AdaBoost [5] and RUSBoost [14] with 100 classifiers. We also compared with Sum Rule [13] in a heterogeneous ensemble where the set of learning algorithms is similar to our method. For the ensemble pruning methods, we selected two high-performance

---

[1] http://archive.ics.uci.edu/ml/datasets.html.
[2] https://www.openml.org.
[3] https://moa.cms.waikato.ac.nz.

Table 1. Information of experimental datasets

#	Datasets	# of samples	# of dimensions	# of class labels
1	Agrawal	1000000	9	2
2	AssetNegotiation-F2	1000000	5	2
3	AssetNegotiation-F3	1000000	5	2
4	AssetNegotiation-F4	1000000	5	2
5	BNG-bridge-v1	1000000	12	6
6	BNG_zoo	1000000	17	7
7	Chess-krvk	28056	6	18
8	DowJones-1985–2003	138166	8	30
9	Electricity-normalized	45312	8	2
10	Hyperplane	1000000	10	2
11	Letter	20000	16	26
12	Penbased	10992	16	10
13	RandomTree	1000000	10	2
14	RBF	1000000	50	4
15	Sine	1000000	4	2
16	Skin_NonSkin	245057	3	2
17	Stagger	1000000	3	2
18	Waveform	1000000	21	3

dynamic ensemble selection methods, namely KNORA Union and KNORA Eliminate (denoted by KNORA-U and KNORA-E) [7] as the benchmark algorithms. The number of nearest neighbors in these dynamic methods was set to 7 [3]. For all methods, we performed the same experimental procedure i.e. run 10-fold cross validation 3 times to obtain 30 test results for each dataset (presented in Table 2). The non-parametric two-tailed Wilcoxon signed-rank test [10] was used to compare the experimental results of the proposed method and a benchmark algorithm on a particular dataset in which p-value < 0.05 deems as difference in experimental results is significant.

4.2   The Influence of Parameters

In this study, we used the mini-batch approach to compute the gradient in updating the credibility threshold through iterations. We examined the influence of the batch size and number of epochs on the performance of the proposed method. Figure 1 presents the relationship between classification error rate and values of batch size $n \in \{1, 16, 32, 64, 128, 256, 521, 1024, 2048, 4096, 8192, All\}$ where *All* means all the number of training data were used as a single batch. Clearly, the line graphs show a common upward trend with different slopes on all datasets as increasing value of batch size can downgrade performance of proposed method. On some datasets like Hyperplane, RBF, Letter and Penbased, the classification error rate increases sharply with the increase of n. Meanwhile,

Table 2. Classification error rates of benchmark algorithms and proposed method (using 3 learning algorithms)

#	KNORA-U		KNORA-E		AdaBoost		Sum Rule		RUSBoost		Proposed Method3	
	Mean	Variance	Mean	Variance	Mean	Variance	Mean	Variance	Mean	Variance	Mean	Variance
1	0.3293▲(4.5)	3.65E-08	0.3440▲(6)	2.53E-07	0.0494▼(1)	3.36E-07	0.3293▲(4.5)	3.04E-08	0.0545▼(2)	3.75E-07	0.3280 (3)	2.50E-11
2	0.0717▲(5)	9.01E-07	0.0558▲(4)	5.55E-07	0.0511▼(1)	4.08E-07	0.1042▲(6)	7.53E-07	0.0553▲(3)	4.02E-07	0.0519 (2)	3.43E-07
3	0.0656▲(5)	7.19E-07	0.0560▲(3)	6.26E-07	0.0531▲(2)	1.01E-06	0.0905▲(6)	7.78E-07	0.0589▲(4)	5.16E-07	0.0525 (1)	3.48E-07
4	0.0740▲(4)	4.34E-07	0.0608▲(3)	2.99E-06	0.0529▼(1)	2.94E-07	0.1139▲(6)	7.08E-07	0.0838▲(5)	5.17E-07	0.0542 (2)	2.52E-06
5	0.3117▲(5)	1.17E-06	0.3017▲(3)	1.51E-06	0.2762▲(1)	1.42E-06	0.3100▲(4)	9.96E-07	0.3376▲(6)	1.42E-06	0.2869 (2)	1.47E-06
6	0.0695▲(4.5)	4.83E-07	0.0585▲(2)	5.55E-07	0.0603▲(3)	6.44E-07	0.0695▲(4.5)	4.34E-07	0.1155▲(6)	1.70E-06	0.0522 (1)	2.50E-07
7	0.3242▲(4)	6.93E-05	0.2955▲(2)	6.11E-05	0.6705▲(5)	3.41E-05	0.3031▲(3)	4.59E-05	0.7433▲(6)	8.09E-05	0.2650 (1)	5.77E-05
8	0.0055▲(5)	2.60E-07	7.0447E-04▲(3)	4.12E-08	2.0024E-04▲(2)	6.52E-09	0.0043▲(4)	3.15E-07	0.6376▲(6)	5.46E-09	0.0000 (1)	0.00E+00
9	0.2192▲(5)	4.93E-05	0.2024▲(3)	2.42E-05	0.1561▼(1)	3.96E-05	0.2126▲(4)	6.07E-05	0.2297▲(6)	9.51E-05	0.1893 (2)	3.22E-05
10	0.0281▲(5)	2.36E-07	0.0128▲(2)	9.55E-08	0.0279▲(4)	3.18E-07	0.0276▲(3)	2.40E-07	0.2542▲(6)	9.62E-06	0.0049 (1)	4.35E-08
11	0.1089▲(3)	5.00E-05	0.0617▲(2)	3.36E-05	0.3460▲(5)	1.15E-04	0.1395▲(4)	6.18E-05	0.7147▲(6)	2.27E-04	0.0507 (1)	3.33E-05
12	0.0435▲(4)	2.58E-05	0.0097▼(1)	1.10E-05	0.0424▲(3)	3.90E-05	0.0895▲(5)	5.46E-05	0.2966▲(6)	9.54E-05	0.0122 (2)	1.41E-05
13	0.1248▲(5)	6.18E-07	0.1179▲(3)	5.43E-07	0.0793▼(1)	1.16E-05	0.1236▲(4)	9.39E-07	0.2529▲(6)	8.36E-06	0.1053 (2)	5.90E-07
14	0.0037▲(3)	4.52E-08	8.7833E-04▲(2)	8.41E-09	0.0475▲(5)	3.01E-02	0.0257▲(4)	1.85E-07	0.1682▲(6)	2.08E-06	2.6667E-06 (1)	1.96E-11
15	0.0119▲(5)	1.40E-07	0.0099▲(3)	6.70E-08	0.0026▼(1)	3.72E-08	0.0115▲(4)	1.05E-07	0.0423▲(6)	7.39E-07	0.0088 (2)	9.35E-08
16	9.0047E-04▲(4)	4.54E-08	5.1145E-04▲(2)	1.64E-08	6.2979E-04▲(3)	2.73E-08	0.0412▲(6)	1.22E-06	0.0265▲(5)	2.05E-06	4.7608E-04 (1)	1.74E-08
17	0.0000 (2.5)	0.00E+00	0.0000 (2.5)	0.00E+00	0.1116▲(5.5)	9.00E-12	0.0000 (2.5)	0.00E+00	0.1116▲(5.5)	9.00E-12	0.0000 (2.5)	0.00E+00
18	0.1591▲(3)	1.06E-06	0.1518▲(2)	1.67E-06	0.1610▲(5)	1.08E-06	0.1594▲(4)	1.18E-06	0.2622▲(6)	1.41E-05	0.1364 (1)	1.38E-06
	Win: 17; Equal:1; Lost:0		Win: 16; Equal: 1; Lost: 1		Win: 11; Equal:0; Lost: 7		Win: 17; Equal:1; Lost: 0		Win:17; Equal:0; Lost:1			
Rk	4.25		2.69		2.75		4.36		5.36		1.58	

* ▲ and ▼ indicate that proposed method is better or worse than benchmark algorithm; (.) indicates the rank of method on the dataset

Rk indicates average ranking of each method

on other datasets such as Agrawal, Electricity-normalized, and BNG-bridge-v1, classification error rates only increase slightly with the increase of the number of batch size. It is also noted that the cost of training is more expensive with smaller value of n. In practice, depending on the training resource and expected performance score, we can choose a suitable value for the batch size parameter. In the next section, we used the batch size $n = 16$ in comparison to the baselines.

Figure 2 presents the classification error rates of the proposed method on experimental datasets where *epochs* parameter was set to 5 and 50. In general, although increasing the number of epochs can improve the ensemble performance, differences in two performance scores on these datasets are not significant. One datasets like Argawal and AssetNegotiation-F2, F3 and F4, the classification error rates only change slightly or remain unchanged with the change of the number of epochs. Only on three datasets Letter, Penbased, and Hyperplane, the differences in classification error rate between two cases are remarkable. In practice, in case of limited resource available, we can choose a small number of epochs for the training process.

4.3   Comparing to the Baselines

Table 2 presents the classification error rates of benchmark algorithms and proposed method in case of using 3 learning algorithms. The following observations can be made:

- Proposed Method$_3$ achieves lowest average rank among all methods (rank value 1.58). On 18 experimental datasets, Proposed Method$_3$ ranks first in 9 cases (50%) and ranks second in 7 cases (38.89%). Our method only performs poorly on Agrawal dataset in which it ranks third.
- Proposed Method$_3$ is better than two DES methods. Comparing to KNORA-U, our method wins in 17 cases and does not lose on any case. Proposed Method$_3$ underperforms KNORA-E on only Penbased datasets (0.0097 vs. 0.0122) while wins this method on 16 datasets.
- Proposed Method$_3$ is significantly better than Sum Rule on 17 datasets. One Stagger dataset where two methods performs equally, both obtain 100% of classification accuracy.
- The performance of Proposed Method$_3$ is better than RUSBoost in 17 cases. Although RUSBoost is special designed for imbalanced data, it significantly underperforms Proposed Method$_3$ on some imbalanced datasets in our experiments such as Chess-krvk and Skin_NonSkin.
- AdaBoost is a high performance ensemble in our experiment in which our method only wins in 11 cases and loses in 7 cases. However, Proposed Method$_3$ is only significantly worse than AdaBoost on Agrawal datasets while our method significantly outperforms on at least 6 datasets Chess-krvk, Hyperplance, Letter, RBF, Stagger, and DowJones-1958–2003.
- The variances of classification error rate of experimental methods on some datasets, especially on synthetic ones such as RBF and Sine, are very small. That means the differences in the classification error rates among 30 results in the test procedure on these datasets are not significant.

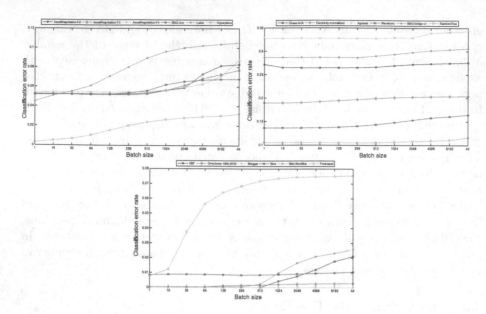

Fig. 1. The influence of number of batch size to classification error rate of proposed method

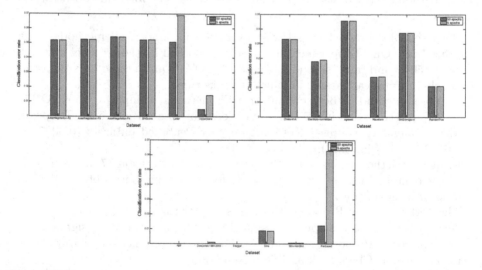

Fig. 2. The performance comparison between proposed method using 5 and 50 epochs

To summarize, Proposed Method$_3$ achieves better performance than two Boosting methods, two DES methods and one simple non-weighted combining method. Proposed Method$_3$ significantly outperforms Sum Rule in all cases which demonstrates the advantage of ensemble weighting technique compared to the simple combining methods.

5   Conclusions and Future Work

We have presented a novel weighted ensemble method for ensemble systems which considers the confidence in the prediction of each classifier. Based on the observation that each classifier provides a different level of confidence in its prediction for each sample, we propose to associate a credibility threshold with each classifier. The confidence in the prediction of each classifier on a sample is compared to the credibility threshold to determine whether the classifier's output should be included in the aggregation. To show the contribution of a classifier in the selected ensemble, we use the difference between the confidence in the prediction and the credibility threshold. This allows us to integrate both the static and dynamic approaches in the proposed method i.e. learning the credibility threshold on the training data by minimizing the entropy loss function and assigning a particular weight associated with each classifier for each test sample. The experiments on diverse data sources show the advantage of the proposed method compared to the benchmark algorithms.

In the future we plan to (1) analyse the convergence of the proposed method, (2) expand the proposed method to handle data stream with concept drift.

References

1. Bakker, B., Heskes, T.: Clustering ensembles of neural network models. Neural Netw. **16**(2), 261–269 (2003)
2. Chen, H., Tiňo, P., Yao, X.: Predictive ensemble pruning by expectation propagation. IEEE Trans. Knowl. Data Eng. **21**(7), 999–1013 (2009)
3. Dang, M.T., Luong, A.V., Vu, T.-T., Nguyen, Q.V.H., Nguyen, T.T., Stantic, B.: An ensemble system with random projection and dynamic ensemble selection. In: Nguyen, N.T., Hoang, D.H., Hong, T.-P., Pham, H., Trawiński, B. (eds.) ACIIDS 2018. LNCS (LNAI), vol. 10751, pp. 576–586. Springer, Cham (2018). https://doi.org/10.1007/978-3-319-75417-8_54
4. Demiriz, A., Bennett, K.P., Shawe-Taylor, J.: Linear programming boosting via column generation. Mach. Learn. **46**(1–3), 225–254 (2002)
5. Freund, Y., Schapire, R.E., et al.: Experiments with a new boosting algorithm. In: ICML, vol. 96, pp. 148–156. Citeseer (1996)
6. Kim, K.J., Cho, S.B.: An evolutionary algorithm approach to optimal ensemble classifiers for DNA microarray data analysis. IEEE Trans. Evol. Comput. **12**(3), 377–388 (2008)
7. Ko, A.H., Sabourin, R., Britto Jr., A.S.: From dynamic classifier selection to dynamic ensemble selection. Pattern Recogn. **41**(5), 1718–1731 (2008)
8. Kuncheva, L.I., Bezdek, J.C., Duin, R.P.: Decision templates for multiple classifier fusion: an experimental comparison. Pattern Recogn. **34**(2), 299–314 (2001)
9. Margineantu, D.D., Dietterich, T.G.: Pruning adaptive boosting. In: ICML. vol. 97, pp. 211–218. Citeseer (1997)
10. Nguyen, T.T., Dang, M.T., Liew, A.W., Bezdek, J.C.: A weighted multiple classifier framework based on random projection. Inf. Sci. **490**, 36–58 (2019)
11. Nguyen, T.T., Nguyen, M.P., Pham, X.C., Liew, A.W.C., Pedrycz, W.: Combining heterogeneous classifiers via granular prototypes. Appl. Soft Comput. **73**, 795–815 (2018)

12. Nguyen, T.T., Nguyen, T.T.T., Pham, X.C., Liew, A.W.C.: A novel combining classifier method based on variational inference. Pattern Recogn. **49**, 198–212 (2016)
13. Nguyen, T.T., Pham, X.C., Liew, A.W.C., Pedrycz, W.: Aggregation of classifiers: a justifiable information granularity approach. IEEE Trans. Cybern. **49**(6), 2168–2177 (2018)
14. Seiffert, C., Khoshgoftaar, T.M., Van Hulse, J., Napolitano, A.: Rusboost: improving classification performance when training data is skewed. In: 2008 19th International Conference on Pattern Recognition, pp. 1–4. IEEE (2008)
15. Ting, K.M., Witten, I.H.: Issues in stacked generalization. J. Artif. Intell. Res. **10**, 271–289 (1999)
16. Woloszynski, T., Kurzynski, M., Podsiadlo, P., Stachowiak, G.W.: A measure of competence based on random classification for dynamic ensemble selection. Inf. Fusion **13**(3), 207–213 (2012)
17. Wu, O.: Classifier ensemble by exploring supplementary ordering information. IEEE Trans. Knowl. Data Eng. **30**(11), 2065–2077 (2018)
18. Yijing, L., Haixiang, G., Xiao, L., Yanan, L., Jinling, L.: Adapted ensemble classification algorithm based on multiple classifier system and feature selection for classifying multi-class imbalanced data. Knowl.-Based Syst. **94**, 88–104 (2016)
19. Zhang, Y., Burer, S., Street, W.N.: Ensemble pruning via semi-definite programming. J. Mach. Learn. Res. **7**, 1315–1338 (2006)

Towards Personalized Radio-Chemotherapy – Learning from Clinical Data vs. Model Optimization

Andrzej Świerniak[1] (iD), Jarosław Śmieja[1(✉)] (iD), Krzysztof Fujarewicz[1] (iD),
and Rafał Suwiński[2] (iD)

[1] Department of Systems Biology and Engineering, Silesian University of Technology,
Akademicka 16, 44-100 Gliwice, Poland
Andrzej.Swierniak@polsl.pl
[2] M. Sklodowska-Curie Memorial Cancer Center and Institute of Oncology,
Gliwice Branch. Wybrzeze Armii Krajowej 15, 44 101 Gliwice, Poland

Abstract. We summarize results of our research studies on models of combined
anticancer radio- and chemotherapy and their comparison with real clinical data.
We use two mathematical techniques, which, to our knowledge, have not been
applied simultaneously: optimal control theory and survival analysis. We recall
results of analytical optimization of combined chemo-radio-therapy for a simple
model of tumor growth with respect to the order, in which these two modes of
treatment should be applied. Then we study both structural and parametric sensitivity of this model and related optimal control problem. Afterwards, we present
results of survival analysis based on the Kaplan-Meier curves for different protocols of chemo-radio-therapy and compare them with real clinical data and results
of optimal treatment protocols.

Keywords: Therapy optimization · Survival analysis

1 Introduction

Despite the progress in novel therapy approaches in fight against cancer, such as, for
example, immunotherapy or gene therapy, chemo- and radiotherapy are most common
in clinical practice. Moreover, as far as adjuvant therapies are concerned, applied before
or after surgery, these two are the most widely used ones. Radiation is then used as a
form of a local treatment, targeting directly the tumor, while chemotherapy supports
killing tumor cells even in distant sites. Both of them are applied in the form of standard
clinical protocols, built upon clinical experience, and hardly take into account the specific
case of an individual patient. Usually, the state of a patient is monitored and the only
treatment personalization consists in discontinuation of drug administration (cancelling
radiotherapy) when it significantly worsens.

Such observation paved the way for numerous attempts to model tumor growth
and treatment. Basically, two approaches have been used. In the first one, the authors
concentrated on arbitrarily chosen protocols and compared their efficacy (e.g. [3, 4])

© Springer Nature Switzerland AG 2020
N. T. Nguyen et al. (Eds.): ACIIDS 2020, LNAI 12033, pp. 371–379, 2020.
https://doi.org/10.1007/978-3-030-41964-6_32

with respect to survival curves [6] that are arguably the main indicator used in clinical practice. That has allowed to compare concurrent and sequential chemo-radio-therapy, but only for those protocols. Therefore, a change in a given protocol might lead to different conclusions. The second approach has been optimal-control oriented, with treatment viewed mathematically as a control variable (see, e.g. [1, 2] and references therein). Though that might lead to important findings, concerning treatment protocols, the optimization problem formulation usually did not take into account specific features of radio- and chemotherapy. Moreover, the optimization goal was not related to survival curves, which made the clinics skeptical about the results obtained.

The goal of this work is to bring the two approaches, described above, closer to each other using clinical data to enhance the quality of models and survival curves, used in clinics, to evaluate the efficacy of treatment in a heterogeneous population of patients. First, the Kaplan-Meier survival curves are analyzed, with patients stratified with respect to various features and types of treatment. We aim at finding features that could be used for better prediction of treatment outcome at the one hand, and on the other hand for creating more realistic pool of virtual patients, for whom treatment is simulated in silico. As far as the former goal is concerned, we concentrate on biomolecular markers that could be easily obtained from standard blood tests. Once that initial phase of clinical data analysis is concluded, we introduce mathematical models of tumor growth and therapy, for which control optimization problem has been stated and necessary conditions derived, indicating the form of optimal treatment protocol. We recall these results, presented in [7] and [8], and subsequently discuss their parametric (i.e., with respect to different model parameter values) and structural (i.e., with respect to different model structure, represented by the equations used to describe model dynamics) sensitivity. It should be noted that existing works, dealing with sensitivity analysis of cancer growth models, concentrate on the former only (see, e.g. [5]) and, in most cases, does not look at the implications of the choice of model structure or parameters for survival curves. Basing on the conclusions drawn at that stage, a pool of virtual patients is created, for whom treatment is simulated and survival curves calculated. That way, a computational framework is created for analysis of treatment efficacy, providing tools for improving it both from the point of view of a general patient population, and an individual patient, represented by a model with parameters based on patient's blood test results.

2   Analysis of Clinical Data

In general, the efficacy of treatment protocols is not based on individual patients' responses but on survival analysis. Kaplan-Meier survival curves are used to show probability of survival in time after diagnosis. While it is relatively easy to test statistical hypotheses about an advantage of one protocol over another, such testing requires a relatively large set of data. For example, our data on lung cancer patients allowed to state that women have overall better prognosis than men (Fig. 1), the advantage of adjuvant radiochemotherapy over chemotherapy is not that clear, due to the differences in observation time (Fig. 2). Even worse, there is no data that would allow to check if sequential treatment (e.g. following an assumption that chemotherapy sensitizes cancer cells to radiation or vice versa) or simultaneous radio- and chemotherapy would yield better prognosis.

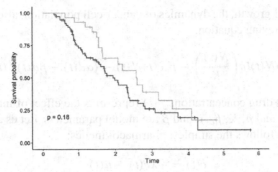

Fig. 1. Comparison of survival data for lung cancer patients for men (*red line*) and women (*blue line*) (Color figure online)

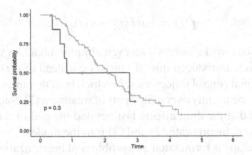

Fig. 2. Comparison of survival data for lung cancer patients that underwent radiochemotherapy (*blue line*) and chemotherapy only (*red line*) (Color figure online)

Comparing efficacy of chemo- to radiochemotherapy is additionally confounded by te fact that the effects of the latter have been observed over shorter time period so far.

Taking into account the above remarks, there is a need for additional means that would support reasoning leading to prognostic conclusions concerning types and protocols of therapy. Mathematical modeling of cancer growth under treatment coupled with application of optimization theory and in silico experiments with a cohort of virtual patients might become such tool.

3 Modeling and Optimization of Radiochemotherapy and Tumor Growth

In this work, the simplest model of tumor growth is considered, in which tumor volume is represented by the size of cancer cells, denoted by $N(t)$. Doses of chemotherapeutic agents and radiation doses are incorporated as control variables $u(t)$ and $d(t)$, respectively. We assume no synergistic or antagonistic effects of one therapy on the other. Following the log-kill hypothesis for chemotherapy [9], in which the rate of killing cancer cells is proportional to the drug concentration, as well as the standard approach to model radiotherapy effects with the linear-quadratic (LQ) term [10], modified by introduction of a term representing DNA damage repair, as well as the most often used

Gompertz model of growth, the dynamics of cancer cell population, affected by therapy is given by the following equation:

$$\dot{N}(t) = -\rho N(t) ln\left(\frac{N(t)}{K}\right) - \beta_c c(t) N(t) - (\alpha d(t) - \beta d(t) f(t)) N(t), \qquad (1)$$

where $c(t)$ denotes drug concentration, $f(t)$ represents the effect of intracellular DNA repair mechanisms and ρ, K, β_c, α and β are model parameters. Let us also assume that drug concentration follows the simplest pharmacokinetics:

$$\dot{c}(t) = -\lambda c(t) + u(t) \qquad (2)$$

where λ corresponds to the half-life of the drug. The effect of DNA repair mechanisms are assumed to follow similar dynamics, i.e.,

$$\dot{f}(t) = -\mu f(t) + d(t). \qquad (3)$$

Some of the previous works assumed an even simpler model, in which control variables $u(t)$ and $d(t)$ were introduced directly into (1), instead of $c(t)$ and $f(t)$, respectively. That led to optimal control trajectories, in which both therapies were to be applied after a delay, always concurrently over a final part of treatment [7]. Such result neglected, among other, the fact that the drug effects last beyond the end of their administration. Therefore, we decided to incorporate (2) and (3) into the model.

The goal of the therapy is formulated as a problem of maximization of tumor control probability (TCP). Assuming constant values of tumor cells density and clonogenic fraction, it can be transformed to the problem of finding measurable functions $[u(t), d(t)]$, representing chemotherapy protocols and irradiation strategy, respectively, that minimize the following performance index:

$$\min_{u(t),d(t)} J = N(T), \qquad (4)$$

where T denotes the fixed end time of treatment. The following constraints are imposed on the control variables:

$$0 \leq u(t) \leq u_{max}, 0 \leq d(t) \leq d_{max}, \qquad (5)$$

$$\int_0^T u(t)dt \leq U, \int_0^T d(t)dt \leq D \qquad (6)$$

Application of the Pontryagin's maximum principle to the control optimization problem, given by (1)–(6) leads to the bang-bang optimal control, whose features (in terms of a sequence or concurrence of control actions) may depend on the model structure [7].

Structural Sensitivity of the Model. Three possible changes in the model structure have been analyzed, in all possible combinations: neglecting pharmacokinetics (excluding (2) from the model and replacing $c(t)$ by $u(t)$ in (1)), neglecting DNA repair mechanisms (excluding (3) from the model and replacing $f(t)$ by $d(t)$ in (1)) and using other, than Gompertzian, growth terms in (1). Taking into account the form optimal control with respect to concurrence or precedence of the two therapies, the solution is not sensitive to incorporation or neglecting of the repair mechanisms. However, the optimal control is sensitive to pharmacokinetics (PK) and this sensitivity has two aspects. First, neglecting PK, regardless of model parameters, leads to the control in the form $0 - u_{max}$, with at least part of the final period of chemotherapy overlapping with radiotherapy. Including PK in the model leads to the form of the optimal control that is dependent on model parameters. Two cases can be distinguished: the optimal control has either $u_{max} - 0$ form or $0 - u_{max} - 0$ form. The latter solution might be attributed to the log-kill hypothesis, while the part $u_{max} - 0$ of the control trajectory is consistent with an observation that the chemotherapeutic effects last beyond the moment the drug dose has been switched to 0 [17].

The last structure change in the sensitivity analysis was related to the tumor growth term. In addition to the Gompertzian one, given in (1), logistic and exponential growth terms have been considered [11]. It has been found out that regardless of the form of the growth term, optimal control structure is always the same, rendering it structurally robust.

Nevertheless, even if the optimal control law has been found to be always in the switching form regardless of the model structure, the time instants, at which the control switches from zero to its maximum value (or the other way) depend on model parameters. It has been a direct conclusion from the formal control optimization results, mentioned in the first paragraph of this section. Parameters, in turn, represent individual properties of cancer in individual patients. Therefore, if these parameters are known, one could apply one of existing computational algorithms to find the best solution for an individual patient (see, e.g. [15, 16] for algorithms that might be implemented to solve various optimization problems).

In the subsequent section, the approach to predict therapy outcomes for an a priori assumed treatment protocols have been presented.

4   Simulation-Based Analysis of Treatment Efficacy

In order to check if the conclusions stemming from formal optimization using Pontryagin's maximum principle hold (with respect to concurrency or precedence of the two forms of treatment) when the treatment results are viewed from another perspective – survival curves (please note that these curves also correspond to maximizing of TCP – but in a different manner), a simulation experiment has been performed. A pool of virtual patients has been created. In each case, cancer growth was assumed to follow the dynamics described by (1)–(3), with parameters sampled for each individual patients from random distributions, as explained farther in the text. Additionally, following the reports on in vitro experiments [12], when sampling the parameter values, ρ has been correlated to α. Since optimization of radio and chemotherapy indicated best protocols in the bang-bang-form, as actually used in clinical practice, such protocols have been used in simulations, with switching times assumed to be defined by clinical standards. In this simulation experiment no protocol optimization has been performed. Instead, three strategies have been compared: sequential chemo- and radiotherapy (CRT), concurrent CRT and radiotherapy only, with respect to their efficacy defined by survival curves. As in clinical oncology, cancer has grown freely for first 14 days, after which the therapy was started (that represented the gap between diagnosis and start of the treatment).

It has been assumed that when $N(t)$ reaches a certain maximum threshold, the virtual patient dies and the time of death was subsequently used for calculation of survival curves. On the other hand, if $N(t)$ reaches a certain minimum threshold, a random number is generated and, depending on its value, either the simulation is continued for that patient or the patient is assumed to be cured.

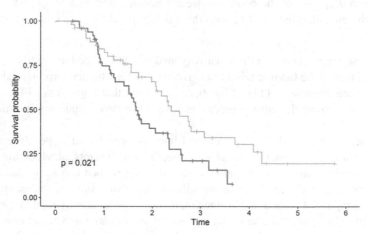

Fig. 3. Comparison of survival data for lung cancer patients with high (*blue line*) and low hemoglobin (*red line*) (Color figure online)

Parameter Sensitivity of the Solutions. First, each virtual patient was characterized by different parameter values, drawn from bivariate normal distribution formed around the values found in the literature [3]. Subsequently, local sensitivity analysis was performed, based on changing distribution parameters for each model parameters and repeating calculations for a new cohort of virtual patients. It has been found that changes growth rate ρ yield largest changes in survival curves. Since growth depends on availability of nutrients and oxygen for cancer cells, and should be negatively regulated by immune cells, survival curves obtained for real patients were checked with respect to their relation to blood morphology parameters (a sample result concerning blood hemoglobin level is shown in Fig. 3). Though initially statistical tests provided positive results, high level of correlation between different parameters of blood morphology (Fig. 4) rendered blood test results not appropriate to determine distributions for sampling model parameters. Therefore, the distribution used to sample model parameters was not related to any of them.

The survival curves obtained in the numerical experiment are shown in Fig. 5. In accordance with clinical data, the show that radiotherapy alone results in significantly worse survival. Moreover, concurrent chemo-radiotherapy should yield better results than a sequential application of chemotherapy and radiotherapy. The same conclusion was reached if a logistic model of growth was used to describe the dynamics of cancer

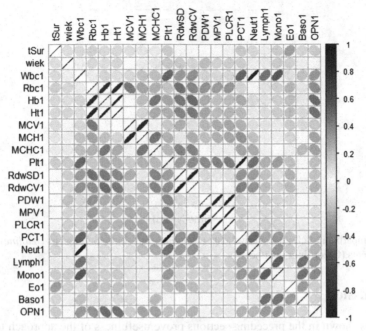

Fig. 4. Correlation map for different blood parameters

cells population, once again proving that these results are not sensitive to the model structure. This is consistent with the results of theoretical optimization, in which optimal control was simultaneously different than zero for some period of time.

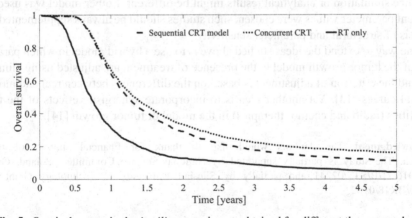

Fig. 5. Survival curves in the *in silico* experiment, obtained for different therapy modes.

With appropriate estimation of parameter values any of presented models could reproduce clinical data with a relatively high accuracy (Fig. 6).

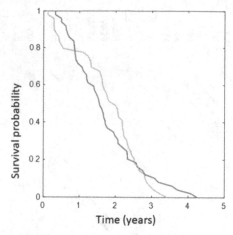

Fig. 6. Comparison of clinical survival data (*blue line*) and simulation results (*red line*) for lung cancer patients (Color figure online)

5　Discussion

The results shown in the preceding sections prove usefulness of the approach to investigate treatment protocols, based on a mathematical formalism. Formal optimization of control for a model of tumor growth, coupled with simulation with a cohort of virtual patients provides additional arguments for claiming higher efficacy of one treatment strategy against another (in the example considered in this paper, it was combined concurrent versus sequential radiochemotherapy). Such conclusion may be not possible to be drawn on the basis of clinical data only (Fig. 2).

Since simulation or analytical results might be different if other model was used or different parameter values were chosen, such studies should be always supplemented by analysis of structural and parameter sensitivity.

One way to extend the ideas studied above is to use a hybrid model in which parameters of the tumor growth model in the presence of treatment are adjusted using clinical data and the criterion of adjustment is based on the differences between real and model survival curves [13]. Yet another idea is to incorporate synergistic effects of the two modalities (radio and chemo- therapies) in the model of tumor growth [14].

Acknowledgment. The Authors would like to thank for financial support of their research. The study is partially supported by National Science Committee, Poland, Grant no. 2016/21/B/ST7/02241 and partially by Silesian University of Technology Grant no. 02/010/BK18/0102.

References

1. Świerniak, A., Kimmel, M., Smieja, J., Puszynski, K., Psiuk-Maksymowicz, K.: System Engineering Approach to Planning Anticancer Therapies. Springer, Cham (2016). https://doi.org/10.1007/978-3-319-28095-0

2. Schättler, H., Ledzewicz, U.: Optimal control for mathematical models of cancer therapies. IAM, vol. 42. Springer, New York (2015). https://doi.org/10.1007/978-1-4939-2972-6

3. Geng, C., Paganetti, H., Grassberger, C.: Prediction of treatment response for combined chemo- and radiation therapy for non-small cell lung cancer patients using a bio-mathematical model. Sci. Rep. **7**, 13542 (2017)

4. Curran, W.J., et al.: Sequential vs concurrent chemoradiation for stage III nonsmall cell lung cancer: randomized phase III trial RTOG 9410. JNCI J. Natl Cancer Inst. **103**(19), 1452–1460 (2011)

5. Dolbniak, M., Kardynska, M., Smieja, J.: Sensitivity of combined chemo-and antiangiogenic therapy results in different models describing cancer growth. Discr. Continuous Dyn. Syst. Ser. B **23**, 145–160 (2018)

6. Dudley, W.N., Wickham, R., Coombs, N.: An introduction to survival statistics: kaplan-meier analysis. J. Adv. Pract. Oncol. **7**(1), 91–100 (2016)

7. Bajgier, P., Fujarewicz, K., Swierniak, A.: Effects of pharmacokinetics and DNA repair on the structure of optimal controls in a simple model of radio-chemotherapy. In: Proceedings of the MMAR Conference, pp. 686–691 (2018)

8. Dolbniak, M., Smieja, J., Swierniak, A.: Structural sensitivity of control models arising in combined chemo-radiotherapy. In: Proceedings of the MMAR Conference, pp. 339–344 (2018)

9. Skipper, H.E., Schabel, F., Wilcox, W.: Experimental evaluation of potential anticancer agents. XIII. on the criteria and kinetics associated with curability of experimental leukemia. Cancer Chemother. Rep. **35**, 1–111 (1964)

10. Fowler, J.F.: The linear-quadratic formula and progress in fractionated radiotherapy. Br. J. Radiol. **62**, 679–694 (1989)

11. Gerlee, P.: The model muddle in search of tumor growth laws. Cancer Res. **73**(8), 2407–2411 (2013)

12. Lee, J.Y., Kim, M.-S., Kim, E.H., Chung, N., Jeong, Y.K.: Retrospective growth kinetics and radiosensitivity analysis of various human xenograft models. Lab. Anim. Res. **32**(4), 187–193 (2016). https://doi.org/10.5625/lar.2016.32.4.187

13. Wolkowicz, S., et al.: Prediction of lung cancer patients' response to combined chemo-radiotherapy using a personalized hybrid model. Mathematica Applicanda **47**(2), 219–229 (2019)

14. Bajger, P., Fujarewicz, K., Swierniak, A.: Optimal control in a model of chemotherapy-induced radiosensilization. Mathematica Applicanda **47**(1), 81–91 (2019)

15. Radu-Emil, P., Radu-Codrut, D.: Nature-Inspired Optimization Algorithms for Fuzzy Controlled Servo Systems. Butterworth-Heinemann, Oxford (2019)

16. Król, D., Lasota, T., Trawiński, B., Trawiński, K.: Investigation of evolutionary optimization methods of TSK fuzzy model for real estate appraisal. Int. J. Hybrid Intell. Syst. **5**(3), 111–128 (2008)

17. Swierniak, A., Smieja, J., Mura, M., Bajger, P.: Modeling and optimization of radio-chemotherapy. Adv. Intell. Syst. Comput. **1033**, 223–233 (2020)

Inference Ability Assessment of Modified Differential Neural Computer

Urszula Markowska-Kaczmar$^{(\boxtimes)}$ⓘ and Grzegorz Kułakowskiⓘ

Department of Computational Intelligence,
Wroclaw University of Science and Technology,
Wyb. Wyspiańskiego 27, 50-370 Wrocław, Poland
urszula.markowska-kaczmar@pwr.edu.pl

Abstract. In this paper we propose three modifications of Differential Neural Computer aiming at the improvement of convergence and speed of the network training. The first one relies on simplifying the mechanism of erasing old information; the second increases the influence of the link between data, and the last one increases the utility of the memory module. We evaluate the proposed modifications using five bAbI tasks. The results analysis gave some insights into modification effects, and the most promising results achieved the DNC modification without erasing vector.

Keywords: Differential neural computer · Modification · Inference · Evaluation

1 Introduction

Neural networks are applied in many domains. They allow efficient image, text, and video recognition [3]. One of the most challenging tasks is text understanding and question answering (Q&A). Application of neural networks to solve these tasks is difficult when they do not have additional memory. In this case, the need to reason causes the networks memorize the whole information in networks' weights and neuron's activities. Still, in this situation, it is not possible to know the relationships between old and new data that are crucial in the case of reasoning or Q&A. In response to this problem arose recurrent neural networks (RNN) that can use information from the past. Still, the classical RNNs fail with distorted information caused by processing long sequences of information. Therefore the networks with long memory have been designed. Two papers presented this idea – [4] and [8] – almost at the same time. The first one describes the additional memory, comparing it to the working memory in the people's brain, because it is possible to store and operate variables using trained rules. The latter gave the name memory networks. These networks allow putting aside information for further recall and use it if necessary to achieve assumed results. The second paper compares it to the standard computer where calculations are separated from data. Neural networks with the memory store necessary data in

ⓒ Springer Nature Switzerland AG 2020
N. T. Nguyen et al. (Eds.): ACIIDS 2020, LNAI 12033, pp. 380–391, 2020.
https://doi.org/10.1007/978-3-030-41964-6_33

memory and only learn, how to operate on them, based on the backpropagation algorithm.

Neural networks with memory are very desired in Q&A, so this task became an indicator of how well these networks can reason. They offer the possibility of reasoning not only from sequential data but also using more complex data, although the ability to handle long-term dependencies in data is still challenging. Differentiable Neural Computer (DNC) [5] is an example of a network with memory. It is possible to use it as a module in different systems, and it can be applied simultaneously to various tasks, like question answering, searching for the shortest path, and reinforcement learning. In spite of its elasticity, the current architecture of DNC is not free from some drawbacks, for instance - high variance of the results, weak convergence, and the long-time needed to train the model.

Researchers try to fulfill this gap by modifying DNC. Franke et al. [2] proposed robust DNC (rsDNC) that contains a slim memory unit (only content-based memory unit) and a bidirectional architecture minimizing the performance variance between different initializations. The paper [1] describes other proposals of DNC modification introducing key-value separation, DNC's deallocation of memory, and link distribution sharpness control. The last-mentioned changes allow increase performance on arithmetic tasks and also improve the mean error rate on the bAbI dataset.

In our work, we focus on the improvement of convergence and speed of training the DNC network by simplifying the mechanism of erasing old information, increasing influence of old links between data and increasing utility of the memory module to obtain the result.

The paper consists of five sections. In the next one, we describe the details of the DNC to give a background for the description of the modifications presented in Sect. 3. Section 4 introduces a course of experiments and obtained results. Conclusions end up the paper. They present a summary and further research plans.

2   Differential Neural Computer Overview

In comparison to other memory networks, the Differential Neural Computer (DNC) [4] characterizes by separated memory that enables storing data without modification for a long time. The network makes use of it if it needs to produce the answer.

The architecture differs from other NNs in the possibility of selective reading and writing data. Thanks to these features, there is no need for frequent information modifications in the next time-step as in LSTM or GRU networks [3]. Additional strength is the generalization improvement of the tasks solved by the network.

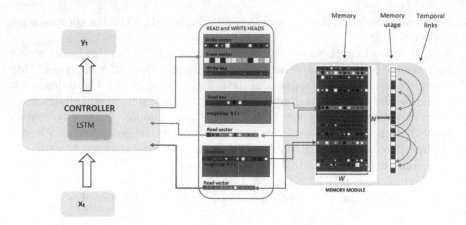

Fig. 1. DNC architecture

2.1   DNC Architecture

The network consists of 2 main components – the controller and DNC memory module, which is the primary concept of the network. The memory is composed of

▲ *reading and writing heads* – these units produce weightings. They are applied to communicate with memory.
▲ *memory* – it is the place of recording information obtained from the controller for further use.
▲ *the associative temporal links* – they can be used to recall the order, in which data were written to the memory (in sequence or reverse).
▲ *vector of memory usage* – it shows the current usage level of each memory location.

The model schema is shown in Fig. 1. It shows the memory size consisting of N locations of size W.

There are three attentional mechanisms implemented in the model:

▲ *reading by content lookup* – the controller emits key vectors that are then compared with the content of each memory row using a similarity measure. The cosine similarity is the most popular one. The row similarities are used by read head to recall association and by write head to modify memory.
▲ *reading by temporal link matrix* – this mechanism allows for writing and then reading the information related to the location dependency of recorded data. The matrix contains order and dependencies between localizations of data.
▲ *dynamic memory allocation* – this mechanism takes responsibility to allocate and free memory locations. It is implemented as a vector of values (between 0 and 1) that reflects the usage of each memory row. The parameter value is increased after each reading and decreased after writing operations.

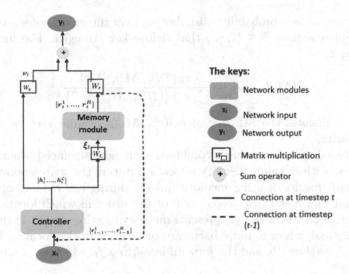

Fig. 2. DNC flowchart

2.2 DNC Performance

DNC's memory M_t stores a collection of N real-valued vectors of size W. Reading is performed by R read heads and the current vector read by the i-th reading head at time t is expressed as follows: $r_t^i = M_t^T w_t^{r,i}$, where $w_t^{r,i}$ is a read weighting vector with components in the range of [0;1]. Write operation is performed by write head. The memory is updated as follows:

$$M_t = M_{t-1} \circ \left(E - w_t^w e_t^T \right) + w_t^w v_t^T \tag{1}$$

where E is a matrix of ones, $\circ$ denotes the element-wise matrix product, w_t^w is a normalized vector of the write weighting vector, v_t is a vector to be written to the memory, e_t is the erase vector that defines how much the elements in the memory have to be erased before the update.

Our DNC's controller is based on LSTM network (Fig. 1). Figure 2 presents DNC flow. The controller takes as an input a vector x_t and a set of read vectors $r_{t-1}^1,...,r_{t-1}^R$. Concatenated input and hidden vectors from both previous timestep h_{t-1}^l and from previous layer h_t^{l-1} are used as an input for each LSTM layer to produce next hidden vector h_t^l. Hidden vectors from all layers at a given timestep are concatenated to emit an output vector ν_t and an interface vector ξ_t. To emit both vectors, the controller computes a function $\mathcal{N}$. The vectors are calculated as follows $[\nu_t, \xi_t] = \mathcal{N}\left(x_t, r_{t-1}^1, \ldots, r_{t-1}^R\right)$. The global output vector of DNC y_t is obtained by another neural network that models function $\mathcal{F}$ of ν_t and read vectors of the current timestep, i.e. $y_t = \mathcal{F}\left(\nu_t, r_t^1, \ldots, r_t^R\right)$.

The simplest memory addressing schemes – content-based addressing – allows to refer to the memory locations where vectors are more similar to a given lookup key k (Fig. 1). To produce weights, a vector-valued function $\mathcal{C}(\cdot)$ is used, which can be interpreted as a probability distribution over the memory locations. The function applies scalars $\beta \in [1, \infty)$ that define key strength. The function is shown in Eq. 2.

$$\mathcal{C}(M, k, \beta)[i] = \frac{exp\{\mathcal{D}(k, M[i, \cdot]))^\beta\}}{\sum_j exp\{(\mathcal{D}(k, M[j, \cdot]))^\beta\}} \tag{2}$$

where: $M[i, \cdot]$ denotes the i-th row of matrix M (a column vector), $\mathcal{D}$ is the cosine similarity.

The content-based addressing is combined with more advanced schemes. Usually, it is used with dynamic memory allocation during the write operation and combined with results of using memory linkage during the reading operation. The temporal link matrix L_t keeps track of the order in which locations have been written. Let a vector w_{t-1} represents memory locations at $(t-1)$ time-step. We can easily assign how to move backward or forward to this location by using the *backward weighting* b_t and the *forward weighting* f_t, which are computed as follows:

$$b_t^i = L_t^T w_{t-1}^r \tag{3}$$

$$f_t^i = L_t w_{t-1}^r \tag{4}$$

The controller uses them and a content weight vector $c_t^r = \mathcal{C}(M_t, k_t^r, \beta_t^r)$ to emit interpolated final read weighting w_t^r by applying three scalar coefficients $\pi_t[1]$, $\pi_t[2]$ and $\pi_t[3]$. $w_t^r = \pi_t[1]b_t^r + \pi_t[2]c_t^r + \pi_t[3]f_t^r$

It is important to record the degree to which memory locations have been written recently. This objective is attained by introducing a new weight vector p_t, which is called the *precedence weight vector*. It is calculated recursively. Index i refers to i-th row of the memory.

$$p_t = \left(1 - \sum_{i=1}^{N} w_t^w[i]\right) p_{t-1} + w_t^w \tag{5}$$

It takes a part in computation of the temporal link matrix L_t. The element $L_t[i, j]$ indicates to what degree memory location i was written after location j.

$$L_t[i, j] = (1 - w_t^w[i] - w_t^w[j]) L_{t-1}[i, j] + w_t^w[i]p_{t-1}[j] \tag{6}$$

The write content vector of weights c_t^w is calculated in a similar way as read content vector of weights c_t^r i.e. $c_t^w = \mathcal{C}(M_t, k_t^w, \beta_t^w)$. Storing information may be made by dynamic memory allocation using the allocation weight vector a_t or by content-based addressing into the locations specified by the *write content weighting* c_t^w. In a given time step, it is also possible that writing is not performed. Therefore the final vector of weights for writing is calculated as:

$$w_t^w = g_t^w [g_t^a a_t + (1 - g_t^a) c_t^w] \tag{7}$$

where g_t^a is a scalar allocation gate which makes the choice between the first two options (a_t and c_t^w). A scalar write gate g_t^w determines to what degree the memory is allocable at this time step.

3   Proposed Modifications of DNC

Despite elasticity, DNC characterizes some drawbacks as a huge variance of results, weak convergence, long time of learning. Our proposed modifications aim at the improvement of convergence and speed of learning by simplification of erasing old information, increasing the impact of linkage between older and newer data, and the increase of the memory module use.

In these modifications, we were inspired by changes introduced in the GRU network that is a modification of the LSTM network. The GRU network differs from LSTM in that it does not hand over additional cell state c_t. We use only the network output being also the cell state h_t sent to the next steps. Another change is a simplified way of adding new and erasing old information. This model joins the forget gate f_t with the input gate in one update gate. In the GRU model, we assume that as much information we erase, as much we can write. Therefore there is one parameters vector less comparing to the LSTM network and the accuracy is similar as using two gates in LSTM [6]. The research, described in [7], where authors conducted random modifications of recurrent neural networks also confirms this conclusion. That is the reason why our first modification of DNC is based on a similar change in the equation describing memory M_t.

The second modification relies on the precedence weight vector change p_t. This change comes from the observation that in the case when write gate g_t^w has a high value, an update of p_t is based on the write vector from the current state. The written data from the previous step has a small impact on the final result. This approach is safe in the case of value calculation (we ensure that $p_t \in \Delta_N$; i.e., Δ_N limits values of the vector to satisfy that the sum of its elements is not higher than 1). In the LSTM and GRU networks, this problem is solved by using gates that, based on another input, a network state and a weight matrix, decide to store or not a further relationship between data. We propose a solution based on maximal value and scaling.

The last proposed modification relies on introducing such a mechanism that the network focuses on using the memory module. The motivation behind this approach lies in the observation that the DNC model is very complicated. In the initial phase, it is easier for the network to rely on ν_t value produced by the controller. Still, to recognize a more complex relationship, the network has to use the memory module. This last modification aims to enforce the network makes correct predictions, basing on previous interaction with the memory module. This modification is the result of some promising changes made in [2], where authors proposed using dropout in the controller's network, and as well as our own observation from the second modification.

Below the details referring to the proposed modifications are presented.

▲ **Simplification of Erasing Old Information** (it will be further abbreviated to *DNC-NEV*; the shortcut from *Not Erasing Vector*). Analogously to LSTM, DNC uses separated mechanisms to erase and store information in memory. Being inspired by simplification of LSTM applied in GRU network, we propose to use similar simplification in DNC. In the case of GRU, combining these both mechanisms does not result in decrease of efficiency. We hope that it will be also true in the case of DNC. This change refers to the way the memory is updated. We proposed modification of Eq. 1 to the form given in Eq. 8.

$$M_t = M_{t-1} \circ (E - w_t^w 1^T) + w_t^w v_t^T \tag{8}$$

where 1 assigns the vector of ones; other elements are described in Eq. 1. Substitution of the erase vector by the vector of ones causes erasing as much information from memory M_t as much we want to write. In this way, there is no need to train the network separately to know how much information should be erased and how much should be written in the memory.

▲ **Increasing the Impact of Linkage Between Data** (it will be further abbreviated to *DNC-SPW*; the shortcut from *Spread Precedence Weighting*). It is easy to notice that in Eq. 5 value of the precedence weight vector p_t is strongly decreased by multiplying it by 1 minus sum of weights w_t^w. This operation satisfies the limitation $\sum_i p_t[i] \leqslant 1$. Calculating a vector p_t element by element, i.e. $p_t = (1 - w_t^w)p_{t-1} + w_t^w$, does not satisfy this condition. Instead the original Eq. 5, we propose to use the following modification: $\hat{p}_t = (1 - max(w_t^w))p_{t-1} + max(w_t^w)w_t^w$. Scaling allows new values to gradually forget old connections. Additionally, the modification decreases the risk that the vector p_t is almost equal to vector w_t^w. In the case when there is a small components number of vector w_t^w with high values, this modification is not essential. In the opposite situation, it offers a longer relationship to store write locations. To ensure that the sum is always exactly 1, we propose additional scaling which is implemented as dividing by the sum of candidate vector $\hat{p}_t$

$$p_t = \frac{\hat{p}_t}{\sum_i(\hat{p}_{t,i})} \tag{9}$$

▲ **Increase of Memory Module Use** (it will be abbreviated to *DNC-ERH*; shortcut from *Enhancement Read Heads*). This modification refers to read vectors. To assign the output y_t DNC uses two elements – the first one is produced by the controller, the second comes from memory module:

$$y_t = W_r[r_t^1, ..., r_t^R] + W_y[h_t^1, ..., h_t^L] \tag{10}$$

Read vectors from the previous time-step $r_{t-1}^1, ..., r_{t-1}^R$ also arrive as an input to the DNC controller. Authors in [6] suggest that read vectors play a significant role in the network efficiency. Following this suggestion, we propose to boost the impact of the memory module of DNC by introducing the gain parameter $\alpha_i \in (1, +\infty)$ for reading vector. Its role is shown in the following equation:

$$r_t^i = \alpha_r(M_t^T w_t^{r,i}) \tag{11}$$

This change influences the output of the network but also the input. It increases the impact of reading vectors in the memory module responses.

4   Experiments

Our goal was to analyse the training efficiency of modified DNC. We observed mean average error and convergence speed of mean average training error. We were also interested whether the modifications cause a decrease in standard deviation. The results should verify in reality our intuitions standing behind the modifications.

Dataset. The experiments were conducted using the benchmark dataset bAbI [8]. It is composed of 20 tasks that allow testing the inference ability of the model during question answering based on a source text. The dataset is commonly used [2,5], because it evaluates many aspects of text understanding, as answer based on a combination of many facts inferred from counting, deduction, induction, reasoning based on time, and many others. It contains data specifically generated to asses the models. The ability to understand similar tasks is the primary condition for deep natural language understanding and reasoning. The description of bAbI construction and the generator of data we have used in the experiments can be accessed from [8].

In our work, we have chosen the English version of the dataset. It contains 10 thousands of training patterns per task with a validation set. Considering the time needed to train the model, we decided to conduct experiments using a subset of bAbI data sets composed of 5 tasks. To preserve the level of generality and complexity of the problem, we assumed training the model based on the overall dataset (simultaneous training the network on various types of tasks included in bAbI). The subset of bAbI contains the following tasks:

Task 2 - Two Supporting Facts. It relies on reasoning using two facts hidden in other sentences that are not connected with an answer. To answer the question, a combination of these two facts is needed.

Task 5 - Three Argument Relations. The task needs a subject and verb recognition in the sentence based on three-argument relations. It verifies whether the model distinguishes who is an object and what it performs.

Task 11 - Basic Coreferences. Using this task we check whether a model can reason the same addressee of action when the person was described by various phrases.

Task 12 - Conjunctions. It verifies the correct assignment of people combined in the sentence by conjunctions to actions.

Task 14 - Time Reasoning. The ability of time reasoning allows for indicating facts on the time axis. During training, the models receive the data sequentially. The events have not to be described chronologically. A good model should be able to store a time relationship to give an answer referring to the event sequence.

Experimental Setup. To check what is the influence of the proposed modification on the DNC performance, we conducted experimental research. The tuning process of the model is highly time-consuming, particularly for a model with high variance. Therefore we decided to base on the hyperparameter values assigned by authors of original paper introducing DNC [5]. They also tested the model on bAbI dataset. As a controller, one LSTM with 256 units with initialized forgetting bias equal to 1 was used. The size N of memory (the number of locations) was decreased from 256 to 48. This change does not change training efficiency but increase the chance to join the content by reusing the network. The size of the memory row was set to 64. Four read heads were used.

The networks were trained using RMSprop with a learning coefficient equal to 0.0001 and a momentum equal to 0.9. Momentum is based on the training process smoothening. It is implemented as following to direction assigned by the accumulation of past average gradients. The batch is equal to 1 because its increase caused a significant decrease in training efficiency. Similarly to [5], the training patterns were randomly assigned from the training data set. The number of all training examples for 5 tasks is 44 5000, 5 000 for validation and 5 000 for testing datasets. Words encoded as one hot were sequentially delivered as inputs. A special symbol "-" indicates the place when the answer is expected. Table 1 shows all hyperparameters values considered in the experiments.

Table 1. Set of parameter values used in experiments.

Element	Hyperparameter	Value
Controller	Recurrent model	LSTM
	Number of units	256
	Initialization b_f	1
	Number of layers	1
DNC	Number of read heads R	4
	The size of row in memory W	64
	Number of locations in memory N	48
Optimization	Method	RMSprop
	Learning coefficient	0.0001
	Momentum	0.9
	Paring back gradients	in$[-10, 10]$
Stop condition	Max number of steps	550 000
	Lack of improvement	67 500 steps
	Min improvement	0.1%
Patterns	Size of mini-batch	1
	Patterns	Randomly
	Input encoding	1 of K (One hot)
	Question separated from text?	No

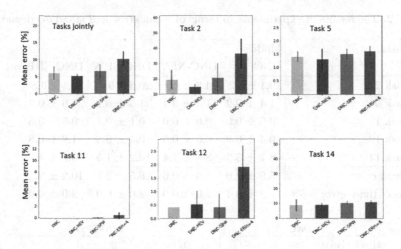

Fig. 3. Mean error and standard deviation on testing set, for all 5 tasks from bAbI used in DNC training

Evaluation Measures. To be consistent with other works referring to the evaluation of DNC, we calculated two measures: (1) the error defined as $(1 - accuracy)$ achieved by the model on all tasks, (2) the number of tasks when the model achieved 95% of accuracy.

Performed Experiments. We performed a series of 4 experiments in each of them we trained and tested one of the following models: DNC, DNC without erase vector – DNC-NEV, DNC with spread precedence weighting – DNC-SPW, DNC with enhancement read heads for $\alpha_r = 4$ – DNC-ERH $= 4$. Each experiment was repeated 5 times using a single set composed of the 5 tasks from bAbI (tasks: 2, 5, 11, 12 i 14).

Analysis of the Mean Error. First, we will focus on the mean error. The results of the experiments are summarized in Table 2 and also shown in Fig. 4. Considering all three modification, DNC-NEV is the best solution. It has significantly improved the result in task 2 and decreased its variance in task 2 and task 14. This fact is worth underlying because one of the DNC drawbacks is the high variance of the results. In the case of the DNC-SPW model (spread precedence weighting), we observe slight falling-off the results. This modification caused an increase of variance in all tasks except 14. Its aim was to enable finding further relationships, but its impact is small. Definitely, the model DNC-ERH $= 4$ worsened the results the most. The strongest decrease can be noticed in task 2, but also in 12. In other tasks, a small regress is observed.

Training Process Analysis. Analysis of mean error (jointly from 5 tasks), presented in the diagrams in Fig. 4 enables the complete convergence comparison during training. We can observe significant improvement in the case of DNC-NEV. The results obtained in each experiment achieved low values. In comparison to the original DNC model, this modification remarkably decreased the

Table 2. The results of experiments in terms of mean error and standard deviation

Tasks	Models			
	DNC	DNC-NEV	DNC-SPW	DNC-ERH $=4$
Task 2	19.1 ± 6.2	14.3 ± 1.9	20.3 ± 9.6	36.1 ± 9.8
Task 5	1.4 ± 0.2	1.3 ± 0.4	1.5 ± 0.2	1.6 ± 0.2
Task 11	0.0 ± 0.0	0.0 ± 0.0	0.1 ± 0.1	0.5 ± 0.5
Task 12	0.4 ± 0.2	0.5 ± 0.5	0.4 ± 0.5	1.9 ± 0.8
Task 14	8.7 ± 3.9	8.8 ± 1.4	10.2 ± 1.5	10.8 ± 1.2
Mean error	5.9 ± 2.0	5.0 ± 0.6	6.5 ± 2.1	10.2 ± 2.2
No of times error >5%	1.8 ± 0.4	2.0 ± 0.0	2.0 ± 0.0	2.0 ± 0.0

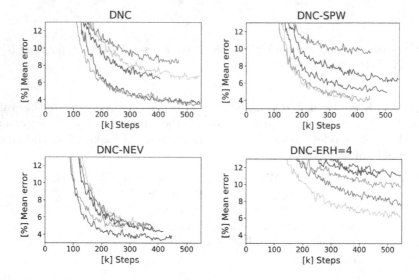

Fig. 4. Error curves on the validation set calculated based on the mean of all 5 chosen tasks from the bAbI dataset. Each curve is presented one experiment trial. The curves are smoothed with a coefficient equal to 0.3.

dispersion of the results between experiments and simultaneously placed the final error values between the lowest values. It also caused quick reaching low error values. In each of 5 experiments, early stopping mechanism was used. Thanks to this fact, the model accomplished values closed to its minimum the quickest on average. In comparison to the original DNC model, the results of DNC-SPW in Table 2 do not differ much but diagram in Fig. 4 shows that on average, these changes should be near 0. The reason may come from the small number of runs in the experiments (there were 5 runs). Modification DNC-ERH $=4$ has not improved results. It caused significant worsening training quality. Although, it is worth mentioning that in comparison to other models, it characterizes by more even deviation distribution between errors of particular experiments.

5   Conclusion

Conducted experiments give some insights which of the proposed modifications are a good choice for further development. From this experiments, we can conclude that DNC-NEV modification is the best from the proposed ones. It caused significant improvement of the network performance and its training convergence. Essential is also a decrease of variance. Our intuitions behind this modification work in practice. It is a good prognostic. The worst results were observed for DNC-ERH $= 4$ model. Probably it was the result of insufficient tuning of the parameter α_r. To fully evaluate the proposed modifications further experimental research is needed with the whole bAbI dataset with more runs. It is our plan for nearest future.

References

1. Csordás, R., Schmidhuber, J.: Improving differentiable neural computers through memory masking, de-allocation, and link distribution sharpness control. CoRR abs/1904.10278 (2019). http://arxiv.org/abs/1904.10278
2. Franke, J., Niehues, J., Waibel, A.: Robust and scalable differentiable neural computer for question answering. CoRR abs/1807.02658 (2018). http://arxiv.org/abs/1807.02658
3. Goodfellow, I., Bengio, Y., Courville, A.: Deep Learning. MIT (2016). http://www.deeplearningbook.org
4. Graves, A., Wayne, G., Danihelka, I.: Neural turing machines. CoRR abs/1410.5401 (2014). http://arxiv.org/abs/1410.5401
5. Graves, A., et al.: Hybrid computing using a neural network with dynamic external memory. Nature, 471–476 (2016). https://doi.org/10.1038/nature20101
6. Greff, K., Srivastava, R.K., Koutník, J., Steunebrink, B.R., Schmidhuber, J.: LSTM: a search space odyssey. CoRR abs/1503.04069 (2015). http://arxiv.org/abs/1503.04069
7. Jozefowicz, R., Zaremba, W., Sutskever, I.: An empirical exploration of recurrent network architectures. In: Proceedings of the 32nd International Conference on International Conference on Machine Learning ICML2015, vol. 37, pp. 2342–2350. JMLR.org (2015). http://dl.acm.org/citation.cfm?id=3045118.3045367
8. Weston, J., Chopra, S., Bordes, A.: Memory networks (2014). https://arxiv.org/abs/1410.3916

Simple Quantum Circuits
for Data Classification

Joanna Wiśniewska[1][✉] and Marek Sawerwain[2]

[1] Institute of Information Systems, Faculty of Cybernetics,
Military University of Technology, Gen. S. Kaliskiego 2, 00-908 Warsaw, Poland
jwisniewska@wat.edu.pl
[2] Institute of Control and Computation Engineering,
University of Zielona Góra, Licealna 9, 65-417 Zielona Góra, Poland
M.Sawerwain@issi.uz.zgora.pl

Abstract. The paper is dedicated to the problem of supervised learning in quantum circuits. We present two solutions: SWAP-test and Simple Quantum Circuits (SQCs) based on the tree tensor networks which are able to properly classify samples from Moons, Circles, Blobs and Iris sets. Moreover, the mentioned circuits were constructed not only for qubits, but also for the units of quantum information with higher freedom level. The SWAP-test, prepared as a part of this paper, works for units like qutrits and ququads – so far this solution has been only discussed in the context of qubits. We present the procedure of data preparation which is important in further data classification with high success rate. It should be emphasized that the shown circuits are effective in pattern recognition in spite of a low level of their complexity.

Keywords: Classification · Quantum circuits · SWAP-test · Qudits

1 Introduction

The machine learning methods are very popular nowadays. Even people who are not particularly interested in science, know the basic facts concerning the artificial intelligence. Thanks to the development in materials engineering, the computers offer amazing computational powers, immense data storage, and swift access to data. Because the idea of artificial intelligence is very attractive, it is not surprising that the potential of modern computers is often utilized to project new experiments and implement new ideas in this field.

Another favorable element in the development of the artificial intelligence seems to be the quantum computing. This new outlook gives exponential speed-up of calculations (e.g. famous Shor's algorithm for prime factorization). Even if this new technology is still unstable and extremely expensive, we regularly receive the press news regarding to achievements in this field (like obtaining quantum supremacy by Google [7]).

One can easily find publications referring to quantum versions [2,13] of methods known in machine learning. There are quantum neural networks [8], quantum

© Springer Nature Switzerland AG 2020
N. T. Nguyen et al. (Eds.): ACIIDS 2020, LNAI 12033, pp. 392–403, 2020.
https://doi.org/10.1007/978-3-030-41964-6_34

kNN methods [17], quantum self-organized maps [16], quantum k-means method [14], etc. In this work, we would like to propose a supervised learning method based on SWAP-test [1] and quantum tensor trees [5]. The novelty of our approach is that we use not only qubits but also units of quantum information with higher freedom level. Our solution is based on the set of elementary gates [10] what allows its implementation on experimental quantum devices [19,21].

The paper is organized in the following way: in Sect. 2 introductory information relating to quantum computing is presented. Sec. 3 contains a description of a classification procedure considered in this work (including the conversion of data to quantum states). In Sect. 4, the results of numerical experiments, pertaining the recognition of data samples by quantum circuits, are shown. Conclusions and a summary may be found in Sect. 5 which is followed by acknowledgements and references to literature.

2   Quantum Computing – Preliminaries

In this section, we would like to briefly present the definitions, related to the field of quantum computing, which are utilized in the next parts of the paper. Let us introduce the following denotations: $\mathbb{R}$ represents the real numbers, $\mathbb{C}$ stands for the complex numbers, i is the imaginary unit. The set of integers is $\mathbb{Z}$, and $\mathbb{Z}^d$ denotes the set of reminders after the division by d (so-called modulo d operation).

A qubit [10] is the most popular unit of the quantum information. It is a natural consequence of the fact that the qubit is the equivalent of the classical bit. The value of a bit is scalar, and it may be zero or one. A state of a qubit is represented as a two-element normalized vector. Usually, the state of qubit is expressed with the use of so-called computational basis which is a set of orthonormal vectors. The most popular basis is the standard basis which for qubits consists of two vectors:

$$|0\rangle = \begin{bmatrix} 1 \\ 0 \end{bmatrix}, |1\rangle = \begin{bmatrix} 0 \\ 1 \end{bmatrix}. \tag{1}$$

The state of a qubit may be expressed as:

$$|\phi\rangle = \alpha_0|0\rangle + \alpha_1|1\rangle, \tag{2}$$

where $\alpha_i \in \mathbb{C}$ are termed as probability amplitudes, under the normalization condition: $|\alpha_0|^2 + |\alpha_1|^2 = 1$.

Just like the bit may be generalized to the dit, the qubit may be generalized to the qudit. The freedom level for bit/qudit equals two. In case of dit or qudit, the freedom level is a natural number $d > 1$. The state of a single qudit is a d-element normalized vector, and the computational basis contains d vectors. The state of a qudit with the use of standard basis is given as:

$$|\psi\rangle = \alpha_0|0\rangle + \alpha_1|1\rangle + \ldots + \alpha_{d-1}|d-1\rangle, \tag{3}$$

where $\alpha_i \in \mathbb{C}$, and $\sum_{i=0}^{d-1} |\alpha_i|^2 = 1$. If we deal with quantum states of n qudits, the number of vectors in basis and the number of vector's entries is d^n.

Obtaining the state of n-qudit register may be calculated as:

$$|\Psi\rangle = |\psi_0\rangle \otimes |\psi_1\rangle \otimes \ldots |\psi_{n-1}\rangle, \tag{4}$$

where $\otimes$ represents the tensor product. Generally, quantum registers may contain qudits of different freedom levels, but, in this work, we take only quantum states of the same d into consideration.

We can name two groups of operations which may be performed on a quantum register: a unitary operation and a measurement. A unitary operation may be understood as applying quantum gates on a register's state. Basic quantum gates, used in this work, are: the negation gate X, the Z gate realizing qubit's rotation through π radians around the z-axis, and the Hadamard gate H. Let us present the qudit versions of these gates [12]. The single qudit negation gate:

$$X|k\rangle = |(k+1) \bmod d\rangle \tag{5}$$

shifts circularly the values of probability amplitudes. The Z gate changes a qudit's phase in the following way:

$$Z|k\rangle = \exp((2\pi \mathrm{i}k)/d)|k\rangle = \xi_d^k|k\rangle \tag{6}$$

where $\xi_d^k = \exp((2\pi \mathrm{i}k)/d)$ are roots of unity ($k = 0, 1, 2, \ldots, d-1$). The gates/operators X, Z come under the generalized Pauli group $\mathcal{P}_d$. These operators are marked as $\Gamma^{j,k}$. They do not commute but meet the following relations:

$$\Gamma^{j,k} = Z^j X^k = \xi_d^{jk} X^k Z^j, \text{ and } X^d = Z^d = I, \tag{7}$$

where I represents d-dimensional identity operator (eye matrix).

The Hadamard gate for qudits is denoted by F:

$$F|k\rangle = \frac{1}{\sqrt{d}} \sum_{l=0}^{d-1} \xi_d^{kl}|l\rangle. \tag{8}$$

The symbol F derives from the Fourier computational basis [18], because the gate F changes the computational basis to the dual form of the Fourier basis. The gate F fulfills: $F^d = I$.

The definition of the Z gate, given above, needs the Lie algebra [3] generator for $SU(d)$ group to be implemented ($d \geq 2$). Let us recollect the procedure of constructing $SU(d)$ generators. The first step is determining the set of projectors:

$$(P^{k,j})_{v,\mu} = |k\rangle\langle j| = \delta_{v,j}\delta_{\mu,k}, \quad 1 \leq v, \mu \leq d. \tag{9}$$

Now, the package of $d(d-1)$ generators, based on the group $SU(d)$, is:

$$\Theta^{k,j} = P^{k,j} + P^{j,k}, \quad \beta^{k,j} = -i(P^{k,j} - P^{j,k}), \tag{10}$$

where k and j fulfill the relation $1 \leq k < j \leq d$.

Next, the set of $(d-1)$ generators is given as:

$$\eta^{r,r} = \sqrt{\frac{2}{r(r+1)}} \left[\left(\sum_{j=1}^{r} P^{j,j} \right) - r P^{r+1,r+1} \right],$$
(11)

where $1 \leq r \leq (d-1)$. As a result of the procedure, the $d^2 - 1$ operators can be formed utilizing the $SU(d)$ group.

We start the definition of the qudit-rotating gates with describing operators $P^{k,j}$, $\Theta^{k,j}$, and $\eta^{r,r}$ as λ_γ (where $\gamma = 1, \ldots, d^2 - 1$). Particular rotations may be realized as an R operator:

$$R_{\lambda_\gamma}(\theta) = \exp\left(\frac{\mathrm{i}\lambda_\gamma \theta}{d}\right),$$
(12)

where $\theta \in \mathbb{R}$. The operators of the rotation are unitary – this can be easily proof because $R_{\lambda_\gamma}(\theta) R_{\lambda_\gamma}^\dagger(\theta) = I$.

In the next sections of this paper, we use two-qudit gates which introduce the entanglement to a quantum state. The $CNOT$ gate for qubits may be generalized for qudits in the following way:

$$CX|x\rangle|y\rangle = |x\rangle|(-x-y) \bmod d\rangle.$$
(13)

Another useful operator is CZ:

$$CZ|x\rangle|y\rangle = \exp\left(\frac{2\pi \mathrm{i} x y}{d}\right) |x\rangle|y\rangle$$
(14)

which changes the phase of a qudit.

Remark 1. Computation based on qudits (instead of qubits) may reduce the size of the quantum gates circuit.

The second group of operations, performed on quantum registers, is a measurement. There are different types of measurement (e.g. Positive Operator-Valued Measure (termed as POVM)). In this work, we focus on the von Neumann measurement [9] which is a projective measurement. Analyzed type of measurement may realized on the whole quantum register or only on some particular qudits. Any projective measurement must refer to a specified computational basis. Let us measure one-qudit state $|\psi\rangle$ in a basis constructed of orthonormal vectors: $\{|u_0\rangle, |u_1\rangle, \ldots, |u_{d-1}\rangle\}$. The spectral decomposition of an observable M is:

$$M = \sum_{i=0}^{d-1} \lambda_i P_i$$
(15)

where λ_i are the eigenvalues of projectors P_i, and each $P_i = |u_i\rangle\langle u_i|$. The results of the measurement are values λ_i. The probability of obtaining λ_i is $p(\lambda_i) = \langle\psi|P_i|\psi\rangle$. A state of the qudit after the measurement is:

$$|\psi'\rangle = \frac{P_i|\psi\rangle}{\sqrt{\langle\psi|P_i|\psi\rangle}} = \frac{P_i|\psi\rangle}{\sqrt{p(\lambda_i)}}.$$
(16)

Remark 2. The character of quantum computations is probabilistic. This means that each experiment must be performed many times and each time the final quantum state is measured. A result of computation is obtained as a probability distribution from measurements outcome.

3   Data Classification with SWAP-Test and Simple Quantum Circuits

The first issue to analyze is the conversion of data. The learning samples are given as classical observations, and they must be written as correct quantum states. Figure 1 depicts the change of points coordinates (real numbers) to the coordinates of points lying on an arc of radius 1. Just like in classical approaches, we obtain the best results of the classification if observations form well separated classes (e.g. points from two different classes are placed in opposite parts of an arc).

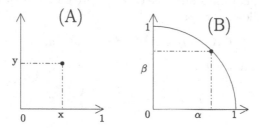

Fig. 1. The conversion of classical data to the amplitudes of quantum states. On the chart (A) a point after the normalization of data is presented. The chart (B) shows a point which coordinates meet the condition $|\alpha|^2 + |\beta|^2 = 1$ – this allows obtaining a quantum state with probability amplitudes like in Eq. (2)

We start the conversion of data with the classical procedure of data normalization [4], known in machine learning and data mining. If the learning set contains n observations with d attributes, then for each attribute l ($l = 1, ..., d$) extreme values must be found: minimal and maximal. The extreme values are utilized to calculate the range of accepted values for every attribute. Next, the data is normalized for attribute l observation by observation. The computed values are obtained in the following way:

$$\bar{x}_{l,m} = \frac{x_{l,m} - \min_l}{\max_l - \min_l} \tag{17}$$

where $x_{l,m}$ denotes the original value of l-th attribute in m-th observation ($m = 1, ..., n$). After this procedure all normalized values $\bar{x}_{l,m} \in [0, 1]$. We would like to treat all attributes values within one observation as amplitudes of one-qudit state. Thus, the second part of normalization is needed – to guarantee that the sum of squared modulus of amplitudes values equals 1.

The symbol d stands for the number of attributes and the freedom level of a qudit. Now, for each d-attribute observation quantum state is calculated. The l-th amplitude of one-qudit state is:

$$\alpha_l^m = \sqrt{\frac{\bar{x}_{l,m}}{\sum_{i=0}^{d-1} \bar{x}_{i,m}}}. \tag{18}$$

This allows building n-element set of quantum states which may be processed by a circuit of quantum gates.

Since we deal with supervised learning, it is necessary to pass the information to which class the sample belongs. We assume that the number of classes is not greater than d in a particular example, and the samples from one class may be accumulated in a proximity of relevant basis state. The quantum state in two-value classifications is calculated as:

$$|\psi_m\rangle = \sin(\gamma + \theta/2) + \cos(\gamma + \theta/2) \tag{19}$$

where $\theta = \alpha_0^m + \alpha_1^m$ and γ represents the additional angle to encode a class for observation m represented by α_r^m amplitudes.

The classification of data, in this paper, is realized by Simple Quantum Circuits (SQC), which are based on the notion of tensor trees [5], and SWAP-test [1]. The data is coded as states of qudits: qubits ($d = 2$), qutrits ($d = 3$), and ququads ($d = 4$). For both the SQC and SWAP-test, the result of the classification is read with the use of quantum measurement. The final state of the computational register is measured, and after a series of experiments we obtain the probability distribution which determines samples classification.

The process of data classification with the use of a SQC, may be presented as three distinctive steps:

(I) circuit's structure defining, i.e. setting the number of qudits, pointing additional rotating gates, and generating entanglement between qudits (towards the qudit/qudits representing the classifier's output),

(II) learning process with the learning set Υ, i.e. utilizing the optimization methods to calculate additional parameters (more precisely: rotation gates parameters) of the quantum circuit which are denoted as $\{\chi_i\}$,

(III) data classification performed by a simulation or a physical realization of the quantum circuit with parameters $\{\chi_i\}$.

The parameters $\{\chi_i\}$ refer to the additional arguments of rotation gates.

We define a structure of the mentioned above circuit as three main tasks. The first one is describing the classifier's input what is presented by the initial state $|\psi\rangle$ of qudits placed in the computational register. The second task is transforming the input data by two kinds of operations – the rotation gates $R_{\lambda_\gamma}(\theta)$, and the gates like CNOT, CPhase which cause the phenomenon of entanglement in the register. The last step is to measure the state $|\psi^c\rangle$ of one or many qudits. A result of the classification process is calculated upon a probability distribution obtained during the measurement.

The initial quantum state's transformation may be written as series of unitary operations:

$$|\psi^c\rangle = U_{r_0}(\chi_0)U_{e_0}U_{r_1}(\chi_1)U_{e_1}\ldots U_{r_{n-1}}(\chi_{n-1})U_{e_{n-1}}|\psi\rangle, \tag{20}$$

where U_{r_i} represents the rotation gates, and U_{e_i} gates introducing entanglement. Operations $U_{r_i}(\chi_0)U_{e_i}$ may be read together because they form one layer in the quantum circuit. In general, we denote by $U(\cdot)$ all unitary operations performed on the state $|\psi\rangle$.

The exemplary initial state for two qubits system may be expressed as:

$$|\psi\rangle = |q_0 q_1\rangle. \tag{21}$$

The qubits q_0 and q_1 may represent observations from the first class and the second class, respectively. In this case, the observations are treated as separate ones. During the classifier's operation, one can decide if the result of computation concerns one or both observations.

Another way to describe the initial state is:

$$|\psi^0\rangle = \alpha_0|00\rangle + \alpha_1|10\rangle, \text{ and } |\psi^1\rangle = \alpha_0|00\rangle + \alpha_1|01\rangle. \tag{22}$$

Here, one of the probability amplitudes is shared for both classes. This approach allows improving the quality of classification because the data coordinates may be distinguished by their values.

A crucial issue is the process of learning, i.e. a proper selection of values $\{\chi_i\}$. To achieve this goal, a learning set with observations, described by the state $|\psi_c\rangle$, and their correct classification was utilized. Let $\Upsilon = \{|\psi\rangle_0, |\psi\rangle_1, |\psi\rangle_2, \ldots, |\psi\rangle_{N-1}\}$ be a learning set. The process of parameters $\{\chi_i\}$ selection may be treated as an optimization problem:

$$\forall_{\Upsilon_k} \quad \theta_j \leq |\langle U(\cdot)\Upsilon_k|\psi_j^c\rangle|^2 \leq \eta_j, \tag{23}$$

where j represents the class for every observation, Υ_k is k-th state from the learning set, and $|\psi_j^c\rangle$ stand for the final state.

An interpretation of Eq. (23) is following: it is expected that for each observation the square of the normalized value of the scalar product lies within the range defined by the values θ_j and η_j. Observations from the learning set are pure quantum states and this implies that $|\langle \Upsilon_k|\psi_j^c\rangle|^2$ represents the probability of measuring the state $|\psi_j^c\rangle$, i.e. the state describing a class j to which we want to classify an i-th observation. The unitary operation U denotes a SQC. We perform the L-BFGS-B method – available in the SciPy package [15,20] – to optimize the value of expression $|\langle U(\cdot)\Upsilon_k|\psi_j^c\rangle|^2$ by adjusting $\{\chi_i\}$ parameters.

The classification of samples requires the information about the probability distribution of results obtained during the measurement. The final states, which are measured as an output of utilized circuits, are denoted as:

$$|\psi_{c0}\rangle = |00\rangle, \ |\psi_{c1}\rangle = |10\rangle. \tag{24}$$

Now, the Fidelity measure [6] may be utilized to assess, if quantum states are similar. In this case, the measure is calculated as:

$$F_{c0} = |\langle\psi|\psi_{c0}\rangle|^2, \quad F_{c1} = |\langle\psi|\psi_{c1}\rangle|^2. \tag{25}$$

After the proper selection of parameters $\{\chi_i\}$, the values of Fidelity should explicitly point out one class, i.e. the probability of measuring $|0\rangle$ on the first qudit should be significantly greater than the probability of measuring $|1\rangle$.

In the case of SWAP-test, the probability of measuring $|0\rangle$ on the first qudit (see circuit (A) on Fig. 2) is very important. If the states processed by the circuit are identical, then $P(|0\rangle) = 1$. Otherwise, the value of $P(|0\rangle)$ is the approximation of the Fidelity measure for analysed states. The probability of measuring $|0\rangle$ in a circuit designed for qubits $(d=2)$ is:

$$P_2(|0\rangle) = \frac{1}{2} + \frac{1}{2} \cdot |\langle\psi|\varphi\rangle|^2. \tag{26}$$

If the obtained probability is $\frac{1}{2}$, the quantum states are orthogonal. The probabilities of measuring $|0\rangle$ on the first qudit for qutrits $(d=3)$ and ququads $(d=4)$ are:

$$P_3(|0\rangle) = \frac{5}{9} + \frac{4}{9} \cdot |\langle\psi|\varphi\rangle|^2, \quad P_4(|0\rangle) = \frac{5}{8} + \frac{3}{8} \cdot |\langle\psi|\varphi\rangle|^2. \tag{27}$$

The orthogonal states for qutrits are described by the probability $\frac{5}{9}$, and for ququads by the probability $\frac{5}{8}$.

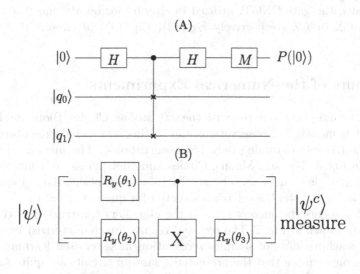

Fig. 2. Two types of quantum circuits (SWAP-test (A), SQC (B)) for the classification of elements from sets Moons, Circles, Blobs and Iris

The schemes of circuits for SQCs and SWAP-test are presented in Fig. 2. The circuits are dedicated to particular data sets which are shown in Sect. 4. It is easy to observe that both types of circuits are characterized by a low level of complexity, e.g. the SQC contains only two layers and needs only two qudits for the computational process. This allows implementing the circuits with the use of cloud-available hardware platforms like IBM Q Experience [19] or Rigetti QCS [21].

It should be emphasized that in SQC for qubits three additional gates R_y, realizing rotations around the Y axis, are used. These gates may be generated by $SU(2)$ group, and the form of the operator is:

$$R_y(\theta) = \exp\left(\frac{i\beta^{0,1}\theta}{2}\right). \tag{28}$$

In case of qudits, equivalents of the gate R_y are constructed with the use of $\beta^{k,j}$ operator. For example, R_y for qutrits are defined as:

$$R_y(\lambda_4) = \exp\left(\frac{i\beta^{0,1}\theta_1}{3}\right), \ R_y(\lambda_5) = \exp\left(\frac{i\beta^{0,2}\theta_2}{2}\right), \ R_y(\lambda_6) = \exp\left(\frac{i\beta^{1,2}\theta_3}{3}\right). \tag{29}$$

All the rotation angles are calculated by solving optimization problem, given in Eq. (23), for a particular learning set.

We obtain a circuit for qutrits as a modification of the qubit circuit. The gates are replaced by their qutrit versions. The R_y gate must be displaced by the sequence of three gates given in Eq. (29). The analogical procedure takes place for ququads.

Naturally, the gate CNOT, utilized in circuits for qubits, must be replaced with gate CX or CZ (respectively Eg. (13), Eq. (14)) in circuits designed for qudits.

4   Results of the Numerical Experiments

Numerical experiments were run with the sets Moons, Circles, Blobs and Iris [11]. Learning sets include 512 observations and testing sets include 256 observations (except the Iris set containing only 150 observations). The mentioned sets are depicted in Fig. 3. The sets Moons, Circles and Blob-2class are constructed for qubits (freedom level $d = 2$). The set Blob-3class contains data prepared for qutrits ($d = 3$), and the Iris set is constructed for ququads ($d = 4$).

Table 1 presents the success rates of the classifiers constructed as the quantum circuits shown in Fig. 2. The received results are characterized by the high efficiency, reaching 99%, of samples recognition in supervised learning. At the same time, one can see that the presented quantum circuits are quite simple.

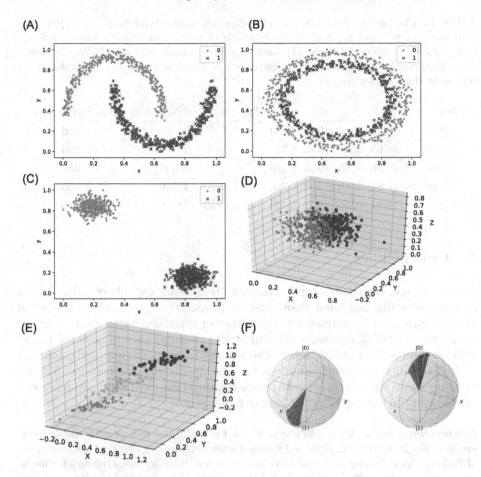

Fig. 3. The data sets after the normalization process: (A) set Moons, (B) set Circles, (C) set Blobs for samples with two features, (D) set Blobs for samples with three features, (E) set Iris presented only by the first three features. In the case (F) quantum states as vectors on the Bloch sphere are shown – the samples come from the Moons set, and it can be observed that they are easy to separate (the values of $P(|0\rangle)$ and $P(|1\rangle)$), after the measurement, will be significantly different

Some observations were incorrectly classified – these are cases when after the normalization and conversion to quantum states the samples became extremely different from the original observations. There are also cases when the quantum states are similar to states from at least two classes.

Table 1. The success ratios for the classification with SWAP-test and SQCs. The columns F_{c0}, F_{c1} and F_{c2} contain the intervals of calculated Fidelity values in SQCs (sets Moons, Circles and Blobs (1) have two final classes, and sets Blobs (2), Iris – three final classes). Two last columns indicate how many samples (of all samples in the set) were classified incorrectly

Set	F_{c0}	F_{c1}	F_{c2}	SWAP-test error	SQC error
Moons	(0.89, 0.98)	(0.02, 0.05)	–	0	4/256
Circles	(0.69, 0.95)	(0.01, 0.11)	–	0	6/256
Blobs (1)	(0.75, 0.92)	(0.01, 0.20)	–	2/256	5/256
Blobs (2)	(0.83, 0.92)	(0.47, 0.56)	(0.01, 0.20)	4/256	2/256
Iris	(0.87, 0.98)	(0.44, 0.59)	(0.01, 0.30)	5/150	15/150

5   Conclusions

As we can see, even very simple quantum circuits are capable of data classification. Naturally, the input data must be properly prepared. First, the set of observations must be normalized. Then, correct quantum states must be generated – the sum of all squares of the amplitudes modulus have to be equal 1. The utilized approach of transforming observations to quantum state, assures that each data sample is unique, and it is easy to classify the observations. Thanks to using qudits for sample sets with more features than two, the circuits are still simple and the number of utilized gates in low.

Acknowledgments. We would like to thank for useful discussions with the *Q-INFO* group at the Institute of Control and Computation Engineering (ISSI) of the University of Zielona Góra, Poland. We would like also to thank to anonymous referees for useful comments on the preliminary version of this paper. The numerical results were done using the hardware and software available at the "GPU μ-Lab" located at the Institute of Control and Computation Engineering of the University of Zielona Góra, Poland.

References

1. Aïmeur, E., Brassard, G., Gambs, S.: Machine learning in a quantum world. In: Lamontagne, L., Marchand, M. (eds.) AI 2006. LNCS (LNAI), vol. 4013, pp. 431–442. Springer, Heidelberg (2006). https://doi.org/10.1007/11766247_37
2. Biamonte, J., Wittek, P., Pancotti, N., Rebentrost, P., Wiebe, N., Lloyd, S.: Quantum machine learning. Nature **549**, 195–202 (2017)
3. Hall, B.C.: Lie Groups, Lie Algebras, and Representations: An Elementary Introduction. Springer, New York (2003)
4. Li, Z., Li, P.: Clustering algorithm of quantum self-organization network. Open J. Appl. Sci. **5**, 270–278 (2015)
5. Liu, D., et al.: Machine learning by unitary tensor network of hierarchical tree structure. New J. Phys. **21**, 073059 (2019)
6. MacMahon, D.: Quantum Computing Explained. Wiley, Hoboken (2007)

7. Murgia, M., Waters, R.: Google claims to have reached quantum supremacy. Financial Times, 20 September 2019
8. Narayanan, A., Menneer, T.: Quantum artificial neural network architectures and components. Inf. Sci. **128**(3–4), 231–255 (2000)
9. von Neumann, J.: Mathematical Foundations of Quantum Mechanics. Princeton University Press, Princeton (1955)
10. Nielsen, M.A., Chuang, I.L.: Quantum Computation and Quantum Information, 10 Anniversary edn. Cambridge University Press, Cambridge (2010)
11. Pedregosa, F., et al.: Scikit-learn: machine learning in Python. J. Mach. Learn. Res. **12**, 2825–2830 (2011)
12. Sawicki, A., Karnas, K.: Universality of single qudit gates. Annales Henri Poincaré **18**, 3515 (2017). https://doi.org/10.1007/s00023-017-0604-z
13. Schuld, M., Sinayskiy, I., Petruccione, F.: An introduction to quantum machine learning. Contemp. Phys. **56**, 172–185 (2015)
14. Veenman, C.J., Reinders, M.J.T.: The nearest sub-class classifier: a compromise between the nearest mean and nearest neighbor classifier. IEEE Trans. PAMI **27**(9), 1417–1429 (2005)
15. Virtanen, P., Gommers, R., Oliphant T.E.: SciPy 1.0-fundamental algorithms for scientific computing in Python. arXiv:1907.10121 (2019)
16. Weigang, L.: A study of parallel self-organizing map. arXiv:quant-ph/9808025v3 (1998)
17. Wiebe, N., Kapoor, A., Svore, K.M.: Quantum nearest-neighbor algorithms for machine learning. Quantum Inf. Comput. **15**(3–4), 318–358 (2015)
18. Zhou, S.S., Loke, T., Izaac, J.A., Wang, J.B.: Quantum fourier transform in computational basis. arXiv:quant-ph/1511.04818v2 (2016)
19. IBM Q experience. https://quantum-computing.ibm.com/. Accessed 28 Sept 2019
20. Jones, E., Oliphant, T., Peterson, P., et al.: SciPy: open source scientific tools for Python. https://www.scipy.org/. Accessed 28 Sept 2019
21. Rigetti QCS. https://www.rigetti.com/qcs. Accessed 28 Sept 2019

A DCA Based Algorithm for Feature Selection in Model-Based Clustering

Viet Anh Nguyen⊙, Hoai An Le Thi⊙, and Hoai Minh Le$^{(\boxtimes)}$⊙

Computer Science and Application Department, LGIPM,
University of Lorraine, Metz, France
{viet-anh.nguyen,hoai-an.le-thi,minh.le}@univ-lorraine.fr

Abstract. Gaussian Mixture Models (GMM) is a model-based clustering approach which has been used in many applications thanks to its flexibility and effectiveness. However, in high dimension data, GMM based clustering lost its advantages due to over-parameterization and noise features. To deal with this issue, we incorporate feature selection into GMM clustering. For the first time, a non-convex sparse inducing regularization is considered for feature selection in GMM clustering. The resulting optimization problem is nonconvex for which we develop a DCA (Difference of Convex functions Algorithm) to solve. Numerical experiments on several benchmark and synthetic datasets illustrate the efficiency of our algorithm and its superiority over an EM method for solving the GMM clustering using l_1 regularization.

Keywords: Model-based clustering · Gaussian Mixture Models · Variable selection · Non-convex regularization · DC programming · DCA

1 Introduction

Clustering (unsupervised classification) is a fundamental unsupervised learning problem and has numerous applications in various domains. In a general clustering problem, given a dataset $X = \left\{ x_i \in \mathbb{R}^D : i \in \{1, ..., N\} \right\}$ of N data points, one aims to divide those data points into K homogeneous clusters such that the similarity of objects within a cluster is high while the similarity of two objects in different clusters is small. Clustering is often referred to as an ill-posed problem due to the difficulty to define in a formal way what the true cluster are [5]. For a given dataset, there does not exist a unique clustering solution and the definition of what the true cluster are should depend on the context of clustering [5,7].

In this work, we are interested in model-based clustering which allows to define the clusters through the probability distribution. Therefore, the clustering result can be interpreted and analysed from a statistical point of view. The advantages of model-based clustering methods lie in the solid probabilistic foundations along with their flexibility compared to other classical clustering methods

© Springer Nature Switzerland AG 2020
N. T. Nguyen et al. (Eds.): ACIIDS 2020, LNAI 12033, pp. 404–415, 2020.
https://doi.org/10.1007/978-3-030-41964-6_35

such as k-means. The first work using mixture model to define clusters was introduced by Wolfe in 1963 [25]. Since then, model-based clustering methods have largely been developed in several application fields. The readers are referred to [18] or [23] for a deep review on existing works of model-based clustering.

In model-based clustering, data is assumed to be generated from a finite mixture of K probability distributions:

$$p(x_i \mid \theta_1, ..., \theta_K) = \sum_{k=1}^{K} \tau_k p(x_i \mid \theta_k), \tag{1}$$

where θ_k, τ_k are respectively the parameter and the mixing proportion of the k-th probability distribution $p(\cdot \mid \theta_k)$, with $\sum_{k=1}^{K} \tau_k = 1$ and $\tau_k > 0$ for all $k \in \{1, ..., K\}$. Among several probability distributions, the Gaussian distribution is certainly the most developed thanks to its interesting properties on both theoretical and computational aspect [3]. In Gaussian mixture models (GMM), each component probability distribution is a Gaussian distribution with mean μ_k and inverse covariance matrix W_k:

$$p(x_i \mid \theta_k = (\mu_k, W_k)) = \sqrt{\frac{\det(W_k)}{(2\tau)^D}} \exp\left(-\frac{1}{2}(x_i - \mu_k)^T W_k (x_i - \mu_k)\right). \tag{2}$$

Then, the parameters of GMM can be estimated by solving the following maximum log-likelihood problem:

$$\max_{\Theta \in D} L(\Theta) := \sum_{i=1}^{N} \log\left(\sum_{k=1}^{K} \tau_k p(x_i \mid \theta_k)\right), \tag{3}$$

where

$$\Theta = (\tau, \theta) = (\tau_1, ..., \tau_K, \mu_1, ..., \mu_K, W_1, ... W_K), \tag{4}$$

and

$$D = \left\{ (\tau, \mu, W) : \sum_{k=1}^{K} \tau_k = 1, \ W_k \succ 0, \ \forall k \in \{1, ..., K\} \right\}. \tag{5}$$

The above optimization problem can be efficiently solved by the Expectation-Maximisation (EM) algorithm. Although the development of model-based clustering methods has been active in last decades, dealing with the high dimensionality of data remains a challenging issue. On the one hand, when the number of dimensions D is high, the model-based clustering methods badly suffer from the well-known curse of dimensionality problem, i.e., a very large number of data is needed to correctly estimate model's parameters [3]. Note that the number of parameters in GMM is a quadratic function of D. On the other hand, it is well-known that high-dimensional data potentially contains non-informative features which can be harmful to clustering result. To deal with the high-dimensional data issue in model-based clustering, several methods have been proposed. A review of existing approaches can be found in [3]. In the first approach, namely dimension reduction, the high-dimensional data is projected into a low-dimensional space then traditional model-based clustering methods can be applied on the projected data. Principal Component Analysis (PCA), Factor Analysis, etc. belong to this first approach. The second approach, feature selection, consists in removing non-informative features and using only a subset of relevant features for the clustering. From a mathematical point of view, the feature selection problem can be formulated as a sparse optimization problem which involves the minimization of the l_0-norm ($\|.\|_0$). The minimization of zero-norm is known to be NP-hard for which several methods have been developed in the literature. The readers are referred to [15] for an extensive overview of existing methods in sparse optimization. The feature selection methods have been intensively developed for numerous classification methods such as SVM [4,12], Semi-Supervised SVM [11], Multi-class Logistic Regression, etc. Motivated by the success of feature selection in classification methods, in this work, we will incorporate it into GMM. The GMM with feature selection is formulated as

$$\max_{\Theta \in \mathcal{D}} L(\Theta) - \lambda \mathcal{P}_{l_0}(\Theta), \tag{6}$$

where $\mathcal{P}_{l_0} := \sum_{k=1}^{K} \sum_{d=1}^{D} |\mu_{kd}|_0$ is the regularization term and $\lambda > 0$ is the trade-off parameters between the two terms. As we have mentioned, the discontinuity of l_0-norm function makes the optimization problem (6) hard to solve. Hence, one usually replaces the l_0-norm by a convex approximation. In [19,26], the authors replaced the l_0 on the variable μ by the l_1 regularization defined as:

$$\mathcal{P}_{l_1}(\mu) = \sum_{k=1}^{K} \sum_{d=1}^{D} |\mu_{kd}|.$$

After the convergence, if $\mu_{kd} = 0$ for all $k \in \{1, ..., K\}$, the dimension d can be considered as non-informative, thus can be removed. In the same manner, Bhattacharya and McNicholas [2] introduced a variant of l_1 regularization function with the presence of proportion of each cluster:

$$\mathcal{P}_{\lambda}(\mu) = N \sum_{k=1}^{K} \tau_k \sum_{d=1}^{D} |\mu_{kd}|.$$

In another work, Wang and Zhu [24] replaced the l_0 norm by the l_∞ norm. Later, Guo et al. [6] introduced the pairwise fusion penalty defined as follows

$$\mathcal{P}_{fusion}(\mu) = \sum_{d=1}^{D} \sum_{1 \leq k < k' \leq K} |\mu_{kd} - \mu_{k'd}|.$$

Instead of shrinking the means to 0, the pairwise fusion penalty shrinks the means towards each other. Thus, dimension d is considered non-informative if all the values of means at that dimension are equal, i.e. $\mu_{kd} = \mu_{k'd}$ for all

$k \neq k'$. It is important to note that, in [6,19,24], the authors assumed further a common diagonal co-variance matrix across clusters. This assumption restricts the correlations between features, therefore reduces the flexibility of the model.

In this work, we will approximate the l_0-norm by a nonconvex function. This choice is motivated by the fact non-convex approximations have been proved to be more efficient than convex approximation and can produce good sparsity [15]. However, the nonconvex approximations make the corresponding optimization problem harder to solve. To the best of our knowledge, this is the first work using a non-convex sparse inducing regularization for feature selection in model-based clustering. Among the well-known existing non-convex approximations, we use the exponential concave function approximation function which has been successfully applied in several machine learning problems such as feature selection in SVM [4,12] and S3VM [11], sparse signal recovery [13], sparse scoring [16], etc. In [12], the authors reformulated the exponential concave approximation as a DC (Difference of Convex functions) and developed an efficient DCA (DC Algorithm) to solve it. Using the same DC decomposition, we will prove that the GMM clustering with feature selection (6) can be written as a DC program and then investigate a DCA based algorithm to solve the resulting optimization problem.

The remainder of the paper is organized as follows. The outline of DC programming and DCA is presented in Sect. 2 then the proposed DCA will be developed in Sect. 3. Computational experiments are reported in Sect. 4.

2   DC Programming and DCA

DC programming and DCA constitute the backbone of smooth/non-smooth nonconvex programming and global optimization [14,21,22]. They address the problem of minimizing a DC function on the whole space $\mathbb{R}^n$ or on a closed convex set $\Omega \subset \mathbb{R}^n$. Generally speaking, a standard DC program takes the form:

$$\alpha = \inf\{F(x) := G(x) - H(x) \,|\, x \in \mathbb{R}^n\} \quad (P_{dc}),$$

where G, H are lower semi-continuous proper convex functions on $\mathbb{R}^n$. Such a function F is called a DC function, and $G - H$ is a DC decomposition of F while G and H are the DC components of F. A DC program with convex constraint $x \in \Omega$ can be equivalently expressed as an unconstrained DC program by adding the indicator function χ_Ω ($\chi_\Omega(x) = 0$ if $x \in \Omega$ and $+\infty$ otherwise) to the first DC component G.

The main idea of DCA is simple: each iteration k of DCA approximates the concave part $-H$ by its affine majorization (that corresponds to taking $y^k \in \partial H(x^k)$) and computes x^{k+1} by solving the resulting convex problem,

$$\min\{G(x) - \langle x, y^k \rangle : x \in \mathbb{R}^n\} \quad (P_k).$$

The sequence $\{x^k\}$ generated by DCA enjoys the following properties [14, 21]:

(i) The sequence $\{F(x^k)\}$ is decreasing.
(ii) If $F(x^{k+1}) = F(x^k)$, then x^k is a critical point of (P_{dc}) and DCA terminates at the k-th iteration.
(iii) If $\mu(G) + \mu(H) > 0$ then the series $\{\|x^{k+1} - x^k\|^2\}$ converges.
(iv) If the optimal value α of (P_{dc}) is finite and the infinite sequence $\{x^k\}$ is bounded then every limit point of the sequence $\{x^k\}$ is a critical point of $G - H$.

DCA is well-known as an efficient approach in the nonconvex programming framework thanks to its versatility, flexibility, robustness, inexpensiveness and their adaptation to the specific structure of considered problems. Numerous DCA-based algorithms have been developed for successfully solving large-scale nons-mooth/nonconvex programs arising in several application areas (see the list of references in [17]). For a comprehensible survey on thirty years of development of DCA, the reader is referred to the recent paper [17].

3   DCA for Solving GMM with l_0-norm Regularization

We approximate the l_0 norm of $s \in \mathbb{R}$ by the exponential concave function [12] defined by $r_\alpha(s) = 1 - \exp(-\alpha|s|)$, where the parameter $\alpha > 0$ controls the tightness of the approximation. By using the exponential concave function $r_\alpha(s)$, the corresponding approximate problem of (6) takes the form:

$$\min_{\Theta \in \mathcal{D}} \mathcal{F}(\Theta) := -L(\Theta) + \lambda \mathcal{P}_\alpha(\Theta), \tag{7}$$

where

$$\mathcal{P}_\alpha(\Theta) = \sum_{k=1}^{K} \sum_{d=1}^{D} r_\alpha(\mu_{kd}). \tag{8}$$

We now prove that $F(\Theta)$ can be rewritten as a DC function. On the one hand, since $-L$ is at least twice differentiable, a natural approach is to use ρ-decomposition, of which $-L(\Theta)$ is written as $-L(\Theta) = G_1(\Theta) - H_1(\Theta)$ where:

$$G_1(\Theta) = \frac{\rho}{2}\|\Theta\|^2,$$

$$H_1(\Theta) = \frac{\rho}{2}\|\Theta\|^2 + L(\Theta).$$

There exists a positive value ρ_0 such that for all $\rho > \rho_0$ the function $H_1(\Theta)$ is convex. In deed, denote by $\lambda_n(\Theta)$ the largest eigenvalue of the Hessian matrix of $-L$ at Θ. It is easy to prove that for all $\rho > max\{0, \lambda_n(\Theta)\}$ the function $H_1(\Theta)$ is convex.

On the other hand, $r_\alpha(s)$ is a DC function with the following DC decomposition [12]:

$$r_\alpha(s) = \alpha|s| - [\alpha|s| - r_\alpha(s)]. \tag{9}$$

Hence $\mathcal{P}_\alpha(\Theta)$ is also a DC function

$$\mathcal{P}_\alpha(\Theta) = G_2(\Theta) - H_2(\Theta),$$
$$G_2(\Theta) = \alpha\|\mu\|_1,$$

$$H_2(\Theta) = \alpha\|\mu\|_1 - \sum_{k=1}^{K}\sum_{d=1}^{D}\left[1 - \exp\left(-\alpha|\mu_{kd}|\right)\right].$$

Consequently, $\mathcal{F}(\Theta) = \mathcal{G}(\Theta) - \mathcal{H}(\Theta)$ is a DC function with $\mathcal{G}(\Theta) := G_1(\Theta) + \lambda G_2(\Theta)$ and $\mathcal{H}(\Theta) := H_1(\Theta) + \lambda H_2(\Theta)$. Thus, DCA can be developed for solving the optimization problem (7). According to general DCA scheme, at each iteration t we compute $\bar{\Theta}^t = (\bar{\tau}^t, \bar{\mu}^t, \bar{W}^t) \in \partial\mathcal{H}(\tau^t, \mu^t, W^t)$ and then compute $(\tau^{t+1}, \mu^{t+1}, W^{t+1})$ as the solution to the following convex problem

$$\min_{\tau,\mu,W \in \mathcal{D}} \left\{\mathcal{G}(\tau,\mu,\mathcal{W}) - \langle(\bar{\tau}^t, \bar{\mu}^t, \bar{W}^t), (\tau,\mu,W)\rangle\right\}. \tag{10}$$

Computation of $(\bar{\tau}^t, \bar{\mu}^t, \bar{W}^t) \in \partial\mathcal{H}(\tau^t, \mu^t, W^t)$
The function $\mathcal{H}$ is differentiable and its gradient can be computed as

$$\bar{\tau}^t = \nabla_\tau\mathcal{H}(\tau^t, \mu^t, W^t) = \rho\tau^t + \nabla_\tau L(\tau^t, \mu^t, W^t), \tag{11}$$
$$\bar{\mu}^t = \nabla_\mu\mathcal{H}(\tau^t, \mu^t, W^t) = \rho\mu^t + \nabla_\tau L(\tau^t, \mu^t, W^t) + \lambda\nabla_\tau H_2(\mu^t), \tag{12}$$
$$\bar{\tau}^t = \nabla_W\mathcal{H}(\tau^t, \mu^t, W^t) = \rho W^t + \nabla_W L(\tau^t, \mu^t, W^t). \tag{13}$$

The gradient of $H_2(\mu^t)$ is given by

$$\frac{\partial H_2}{\partial \mu_{kd}}(\mu^t) = \begin{cases} \alpha(1 - exp(-\alpha\mu_{kd}^t)) & \text{if } mu_{kd}^t \geq 0 \\ -\alpha(1 - exp(\alpha\mu_{kd}^t)) & \text{if } mu_{kd}^t < 0 \end{cases} \quad \forall k = 1..K, d = 1..D. \tag{14}$$

Let us now consider the function

$$p(\theta; x) = p(x \mid \theta = (\mu, W)) = \sqrt{\frac{\det(W)}{(2\pi)^D}}\exp\left(-\frac{1}{2}(x-\mu)^T W(x-\mu)\right).$$

We have

$$\frac{\partial p(\theta; x)}{\partial \mu} = \sqrt{\frac{\det(W)}{(2\pi)^D}}\exp\left(-\frac{1}{2}(x-\mu)^T W(x-\mu)\right)W(x-\mu). \tag{15}$$

and

$$\frac{\partial p(\theta; x)}{\partial W} = \frac{1}{(2\pi)^{D/2}}\frac{1}{2\sqrt{\det(W)}}\frac{\partial \det(W)}{\partial W}\exp\left(-\frac{1}{2}(x-\mu)^T W(x-\mu)\right)$$

$$= +\sqrt{\frac{\det(W)}{(2\pi)^D}}\exp\left(-\frac{1}{2}(x-\mu)^T W(x-\mu)\right)\left[-\frac{1}{2}(x-\mu)(x-\mu)^T\right]. \tag{16}$$

Since $W \succ 0$, Jacobi's formula implies

$$\frac{\partial \det(W)}{\partial W} = \text{adj}^T(W) = \text{adj}(W) = \det(W)W^{-1}.$$

Thus, we obtain

$$\begin{aligned}
\frac{\partial p(\theta; x)}{\partial W} &= \frac{1}{(2\pi)^{D/2}} \frac{\sqrt{\det(W)}}{2} \exp\left(-\frac{1}{2}(x-\mu)^T W(x-\mu)\right) W^{-1} \\
&= -\frac{1}{2}\sqrt{\frac{\det(W)}{(2\pi)^D}} \exp\left(-\frac{1}{2}(x-\mu)^T W(x-\mu)\right)(x-\mu)(x-\mu)^T \\
&= \frac{1}{2}\sqrt{\frac{\det(W)}{(2\pi)^D}} \exp\left(-\frac{1}{2}(x-\mu)^T W(x-\mu)\right)\left[W^{-1} - (x-\mu)(x-\mu)^T\right].
\end{aligned}$$

$$\tag{17}$$

Denotes

$$A_i^t = \sum_{k=1}^{K} \tau_k^t\, p(x_i \mid \theta_k^t), \qquad \forall i = 1, \ldots, N. \tag{18}$$

$$\frac{\partial L}{\partial \tau_k}(\Theta^t) = \sum_{i=1}^{N} \frac{p(x_i \mid \theta_k^t)}{A_i^t}, \qquad \forall k = 1, \ldots, K. \tag{19}$$

$\nabla L(\Theta^t)$ can be computed as follows

$$\nabla_{\tau_k} L(\Theta^t) = \sum_{i=1}^{N} \frac{p(x_i \mid (\mu_k^t, W_k^t))}{\sum_{l=1}^{K} p(x_i \mid (\mu_l^t, W_l^t))}, \tag{20}$$

$$\begin{aligned}
\nabla_{\mu_k} L(\Theta^t) &= \sum_{i=1}^{N} \frac{\tau_k^t}{A_i^t} \frac{\partial p(\theta_k^t; x_i)}{\partial \mu_k} \\
&= \sum_{i=1}^{N} \frac{\tau_k^t}{A_i^t} \sqrt{\frac{\det(W_k^t)}{(2\pi)^D}} \exp\left(-\frac{1}{2}(x_i - \mu_k^t)^T W_k^t(x_i - \mu_k^t)\right) \times \\
&\qquad\qquad\qquad\qquad\qquad\qquad\qquad\qquad W_k^t(x_i - \mu_k^t), \tag{21}
\end{aligned}$$

$$\begin{aligned}
\nabla_{W_k} L(\Theta^t) &= \sum_{i=1}^{N} \frac{\tau_k^t}{A_i^t} \frac{\partial p(\theta_k^t; x_i)}{\partial W_k} \\
&= \sum_{i=1}^{N} \frac{\tau_k^t}{A_i^t} \frac{1}{2} \sqrt{\frac{\det(W_k^t)}{(2\pi)^D}} \exp\left(-\frac{1}{2}(x_i - \mu_k^t)^T W_k^t(x_i - \mu_k^t)\right) \times \\
&\qquad\qquad\qquad\qquad\qquad \left[(W_k^t)^{-1} - (x_i - \mu_k^t)(x_i - \mu_k^t)^T\right]. \tag{22}
\end{aligned}$$

Remark. The computation of $\nabla L(\Theta)$ requires to compute the determinant of W_k and the inverse matrix of W_k which can be time consuming. In practice, we can use an automatic differentiation method to numerically compute an approximate value of $\nabla L(\Theta)$. Note that, in this case, DCA still converges to a critical point [17].

Compute $(\tau^{t+1}, \mu^{t+1}, W^{t+1})$ by solving the convex sub-problem (10)

Fortunately, the optimization problem (10) is separable on all three variables. Thus, we have

$$\tau^{t+1} \in \arg\min \left\{ \frac{\rho}{2}\|\tau\|^2 - \langle \bar{\tau}^t, \tau \rangle : \sum_{k=1}^{K} \tau_k = 1 \right\}, \tag{23}$$

$$\mu^{t+1} \in \arg\min \left\{ \frac{\rho}{2}\|\mu\|^2 + \alpha\|\mu\|_1 - \langle \bar{\mu}^t, \mu \rangle \right\}, \tag{24}$$

$$W^{t+1} \in \arg\min \left\{ \frac{\rho}{2}\|W\|^2 - \langle \bar{W}^t, W \rangle : W_k \succ 0, \ \forall k \in \{1, ..., K\} \right\}. \tag{25}$$

The solution of (23) is nothing else but the projection of the point $\dfrac{\bar{\tau}^t}{\rho}$ onto a simplex $\Delta := \left\{ \tau \in (0,1)^K : \sum_{k=1}^{K} \tau_k = 1 \right\}$. The projection of points into a simplex can be efficiently computed. For instance, we can use a very inexpensive algorithm developed in [10]. Similarly, W_k^{t+1}, for all $k = 1..K$, is the projection of $\dfrac{\bar{W}_k^t}{\rho}$ onto the positive definite cone $\mathcal{W}_k := \{W_k : W_k \succ 0\}$ for which some efficient algorithms are available, e.g., QUIC [8]. As for the problem (24), it can be efficiently solved using the soft thresholding operator [1].

Finally, the DCA scheme for solving the problem (7) can be described as follows.

DCA-GMM

Initialization: Let $(\tau^{(0)}, \mu^{(0)}, W^{(0)}) \in \mathcal{D}$ be a initial point, $\alpha > 0$, $\lambda \geq 0$ and $\rho > L$. $t \leftarrow 0$.

repeat

   Compute $\bar{\Theta}^t = (\bar{\tau}^t, \bar{\mu}^t, \bar{W}^t) \in \partial \mathcal{H}(\tau^t, \mu^t, W^t)$ using (11), (12) and (13).

   Compute $(\tau^{t+1}, \mu^{t+1}, W^{t+1})$ by

$$\tau^{t+1} = \text{Proj}_\Delta(\frac{\bar{\tau}^t}{\rho}),$$

   Solve (24) using soft thresholding operator to obtain μ^{t+1},

$$W_k^{t+1} = \text{Proj}_{\mathcal{W}_k}(\frac{\bar{W}_k^t}{\rho}), \forall k = 1..K.$$

until stopping criterion.

---

4   Numerical Experiment

Comparative Algorithm. We compare our algorithm with a modified EM algorithm presented in the work of Zhou et al. [26] for solving the GMM with l_1 regularization, namely

$$\max_{\Theta \in \mathcal{D}} L\left(\Theta\right) - \lambda \mathcal{P}_{l_1}\left(\Theta\right), \tag{26}$$

with $\mathcal{P}_{l_1}\left(\Theta\right) = \sum_{k=1}^{K} \sum_{d=1}^{D} |\mu_{kd}|$.

Comparative Criteria. To evaluate the performance of algorithms, we consider the following criteria

- Adjusted Rand Index (ARI): a well-known measure of the similarity between two clusterings [9].
- Selected Feature percentage (SF): the percentage of selected features over the total number of features.
- True Positive Rate (TPR): the percentage of features selected which is informative over all informative features.
- False Positive Rate (FPR): the percenrage of non-informative features selected over all non-informative features.
- Computation time.

Experiment Setting. The parameter α is chosen from the set $\{1, 5, 10\}$ while λ belongs to $\{0.1, 0.2, \ldots, 10\}$. We use a grid search procedure for choosing the besst value of α and λ. To reduce the effects of initial values and local minima, we repeat the search multiple times. The grid point with highest average BIC value, as defined below, is chosen as the optimal tuning parameters.

$$BIC(\alpha, \lambda) = \sum_{i=1}^{N} \log \left(\sum_{k=1}^{K_e} \widetilde{\pi_k} p\left(x_i \mid \widetilde{\theta}_k\right) \right) - \frac{1}{2} \log(N) \sum_{k=1}^{K_e} P_k.$$

We run each algorithm 100 times with the optima parameters and report the mean and standard deviation of each criteria. K-means is used for finding an initial point for all algorithms. Convex subproblems on sparse covariance selection are solved by QUIC software [8]. We use an automatic differentiation algorithm [20] to compute an approximate value of $\nabla L(\Theta)$.

The experiments are performed on a Intel Core i7 3.60 GHz PC with 16 GB of RAM and the codes were written in MATLAB.

Experiment 1. The purpose of this experiment is to evaluate the algorithm's capacity to select informative features to provide high classification accuracy. Hence, the experiment is performed on a synthetic dataset whose the informative features are known a priori. The synthetic dataset is generated as follows. First, we create three vectors of dimension 20, namely m_1, m_2, m_3. Three first elements of m_1 are set to 1 and the rest are 0. Similarly, the 4th, 5th and 6th elements of m_2 are 1 and the rest are 0. The element 7th, 8th and 9th of m_3 are equal to 1 and the rest are 0. Using these three vectors, we generate three corresponding multivariate Gaussians, each possesses m_i as its mean and the co-variance matrix is unit diagonal matrix. From each Gaussian, we randomly generate 100 data points. Thus, the synthetic dataset contains 300 data points evenly divided into 3 classes; each data point is represented by 20 features but only first 9 features are informative.

Table 1. Comparative results on synthetic dataset

	DCA-GMM	EM-GMM [26]
SF(%)	55 ± 8.5	**54 ± 15.57**
TPR(%)	**91.11 ± 8.76**	71.11 ± 18.59
FPR(%)	**25.45 ± 11.18**	40 ± 13.79
ARI	**0.66 ± 0.02**	0.07 ± 0.02
Time(s)	**6 ± 0.029**	106 ± 1.4935

We observe from the results (Table 1) that *EM-GMM* selects slightly less features than *DCA-GMM* (54% vs 55%). However, *DCA-GMM* selects more informative features (91.11%) than *EM-GMM* (71.11%). Consequently, *DCA-GMM* selects less non-informative than *EM-GMM* (25.45% vs 40%). As for the ARI criterion, *DCA-GMM* obtains 0.6595 which is significantly higher than 0.0663 of *EM-GMM*. The computation time of *DCA-GMM* is only 6 s while *EM-GMM* needs 106 s to furnish a result.

Experiment 2. In the second experiment, we compare the performance of *DCA-GMM* and *EM-GMM* on several Benchmark datasets. The Benchmark datasets are taken from the well-know UCI Machine Learning Repository. We summarize the information of used Benchmark datasets in Table 2 and report the comparative results in Table 3.

Table 2. Datasets

Dataset	Instances (N)	Dimension (D)	Classes (K)
comp	3891	10	3
ionosphere	351	32	2
iris	150	4	3
thyroid	215	5	3

We observe that *DCA-GMM* has smaller SF in all datasets compared to *EM-GMM*, which selects all features in all datasets. In *iris* dataset, *EM-GMM* has higher ARI than *DCA-GMM* (0.9039 vs 0.7555) while for the other datasets (*comp*, *ionosphere* and *thyroid*), the ARIs of *DCA-GMM* are higher. As for the computation time, *DCA-GMM* is significantly faster than *EM-GMM*, e.g. up to 71 times faster on *comp* dataset.

Table 3. Comparative results on Benchmark datasets

Datasets		DCA-GMM	EM-GMM [26]
comp	SF(%)	**80 ± 9.01**	100 ± 0
	ARI	**0.966 ± 0.0037**	0.9279 ± 0.0015
	Time(s)	**9.1 ± 1.7**	641.8 ± 185.2
ionosphere	SF(%)	**43.75 ± 3.03**	100 ± 0
	ARI	**0.7716 ± 0.0099**	0.4089 ± 0
	Time(s)	**5.8 ± 0.09**	78 ± 0.04
iris	SF(%)	**75 ± 0**	100 ± 0
	ARI	0.7555 ± 0.0091	**0.9039 ± 0**
	Time(s)	**0.39 ± 0.09**	24.6 ± 0.06
thyroid	SF(%)	**60 ± 0**	100 ± 0
	ARI	**0.9357 ± 0.0028**	0.8933 ± 0
	Time(s)	**0.3 ± 0.009**	18.1 ± 0.023

Overall, *DCA-GMM* furnishes better result than *EM-GMM* on all three comparative criteria.

5   Conclusion

We have studied the Gaussian mixture model clustering. Furthermore, to deal with the high-dimensional data, we use feature selection which involves the minimization of l_0-norm. The purpose of feature selection is double. Firstly, by removing non-informative features, we could improve the clustering results. Secondly, it also decreases the number of parameters to estimate in the GMM clustering. We approximate the l_0 norm by a non-convex function, namely the exponential concave function. This is the first work using a non-convex approximation for feature selection in model-based clustering. The resulting problem is then reformulated as a DC function and we developed a DCA to solve it.

We have carefully conducted numerical experiments on both synthetic and Benchmark datasets. The results on synthetic dataset show that our algorithm *DCA-GMM* selects more informative features than *EM-GMM* and consequently gives better classification accuracy. Furthermore, *DCA-GMM* is 17.6 times faster than *EM-GMM* on synthetic dataset. Similarly, *DCA-GMM* outperforms *EM-GMM* on several Benchmark datasets, with respect to all three comparative criteria.

References

1. Beck, A., Teboulle, M.: A fast iterative shrinkage-thresholding algorithm for linear inverse problems. SIAM J. Imaging Sci. **2**, 183–202 (2009)
2. Bhattacharya, S., McNicholas, P.D.: A LASSO-penalized BIC for mixture model selection. Adv. Data Anal. Classif. **8**, 45–61 (2014)

3. Bouveyron, C., Brunet, C.: Model-based clustering of high-dimensional data: a review. Comput. Stat. Data Anal. **71**, 52–78 (2013)
4. Bradley, P.S., Mangasarian, O.L.: Feature selection via concave minimization and support vector machines. In: Proceedings of the Fifteenth International Conference on Machine Learning ICML 1998, pp. 82–90 (1998)
5. Grun, B.: Model-based clustering. In: Fruhwirth-Schnatter, S., Celeux, G., Robert, C.P. (eds.) Handbook of Mixture Analysis. Taylor and Francis, New York (2019)
6. Guo, J., Levina, E., Michailidis, G., Zhu, J.: Pairwise variable selection for high-dimensional model-based clustering. Biometrics **66**, 793–804 (2009)
7. Hennig, C.: What are the true clusters? Pattern Recogn. Lett. **64**, 53–62 (2015)
8. Hsieh, C.J., Sustik, M.A., Dhillon, I.S., Ravikumar, P.: QUIC: quadratic approximation for sparse inverse covariance estimation. J. Mach. Learn. Res. **15**, 2911–2947 (2014)
9. Hubert, L., Arabie, P.: Comparing partitions. J. Classif. **2**, 193–218 (1985)
10. Judice, J., Raydan, M., Rosa, S.: On the solution of the symmetric eigenvalue complementarity problem by the spectral projected gradient algorithm. Numer. Algorithms **47**, 391–407 (2008)
11. Le, H.M., Le Thi, H.A., Nguyen, M.C.: Sparse semi-supervised support vector machines by DC programming and DCA. Neurocomputing **153**, 62–76 (2015)
12. Le Thi, H.A., Le, H.M., Nguyen, V.V., Pham Dinh, T.: A DC programming approach for feature selection in support vector machines learning. J. Adv. Data Anal. Classif. **2**, 259–278 (2013)
13. Le Thi, H.A., Nguyen Thi, B.T., Le, H.M.: Sparse signal recovery by difference of convex functions algorithms. In: Selamat, A., Nguyen, N.T., Haron, H. (eds.) ACIIDS 2013. LNCS (LNAI), vol. 7803, pp. 387–397. Springer, Heidelberg (2013). https://doi.org/10.1007/978-3-642-36543-0_40
14. Le Thi, H.A., Pham Dinh, T.: The DC (difference of convex functions) programming and DCA revisited with DC models of real world nonconvex optimization problems. Ann. Oper. Res. **133**(1–4), 23–46 (2005)
15. Le Thi, H.A., Pham Dinh, T., Le, H.M., Vo, X.T.: DC approximation approaches for sparse optimization. Eur. J. Oper. Res. **244**(1), 26–46 (2015)
16. Le Thi, H.A., Phan, D.N.: DC programming and DCA for sparse optimal scoring problem. Neurocomput. **186**(C), 170–181 (2016)
17. Le Thi, H.A., Pham Dinh, T.: DC programming and DCA: thirty years of developments. Math. Program. **169**(1), 5–68 (2018)
18. McNicholas, P.: Model-based clustering. J. Classif. **33**, 331–373 (2016)
19. Pan, W., Shen, X.: Penalized model-based clustering with application to variable selection. J. Mach. Learn. Res **8**, 1145–1164 (2007)
20. Paszke, A., et al.: Automatic differentiation in PyTorch. In: NIPS Autodiff Workshop (2017)
21. Pham Dinh, T., Le Thi, H.A.: Convex analysis approach to DC programming: theory, algorithms and applications. Acta Math. Vietnamica **22**(1), 289–355 (1997)
22. Pham Dinh, T., Le Thi, H.A.: A D.C. optimization algorithm for solving the trust-region subproblem. SIAM J. Optim. **8**(2), 476–505 (1998)
23. Stahl, D., Sallis, H.: Model-based cluster analysis. Comput. Stat. **4**, 341–358 (2015)
24. Wang, S., Zhu, J.: Model-based high-dimensional clustering and its application to microarray data. Biometrics **64**, 440–448 (2008)
25. Wolfe, J.: Object cluster analysis of social areas. Master's thesis, Ph.D. thesis, California, Berkeley (1963)
26. Zhou, H., Pan, W., Shen, X.: Penalized model-based clustering with un-constrained covariance matrices. Electron. J. Stat. **3**, 1473–1496 (2007)

PRTNets: Cold-Start Recommendations Using Pairwise Ranking and Transfer Networks

Dylan M. Valerio and Prospero C. Naval Jr.[✉]

Computer Vision and Machine Intelligence Group,
Department of Computer Science,
University of the Philippines,
Quezon City, Philippines
dylan_kemuel.valerio@upd.edu.ph, pcnaval@up.edu.ph

Abstract. In collaborative filtering, matrix factorization, which decomposes the ratings matrix into low rank user and item latent matrices is widely used. The decomposition is based on the rating scores of users to item, with the user and item latent matrices sharing a common embedding space. A similarity function between the two represents the predicted rating of a user to an item. However, this matrix factorization approach falls short for cold-start recommendation where items have very few or no ratings. This paper puts forward a novel approach of doing cold-start recommendation by using a neural network, the Transfer Network, to learn a nonlinear mapping from item features to the item latent matrix. The item latent matrix is produced by another network, the Pairwise Ranking Network, which utilizes pairwise ranking functions. The Pairwise Ranking Network efficiently utilizes implicit feedback by optimizing the ranking of the recommendation list. We find the optimal architecture for the Pairwise Network and the Transfer Network through warm-start and cold-start evaluation. With the Transfer Network, we map the Tag Genome dataset to the item latent matrix and produce cold-start recommendations for a test set derived from the MovieLens 20M dataset. Our approach yielded a significant margin of improvement of 0.276 and 0.089 average precision at $k = 10$ over the baseline LightFM and neighborhood averaging methods respectively.

Keywords: Machine learning · Recommender systems · Neural networks · Transfer learning

1 Introduction

Recommendation algorithms are a key component in many commercial applications deployed in the web. Due to the widespread use of the internet, websites have been serving thousands to hundreds of millions of items to millions of users. Each user has unique preferences and limited attention span. Recommender systems address this issue by enabling personalization through matching users with

© Springer Nature Switzerland AG 2020
N. T. Nguyen et al. (Eds.): ACIIDS 2020, LNAI 12033, pp. 416–428, 2020.
https://doi.org/10.1007/978-3-030-41964-6_36

items that they are most likely to engage, watch or purchase. Hence, the task of a recommendation algorithm is to provide users a ranked list of interesting items. This *ranking problem* is a difficult task since searching for the possible ranking lists of size k over n items for all m users is of $\mathcal{O}(m\binom{n}{k})$, a significant combinatorial challenge. Another challenge that recommendation algorithms must take into account is the fact that items and users are added continuously. This means that new items or users that do not have interactions yet cannot be recommended using collaborative filtering. This is called the *cold-start problem*.

In this research, we improve existing algorithms that tackle the cold-start problem. We use PRTNets, Pairwise Ranking and Transfer Networks, that can create cold-start recommendations and within its framework also handles the ranking problem. First, we develop neural networks that use pairwise loss functions to create a ranked list of items. The Pairwise Ranking Network factorizes the ratings data into the user and item latent matrices through matrix factorization with pairwise loss functions. Secondly, we use another neural network, the Transfer Network, to transform item features to the item latent matrix created by the Pairwise Ranking Networks. Through item features, items that do not yet have interactions form a ranked list of items that is personalized for each user. Both networks are capable of scaling to large datasets through online learning.

2   Related Literature

2.1   Matrix Factorization Using Pairwise Loss Functions

Collaborative filtering algorithms create recommendations by utilizing the ratings matrix $R^{m \times n}$ of m users to n items, where each element of this matrix, r_{ui}, denotes the rating of user $u \in U$ to an item $i \in I$. Matrix factorization, a popular collaborative filtering algorithm, utilizes the latent structure of the ratings matrix by decomposing it into two sets of lower rank dense matrices $P \in R^{m \times k}$ and $Q \in R^{k \times n}$, where k is the rank. These two low rank dense matrices are called the user and item latent matrices. The dot product of the row and column vectors is the predicted rating.

Factorization of the ratings matrix involves the minimization of various loss functions for different recommendation tasks. Pointwise loss functions such as weighted regularized matrix factorization, Singular Value Decomposition (SVD), and alternating least squares, involve the minimization of a squared loss objective between the predicted rating and the original rating [4, 6, 13]. Another set of methods make use of pairwise loss functions which predict the optimal ranked lists for each user rather than minimizing the difference between the predicted rating and the original rating [7, 8, 11, 15]. These functions have also the advantage of being able to use implicit feedback (user likes, item views, music plays) which involve a vast majority of user interactions.

For a specific user u and item i, the dot product of the user latent vector p_{uf} and the item latent vector q_i denotes the score $\hat{y}_{ui}$ (Eq. 1). In pairwise loss functions, this does not represent a predicted rating but rather denotes a numerical value to base the ranking of recommended items.

$$\hat{y}_{ui} = p_u \cdot q_i \tag{1}$$

Optimizing the ranking for all users and all items is inefficient especially for large datasets so pairwise loss functions are proposed as a solution to create an approximate ranking function. These are differentiable functions that can be solved via gradient descent. Pairwise ranking loss functions use triplets – a user u, a positive item $i \in I_u^+$ by that user and another item that is not rated by that user, $j \in I_u^-$. I_u^+ denotes the positive items by user u and I_u^- denotes the irrelevant item set for the same user. Pairwise ranking models optimize its parameters, the latent matrices P and Q, such that the difference of the positive and negative item scores for all users are maximized.

Rendle et al. proposed Bayesian Personalized Ranking (BPR) [11]. Several gradient descent passes over the dataset is required for convergence. The BPR loss function is described by

$$\text{BPR} = \sum_{(u,i,j) \in D_s} \ln \sigma(\hat{x}_{uij}) \tag{2}$$

where $\hat{x}_{uij} := \hat{y}_{ui} - \hat{y}_{uj}$ and $D_s = \{(u,i,j)\} | u \in U, i \in I_u^+, j \in I_u^-\}$. σ denotes the logistic function.

Weighted Approximate-Rank Pairwise (WARP) loss is another pairwise loss function that balances the need for an appropriate ranking function and training time. WARP uses repeated sampling of negative items. To make WARP amenable to parallel updates, Kula proposed an approximation to WARP using the adaptive hinge loss function [7]. Several negative items are sampled and the hinge function is computed using the maximum negative score incurred (Eq. 3). This introduces the same mechanism as the original algorithm, with a large gradient update if at least one of the sampled negative items have a significantly larger score than the positive item.

$$\text{WARP-AHL} = \sum_{(u,i,j) \in D_s} max(0, 1 + \hat{y}_{uj}^{max} - \hat{y}_{ui}) \tag{3}$$

The advantage of using this method is apparent in modern GPU systems where a large batch of updates could be done in parallel. A similar reasoning of leveraging on modern GPU computation was proposed by Liu and Natarajan [8]. They introduced the following Weighted Margin-Rank Batch Loss (WMRB) which extends WARP by utilizing batches to estimate the ranking function:

$$\text{WMRB} = \log\left(1 + \frac{|I|}{|Z|} \sum_{(u,i,j) \in D_s} \sum_{j \in Z} max(0, 1 + \hat{y}_{uj} - \hat{y}_{ui})\right) \tag{4}$$

where Z denotes the subset of I randomly sampled items without replacement.

2.2   Mapping Item Features for Cold-Start Recommendation   •

Mapping item features to the item latent matrix is one of the existing solutions to the cold-start problem of collaborative filtering algorithms. This new representation enables items with no ratings to be embedded in the same space as the item latent matrix.

The work of Gantner et al. is perhaps the first to study mapping item features to the collaborative filtering domain [1]. They used a linear mapping from the item features to the item latent matrix by using the BPR criterion. They found that optimizing the BPR criterion produces better results than the least squares difference of the mapped item latent matrix to the original item latent matrix. On the other hand, Oord et al. have shown better results using the least squares difference in learning the mapping function [9]. Oord et al. use raw audio features as their item features for cold-start recommendation. They trained a deep convolutional neural network to map mel-frequency cepstral coefficients to the item latent matrix created by WRMF [4].

Kula has proposed a hybrid recommender, LightFM, where content metadata is also used in conjunction with transaction data. As a special type of Factorization Machine [10], LightFM uses the second-order interactions between users, items and their features. With this method, the metadata features explain part of the structure of the user interactions. Kula experimented over MovieLens 10M combined with the Tag Genome dataset, as well as the CrossValidated dataset from StackExchange. He has found improvements using LightFM over ALS-like matrix factorization and latent semantic indexing, particularly in cold-start scenarios. For this reason we compare our proposed approach with LightFM, since the later has shown state of the art results for a large dataset [7].

Our work differs from [1] and [7] in terms of the mapping algorithm used. We use a neural network, the Transfer Network to learn a nonlinear mapping from the item features to the item latent matrix. Transfer Networks benefit from the growing literature around neural network architectures such as dropout [12], batch norm [5] and adaptive learning [3], to name a few. This is a clear improvement over the use linear transformations similar to the work of [9].

3   Description of the Algorithm

We use a two-step approach. First, we implement standard collaborative filtering using Pairwise Ranking Networks to factorize the ratings matrix. This step is needed to create user and item latent matrices that produce relevant recommendation lists for each user. Second, we use the Transfer Network to map item features into the item latent matrix. At evaluation time of cold-start items, we use its item features as input and produce a ranking score $\hat{y}_{ui}$.

The Pairwise Ranking Network algorithm is described in Algorithm 1. To produce a ranking score $\hat{y}_{ui}$, a forward pass is performed via a multiplication operator across the latent feature axis. To optimize for the user and item latent matrix, we use pairwise loss functions Ω that approximate the ranking function. We use the gradient of these loss functions in the backward pass to update

---

Algorithm 1. Pairwise Ranking Network Algorithm

Result: User latent matrix P and item latent matrix Q
Input: Neural network parameters
　　　　　Loss function $\Omega \in \{$BPR, WARP-AHL, WMRB$\}$
　　　　　Optimizer $\eta \in \{$SGD, rmsprop, adagrad$\}$
　　　　　Batch size n
　　　　　Training and validation set D_{tr} and D_v
　　　　　$D_{tr} := \{(u, i, j) \,|\, u \in U, i \in I_u^+, j \in I_u^-\}$
repeat
　　│　Sample $D_s := \{d \in D_{tr} \ \wedge \ d \notin D_v\}, |D_s| = n\}$
　　│　Perform forward pass and compute minibatch loss:
　　│　　　$J(P, Q, nD_s) := \frac{1}{n} \sum_{z=0}^{n} \Omega(P, Q, D_s)$
　　│　Perform backward pass:
　　│　　　Update the user latent matrix: $P := P - \eta(\nabla J(P, D_s))$
　　│　　　Update the item latent matrix: $Q := Q - \eta(\nabla J(Q, D_s))$
　　│　Compute validation error of parameters P and Q using D_v
until *validation AP@k does not improve*;

---

the parameters. To train pairwise loss functions, triplets are used of the form $D = \{(u, i, j) | u \in U, i \in I_u^+, j \in I_u^-\}$ where I_u^+ is the set of items rated by the user u and I_u^- is the set of items not rated by user u.

The Transfer Network algorithm is described in Algorithm 2 where we adopt the notation of [1]. The item feature mapping is described by

$$\hat{y}_{ui} = \sum_{f=1}^{k} \sum_{l=1}^{n} p_{uf} \phi_f(a_{li}) \tag{5}$$

where a_{li} is the one-hot encoded feature set of item i and $\phi : R^n \rightarrow R^k$ denotes the vector-valued function that maps a_i to the predicted item latent matrix $\hat{q}_k$. We use the squared error loss function to optimize the item feature mapping ϕ. The output, $\hat{y}_{ui}$, is then used as the ranking score of the input item.

3.1   PRTNet Architectures

The Pairwise Networks in Fig. 1 take in as input triplets, the user, the positive items and the negative items. The Fully Connected Layer (embeddings) is a layer with linear activations which represents the user and item latent matrices. The positive and negative items share a single item latent representation. Recall that the dot product of the user and item latent matrix is the score which the ranking of the items is based – a higher score represents a higher ranking. The merge layer implements the BPR, WARP-AHL and WMRB losses. Dropout was used as a hyperparameter in our experiments.

The Transfer Network is a feedforward neural network with the number of output units equal to the dimensionality of the item latent matrix. It maps the

---

Algorithm 2. Transfer Network Algorithm

Result: Mapping function ϕ
Input: Neural network parameters
          Optimizer $\eta \in \{\text{SGD, rmsprop, adagrad}\}$
          Batch size n
          Item features $a \in A$
          Training and validation set A_{tr} and A_v
repeat
    Sample $A_s := \{a \in A_{tr} \wedge a \notin A_v\}, |A_s| = n$
    Perform forward pass and compute minibatch squared error loss:
        $J(\phi, n, A_s) := \frac{1}{n} \sum_{i=1}^{n} \frac{1}{2} (\hat{q}_i - q_i)^2$ where $\hat{q}_i := \phi(a_i)$ for all A_s
    Perform backward pass:
        Update the k-th layer: $\phi_k := \phi_k - \eta(\nabla J(\phi_k, A_s))$
    Compute validation error of parameters ϕ using A_v
until *validation squared error does not improve;*

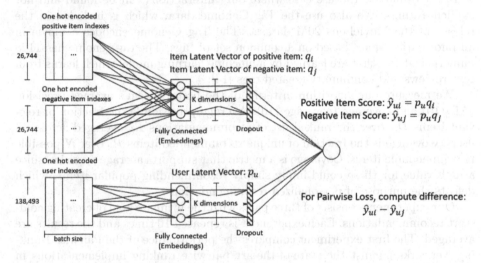

Fig. 1. BPR, WARP-AHL and WMRB Network Implementations. The number of input neurons are equal to the number of the items and users in the dataset, which are 26,744 and 138,493, respectively.

item features of a specific item to its item latent matrix. For this research, the optimization criterion used is least squares. The ReLU activation is used in all layers of the network (Fig. 2).

4 Experiments

We use the MovieLens 20M dataset which contains ratings of movies from 1990 to 2017 [2]. In this research, we use the positive feedback dataset, which consider the ratings of 4 and above to be marked as one and the rest are marked zero.

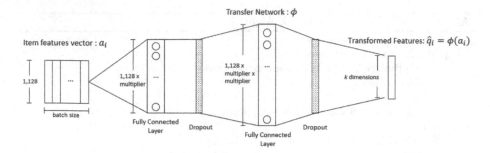

Fig. 2. Transfer Network Implementation. The number of input neurons are equal to the input dimensionality of the tag genome dataset which are 1,128 numeric features. The multipliers are hyperparameters to be optimized in the experiments.

This is to simulate the use case where only interactions can be found and not explicit ratings. We also use the Tag Genome data, which is included in the release of the MovieLens 20M dataset. The Tag Genome encodes a movie in an information space based on a common set of tags. The tag genome has 1128 numeric features that are learned by a machine learning model which learns from text reviews and community-created tags [14].

We measure our algorithm in terms of the first 10 item's average precision (AP@10) and coverage (coverage@10). Average precision is the fraction of relevant items D_+ over the ranked list of recommendations R_k averaged across all users. Coverage is the fraction of unique recommended items $\hat{R}_k$ over N possible recommendable items. Coverage is a metric that supports average precision since a high value for these could imply simply recommending popular items, which defeats the purpose of personalized recommendations.

Our experiment consists of three parts with the ultimate goal of creating cold-start recommendations. Each experiment is repeated 10 times and the results are averaged. The first experiment compares the performance of the Pairwise Ranking Networks against the state-of-the-art pairwise ranking implementations in LightFM [7] using AP@10. We evaluate and test the metrics using the warm-start scenario, which are evaluated by testing and recommending items seen in training. The best resulting models are saved to use as targets for the Transfer Networks. The second experiment aims to find the best Transfer Network architecture. We use the created item latent matrix of the best resulting model configuration of the first experiment as the target. This is a similar scenario as the first experiment, except that the mapped item latent matrix from the Transfer Network is used to create recommendations. The third experiment compares the cold-start recommendations of the Transfer Network against other methods. The best configurations for the Transfer Networks are used in the cold-start recommendations. The Pairwise Ranking Networks were trained on the cold-start training set and evaluated on its corresponding test set, which comprises of items not seen in training. To create the item feature mapping, the item latent matrix from the Pairwise Ranking Network is used. The warm-start and cold-start train-test splits are illustrated in Fig. 3.

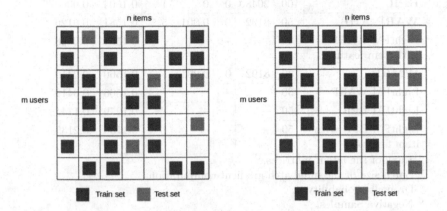

Fig. 3. Warm-start train-test split (left), cold-start train-test split (right). The warm-start train-test splits includes splitting the ratings of 25% of the users that have watches greater than 18, the median of the number of rated items per user. 10% of their ratings are put into the test set. The cold-start train-test splits include putting all the ratings of movies of the year 2014 and beyond to be part of the test set.

5   Results

5.1   Experiment on the Best Architecture for the Pairwise Ranking Networks

In Table 1, the best AP@10 results for the Pairwise Ranking Network implementations and their hyperparameters are shown. BPR and WARP-AHL performed better in terms of AP@10 than LightFM WARP with item features. On the other hand, LightFM results outputted better coverage than all Pairwise Ranking Networks. The Pairwise Ranking Networks have emphasized more on optimizing AP@10 than diversifying the recommendation list.

BPR has the largest number of latent factors. On the other hand, WMRB requires the least number of latent factors. WARP-AHL and WMRB seem to require learning rate decay to stabilize its training. We have found that applying a decay in the learning rate every epoch leads to significantly better results. Lastly, adding dropout did not seem to help the Pairwise Ranking Networks.

Table 1. The best BPR, WARP-AHL and WMRB results on the warm-start scenario

Loss	LF[a]	BS[b]	P[c]	Decay	NS[d]	ap@10[e]	cov@10 [f]
BPR	100	20480	0	0	1	0.1604	0.0069
WARP-AHL	50	8192	0	0.001	5	0.1581	0.0126
LightFM WARP with item features	30	–	–	–	–	0.1561	0.0068
WMRB	30	8192	0	0.001	5	0.1560	0.0419
LightFM WARP	30	–	–	–	–	0.1506	0.0419
LightFM BPR	50	–	–	–	–	0.1279	0.1344
LightFM BPR with item features	50	–	–	–	–	0.1136	0.2135

[a] Rank of the latent matrices.
[b] Batch size of the minibatch gradient computation.
[c] Dropout probability.
[d] Negative Samples.
[e] Average Precision@10.
[f] Coverage@10.

5.2 Experiment on the Transfer-Networks' Capability on the Warm-Start Data

In Table 2, the best AP@10 results for the Transfer Networks are shown. All these results are worse than each of the implementations in Experiment 1 since the Transfer-Network use only the item features as input variables. This comes at a factor of ~0.09 average precision lost. This loss in performance may be attributable to the degree of feature expressiveness and the complexity of mature users. As an example, there are still nuances between users who like Oscar-winning dramas that feature complex human experiences. These are nuances that may not be captured by item features alone. Mature users would tend to exhibit these diverse preferences as well, which adds difficulty to the recommendation problem.

For BPR and WMRB, a deeper network performs best while for WARP-AHL, a wider network is the best. All three configurations use ReLU activations and require normalization by dropout. For the wider source networks, BPR and WMRB, it is interesting that batch normalization is required since deeper networks run the problem of smaller gradients. Batch normalization makes the training faster and helps in convergence for deeper networks. It is also noteworthy than a smaller batch size is required for BPR and WMRB, since a deeper architecture requires better regularization.

Table 2. The best BPR, WARP-AHL and WMRB results of the Transfer-Networks on the warm-start scenario

Loss	Arch[a]	BS	Optimizer	p	BN[b]	AP@10	coverage@10
BPR	[3,5]	256	adagrad	0.01	TRUE	0.07743	0.0110
WARP-AHL	[10]	512	adagrad	0.01	FALSE	0.075119	0.0056
WMRB	[3,5]	256	adagrad	0.01	TRUE	0.086873	0.0042

[a] Architecture refer to the neural network architecture, where [3,5] means that the network has two fully-connected layers. The first number denotes that the input layer's dimensionality is multiplied by 5. The second number denotes that the output dimensionality of the second layer is multiplied by 3. With the tag genome features, this amounts to 5640 and 16920 units for the second and third layers.

[b] If batch norm was applied.

5.3 Experiment on the Transfer-Networks' Capability on the Cold-Start Data

In Table 3, it is clear that each of the Transfer-Networks perform better than their LightFM counterparts in creating recommendations for the cold-start scenario. The results of the Transfer-Networks vary significantly, although at a level still superior to the LightFM implementations. WARP-AHL is significantly higher in AP@10 than neighborhood averaging. Among the other Transfer Networks, it has the highest coverage@10 where 27% of the catalog of 549 cold-start movies is being recommended. Together with the AP@10 results, this shows the degree of personalization offered by PRTNet.

Neighborhood averaging, a very simple content-based filtering algorithm achieved second. It is a technique that averages over the tag genome values and the closest 10 cold-start items to this averaged vector is recommended[1]. This simple algorithm outputted slightly better results than BPR, WMRB and even the LightFM implementations. It also has the highest coverage@10, where 91% of the cold-start items catalog is already being recommended at the first 10. We think that its success is because of the "denseness" of the MovieLens dataset, where ratings for each movie is numerous. It may also be attributable to the richness of the tag genome, where each tag describes movies in a fine detail. For sparser datasets and coarser item features, this averaging technique may not be suited. This is a point for further study.

The statistics for the training time for the best Pairwise Ranking and Transfer Network implementation is shown in Table 4. For the Pairwise Ranking Networks, it is clear that WMRB is significantly faster than BPR and WARP training. This is attributable to its smaller number of latent factors to update. On the other hand, BPR is both the slowest and the most varied in its training time

---

[1] Neighborhood averaging was ran using the entire training set and is a deterministic algorithm.

Table 3. Results for experiment 3: metrics on cold-start data

Algorithm	Mean AP@10	Std AP@10	coverage@10
WARP-AHL	0.294	0.060	0.2716
Neighborhood averaging	0.205	-	0.91
BPR	0.196	0.089	0.0905
WMRB	0.193	0.144	0.0512
LightFM BPR	0.0183	0.010	0.7967
LightFM WARP	0.0170	0.004	0.7581
LightFM BPR with Item Features	0.005	0.015	0.7967
LightFM WARP with Item Features	0.0144	0.005	0.7490

convergence. Recall that in our experiments BPR required the largest number of latent factors. For the Transfer Networks, WARP-AHL is the fastest algorithm in terms of the total training time since it has the least number of training epochs. This is because the best architecture found for WARP-AHL is a single wide layer as opposed to BPR and WMRB. Convergence for the weights was achieved in fewer epochs and it also resulted in being the best network for cold-start recommendation. In other words, the less complex WARP-AHL Transfer Network achieved better generalization and faster training time than its more complex counterparts.

Table 4. Mean training time in seconds of the best networks

	Pairwise Networks		Transfer Networks	
	Train time	Time/epoch	Train time	Time/epoch
BPR	4130.31 ± 849.88	96.47 ± 0.30	364.11 ± 89.03	1.30 ± 0.19
WARP-AHL	1145.97 ± 102.83	54.17 ± 0.52	313.53 ± 96.08	1.31 ± 0.12
WMRB	592.24 ± 121.08	17.89 ± 0.38	385.89 ± 95.52	1.30 ± 0.21

5.4   Conclusion

In this paper, we used neural networks to map item features to the collaborative filtering domain represented by item latent matrices. The item latent matrices are trained by neural networks of pairwise ranking functions to enable training on implicit feedback. These neural networks are nonlinear mapping functions that enable recommendation of items without ratings data.

We have shown success with using Pairwise Ranking and Transfer Networks for cold start recommendations. Our Transfer-Networks yielded significantly better results than LightFM methods. Overall, our proposed WARP-AHL function is the best function for the Pairwise Ranking Network and Transfer Network. It has achieved comparable performance with BPR in two experiments.

In a third experiment, it significantly outperforms BPR and WMRB. WARP-AHL also achieved the best coverage@k among these methods.

For future work, more research is needed using other datasets. The MovieLens dataset is a dense type of collaborative filtering dataset, that is, there are many more users with many ratings than users with fewer ratings. It is interesting to study the efficiency of Pairwise Ranking Networks and Transfer Networks for cold-start recommendation in sparser datasets, that is, has a less than 0.34% non-zero entries. We also plan to implement deeper architectures for the Pairwise Ranking Network instead of the single layer decomposition in this research. This can introduce a hierarchy of latent features in a way similar to the advances in deep learning. A multi-layer decomposition of the ratings matrix, as well as a simultaneous nonlinear mapping function, could be computed efficiently with modern GPU architectures. Lastly, we also plan to create networks that simultaneously learn the optimal ranked list and the mapping function. This may introduce a tighter coupling of the item feature mapping and the item latent matrix, making the model generalize well to both scenarios of warm-start and cold-start recommendation.

References

1. Gantner, Z., Drumond, L., Freudenthaler, C., Rendle, S., Schmidt-Thieme, L.: Learning attribute-to-feature mappings for cold-start recommendations. In: Proceedings of the IEEE International Conference on Data Mining (ICDM), pp. 176–185 (2010). https://doi.org/10.1109/ICDM.2010.129
2. Harper, F.M., Konstan, J.A.: The MovieLens datasets: history and context. ACM Trans. Interact. Intell. Syst. 5(4), 19:1–19:19 (2015). https://doi.org/10.1145/2827872
3. Hinton, G.E., Srivastava, N., Swersky, K.: Lecture 6a- overview of mini-batch gradient descent. In: COURSERA: Neural Networks for Machine Learning, p. 31 (2012). http://www.cs.toronto.edu/~tijmen/csc321/slides/lecture_slides_lec6.pdf
4. Hu, Y., Koren, Y., Volinsky, C.: Collaborative filtering for implicit feedback. In: IEEE International Conference on Data Mining, pp. 263–272 (2008). https://doi.org/10.1109/ICDM.2008.22
5. Ioffe, S., Szegedy, C.: Batch normalization: accelerating deep network training by reducing internal covariate shift (2015). http://arxiv.org/abs/1502.03167
6. Koren, Y.: The BellKor solution to the Netflix grand prize. Netflix prize documentation, 1–10 August 2009. http://www.stat.osu.edu/~dmsl/GrandPrize2009_BPC_BellKor.pdf
7. Kula, M.: Metadata embeddings for user and item cold-start recommendations. In: CEUR Workshop Proceedings (2015)
8. Liu, K., Natarajan, P.: WMRB: learning to rank in a scalable batch training approach (2017)
9. van den Oord, A., Dieleman, S., Schrauwen, B.: Deep content-based music recommendation. In: Electronics and Information Systems department (ELIS), p. 9 (2013). https://doi.org/10.1109/MMUL.2011.34.van. http://papers.nips.cc/paper/5004-deep-content-based-music-recommendation.pdf

10. Rendle, S.: Factorization machines. In: Proceedings of the 2010 IEEE International Conference on Data Mining (ICDM 2010), pp. 995–1000. IEEE Computer Society, Washington, DC (2010). https://doi.org/10.1109/ICDM.2010.127
11. Rendle, S., Freudenthaler, C., Gantner, Z., Schmidt-Thieme, L.: BPR: Bayesian personalized ranking from implicit feedback (2012)
12. Srivastava, N., Hinton, G., Krizhevsky, A., Sutskever, I., Salakhutdinov, R.: Dropout: a simple way to prevent neural networks from overfitting. J. Mach. Learn. Res. **15**, 1929–1958 (2014)
13. Takács, G.: Alternating least squares for personalized ranking (2012). https://doi.org/10.1145/2365952.2365972
14. Vig, J., Riedl, J.: The tag genome: encoding community knowledge to support novel interaction. ACM Trans. Interact. Intell. Syst. **2**(44) (2012). https://doi.org/10.1145/2362394.2362395
15. Weston, J., Bengio, S., Usunier, N.: WSABIE: scaling up to large vocabulary image annotation (2011)

Deep Learning Models

Dynamic Prototype Selection by Fusing Attention Mechanism for Few-Shot Relation Classification

Linfang Wu[1,2], Hua-Ping Zhang[1(✉)], Yaofei Yang[1,3], Xin Liu[4],
and Kai Gao[2(✉)]

[1] Lab of NLPIR Big Data Search and Mining, Beijing Institute of Technology,
Beijing 101300, China
linfangwu0112@163.com, kevinzhang@bit.edu.cn, yangyaofei@gmail.com
[2] School of Information Science and Engineering,
Hebei University of Science and Technology, Shijiazhuang 050018, China
gaokai@hebust.edu.cn
[3] Beijing Information Science and Technology University,
Beijing 101300, China
[4] Beijing Institute of Information Technology, Beijing 101300, China
jfz97@163.com

Abstract. In a relation classification task, few-shot learning is an effective method when the number of training instances decreases. The proto typical network is a few-shot classification model that generates a point to represent each class, and this point is called a prototype. The mean is used to select prototypes for each class from a support set in a prototypical network. This method is fixed and static, and will lose some information at the sentence level. Therefore, we treat the mean selection as a special attention mechanism, then we expand the mean selection to dynamic prototype selection by fusing a self-attention mechanism. We also propose a query-attention mechanism to more accurately select prototypes. Experimental results on the FewRel dataset show that our model achieves significant and consistent improvements to baselines on few-shot relation classification.

Keywords: Relation classification · Few-shot learning · Attention mechanism

1 Introduction

Relation classification (**RC**) is an important task in *natural language processing* (**NLP**). It is applied to many downstream tasks, such as knowledge graphs (**KGs**) and *question answering* (**QA**), aiming to determine the correct relation between two entities in a given sentence.

Conventional supervised models [6,7] achieve successful results at this task. But conventional supervised learning methods need annotated data, which is

N. T. Nguyen et al. (Eds.): ACIIDS 2020, LNAI 12033, pp. 431–441, 2020.
https://doi.org/10.1007/978-3-030-41964-6_37

time-consuming and labor-intensive. A distant supervision (DS) mechanism is a primary approach to alleviate this problem, and it automatically annotates adequate amounts of training instances [8].

Some models achieve promising results on common relations, but the performance of a relation drops dramatically when its number of training instances decreases. About 58% of the relations in NYT-10[1] are long-tail, with fewer than 100 instances [5]. In a relation classification task, few-shot learning is an effective method when the number of training instances decreases. Hence many few-shot learning methods have been studied in recent years which contain some metric learning methods such as the *Prototypical Network* (**PN**) [9].

The core of PN is the prototype, which is a representation of each class encoded by a *convolutional neural network* (**CNN**). Therefore, the distance between the instances of a query set and the prototype is calculated by the square of L2 distance. This method averages all instances of each class as prototypes, and it has some disadvantages. First, the mean selection method is a static process, and it may lose some information at the sentence level. Second, each instance should have its own weight. A different confidence should be applied to each instance for each relationship, but this method treats them equally.

We can generalize the mean selection as a special attention mechanism, and to better select prototypes and enhance accuracy, we expand the mean selection as a dynamic selection by fusing a self-attention mechanism. By this method, we can assign different weights to each instance and update weights dynamically. We also propose a query-attention mechanism that considers the instance information of a query set, gets better semantic information, and selects prototypes more accurately. In our model, we use self-attention and query-attention mechanisms to construct a weight matrix. It reflects the weight of each instance in each class. We build sentence-level attention over multiple instances, which is expected to dynamically select the prototype. Compared to the PN model, the model fusing an attention mechanism can select prototypes according to the weight of each sentence and update weights dynamically. In particular, the attention mechanism that consider the sentence information in the query set can match with prototypes more accurately. The main contributions of this work is twofold:

1. We expand the mean selection as a dynamic selection by fusing a self-attention mechanism.
2. We propose a query-attention mechanism that considers the instance information of a query set.

2   Related Work

2.1   Few-Shot Learning in RC

Many methods have been proposed for RC, such as embedding, kernel, and neural methods. Neural networks have shown great power in supervised tasks

---

[1] This dataset is generated by a distant supervision mechanism, and it contains much noise.

and have been widely used in NLP tasks in recent years. Gormley et al. [4] used word embedding and position embedding as inputs of CNN to obtain sentence representation.

Many models use the distant supervision dataset and obtain good results. Nevertheless, these models are mainly devoted to reducing the noise caused by distant supervision without considering the influence of long-tail relations. Therefore, the small sample learning method becomes a point of focus.

Much effort has been devoted to few-shot learning. Caruana [2] fine-tuned pre-trained models from classes containing many instances to those with few instances using transfer learning methods. Metric learning methods [11] have been proposed to learn the distance distributions among classes. Garcia and Bruna [3] proposed the architecture of the graph neural network, which propagates label information from labeled samples toward unlabeled query samples. Meta-learning has been recently proposed, and such models [9] have achieved significant results on several few-shot benchmarks. Much work in NLP focuses on the zero-shot/semi-supervised scenario [12]. Regarding the RC task, Han et al. [5] proposed a new large supervised few-shot RC dataset and conducted a comprehensive evaluation of few-shot learning methods on this dataset.

2.2 Attention Mechanism

Bahdanau et al. [1] used an attention mechanism on machine translation tasks, which was the first research to apply an attention mechanism to NLP. Vaswani et al. [10] summarized the attention mechanism to a query action from sets of key-value pairs.

3 Few-Shot with Prototypical Network

In this section, we introduce the Prototypical Network (PN). It is first necessary to describe the formulation of few-shot because the dataset usage and training steps differ from other machine learning methods.

3.1 Task Formulation

Given a sentence with the annotated entity pairs e_1 and e_2, we aim to identify the relation between e_1 and e_2. Here, $\mathcal{R} = \{r_1, ..., r_C\}$ defines all relations in the dataset. In training, we need to select some instances as support and query sets. We randomly select $N * K$ instances as the *support set*, which has N classes. Then we select some instances, called the *query set*, which do not overlap with those in the support set.

Using the support set S consists of labeled instances to predict the unlabeled instances in query set $\mathcal{Q}$. The support set contains $\mathcal{N}$ classes randomly selected from $\mathcal{R}$, and each class contains $\mathcal{K}$ samples. The support set is defined as

$$S = \{(x_1^1, r_1), \dots, (x_1^K, r_1)$$
$$\dots,$$
$$(x_N^1, r_N), \dots, (x_N^K, r_N)\}$$

(1)

$x_j^i = [w_1, w_2, \ldots, w_l]$ denotes an instance where $j = r_j$, and it is a sequence of tokens $[w_1, w_2, \ldots, w_l]$, where l denotes the maximum length. Each sentence has a different length, so we need to set a value to pad them to the same length.

3.2 Prototype Selection and Model Training

The structure of PN is shown in Fig. 1.

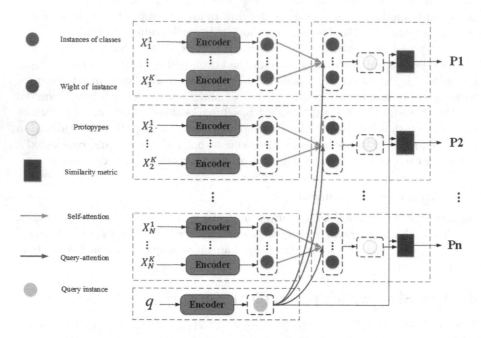

Fig. 1. Architecture of our models in an N-way K-shot learning scenario. First, the sentences are encoded by CNN. Then, through the blue or red array, a prototype of each class is selected by self-attention or query-attention respectively. Finally, prototypes will be used for classification when comparing to the encoded instance from a query set. (Color figure online)

Given an instance $x = [w_1, w_2, \ldots, w_l]$, the pre-trained model $GloVe^2$ maps each word to a vector $v_{word} \in \mathbb{R}^{d_w}$. To highlight the role of entities, we introduce position features that map the relative distance between each word and two entities to two vectors $v_{position1} \in \mathbb{R}^{d_p}$, $v_{position2} \in \mathbb{R}^{d_p}$. Finally, the dimension of each word is $d_w + d_p * 2$. After embedding, we use CNN as an encoder to extract the high-level features of each sentence and get the semantic information of instances. The process of embedding and encoding is defined as

$$h = CNN(V_{word} \bigoplus V_{position1} \bigoplus V_{position2}). \tag{2}$$

---

[2] https://nlp.stanford.edu/projects/glove/.

After extracting semantic and positional information of each instance by the encoder, we take the mean of K instances to be prototypes, and we get N prototypes. These are representations of each class. Then we calculate the distance between each query instance and each prototype. The closest prototype to the query instance is the prediction of the query instance. Finally, we compute the loss between truth labels and predictions and train with backpropagation.

The support set is used to calculate prototypes. For each class of the support set, PN takes prototypes by averaging all instances in each class. The process of getting prototypes c_n can be expressed as

$$c_n = \frac{1}{|S_n|} \sum_{(x_i, y_i) \in S_n} h_i, \tag{3}$$

where n is a class whose prototype is c_n. The square of L2 distance is used to classify the class of the support set:

$$d(h, c_n) = \|h_q - c_n\|^2 \tag{4}$$

$$P_h = \frac{exp(-d(h_q, c_n))}{\sum_{n'} exp(-d(h_q, c_{n'}))}. \tag{5}$$

Equation 5 is a softmax, where h_q is the encoded instance in the query set. From the above formula, we can get the probability of each class and predict instances in a query set.

The averaging method in prototype selection has some disadvantages. First, to get a prototype by the mean selection method is a static process. If an instance contains rich information, mean selection can only get a part of the information, and it may lose some information at the sentence level. Second, each instance should have its own weight because a different confidence should be applied to each instance for each relationship, but this averaging method treats them equally. If we can set the weight of each instance dynamically, then a better prototype will be selected, and this will result in better performance.

4   Dynamic Prototype Selection

We employ dynamic prototype selection to select the prototype in PN. The mean selection is shown in Fig. 2(a), and our dynamic selection is shown in Fig. 2(b). The difference in these selections is the weight. The mean selection uses a fixed $\frac{1}{k}$, and dynamic selection uses w_i^j, which is dynamic. To unify those two selections, the mean selection is generalized as a special attention mechanism, as described below.

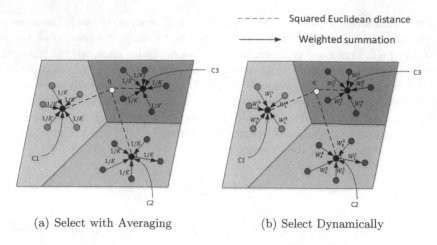

(a) Select with Averaging   (b) Select Dynamically

Fig. 2. Comparison of different selections

4.1 Attention and Mean Selection

The attention function used in this paper can be defined as follows, and the prototype can be generated using this section:

$$attention = W$$

$$\text{where } \sum_{i=1}^{k} W_i = 1. \tag{6}$$

$$c_n = \sum_{i=1}^{k} W_i h_i. \tag{7}$$

The output of the *attention* function is a vector, which has k dimensions in an N-way k-shot model. W_i is the weight of ith instance in each class. The c_n is the selected point from support set.

In PN, the mean is used for all instances of each task, and we use the results as prototypes to represent each class. The mean can be represented as an attention that all instances have the same weight,

$$attention = [\frac{1}{k}, \frac{1}{k}, \ldots, \frac{1}{k}]. \tag{8}$$

Now that the mean is connected with attention, we will show several attention implementations.

4.2 Selection with Self-attention Mechanism

Here, we only consider the support set information, so we use self-attention mechanism to build the weight matrix. As described by Vaswani et al. [10], self-attention mechanism will be used due to its good performance.

$$W^{co} = softmax(\frac{S^* S^{*T}}{\sqrt{dim}}),$$ (9)

$$where \ S^* = SW_s + b_s$$

and $W_s \in \mathbb{R}^{dim \times dim}$, dim is the dimension of the encoded instance, b_s is the bias, and W^{co} is a matrix which we called co-attention. But what we want is an attention vector W. We use W^{co} to generate the attention vector, whose every element is generated by averaging every column:

$$Attention_{self}(S) = [a_1, a_2, \ldots, a_k]$$

$$a_i = \frac{\sum_{j=1}^{k} W_{ij}^{co}}{k}.$$ (10)

4.3 Selection with Query-Attention Mechanism

We want to compute attention with more information. Not only the support set but the query set is taken into account to gather more semantic information. Therefore, we employ query-attention. We use the same method to generate the attention vector from the co-attention matrix in the self-attention mechanism. We generate the co-attention matrix as follows, where $W_Q \in \mathbb{R}^{dim \times 4dim}$ and $W_{atten} \in \mathbb{R}^{4dim \times k}$:

$$Attention_{query}(S, Q) = softmax((|S - Q|W_Q) * W_{atten} + b_{atten}).$$ (11)

5 Experiment

5.1 Dataset

The FewRel dataset consists of 100 relations, each with 700 instances. The average number of tokens in each sentence is 24.99, and there are 124,577 unique tokens in total. The 100 relations are split into 64, 16 and 20 for training, validation and test respectively. It is a large-scale supervised dataset for few-shot relation classification tasks. To address the wrong-labeling problem in most distantly supervised RC datasets, researchers apply crowd-sourcing to manually remove the noise[3].

5.2 Parameter Settings

For the model setup, we set the character embedding d_w to 50 and the position embedding size to 10.[4] The kernel size of CNN is 3 and the number of kernels is

---

[3] Many previous works have worked on automatically removing noise from distantly supervision. Instead, they use crowd-sourcing methods to achieve a high accuracy. https://github.com/thunlp/FewRel.

[4] Both position embedding relative to entity 1 and entity 2 are 5. The position embedding size is combined them together.

230. For regularization, the dropout is used in the model with probability 0.1. Settings for most parameters are the same as for the prototypical network.

We let stochastic gradient descent (SGD) be the optimizer and set the initial learning rate to 0.1, decreasing by 10^{-5} after 10000 steps. We use a batch size of 4 and train all models with 30000 steps. Based on previous work, providing more classes during training can obtain better results, so we use 20 classes to train the model. During training, 20 classes are randomly selected as N, and five instances are randomly selected as K in each batch. The parameters in this model are randomly initialized from a uniform distribution over $[-0.05, 0.05]$.

5.3   Results

Our experiments try out four few-shot learning configurations, 5-way 1-shot, 5-way-5 shot, 10-way 1-shot, and 10-way 5-shot, which were the same as Han et al. [5] Table 1 shows that the dynamic selection method fusing attention mechanism achieves a better result compared to the other models.

Table 1. Accuracies (%) of different models on FewRel test set.

Model	5-way 1-shot	5-way 5-shot	10-way 1-shot	10-way 5-shot
Meta*	64.46 ± 0.54	80.57 ± 0.48	53.96 ± 0.56	69.23 ± 0.52
GNN*	66.23 ± 0.75	81.28 ± 0.62	46.27 ± 0.80	64.02 ± 0.77
SNAIL*	67.29 ± 0.26	79.40 ± 0.22	53.28 ± 0.27	68.33 ± 0.25
PN**	69.20 ± 0.20	84.79 ± 0.16	56.44 ± 0.22	75.55 ± 0.19
PN-Self[†]	$\mathbf{72.39 \pm 0.20}$	$\mathbf{89.44 \pm 0.37}$	$\mathbf{60.01 \pm 0.22}$	$\mathbf{81.51 \pm 0.12}$
PN-Query[‡]	$\mathbf{72.58 \pm 0.20}$	$\mathbf{90.30 \pm 0.17}$	$\mathbf{60.97 \pm 0.22}$	$\mathbf{83.19 \pm 0.10}$

* The scores of these three models are from Han et al. [5] which systematically adopts the most recent state-of-the-art few-shot learning methods for RC.
** PN is the prototypical network, which is the baseline in these experiments.
† This is our model PN with self-attention, as described in Sect. 4.2.
‡ This is our model PN with query-attention, as described in Sect. 4.3.

5.4   Analysis

Experimental results show that the model fusing attention mechanism can enhance the accuracy of prediction. The dynamic selection method can deal with the problem that treat all instances as a same weight and update the weight of each instance dynamically. In addition, our proposed query-attention mechanism considers semantic information of query sets, and therefore can achieve higher accuracy.

Because we use a neural network to replace the mean method so we want to know the increase of accuracy come from attention mechanism or just because it's a neural. Therefore, we further experiment with several models that replace our attention mechanism based dynamic selection with a multi-layer fully connected neural network. We employ several types of fully connected networks with different action functions and dropout rates, and the results are shown in Table 2. The multi-layer perception (MLP) method is worse than our model, and most of them are even worse than the PN. This shows that not every neural network can improve the performance. The model fusing attention mechanism make an improvement because the attention mechanism is a generalization of the mean and it can dynamically select prototypes more accurately.

Table 2. Accuracies (%) of using MLP selection method and attention selection method on FewRel

Model	Layers*	Activation**	Dropout	Accuracy†
PN+MLP	1	Relu	0.5	83.26 ± 0.22
			0.3	83.67 ± 0.16
			0.1	84.09 ± 0.19
		tan	0.5	75.22 ± 0.20
			0.3	82.71 ± 0.18
			0.1	84.49 ± 0.21
	2	Relu	0.5	82.71 ± 0.18
			0.3	83.44 ± 0.21
			0.1	82.90 ± 0.32
		tan	0.5	19.89 ± 0.21
			0.3	84.40 ± 0.13
			0.1	85.34 ± 0.28
PN			0.1	84.79 ± 0.16
PN+Self			0.1	**89.44 ± 0.37**
PN+Query			0.1	**90.30 ± 0.17**

* This parameter layers describe as the hidden layer of MLP.
** This parameter layers describe as the activation function of MLP. We use two nonlinear activation functions.
† We used the accuracy of 5 way 5 shot task as the basis for our evaluation.

6 Conclusion

In original PN, the mean is used to select the prototype, and it is static and fixed. This method may loss some information at the sentence level and reduce the performance of the entire model. We treat the mean as a special attention mechanism that is static. According to this, we generalize the mean selection to a real attention mechanism. We experiment with the two attention mechanisms of self-attention and query-attention. Those dynamic selection methods obtain better results than the original PN. Further analysis shows that the increased performance is due not to a neural network but to an attention mechanism. Our attention mechanism is a generalization of the mean, and it can dynamically and accurately select the prototype and enhance the accuracy.

Treating the attention mechanism as a generalization of the mean is a novel strategy to dynamically select the prototype. It is a new way to find a more accurate method to select prototypes.

Acknowledgement. This paper is sponsored by National Science Foundation of China (61772075) and National Science Foundation of Hebei Province (F2017208012).

References

1. Bahdanau, D., Cho, K., Bengio, Y.: Neural machine translation by jointly learning to align and translate. arXiv:1409.0473 [cs, stat], September 2014
2. Caruana, R.: Learning many related tasks at the same time with backpropagation. In: Tesauro, G., Touretzky, D.S., Leen, T.K. (eds.) Advances in Neural Information Processing Systems 7, pp. 657–664. MIT Press, Cambridge (1995)
3. Garcia, V., Bruna, J.: Few-shot learning with graph neural networks. arXiv:1711.04043 [cs, stat], November 2017
4. Gormley, M.R., Yu, M., Dredze, M.: Improved relation extraction with feature-rich compositional embedding models. arXiv:1505.02419 [cs], May 2015
5. Han, X., et al.: FewRel: a large-scale supervised few-shot relation classification dataset with state-of-the-art evaluation. arXiv:1810.10147 [cs, stat], October 2018
6. Huang, Y.Y., Wang, W.Y.: Deep residual learning for weakly-supervised relation extraction. arXiv:1707.08866 [cs], July 2017
7. Ji, G., Liu, K., He, S., Zhao, J.: Distant supervision for relation extraction with sentence-level attention and entity descriptions. In: Thirty-First AAAI Conference on Artificial Intelligence, February 2017. https://www.aaai.org/ocs/index.php/AAAI/AAAI17/paper/view/14491
8. Mintz, M., Bills, S., Snow, R., Jurafsky, D.: Distant supervision for relation extraction without labeled data. In: Proceedings of the Joint Conference of the 47th Annual Meeting of the ACL and the 4th International Joint Conference on Natural Language Processing of the AFNLP: Volume 2 (ACL-IJCNLP 2009), vol. 2, p. 1003. Association for Computational Linguistics, Suntec, Singapore (2009). http://portal.acm.org/citation.cfm?doid=1690219.1690287
9. Snell, J., Swersky, K., Zemel, R.: Prototypical networks for few-shot learning. In: Guyon, I., et al. (eds.) Advances in Neural Information Processing Systems 30, pp. 4077–4087. Curran Associates, Inc. (2017). http://papers.nips.cc/paper/6996-prototypical-networks-for-few-shot-learning.pdf
10. Vaswani, A., et al.: Attention is all you need. In: Guyon, I., et al. (eds.) Advances in Neural Information Processing Systems 30, pp. 5998–6008. Curran Associates, Inc. (2017). http://papers.nips.cc/paper/7181-attention-is-all-you-need.pdf

11. Vinyals, O., Blundell, C., Lillicrap, T., Kavukcuoglu, k., Wierstra, D.: Matching networks for one shot learning. In: Lee, D.D., Sugiyama, M., Luxburg, U.V., Guyon, I., Garnett, R. (eds.) Advances in Neural Information Processing Systems 29, pp. 3630–3638. Curran Associates, Inc. (2016). http://papers.nips.cc/paper/6385-matching-networks-for-one-shot-learning.pdf
12. Xie, R., Liu, Z., Jia, J., Luan, H., Sun, M.: Representation learning of knowledge graphs with entity descriptions. In: Thirtieth AAAI Conference on Artificial Intelligence, March 2016. https://www.aaai.org/ocs/index.php/AAAI/AAAI16/paper/view/12216

An Enhanced CNN Model on Temporal Educational Data for Program-Level Student Classification

Chau Vo[✉] [iD] and Hua Phung Nguyen[✉] [iD]

Ho Chi Minh City University of Technology,
Vietnam National University – HCMC, Ho Chi Minh City, Vietnam
{chauvtn,nhphung}@hcmut.edu.vn

Abstract. In educational data mining, study performance prediction is one of the most popular tasks to forecast final study status of students. Via these predictions, in-trouble students can be identified and supported appropriately. In the existing works, this task has been considered in many various contexts at both course and program levels with different learning approaches. However, the real-world characteristics of the task's inputs and output such as temporal aspects, data imbalance, and data shortage with sparseness have not yet been fully investigated. Making the most of deep learning, our work is the first one handling those challenges for the program-level student classification task on temporal educational data. In a simple but effective manner, a novel solution is proposed with convolutional neural networks (CNNs) to exploit their well-known advantages on images for temporal educational data. Moreover, image augmentation is done in different ways so that data shortage with sparseness can be overcome. In addition, we adapt new loss functions (Mean False Error and Mean Squared False Error) to make CNN models tackle data imbalance better. As a result, the task is resolved by our enhanced CNN models with more effectiveness and practicability. Indeed, in an empirical study on three real temporal educational datasets, our models outperform other traditional models and original CNN variants on a consistent basis with Accuracy of about 85%–95%.

Keywords: Program-level student classification · Deep learning · Convolutional neural network · Data imbalance · Data sparseness

1 Introduction

In educational data mining, study performance prediction of each student is one of the most popular tasks. It has been supported in many various contexts in the existing works which are reviewed as follows. Classified with respect to the performance level, [4, 9–11] conducted the task at the course level to predict a final study status or a final grade of each student after the end of the course, while few works defined the task at the program level to predict a final study status of each student after the period of time completing the curriculum. Classified with respect to the learning approaches, [10] and many existing works in the past used traditional supervised learning models, while [2, 9, 11, 12] used

© Springer Nature Switzerland AG 2020
N. T. Nguyen et al. (Eds.): ACIIDS 2020, LNAI 12033, pp. 442–454, 2020.
https://doi.org/10.1007/978-3-030-41964-6_38

deep learning models. As an advanced approach, deep learning has been examined in [17] and reviewed in [5]. In particular, [17] discussed the use of deep learning for student proficiency estimation in an e-tutoring system. It was realized that deep learning had their own merits such as encoding learning dynamics and discovering novel representations. However, it is argued that deep learning might not have been the panacea, but promising. In addition, datasets play an important role in selecting a prediction model. Deep learning might be needed if datasets are large enough.

As compared to the existing works, in this paper, our work supports study performance prediction as a student classification task in the supervised learning approach with deep learning models. We consider the task at the program level based on temporal educational data of regular students in an academic credit system.

Moreover, our work takes into account the real-world characteristics of the task's inputs and output to make its context more practical. For the inputs, temporal aspects, data imbalance, and data shortage with sparseness are supported. For the output, we use the training dataset of the study results of the past students in the previous generation to build a model for predictions of each student in the current generation at the program level. These characteristics make the task more challenging, especially with deep learning which often requires (extremely) large datasets of images and texts.

Nowadays, [2, 4, 8, 9, 12, 16, 18] and the other existing works with deep learning on educational data reviewed in [5] have not yet taken into consideration the aforementioned task and its context. Indeed, [2] utilized three variants of recurrent networks: traditional recurrent neural network (RNN), a Gated Recurrent Unit (GRU) neural network, and a Long-Short Term Memory network (LSTM) to build sensor-free affect detectors on temporal data from ASSISTments, a free web-based learning platform in a classroom environment. The authors did not face the challenges stated in our work. Nevertheless, their results showed great improvement with deep learning models as compared to the previous ones. In [9], the authors proposed a prediction model, GritNet, based on the bidirectional long short-term memory (BiLSTM) model to predict the study performance of students according to their interactions with online coursework on Udacity. In [8], GritNet was later introduced with domain adaptation to transfer a GritNet model trained on a past course to a new one. The task supported in [8, 9] was defined at the course level. In their models, globally max-pool representation of the hidden states over the entire sequence was defined to tackle data imbalance. Nonetheless, no detail about sparseness was mentioned. In [12], the authors proposed an Exercise-Enhanced Recurrent Neural Network framework to model student exercising process with a BiLSTM model for exercise semantic representations from texts and then with a LSTM to trace student states. After that student performance predictions were made with Markov property or Attention mechanism. Their models were evaluated on real-world mathematic datasets from an online learning system for senior high-school students with exercise resources. In [16], the authors proposed a dropout prediction model for students in massive open online courses (MOOCs), based on a deep, fully-connected, feed-forward neural network using clickstream data from 40 MOOCs from HarvardX. The authors also considered data imbalance and used reweighting the loss of the classes, but the problem was not completely solved. In [18], the authors defined a Deep Belief Network for Automatic Short Answer Grading, using data from Cordillera, a natural language intelligent tutoring system for teaching introductory college physics. Their model provided better results than the traditional classifiers. Like [11], [4] performed a dropout prediction task at the course level. Unlike [11], [4] aimed at MOOCs, particularly using time series data from the Coursera and edX courses. The authors used an RNN model with LSTM cells to obtain better dropout prediction results as compared to the others. Unfortunately, no challenge like those in our task was detailed in the aforesaid works.

As the first work handling those challenges for the program-level student classification task on temporal educational data, our work constructs a deep learning-based solution in a simple but effective manner. In this solution, convolutional neural networks (CNNs) are used so that their well-known advantages on images can be applied to temporal educational data. Therefore, temporal educational data are first transformed into color images. Those resulting images can also preserve the temporal aspects of the data. Moreover, image augmentation can be done on those images in different ways so that data shortage with sparseness can be overcome. Thanks to the today's deep learning framework, customized loss functions can be implemented for deep learning models. With that facility, we adapt new loss functions (Mean False Error and Mean Squared False Error) in [14] to handle data imbalance with CNN models. As a result, our task can be resolved with such enhanced CNN models for more effectiveness and practicability. Indeed, an empirical study on three real temporal educational datasets has shown that our enhanced CNN models outperform other traditional models and original CNN variants on a consistent basis. They can provide more correct predictions with Accuracy of about 85%–95%. These correct predictions are helpful for identifying in-trouble students so that more appropriate support can be prepared to enable them to graduate from a university ultimately.

2 A Student Classification Task on Temporal Educational Data

Program-level student classification is an educational data mining task to classify a current student into an appropriate final study status group after the end point in time of the program. When the task is conducted on temporal educational data, all the historical study results of each student are examined. At that moment, knowledge chains accumulated so far can be analyzed towards the final success of each student. Nevertheless, such a task has not yet been investigated in the existing works.

Given a generation of u students with all the historical study results of each student, the task is to classify u students into two groups: the first including the students who will graduate from university and the second including the students who will never. It is regarded as a prediction task that predicts the final study status of a student with two labels: "graduating" and "study_stop", respectively.

To solve the aforementioned task, the historical data of the previous generation are used in a supervised learning approach to support a current generation. This forms a practical context of the task in our real world. The following are its formal definitions.

Given a dataset D_m^t about m students at a point t in time for $t = 1 \ldots T$, D_m^t is defined as follows: $D_m^t = \left\{ X_1^t, X_2^t, \ldots, X_m^t \right\}$. In our work, a point t in time is a semester.

In D_m^t, X_i^t is a vector representing a student at a point t in time as follows:

$$X_i^t = \left(x_{i1}^t, x_{i2}^t, \ldots, x_{ip}^t \right) \text{ for } i = 1 \ldots m$$

Where p is the dimensionality of a vector which is the number of courses required for graduation and each x_{ij}^t with $j = 1 \ldots p$ is a positive real number in a grading scale, e.g. [0, 10], representing a grade that a student has got for the course at the dimension j. The total data set D_m about m students in T semesters is defined as follows:

$$D_m = \left\{ D_m^1, D_m^2, \ldots, D_m^t, \ldots, D_m^T \right\}$$

Regarding the study status labels, Y is a set of labels defined: $Y = \{0, 1\}$. In our experiments, 1 is used for "study-stop" and 0 for "graduating" to imply unsuccessful and successful students, respectively.

Using these definitions, the inputs, output, and processing of the task are formally stated below, using U as an unlabeled dataset of u current students in T semesters at present, denoted as $T_{current}$, and L as a labeled dataset of n past students in T semesters in the past, denoted as T_{past}.

$$U = (D_u, Y_u) \text{ in } T_{current} \text{ where } Y_u = \{unknown_i\} \text{ for } i = 1 \ldots u$$
$$L = (D_n, Y_n) \text{ in } T_{past} \text{ where } Y_n = \{y_i\} \text{ for } y_i \in Y \wedge i = 1 \ldots n$$

D_n and D_u are defined like D_m for n past students and u current students, respectively. Y_n in L is a set of n **known** study status labels each of which shows the final study status of a corresponding past student; while Y_u in U is a set of u **unknown** study status labels for u corresponding current students to be predicted.

The task is now resolved with two phases as follows:

(i).   Construct a classifier: $H = learning(L)$, where H is our model obtained from the process of a supervised learning algorithm *learning* on a labeled dataset L.
(ii). Classify unlabeled data: $Y_u = H(D_u)$ such that $Y_u = \{y_i\}$ for $y_i \in Y \wedge i = 1 \ldots u$.

3   The Proposed CNN Model with Additional Enhancements

In this section, a novel solution to our program-level student classification task on temporal educational data is proposed with enhanced CNN models. CNN models are chosen due to two main facts. The first is their great success in image and text processing. This success has been shown with textual data in the course forum in [11]. The second is their capability of examining spatial-temporal associations layer by layer in inputs. Such a capability is needed when a certain number of courses in a few semesters of each student must be examined. Moreover, additional enhancements like including new loss functions and data augmentation are available with CNN models so that they can handle small-sized datasets with imbalance and sparseness well.

3.1   Model Design

a. From temporal educational data to images for a CNN model. In this transformation, temporal study results of each student are processed into a color image. At that moment, differences between the students can be obtained by means of visual features layer by layer with a representation learning process of a deep learning model. Such differences can then be fed to the final learning process for classification. An illustration is shown in Fig. 1 to represent "graduating" and "study_stop" students.

(a). Representation of a "graduating" student   (b). Representation of a "study_stop" student

Fig. 1. Illustration on image representations of "graduating" and "study_stop" students

From each X_i^t vector in D^t representing a student at a time point t for $t = 1 \dots T$, a p-by-T matrix is generated and fed to a color (*Red, Green, Blue*) channel of an image. Before transformation, each student i is represented over T time points:

$$X_i = \left(\left(x_{i1}^1, x_{i2}^1, \dots, x_{ip}^1 \right), \dots, \left(x_{i1}^t, x_{i2}^t, \dots, x_{ip}^t \right), \dots, \left(x_{i1}^T, x_{i2}^T, \dots, x_{ip}^T \right) \right)$$

After transformation, each student i is represented over T time points as an image:

$$X_i = \begin{pmatrix} \left(x_{i1}^1, x_{i2}^1, \dots, x_{ip}^1 \right) \\ \vdots \\ \left(x_{i1}^t, x_{i2}^t, \dots, x_{ip}^t \right) \\ \vdots \\ \left(x_{i1}^T, x_{i2}^T, \dots, x_{ip}^T \right) \end{pmatrix} \quad \text{for a color channel}$$

In this representation, each window of a certain size can be used to slide over such an image so that visual features can be captured via color, brightness, shading, region growing, and so on. Those features imply that the temporal connections between study results from different courses over time can be extracted. They become more helpful when not only local but also global ones can be generated in deep learning.

b. New loss functions for a CNN model. Nowadays, there are several different approaches handling data imbalance in deep learning as introduced in [3, 7, 14]. For educational data, [11] has examined a cost-sensitive learning approach in [7] and a loss function-based approach in [14]. It is found that a loss function-based approach was more appropriate for educational data. Therefore, mean false error (MFE) and mean squared false error (MSFE) which are loss functions minimized for the deep neural networks handling data imbalance proposed in [14] are presented as follows:

$$MFE = FPE + FNE$$

$$MSFE = FPE^2 + FNE^2$$

Where FPE is mean false positive error, FNE is mean false negative error, N is the number of instances in the negative class (i.e. majority class), and P is the number of instances in the positive class (i.e. minority class). FPE and FNE are given below:

$$FPE = \frac{1}{N} \sum_{i=1..N} \frac{1}{2}(d^{(i)} - y^{(i)})^2$$

$$FNE = \frac{1}{P} \sum_{i=1..P} \frac{1}{2}(d^{(i)} - y^{(i)})^2$$

Compared to each other, MFE is less complex than MSFE although [11] has confirmed the more effectiveness of MSFE. In the context of our task, data shortage makes us hard to anticipate if MSFE is still more effective. Therefore, the traditional loss functions like mean squared error (MSE) and cross-entropy along with MFE and MSFE need to be examined for our CNN model on temporal educational data.

c. Data augmentation. Once our temporal educational data become images, image preprocessing techniques can be used to obtain different versions of each image in the original dataset. By doing that, a small-sized training dataset can be enlarged and well fit for deep learning. Meanwhile, sparseness can be resolved when zeros play a role of image background and only non-zero values are considered with the image preprocessing techniques. Among many various techniques such as flipping, rotation, shear, brightness, zoom, and shifting, we select three of them as follows for augmentation according to the characteristics of our temporal educational data.

- *Shear*: given a degree value d, this technique shears the angle in counter-clockwise direction in d degrees. Applying to temporal educational data, the study results of some courses in some semesters will move from the bottom to the top in counter-clockwise direction. At the same time, the regions at the top on the left and at the bottom on the right are removed, leading to a corresponding removal of the study results of the general courses in the first year and those of the courses in the latest year. These data have less impact on classification because for most students, the first ones are nearly the same and the second ones are almost zeros.
- *Zoom*: given a zoom value z, this technique makes a random zoom between $(1 - z)$ and $(1 + z)$ corresponding to zoom-in and zoom-out. Such zooms help us stress the available study results and skip zeros, leading to sparseness reduction.
- *Horizontally flipping*: this technique randomly flips an image horizontally. Consequently, the study results in the T-th semester become those in the 1st semester, the study results in the $(T - 1)$-th semester become those in the 2nd semester, and so on. A normal student will have a quite full image along the study path of knowledge accumulation with respect to the curriculum and thus, more differences between the original image and the transformed one. By contrast, an in-trouble student will have a less full image with more sparseness if he/she has studied just for a while. In another case, an in-trouble student will have a quite full image, but containing several study paths because of the on-time courses and the retaken courses. Such discriminations are helpful for class separation in the learning process.

Using these techniques, our resulting model will get more effective than the one on original data with no augmentation. This hypothesis is tested in our empirical study.

d. An enhanced CNN model for program-level student classification on temporal educational data. Putting them altogether, an enhanced CNN model is achieved. This model is then configured with an appropriate number of convolutional layers, pooling layers, and fully connected layers for classification.

- *Convolutional layers*: due to small image sizes of $p \times T$ where p is the number of features and T is the number of time points, our CNN model is designed with two convolutional layers. The first one helps us extract local study trends in study results over time. The latter generates global ones by combining the local study trends which have just been derived. In particular, each filter of the first convolutional layer forms a view of study results corresponding to a group of courses over some contiguous points in time. After that, each filter of the second convolutional layer combines those detailed views into a larger group of courses over time to reflect the study characteristics of each student via all the courses he/she has studied so far with respect to the entire curriculum of the program.
- *Pooling layers*: after each convolutional layer, there is a max pooling layer selected according to the sparseness properties of temporal educational data. Indeed, averages might include extremes like zeros into the next layers.
- *Fully connected layers*: after the second max pooling layer, several fully connected layers are added to perform classification. The last layer among the fully connected layers is the output layer of our model with only one node to support binary classification. Only this layer uses "Softmax" activation function while the others use "ReLU". Besides, no dropout layer is used because of small-sized datasets.

For enhancements, our CNN model is equipped with data augmentation in its preprocessing phase and with a new loss function in its learning phase.

3.2 Model Characteristics

In this subsection, the main characteristics of our enhanced CNN model are discussed.

Firstly, the novelty of our model is highlighted. Compared to deep learning models in the existing works like [2, 8, 9, 11, 16–18] and those reviewed in [5], our model is the first deep learning model for program-level student classification on temporal educational data with data imbalance and sparseness. An entire solution is novel for the task although its design is based on the existing deep learning supports.

Secondly, not only effectiveness but also practicability with data imbalance is examined for the proposed model. In [8, 9, 11], data imbalance is considered. However, the context of each solution in these works is different from ours when our deep learning model works with the images transformed from temporal numerical data. Resolving such data characteristics is our first contribution to the task.

Thirdly, the practicability of our model is expressed via how it deals with data shortage: small-sized training dataset and sparseness. Normally, deep learning requires large to extremely large training datasets so that generality can be reached effectively.

Such large datasets for a program-level supervised learning task are hardly available because of the dynamics with often changes in the educational domain as compared to other domains. In this situation, we still desire a deep learning model for our task and thus need to overcome this challenge with data augmentation. Compared to the works like [2, 8, 9, 11, 16–18] and those discussed in [5], our work is the first one that takes this challenge into account and tackles it effectively as shown next.

Generally speaking, by means of deep learning, the learning process on temporal educational data structured in images now conducts both representation learning for visual features and supervised learning for classification. It exploits the dynamics in our temporal educational data as explained previously for more effectiveness.

4   An Empirical Evaluation

4.1   Experiment Settings

In order to evaluate the enhanced CNN models, an empirical study is conducted.

The study used four real datasets of the students who took Computer Science program in 2005–2008 at Faculty of Computer Science and Engineering, Ho Chi Minh City University of Technology, Vietnam National University – Ho Chi Minh City [1]. Details of each dataset is given in Table 1. Each of them is named according to the year of its corresponding generation. For each generation, the study results of the students with respect to their curriculum in six contiguous semesters were used. For our datasets, there are 43 dimensions (features, attributes) corresponding to 43 subjects in the curriculum. For labeling each vector representing each student, two class labels, "study_stop" and "graduating", are used. "study_stop" shows that the student did not study any longer and thus, never obtains a degree; while "graduating" implies that the student completed the program successfully after a full period of study time. Regarding distribution over each class, there is a high data imbalance in each dataset. This is a challenge that needs to be considered in the learning process appropriately. Besides, our datasets are sparse due to early prediction. The number of unknown values is high corresponding to the unknown grades of the courses that the students have not yet taken. In these cases, zeros (0 s) are used with the percentages in Table 1.

Table 1. Data descriptions

Dataset	Total		"study_stop"		"graduating"		Number of values	Sparseness with 0 s	
	#	%	#	%	#	%		#	%
2005	214	100	28	13.08	186	86.92	55,212	30,226	54.75
2006	209	100	17	8.13	192	91.87	53,922	31,789	58.95
2007	229	100	62	27.07	167	72.93	59,082	34,156	57.81
2008	243	100	75	30.86	168	69.14	62,694	34,974	55.79

Furthermore, we consider their temporal aspects of educational data at two levels. The first one is the student level examining six contiguous semesters when each student is taken into account. This level enables to include the study performance history of each student in the learning process. The latter is the generation level associated with each generation which each dataset corresponds to. At the generation level, the previous dataset is used to support the task on the current one in such a way that a practical context can be established for the task. Therefore, our datasets are temporal and different from those used in the existing works. In connection with these temporal aspects, our task needs to be tackled with a more effective new model.

For traditional models, we used the existing ones in Weka library [15]. They are k-nearest neighbors (k-NN) with $k = 1$ and Euclidean metric, Logistic Regression (LR), Naïve Bayes, Neural Networks (NN), and Support Vector Machine (SVM) which is an SMO model with the Radial Basis Function kernel. The choice of these models stems from their popularity and relatedness to CNN models. Default parameter settings were used to simplify the experiments and avoid bias implementation.

For the implementation of our enhanced CNN models, Keras [6] and Theano [13] were used. Their configurations were made according to data characteristics. Particularly, due to small-sized images, only two convolutional layers with 256 and 128 filters using kernel size $= (6, 6)$ and $(3, 3)$, strides $= (6, 6)$ and $(3, 3)$, respectively, and two corresponding MAX pooling layers using pool size $= (2, 2)$ can be included in our models. After that, a flatten layer is used to convert 3D feature maps to 1D feature vectors. Next, five fully connected layers are set for training on dataset 2005, while 7 fully connected layers on datasets 2006 and 2007. No dropout layer is needed because of data sparseness. For all the models, "ADAM" optimizer was used with batch size $= 64$ and epoch $= 10$, while the activation function was "ReLU" for the internal layers and "Softmax" for the output layer. Their summaries are given in Table 2.

Table 2. Summaries of our CNN models.

Training	Trainable params	Layers (nodes)
2005	501,441	Conv2D(256)-MaxPooling-Conv2D(128)-MaxPooling-Flatten-Dense(128)-Dense(64)-Dense(64)-Dense(32)-Dense(1)
2006	536,577	Conv2D(256)-MaxPooling-Conv2D(128)-MaxPooling-Flatten-Dense(128)-Dense(128)-Dense(128)-Dense(64)-Dense(64)-Dense(64)-Dense(1)
2007	513,729	Conv2D(256)-MaxPooling-Conv2D(128)-MaxPooling-Flatten-Dense(128)-Dense(128)-Dense(128)-Dense(64)-Dense(64)-Dense(32)-Dense(1)

In the learning process, we consider four loss functions as previously presented. They are binary cross-entropy (CE), mean squared error (MSE), mean false error (MFE), and mean squared false error (MSFE). The first two are well-known for a binary classification task; while the last two were introduced in [14]. They were also examined in [11] for effectiveness confirmation on temporal educational data with data imbalance at the course level. Nevertheless, it is questionable to pick a proper loss function for deep learning on temporal educational data at the program level.

Moreover, thanks to *ImageDataGenerator*(), image augmentation has been done with *shear_range*, *zoom_range*, and *horizontal_flip*. These techniques were selected because of the temporal aspect of our datasets on the horizontal line. They generated more images to enlarge our small-sized training sets from these various perspectives.

For comparison, Accuracy (%) is used for correct predictions. The higher values imply the better model. Also, the best results are shown in bold in Tables 3, 4 and 5.

Table 3. Accuracy values from different classification models.

Dataset	k-NN	LR	Naïve Bayes	NN	SVM	Enhanced CNN
2006	67.46	23.45	56.46	10.05	68.90	**94.74**
2007	82.10	79.48	79.91	82.97	82.10	**85.15**
2008	79.42	80.66	83.13	84.36	83.54	**85.19**

Table 4. Accuracy values from our enhanced CNN model with different loss functions.

Dataset	CE	MSE	MFE	MSFE
2006	91.87	93.78	**94.74**	94.26
2007	72.93	72.93	**85.15**	83.84
2008	83.54	84.77	**85.19**	**85.19**

Table 5. Accuracy values from our enhanced CNN model before and after augmentation.

Dataset	Augmentation	CE	MSE	MFE	MSFE
2006	*Before*	91.87	91.87	92.34	92.34
	After	91.87	**93.78**	**94.74**	**94.26**
2007	*Before*	72.93	72.93	75.11	74.24
	After	72.93	72.93	**85.15**	**83.84**
2008	*Before*	80.25	84.77	84.36	83.13
	After	**83.54**	84.77	**85.19**	**85.19**

For evaluation, we raise three questions in this empirical study as follows:

– Do our enhanced CNN models outperform the traditional models?
– Are our enhanced CNN models more effective with an appropriate loss function?
– Are our enhanced CNN models more effective with image augmentation?

4.2   Experimental Results and Discussions

For the first question, Table 3 shows that the best Accuracy values are from our models. There are significant differences between our models and the others because our Accuracy values are higher than not only those of the others but also the percentages of the majority classes in the datasets. It is worth noting that the highest improvement was obtained for dataset 2006 on which NN failed to make correct predictions. On this dataset, most of the models tend to assign a minority class label to any predicted instance, leading to the lower Accuracy values. By contrast, our model can distinguish well between the instances of the minority class and those of the majority class. On the other datasets, NN gives the second best Accuracy values. In these cases, our models outperform NN when more instances can be used from image augmentation and deep representation on the instances can be conducted. This is understandable as the fully connected layers in our models play a role of a neural network conducting a classification task. Besides, in our models, a new loss function can be defined for a better learning process; while the MSE function was fixed with NN. On a consistent basis, our models yield more correct predictions on these three datasets.

To answer the second question, the results in Table 4 indicate the appropriateness of MFE loss function for our enhanced CNN models in all the cases. Among the loss functions, MFE loss function seems to be comparable to MSFE loss function. However, MFE is more suitable. This selection at the program level is different from that in [11] at the course level because of data shortage in our task, leading to the fact that it is hard to achieve better convergences with a more complex loss function like MSFE. In contrast, more effectiveness can be attained with MFE loss function to tackle data imbalance in our datasets. Nevertheless, such better results with MFE loss function need more investigation with a multiclass classification task in the future.

Next, the third question is answered with the results in Table 5 which reflect the significant contribution of image augmentation to the effectiveness of our models. Although our models using either MFE or MSFE with no augmentation are better than or comparable to their variants using either CE or MSE, their performance can be well enhanced with image augmentation. This is because MFE and MSFE are more complex loss functions as compared to CE and MSE, leading to a need of more training instances. Using image augmentation, our training datasets have been enlarged. The resulting larger training datasets help the learning process perform better. For CE and MSE, augmentation also has a certain impact on the effectiveness of the models using these loss functions. Generally speaking, deep learning models on our small-sized training datasets have been successfully enhanced by image augmentation techniques, which cannot be obtained with our original numerical datasets.

In short, this empirical evaluation has confirmed the appropriateness of our enhanced CNN models for student classification on temporal educational data. They also make more correct predictions than the traditional ones and their original variants. As a result, our models help forecasting the final study status of each student so that stakeholders can make suitable changes for more successful students.

5   Conclusions

In this paper, we have built an enhanced convolutional neural network model to classify university students at the program level using their temporal study results in the first few years. Although numerical, such temporal educational data have been restructured to become images. Those images can be then processed effectively by deep learning because the inherent temporal relationships in temporal educational data can be captured easily from the pixel locations of the resulting images. Besides, making the most of a CNN model on images, two enhancements that have been made on a CNN model on temporal educational data are a choice of a loss function and image augmentation. The first enhancement is related to the optimization of the learning process, while the second one is connected to the dataset size. In particular, among the existing loss functions, the loss functions that can handle data imbalance like mean false error and mean squared false error seem to be more appropriate for the learning process on our very much imbalanced educational datasets. For image augmentation, image processing techniques have helped us tackle the small size of a training dataset with a deep learning model. Indeed, once images are obtained from temporal educational data, more images can be generated from those so that our small dataset can get larger for the learning process and thus, the learning process can return a more effective model. Consequently, with Accuracy of about 85%–95%, our enhanced CNN models outperform the other traditional models and unenhanced original variants through a number of experiments on three various real temporal educational datasets.

As of this moment, our enhanced CNN model performs binary classification on temporal educational data. Therefore, we plan to extend the proposed model for multi-class classification. This extension is nontrivial due to data imbalance and overlapping of temporal educational data. In addition, integrating these models into the educational decision support system is also of our interest for decision making support.

References

1. Academic Affairs Office, Ho Chi Minh City University of Technology, Vietnam, 29 June 2017. http://www.aao.hcmut.edu.vn
2. Botelho, A.F., Baker, R.S., Heffernan, N.T.: Improving sensor-free affect detection using deep learning. In: Proceedings of the 18th International Conference on Artificial Intelligence in Education, pp. 40–51 (2017)
3. Buda, M., Maki, A., Mazurowski, M. A.: A systematic study of the class imbalance problem in convolutional neural networks. arXiv preprint arXiv:1710.05381 (2017)
4. Fei, M., Yeung, D-Y.: Temporal models for predicting student dropout in massive open online courses. In: Proceedings of the IEEE International Conference on Data Mining Workshop, pp. 256–263. IEEE (2015)
5. Hernández-Blanco, A., Herrera-Flores, B., Tomás, D., Navarro-Colorado, B.: A systematic review of deep learning approaches to educational data mining. Complexity **2019**, 1–22 (2019)
6. Keras: the Python deep learning library. keras.io. Accessed 24 Sept 2019
7. Khan, S.H., Hayat, M., Bennamoun, M., Sohel, F., Togneri, R.: Cost-sensitive learning of deep feature representations from imbalanced data. IEEE Trans. Neural Netw. Learn. Syst. **PP**(99), 1–15 (2017)

8. Kim, B.-H., Vizitei, E., Ganapathi, V.: GritNet 2: real-time student performance prediction with domain adaptation. arXiv:1809.06686v3 [cs.CY], pp. 1–8 (2019)

9. Kim, B.-H., Vizitei, E., Ganapathi, V.: GritNet: student performance prediction with deep learning. In: Proceedings of the 11th International Conference on Educational Data Mining, pp. 1–5 (2018)

10. Kravvaris, D., Kermanidis, K.L., Thanou, E.: Success is hidden in the students' data. Artif. Intell. Appl. Innov. **382**, 401–410 (2012)

11. Nguyen, P.H.G., Vo, C.T.N.: A CNN model with data imbalance handling for course-level student prediction based on forum texts. In: Nguyen, N.T., Pimenidis, E., Khan, Z., Trawiński, B. (eds.) ICCCI 2018. LNCS (LNAI), vol. 11055, pp. 479–490. Springer, Cham (2018). https://doi.org/10.1007/978-3-319-98443-8_44

12. Su, Y., et al.: Exercise-enhanced sequential modeling for student performance prediction. In: Proceedings of the 32nd AAAI Conference on Artificial Intelligence, pp. 2435–2443 (2018)

13. Theano: a Python library to define, optimize, and evaluate mathematical expressions. http://www.deeplearning.net/software/theano. Accessed 24 Sept 2019

14. Wang, S., Liu, W., Wu, J., Cao, L., Meng, Q., Kennedy, P.J.: Training deep neural networks on imbalanced data sets. In: Proceedings of the 2016 International Joint Conference on Neural Networks (IJCNN), pp. 4368–4374. IEEE (2016)

15. Weka 3. http://www.cs.waikato.ac.nz/ml/weka. Accessed 28 June 2017

16. Whitehill, J., Mohan, K., Seaton, D., Rosen, Y., Tingley, D.: Delving deeper into MOOC student dropout prediction. arXiv:1702.06404v1 [cs.AI], pp. 1–9 (2017)

17. Wilson, K.H., et al.: Estimating student proficiency: deep learning is not the panacea. In: Proceedings of the Neural Information Processing Systems Workshop on Machine Learning for Education, pp. 1–8 (2016)

18. Zhang, Y., Shah, R., Chi, M.: Deep learning + student modeling + clustering: a recipe for effective automatic short answer grading. In: Proceedings of the 9th International Conference on Educational Data Mining, pp. 562–567 (2016)

Ensemble of Multi-channel CNNs for Multi-class Time-Series Classification. Depth-Based Human Activity Recognition

Jacek Treliński and Bogdan Kwolek[✉]

AGH University of Science and Technology, 30 Mickiewicza, 30-059 Krakow, Poland
{tjacek,bkw}@agh.edu.pl
http://home.agh.edu.pl/~bkw/contact.html

Abstract. In this work we present a new algorithm for multivariate time-series classification. On multivariate time-series of features we train multi-class, multi-channel CNNs to model sequential data. The multi-channel CNNs are trained on time-series drawn with replacement from a pool of augmented time-series. The features extracted by such bagging meta-estimators are used to train SVM classifiers focusing on hard samples that are close to the decision boundary and multi-class logistic regression classifiers returning well calibrated predictions by default. The recognition is done by a soft voting-based ensemble, built on SVM and logistic regression classifiers. We demonstrate that despite limited amount of training data, it is possible to learn sequential features with highly discriminative power. The time-series were extracted in tasks including classification of human actions on depth maps only. The experimental results demonstrate that on MSR-Action3D dataset the proposed algorithm outperforms state-of-the-art depth-based algorithms and attains promising results on UTD-MHAD dataset.

Keywords: Convolutional neural networks · Multivariate time-series · Depth-based human action recognition

1 Introduction

Automatic recognition of human activities – commonly referred to as Human Activity Recognition (HAR) – has emerged as a key research area. The aim of visual activity recognition is to determine whether a given action occurred in image or depth sequence. HAR has gained importance in recent years due to its applications in various areas such as health-care services, security and surveillance, entertainment, smart home, and thus there is a rapidly increasing demand for systems that allow recognizing human activities [1]. This is a challenging problem because of high complexity of human actions, complex motion patterns, variation in motion patterns, occlusions, variation of appearance, etc.

© Springer Nature Switzerland AG 2020
N. T. Nguyen et al. (Eds.): ACIIDS 2020, LNAI 12033, pp. 455–466, 2020.
https://doi.org/10.1007/978-3-030-41964-6_39

One of the main issues is that the same action can be performed in many different ways, even by the same person. Thus, the main research challenge is to develop a proper representation of actions, which is both discriminative and general. With regard to data used for representing human behavior, the approaches can be divided into visual sensor-based, non-visual sensor-based, and multi-modal categories. Sensor-based activity recognition is a difficult task due to the inherent noisy nature of measurements or observations. Visual sensors deliver 2D or 3D images, whereas other sensors delver one-dimensional/multi-channel signals. Vision-based activity recognition has been a research focus for a long period of time due to unobtrusiveness, big potential in surveillance [2] as well as ability to cover the subject and the context in which the activity took place [1]. Wearable sensor-based approaches, on the other hand, does not suffer from occlusion, lighting conditions and other visual constraints.

In recent years, substantial progress has been made in research on behavior identification and understanding. Even though significant progress that has been achieved, its associated tasks are far from being solved and many unsolved problems remain. Due to non-rigid shape of the humans, viewpoint variations, occlusions, intra-class variations, and plenty relevant challenges and environmental complexities, current state-of-the-art algorithms have poor performance in comparison to human capabilities of recognizing and understanding human motions and actions. In order to facilitate development and evaluation of new algorithms as well as to facilitate development and evaluation of new algorithms, several benchmark datasets have been recorded and made publicly available in the last decades [3,4]. The release of consumer depth cameras, like Microsoft Kinect, has significantly lighten many of difficulties that lower the action recognition performance on the basis of traditional RGB images. These sensors provide in addition to the RGB image a depth map allowing to cope with viewpoints as well as illumination and color changes. For the same reason, 3D-based approaches provide higher accuracy than 2D-based approaches.

The MSR-Action3D [3] is one of the most frequently used datasets in the research as well as in evaluation of algorithms using 3D information. The recently introduced UTD-MHAD dataset [4] has four types of data modalities: RGB, depth, skeleton joint positions, and the inertial sensor signals. Most approaches to depth-based action recognition rely on 3D positions of body joints, which can be determined for instance by the MS Kinect sensor [5]. As mentioned in a recent work [6], there are only few researches on depth-based human action recognition using CNN, mainly because datasets with depth modality are relatively small-scale. The recognition performances that are achieved by such algorithms are generally lower in comparison to recognition performances of skeleton-based methods. However, presently the choice of cameras that estimate locations of body joints with sufficient 3D accuracy is quite limited.

In the past years, traditional pattern recognition approaches allowed us to achieve remarkable progress in depth-based action recognition [7,8]. However, since these methods often rely heavily on manual extraction of features, where only shallow features can be learned, their generalization capabilities and recognition efficiencies are not as high as they should be. The recent advancement of deep learning makes it possible to perform automatic high-level feature extraction and thus allows to achieve promising results in many areas, including HAR. These methods employ trainable feature extractors and computational models with multiple processing layers for action representation and recognition. This means that HAR can be achieved through end-to-end learning. However, deep learning-based models for human action recognition require a huge amount of image or depth map sequences for training. Since collecting and annotating huge amounts of data is immensely laborious, the current datasets for 3D action recognition typically have 10, 20, 27 or a little more types of actions, which were performed by a dozen or dozen actors. Due limitations mentioned above and also that the number of sequences in the currently available datasets with the 3D data is typically smaller than one thousand, recognition of actions on the basis of only 3D depth maps is very challenging.

In this work we present a new algorithm for multivariate time-series classification. On multivariate time-series of features we train multi-class, multi-channel CNNs to model sequential data. The multi-channel CNNs are trained using time-series drawn with replacement. The features extracted by such bagging meta-estimators are used to train SVM classifiers focusing on hard samples that are close to the decision boundary (the support vectors) and multi-class logistic regression classifiers returning well calibrated predictions by default. The recognition is done by a soft voting-based ensemble, built on SVM and logistic regression classifiers. We demonstrate that even despite limited number of training data it is possible to learn sequential features with highly discriminative power. The time-series were extracted in tasks including classification of human actions on depth maps only. In each depth map, on the basis the non-zero pixels representing the person shape the frame-features were calculated. The frame-features evolving over time make multivariate time-series representing shape motion.

As far as we know, the multi-channel convolutional neural networks [9,10] have not been utilized in human action recognition until now. We demonstrate that this new algorithm has a remarkable potential. We demonstrate experimentally that on MSR-Action3D dataset the proposed algorithm outperforms state-of-the-art depth-based algorithms and attains promising results on challenging UTD-MHAD dataset. One of the most important features of our algorithm is that it needs no skeleton detection. We demonstrate experimentally that despite not utilizing the skeleton modality, the proposed algorithm attains better classification performance than several skeleton-based algorithms, which usually achieve better results in comparison to depth-based only algorithms. It is worth noting that several RGB-D cameras, including most stereovision ones delivers no skeleton modality.

2 The Algorithm

At the beginning of this Section we explain how frame-features are determined. Afterwards, in Subsect. 2.2 we discuss the extraction if time-series features using multi-channel, temporal CNNs. We describe augmentation of time-series and the ensemble consisting of meta-estimators.

2.1 Frame-Features

For each depth frame we calculate features describing the person's shape. We project the acquired depth maps onto three orthogonal Cartesian views to capture the 3D shape and motion information of human actions [4]. This indicates that we determine side-view and top projections of depth maps. Specifically, a depth map acquired by the depth sensor is projected onto three 2D orthogonal Cartesian planes, where xy plane represents frontal view, yz plane exemplifies the side view and xz plane represents top view. Only pixels representing the extracted person in depth maps are employed to calculate the features. The following frame-features were calculated on such depth maps:

- area ratio (calculated only for frontal depth map in axes x, y, expressing the area occupied by the person to total number of pixels in the depth map),
- standard deviation (axes x, y, z), skewness (axes x, y, z),
- correlation (xy, xz and zy axes),
- x–coordinate for which the corresponding depth value represents the closest pixel to the camera,
- y–coordinate for which the corresponding depth value represents the closest pixel to the camera.

This means that the person shape in each depth maps is described by twelve features. A human action represented by a number of depth maps is described by a multivariate time-series of length equal to number of frames and dimension equal to twelve.

2.2 Extracting Time-Series Features Using Multi-channel, Temporal CNNs

Multi-channel, Temporal CNN for Feature Extraction. In multi-channel, temporal CNNs the 1D convolutions are applied in the temporal domain. In this work, the time-series (TS) of frame-features were used to train multi-channel CNNs. The number of channels is equal to the number of frame-features, i.e. to twelve. The multivariate time-series were interpolated to the length equal to 128. This means that regardless of the length of the multivariate time-series, the length of time-series representing any action is equal to 128. Cubic-spline algorithm has been used to interpolate the TS to the common length. The first layer of the MC CNN is a filter (feature detector) operating in time domain. Having on regard that the amount of the training data in current datasets for depth-based action recognition is quite small, the neural network consists of two convolutional layers, each with 8×1 filter and 4×1 max pool, see Fig. 1. The number of neurons in the dense layer is equal to one hundred. The neural networks have been trained on time-series of all training data sequences. The number of output neurons is equal to number of the classes. Nesterov Accelerated Gradient (Nesterov Momentum) has been used to train the network, in 1500 iterations, with momentum set to 0.9, learning rate equal to 0.001, L1 parameter set to 0.001 and dropout set to 0.5. The neural networks trained in such a way have been utilized to classify the actions as well as to extract the features. After training, the output of the dense layer of each trained neural network has been used to extract the features. The size of the feature vector is equal to one hundred, see Fig. 1.

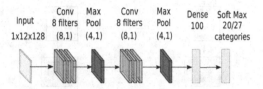

Fig. 1. Flowchart of the multi-channel CNN for multivariate time-series modeling and classification.

Time-Series Augmentation. The multi-channel CNNs have been trained on augmented time-series. Additional four time-series were generated for each input data sequence through extracting:

- (i) the first sixteen data and then interpolating to 8,
- (ii) the first sixteen data and then interpolating to 32,
- (iii) the last sixteen data and then interpolating to 8,
- (iv) the last sixteen data and then interpolating to 32,

and then adding such interpolated subsequences to the input data sequence, and finally interpolating the concatenated TS to the length equal to 128. Afterwards, two additional time-series were generated by scaling the data sequences in time domain by 2 and 0.5.

Ensemble for Action Classification. Two multi-channel, temporal CNNs were trained using two differently augmented datasets. The features extracted by the neural networks were then used to train two multi-class classifiers. We trained a multi-class linear SVM and a Logistic Regression (LR) classifier. The LR returns well calibrated predictions by default as it directly optimizes the log-loss, and thus it has been selected to be used in the ensemble. The SVM focuses on hard samples that are close to the decision boundary (the support vectors), and this was the main motivation of using it in the ensemble. Weighted average probabilities (soft voting) was used in classification of human actions. Figure 2 shows a basic ensemble consisting of two classifiers, which has been used in evaluations, c.f. experimental results in Sect. 3 A grid search has been utilized in order to tune the hyperparameters of the individual estimators.

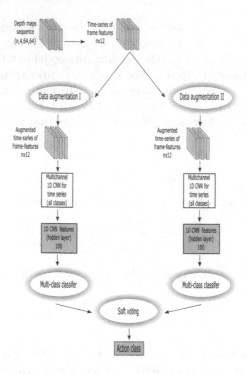

Fig. 2. Flowchart of the base ensemble.

In ensemble algorithms, bootstrap aggregating, also called bagging [11], consists in building several instances of the classifier on random subsets of the original training set and then aggregating their individual predictions to form a final decision. By introducing randomization into the training procedure and then making an ensemble on classifiers trained on randomized subsets the variance of the predictor is reduced. Bagging is also a way to reduce overfitting, without making it necessary to adapt the underlying base classifier. When random subsets of the dataset are drawn as random feature subsets, then the ensemble method is known as Random Subspaces [12]. Panov and Džeroski [13] demonstrated that combining the bagging and random subspaces permits creating better ensembles.

Figure 3 depicts a flowchart of the ensemble with multi-channel CNNs that were trained using both all time-series and time-series drawn with replacement. At the beginning the time-series were augmented using techniques described in Subsect. 2.2. Afterwards, six multi-channel CNNs were trained on time-series drawn with replacement, and one multi-channel CNNs was trained on all augmented time-series. Finally, the features extracted by such bagging meta-estimators have been employed to train multi-class SVM classifiers and multi-class logistic regression classifiers.

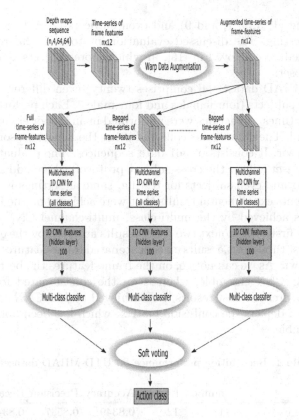

Fig. 3. Flowchart of the algorithm for human action classification using multivariate time-series.

3   Experimental Results and Discussion

The proposed algorithms have been evaluated on two publicly available benchmark datasets: MSR Action3D dataset [3] and UTD-MHAD dataset [4]. The datasets were selected having on regard their frequent use by action recognition community in the evaluations and algorithm comparisons. In all experiments and evaluations, 557 sequences of MSR Action3D dataset were investigated. Half of the subjects were utilized to provide the training data and the rest of the subjects has been employed to get the test subset. In the discussed classification setting, half of the subjects are used for the training, and the rest for the testing, which is different from evaluation protocols based on AS1, AS2 and AS3 data splits and averaging the classification accuracies over such data splits. Another aspect of this is that the classification performances achieved in the utilized setting are lower in comparison to classification performances, which are achieved on AS1, AS2, AS3 setting due to bigger variations across the same actions performed by different subjects. The cross-subject evaluation protocol [8,14] has been applied in all evaluations. Specifically, in the cross-subject protocol, odd subjects are used for training (1, 3, 5, 7, and 9) and even subjects (2, 4, 6, 8, and 10) are employed for testing. The discussed evaluation protocol is different from the evaluation procedure employed in [15], in which more subjects were utilized in the training subset.

The UTD-MHAD dataset [4] comprises twenty seven different actions performed by eight subjects (four females and four males). Each performer repeated each action four times. All actions were performed in an indoor environment with fixed background. The dataset was collected using the Kinect sensor and a wearable inertial sensor. It consists of 861 data sequences. The evaluation protocol used for this dataset follows the cross-subject protocol, where odd subjects were used for training and even subjects for testing, same as settings in [4].

Table 1 presents experimental results that were achieved on the UTD-MHAD dataset. Results achieved by the multi-class, multi-channel CNN classifier are presented in the first row. In next two rows, results achieved by the ensembles are shown. In the last three rows, results on concatenated frame-features with inertial features are shown. As we can notice, on the frame-features the best results were achieved by MC CNN ensemble, whereas on the concatenated frame-features with inertial features the best results were achieved by MC CNN ensemble with bagging. Figure 4 depicts the confusion matrix, which has been obtained by the MC CNN ensemble.

Table 1. Recognition performance on UTD-MHAD dataset.

Prediction	num cl.	f. num.	Accuracy	Precision	Recall	F1-score
MC CNN	1	12	0.8349	0.8537	0.8349	0.8319
MC CNN ensemble	2	12	0.8651	0.8788	0.8651	0.8624
MC CNN ens.+bag.	7	12	0.8535	0.8648	0.8535	0.8496
MC CNN (inert)	1	18	0.9256	0.9343	0.9256	0.9239
MC CNN ensemble (inert)	2	18	0.9186	0.9298	0.9186	0.9171
MC CNN ens.+bag. (inert)	7	18	0.9302	0.9410	0.9302	0.9286

Table 2 presents experimental results that were achieved on the MSR Action 3D dataset. As we can observe, the best results were achieved by MC CNN ensemble with bagging. Comparing the results achieved by the MC CNN ensemble and the MC CNN ensemble with bagging we can observe that bagging improved the classification precision. Figure 5 depicts the confusion matrix, which has been obtained by the MC CNN ensemble.

Table 3 presents the recognition performance of the proposed method compared with previous methods. Most of current methods for action recognition on UTD-MHAD dataset are based on 3D positions of body joints. These methods usually achieve better results than methods relying on depth data only. Despite the fact that our method employs only depth modality, it outperforms many of them. Methods based on depth data only have wider range of applications since not all depth cameras have support for skeleton extraction. Our

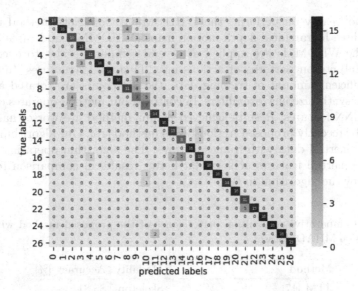

Fig. 4. Confusion matrix on UTD-MHAD dataset, obtained by the MC CNN ensemble.

Table 2. Recognition performance on MSR Action 3D dataset.

Prediction	num cl.	f. num.	Accuracy	Precision	Recall	F1-score
MC CNN	1	12	0.9345	0.9419	0.9345	0.9344
MC CNN ensemble	2	12	0.9455	0.9471	0.9455	0.9438
MC CNN ens.+bag.	7	12	0.9455	0.9517	0.9455	0.9435

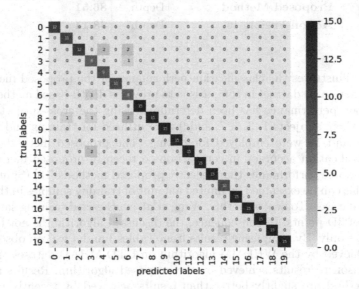

Fig. 5. Confusion matrix on MSR-Action3D dataset, obtained by the MC CNN ensemble.

method noticeably outperforms the WHDMM+3DConvNets method that utilizes weighted hierarchical depth motion maps (WHDMMs) and three 3D ConvNets. The WHDMMs are applied at several temporal scales to encode spatiotemporal motion patterns of actions into 2D spatial structures. In order to collect sufficient amount of training data, the 3D points are rotated and then utilized to synthesize new exemplars. In contrast, our method operates on multichannel CNN features. Results achieved by our method are worse than results achieved by recently proposed Action-fusion method [16]. Our algorithm permits fusing the inertial data with frames-features. The algorithm operating on both depth and inertial features achieves far better results in comparison to results obtained by the algorithm in [16].

Table 3. Comparative recognition performance of the proposed method with recent algorithms on MHAD dataset.

Method	Modality	Accuracy [%]
JTM [17]	Skeleton	85.81
SOS [18]	Skeleton	86.97
Kinect & inertial [4]	Skeleton	79.10
Struct. body [19]	Skeleton	66.05
Struct. part [19]	Skeleton	78.70
Struct. joint [19]	Skeleton	86.81
Struct. SzDDI [19]	Skeleton	89.04
WHDMMs+ConvNets [14,19]	Depth	73.95
Proposed Method	Depth	**86.51**
Action-Fusion [16]	Depth	**88.14**

Table 4 illustrates the classification performance of the proposed method on the MSR-Action3D dataset in comparison to previous depth-based methods. The classification performance has been determined using the cross-subject evaluation [20], where subjects 1, 3, 5, 7, and 9 were utilized for training and subjects 2, 4, 6, 8, and 10 were employed for testing. The proposed method achieves better classification accuracy in comparison to recently proposed method [17], and it has worse performance than recently proposed methods [18] (Split II) and [14,19]. This can be explained by limited amount of training samples in the MSR-Action3D dataset. To deal with this, Wang et al. synthesized training samples on the basis of 3D points. In consequence, the discussed algorithm is not based on depth maps only. By comparing results from Tables 3 and 4 we can observe that results achieved by the WHDMM algorithm on UTD-MHAD dataset are worse in comparison to results achieved by the proposed algorithm. Results attained by our method are slightly better than results achieved by recently proposed Action-fusion method [16].

Table 4. Comparative recognition performance of the proposed method with recent algorithms on MSR Action 3D dataset.

Method	Split	Modality	Acc. [%]
3DCNN [17]	Split II	Depth	84.07
DMMs [7]	Split II	Depth	88.73
PRNN [18]	Split II	Depth	94.90
WHDMM+CNN [14]	Split I	Depth	100.00
S DDI [19]	Split I	Depth	100.00
Action-Feusion [16]	Split I	Depth	**94.51**
Proposed Method	Split I	Depth	**94.55**

4 Conclusions

In this paper we presented a novel algorithm for classification of multivariate time-series using multi-channel CNNs. The multi-channel CNNs are learned on time-series drawn with replacement from a pool of augmented time-series. The features extracted by bagging meta-estimators are utilized to train SVM multi-class classifiers and multi-class logistic regression classifiers. The recognition is done by a soft voting-based ensemble, built on such multi-class classifiers. We demonstrated that the presented algorithm achieves promising results in tasks comprising human action recognition on depth maps We demonstrated experimentally that our algorithm outperforms recent skeleton-based methods, which typically achieve better results than methods based on depth maps only. It outperforms state-of-the-art algorithms on MSR-Action3D dataset and attains promising results on challenging UTD-MHAD dataset. The source code of the presented algorithms is available at https://github.com/tjacek/DeepActionLearning.

Acknowledgment. This work was supported by Polish National Science Center (NCN) under a research grant 2017/27/B/ST6/01743.

References

1. Liang, B., Zheng, L.: A survey on human action recognition using depth sensors (2015)
2. Guo, K., Ishwar, P., Konrad, J.: Action recognition from video using feature covariance matrices. IEEE Trans. Image Process. **22**(6), 2479–2494 (2013)
3. Li, W., Zhang, Z., Liu, Z.: Action recognition based on a bag of 3D points (2010)
4. Chen, C., Jafari, R., Kehtarnavaz, N.: UTD-MHAD: a multimodal dataset for human action recognition utilizing a depth camera and a wearable inertial sensor, September 2015
5. Shotton, J., et al.: Real-time human pose recognition in parts from single depth images. Commun. ACM **56**(1), 116–124 (2013)

6. Wu, H., Ma, X., Li, Y.: Hierarchical dynamic depth projected difference images-based action recognition in videos with convolutional neural networks. Int. J. Adv. Robot. Syst. **16**(1) (2019)

7. Yang, X., Zhang, C., Tian, Y.L.: Recognizing actions using depth motion maps-based histograms of oriented gradients (2012)

8. Xia, L., Aggarwal, J.: Spatio-temporal depth cuboid similarity feature for activity recognition using depth camera (2013)

9. Zheng, Y., Liu, Q., Chen, E., Ge, Y., Zhao, J.L.: Time series classification using multi-channels deep convolutional neural networks. In: Li, F., Li, G., Hwang, S., Yao, B., Zhang, Z. (eds.) WAIM 2014. LNCS, vol. 8485, pp. 298–310. Springer, Cham (2014). https://doi.org/10.1007/978-3-319-08010-9_33

10. Zheng, Y., Liu, Q., Chen, E., Ge, Y., Zhao, J.L.: Exploiting multi-channels deep convolutional neural networks for multivariate time series classification. Front. Comput. Sci. **10**(1), 96–112 (2016)

11. Breiman, L.: Bagging predictors. Mach. Learn. **24**(2), 123–140 (1996)

12. Ho, T.K.: The random subspace method for constructing decision forests. IEEE Trans. Pattern Anal. Mach. Intell. **20**(8), 832–844 (1998)

13. Panov, P., Džeroski, S.: Combining bagging and random subspaces to create better ensembles. In: R. Berthold, M., Shawe-Taylor, J., Lavrač, N. (eds.) IDA 2007. LNCS, vol. 4723, pp. 118–129. Springer, Heidelberg (2007). https://doi.org/10.1007/978-3-540-74825-0_11

14. Wang, P., Li, W., Gao, Z., Zhang, J., Tang, C., Ogunbona, P.: Action recognition from depth maps using deep convolutional neural networks. IEEE Trans. Hum.-Mach. Syst. **46**(4), 498–509 (2016)

15. Xia, L., Chen, C.C., Aggarwal, J.: View invariant human action recognition using histograms of 3D joints (2012)

16. Kamel, A., Sheng, B., Yang, P., Li, P., Shen, R., Feng, D.: Deep convolutional neural networks for human action recognition using depth maps and postures. IEEE Trans. Syst. Man Cybern.: Syst. **49**(9), 1806–1819 (2019)

17. Wang, P., Li, W., Li, C., Hou, Y.: Action recognition based on joint trajectory maps with convolutional neural networks. Knowl.-Based Syst. **158**, 43–53 (2018)

18. Hou, Y., Li, Z., Wang, P., Li, W.: Skeleton optical spectra-based action recognition using convolutional neural networks. IEEE Trans. Circuits Syst. Video Technol. **28**(3), 807–811 (2018)

19. Wang, P., Wang, S., Gao, Z., Hou, Y., Li, W.: Structured images for RGB-D action recognition (2017)

20. Wu, Y.: Mining actionlet ensemble for action recognition with depth cameras (2012)

Post-training Quantization Methods
for Deep Learning Models

Piotr Kluska[1] and Maciej Zięba[2,3(✉)]

[1] AXA XL, Wrocław, Poland
piotr.kluska@axaxl.com
[2] Faculty of Computer Science and Managment,
Wroclaw University of Science and Technology, Wrocław, Poland
maciej.zieba@pwr.edu.pl
[3] Tooploox Ltd., Wrocław, Poland

Abstract. In this work, we perform a comprehensive study on post-training quantization methods for convolutional neural networks in two challenging tasks: classification and object detection. Furthermore, we introduce a novel method that quantizes every single layer to the smallest bit width, which does not introduce accuracy degradation. As a result, the model layers are compressed to a variable number of bit widths preserving the quality of the model. We provide experiments on object detection and classification task and show, that our method compresses convolutional neural networks up to 87% and 49% in comparison to 32 bits floating-point and naively quantized INT8 baselines respectively while maintaining desired accuracy level.

Keywords: Quantization · Convolutional neural networks · Network compression

1 Introduction

Nowadays, deep learning models are capable of solving challenging machine learning tasks in natural language processing and computer vision like classification, object detection, segmentation, and many others [6,12,18]. While most of the convolutional neural networks (CNNs) are designed to achieve better accuracy, their complexity and size increase [10,14,28]. Furthermore, they require a graphics processing unit (GPU) or tensor processing unit (TPU) with an enormous amount of computing power for training and even to run inference in real-time. Therefore, lack of accelerators in an environment with constraints on energy, computing power or memory usage makes state of the art CNNs hard to deploy onto autonomous vehicles, embedded or mobile devices. To tackle this challenge, the family of mobile-friendly networks had been proposed [1,11,29]. However, usually one has to sacrifice accuracy for inference time and memory consumption [1].

© Springer Nature Switzerland AG 2020
N. T. Nguyen et al. (Eds.): ACIIDS 2020, LNAI 12033, pp. 467–479, 2020.
https://doi.org/10.1007/978-3-030-41964-6_40

Additionally, there is ongoing research in the model compression area with usage of pruning and quantization [8,9,32,34]. First effectively introduces sparsity in weights and activations of the artificial neural network, usually followed by retraining to recover reduced loss of accuracy [8,16]. Latter focuses on converting floating point (FP32) weights and activations to lower bit precision integers [7,9,17]. It was shown that an artificial neural network can effectively be compressed to binary or ternary weights [15,21,34]. However, to achieve such compression, CNN has to be trained from scratch [3,32,34]. On the other hand, post-training quantization allows compressing already trained models. Nevertheless, compressing weights and activation may result in accuracy loss and may also require retraining to achieve similar performance as the original CNN. Compression to integer precision reduces computation, memory and energy usage making CNNs attractive to run in constrained environments [5,9].

In this study, we focus on post-training quantization and its effects on image classification and object detection tasks. Through our research, we have used two deep object detectors that differ in their architecture and prediction procedure - YOLOv3 [23] and Faster-RCNN [24]. Moreover, we also consider two CNNs classifiers - Darknet53 [23] and ResNet50 [10] that are backbones of the aforementioned detectors. In order to assess the quality of the post-training quantized CNNs we evaluate them on well known public datasets - Microsoft Common Objects in Context [18] and ImageNet [12] for object detectors and classifiers respectively. Additionally, we proposed a method that iteratively quantizes each layer in CNN to the smallest possible integer precision that does not introduce accuracy degradation.

We can summarize the contribution of this study to the following points:

- In Sect. 3.2 we evaluate the state of the art object detection models performance - YOLOv3 and Faster-RCNN - after post-training quantization. Additionally, we examine classification CNNs: Darknet53 and ResNet50.
- We compress the model to mixed-integer precision in order to achieve the best compression with small accuracy loss. To best of our knowledge, this is the first such approach, especially for the object detection task.

2   Related Work

2.1   Post-training Quantization

While quantization to binary or ternary weights requires quant-aware training one might find post-training quantization sufficient. Furthermore, with post-training quantization, one can compress already trained CNN rather than training from scratch. However, quantization might introduce accuracy degradation also called quantization error. In work by Sung et al. [27] the authors compared the effects of post-training quantization CNNs with and without retraining on different quantization levels, showing that retraining procedure can recover accuracy loss.

NVIDIA provides a platform to deploy deep learning models for inference - TensorRT. It optimizes the deep CNN operation graph by fusing and removing unused layers. Furthermore, it allows reducing precision to 16-bit floating-point or quantization to 8-bit integer precision [19]. It requires calibration dataset, on which it collects activations histograms. Those histograms are used to minimize loss of information that is measured by Kullback-Leibler divergence between reference and quantized weights.

In their work Banner et al. [2] empirically show that activations of CNNs follow either Gaussian or Laplace distributions. Based on that they introduce a method that post-train quantizes activations to a 4-bit integer. Furthermore, they utilize vector quantization by building a lookup table with K-Means algorithm to compress weights to INT4 or INT8.

Finally, Ristretto [7] is a framework developed in Caffe that converts all layers of CNN uniformly to the dynamic fixed point, minifloat or integer precision.

2.2 Quantization of Object Detection CNNs

Object detection models can be divided into two categories: one-stage (YOLOv3) and two-stage (Faster-RCNN) networks [26]. Both types are trained to classify and localize objects in the images, but they take different approaches to do it.

Nakahara et al. [20] took YOLOv2 [22] architecture and performed network surgery to excise feature extractor. Next, they binarized and trained the network with a parallel support vector regressor for classification and localization tasks. They have achieved almost 41 frames per second at 67.6 mean average precision (mAP) on PASCAL VOC 2007 dataset [6] while running on FPGA.

Moreover, Jacob et al. [3] evaluated MobileNet Single Shot Detector [25] after quantization-aware training utilizing "fake quantization" to INT8. They reported a 50% faster inference time with 1.8% loss in overall accuracy.

Quantization Mimic [31] builds on knowledge distillation and adds quantization to the process. They employ quantization-aware training both on teacher and student networks and show that on Pascal VOC 2007 dataset achieve 47 mAP with quantized Faster-RCNN student network.

Last but not least, training of low bit-width networks (LBW-Net) was proposed by Yin et al. [33] which during training minimizes Euclidian distance between full-recision and quantized weights and show that their 6-bit integer network achieves 78.24% mAP compared to 78.94% FP32 reference network.

3   Methods

In this section, we present two of the well-known post-training compression techniques: batch normalization folding and post-training.quantization. Next, we introduce our novel approach that is capable to compress deep neural network models to a variable number of bits obtaining better compression than naive quantization. The models used within this study were downloaded from the Internet. For object classification we used: ResNet-50[1] and Darknet53[2] deep neural networks. As for the object detection task, we used Faster-RCNN[3] and YOLOv3[4] architectures.

3.1 Batch Normalization Folding

During the inference, the batch normalization layer transforms input by moving average ($E[x]$) and variance ($Var[x]$) (Eq. (1)) that were calculated during the training. Therefore, we can treat it as a linear transformation and fold the parameters of the batch normalization layer into the preceding convolution layer [13] (Eqs. (2, 3)). As a result, we can reduce the total number of the parameters and the number of multiply-add calculations.

$$\hat{x}_i = \gamma \frac{x_i - E[x]}{\sqrt{Var[x] + \epsilon}} + \beta \tag{1}$$

$$\hat{W} = \frac{\gamma W}{\sqrt{Var[x] + \epsilon}} \tag{2}$$

$$b = \beta - \frac{\gamma E[x]}{\sqrt{Var[x] + \epsilon}} \tag{3}$$

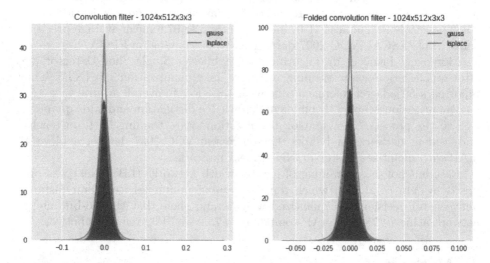

Fig. 1. Histogram of YOLOv3 convolution filter weights before and after folding batch normalization.

---

where: x_i and $\hat{x}_i$ are the inputs and outputs values of the batch normalization layer. γ and β are a scale factor and shift respectively. W is the layer weights before folding and $\hat{W}$ is after. With b we denote layer's bias. Example effect of this operation can be observed in the Fig. 1.

3.2   Post Training Quantization

In our research we perform range-based linear post-training - also called "naive" - quantization: symmetric and asymmetric (signed and unsigned) as seen in Fig. 2. Asymmetric quantization can fully utilize integer range in comparison to symmetric which may leave unused bits in its range. To convert floating-point value to an integer, we calculate scale factor (Δ) based on statistics gathered from N (N = 1000) samples taken from the training dataset.

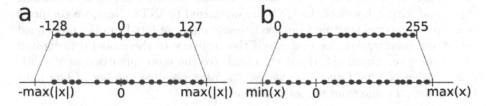

Fig. 2. (a) Example of symmetric quantization of floating-point range x to INT8 with respect to 0, where the range is based on the maximum absolute value of minimum or maximum value in FP32. (b) On the contrary, in the asymmetric mode we map floating-point range to integer range with zero-point offset.

In asymmetric quantization, we have to also consider to represent zero within the quantization range. This can be done by introducing quantization bias called zero points (z_p). We use z_p to exactly match zero from floating-point range to integer range. Symmetric quantization differs from asymmetric by introducing aforementioned bias (Eqs. (4, 5)).

To determine scale factor and zero-point following statistics were collected on inputs and outputs of layers and activations: minimum and maximum, average minimum and maximum and standard deviation.

$$W_s = round(\Delta * W) \tag{4}$$

$$W_a = round(\Delta * W) - z_p \tag{5}$$

where: with Δ we represent quantization scale factor. W_s (symmetric) and W_a (asymmetric) symbolize layers' weights after quantization.

Scale factor (Δ) can be calculated once for the whole layer or multiple per-channel. When calculated per-channel we assume that distributions of weights can differ in every single channel, in result giving a more calibrated range of integer precision. We calculate scale factor per channel for convolution filters to embrace that weights across channels can be varied. Furthermore, to avoid over-flow due to low precision we include INT32 accumulator for results of additions and multiplications. Moreover, all biases in layers are also quantized to INT32.

3.3   Iterative Mixed-Precision Quantization

We propose a method that iteratively quantizes CNN with the aim to find a minimum number of bits per layer from the predefined integer set - $\{N_{min} \leq n \leq N_{max}\}$, where N_{min} and N_{max} define quantization search space - that does not introduce degradation loss within the given threshold. This method compared to naive quantization can utilize variable bit-width compression across the layers effectively reducing memory usage by the model. The method's algorithm is described in the algorithm listing 1. To verify CNN performance after quantization of each layer, this method requires the validation dataset, baseline metric calculated on this dataset and acceptable accuracy decrease (Δ). First, input and output layers of the CNN are quantized to INT8. Next, we compress one layer at a time, starting from the lowest number of bits from the predefined set. After quantization, we evaluate if the accuracy of the model deteriorated below the given threshold. If not then we leave the layer quantized at the current number of bits. Otherwise, we get the next possible number of bits until the maximum value from the set is reached.

> **Input:** CNN, validation dataset, baseline accuracy (acc), threshold (Δ),
> N_{min} , N_{max}
> **Output:** Quantized CNN, accuracy of quantized model $a\hat{c}c$
> 1 **begin**
> 2   Quantize input and output layers weights and activations of CNN to INT8;
> 3   **for** *layer in CNN.layers* **do**
> 4   **for** *n_bits in [$N_{min}...N_{max}$]* **do**
> 5   Quantize(layer, n_bits);
> 6 $a\hat{c}c$ = evaluate(CNN, validation_dataset);
> 7   **if** *acc - $a\hat{c}c$ ¡= Δ* **then**
> 8   break;
> 9   **end**
> 10   **end**
> 11   **end**
> 12 **end**

Algorithm 1. Iterative mixed precision procedure. We quantize layers sequentially starting

4   Experiments

Models. In this study, we performed experiments on image classification and object detection. During the experiments, we used architectures listed in Sect. 3. All the models were downloaded from the Internet. Before performing post-training quantization on the models the batch normalization layer was folded in the preceding layer as described in Sect. 3.1.

Datasets. The models were evaluated on following validation datasets: ImageNet [12] and MS COCO [18] for image classification and object detection respectively. ImageNet validation dataset contains 50000 images with 1000 unique classes. MS COCO validation dataset consists of 5000 images with 80 distinct object classes with bounding box annotations.

Metrics. Image classification evaluation was done with Top-1 and Top-5 metrics. For object classification task we used average precision (AP) and average precision at the intersection over union equal to 0.5 (AP^{50}). Finally, we calculated the size of the network parameters after quantization and reported it in megabytes (MB).

Iterative Mixed-Precision Quantization. Quantization search space was conducted with the following set $\{4, 6, 8\}$. The models were compared against baseline (FP32) models and naive (INT8) post-training asymmetric quantization.

5   Results

5.1   Post-training Quantization

First, we employ one-shot full per-channel quantization of the weights of the CNNs layers to 4, 6 or 8 bits. Based on results observation we can find that YOLOv3 and DarkNet architectures lose most of the accuracy during symmetric quantization (Table 1) in comparison to Faster-RCNN and ResNet. We hypothesize, that the quantization to lower bits may impact LeakyReLU activation function due to its nature. We also observe that all CNNs - except ResNet50 - perform better when we utilize asymmetric quantization (Tables 2 and 3) rather than symmetric (Table 1). We can observe, that YOLOv3, when quantized to INT8, achieves better AP and AP^{50} than baseline FP32, while in the other CNNs we can observe accuracy degradation. Our results are in line with Banner et al. [2] that post-training quantization to INT4 leads to an accuracy equal to zero. Finally, it can be seen that object detection CNNs loose relatively less precision than classification neural network accuracy.

Table 1. Effects of per-channel symmetric quantization on object detection and classification tasks. Weights and activations represent number of bits used. Abbreviation Acts - Activations.

Weights	Acts	YOLOv3		Faster-RCNN		Darknet53		ResNet50	
		AP	AP^{50}	AP	AP^{50}	Top-1	Top-5	Top-1	Top-5
FP32	FP32	33.6	57.1	36.9	58.6	77.2	93.8	76.1	92.9
4	4	0	0	0.2	0.4	0.1	0.5	0.2	0.9
4	6	2	6.1	2.08	3.63	0.1	0.5	0.2	0.9
6	6	5.4	16.8	30.3	50.4	0.6	1.9	60.1	83.3
6	8	28.7	5.4	35.2	56.8	61.1	83.7	69.1	90.6
8	8	30.2	55.6	**36.2**	**58.1**	62.2	84.5	**69.8**	91.1
8	10	**33.9**	**58.2**	**36.2**	58	**75**	**93.4**	68.6	**91.4**

Table 2. Effects of per-channel signed asymmetric quantization on object detection and classification tasks.

Weights	Acts.	YOLOv3		Faster-RCNN		Darknet53		ResNet50	
		AP	AP^{50}	AP	AP^{50}	Top-1	Top-5	Top-1	Top-5
FP32	FP32	33.6	57.1	36.9	58.6	77.2	93.8	76.1	92.9
4	4	0	0	2.3	4.7	0.1	0.5	0.1	0.8
4	6	10.7	27.8	23.5	40.7	0.5	1.5	35.3	58.9
6	6	16.7	41.9	31.9	52.4	0.6	1.9	60.1	83.3
6	8	31.2	56.8	36	57.9	66.8	87.9	68.3	90.1
8	8	32.7	57.4	36.1	58.1	66.7	87.8	**68.1**	**90.2**
8	10	**34.1**	**58.2**	**36.2**	**58.2**	**75.4**	**93.5**	67.4	90

Table 3. Effects of per-channel unsigned asymmetric quantization on object detection and classification tasks.

Weights	Acts.	YOLOv3		Faster-RCNN		Darknet53		ResNet50	
		AP	AP^{50}	AP	AP^{50}	Top-1	Top-5	Top-1	Top-5
FP32	FP32	33.6	57.1	36.9	58.6	77.2	93.8	76.1	92.9
4	4	0	0	2.3	4.6	0.1	0.5	0.1	0.8
4	6	10.8	27.9	23.5	40.6	0.5	1.6	35.5	59.1
6	6	16.6	42	31.9	52.6	0.6	1.9	60.3	83.3
6	8	31.3	56.7	36	57.8	66.8	87.9	60.3	83.3
8	8	32.8	57.6	**36.2**	58	66.8	87.7	**68.2**	90.3
8	10	**34.2**	**58.3**	**36.2**	**58.1**	**75.4**	**93.5**	67.4	**90.5**

5.2 Iterative Mixed-Precision Quantization

We examine our method on object detection task with a threshold within 1–2% overall AP^{50} loss. We can observe in Table 4 that both YOLOv3 and Faster-RCNN achieve the best trade-off of accuracy for size when the threshold is set to 1.5% or 1.75%. Our method achieves compression up to 82% and 87% of YOLOv3 and Faster-RCNN respectively in comparison with their FP32 baselines. When compared to INT8 our method compresses up to 34% and 49% YOLOv3 and Faster-RCNN respectively within an acceptable threshold.

Table 4. Evaluation of iterative mixed quantization on object detection CNNs. We report FP32 baseline and naive signed asymmetric quantization to INT8 and compare their performance and size (in megabytes) to compressed models by our method.

Threshold	YOLOv3				Faster-RCNN			
	AP	AP^{50}	Size	AVG Bits	AP	AP^{50}	Size	AVG Bits
Baseline	33.6	57.1	236.2	32	36.9	58.6	159.2	32
Naive	34.1	58.2	59	8	36.2	58.2	39.8	8
1	31.9	56	48.1	6.13	**35.7**	**57.6**	33.3	7.64
1.5	**31.9**	**56.2**	41	5.51	35.6	**57.6**	21.2	5.23
1.75	31	55.3	40.4	5.91	35.1	57.3	**20.3**	**4.77**
2	31	55.2	**39**	**5.36**	34.8	56.6	25.8	5.7

Last but not least, we present in Table 5 that iterative mixed-precision is also applicable to the classification task. Based on results from Sect. 3.2 we set acceptable top-1 degradation threshold at 5, 8, 9 and 10 percent. Our best ResNet50 is better than naively quantized to INT8 by 0.7% top-1 score while taking almost 18% less of memory. Moreover, our smallest ResNet50 parameters require 43%

Table 5. Evaluation of iterative mixed quantization on classification CNNs. Similarly, we report FP32 baseline and naive signed asymmetric quantization to INT8 and compare their performance and size (in megabytes) to compressed models by our method.

Threshold	Darknet53				ResNet50			
	Top-1	Top-5	Size	AVG Bits	Top-1	Top-5	Size	AVG Bits
Baseline	77.2	93.8	158.6	32	76.1	92.9	97.4	32
Naive	75.4	93.5	39.65	8	68.1	90.2	24.35	8
5	**73.9**	**92.4**	25.7	4.94	**68.8**	**91**	20	7.2
8	70.8	90.8	**23.1**	**4.34**	66.4	88.9	17.9	5.34
9	69.3	89.3	27.2	4.56	68.1	90.2	**14**	**4.79**
10	70.9	90.4	24.9	4.41	64.7	87.2	18.4	5.62

less memory than INT8 reference with the same level of accuracy and 85% than FP32 baseline. Similarly, Darknet53 CNN was compressed up to 85% compared to FP32 baseline and 46.7% to post-training quantization INT8. We observe that with the threshold set to 8% almost all of the artificial neural network weights are quantized to INT4 with just 6.4% degradation of accuracy. Contrary to naive INT4 quantization which accuracy was diminished to almost zero as reported in the signed asymmetric results Table 2. Yet, only taking 4MB more of space. It is worth remarking that only in case of ResNet50 the selected threshold was exceeded while remaining CNNs did meet given criteria.

6 Discussion

When deploying state of the art CNNs to embedded devices, one has to consider memory usage, energy consumption and inference time. Especially in systems that require real-time analysis of images like in autonomous vehicles. To make it feasible while including all limitations we consider quantization of the CNN to integer precision. In our work, we focused on memory usage. Nevertheless, we assume that on the application-specific integrated circuit the prediction time should decrease. In the literature, the focus is placed on a classification task [2,3,7,9,17] rather than on object detectors. In this research, we have conducted comprehensive experiments of post-train quantization on object detection and classification tasks. Our results were presented on widely used CNN architectures that can be used in the future as a reference point. We found that object detection artificial neural networks were less prone to precision degradation than classification CNNs. Last but not least, we found that increasing the bit width of the layer's activation by 2 had a positive impact on CNN performance, especially YOLOv3 which average precision on the validation dataset was higher than a baseline FP32.

Most of the quantization methods - with and without training - leave input and output layers at FP32 and quantize the whole network to INT8 or INT4 [4]. Iterative mixed quantization is a method that builds on the layer's sensitivity and embraces that layers can operate on different bit widths while preserving the desired accuracy. Our method can compress the baseline network by up to 87% which is nearly the size of quantized CNN to INT4, yet maintaining precision within the given threshold. After quantization, the storage and memory requirements of CNNs make them viable to be deployed on embedded devices. Furthermore, our method does not require retraining the CNN after quantization and allows for the practical trade-off between accuracy and memory storage. Nevertheless, we observed that setting a higher threshold for Faster-RCNN, ResNet50 or Darknet53 had an inverse effect of anticipated. The method aggressively quantized the stem of CNN, which in result had to promote higher precision in following layers to meet the given threshold. To avoid this in the future, one could introduce rules that would enforce the quantization of initial layers to higher precision. Alternatively, iterative mixed-precision quantization could benefit from minimizing loss like Kullback-Leiber divergence (NVIDIA TensorRT

[19]) or Euclidean distance [33] between quantized and baseline weights or monitoring if the objective loss does not increase.

Recently Wang et al. [30] have proposed the Hardware-Aware Automatization Quantization method which utilizes reinforcement learning to find the optimal number of bits per layer and activation for given hardware.

7 Summary

In our study, we have carried thorough examinations on the effects of post-training quantization of object detection and classification CNNs. We have shown, that naively quantized object detector to INT8 can perform as good as reference FP32 CNN. Furthermore, we have shown that increasing activation's bit width by 2 relative to the weight's bit width even boosted YOLOv3 performance compared to the baseline FP32 model.

We introduced and evaluated method - iterative mixed-precision quantization. We provide comprehensive experiments on object detection (one stage and two-stage detectors) and classification tasks and show that our method achieves up to 87% and 49% compression compared to 32 floating-point and 8 integer bits baselines respectively while keeping degradation of accuracy within an acceptable threshold.

References

1. Andrew, H., et al.: Searching for MobileNetV3, Technical report (2019). http://arxiv.org/abs/1905.02244
2. Banner, R., Nahshan, Y., Hoffer, E., Soudry, D.: Post training 4-bit quantization of convolution networks for rapid-deployment. Technical report (2018). https://arxiv.org/pdf/1810.05723.pdf
3. Benoit, J., et al.: Quantization and training of neural networks for efficient integer-arithmetic-only inference. Technical report (2017). https://arxiv.org/pdf/1712.05877.pdf
4. Choi, J., Wang, Z., Venkataramani, S., Chuang, P.I.J., Srinivasan, V., Gopalakrishnan, K.: PACT: parameterized clipping activation for quantized neural networks. Technical report (2018). http://arxiv.org/abs/1805.06085
5. Dally, W.J., et al.: Hardware-enabled artificial intelligence. https://research.nvidia.com/sites/default/files/pubs/2018-06_Hardware-Enabled-Artificial-Intelligence/VLSI2018_HardwareAI.pdf.PDF
6. Everingham, M., Van Gool, L., Williams, C.K.I., Winn, J., Zisserman, A.: The pascal visual object classes (VOC) challenge. Int. J. Comput. Vis. **88**(2), 303–338 (2010)
7. Gysel, P., Pimentel, J., Motamedi, M., Ghiasi, S.: Ristretto: a framework for empirical study of resource-efficient inference in convolutional neural networks. IEEE Trans. Neural Netw. Learn. Syst. **29**(11), 5784–5789 (2018). https://doi.org/10.1109/TNNLS.2018.2808319
8. Hacene, G.B., Gripon, V., Arzel, M., Farrugia, N., Bengio, Y.: Quantized guided pruning for efficient hardware implementations of convolutional neural networks. Technical report (2018). https://arxiv.org/pdf/1812.11337.pdf

9. Han, S., Dally, W.J.: Bandwidth-efficient deep learning. In: 2018 55th ACM/ESDA/IEEE Design Automation Conference (DAC), pp. 1–6. IEEE (2018)

10. He, K., Zhang, X., Ren, S., Sun, J.: Deep residual learning for image recognition. Technical report (2016). https://doi.org/10.1109/CVPR.2016.90. http://image-net.org/challenges/LSVRC/2015/

11. Iandola, F.N., Han, S., Moskewicz, M.W., Ashraf, K., Dally, W.J., Keutzer, K.: SqueezeNet: alexnet-level accuracy with 50x fewer parameters and <0.5MB model size. Technical report (2016). http://arxiv.org/abs/1602.07360

12. Deng, J., Dong, W., Socher, R., Li, L.-J., Li, K., Fei-Fei, L.: ImageNet: a large-scale hierarchical image database. In: 2009 IEEE Conference Computer Vision Pattern Recognition, pp. 248–255 (2009). https://doi.org/10.1109/CVPRW.2009.5206848. http://ieeexplore.ieee.org/lpdocs/epic03/wrapper.htm?arnumber=5206848

13. Krishnamoorthi, R.: Quantizing deep convolutional networks for efficient inference: a whitepaper. arXiv preprint arXiv:1806.08342 (2018)

14. Krizhevsky, A., Sutskever, I., Hinton, G.E.: ImageNet classification with deep convolutional neural networks. Technical report (2012)

15. Li, F., Zhang, B., Liu, B.: Ternary weight networks (2016). https://github.com/fengfu-chris/caffe-twns. http://arxiv.org/abs/1605.04711

16. Li, H., Kadav, A., Durdanovic, I., Samet, H., Graf, H.P.: Pruning filters for efficient ConvNets. Technical report (2016). http://arxiv.org/abs/1608.08710

17. Lin, D.D., Talathi, S.S., Annapureddy, V.S.: Fixed point quantization of deep convolutional networks. Technical reports (2016). https://arxiv.org/pdf/1511.06393.pdf

18. Lin, T.Y., et al.: Microsoft COCO: common objects in context. In: Proceedings of IEEE Computer Society Conference on Computer Vision and Pattern Recognition, pp. 3686–3693 (2015). https://doi.org/10.1109/CVPR.2014.471. http://arxiv.org/abs/1405.0312

19. Migacz, S.: 8-bit inference with TensorRT. In: GPU Technology Conference (2017). http://on-demand.gputechconf.com/gtc/2017/presentation/s7310-8-bit-inference-with-tensorrt.pdf

20. Nakahara, H., Yonekawa, H., Fujii, T., Sato, S.: A lightweight yolov2: a binarized CNN with a parallel support vector regression for an FPGA, pp. 31–40 (2018). https://doi.org/10.1145/3174243.3174266

21. Rastegari, M., Ordonez, V., Redmon, J., Farhadi, A.: XNOR-net: imagenet classification using binary convolutional neural networks. In: Leibe, B., Matas, J., Sebe, N., Welling, M. (eds.) ECCV 2016. LNCS, vol. 9908, pp. 525–542. Springer, Cham (2016). https://doi.org/10.1007/978-3-319-46493-0_32

22. Redmon, J., Farhadi, A.: YOLO9000: better, faster, stronger. In: Proceedings of 30th IEEE Conference on Computer Vision on Pattern Recognition, CVPR 2017, pp. 6517–6525 (2017). https://doi.org/10.1109/CVPR.2017.690. http://pjreddie.com/yolo9000/

23. Redmon, J., Farhadi, A.: YOLOv3: an incremental improvement. Technical report (2018). http://arxiv.org/abs/1804.02767

24. Ren, S., He, K., Girshick, R., Sun, J.: Faster R-CNN: towards real-time object detection with region proposal networks. IEEE Trans. Pattern Anal. Mach. Intell. **39**(6), 1137–1149 (2017). https://doi.org/10.1109/TPAMI.2016.2577031. http://image-net.org/challenges/LSVRC/2015/results. http://www.ncbi.nlm.nih.gov/pubmed/27295650

25. Sandler, M., Howard, A., Zhu, M., Zhmoginov, A., Chen, L.C.: MobileNetV2: inverted residuals and linear bottlenecks. In: Proceedings of IEEE Computer Society Conference on Computer Vision and Pattern Recognition, pp. 4510–4520 (2018). https://doi.org/10.1109/CVPR.2018.00474. https://arxiv.org/pdf/1801.04381.pdf

26. Sultana, F., Sufian, A., Dutta, P.: A review of object detection models based on convolutional neural network. Technical report (2019). http://arxiv.org/abs/1905.01614

27. Sung, W., Shin, S., Hwang, K.: Resiliency of deep neural networks under quantization (2015). https://arxiv.org/pdf/1511.06488.pdf

28. Szegedy, C., Vanhoucke, V., Ioffe, S., Shlens, J., Wojna, Z.: Rethinking the inception architecture for computer vision. Technical report (2016). https://doi.org/10.1109/CVPR.2016.308. https://arxiv.org/pdf/1512.00567.pdf

29. Tan, M., et al.: MnasNet: platform-aware neural architecture search for mobile (2018). https://github.com/tensorflow/tpu/. http://arxiv.org/abs/1807.11626

30. Wang, K., Liu, Z., Lin, Y., Lin, J., Han, S.: HAQ: hardware-aware automated quantization with mixed precision. In: Proceedings of the IEEE Conference on Computer Vision and Pattern Recognition, pp. 8612–8620 (2019)

31. Wei, Y., Pan, X., Qin, H., Ouyang, W., Yan, J.: Quantization mimic: towards very tiny CNN for object detection. Technical report. https://arxiv.org/pdf/1805.02152v3.pdf

32. Ye, S., et al.: Progressive DNN compression: a key to achieve ultra-high weight pruning and quantization rates using ADMM (2019). https://bit.ly/2TYx7Za. http://arxiv.org/abs/1903.09769

33. Yin, P., Zhang, S., Qi, Y., Xin, J.: Quantization and training of low bit-width convolutional neural networks for object detection (2016). https://arxiv.org/pdf/1612.06052v2.pdf. http://arxiv.org/abs/1612.06052

34. Zhou, S., Wu, Y., Ni, Z., Zhou, X., Wen, H., Zou, Y.: DoReFa-Net: training low bitwidth convolutional neural networks with low bitwidth gradients (2016). http://arxiv.org/abs/1606.06160

Semi-supervised Representation Learning for 3D Point Clouds

Adrian Zdobylak[1] and Maciej Zieba[1,2]($\boxtimes$)

[1] Wroclaw University of Science and Technology, Wrocław, Poland
zdobylak.adrian@gmail.com, maciej.zieba@pwr.edu.pl
[2] Tooploox, Wrocław, Poland

Abstract. The recent development in the fields of autonomous vehicles, robot vision and virtual reality caused a shift in the research focus - more attention is paid to 3D data representation. In this work, we introduce a novel approach for learning representations for 3D point clouds in semi-supervised mode. The main idea of the approach is to combine the benefits of training autoencoders designed for 3D point clouds in unsupervised mode together with the triplet loss utilized for supervised examples. The proposed method was evaluated considering the classification task and using a challenging benchmark dataset for 3D point clouds.

Keywords: Representation learning · Point cloud · Semi-supervised learning · Triplet networks

1 Introduction

Deep Convolutional Neural Networks (CNNs) [3,6,7,14,16] are responsible for numerous breakthroughs in the pattern recognition field. They have become a standard choice if input data is an image, represented either as two-dimensional or three-dimensional tensor (e.g. color as a third dimension). In other words, these architectures require regular input data. It seems reasonable to conclude that the mentioned methods might produce satisfactory results for 3D data in the form of voxel grids. In practice, 3D objects are sparse, which causes input volume to be relatively big. This, in turn, occurs to be an inefficient representation. Feeding a deep convolutional network with sparse input data significantly slows down the learning procedure. This motivates exploration of learning capabilities of other data representations and novel architectures design for learning interesting latent representations.

The aim of the work is to develop a novel algorithm for learning representations for 3D point clouds trained in a semi-supervised manner. The proposed model utilizes triplet learning together with the specific autoencoding structure that is trained in an end-to-end framework. The encoding network in the considered autoencoder is represented by PointNet [11] that can be easily used for 3D objects represented by the sets of points (point clouds) due to the robustness to

© Springer Nature Switzerland AG 2020
N. T. Nguyen et al. (Eds.): ACIIDS 2020, LNAI 12033, pp. 480–491, 2020.
https://doi.org/10.1007/978-3-030-41964-6_41

the permutations of the points delivered on the input. PointNet returns a single point cloud representation that is invariant to the input order by utilizing a specific version of the max-pooling strategy. On the other hand, triplet learning is an effective method that can be applied to discover some discriminative latent representations even from a small number of samples [22]. Combining those two ideas is beneficial in terms of the quality of 3D points representation as well as reconstruction capabilities of the model. The paper delivers a quantitive evaluation of that approach for classification retrieval and reconstruction tasks. For evaluation purposes, the ShapeNet [2] dataset is used.

The Sect. 2 gives a brief overview of the development of the machine learning methods for 3D data processing and representation learning. The Sect. 3 introduces theoretical aspects of the employed model and describes loss functions used for training purposes. Conducted experiments and the conclusions are extensively discussed in Sect. 4. The last Sect. 5 provides a summary of the article.

2   Related Work

One of the first attempts to utilize machine learning models for 3D point clouds were made in [9,18], where CNNs have been successfully applied to 3D volumetric data. In [15] authors proposed Multi-view CNN (MVCNN) to render multiple 2D images from every single 3D object, from different angles. The state of the art accuracy was achieved with a variation of Multi-view CNN - RotationNet [5], where authors enrich the existing MVCNN solution with an additional rotation network.

Typical neural network architectures are incapable of processing points represented with sets because they are not invariant to the given points ordering. That issue was solved in [11]. The authors proposed PointNet, where the order invariance is obtained by a single symmetric function, max pooling. Effectively, the network learns a set of optimization functions/criteria that select interesting or informative points of the point cloud and encode the reason for their selection. The idea of PointNet was extended in [12] by proposing PointNet++ that utilizes hierarchical point set feature learning in order to improve the ability to recognize fine-grained patterns and complex scenes.

Another approach for 3D point data is an octree-based convolutional neural network (O-CNN) [17]. The key idea of their method is to represent the 3D shapes with octrees and perform 3D CNN operations only on the sparse octants occupied by the boundary surfaces of 3D shapes. An interesting approach was presented in [8] where authors propose to use hierarchical feature extraction with permutation invariant network and Self Organized Maps (SOM) to produce two-dimensional representation of the point cloud. The most recent works, VoxelNet [21], and the further extension F-PointNet [10], are complex end-to-end models capable to be trained on large numbers of point clouds.

Besides numerous supervised approaches proposed for 3D point clouds, there are also a couple of attempts to train informative representations in an unsupervised manner. In [1] authors propose an autoencoder for representing 3D point clouds that takes point cloud as an encoding network and recommend to apply one of two possible distances as reconstruction losses: Chamfer and Earth-Mover. The latent representation can be further mapped to the arbitrary given distribution with the specific GAN model trained on the latent space in the stacked mode. In [20] authors propose adversarial autoencoder for 3D point cloud that enforces assumed prior distribution on the latent space and is trained in the end-to-end framework. One of the recent papers [19] utilizes VAE together with continuous normalizing flows to generate 3D point clouds.

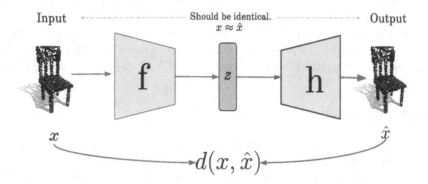

Fig. 1. The architecture for autoencoder designed for 3D point clouds.

The presented approaches are focused either on supervised or unsupervised learning, while our method is directed on a semi-supervised training framework. Moreover, we show in the experimental studies that training autoencoder together with triplet loss increase the overall quality of the bottleneck representation and does not decrease the reconstruction capabilities of the model.

3   Methods

In this section, we introduce the model for learning representations of 3D point clouds in semi-supervised mode. First, we provide the details about the autoencoders for 3D objects together with the losses that operate directly on point clouds. Next, we describe triplet neural networks that are effectively used to train representations in supervised mode. Finally, we present our variation of autoencoder trained in a semi-supervised mode for learning 3D point clouds.

3.1   Autoencoder for 3D Point Clouds

The typical autoencoder (see Fig. 1) is composed of two parts: encoder $f(\cdot)$ that maps an object $\mathbf{x}$ from data space to some latent space $\mathbf{z}$ ($\mathbf{z} = f(\mathbf{x})$) and decoder $g(\cdot)$ that maps objects from latent to data space, and tries to reconstruct

$\mathbf{x}$ from $\mathbf{z}$ by $\hat{\mathbf{x}} = g(\mathbf{z})$. The usefulness of the autoencoder comes from the *bottleneck* layer $\mathbf{z}$, which stores data encoding using compact data representation.

Assuming given dataset $D = \{\mathbf{x}_n\}_{n=1}^{N}$ the autoencoder is trained with the following reconstruction loss:

$$L_{rec} = \frac{1}{N} \sum_{n=1}^{N} d(\mathbf{x}_n, g(f(\mathbf{x}_n))), \tag{1}$$

where $d(\cdot, \cdot)$ is assumed distance function, usually defined as $d(\mathbf{x}_n, \bar{\mathbf{x}}_n) = \frac{1}{2}\|\mathbf{x}_n - \bar{\mathbf{x}}_n\|_2^2$.

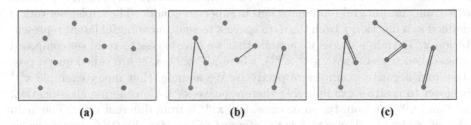

(a)   (b)   (c)

Fig. 2. Visualization of matching nearest points in Chamfer distance. An autoencoder takes input points (green dots) and outputs the same number of reconstructed points (red dots, see a). In the first step, every input point is compared to the closest output point (b). The next step is to repeat the operation by comparing output points to input points (c) (Color figure online).

The provided structure of the autoencoder can be applied for the objects $\mathbf{x}$ that can be represented in vectored form. However, point cloud is defined as a set of data points $S = \{\mathbf{p}_1, \ldots, \mathbf{p}_K\}$, where each point $\mathbf{p}_k$ lies in the D dimensional space, which usually equals 3. In order to adjust the autoencoder for 3D point clouds, relating to the model proposed in [1], we make the following assumptions:

- We represent encoder $f(\cdot)$ using PointNet model (see [11] for details). In other words, it means that for all of possible permutations of the points stored in point cloud S we will obtain the same $\mathbf{z}$ representation returned by $f(\cdot)$. This phenomenon is obtained by the application of max pooling as a symmetric function.
- Simple neural network with fully connected layers is used as the decoding function $g(\cdot)$. The function returns reconstructed point cloud with some trained ordering of the points.
- Chamfer distance is used for reconstruction in order to match the points from the input point cloud to the reconstructions returned by $g(\cdot)$ model.

Assuming two point clouds S_1 and S_2 the Chamfer distance function can be defined as:

$$d(S_1, S_2) = \sum_{\mathbf{p}_1 \in S_1} \min_{\mathbf{p}_2 \in S_2} \|\mathbf{p}_1 - \mathbf{p}_2\|_2^2 + \sum_{\mathbf{p}_2 \in S_2} \min_{\mathbf{p}_1 \in S_1} \|\mathbf{p}_1 - \mathbf{p}_2\|_2^2. \tag{2}$$

For a given training set composed of the point clouds $\mathcal{D} = \{S_1, \ldots, S_N\}$ the autoencoder is trained by minimizing the reconstruction loss given by Eq. (1) with Chamfer distance function.

Chamfer loss computes distances between each point of set S_1 and its closest counterpart in set S_2, and then computes the same distance from point of view of set S_2. As equation may not be straightforward, the example procedure of matching points is visualised and explained in Fig. 2.

3.2   Supervised Training with Triplet Loss

It was shown in previous works [4, 13] that triplet networks are effective in terms of learning meaningful representations in supervised mode. The triplet network is defined as a function f from the data space $\mathbf{x}$ to some meaningful latent representation $\mathbf{z}$. To train a model we assume that we have access to train set composed of so-called triplets $(\mathbf{x}^{(q)}, \mathbf{x}^{(p)}, \mathbf{x}^{(n)})$, where $\mathbf{x}^{(q)}$ $\mathbf{x}^{(p)}$, $\mathbf{x}^{(n)}$ are called query, positive and negative examples respectively. We assume, that query example $\mathbf{x}^{(q)}$ is closer to positive example $\mathbf{x}^{(p)}$ than negative $\mathbf{x}^{(n)}$. Considering classification, $\mathbf{x}^{(q)}$ and $\mathbf{x}^{(p)}$ are from the same class, and $\mathbf{x}^{(n)}$ is from different class. The main idea of triplet-based approach is to enforce f to transfer objects from data space to the latent representation, where similar objects (objects from the same class) are represented by similar vectors $\mathbf{z}$. This result can be obtained by minimizing the following loss function:

$$L_t = \frac{1}{N_t} \sum_{n=1}^{N_t} [d_t(f(\mathbf{x}_n^{(q)}), f(\mathbf{x}_n^{(p)})) - d_t(f(\mathbf{x}_n^{(q)}), f(\mathbf{x}_n^{(n)})) + \alpha]_+, \qquad (3)$$

where N_t is number of triplets in training data, $(\cdot)_+$ denotes a hinge loss $max(\cdot, 0)$, if the distance between examples in latent space is smaller than a margin α (α is a hyperparameter to be tuned), then the loss for such a case equals 0. Various distance functions $d_t(\cdot, \cdot)$ can be applied for the problem, however ℓ_1 and ℓ_2 are sufficient for practical applications. The triplet training framework can be easily applied to the point cloud case, simply by representing $f(\cdot)$ with PointNet architecture and taking some intermediate layer of the model (after max pooling stage) to obtain some order invariant representation $\mathbf{z}$.

3.3   Semi-supervised Approach for Training Point Clouds

In previous subsections, we presented two models that can be applied to learn meaningful representations for 3D point clouds. The autoencoder is capable of representing point cloud data on bottleneck space and is trained in an unsupervised manner. The triplet model can be easily used to train robust data representation with supervised data. In this work we aim at designing a model that can be easily trained in a semi-supervised way, assuming that we have limited access to the labeled data. Data annotation is often a costly and error-prone procedure. In order to reduce the effort, one might decide to annotate only a part of the collected data. Using only a small supervised part for training may lead to poor generalization ability. One way of overcoming the issue is to utilize unlabeled data in a semi-supervised manner.

The main idea of our approach is to enforce a PointNet model $f(\cdot)$ to act as an encoder for unsupervised data and as a triplet network for supervised examples. In order to achieve this we propose to train the autoencoder by minimizing the global loss that is composed of reconstruction loss fed by unsupervised data and triplet loss trained on labeled part of the dataset. Assuming given unlabeled data $D = \{S_n\}_{n=1}^{N}$, where some subset is supervised, what in fact means that it can be used to form set of triplets $D_t = \{(S_n^{(q)}, S_n^{(p)}, S_n^{(n)})\}_{n=1}^{N_t}$. We aim to optimize the following global objective:

$$L = L_{rec} + \lambda \cdot L_t, \tag{4}$$

where L_{rec} is reconstruction fed by D and represented by Eq. (1), L_t is triplet network trained with D_t, λ is a hyperparameter that controls trade-off between component losses.

The network is now trained simultaneously to reconstruct input points (autoencoder), as well as learn representation embedding $\mathbf{z}$ using the supervised part of the dataset. This indicates that L_{rec} is computed for a pair of an input-output point cloud, while L_t operates only on the embedding layer.

In practical implementation, the single batch consists of a mixture of labeled and unlabeled data. For the batch of size B, $\frac{1}{2}B$ of examples have the label assigned, the other half does not. As the unlabeled part of the dataset is significantly more numerous, for every epoch, unsupervised examples will be sampled to fill the batches.

4   Experiments

The main goal of the experiments is to evaluate the quality of the representations obtained with the semi-supervised approach proposed in this paper in comparison to a simple triplet model trained entirely on supervised data and simple autoencoder trained in purely unsupervised mode. We are going to evaluate our semi-supervised approach taking into consideration the classification, retrieval and reconstruction capabilities of the model.

4.1   Metrics

Two metrics have been selected to evaluate the quality of the embeddings returned by the encoder: mean average precision and classification accuracy. The first one measures how well (on average) the test examples are sorted (according to the Euclidean distance) to the query example assuming that examples are represented by their embeddings in latent space $\mathbf{z}$. The average precision for a single example is maximal and equal 1 if all of the examples from the same class in test set have the shortest distances to the query example which is from the same class. The classification accuracy refers to the accuracy of linear SVM trained in stacked mode directly on the embeddings using the same part of the dataset which was previously used to learn representations.

4.2   Dataset

The experiments were conducted on the ShapeNetCore, the subset of ShapeNet [2] dataset. It consists of 57 different object categories with 57449 unique objects, every object is represented by 2048 points. The point clouds have been generated by sampling the points from 3D CAD models. The dataset is highly imbalanced, the four most numerous categories cover more than half of the dataset. Obviously, the disproportion of that magnitude may cause domination over underrepresented classes, therefore, for the training set the maximum number of N examples per class is used. This setup does not only compensate for the data imbalance problem - the triplets are data-hungry and will perform significantly better with larger supervision, but it also allows us to examine the difference against the baseline PointNet autoencoder.

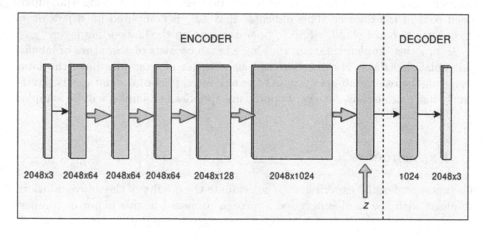

Fig. 3. Architecture autoencoder with PointNet as an encoder.

4.3   Model Architecture and Training Details

The model used for conducting experiments, autoencoder with PointNet encoder, is presented in Fig. 3. Depending on the target different parts of the schema are relevant: unsupervised training uses both encoder and decoder, supervised triplet learning model utilizes layer z for triplet loss.

The encoder structure is composed of five consecutive convolutional layers, after each layer batch normalization is applied. Between the last convolutional and embedding layer, we use dropout (not shown in Fig. 3) with a probability of 0.8. The convolutional layers are two dimensional, the filters have size $[1, C]$, where C is the channel size. This kind of operation is equivalent to applying shared MLP to every single point. The latent representation **z** is simply obtained by taking max pooling operation with respect to the point size dimension. The decoding network is composed of one hidden fully connected layer.

To determine the best size of z we have tested various embedding sizes, such as: 8, 16, 32, 64, 128, 256, 512. We selected the desired size of embedding by choosing possibly the least dimensional one, simultaneously maximizing mean average precision score. The first three, on average, yielded lower scores (0.58–0.67 mAP), having relatively high variance. We decided to go with the hidden layer of size 64, which often outperformed larger embeddings.

4.4   Experimental Results

Semi-supervised learning may be beneficial when the availability of annotated data is somehow limited. It can be observed in Fig. 4, where the influence of the size of labeled data on the accuracy of linear SVM is examined for two models: simple triplet model trained on limited data and semi-supervised approach, where the supervised part of training is conducted on the same subset of labeled examples. In each experiment, we construct the supervised part of the training set independently by drawing no more then 30/50/100/200 examples of every class. For the least numerous datasets, we noted a significant improvement in classification accuracy for the training conducted in semi-supervised mode, 5.7%, and 3.9% respectively. In the third case our semi-supervised approach also turned out to be slightly better than the triplet network, we observed

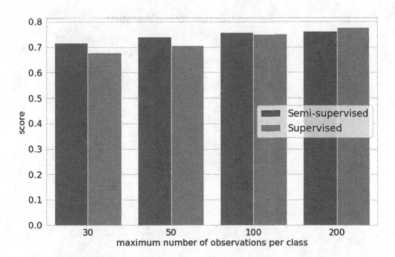

Fig. 4. Classification accuracy of linear SVM (trained on embeddings) with various numbers of supervised examples. Two models are considered in this experiment: our semi-supervised approach that takes the given number of supervised examples for triplet loss and entire unlabeled training data for reconstruction loss (blue bars) and simple triplet model trained on supervised data (orange bars). (Color figure online)

an increase below 1%. The last experiment ended with a supervised method advantage. The analysis has shown the profit from utilizing an additional unsupervised portion of training data, especially when the majority of the dataset is not yet labeled.

Table 1. Minimal reconstruction loss values of our jointly objective (a), and baseline PointNet autoencoder (b).

	Reconstruction loss
(a) Ours	0.000992
(b) PointNet AE	0.000989

In further experiments, we examined the reconstruction ability of our method in comparison to PointNet AE that was designed for this task. From now on, we focus on the setup with a maximum of 50 examples per class. Figure 5 presents validation losses. For the semi-supervised method, we focus only on the reconstruction part, excluding the second objective. The first thing to explain is the difference in curves shape. In our method we mix labeled and unlabeled examples evenly, then we iterate over the annotated dataset and for each batch we fill the rest with samples without class assigned. This indicates that if we use a small portion of the dataset for the supervised part, then also an unlabeled part will be small. As we sample from the whole dataset, it will eventually converge,

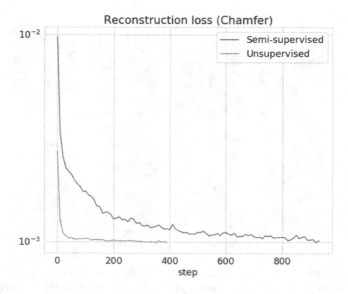

Fig. 5. Reconstruction losses of PointNet AE (unsupervised) and our semi-supervised method. For unsupervised training the whole dataset was used, where in semi-supervised labeled part was limited to maximum of 50 examples per class.

but as the semi-supervised epoch contains fewer examples than unsupervised one (whole dataset), reconstruction loss of AE converges quicker. At the end, loss values are similar, which is highlighted in Table 1.

Table 2. Comparison between our semi-supervised approach (a), baseline unsupervised PointNet autoencoder (b), and supervised triplet learning (c).

	mAP score	SVM accuracy
(a) Ours	0.708 ± 0.003	**0.735 ± 0.004**
(c) PointNet AE	0.575 ± 0.006	–
(b) Triplets	**0.718 ± 0.013**	0.707 ± 0.004

4.5   Discussion of the Results

Conducted experiments (Table 2) have shown that despite worse results in comparison to triplet learning, our method produced embeddings which significantly improved classification score. At the same time, our network was also able to learn to reconstruct point clouds, with the same efficiency as PointNet's autoencoder (Table 1). We observed that our semi-supervised approach reduces overfitting, acting as an additional regularizer.

Indeed our semi-supervised method does not surpass the triplet learning if the whole supervised dataset is used, however, we would like to note two advantages of our approach:

- similarly to triplets, it produces decent embeddings (with respect to mAP score), but also an additional byproduct in the form of reconstruction network; by jointly training, the reconstruction term converges quicker than in vanilla AE,
- semi-supervised approach achieves higher classification accuracy if only a subset of an annotated dataset is used.

5   Conclusion

In this paper we introduce the novel end-to-end semi-supervised training model for 3D point clouds. The main contribution of the model is combining the benefits of using autoencoders together triplet neural networks to obtain robust data representation together with the good reconstruction capabilities.

The conducted experiments using the embedded space show some benefits of the semi-supervised approach comparing to the standard triplet model trained on the limited portion of data. The classification accuracy of semi-supervised model trained on the limited number of examples was significantly higher than for the embeddings produced by a simple triplet approach. However, no significant difference was observed, when mAP calculated with simple Euclidean distance was taken under consideration. Despite this, we can observe the great additional asset of combining the triplet and the reconstruction targets - the decoder network. The embedding can now be inverted to obtain a point cloud. It is also worth mentioning, that the new method evinces the ability to act as a regularizer during the representation learning.

For the future works, to better measure the differences in the accuracy score, it might be a good idea to train the classifier on the dataset with a constant number of instances. Another possible way of exploring capabilities of the presented method is using different proportion of labeled and unlabeled examples. We also consider utilizing Earth-Mover instead of Chamfer distance to achieve better reconstruction capabilities of the model.

References

1. Achlioptas, P., Diamanti, O., Mitliagkas, I., Guibas, L.: Learning representations and generative models for 3D point clouds. In: International Conference on Machine Learning, pp. 40–49 (2018)
2. Chang, A.X., et al.: ShapeNet: an information-rich 3D model repository. CoRR abs/1512.03012v1 (2015). http://arxiv.org/abs/1512.03012v1
3. He, K., Zhang, X., Ren, S., Sun, J.: Deep residual learning for image recognition. In: Proceedings of the IEEE Conference on Computer Vision and Pattern Recognition, pp. 770–778 (2016)
4. Hoffer, E., Ailon, N.: Deep metric learning using triplet network. In: Feragen, A., Pelillo, M., Loog, M. (eds.) SIMBAD 2015. LNCS, vol. 9370, pp. 84–92. Springer, Cham (2015). https://doi.org/10.1007/978-3-319-24261-3_7
5. Kanezaki, A., Matsushita, Y., Nishida, Y.: RotationNet: joint object categorization and pose estimation using multiviews from unsupervised viewpoints. In: Proceedings of the IEEE Conference on Computer Vision and Pattern Recognition, pp. 5010–5019 (2018)
6. Krizhevsky, A., Sutskever, I., Hinton, G.E.: ImageNet classification with deep convolutional neural networks. In: Advances in Neural Information Processing Systems, pp. 1097–1105 (2012)
7. LeCun, Y., Bottou, L., Bengio, Y., Haffner, P., et al.: Gradient-based learning applied to document recognition. Proc. IEEE **86**(11), 2278–2324 (1998)
8. Li, J., Chen, B.M., Hee Lee, G.: SO-net: self-organizing network for point cloud analysis. In: The IEEE Conference on Computer Vision and Pattern Recognition (CVPR), June 2018
9. Maturana, D., Scherer, S.: VoxNet: a 3D convolutional neural network for real-time object recognition. In: 2015 IEEE/RSJ International Conference on Intelligent Robots and Systems (IROS), pp. 922–928. IEEE (2015)
10. Qi, C.R., Liu, W., Wu, C., Su, H., Guibas, L.J.: Frustum pointnets for 3D object detection from RGB-D data. In: Proceedings of the IEEE Conference on Computer Vision and Pattern Recognition, pp. 918–927 (2018)
11. Qi, C.R., Su, H., Mo, K., Guibas, L.J.: PointNet: deep learning on point sets for 3D classification and segmentation. CoRR abs/1612.00593v2 (2016). http://arxiv.org/abs/1612.00593v2
12. Qi, C.R., Yi, L., Su, H., Guibas, L.J.: Pointnet++: deep hierarchical feature learning on point sets in a metric space. CoRR abs/1706.02413v1 (2017). http://arxiv.org/abs/1706.02413v1

13. Schroff, F., Kalenichenko, D., Philbin, J.: FaceNet: a unified embedding for face recognition and clustering. In: Proceedings of the IEEE Conference on Computer Vision and Pattern Recognition, pp. 815–823 (2015)
14. Simonyan, K., Zisserman, A.: Very deep convolutional networks for large-scale image recognition. arXiv preprint arXiv:1409.1556 (2014)
15. Su, H., Maji, S., Kalogerakis, E., Learned-Miller, E.: Multi-view convolutional neural networks for 3D shape recognition. In: Proceedings of the 2015 IEEE International Conference on Computer Vision (ICCV), ICCV 2015, pp. 945–953. IEEE Computer Society, Washington (2015). https://doi.org/10.1109/ICCV.2015.114. http://dx.doi.org/10.1109/ICCV.2015.114
16. Szegedy, C., et al.: Going deeper with convolutions. In: Proceedings of the IEEE Conference on Computer Vision and Pattern Recognition, pp. 1–9 (2015)
17. Wang, P.S., Liu, Y., Guo, Y.X., Sun, C.Y., Tong, X.: O-CNN: octree-based convolutional neural networks for 3D shape analysis. ACM Trans. Graph. (TOG) 36(4), 72 (2017)
18. Wu, Z., et al.: 3D ShapeNets: a deep representation for volumetric shapes. In: Proceedings of the IEEE Conference on Computer Vision and Pattern Recognition, pp. 1912–1920 (2015)
19. Yang, G., Huang, X., Hao, Z., Liu, M.Y., Belongie, S., Hariharan, B.: PointFlow: 3D point cloud generation with continuous normalizing flows. arXiv preprint arXiv:1906.12320 (2019)
20. Zamorski, M., et al.: Adversarial autoencoders for compact representations of 3D point clouds (2018)
21. Zhou, Y., Tuzel, O.: VoxelNet: end-to-end learning for point cloud based 3D object detection. In: The IEEE Conference on Computer Vision and Pattern Recognition (CVPR), June 2018
22. Zieba, M., Wang, L.: Training triplet networks with GAN. arXiv preprint arXiv:1704.02227 (2017)

Stock Return Prediction Using Dual-Stage Attention Model with Stock Relation Inference

Tanawat Chiewhawan[iD] and Peerapon Vateekul[(✉)][iD]

Chulalongkorn University Big Data Analytics and IoT Center (CUBIC),
Department of Computer Engineering, Faculty of Engineering, Chulalongkorn University,
Bangkok 10330, Thailand
6170929321@student.chula.ac.th, peerapon.v@chula.ac.th

Abstract. Deep learning models have become widely accessible for stock prediction tasks. However, most of the research in this area focuses on only a single stock or an index and often formulates the problem to optimize only on the accuracy. Our paper proposed a more profit-oriented framework by formulating the problem into multiple stock returns prediction as well as introducing a relation inference for stock ranking. This setup can diversify investment and eventually enhance trading profits while maintaining the regression accuracy. Moreover, it is become more challenging to process multiple time-series features simultaneously because of the great variety of available information in the financial market. We mitigate this with the state-of-the-art model for time-series forecasting, the Dual-stage attention recurrent neural networks (DA-RNN), and train them with the shared-parameter model setting. The attention layer within DA-RNN helps the model captures the relevance insight among the features. We conducted experiments on major 64 target stocks from the SET market with RMSE, mean reciprocal rank, and annualized profit returns as evaluation metrics. The results show that our proposed model framework (DA-RANK) can predict multiple stock returns in ranking order and able to produce a desirable improvement in profitability over other baseline models.

Keywords: Deep learning · Ranking-aware loss function · Long Short-term memory model · Dual-stage attention · Stock prediction · Stock ranking

1 Introduction

Stock prediction is notoriously a challenging subject because of the high volatility and the influence of dynamic external factors such as the global economy and investor's behavior. This predictability of stocks has long been controversial. The earlier works on the efficient-market hypothesis (EMH) [1, 2] suggest that the price reflects all information suddenly, and the movement is random processes. However, various studies from many fields attempt to explore this challenge. Recently, machine learning and deep learning are emerging with promising results.

In this paper, we propose a framework for multiple stock returns prediction to handle various stock time series features, and to capture stock relations. We modified the Dual-Stage-attention model (DA-RNN), the original work of Qin [3], to tackle the features

© Springer Nature Switzerland AG 2020
N. T. Nguyen et al. (Eds.): ACIIDS 2020, LNAI 12033, pp. 492–503, 2020.
https://doi.org/10.1007/978-3-030-41964-6_42

and the temporal relations. Next, we transformed a set of stock features to allow fixed batch size training. This set up allows us to infer stock relations with a combination of loss functions: regression loss and ranking loss. As first mentioned in [4], a model with high accuracy does not always lead to the optimum profit when trading, Table 1 demonstrates this discrepancy. Thus, our framework can focus more on profit using the relational ranking. Our prediction target is the next day returns; we describe this labeling process in Sect. 3.1. Finally, we conducted the experiments on 64 targets stock of SET. The results show that our model could improve annualized returns over the baseline while maintaining predicted returns accuracy.

The remainder of this paper is organized into sections as follows: Sect. 2 discusses related works. Section 3 describes the proposed framework. The experimental setting is in Sect. 4, and the results from our proposed method are in Sect. 5. Finally, we present the conclusion at the end of the paper.

Table 1. An intuitive explanation for the discrepancy[a] in prediction accuracy and actual profit, reference from Table 2 of [4]

Method	Ground truth			(1) Ranking-aware prediction					(2) MSE optimized prediction				
Stocks	A	B	C	A	B	C	MSE	Profit	A	B	C	MSE	Profit
Returns	+30	+10	−50	**+50**	−10	−50	266	+30	+20	**+30**	−40	200	+10

[a]Method (1) suggests a higher profit stock for trading while method (2) with higher regression accuracy (less mean square loss) suggest a lower profit stock

2   Related Works

This section refers to previous related studies for our research. We group the works as described below.

2.1   Machine Learning and Deep Learning in Stock Prediction

Machine learning has become increasingly popular in stock prediction research. For example, the works from [5, 6] conduct comparative experiments using algorithms such as Random Forest (RF), and Artificial Neural Network (ANN). The results show that RF outperforms other baseline models in the metrics of accuracy on stock trend classification as well as on the trading profit. Recently, more modern approaches start utilizing a deep learning model in their studies. References [7–9] implemented the Long-Short Term Memory recurrent neural network (LSTM) [10] with successful results. In [9] used the LSTM with numerous generated technical indicators as features to predict stock trends successfully. While [11] explored a modified LSTM to enhance the model's feature extraction.

However, with a various selection of features available in the financial market, the problem of stock prediction becomes more challenging. Hence, recent deep learning researches aim to implement new techniques to enhance models on those challenges. Hollis, Yi, and Viscardi [12] investigated an LSTM model with an attention mechanism. Their results align with other researches showing time-series forecast improvement [3, 13]. This paper will also utilize the attention mechanism to boost model prediction results with numerous time-series features.

2.2 Stock Prediction with Multi-variate Inputs

In 2017, Nelson et al. [9] explored the feasibility of the Long-Short-term memory model on the stock trend prediction task. The LSTM model is best-known for sequential prediction [10]. The experiment was on five individual stocks with 180 features, including generated technical indicators and price information. The results show that, on average, the LSTM outperforms other baselines. However, in some particular stock, Random forest shows better results.

In the same year, Qin et al. [3] proposed a nonlinear autoregressive exogenous model called "Dual-Stage Attention Recurrent Neural Network (DA-RNN)" for time-series task. DA-RNN is an attention-based recurrent neural network with two attention mechanisms, input attention, and temporal attention. First, the input attention layer enhances a recurrent model that handles multiple time series inputs. It captures and differentiates importance among feature inputs then applies attention weights to them. Second, a temporal attention layer processes the encoded information from the input attention layer and grasps the significance of them at each time step then applies the temporal attention weight ahead of the LSTM prediction layer. Their experiment tests DA-RNN on the NASDAQ100 dataset with more than 80 inputs for driving features. Even though the DA-RNN shows promising prediction performance but the author designed DA-RNN for a single time series prediction, which does not fulfill our multiple stocks prediction objectives. Figure 2 shows a simplified diagram of modified DA-RNN from Figure 1 of [3].

2.3 Multiple Stocks Prediction and Stock Ranking

In 2017, Fischer and Krauss [7] conducted a comprehensive study on stock trend prediction using LSTM. Their works focus on 500 stocks prediction in the S&P 500 index. They select top-k stocks from multiple stocks prediction for long positions and bottom-k stocks for short positions. This setting with LSTM shows better results than other models such as the Random Forest and vanilla Deep Neural Networks. However, their numerical feature was only a single sequence of the stock return, neglecting other possible relevance features.

In 2019, Feng et al. [4] proposed a model framework called "Temporal Relational Ranking" for stock prediction. Their model proposed a temporal graph convolution, which processes stock relations such as ownership, partnership (sparse binary features). They also introduce stock relation ranking frameworks that utilized ranking loss in their model. However, they conduct the model experiments on only five feature series, which are the closed price and four periods of a simple moving average of the close price. We apply their proposed relation ranking framework to our model by adding multi-features processing of the attention mechanism.

There are three main benefits of multiple stocks prediction over a single stock prediction. Firstly, we could capture the relations between stocks. Akitas [14] shows that grouping stocks within the same industry could benefit model predictions. For example, companies that are competitors could have an opposite trend, while companies that are trading partners could have similar trends. Section 3.2 will describes how we capture stock relations. Secondly, an optimum buying or selling signal for a single stock does not occur very often. Exploring multiple stocks increases trading opportunities and reducing risk. Finally, with multiple stock returns prediction, we can train the model to rank among those stocks and suggest only the top expected return for trading. This ranking method can help the model to improve profits. As mentioned earlier in [4], there is a discrepancy between optimizing the model accuracy and maximizing profit.

3 Proposed Framework

Our proposed framework aims to improve the performance of multiple stock returns prediction using various time-series inputs. The framework, as shown in Fig. 1, starts with data pre-processing and normalization for time-series inputs. Next, we apply the fixed batch data transformation before moving on to the prediction model, and we will discuss the detail of this transformation in Sect. 3.2.

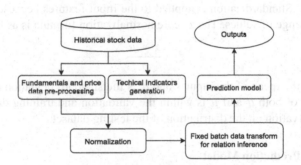

Fig. 1. Proposed framework

3.1 Data Pre-processing

Fundamentals and Price Data. The fundamentals data for each stock are transformed with forward-filling if they are quarterly updated to be consistent with other daily frequency data. There are seven fundamentals attributes that were represented in other forms, namely, in the percentage of changes from last quarter, the percentage of changes from last year, and cumulative value since the beginning of the year. Table 6 the Appendix describes details of all of the fundamental and price data used in this study.

Technical Indicators. We adopt a list of indicators proposed in [11] then generated them with our proposed period from short to long terms: (5, 7, 10, 14, 20, 30, 50, 75, 100) days. The total number of generated indicators is 9-periods multiply by 17 time-series from 15-indicators (MACD provide three series) equals 153 features. Table 2 shows the full list of our technical indicator features. We applied this indicator generation for every single stock in our target stocks.

Table 2. List of 15 technical indicators generated from historical price data [11]

RSI	EMA	TripleEMA	MACD	CMFI
William%R	SMA	CCI	PPO	DMI
WMA	HMA	CMO	ROC	PSI

Data Labeling. The ground truth for next step prediction is a 1-day return ratio for the following day, as shown in Eq. (1).

$$y_t^i = \left(p_{t+1}^i - p_t^i\right)/p_t^i \tag{1}$$

Where y_t^i is the return ratio for the stock i at time step t and p_t^i is the close price of the day t while p_{t+1}^i is the close price of the next day.

Normalization. Standardization is applied to the input features because each of them has a different range of values. The z-score normalization formula is as follows.

$$z = \frac{x - \mu}{\sigma} \tag{2}$$

Where μ is the mean of the input x and σ is the standard deviation of the input x. The calculation of both σ and μ is within the validation and training dataset to avoid our model observation on the distribution of the testing dataset.

3.2 Proposed Prediction Model

Our proposed model, Dual Attentional Ranking model (DA-RANK), aims to simultaneously predict sets of stock returns with relation inferences between them. The model structure consisted of two parts (i) Features relevance and temporal attentional recurrent neural network (ii) Stock relation inference framework.

Features Relevance and Temporal Attention. We select the state-of-the-art attention model for time series predictions called "Dual-Stage Attention-Based Recurrent Neural Network (DA-RNN)" to enhance feature and temporal relevance. Our core deep learning network is a modification from the original work of Qin [3]. We added a batch normalization [15] layer before the Softmax layer in the input attention layer, as shown in Fig. 2 to enhance attention weights calculation. We omitted the full detail of DA-RNN is from our paper as we introduce minor changes to the original work.

Stock Relation Inference. This second part of the model's purpose is to integrate stock relations during model training. We impose two methods (i) Fixed-batch training for a shared-parameter model and (ii) Pair-wise ranking loss.

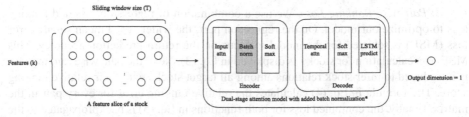

Fig. 2. A simplified diagram of dual-stage-attention recurrent neural networks from Figure 1 in [3], with added batch normalization layer [15]

(i) Fixed-batch Training for a Shared-parameter Model. A fixed-batch training is a designed we adopted from [4] to achieve ranking loss while training a model. This design fixed the size of a training data batch to equal the number of target stocks. Thus, the model can simultaneously train stock data within the same period and calculate a ranking loss. Also, with this setting, the weights of models are shared among all target stock rather than we construct multiple models separately per stock. We called this a shared-parameter model. Figure 3. shows the proposed fixed batch transformation. A slice of single stock's features has a dimension of T × k, where k is the number of time-series features for each stock (e.g., technical indicators, financial parameter series), and T is the sliding window for those features. We prepare this slice for each stock within our target N stocks. This collection of N feature slices is size-equivalent to the training batch size and represents multiple stock information during the same period. All N stocks share the same weight in the modified DA-RNN model Fig. 3(b) as a result of our fixed batch size setting. These model's shared weights are updated when the model observes all N slices of stock features in a batch during training.

There are three benefits to this design. First, it favors the ranking loss calculation, which we will cover in the next section. Second, the model becomes universal from the shared-parameter among target stocks concept, with the ability to predict particular stock independently. To be more specific, the model treats individual stock as one separate set of features within a training batch. The trained model could still predict any stock without the need to retraining the whole model again when any stock ceases to trade in the market. Finally, it reduces the model weights per data by a factor of target stocks. The lower model weights imply faster training and more straightforward to converge for the solution.

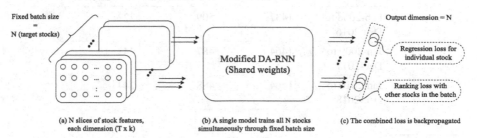

Fig. 3. Diagram of the proposed model DA-RANK: (a) input features slices (b) a modified DA-RNN unit, the model's weights are shared among all stocks (c) combination of loss functions

(ii) Pair-wise Ranking Loss. We use a combination of regression loss and ranking loss to optimize our model. On the regression part, the widely used mean square error loss (MSE) is selected for the model to focus on the return prediction accuracy. This MSE loss calculation for stock i is displayed in Eq. (3). Next, the pair-wise ranking loss is introduced to infer stock relations among all target stocks with their relative ranking score. The formula in Eq. (4) calculates the relative ranking error for every pair in the matrix. Finally, the combined loss for both functions in Eq. (5) is backpropagated to the model when learning a fixed batch size data Fig. 3(c).

$$MSE\ loss^i = \left(\hat{y}_t^i - y_t^i\right)^2 \tag{3}$$

$$Pairwise - Ranking\ loss = \sum_{i=0}^{N} \sum_{j=0}^{N} \max(0, -\left(\hat{y}_t^i - \hat{y}_t^j\right)\left(y_t^i - y_t^j\right)) \tag{4}$$

$$Combined\ Loss = \frac{\sum_{i=0}^{N} MSE\ loss^i}{N} + \alpha(Pairiwise - Rankingloss) \tag{5}$$

Where N is the number of target stocks to be predicted simultaneously, the $\hat{y}_t^i$ is the predicted return for stock i at time step t, the y_t^i is the label describes in the Eq. (1), and α (alpha) is a weighting ratio tradeoff between the regression accuracy and the ranking accuracy, which is one of the hyperparameters to be tuned.

4   Experimental Setting

4.1   Dataset

The stock's end of the day (EOD) data[1] used in our research is from The Stock Exchange of Thailand (SET) Market, corresponding to the period from 12th February 2008 to 28th December 2018. We use daily frequency data due to computational and data limitations. Data are split into three sets for each stock, as summarized in Table 3. The training,

Table 3. Dataset records summary

No.	Data period	Training records/stock	Validating records/stock	Testing records/stock
1	Feb-2008 to Dec-2016	1437	509	222
2	Jan-2009 to Dec-2017	1464	487	243
3	Jan-2010 to Dec-2018	1464	488	244

---

[1] The provided information was retrieved via the SET SMART portal with permission granted for academic purposes.

validating, and testing records for the model are approximately 92,000/31,500/15,000 per period, respectively, when considering all target stocks.

As suggested by [11, 16], we split the data into training, validating, and testing period, as shown in Fig. 4. This setting is to verify the robustness of the model over time.

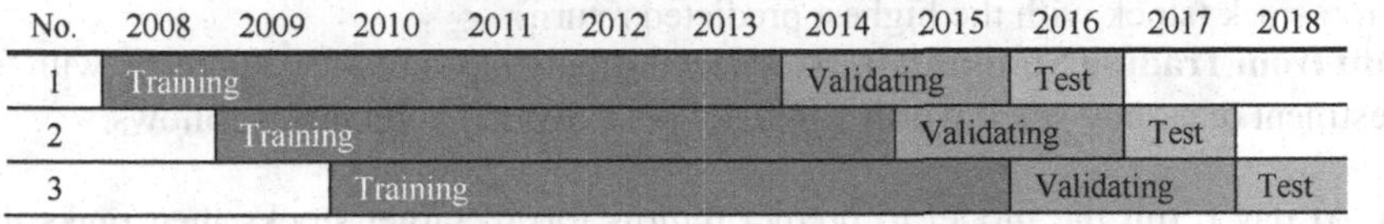

No.	2008	2009	2010	2011	2012	2013	2014	2015	2016	2017	2018
1	Training						Validating		Test		
2		Training						Validating		Test	
3			Training						Validating		Test

Fig. 4. Illustration of the sliding-splitting period for stock performance training and evaluation

4.2 Target Stocks Pre-selection Criteria

There are three considerations for our selection of target stock principles. With the below criteria, we pre-select 64 target stocks out of the SET100 index[2].

1. Stock information availability through training to testing periods
2. Sufficient liquidity to assume order always get filled
3. Sufficient volume and big market cap, to avoid price manipulation and to assume that our trading effect on the price can be neglect

4.3 Compared Methods

In this section, we define the multiple methods for performance comparison. The baselines include simple investment, machine learning, and other deep learning models.

1. Index performance – buy and hold (buy at the start of the period, sell at the end)

   • SET, SET100, and SET64[3]: buy and hold

2. Random Forest: represent a non-deep learning model (tree size: 200, max depths: 5)
3. LSTM[4] – as a general deep learning baseline (1 layer, vanilla LSTM)
4. LSTM-RANK – to compare the effectiveness of relation inference over LSTM
5. DA-RNN (see footnote 4) - to compare the effect of attention mechanisms
6. DA-RANK (proposed model) to compare both dual attention and relation inference

---

[2] SET100 index includes top stock with high market capitalization and trading volumes. We select SET100 Thai market capital as of 7 February 2018. These 64 stocks started trading before 2008 and still active in 2018.

[3] SET 64 refers to the 64 target stocks, we invested in equally size (e.g. 1000 dollars per stock).

[4] The outputs from LSTM/ DA-RNN are transformed from price to return using Eq. 1.

4.4 Evaluation Metrics

We compare the performance of each model with three measures:

Root Mean Square Error (RMSE). standard evaluation to the predicted return
Mean Reciprocal Ranking (MRR). evaluate the model on the ranking performance of the top stock (stock with the highest predicted return)
Profit from Trading Simulation. We select the daily buy-hold-sell strategy[5] with fixed investment (e.g., buy stock worth 1,000 dollars daily). The details as follows:

- At day t, run the model to predict returns for all target stocks then ranks those predicted return to select only the top stock to buy with fixed investment
- At the day t + 1, sell the stock bought from day t at the close price of day t + 1.

4.5 Hyperparameter Tuning

We optimized our model with the Adaptive Moment Estimation (Adam) algorithm with an initial learning rate of 0.001. Next, a grid search for hyperparameter was applied to the range of parameters as follows: hidden unit (16, 32), window size - T equal 5, and regression-ranking tradeoff: Alpha - α (0, 10, 100, 1000). We choose this alpha range because we observed that the average magnitude of the MSE loss on the first model epoch is around 95 times larger than the ranking-aware loss.

5   Experimental Results

We select the model with the best MRR in the validation dataset to evaluate on the test dataset, the performance on four metrics is shown in Table 4. The results show that our DA-RANK model, on the three-year average, top performs other models on the MRR and the % average annual profit at 0.1193 and 71.97% respectively. However, on the RMSE, the RF consistently top performs over three years at 0.019, while our model ranks the third at 0.030. On the effectiveness of relation inference, it can improve both DA-RNN and LSTM model's performance in all metrics except for DA-RNN's RMSE.

The standards deviation of the tested results as shown in Table 5 is to illustrate the robustness of the models. We found that the DA-RANK ($\pm41.3\%$) and DA-RNN ($\pm39.9\%$) are not as robust as LSTM ($\pm12.2\%$) and LSTM-Rank (±21.6) on % profit. While on RSME for models with DA-RNN show significantly more robust at around ±0.005.

Another benefit of using the attention mechanism over dimensionality reduction is the model interpretability. We can extract the attention weights to learn which features are relevance among all features for a stock at any particular time. For example, during the 2018 period, the top 3 relevance features for PTTEP to the predicted returns are 1. Close price, 2. TEMA (5 periods) 3. HMA (5 periods). This interpretability would be difficult if we performed dimensionality reduction to the input features.

---

[5] This strategy assumes that the trading volume always sufficient to satisfy buying or selling at close price. And the fee is neglected but we can recalculate percent profit after fee with:
$\%Return_{after\ fee} = \%Return_{before\ fee} \times (1 - fee)^2$.

Table 4. Profit, MRR, and RMSE on test data set comparison with baseline models

Model	Profit %				MRR top stock			
	2016	2017	2018	Avg.	2016	2017	2018	Avg.
SET	20.00%	12.20%	-12.10%	6.70%				
SET100	20.20%	14.90%	-11.40%	7.90%				
SET64	25.80%	19.10%	-16.00%	9.60%				
RF	80.83%	7.96%	88.72%	59.17%	0.112	0.075	0.134	0.1070
LSTM	66.51%	-22.52%	0.80%	14.93%	0.093	0.068	0.122	0.0942
LSTM-RANK	75.86%	5.94%	64.65%	48.81%	0.117	0.088	**0.150**	0.1180
DA-RNN	**122.87%**	**27.64%**	-20.42%	43.36%	**0.148**	0.108	0.100	0.1188
DA-RANK	77.76%	27.01%	**111.13%**	**71.97%**	0.111	**0.113**	0.134	**0.1193**

Model	RMSE				Sharpe Ratio			
	2016	2017	2018	Avg.	2016	2017	2018	Avg.
RF	**0.020**	**0.017**	**0.019**	**0.019**	0.141	0.022	0.117	0.093
LSTM	0.133	0.039	0.048	0.073	0.130	-0.044	0.001	0.029
LSTM-RANK	0.089	0.041	0.032	0.054	0.138	0.009	0.069	0.072
DA-RNN	0.023	0.021	0.038	0.027	**0.196**	**0.049**	-0.026	0.073
DA-RANK	0.031	0.021	0.038	0.030	0.124	0.046	**0.142**	**0.104**

Table 5. The average and standard deviation for the test dataset for all hyperparameter

Model	% Profit	MRR	RMSE	Sharpe ratio
LSTM	16.62% ± 12.18%	0.096 ± 0.004	0.09 ± 0.022	0.03 ± 0.016
LSTM-RANK	44.94% ± 21.16%	0.101 ± 0.016	0.15 ± 0.126	0.08 ± 0.032
DA-RNN	45.49% ± 39.93%	0.114 ± 0.020	0.03 ± 0.004	0.07 ± 0.056
DA-RANK	44.46% ± 41.73%	0.111 ± 0.012	0.03 ± 0.005	0.06 ± 0.062

6 Conclusion

In this paper, we introduce a new framework for multiple stock returns prediction called "DA-RANK". We tailored the framework to enhance model profits with dual-stage-attention and relation inference. Our model could capture stock relations (relation inference) with three implementations, shared-parameter model, fixed batch size training, and ranking loss. More importantly, the model able to process a large number of features input through our modified dual attention mechanism. We evaluate the performance with RMSE, MRR, and annualized profit from trading simulation on 64 target stocks within the SET market using three annual-sliding periods of test data.

The results show that our proposed model DA-RANK overtakes baseline models on profitability while preserving satisfied regression accuracy. This profit enhancement was due to the ability to ranking stocking during the prediction and capability to process various feature inputs. The DA-RANK was unable to outperform the Random forest on the RMSE metric. However, this downfall is supported by the initial assumption that there is a discrepancy between optimized accuracy and maximizing profits.

In the future work, the research can be extended to the high-frequency trading or intra-day data; however, in this paper, were limited on the data and computing power.

Acknowledgment. Our work is partly supported on the research budget by Capital Market Research Institute (CMRI), The Stock Exchange of Thailand (SET), during Capital Market Research Innovation contest 2019. and permission to use the SET SMART dataset for academic purposes from the Financial Laboratory, Chulalongkorn Business School and Asst. Prof. Tanakorn Likitapiwat.

Appendix

Table 6. Fundamentals and price data description, 52 attributes in total

Attribute name	Description			Count
A/P Turnover	Seven attributes presented in 5 forms below • Q - at the quarter data • Cum. Q - cumulative quarter value since the first day of the year • QoQ % - percent change from the previous quarter • YoY % - percent change from the previous year, same quarter • YoY Cum. - percent change from YE data (cumulative)			35
D/E Ratio				
Fixed Asset				
Shareholder Equity				
Total Asset				
Total Liability				
Total Revenue				
Attribute name: Quarter data				6
Cash Cycle Period	Net Profit Margin	Net Profit		
Earnings per Share	Return of Asset	Return of Equity		
Attribute name: Daily data				11
Close Price	Open Price	Stock Trade Volume	P/E Ratio	
High Price	Book Value	Transaction Volume	P/BV Ratio	
Low Price	Market Value	Market Capital		

References

1. Fama, E.F.: Efficient capital markets: a review of theory and empirical work. J. Financ. **25**(2), 383–417 (1970). https://doi.org/10.2307/2325486
2. Malkiel, B.G.: Reflections on the efficient market hypothesis: 30 years later. Financ. Rev. **40**(1), 1–9 (2005)
3. Qin, Y., Song, D., Chen, H., Cheng, W., Jiang, G., Cottrell, G.: A dual-stage attention-based recurrent neural network for time series prediction. arXiv e-prints (2017)
4. Feng, F., He, X., Wang, X., Luo, C., Liu, Y., Chua, T.-S.: Temporal relational ranking for stock prediction. ACM Trans. Inf. Syst. **37**, 1–30 (2019). https://doi.org/10.1145/3309547
5. Ballings, M., Van den Poel, D., Hespeels, N., Gryp, R.: Evaluating multiple classifiers for stock price direction prediction. Expert Syst. Appl. **42**(20), 7046–7056 (2015)
6. Patel, J., Shah, S., Thakkar, P., Kotecha, K.: Predicting stock and stock price index movement using trend deterministic data preparation and machine learning techniques. Expert Syst. Appl. **42**(1), 259–268 (2015)
7. Fischer, T., Krauss, C.: Deep learning with long short-term memory networks for financial market predictions. Eur. J. Oper. Res. **270**(2), 654–669 (2018). https://doi.org/10.1016/j.ejor.2017.11.054
8. Chen, K., Zhou, Y., Dai, F.: A LSTM-based method for stock returns prediction: a case study of china stock market. In: 2015 IEEE International Conference on Big Data (Big Data), pp. 2823–2824. IEEE (2015)
9. Nelson, D.M., Pereira, A.C., de Oliveira, R.A.: Stock market's price movement prediction with LSTM neural networks. In: 2017 International Joint Conference on Neural Networks (IJCNN), pp. 1419–1426. IEEE (2017)
10. Hochreiter, S., Schmidhuber, J.: Long short-term memory. Neural Comput. **9**(8), 1735–1780 (1997). https://doi.org/10.1162/neco.1997.9.8.1735
11. Sezer, O.B., Ozbayoglu, A.M.: Algorithmic financial trading with deep convolutional neural networks: time series to image conversion approach. Appl. Soft Comput. **70**, 525–538 (2018). https://doi.org/10.1016/j.asoc.2018.04.024
12. Hollis, T., Viscardi, A., Yi, S.E.: A comparison of LSTMs and attention mechanisms for forecasting financial time series. CoRR abs/1812.07699 (2018)
13. Guo, T., Lin, T.: Multi-variable LSTM neural network for autoregressive exogenous model. arXiv preprint arXiv:1806.06384 (2018)
14. Akita, R., Yoshihara, A., Matsubara, T., Uehara, K.: Deep learning for stock prediction using numerical and textual information. In: 2016 IEEE/ACIS 15th International Conference on Computer and Information Science (ICIS), 26–29 June 2016, pp. 1–6 (2016)
15. Ioffe, S., Szegedy, C.: Batch normalization: accelerating deep network training by reducing internal covariate shift. arXiv preprint arXiv:1502.03167 (2015)
16. Oncharoen, P., Vateekul, P.: Deep learning using risk-reward function for stock market prediction. Paper Presented at the 2018 2nd International Conference on Computer Science and Artificial Intelligence, Shenzhen, China (2018)

Comparison of Aggregation Functions for 3D Point Clouds Classification

Maciej Zamorski[1,2(✉)], Maciej Zięba[1,2], and Jerzy Świątek[1]

[1] Faculty of Computer Science and Management, Wrocław University of Science and Technology, Wrocław, Poland
maciej.zamorski@pwr.edu.pl
[2] Tooploox, Wrocław, Poland

Abstract. The three-dimensional data is the core tool behind environment aware algorithms used in e.g. SLAM or autonomous driving. As a data format, point clouds are becoming increasingly popular, due to their high-resolution and mapping fidelity. However, representing data as points, rather than voxels, comes with very high processing complexity, as machine learning models need to deal with permutation-invariance within samples. The PointNet architecture provides an easy and efficient way to deal with the point cloud data, by performing feature extraction for each point separately and then computing feature-wise max function. In this work, we present a comparison of different permutation-invariant functions used for this aggregation evaluated on the ShapeNet dataset for the classification task.

Keywords: Representation learning · Point clouds · Deep learning · Input permutation invariance

1 Introduction

Deep learning models are known for their superiority over other machine learning approaches in image and video processing applications. The deep architectures that utilize convolution filters are capable to solve various image processing tasks including segmentation, classification, object detection, and image generation. However, that group of models is ineffective for input data represented by the point clouds - sets of points creating the objects in 3D space because of the sparse input data that significantly slows down the learning procedure. To overcome that issue PointNet model [7] for point cloud representation was introduced. The central idea of the model is that it is invariant to the permutations of the input data, which practically means, that for any order of the points delivered on the input the PointNet returns the same output values. This phenomenon is achieved by the application of a specific max(·) aggregation function that is permutation invariant and is applied to the dimension combined with the number of points.

In this work, we make a deep analysis of aggregation procedures applied to the PointNet. First, we analyze the quality and diversity of key points that

N. T. Nguyen et al. (Eds.): ACIIDS 2020, LNAI 12033, pp. 504–513, 2020.
https://doi.org/10.1007/978-3-030-41964-6_43

are selected within the max(·) aggregation procedure. Next, we propose and evaluate other aggregation functions that are also permutation invariant to the ordering of the input points. We show, that combining aggregation functions may lead to better classification results. For the evaluation purposes, we make use of ShapeNet [2] dataset and compare the various aggregation techniques using the classification accuracy of the PointNet model.

This work is organized as follows. Section 2 gives a brief overview of the development of the machine learning methods for 3D data processing and representation learning. Section 3 introduces the PointNet model and aggregation procedures that we evaluated in this work. Conducted experiments and the conclusions are extensively discussed in Sect. 4. The last Sect. 5 provides a summary of the article.

2   Related Works

In the initial works about models for 3D point clouds authors successfully applied CNNs to 3D volumetric data [5,12]. In [10] authors represent 3D objects by multiple 2D images gathered from different angles and propose a tailored model for multi-view examples - Multi-view CNN (MVCNN). The state of the art accuracy was achieved with a variation of Multi-view CNN - RotationNet [3], where authors enrich the existing MVCNN solution with an additional rotation network.

The voxel representation is rather problematic because of the dimensionality and the sparsity of the input data. It would we beneficial if the model is operating on the set of 3D coefficients. However, typical neural network architectures are incapable of processing sets of points because they are not invariant to the given points ordering. To overcome that issue authors of [7] proposed PointNet, the specific deep learning model for which the order invariance is obtained by application of a single symmetric function, max pooling. Effectively the network learns a set of optimization functions/criteria that select interesting or informative points of the point cloud and encode the reason for their selection. The idea of PointNet was extended in [8] by proposing PointNet++ that utilizes hierarchical point set feature learning in order to improve the ability to recognize fine-grained patterns and complex scenes.

In [11] authors propose another interesting approach for 3D point data, an octree-based convolutional neural network (O-CNN). The key idea of our method is to represent the 3D shapes with octrees and perform 3D CNN operations only on the sparse octants occupied by the boundary surfaces of 3D shapes. An interesting approach was presented in [4] where authors propose to use hierarchical feature extraction with permutation invariant network and Self Organized Maps (SOM) to produce a two-dimensional representation of the point cloud. The most recent works, VoxelNet [15], and the further extension F-PointNet [6] are complex end-to-end models capable to be trained on large numbers of point clouds.

Besides numerous supervised approaches proposed for 3D point clouds, there are also a couple of attempts to train informative representations in an unsupervised manner. In [1] authors propose an autoencoder for representing 3D point clouds that takes point cloud as an encoding network and recommend to apply one of two possible distances as reconstruction losses: Chamfer and Earth-Mover. The latent representation can be further mapped to the arbitrary given distribution with the specific GAN model trained on the latent space in the stacked mode. The model is utilizing simple PointNet architecture in the encoding part of the model. An interesting extension is presented in [14] where authors propose to use a variation of adversarial autoencoder for 3D point clouds that is trained in the end-to-end framework. The key idea of the approach is that it enforces assumed prior distribution on the latent space for generative purposes. One of the recent papers [13] utilizes VAE together with continuous normalizing flows to generate 3D point clouds. Another interesting approach is described in [9], where the authors utilize the variation of a conditional flow-based generative model designed for 3D data.

Most of the current models that are solving various tasks with the point clouds assume some variations of PointNet structure. Therefore it is beneficial to propose and evaluate novel aggregation techniques that can be utilized in various models designed for the point clouds.

3   Method

In this section, we describe the deep learning framework used for processing the point cloud data. First, we state the importance of using the models that are permutation invariant to the passed input, following by describing the PointNet as a permutation invariant model. Lastly, we describe other permutation invariant aggregations as a study of improving information richness embedded in learned representations and its application to a classification task on a ShapeNet dataset.

3.1   Model Invariance Problem for Point Clouds

Contrary to the image data, where pixels are stored in a structured, grid-array order, point cloud samples are given as sets of points, which are (due to the basic mathematical properties of the sets) orderless and contain distinct objects. This property greatly increases the complexity of the point cloud data, as neural networks typically assume an ordered input. To avoid training models for the $n!$ possible permutations of each input example (n being the number of the point clouds in the dataset) it is necessary to construct the machine learning models in a way, that is invariant to the input permutation. The usual approach for dealing with permutation invariant data is processing each element (in our case, each point) of the input (here: point cloud) independently, followed by an aggregation of outputs by a permutation invariant function. We can define such function as follows.

Let's assume a set of point clouds, denoted as $\mathcal{S} = \{S_1, S_2, \ldots, S_n\}$. Each point cloud S_k, is made of points $s_{k,1}, s_{k,2}, \ldots, s_{k,l} \in \mathbb{R}^3$. The function $f : \mathbb{R}^{k \times 3} \to \mathbb{R}^h$ and $f(S_k) = z_k$ is called *a permutation invariant function*, when we want to obtain constant output of the function for the same point cloud, regardless of the order of the points composing that point cloud. This can be stated as:

$$f(S_k) \equiv f(s_{k,1}, s_{k,2}, \ldots, s_{k,l}) = f(s_{k,\pi_1}, s_{k,\pi_2}, \ldots, s_{k,\pi_l}), \tag{1}$$

where $\pi_1, \pi_2, \ldots \pi_l$ is an arbitrary permutation of the range $1, 2, \ldots l$.

3.2   PointNet as a Permutation Invariant Structure

In this work, we use the PointNet model, introduced in [7]. The main characteristic of this architecture is performing feature extraction for each point separately and then aggregating those features point-wise with permutation-invariant function into a single vector of global features. In [7] the authors presented the architecture as shown in Fig. 1, with f being max-pooling function, MLP 1 being two fully-connected layers with 64 units both, MLP 2 being two fully-connected layers with 128 and h_z units, MLP 3 being three fully-connected layers with 512, 256 and h_y units, where h_y- number of classes. The TNet modules are referring to learnable affine transformations in a 3-dimensional space (point space) and a 64-dimensional space (feature space). They are modeled as "mini PointNets", but are returning a transformation matrices of sizes 3×3 and 64×64 respectively.

PointNet achieves permutation invariance with a max-pooling function f (marked as a green region in the Fig. 1). In this work, we replace the max pooling with other types of aggregations in order to study potential benefits for representations expressiveness learned by the model and a subsequent increase in classification accuracy. We compare simple, single-statistic functions such as mean($\cdot$), median($\cdot$) or sum($\cdot$) with combination of those statistics, further aggregated with the use of the neural networks. We also present a qualitative comparison between max($\cdot$) and median($\cdot$) functions, by comparing the critical subsets of point clouds chosen by those methods, consisting of points having the activations, that were returned.

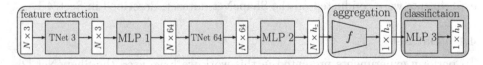

Fig. 1. A PointNet architecture consisting of three parts: (a) feature extraction encoding separately each point of a given point cloud to a feature vector of length h_z, (b) aggregation combining features feature-wise across all points, (c) classification part learning a mapping from aggregated features to class labels. The TNet K refers to learnable affine transforms in K dimensions introduced in [7]. The MLP M networks architecture is described in a Sect. 3.2.

3.3   Current Aggregation Techniques

Currently feature aggregation functions are such functions f (with properties described in Sect. 3.1) that for matrices of local features $F \in \mathbb{R}^{l \times h}$ (where l - number of points in the point cloud, h - number of features for each point in the point cloud) perform feature-wise aggregations across all points, resulting in a global feature vector $\mathbf{z} \in \mathbb{R}^h$. Most widely aggregation function is a $\max(\cdot)$ function (presented in [7]), returning maximum local activation value among all points for specific features.

However, these aggregation techniques come with limitations. Using the max function to construct a feature, takes only the maximum activation value into consideration while ignoring the other values and the shape of the distribution of all activations. By using other aggregation functions, such as sum, mean or median the whole local features distribution is used for creating a global feature.

In this work, we also propose using an approach consisting of a combination of different aggregation functions. First, we calculate the statistics with different methods: sum, median, mean and max functions, obtaining vectors of size h each. Those statistics were later grouped into two groups: (1) sum, mean & max and (2) median, mean & max, resulting in two vectors of size $3h$. In the end, we pass those vectors through a two-layer neural network (with layers of length $2h$ and h) to reduce the global vector size from $3h$ to h. It is done to ensure that those "combined" vectors were comparable with the standard approaches.

Another proposed approach is a modification of described $\max(\cdot)$ function to return k biggest values from the input set, instead of just the maximal one. We call this method a "top-k" function. Those values are later passed through a one-layer neural network to reduce the vector size from $k \cdot h$ to h in order to compare the method with other approaches.

4   Experiment

The goal of the experiment is to evaluate the classification abilities of the classification network, based on different aggregation functions f used to combine features point-wise. In addition to standard permutation-invariant functions max, mean, median, three combinations were also checked: (1) sum-mean-max (SMM), (2) median-mean-max (MMM) and (3) max top-k, concatenated together and passed through the one-layer neural network to match the encoding size. The experiments were conducted on the ShapeNet dataset, described in Sect. 4.1.

4.1   Dataset

To perform experiments presented in this work we have used the ShapeNet [2] dataset. It consists of over 50000 point clouds, each given as 2048 points in 3-dimensional space and belonging to 55 classes. The classes are heavily imbalanced, ranging from 65 to 23 thousand samples per class. Each point cloud is given pre-aligned to a vertical axis (usually denoted in Cartesian coordinates as axis "z"). The dataset was split into 85% train – 5% validation – 10% test part while keeping proportions between the classes.

4.2 Metrics

Three metrics were used for quantitative evaluation:

- accuracy – percent of correctly classified test samples, given as

$$\text{accuracy} = \frac{1}{N} \sum_{n=1}^{N} \mathbb{1}(\hat{y}_n = y_n), \tag{2}$$

where N - number of samples, $\hat{y}_n$ - predicted class of n-th sample, y_n - true class of n-th sample and $\mathbb{1}(\cdot)$ - indicator function
- balanced accuracy – average of recall values calculated per class. As the dataset is very imbalanced it provides better overview of classification ability on underrepresented classes, given as

$$\text{b. accuracy} = \frac{1}{C} \sum_{c=1}^{C} \frac{1}{S_c} \sum_{s}^{S_c} \mathbb{1}(\hat{y}_s = c), \tag{3}$$

where C - number of classes in the dataset, S_c - number of samples that belong to class c, $\hat{y}_s$ - predicted class of sample y_s, belonging to class c and $\mathbb{1}(\cdot)$ - indicator function
- encoding diversity – for the aggregation functions that return a feature value of a specific point (e.g. max, median) we calculate how many unique points were taken into constructing final feature vector, given as

$$\text{diversity} = \frac{1}{N} \sum_{n=1}^{N} \frac{|\text{set}(f(x_n))|}{h_z}, \tag{4}$$

where $\text{set}(A)$ - set consisting of elements of A, $|A|$ - number of elements of A, h_z - length of a feature vector, N - number of samples.

4.3 Results

In this section, we present the results on the classification tasks achieved by the PointNet model with different aggregation functions. The experiments were conducted for five different lengths of the feature vector $h_z \in \{8, 16, 25, 50, 100\}$.

In Table 1 we present the results for model accuracy, without taking the ShapeNet class imbalance into account. We can observe, that combining several types of features (in this case MMM) yields much better results than using any of the other aggregation types in separation. If only a single-statistic feature aggregation is considered, then using mean and median allows higher performance than the max function baseline.

Table 2 presents the model performance with consideration of the class imbalance. First, the accuracy was calculated for every class separately, and then the results were averaged to obtain more informative data about the classification abilities. It can be observed, that using a statistic that takes all points into

Table 1. Comparison of a classification accuracy given different aggregation functions and lengths h_z of the global feature vector.

Aggregation	$h_z = 8$	$h_z = 16$	$h_z = 25$	$h_z = 50$	$h_z = 100$
Max	0.8426	0.8468	0.8494	0.8471	0.8484
Mean	0.8563	0.8610	0.8632	0.8648	0.8703
Median	0.8536	0.8559	0.8623	0.8636	0.8677
MMS	0.8561	0.8580	0.8642	0.8677	0.8717
MMM	**0.8735**	**0.8671**	**0.8701**	**0.8739**	**0.8765**

Table 2. Comparison of a classification balanced accuracy given different aggregation functions and lengths h_z of the global feature vector.

Aggregation	$h_z = 8$	$h_z = 16$	$h_z = 25$	$h_z = 50$	$h_z = 100$
Max	0.6704	0.6785	0.6805	0.6826	0.6879
Mean	**0.6966**	**0.7029**	**0.7091**	0.6938	0.7075
Median	0.6861	0.6824	0.6954	0.6956	0.7094
MMS	0.6682	0.6767	0.7082	0.7056	0.7021
MMM	0.6854	0.6880	0.7035	**0.7072**	**0.7182**

Table 3. Comparison of accuracy and balanced accuracy between top-k of max values. Note that using the top-1 aggregation is the same as using the standard max function.

Top-k	Accuracy	Recall
$k = 1$	0.8468	0.6785
$k = 2$	0.8547	0.6802
$k = 3$	**0.8560**	0.6796
$k = 4$	0.8554	**0.6987**
$k = 5$	0.8555	0.6869

Table 4. Comparison of the diversity of the points that compose the feature vector. A higher value means that a bigger portion of the feature vector was created by taking values of the unique points from the point cloud.

Aggregation	$h_z = 8$	$h_z = 16$	$h_z = 25$	$h_z = 50$	$h_z = 100$
Max	0.9333	0.8829	0.8659	0.7736	0.6409
Median	**0.9821**	**0.9530**	**0.9197**	**0.8063**	**0.7151**

account while calculating the feature value (i.e. mean) allows obtaining the best results when using short feature vectors.

The original PointNet [7] used a max function to aggregate feature activations. In Table 3 we present the results of taking additional activations (k

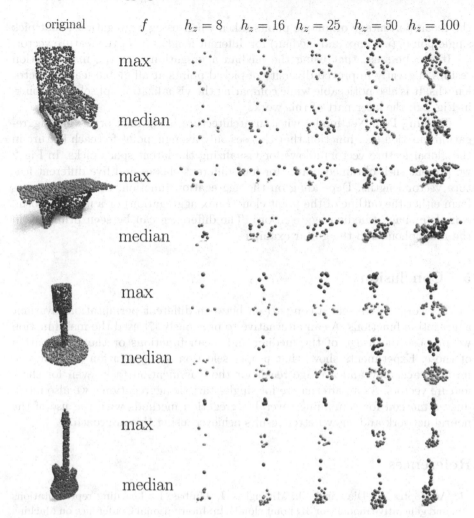

Fig. 2. Visualization of points that were used for global feature vector composition for different types of aggregation function f and feature vector length h_z. Original point cloud presented in the left column.

maximal values ins ted of just 1). The values from 1 to 5 were tested, as getting more points resulted in diminishing or no improvement. Based on these results we conclude, that when using max as an aggregation function, using the top 4 activation values is optimal.

Using the aggregation functions that pick specific feature values out of all given feature values among points in the cloud may result in selecting the same point for more than one feature in the feature vector, potentially reducing the amount of information represented in the final feature vector. Table 4 presents the mean percentage of unique points taken when using aggregations, that pick single points (i.e. max and median) for different lengths h_z of the feature vector.

It can be seen, that using the median aggregation over the max function results in greatly improved diversity of picked points at all global feature vector lengths. It is also noticeable while comparing the visualizations of selected points in Fig. 2 in the later part of this work.

Training PointNet model with an architecture using max or median aggregation function (i.e. function that chooses only a single point for each feature in the global feature vector) allows for visualizing the latent space picks. In Fig. 2 we present such obtained points for four different classes and five different feature vector lengths. Depending on the aggregation function, the picked points form either the outline of the point cloud (max aggregation) or a more uniform, complete shape (median aggregation). The difference can be seen primarily in the microphone and the guitar examples.

5   Conclusion

In this work, we present a comparison between different permutation-invariant aggregation functions. As an alternative to previously [7] used the max function we propose the usage of the median and mean functions or the combination of those. Experiments show, that proper selection of aggregation function can improve accuracy and average recall for the classification task, even for short feature vectors. As an alternative for single-statistic aggregations, we also introduce a method for combining several aggregation methods with the use of the neural network and show better results achieved using this approach.

References

1. Achlioptas, P., Diamanti, O., Mitliagkas, I., Guibas, L.: Learning representations and generative models for 3D point clouds. In: International Conference on Machine Learning, pp. 40–49 (2018)
2. Chang, A.X., et al.: ShapeNet: an information-rich 3D model repository. arXiv preprint arXiv:1512.03012 (2015)
3. Kanezaki, A., Matsushita, Y., Nishida, Y.: Rotationnet: joint object categorization and pose estimation using multiviews from unsupervised viewpoints. In: Proceedings of the IEEE Conference on Computer Vision and Pattern Recognition, pp. 5010–5019 (2018)
4. Li, J., Chen, B.M., Hee Lee, G.: SO-Net: self-organizing network for point cloud analysis. In: The IEEE Conference on Computer Vision and Pattern Recognition (CVPR), June 2018
5. Maturana, D., Scherer, S.: VoxNet: a 3D convolutional neural network for real-time object recognition. In: 2015 IEEE/RSJ International Conference on Intelligent Robots and Systems (IROS), pp. 922–928. IEEE (2015)
6. Qi, C.R., Liu, W., Wu, C., Su, H., Guibas, L.J.: Frustum PointNets for 3D object detection from RGB-D data. In: Proceedings of the IEEE Conference on Computer Vision and Pattern Recognition, pp. 918–927 (2018)

7. Qi, C.R., Su, H., Mo, K., Guibas, L.J.: PointNet: deep learning on point sets for 3D classification and segmentation. In: Proceedings of the IEEE Conference on Computer Vision and Pattern Recognition, pp. 652–660 (2017)
8. Qi, C.R., Yi, L., Su, H., Guibas, L.J.: PointNet++: deep hierarchical feature learning on point sets in a metric space. CoRR abs/1706.02413v1 (2017). http://arxiv.org/abs/1706.02413v1
9. Stypułkowski, M., Zamorski, M., Zięba, M., Chorowski, J.: Conditional invertible flow for point cloud generation. arXiv preprint arXiv:1910.07344 (2019)
10. Su, H., Maji, S., Kalogerakis, E., Learned-Miller, E.: Multi-view convolutional neural networks for 3D shape recognition. In: Proceedings of the 2015 IEEE International Conference on Computer Vision (ICCV), ICCV 2015, pp. 945–953. IEEE Computer Society, Washington (2015). https://doi.org/10.1109/ICCV.2015.114
11. Wang, P.S., Liu, Y., Guo, Y.X., Sun, C.Y., Tong, X.: O-CNN: octree-based convolutional neural networks for 3D shape analysis. ACM Trans. Graph. (TOG) **36**(4), 72 (2017)
12. Wu, Z., et al.: 3D ShapeNets: a deep representation for volumetric shapes. In: Proceedings of the IEEE Conference on Computer Vision and Pattern Recognition, pp. 1912–1920 (2015)
13. Yang, G., Huang, X., Hao, Z., Liu, M.Y., Belongie, S., Hariharan, B.: PointFlow: 3D point cloud generation with continuous normalizing flows. arXiv preprint arXiv:1906.12320 (2019)
14. Zamorski, M., et al.: Adversarial autoencoders for compact representations of 3D point clouds. arXiv preprint arXiv:1811.07605 (2018)
15. Zhou, Y., Tuzel, O.: VoxelNet: end-to-end learning for point cloud based 3D object detection. In: The IEEE Conference on Computer Vision and Pattern Recognition (CVPR), June 2018

Benign and Malignant Skin Lesion Classification Comparison for Three Deep-Learning Architectures

Ercument Yilmaz[1]([✉])(iD) and Maria Trocan[2](iD)

[1] Karadeniz Technical University, Trabzon, Turkey
ercument@ktu.edu.tr
[2] Institut Supérieur d'Électronique de Paris, Paris, France
maria.trocan@isep.fr
http://www.ercumentyilmaz.com

Abstract. Early detection of melanoma, which is a deadly form of skin cancer, is vital for patients. Differential diagnosis of malignant and benign melanoma is a challenging task even for specialist dermatologists. The diagnostic performance of melanoma has significantly improved with the use of images obtained via dermoscopy devices. With the recent advances in medical image processing field, it is possible to improve the dermatological diagnostic performance by using computer-assisted diagnostic systems. For this purpose, various machine learning algorithms are designed and tested to be used in the diagnosis of melanoma. Deep learning models, which have gained popularity in recent years, have been effective in solving image recognition and classification problems. Concurrently with these developments, studies on the classification of dermoscopic images using CNN models are being performed. In this study, the performance of AlexNet, GoogLeNet and Resnet50 CNNs were examined for the classification problem of benign and malignant melanoma cancers on dermoscopic images. Dermoscopic images of 19373 benign and 2197 malignant lesions obtained from ISIC database were used in the experiments. All three CNNs, which were the former winners of the ImageNet competition, have been reconfigured to perform binary classification. In the experiments 80% of the images were used for training and the remaining 20% were used for validation. All experiments were performed with the same parameters for each CNN models. According to the experiments ResNET50 model achieved the best performance with 92.81% classification accuracy and AlexNet was first-ranked in terms of the time complexity measurements. The development of new models based on existing CNN models with a focus on dermoscopic images will be the subject of future studies.

Keywords: Melanoma · Benign and Malignant lesion classification · Deep learning · Convolutional neural networks · AlexNet · GoogLeNet · ResNet

Dr. E. Yilmaz's contribution was supported by The Scientific and Technological Research Council of Turkey (TUBITAK) under Grant 1059B191802000.

© Springer Nature Switzerland AG 2020
N. T. Nguyen et al. (Eds.): ACIIDS 2020, LNAI 12033, pp. 514–524, 2020.
https://doi.org/10.1007/978-3-030-41964-6_44

1   Introduction

Melanoma is a lethal form of skin cancer caused by dark pigment producing cells called melanocytes. This type of cancer develops as a result of the genetic mutation of melanocytes, that produce melanin which determines the skin and hair color. Melanoma is observed at a low percentage of 4% to 5% among dermatological cancers [1]. According to the cancer statistics mentioned in a recent study, it is predicted that 1,762,450 cancer cases will be observed in the United States in 2019 and 96,480 of these cases will be of melanoma type [2]. Melanoma is responsible for 75% to 80% of deaths from such skin cancers.

Aggressive growth and metastasis are among the general characteristics of melanoma. Melanoma can be easily treated when diagnosed in its early stages (stage 1). Melanoma has a survival rate of 97% at stage 1. Disease can progress rapidly to the last stage (stage 4) due to the late diagnosis or improper treatment and becomes untreatable. Only 14% of patients with metastatic melanoma survive for up to 5 years. Surgical excision following early diagnosis of melanoma is the only way to eliminate the disease that will result in death. However, excision of benign lesions will lead to increased morbidity and unnecessary health expenditures [3].

Given the complexity and similarity of skin lesions, it can be challenging to make an accurate diagnosis. Melanotic lesions can be diagnosed through the evaluation of multiple parameters, such as pattern analysis of the lesional region, comparison of the suspicious lesion with the patient body, patient's medical history and evolution of the lesion.

Before 1980, the melanoma was recognized by signs of overgrowth, ulceration, and fungus of the skin lesion in its last stage, and unfortunately it was too late for the patient. After 1985, ABCD [4] criterion was defined by members of a group at New York University. In this abbreviation A stands for asymmetry, B stands for border irregularity, C stands for color variation and D stands for diameter >6 mm. In the following years this criterion was updated as ABCDE in which E stands for evolution of the lesion (skin, size, color changes or other symptoms). The characteristics of the early stages of melanoma is defined according to this criterion.

Although pathological diagnosis is accepted as the gold standard [5], significant advances have been made in the early diagnosis of melanoma as a result of innovations in clinical methods and imaging technologies. In the 1990s, with the newly developed dermatological light-based visual techniques such as dermoscopy [6], clinicians were able to spot the signs that could not be observed with the naked eye where the lesion is located. Later, digital dermoscopy was emerged as a result of the development of computer-assisted methods. More successful diagnostic results were achieved with digital dermoscope than the manual dermoscope [7]. In clinical investigations the accuracy of melanoma diagnoses has increased over the years due to the usage of these light-based visual technologies [8].

Patients who have skin disorders, do not think or delay to see the doctor unless they encounter very serious symptoms [9]. Nowadays, it is possible to perform patient examination on the internet with the method called telederma-tology [10]. In teledermatology, patients can receive pre-diagnosis information about the disease from specialist physicians by transmitting the images taken by mobile phone or professional camera to the clinics. The personal experience of physician is very important in the diagnosis of skin lesions. Physician attempt to make a diagnosis by comparing the local features of the lesions with the rest of the patient's body [11]. On the other hand, some studies have reported that the diagnoses made by the teledermatology method have higher accuracy in skin cancer diagnoses [12].

It is understood that dermatological images will have an important role in the pre-diagnosis and even more advanced stages of disease diagnosis. Digital images of skin lesions can be used in teledermatology or can be used as training data for artificial intelligence models [13]. Later these models will integrated to computer aided diagnostic systems in parallel with today's technological developments. In literature, we have frequently come across studies about comparison of deep learning (DL) performance against clinicians in recent years [14]. Convolutional Neural Network (CNN), which is a class of DL, models are used in most of these studies. When we examine the ones about skin lesion diagnosis, it is understood that the diagnostic accuracy of CNN's are getting close to the performance of dermatologists in each new study [15]. According to some of these new studies, it is also stated that CNN models are more successful than dermatologists [16] and pathologists, [17].

2   Benign - Malignant Skin Lesion Classification Using Deep Learning

Deep learning [18] architectures allow computational models of multiple process-ing layers to learn the representation of data. These methods have significantly improved the latest technology in speech recognition, visual object recognition, object detection and many other areas such as drug discovery and genomics. Deep learning discovers and learns complex information in large datasets using the backpropagation algorithm to show how a machine must change the inter-nal parameters used to calculate the output on each layer of output from the previous layer.

2.1   AlexNet

AlexNet [19] is a convolutional neural network model that was the winner of the 2012 ImageNet contest. The ImageNet has been held since 2010 and is based on the problem of classifying more than 14 million images into one thousand different classes. This model has achieved 15.3% top-5 error rate in the year it participated the competition and attracted all the attention. Alexnet achieved this result by reaching 10% better classification accuracy than its closest com-petitor. Although convolutional neural networks gave their first successful results with the LeNet-5 model developed by Yann LeCun in 1998 [18], it was very difficult to implement CNN's in practice. Because CNN models required high processing power to perform required calculations adn they were computationally expensive at that period. In recent years, where we have benefited from the high computing capabilities of GPU processors, it has been possible to implement CNN architectures.

Although Alexnet has similar architecture to Lenet5, it is a deeper and larger CNN architecture. Lenet 5 was consisted of 3 convolutions, 2 sub-sampling and 1 fully connected layers. In Lenet5, average pooling was used for sub-sampling and tanh and sigmoid activation functions were used for the calculations. Alexnet, on the other hand, consists of a total of 8 main layers, of which the first five are convolution layers and the last three are fully connected layers (see Fig. 5). Max-pooling layers are available to reduce the size between these layers during application. ReLU (Rectified Linear Unit) activation function, which offers better training performance compared to tanh and sigmoid functions used in Lenet 5, was preferred in the calculations. Alexnet receives RGB images to 227×227 as input. In this model, a total of 62,378,344 parameters are calculated in all layers as a result of the operations performed on the image.

2.2   GoogLeNet

GoogleNet [20] is a CNN model developed by Google. This CNN is the winner of the 2014 ImageNet contest and has reached a top-5 error rate of 6.67% in the competition. This ratio is very close to human performance, which is 5.1%. This model has 22 layers and is deeper and wider than previous CNNs (see Fig. 6). Unlike AlexNet, the number of parameters calculated in these layers is about 5 million which means that it will require 12 times less computation than AlexNet. Another difference of GoogLeNet is that it does not have fully connected layers. Instead, there are Inception modules which makes the network much deeper and wider. In the Inception module, 1×1, 3×3, 5×5 convolution and 3×3 maxpooling operations are applied to the data transmitted from the previous layer. Unlike Naive Inception, which has a high calculation cost, 1×1 convolution is applied before 3×3 and 5×5 convolutions in order to reduce the number of calculations. For the same purpose, 1×1 convolution was also added after maxpooling (see Fig. 7). The different features obtained as a result of these operations applied to the data from the previous layer are concatenated and transmitted to the next layer. Unlike previous deep architectures, GoogLeNet uses global average pooling instead of using a fully connected layer at the end of the network. As a result, it was stated that top-1 accuracy improved by 0.6%. Some middle softmax branches are present at the intermediate layers of this network which were designed as auxiliary classifiers. The calculated loss obtained from these layers is added to the total loss with a weight of 0.3. These layers were added for the purpose of regularizing the network and preventing the gradient vanishing.

2.3 ResNet

Resnet [21] is a more advanced version of CNNs. This model, which stands for Residual Neural Network, is different from previous CNN models. Theoretically, it is assumed that the accuracy of classification will increase as the network depth increases in CNNs. However, no matter how deep the model is, it reaches saturation after a point. From this point on, detonation occurs and the classification reports accuracy. Resnet adds shortcuts between layers to resolve this issue. This method prevents corruption while the CNN deepens. Bottleneck blocks are used for acceleration training at ResNet. There are several types of Resnet. Resnet has many varieties, one of which is the winner of the 2015 ImageNet competition. With 152 layers, this Resnet model reached a top 5 error rate of 3.57% in the competition. In Resnet model, instead of using 2 (3×3) convolution, (1×1), (3×3), (1×1) convolution layers are used (see Fig. 8). In this study, a smaller and more frequently used version of Resnet152, Resnet50 model, was examined.

3 Experimental Framework and Results

3.1 Hardware Configuration and Software

We conducted experiments on a device with 2.40 GHz 8 Core Intel 7 4700HQ CPU, 16 GB of RAM and NVIDIA GeForce GT 750M Graphic Processing Unit (GPU). We developed our algorithms using Matlab R2018b with Deep Learning Toolbox. Three CNN models were tested on GPU with Marlab 2018b.

3.2 Dataset Descriptions

We referred to the International Skin Imaging Collaboration: Melanoma Project (ISIC), public archive of clinical and dermoscopic images of skin lesions [22]. This database consists of 23906 skin lesion images. We selected a subset of 21570 dermoscopic images from this database which consisted of 19373 benign and 2197 malignant diagnostic attributes.

3.3 Data Preprocessing

Both pre-trained models are designed to classify images of specific sizes. AlexNet accepts 227×227 images as input. GoogLeNet and ResNet50 are designed to classify 224×224 images. In order to retrain each classifiers for different types of images, the existing dermoscopic images need to be adjusted to the dimensions accepted by the classifier. Accordingly, each image was resized to fit the input layers of the CNNs by applying bicubic interpolation.

3.4   Configuration of CNNs

AlexNet, GoogleNet and ResNet were originally designed for the ImageNet competition to classify millions of images into 1000 different classes. Therefore, the final softmax layers of both pre-trained CNNs are designed to give 1000 different probability values of 1000 different classes. In order to apply binary classification of benign and malignant lesions with those CNNs, we modified the last output layer and set output size equal to 2, weight learn rate factor to 10 and bias learn rate factor to 10.

3.5   Training Setup

For the experiments, 80% of 21570 images were randomly allocated for training and 20% for validation. Both classifiers used the same randomly chosen sets for training and validation. We also configured the minibatch size equal to 10, the learning rate to 0.0003, and the number of epoch to 5 for the training setup.

3.6   Classification Performance

As can be seen, ResNet50 and GoogLeNet have achieved more successful classification accuracy performance results than AlexNet. Table 1 gives detailed information about performances of the CNNs. All three models reached these scores with 8625 iterations in 5 epochs during the training phase. Each epochs lasted in 1725 iterations. As shown in results all three models, a classification accuracy score of which slightly over 90% was achieved at the end of the first epoch (see Figs. 1, 2 and 3). This score was increased in the next 5 epochs with small improvements. According to the results, three CNNs which are capable of classifying images of 1000 categories of different types, may also be useful in the classification of benign and malignant lesions on dermatology images (see Fig. 4).

Table 1. Classification performance of the three CNNs

Classifier	Accuracy	Elapsed time
AlexNet	91.40%	239 min 4 s
GoogLeNet	92.70%	283 min 50 s
ResNet50	92.81%	354 min 46 s

3.7   Time Complexity Comparison of Three CNN Models

The CNNs tested in this study were retained in their original state and the final layers of the models were modified to apply binary classification. All three models were trained on the same platform with the same data and their performances were measured. Among these models, AlexNet calculates approximately 60 million parameters for a picture, GoogleNet calculates 5 million parameters, and ResNet50 calculates 25.6 million parameters. According to the results the

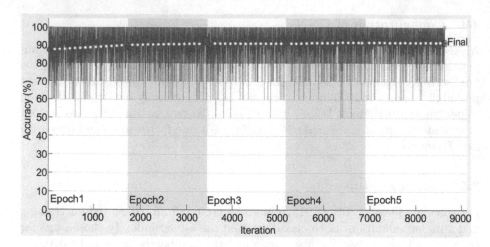

Fig. 1. AlexNet's training accuracy evolution over 5 epochs

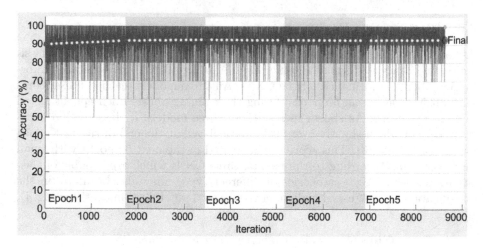

Fig. 2. GoogleNet's training accuracy evolution over 5 epochs

training phase lasted 239 min 4 s for AlexNet, 283 min 50 s for GoogLeNet and 354 min 46 s for ResNet50 on the same 80% sub-data set. The remaining 20% subset of images used for validation on both CNNs. Validation step lasted in 5 min 17 s for AlexNet, 6 min 52 s for GoogleNet and 8 min 18 s for ResNet50.

Although ResNet's classification accuracy score is 1.41% better than AlexNet, it is observed that AlexNet is more advantageous in terms of time complexity. GoogleNet ranked second according to time complexity scores, however it also missed the first place with a small margin in classification accuracy. In this case, it can be concluded that GoogLeNet is more advantageous in terms of costperformance evaluation. ResNet's first-place achievement of classification accuracy score was overshadowed by its long computation time.

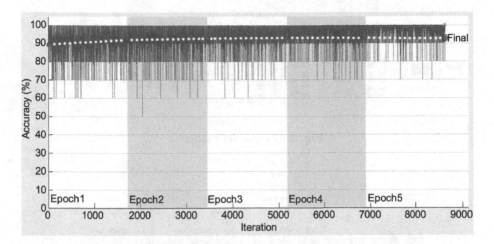

Fig. 3. ResNet50's training accuracy evolution over 5 epochs

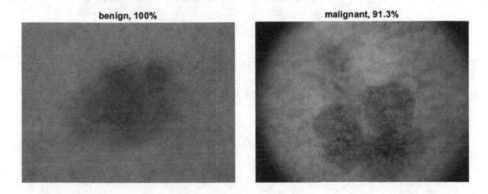

Fig. 4. Benign and malignant lesion classification example with AlexNet

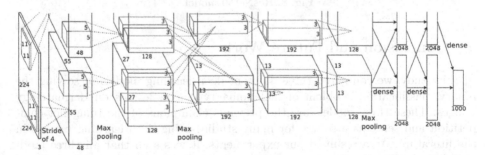

Fig. 5. AlexNet model.

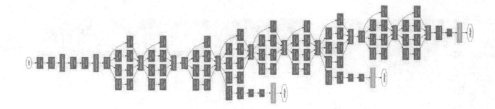

Fig. 6. GoogLeNet model.

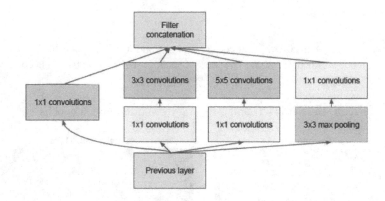

Fig. 7. GoogLeNet inception layer.

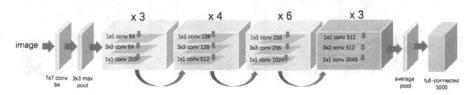

Fig. 8. ResNet50 model.

4   Conclusion and Future Work

In this study, we compared three different deep learning models in solving the
binary classification problem of melanoma type skin lesions on dermatology
images. These three models are the previous champions of the Imagenet competition and serve as a source for many studies using deep learning models in
the literature. As a result of our experiments, it was seen that all three models reached 90% classification accuracy scores. The ResNet50 model, with the
highest classification accuracy, lags behind time complexity measurements. The
AlexNet model, which has the lowest classification success, was first-ranked in
time complexity measurements. The GoogLeNet model, which takes second place
in both measurements, can be said to be more balanced model according to the
performance-cost measurements.

According to some studies carried on pathological diagnosis of melanoma, a second expert dermapathologist review may change the diagnosis. In a study of which 567 patients diagnosis report were evaluated, 3% (19) of these cases diagnosis were changed to benign after a second expert review in contrast to the initial report [23]. In such a case where there are controversies about pathological diagnosis, the effective use of computer-aided systems and consequently CNN algorithms will increase the success of differential diagnosis. Thus, biopsy tests which are costly for health expendituresor and stressful for patients may not be requested in some cases.

In addition, considering the promising results of our study and the previous studies showing that deep learning models give more successful results than dermatologist diagnoses, it can be predicted that computer-aided diagnostic systems will be used more in the detection of skin cancer in the following years. The CNN models, which will be developed in line with this objective, is expected to achieve high classification accuracy with short calculation time. In our future studies, we aim to achieve more successful results by modifying some layers of all three models. The long-term objective is to develop systems that can make a differential diagnosis for different types of skin lesions on images taken with mobile phones or digital cameras.

References

1. Miller, A.J., Mihm, M.C.: Melanoma. N. Engl. J. Med. **355**(1), 51–65 (2006)
2. Argenziano, G., et al.: Accuracy in melanoma detection: a 10-year multicenter survey. J. Am. Acad. Dermatol. **67**(1), 54–59.e1 (2012)
3. Siegel, R.L., Miller, K.D., Jemal, A.: Cancer statistics, 2019. CA Cancer J. Clin. **69**(1), 7–34 (2019)
4. Rigel, D.S., Russak, J., Friedman, R.: The evolution of melanoma diagnosis: 25 years beyond the ABCDs. CA Cancer J. Clin. **60**(5), 301–316 (2010)
5. Darragh, C.T., Clayton, A.S.: Melanoma in situ. In: Hanlon, A. (ed.) A Practical Guide to Skin Cancer, pp. 97–115. Springer, Cham (2018). https://doi.org/10.1007/978-3-319-74903-7_5
6. Errichetti, E., Stinco, G.: Dermoscopy in general dermatology: a practical overview. Dermatol. Ther. **6**(4), 471–507 (2016)
7. Kittler, H., Pehamberger, H., Wolff, K., Binder, M.: Diagnostic accuracy of dermoscopy. Lancet Oncol. **3**(3), 159–165 (2002)
8. Sinz, C., et al.: Accuracy of dermatoscopy for the diagnosis of nonpigmented cancers of the skin. J. Am. Acad. Dermatol. **77**(6), 1100–1109 (2017)
9. Winterbottom, A., Harcourt, D.: Patients' experience of the diagnosis and treatment of skin cancer. J. Adv. Nurs. **48**(3), 226–233 (2004)
10. Lee, J.J., English, J.C.: Teledermatology: a review and update. Am. J. Clin. Dermatol. **19**(2), 253–260 (2018)
11. Marghoob, A.A., Scope, A.: The complexity of diagnosing melanoma. J. Invest. Dermatol. **129**(1), 11–13 (2009)
12. Finnane, A., Dallest, K., Janda, M., Soyer, H.P.: Teledermatology for the diagnosis and management of skin cancer: a systematic review. JAMA Dermatol. **153**(3), 319–327 (2017)

13. Petrie, T., Samatham, R., Witkowski, A.M., Esteva, A., Leachman, S.A.: Melanoma early detection: big data, bigger picture. J. Invest. Dermatol. **139**(1), 25–30 (2019)
14. Liu, X., et al.: A comparison of deep learning performance against health-care professionals in detecting diseases from medical imaging: a systematic review and meta-analysis. Lancet Digit. Health **1**(6), e271–e297 (2019)
15. Haenssle, H.A., et al.: Man against machine: diagnostic performance of a deep learning convolutional neural network for dermoscopic melanoma recognition in comparison to 58 dermatologists. Ann. Oncol. **29**(8), 1836–1842 (2018)
16. Brinker, T.J., et al.: Deep neural networks are superior to dermatologists in melanoma image classification. Eur. J. Cancer **119**, 11–17 (2019)
17. Hekler, A., et al.: Deep learning outperformed 11 pathologists in the classification of histopathological melanoma images. Eur. J. Cancer **118**, 91–96 (2019)
18. LeCun, Y., Bengio, Y., Hinton, G.: Deep learning. Nature **521**(7553), 436–444 (2015)
19. Krizhevsky, A., Sutskever, I., Hinton, G: ImageNet classification with deep convolutional neural networks. In: NIPS (2012)
20. Szegedy, C., et al.: Going deeper with convolutions. In: The IEEE Conference on Computer Vision and Pattern Recognition (CVPR), Boston, pp. 1–9 (2015)
21. He, K., Zhang, X., Ren, S., Sun, J.: Deep residual learning for image recognition. In: Proceedings of the IEEE Conference on Computer Vision and Pattern Recognition, Las Vegas, pp. 770–778 (2016)
22. International Skin Imaging Collaboration (ISIC) Project. https://www.isic-archive.com/. Accessed 1 Oct 2019
23. Suzuki, N.M., Saraiva, M.I.R., Capareli, G.C., Castro, L.G.M.: Histologic review of melanomas by pathologists trained in melanocytic lesions may change therapeutic approach in up to 41.9% of cases. Anais Bras. Dermatol. **93**(5), 752–754 (2018)

Advanced Data Mining Techniques and Applications

Automatic Parameter Setting in Hough Circle Transform

Pei-Yu Huang[1], Chih-Sheng Hsu[2], Tzung-Pei Hong[1,2(✉)] ⓘ, Yan-Zhih Wang[3],
Shin-Feng Huang[4], and Shu-Min Li[1]

[1] Department of Computer Science and Engineering, National Sun Yat-sen University,
Kaohsiung, Taiwan
[2] Department of Computer Science and Information Engineering,
National University of Kaohsiung, Kaohsiung, Taiwan
tphong@nuk.edu.tw
[3] Footprintku Inc., Kaohsiung, Taiwan
[4] Department of Applied Mathematics, National University of Kaohsiung, Kaohsiung, Taiwan

Abstract. The Hough circle transform has been widely used in pattern recognition
to find circles in images. It, however, has many parameters to be adjusted before
usage. This study focuses on the analysis of the circular size and distribution
density for the images of ball maps in electronic parts. The amounts of projection
of the binary conversion quality in different directions are calculated, and the
parameters required in the Hough circle transform are automatically adjusted based
on the amounts of projection. In this way, the time for setting blind tests could be
reduced and the accuracy of pattern recognition could be improved.

Keywords: Hough circle transform · Pattern recognition · Parameter setting ·
Ball map

1 Introduction

In recent years, with the popularization of the information industry, many technical information transmission methods have also been converted from traditional paper materials into digital data. The digitized data is also easier to apply and automate the entire
workflow.

In the past, there have been few studies on digitizing electronic component specifications. Part of scanning the contents of an electric part specification is to convert the
image content in the file into text. For example (Fig. 1), this kind of picture is called
ball map. The ball map contains information about the foot of various electronic parts.
Different images have different sizes of circles and character information. The purpose
of the study was to remove the noise from the round frame and to extract the text [9]. In
this paper, we have found a way to automatically detect the parameters that are suitable
for bringing in, based on Hough Circle Transform, for various images with different
circle sizes. In this way, it is not necessary to manually test each picture to find suitable
parameters [12].

© Springer Nature Switzerland AG 2020
N. T. Nguyen et al. (Eds.): ACIIDS 2020, LNAI 12033, pp. 527–535, 2020.
https://doi.org/10.1007/978-3-030-41964-6_45

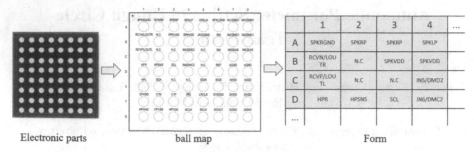

Fig. 1. Converting the pin information map in an electronic part into a form.

The remainder of the paper is organized as follows: In Sect. 2, the related work will be discussed. Section 3 is devoted to the method we proposed. Section 4 presents the experimental results and discussion. The conclusion is reported in Sect. 5.

2   Related Work

2.1   Hough Circle Transform

The Hough transform is one of the feature extraction techniques applied to image recognition and analysis. This purpose is going to find lines, circles, and rectangles in the pictures [11]. The Hough circle transform is a variant of Hough transform for detecting circles in imperfect image inputs. We know that in two-dimensional space, the equation of a circle can be expressed as:

$$(x - a)^2 + (y - b)^2 = r^2,$$

where a, b are the coordinates of the center of the circle, r is the radius of the circle, and (x, y) are all parameters that satisfy this equation on the plane. The steps for Hough Circle Transform are as follows:

(1)   The gradient of the points on the circumference points to the position of the center of the circle. For each point, the count is increased only along the gradient direction, and the range is defined by the maximum and minimum values of the predetermined radius. If the threshold is exceeded, the point is the center of the circle.

(2)   The distance between the center of the circle and the point is calculated. The maximum value is the radius of the circle.

2.2   Fully Convolutional Network

In artificial intelligence and machine learning, the most commonly used Convolutional neural network [10], also known as CNN, CNNs, or ConvNets. CNN's architecture consists of three parts: Convolution, Pooling, and Fully Connection. The feature map describes the corresponding feature data that should be generated for that step.

In this article, parts of the automated process are used by Fully Convolutional Networks. Fully Convolutional Networks (FCN) adjusted the CNN architecture, and the Pooling and Fully Connection were replaced by Convolution. Also, the text detection method based on FCN module construction: An Efficient and Accurate Scene Text Detector (EAST) is a text detector based on deep learning [4]. The main idea comes from U-Net [6], which uses a U-shaped structure to obtain pixel-level segmentation prediction results and pixel-level geometric prediction results. Based on these two results, the coordinate values of the four vertices of each bounding box can be calculated. Then remove the redundant, duplicate bounding box through NMS. EAST is often used for text detection in natural scenes, so it has excellent text detection results [2]. The natural scenes that affect text detection include sensor noise, light, blur and angle [1]. Therefore, we chose the part of the text detection that EAST applied in this study [3].

3   Method

3.1   Flow Diagram

Figure 2 shows the flowchart of automatic conversion. First, the approach imports the image into the EAST neural network. Second, it uses the previously trained neural network model for text area detection to identify the area containing characters as text. Finally, it places the recognized text of the image into the table automatically.

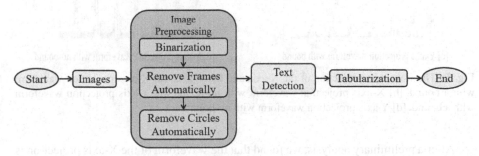

Fig. 2. The flow chart of automatic conversion

Before importing an image into the neural network for text recognition, we use pre-processing on the image to improve the accuracy of the neural network to find the text area in the picture. Preprocessing includes binarization, removing frames and removing circles [8]. Binarization converts the three-channel color map into a black-and-white image to facilitate subsequent calculation of the pixel quality of the picture [5]. Removing frames removes the solid outline of the image that affects the projected waveform, making the numerical analysis more accurate and straightforward. Removing circles removes the rounded frame from the image, including stable and dashed circles. It is because in our previous implementations, we found that the edge of the circle is a severe disturbance for the neural network. Therefore, before the picture is imported into the neural network, the frame line containing the circle is eliminated.

3.2 Pixel Projection and Noise Removal

In Removing Circles, we use Hough circle transform. For Hough circle transform, we need to adjust the appropriate parameters according to the size and density of the circle in the image, such as the upper and lower limits of the circular diameter and the minimum distance between two centers of circles.

To automatically find the appropriate parameters as the input value of Hough circle transform, we first divide the binarized image into the pixel accumulation of the X-axis and the Y-axis. We then obtain the waveform of both the X-axis and the Y-axis projection amounts (Fig. 3).

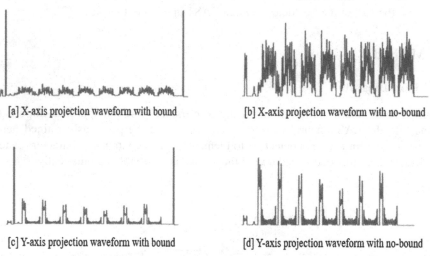

[a] X-axis projection waveform with bound   [b] X-axis projection waveform with no-bound

[c] Y-axis projection waveform with bound   [d] Y-axis projection waveform with no-bound

Fig. 3. The waveform projection of the X-axis and the Y-axis. [a] X-axis projection waveform with a bound. [b] X-axis projection waveform with no-bound. [c] Y-axis projection waveform with a bound. [d] Y-axis projection waveform with no-bound.

After a preliminary analysis, we found that the waveform of the X-axis projection is more correlated with the parameters of the Hough circle transform. Then, the waveform of the X-axis projection amount will be further statistically analyzed. The result is shown in Fig. 4.

The X-axis projection waveform processed by the module of removing frames (Fig. 3[b] and [d]) reduces the outliers of the frame projection on the waveform, and the waveform becomes more regular. We then normalize the waveform, calculate the continuous length set of the maxima of the binary wave, and estimate the approximate diameter of the circle. The constant value set of the minimum value of the binary wave can be used to calculate the distance between the two centers (the median is used as the estimated value currently).

```
RemoveCircle()
```
$$X\_list = [x_{p1}, x_{p2}, \ldots, x_{ph}] \quad //Each\ X - axis\ projection$$

$$x_{pn} = \sum_{i=1}^{h} img[x_n][y_i]$$

$$if\ x_{pn} \geq avg(X_{list}):$$
$$\quad x_{pn} \leftarrow x_{high}$$
$$else\ x_{pn} \leftarrow x_{low}$$

$$for\ n = 1\ to\ h:$$
$$\quad if\ x_{p(n-1)} = x_{high}\ and\ x_p = x_{high}:$$
$$\quad\quad Length_{high} \leftarrow Length_{high} + 1$$
$$\quad else:$$
$$\quad\quad if\ Length_{high} > 0:$$
$$\quad\quad\quad H\_list.\,append(Length_{high})$$
$$\quad if\ x_{p(n-1)} = x_{low}\ and\ x_p = x_{low}:$$
$$\quad\quad Length_{low} \leftarrow Length_{low} + 1$$
$$\quad else:$$
$$\quad\quad if\ Length_{low} > 0:$$
$$\quad\quad\quad L\_list.\,append(Length_{low})$$
$$Diameter \leftarrow median(H\_list)$$
$$CircleDis \leftarrow median(L\_list)$$
$$HoughCircle(Diameter,\ CircleDis)$$

Fig. 4. The algorithm for the projection of the X-axis and Y-axis.

4 Experiment and Discussion

Before importing the image into the neural network for text detection, we will preprocess the image to get better text detection results. The preprocessing of the image includes three parts: binarization, removing frames and removing circles.

4.1 Images Binarization

The original color or grayscale image is uniformly transferred to a picture with only black and white pixel values to calculate the projection amount of the black pixel on the X-axis.

4.2 Removing Frames Automatically

As shown in Fig. 5, it can be found that the waveform of the X-axis projection without the frame is more regular than the figure with the frame.

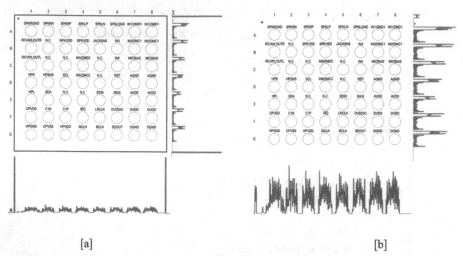

[a]   [b]

Fig. 5. The projection waveform of the X-axis and Y-axis. [a] The picture has the frame. [b] The picture removed the frame

4.3 Removing Circles Automatically

This part normalizes the waveform of the X-axis projection amount (Fig. 6) and automatically calculates the waveform value to obtain the most suitable original diameter and center-to-center spacing [7].

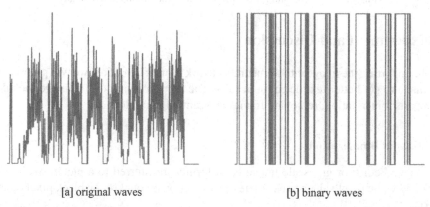

[a] original waves   [b] binary waves

Fig. 6. Converting initial waves to binary waves. [a] original waves [b] binary waves

Then it automatically takes the obtained parameters into the Hedge circle transform to eliminate the circular edges in the picture (Fig. 7).

	1	2	3	4	5	6	7	8
A	SPKRGND	SPKRN	SPKRP	SPKLP	SPKLN	SPKLGND	IN1/DMD1	IN1/DMD1
B	RCVN/LOUTR	N.C.	SPKVDD	SPKVDD	JACKSNS	IN3	IN2/DMC1	IN2/DMC1
C	RCVP/LOUTL	N.C.	N.C.	IN5/DMD2	N.C.	IN4	MICBIAS	MICBIAS
D	HPR	HPSNS	SCL	IN6/DMC2	N.C.	REF	AGND	AGND
E	HPL	SDA	N.C.	N.C.	SDIN	BIAS	AVDD	AVDD
F	CPVDD	C1N	C1P	IRQ	LRCLK	DVDDIO	DVDD	DVDD
G	HPGND	CPVSS	HPVDD	MCLK	BCLK	SDOUT	DGND	DGND

Fig. 7. Removing the frame and circles in the ball map image

4.4 Experimental Result

Compare the results before and after using automatic parameter setting, we can see from the colored part of Fig. 8[a] that the circles found by Hough circle transform is very messy. However, using our proposed method, all the circles can be found out without loss (Fig. 8[b]).

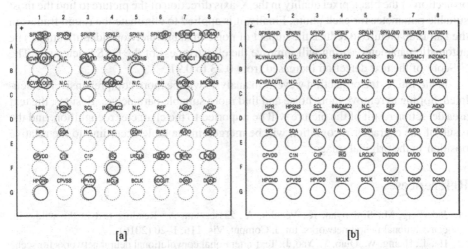

[a]   [b]

Fig. 8. [a] Hough circle transform using present parameters. [b] Hough circle transform using automatic parameter setting.

The preprocessed text detection results (Fig. 9[a]) are significantly better than the original image (Fig. 9[b]). As a result, good image preprocessing can improve the accuracy of image recognition.

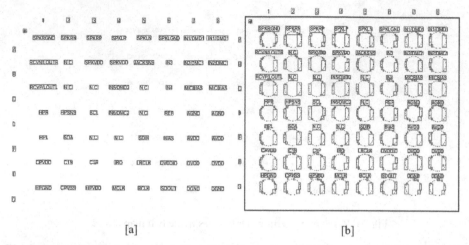

[a]   [b]

Fig. 9. Removing the frame and circles in the ball map image. [a] The text detection results after processing. [b] Original text detection result.

5   Conclusion

This article presents an architecture that automatically adjusts the parameters in the Hough circle transform. The projected waveform is further analyzed by the cumulative projection of the black pixel quality in the X-axis direction of the picture to find the most suitable parameter for each photo. Finally, it is applied to eliminate the round frames in the ball map picture to reduce the impact of noise in neural-network identification. The automatic parameter setting will improve the effectiveness of the EAST detection. The F-Score of the text detection increased from 0.23 to 0.91.

In the future, the projection waveform analysis will be more comprehensively calculated, hoping to remove more noise and find more suitable parameters. Besides, the text area detection result of the picture will be imported to the character recognition, and the result of the character recognition will be merged into a table according to the relative position in the original image.

References

1. Jaderberg, M., Simonyan, K., Vedaldi, A., Zisserman, A.: Reading text in the wild with convolutional neural networks. Int. J. Comput. Vis. **116**, 1–20 (2016)
2. He, T., Huang, W., Qiao, Y., Yao, J.: Text-attentional convolutional neural networks for scene text detection. IEEE Trans. Image Process. **25**, 2529–2541 (2016)
3. Zhou, X., et al.: EAST: an efficient and accurate scene text detector. In: The IEEE International Conference on Computer Vision and Pattern Recognition, pp. 2642–2651 (2017)
4. Zhang, Z., Zhang, C., Shen, W., Yao, C., Liu, W., Bai, X.: Multi-oriented text detection with fully convolutional networks. In: The IEEE International Conference on Computer Vision and Pattern Recognition, pp. 4159–4167 (2016)
5. Long, J., Shelhamer, E., Darrell, T.: Fully convolutional networks for semantic segmentation. In: The IEEE International Conference on Computer Vision and Pattern Recognition, pp. 3431–3440 (2015)

6. Ronneberger, O., Fischer, P., Brox, T.: U-Net: convolutional networks for biomedical image segmentation. In: The 18th International Conference on Medical Image Computing and Computer Assisted Intervention, pp. 234–241 (2015)
7. Yadav, V.K., Batham, S., Acharya, A.K., Paul, R.: Approach to accurate circle detection: circular Hough transform and local maxima concept. In: The International Conference on Electronics and Communication Systems (2014)
8. Bieniecki, W., Grabowski, S., Rozenberg, W.: Image preprocessing for improving OCR accuracy. In: The International Conference on Perspective Technologies and Methods in MEMS Design, pp. 75–80 (2007)
9. Wolf, C., Jolion, J.M.: Object count/area graphs for the evaluation of object detection and segmentation algorithms. Int. J. Doc. Anal. Recogn. **8**, 280–296 (2006)
10. Zhang, G.P.: Neural networks for classification: a survey. IEEE Trans. Syst. Man Cybern. Part C (Appl. Rev.) **30**, 451–462 (2000)
11. Illingworth, J., Kittler, J.: The adaptive Hough transform. IEEE Trans. Pattern Anal. Mach. Intell. **9**, 690–698 (1987)
12. Hardouin, C., Yao, J.F.: Multi-parameter auto-models with applications to cooperative systems. Comptes Rendus Mathematique **345**(6), 349–352 (2007)

Construction of an Intelligent Tennis Coach Based on Kinect and a Sensor-Based Tennis Racket

Chun-Hao Chen[1], Che-Kai Fan[2], and Tzung-Pei Hong[3,4(✉)]

[1] Department of Information and Finance Management,
National Taipei University of Technology, Taipei, Taiwan
chchen6814@gmail.com
[2] Department of Computer Science and Information Engineering,
Tamkang University, Taipei, Taiwan
kevin03058277@gmail.com
[3] Department of Computer Science and Engineering,
National Sun Yat-sen University, Kaohsiung, Taiwan
tphong@nuk.edu.tw
[4] Department of Computer Science and Information Engineering,
National University of Kaohsiung, Kaohsiung, Taiwan

Abstract. The purpose of this paper is to develop an artificial intelligence coach system which can be used to provide beginners a more economical way to learn tennis and to achieve as similar as the coach to guide the training. To reach the goal, artificial intelligence coach system is composed of three phases that are (1) Tennis swing motion data collection, (2) CAST-based swing motion group construction, and (3) Online tennis swing motion analysis. In the first phase, the tennis motions are collected using the Kinect and the built sensor-based tennis racket. Then, the cluster affinity search technique (CAST) and the dynamic time warping (DTW) are utilized to divide the collected swing motion series data into groups to form swing motion groups. In the third phase, using the swing motion groups, the system can provide possible tennis motion improvement suggestions. At last, experiments were also conducted on the real dataset to show the effectiveness of the proposed system.

Keywords: Clustering technique · Tennis swing motion · Dynamic time warping · Kinect · Sensor

1 Introduction

As many Taiwanese professional tennis players have become famous in international tennis competitions, people playing tennis in Taiwan has grown in recent years. As we all know, to learn how to play tennis is a very difficult task for beginners. Basically, they can start to practice with different approaches. For example, firstly, they can be trained by themselves using various assistant tools, e.g., the wall, self-training equipment, or secondly, they can hire coaches to teach them the tennis knowledge and skills. However, when the first approach is adopted, the beginners are easily to practice using the wrong swing postures that will result in suffering multiple injuries to their body. In addition, after having the wrong habits of swing posture, it will increase the difficulty of posture adjustment in the future. On the contrary, hiring coaches to teach tennis knowledge and swing postures, it will decrease the probability of being injured. But the problem is the coach's time is limited and the cost is expensive.

To solve the above mentioned problem, the literature shows that the Kinect is a commonly device using for the identification of human body skeleton, and the application fields include: gesture recognition [3], exergaming [4], human action detection [6], gait analysis [1, 8] and physiotherapy [7]. For tennis swing motion analysis, Fan-Chiang et al. proposed an approach for tennis swing motion recognition based on the Kinect [5]. It first used the Kinect to retrieve skeleton coordinates to build a three-dimensional Cartesian coordinate system. Then, based on the body posture of a player detected by the Kinect, the fore hand or back hand were recognized. The starting and ending timings of swing motions were set. After receiving the swing motion series data, the dynamic time warping (DTW) was utilized to compare the received data with the given coach swing motion data to identify three types of motions that were forehand (backhand) drive, volley and slice. Finally, the suggestions were given based on the comparison results. However, the tennis racket face orientation during a swing which could be an important factor to have a high quality hit is not considered in that approach.

In this paper, the purpose is to provide a more intelligent way for beginners to learn tennis, and as a result, the artificial intelligence tennis coach (AITC) system which can provide suggestions like coaches to beginners to learn tennis skill is proposed. To reach the goal, we first implement a sensor-based tennis racket which can be used to collect racket face orientation data. Then, the Kinect and the designed sensor-based tennis racket are utilized together to collect six tennis swing motions, including the forehand drive, forehand volley, forehand slice, backhand drive, backhand volley and backhand slice. Using the Kinect, the x, y and z axes values of the wrist, elbow and shoulder, totally night attributes, are recorded. For the MPU6050 sensor (G-sensor) in the tennis racket, the x, y and z axes values of the sensor, totally three attributes that are used to identify racket face orientation, are received. Hence, the twelve attributes are employed to represent a tennis swing motion at a time point. By using the collected tennis swing trajectory series, for each swing motion, the clustering algorithm, called the cluster affinity search technique (CAST) [2], is utilized to divide the tennis swing trajectory series into groups to form the swing motion groups. At last, based on the proposed system, the received tennis swing trajectory series from a new user can be compared to the swing motion groups to recognize the swing motion type and provide possible improvement suggestions.

2   Tennis Swing Motion Data Collection

In this section, how to collect tennis swing trajectory series is described. Because the tennis racket face orientation during a swing motion has a critical impact on the swing quality, we took it into consideration and built the sensor-based tennis racket to receive the racket face data. The built sensor-based tennis racket and its related information are shown in Fig. 1.

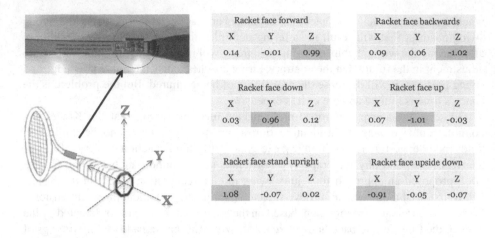

Racket face forward				Racket face backwards		
X	Y	Z		X	Y	Z
0.14	-0.01	0.99		0.09	0.06	-1.02

Racket face down				Racket face up		
X	Y	Z		X	Y	Z
0.03	0.96	0.12		0.07	-1.01	-0.03

Racket face stand upright				Racket face upside down		
X	Y	Z		X	Y	Z
1.08	-0.07	0.02		-0.91	-0.05	-0.07

Fig. 1. The built sensor-based tennis racket.

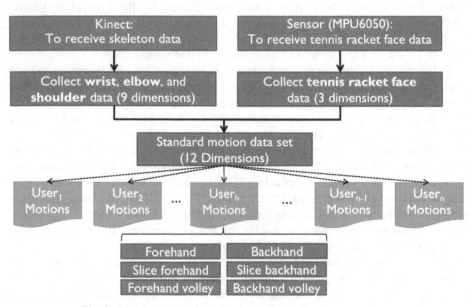

Fig. 2. The flowchart of tennis motion data collection procedure.

Figure 1 shows that the MPU6050 (G-sensor) was embedded on the tennis racket (left side of the figure). Then, the racket face orientation of the six swing motions can be identified via the x, y and z axes values. From the right hand side of the figure, they indicate that when the states of racket face orientation are "forward" or "backward", the values of z axis are very close to 1 or -1. When the states of racket face orientation are "down" or "up", the values of y axis are very close to 1 or -1. When the states of racket face are "stand upright" or "upside down", the values of x axis are very close to 1 or -1. Hence, using the built sensor-based tennis racket, the proposed system can provide beginners possible improvement suggestions based on the data received from users in the online tennis motion analysis phase.

The flowchart of tennis motion data collection procedure by using the Kinect and the built sensor-based tennis racket is shown in Fig. 2.

Figure 2 shows that the Kinect is used to collect a player's skeleton data in a certain period of time according to the predefined start and end conditions. The MPU6050 sensor is employed to receive the tennis racket face data. The twelve dimensions (attributes) are utilized to represent a swing motion. The first night dimensions are used to store the wrist, elbow and shoulder data received by the Kinect. The last three dimensions are utilized to indicate the tennis racket face orientation data received from the MPU6050 sensor. In other words, a tennis swing trajectory series in a period of time is recorded by a t-by-n matrix, where t and n are the number of time points and number of attributes. After the first phase, the swing trajectory series are gathered to form the standard motion dataset.

3   CAST-Based Swing Motion Group Construction

Because swing motions of all participants may difference, it is necessary to divide the swing motion trajectory series into groups for further analysis after having the standard motion dataset. To reach the goal, the CAST clustering algorithm which can automatically find groups without the group numbers is employed in the second phase to divide the trajectory series into groups, and the flowchart of the procedure is shown in Fig. 3.

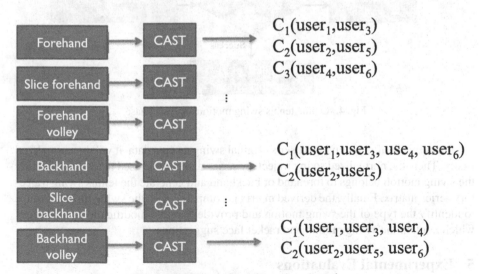

Fig. 3. CAST-based tennis motion clustering

From Fig. 3, we can know that each of the six swing motions is divided into groups using the CAST. Taking backhand volley as example, the motions from all players has been divided into two swing motion groups that mean the $user_1$, $user_3$ and $user_4$ have similar backhand volley motions in cluster C_1 as well as the $user_2$, $user_5$ and $user_6$ in C_2.

4   Online Tennis Swing Motion Analysis

In the online analysis phase, the proposed system will be utilized to receive a new user's swing motion data via the Kinect and the built series-based tennis racket. Then, the received swing motion trajectory series is compares to the derived swing motion groups for identifying swing motion type and finding possible modification suggestions. The flowchart of the online tennis swing motion analysis phase is shown in Fig. 4.

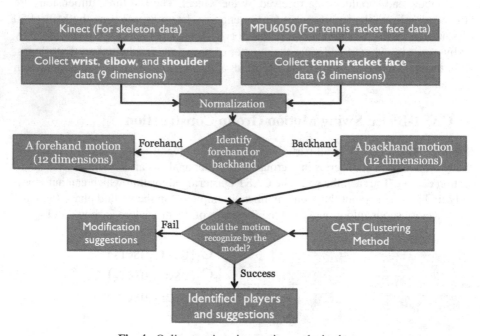

Fig. 4. Online tennis swing motion analysis phase.

From Fig. 4, after receiving a user's original swing motion data, it will be normalized firstly. Then, the normalized swing trajectory series data will be used to identify whether the swing motion belongs to forehand or backhand and generate the tennis swing trajectory series matrix. Finally, the derived matrix is compared with the swing motion groups to identify the type of the swing motion and provide possible modification suggestions which contain the swing motion and racket face suggestions.

5   Experimental Evaluations

5.1   Dataset Description

In this paper, the six swing motions were collected from the eight tennis players of university team. Five of them did ten swings for every swing motion, and 300 (= 10 ×

5 × 6) tennis swing trajectory series matrix were collected. Three of them did twenty swings for every swing motion. Hence, 360 (= 20 × 3 × 6) tennis swing trajectory series matrix were collected. In other words, totally 660 tennis swing trajectory series were used for the experimental evaluations, and each swing motion has 110 tennis swing trajectory series.

5.2 Evaluations of the Derived Swing Motion Groups

Before using the CAST to derive the swing motion groups, the distance matrix of every two swing motions is calculated by the DTW. Then, the distance matrix is converted to a similarity matrix. The range of an element in the similarity matrix is then between [0, 1]. Utilizing the similarity matrix, the swing motion groups for the six tennis swing motions

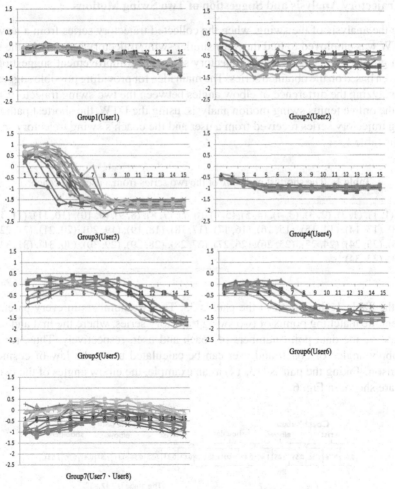

Fig. 5. The swing motion groups of forehand volley.

can be derived using the CAST. Taking forehand volley as an example, the derived swing motion groups are shown in Fig. 5.

From Fig. 5, we can observe that the tennis swing trajectory series of the eight players are divided into seven groups. From the results, they indicate that every swing motion group can map to a specific player, e.g., the representative player of *Group1* is *User1*, etc. And, the interesting part is the group *Group7* which contains tennis swing trajectory series from *User7* and *User8*. In other words, *User7* and *User8* have similar tennis swing trajectory series of forehand volley. Based on the results, in the online tennis swing motion analysis, when a new user's tennis swing trajectory series is similar to *Group7*, the system can not only identify the swing motion type but also provide possible improvement suggestions based on the tennis swing trajectory series of *User7* and *User8*.

5.3   Trajectory Analysis and Suggestion of Two Swing Motions

During the analysis of the swing, when the collected trajectory series from a new user, the DTW was employed to find the most similar swing trajectory series from the swing motion groups as a coach's swing trajectory series. The improvement suggestions are then given based on the obtained series. The main idea for providing possible suggestions is by analyzing the difference of elbow angles between the two swing trajectory series.

In the online tennis swing motion analysis, using the DTW, the shortest path *sph* of a swing trajectory series received from a user and the coach's swing trajectory series is given in Table 1.

Table 1. The shortest path *sph* of the two series from a user and coach.

(0, 0), (1, 1), (1, 2), (2, 3), (3, 4), (4, 5), (5, 6), (6, 7), (7, 8), (8, 9), (9, 10), (10, 11), (11, 12), (12, 13), (13, 14), (14, 15), (15, 16), (16, 17), (17, 18), (18, 19), (19, 20), (20, 21), (21, 22), (22, 23), (23, 24), (24, 25), (25, 26), (26, 27), (27, 28), (28, 29), (29, 30), (30, 31), (31, 32), (32, 33), (33, 34)

In the Table 1, the length of the path is 35. The two numbers in every parenthesis represent the matching points of two swing trajectory series, where the first and second numbers are the data point numbers of coach and user, respectively. Thus, for every pair, elbow angles of coach and user can be calculated using the law of cosines for comparison. Taking the pair is (13, 14) as an example, the elbow angles of the two data points are shown in Fig. 6.

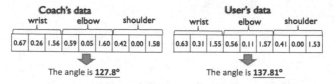

Fig. 6. The elbow angles of the coach and user.

From Fig. 6, we can know that the user's angle is larger than coach's. Because there are 35 data points, a measurement is defined to check whether user's swing motion is correct via the shortest path sph as follows:

$$\text{angleJudgment}(sph) = \text{CorrectNumberOfPoints}(sph)/|sph|.$$

Utilizing the formula, if the value of the formula is larger than a given threshold, the system will report there is no significant difference between the trajectory series of user and coach. In order to make the suggestion much easier for the user to understand, a swing motion is divided into three periods, including the start, during swing and end periods. When the value of the angle judgment measurement is smaller than the threshold, the suggestion for during swing period will be provided. When the threshold was set at 0.8, the suggestions for the shortest path received from another swing are illustrated in Table 2.

Table 2. The suggestions for the user in during swing period.

Data points in during swing period	Elbow angle	Suggestions
(5, 4)	(123.96, 116.36)	Correct
(6, 4)	(120.57, 116.36)	Correct
(7, 5)	(122.10, 139.80)	Elbow angle: too big
(8, 6)	(125.12, 139.19)	Elbow angle: too big
(9, 7)	(130.44, 128.73)	Correct

From Table 2, we can observe that the proposed system indicated that elbow angles of two pairs (7, 5) and (8, 6) are too big. Thus, based on the reported suggestions, the user can understand the problem and to improve the swing motion.

6 Conclusions and Future Work

This paper has designed a tennis practice system as the artificial intelligence tennis coach for beginners to learn tennis swing motions. Using the system, beginners can have a more economical way to learn tennis skills in a more effective way and to reduce the probability of body to be injured. To reach the purpose, the tennis swing motion data collection module was first implemented to collect six swing motions using the Kinect and the built sensor-based tennis racket. Then, each of the swing motions is divided into groups by the CAST to form swing motion groups. At last, the received tennis swing trajectory series from a new user can be compared with the swing motion groups to get suggestions for improving the swing motion. Experiments were also made on the real dataset to show the effectiveness of the proposed system in terms of providing useful suggestions in the middle period of a swing motion. In the future, we will attempt to find more tennis players to collect their swing motion data to make the swing motion groups more reliable and tune the conditions for collecting the tennis swing trajectory series more easily.

Acknowledgments. This research was supported by the Ministry of Science and Technology of the Republic of China under grant MOST 108-2221-E-032-037.

References

1. Abiddin, W.Z.W.Z., Jailani, R., Omar, A.R., Yassin, I.M.: Development of Matlab Kinect skeletal tracking system for gait analysis. In: IEEE Symposium on Computer Applications & Industrial Electronics (2016)
2. Ben-Dor, A., Friedman, N., Yakhini, Z.: Clustering gene expression patterns. J. Comput. Biol. **6**(3/4), 281–297 (1999)
3. Chaudhary, A., Prashanth, D., Raheja, J.L., Minhas, M., Shah, T.: Robust gesture recognition using Kinect: a comparison between DTW and HMM. Optik **126**, 1098–1104 (2015)
4. Diest, M.V., Lamoth, C.J.C., Postema, K., Stegenga, J., Verkerke, G.J., Wörtche, H.J.: Suitability of Kinect for measuring whole body movement patterns during exergaming. J. Biomech. **47**, 2925–2932 (2014)
5. Fan-Chiang, Y.M., Chang, C.H., Hsieh, C.T., Lin, C.W.: Kinect-based tennis swing motion analysis and its application. In: Taiwan E-Learning Forum, pp. 1–10 (2014)
6. Goudelis, G., Gourgari, S., Karpouzis, K., Kollias, S.: THETIS: three dimensional tennis shots a human action dataset. In: IEEE Conference on Computer Vision and Pattern Recognition Workshops (2013)
7. Huber, M.E., Leeser, M., Seitz, A.L., Sternad, D.: Validity and reliability of Kinect skeleton for measuring shoulder joint angles: a feasibility study. Physiotherapy **101**, 389–393 (2015)
8. Li, J., Wang, Y., Sun, J., Zhao, D.: Gait recognition based on 3D skeleton joints captured by Kinect. In: International Conference on Image Processing (2016)

Data Pre-processing Based on Convolutional Neural Network for Improving Precision of Indoor Positioning

Eric Hsueh-Chan Lu(✉) ⓘ, Kuei-Hua Chang, and Jing-Mei Ciou

Department of Geomatics, National Cheng Kung University, No. 1, University Road, Tainan City 701, Taiwan (R.O.C.)
luhc@mail.ncku.edu.tw, goeatsmall@gmail.com, s0918038832@gmail.com

Abstract. In the past, indoor positioning technology was mainly based on pedestrian dead reckoning and wireless signal positioning methods, but it was easy to cause some problems such as error accumulation and signal interference. Positioning accuracy still needs to be improved. With the development of neural networks in recent years, many researchers have successfully applied the neural network to the indoor positioning problem based on the Convolutional Neural Network (CNN). This technique mainly determines the position of the image by matching the image features. CNN faces the same challenges as other supervised learning. If the "clean" data cannot be collected, the trained model will not achieve good positioning accuracy. For CNN used for indoor positioning, if someone passes through in the training data, causing the person to appear in different positions of the images, the model may think that the images are the same location. To solve this problem, we propose a data pre-processing method to improve the accuracy of indoor positioning based on CNN. In this method, the moving objects recognized in training and testing data are modified in different ways. We perform data pre-processing method based on Mask R-CNN and YOLO, and then integrate the pre-processing method to PoseNet the famous CNN indoor positioning architecture. Through real experimental analysis, removing moving objects can effectively improve indoor positioning accuracy about 46%.

Keywords: Convolutional Neural Network · Indoor positioning · Object detection · Data pre-processing

1 Introduction

The development of spatial information technology has become more and more rapid in recent years, and the research and application on Location-Based Services (LBSs) [10] and route planning [5] related to positioning have attracted more and more attention. In the outdoor environment, people can obtain accurate surface location information through GPS and GNSS, and many areas also generate more convenient and appropriate services due to the emergence of global satellite positioning systems, such as satellite

© Springer Nature Switzerland AG 2020
N. T. Nguyen et al. (Eds.): ACIIDS 2020, LNAI 12033, pp. 545–552, 2020.
https://doi.org/10.1007/978-3-030-41964-6_47

navigation systems, smart parking systems, various geodesy, etc. Although the global satellite positioning system brings convenience to people's livelihood applications, when satellite signals are blocked, the positioning application will also fail. For example, indoor or basement is the most vulnerable to missing signals. Once the signal is blocked, GPS will not be able to continue to provide location services. Therefore, how to continuously calculate the position after the satellite signal fails, making indoor positioning technology a hot spot. Research topics. Indoor positioning is widely used and has high commercial value. Common application fields include route guidance of stations, AR interaction of art museums, smart navigation of department stores, and cargo monitoring of factories. Therefore, more and more scholars explore indoor positioning.

In the past, indoor positioning technology can be mainly divided into Pedestrian Dead Reckoning (PDR) [4] and wireless signals [2, 8], but it is easy to cause problems such as error accumulation and signal interference, and there is still much room for improvement in positioning accuracy. With the development of neural networks in recent years, in addition to predicting data and identifying images, many researchers have successfully applied Convolutional Neural Network (CNN) to indoor positioning technology. It is regarded as the concept of the human eye, analyzes the environmental characteristics in the image, and judges the position of the photographer itself by matching the feature values. CNN faces the same challenges as other supervised learning. If a large amount of "clean" data cannot be collected, the trained model will not achieve the desired positioning accuracy. It is easier to collect a large amount of data, but it cannot ensure that there is no interference in the data. Therefore, it is very important to pre-process the data.

This paper makes effort on precision improvement of indoor positioning and is intended to solve the situation of interference in CNN training data. If there is a passerby when we collect the training data, the passerby will be repeatedly displayed in images in different positions. So model may consider these images of passerby as the same position. Such a model cannot achieve good accuracy. Therefore, we propose a method for pre-processing data to improve the accuracy of CNN indoor positioning. The moving objects detected in the training and testing data are modified in different ways. In the experiment, we are based on Mask R-CNN [7] and YOLO [6] two kinds of CNN networks for data pre-processing and integration into the famous CNN indoor positioning architecture – PoseNet [1, 3] improved positioning accuracy.

The remaining of this research is organized as follows. Section 2 reviews related work on indoor positioning and CNN issues. In Sect. 3, we explain the details of data pre-processing and CNN-based indoor positioning models. The experimental evaluations are shown in Sect. 4. Finally, the conclusions and future work are mentioned in Sect. 5.

2 Related Work

In this section, we review some important studies related to indoor positioning issues. To collect available information for indoor positioning, Lan *et al.* proposed Pedestrian Dead Reckoning (PDR) [4] based on IMU technology to track the user's trajectory and detect when the user leaves the parking space so that the next user can use the mobile phone to get the location service for the inquiring parking space. Wireless signal positioning is another indoor positioning solution. The common signal type includes infrared, WiFi and Bluetooth. The calculation method includes proximity positioning, intersection method and feature comparison. Grossmann *et al.* set up a wireless network in the museum's exhibition hall [2]. The Access Point (AP) of the road, and uses the Received Signal Strength Index (RSI) to obtain location information. Subhan *et al.* based on the use of feature matching in Bluetooth to perform position estimation [8]. The accuracy of the indoor positioning system depends largely on the parameter values of the comparison and the measurement results of the surrounding environment.

Convolution Neural Network (CNN) is an effective recognition algorithm that has been widely used in image recognition, object detection and localization in recent years. To accurately estimate the attitude of a monocular camera, Kendall *et al.* proposed the CNN model PoseNet [1, 3] for regression pose estimation. PoseNet is a CNN indoor positioning architecture for regression pose estimation. Its network architecture is based on GoogleNet [9] proposed by Szegedy *et al.* The input is a color image and the output is changed to a seven-dimensional pose vector. The paper indicates that PoseNet can be used for both outdoor and indoor positioning. Ren *et al.* proposed Mask R-CNN [7] that is to mark objects identified in the image as masks close to the pixel level. In the past, the practice of masking was called RoIPooling. When the value was processed, the nearest interpolation method was used, and the output pixel value was the nearest pixel value, so that the resulting mask would be offset. The obtained area size is not an integer, and the mask after taking the integer cannot reach the pixel level. Therefore, the Mask R-CNN changes the RoIPooling to use the bilinear interpolation method, and performs lincar interpolation in two directions. The output pixel value is a weighted average of the surrounding pixel intensities, the pixel values are relatively continuous, and the mask position is more precise, and this method is also referred to as RoIAlign. In Mask R-CNN, the previous numerical processing method was changed and the pixel-level mask was achieved. YOLO [6] published by Facebook AI Research (FAIR) emphasizes not the pixel-level mask, but the actual recognition speed. The main method is to produce S*S squares, each square predicts the confidence score and type of the contained object by itself, and finally outputs the highest scored square. YOLO designed the network as end-to-end, which not only makes training easier, but also speeds up overall.

3 Proposed Method

In this section, we introduce our proposed indoor positioning method that can be divided into two phases including data pre-processing and CNN-based indoor positioning model.

3.1 Data Pre-processing

Training data are very important for supervised learning. The deep learning prediction model used for indoor positioning requires "clean" data. If there is a passerby when we collect the training data, the passerby will repeat in images in different positions. So model may consider these images of passerby as the same position. Such a model cannot achieve good accuracy. Therefore, we try to identify the moving object in the training and testing data and modify it in different ways for data pre-processing.

Data pre-processing is divided into two steps. The first step is to detect the object in the image. The detected object is a moving object. The techniques used in this step are YOLO and Mask R-CNN. The difference is that YOLO marks the object in a square, and the Mask R-CNN is marked according to the shape of the object. Then we paint the marked area. As shown in Fig. 1, using YOLO and Mask R-CNN to change the moving object to white, test whether the modified moving object will increase the accuracy of positioning. Since we found that the results of Mask R-CNN could not completely cover moving objects. We increased the marked area by ourselves, then test whether the accuracy was affected.

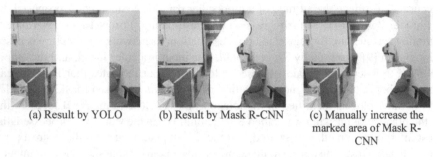

(a) Result by YOLO   (b) Result by Mask R-CNN   (c) Manually increase the marked area of Mask R-CNN

Fig. 1. Paint the marked area to white.

The second step is to use different strategies to paint the marked area of detected moving objects. The difference in strategy is mainly the difference in color. Since white is a subjective setting, it is assumed that white should be replaced with the most color in the image, as shown in Fig. 2, and whether the altered color affects the accuracy.

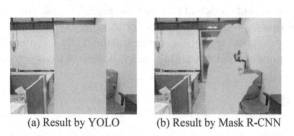

(a) Result by YOLO   (b) Result by Mask R-CNN

Fig. 2. Paint the marked area to the most color in the image.

We also tried to adjust the color of the painting to the average color of the image. From the experiment in the previous step, we found that if the moving object occupies too much area of the image, it would cause the painting color to approach the color of the moving object. Therefore, we changed it by calculating the average of all the colors in the image without the moving object, as shown in Fig. 3.

Fig. 3. Paint the marked area to the average color in the image.

3.2 CNN-Based Indoor Positioning Model

This article refers to the PoseNet architecture and makes some minor adjustments, using CNN to directly estimate the camera position from the image training model. Figure 4 shows the PoseNet adjusted by GoogLeNet-based architecture. The main adjustments are in two parts:

1. Replaces the three multi-classifiers with a regression amp, and each final fully-connected layer outputs a seven-dimensional pose containing the three-dimensional position and the quaternion.
2. Before the final affine regression, insert a fully connected layer with a feature size of 2048 to form a 23-layer PoseNet architecture. This is to generate a positioning vector that can be explored by PoseNet.

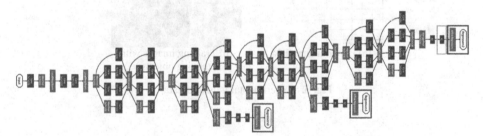

Fig. 4. The network architecture of PoseNet.

We also made some adjustments to the PoseNet architecture. The PoseNet model finally outputs the location and direction information, but the direction is not needed for our goal. Our ultimate goal is to let the user use the mobile phone for image positioning. The focus is on positional accuracy. The mobile phone itself is equipped with an accelerometer, gyroscope and magnetometer triaxial sensor. We only need to grab the information of the gyroscope in the mobile phone to know the direction angle, so we don't need to use CNN to estimate the direction. Thus, the loss function also needs to be adjusted together. The loss function is as shown in Formula (1), where $\widehat{P}$ and P are the position prediction value and the position true value, respectively. In the classification problem, each output tag contains at least one training sample, but for regression problems, the output tags are mostly continuous or infinite.

$$\text{loss(I)} = \widehat{P} - P \tag{1}$$

4 Experimental Evaluation

To evaluate the performance of indoor positioning by integrating data pre-processing and CNN model, a series of experiment are conducted by using the real data. All the experiments in this thesis are carried out on the Tensorflow platform. The hardware device used is Geforce GTX 1080 TI, the operating system is Linux Ubuntu 16.04, and the image source is Samsung Galaxy S7 edge.

4.1 Experimental Data and Setting

The experimental field used in this thesis is our laboratory, as shown in Fig. 5(a). In the experiment, we define people as moving objects, Fig. 5(b) shows one of the image samples. There are total 48 images for training, 11 have been painted. And total 24 images for testing, 8 have been painted.

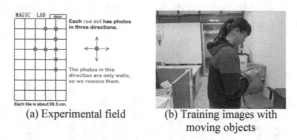

(a) Experimental field   (b) Training images with
                                 moving objects

Fig. 5. Experimental field and data.

4.2 Experimental Results

In the experiment, the images were altered in different ways and colors, so we combined different training and testing data to test the positioning error. The total experimental results are shown in Table 1 (unit: meter). If you use an image without any pre-processing during training, the average error of positioning will be nearly 4 m, and using YOLO and Mask R-CNN to change the moving object to white can improve the error, and YOLO is the most effective. Ok, it can improve nearly 1 m. Since the results of the Mask R-CNN coating did not completely cover the moving object, the experiment was manually added to the modification range, and the experimental results were still not as good as the YOLO correction. The reason may be because YOLO's modification method is a complete box, and Mask R-CNN's method of retouching preserves the contour of the object so that CNN still regards it as an important feature value.

Table 1. Comparison of the effects of different data pretreatment methods. Note that the parentheses represent the colors we used to paint on the moving object. '*W*' denotes white; '*M*' denotes the most color in the image; '*A*' denotes the average color of the image; and '*Mask+*' denotes we increased the marked area of Mask R-CNN.

Testing	Training						
	Original image	YOLO (W)	Mask (W)	Mask+ (W)	Mask+ (M)	YOLO (M)	YOLO (A)
Original image	**3.8039**	2.9908	3.2976	3.0704	2.8247	2.8221	2.0480
YOLO (W)	4.0789	2.8166	2.5835	3.1854	3.3238	2.5716	2.0548
Mask (W)	3.6532	2.7910	3.0055	3.2715	2.9972	2.6301	2.0241
Mask+ (W)	3.5441	2.6780	2.9916	3.2618	3.1098	2.6438	2.0993
Mask+ (M)	3.5024	3.0143	3.1918	3.2517	3.1849	2.5623	2.0638
YOLO (M)	3.5914	2.7731	2.6746	3.2002	3.4788	2.5586	**1.9683**
YOLO (A)	3.5750	2.7714	2.6850	3.1673	3.4581	2.5770	1.9799

Painting a moving object to white is personally subjective and unfounded, so the painted color is adjusted to the most frequently appearing color in the image, but in some images the moving object occupies too much area and the resulting color is approach to the object. Therefore, such color did not greatly improve the positioning error. Finally, the color calculation method is adjusted to the average color in the image except the moving object. It can be seen from the results that use average color to pre-process training data can reduce the positioning error in different testing data and can reduce the error of nearly 2 m compared with the data without any pre-processing. From all the above experimental results, it can be seen that regardless of the pre-processing method of the training and testing data, the positioning accuracy can be better than that of the original data. The best result is to use YOLO to detect moving objects and then paint them according to the average color of the image.

5   Conclusion and Future Work

This paper verifies the influence of moving objects on Convolutional Neural Network (CNN) indoor positioning accuracy through experiments. It also proposes to use Mack R-CNN and YOLO to pre-process the data with different strategies. The results show that the modified moving objects can effectively improve the positioning accuracy, regardless of what kind of strategy can improve accuracy. In addition to the people we have modified in the experiment, similar indoor positioning methods also mentioned that the change of furniture position will seriously affect the positioning accuracy. In the future work, we will consider the furniture that change position very often as moving objects too, or try some different ways of data pre-processing. In addition, we plan to test in a wider indoor field and try to develop a fully automated indoor field data collection mechanism, such as the use of drones or customized devices, to improve the overall efficiency of the experiment.

Acknowledgment. This research was supported by Ministry of Science and Technology, Taiwan, R.O.C. under grant no. MOST 108-2621-M-006-008 -.

References

1. Ciou, J.-M., Lu, E.H.-C.: Indoor positioning using convolution neural network to regress camera pose, ISPRS Geospatial Week, 1289–1294 (2019)
2. Grossmann, U., Gansemer, S., Suttorp, O.: RSSI-based WLAN indoor positioning used within a digital museum guide. Int. J. Comput. **7**(2), 66–72 (2014)
3. Kendall, A., Grimes, M., Cipolla, R.: PoseNet: a convolutional network for real-time 6-DOF camera relocalization. In: IEEE International Conference on Computer Vision, pp. 2938–2946 (2015)
4. Lan, K.-C., Shih, W.-Y.: An indoor locationtracking system for smart parking. Int. J. Parallel Emergent Distrib. Syst. **29**(3), 215–238 (2014)
5. Lu, E.H.-C., Chen, H.-S., Tseng, V.S.: An efficient framework for multirequest route planning in urban environments. IEEE Trans. Intell. Transp. Syst. **18**(4), 869–879 (2017)
6. Redmon, J., Divvala, S., Girshick, R., Farhadi, A.: You only look once: unified, real-time object detection. In: IEEE International Conference on Computer Vision and Pattern Recognition, pp. 779–788 (2016)
7. Ren, S., He, K., Girshick, R., Sun, J.: Faster R-CNN: towards real-time object detection with region proposal networks. Adv. Neural Inf. Process. Syst. 91–99 (2015)
8. Subhan, F., Hasbullah, H., Rozyyev, A., Bakhsh, S.T.: Indoor positioning in bluetooth networks using fingerprinting and lateration approach. In: IEEE International Conference on Information Science and Applications, pp. 1–9 (2011)
9. Szegedy, C., et al.: Going deeper with convolutions. In: IEEE International Conference on Computer Vision and Pattern Recognition, pp. 1–9 (2015)
10. Tseng, V.S., Lu, E.H.-C., Huang, C.-H.: Mining temporal mobile sequential patterns in location-based service environments. In: IEEE International Conference on Parallel and Distributed Systems, pp. 1–8 (2007)

Efficient Approaches for Updating Sequential Patterns

Show-Jane Yen and Yue-Shi Lee[✉]

Department of Computer Science and Information Engineering, Ming Chuan University,
Taoyuan, Taiwan
{sjyen,leeys}@mail.mcu.edu.tw

Abstract. Mining sequential patterns is to find the sequential purchasing behaviors for most of the customers in a transaction database. By using sequential patterns, it is possible to predict which products will be purchased in the future after the customer purchases certain commodities. Nowadays, transaction data is continuously added to the database. It is an important issue to update the sequence pattern efficiently in this environment. The previous efficient approach is to store the transactions in a tree structure. When the transactions were added, the tree structure could be updated according to the newly added items. It still needs to re-find the sequential patterns from the updated tree structure and re-scan the original transactions, without using the previous patterns. Therefore, we propose two algorithms for mining and maintaining the discovered sequential patterns when the transactions are added into the database. Our algorithms do not need to re-scan the original transactions and re-generate the existing sequential patterns, which just need to process the added transactions to update the existing sequential patterns. The experimental results also show that our algorithms outperforms the previous approaches.

Keywords: Data mining · Sequential pattern · Data stream · Transaction database

1 Introduction

Because the demand of storing data is growing rapidly and the capacity of the storage is getting larger, data stored in a database become larger in a daily basis. How to discover the potentially useful information existing in large databases becomes a popular research field in the computer science. *Data mining* [1] is the promising technology in discovering such useful information from large databases in order to enhance the quality of decision makings for the decision makers in all kinds of practical applications.

First of all, we give some terminologies about sequential pattern. A transaction database consists of a set of records. A record typically consists of the transaction identifier, the customer (buyer) identifier, the transaction date (or transaction time), and the items purchased in this transaction. That is, a transaction contains a set of items which are bought by a customer at the same time. Mining sequential patterns [2, 9–13] is to find the sequential purchasing behaviors for most customers from a transaction database. For

© Springer Nature Switzerland AG 2020
N. T. Nguyen et al. (Eds.): ACIIDS 2020, LNAI 12033, pp. 553–564, 2020.
https://doi.org/10.1007/978-3-030-41964-6_48

example, there may be a sequential pattern "<{shirts, necktie}{jacket}{shoes}> 70%" discovered from a transaction database in a department store, which means that seventy percent of the customers buy jacket after buying shirts and necktie, and then they buy shoes after buying jacket. We can use this information to predict what products the users will purchase when they bought some items. Therefore, when a user bought some items, the seller can recommend or promote the other products which the user is interested in according to the discovered sequential patterns.

We define more terminologies which are adopted in the process of mining sequential patterns: An itemset is a non-empty set of items. We say that an *itemset* X is contained in a transaction T. A *sequence* is an ordered list of the itemsets. A sequence s is denoted as $<s_1, s_2, ..., s_n>$, where s_i is an itemset. A sequence $<a_1, a_2, ..., a_n>$ is contained in another sequence $<b_1, b_2, ..., b_m>$ if there exist n integers $i_1 < i_2 < ... < i_n, 1 \leq i_k \leq m$ such that $a_1 \subseteq b_{i_1}, a_2 \subseteq b_{i_2}, ..., a_n \subseteq b_{i_n}$. A *customer sequence* is a list of all the transactions for a customer, which is ordered by increasing transaction-time. A *customer sequence database* includes a set of customer sequences. For example, Table 1 is a transaction database, and Table 2 is a customer sequence database which is transformed from Table 1.

Table 1. A transaction database

CID	Time	Item
10	2016/1/1, 10:30	AB
10	2016/1/1, 11:36	ABD
10	2016/1/1, 11:40	BCD
20	2016/2/1, 12:20	AB
20	2016/3/2, 15:36	BD
30	2016/2/2, 09:18	CD
40	2016/1/23, 15:11	BD

Table 2. A customer sequence database

CID	Customer sequence
10	<(AB)(ABD)(BCD)>
20	<(AB)(BD)>
30	<(CD)>
40	<(BD)>

The *support for a sequence* s is the ratio of the number of customer sequences that contain s to the total number of the customer sequences in a customer sequence database.

The *support count for a sequence* is the number of the customer sequences that support this sequence. If the support for a sequence s satisfies a user-specified *minimum support* threshold, then s is called *frequent sequence* or *sequential pattern*. The *length* of a sequence s is the number of items in the sequence. A sequence of length k is called a *k-sequence*, and a frequent sequence of length k a *frequent k-sequence*. In general, before generating frequent sequences, we need to generate *candidate sequences*, and scan the database to count the support for each candidate sequence to determine if it is a frequent sequence. A candidate sequence of length k is called a *candidate k-sequence*.

Since the customer transactions will grow rapidly, the user behaviours will change from time to time. Some approaches have been proposed for mining sequential patterns [2, 11–13], but they all assume that the database is static and it requires re-mining the entire database to obtain up-to-date sequential patterns. Therefore, it is very time-consuming to re-find the sequential patterns.

In 2004, Han et al. proposed the IncSpan algorithm [3] for mining sequential patterns when the transactions are added to the database. The authors propose a semi-frequent sequences which are sequences with support no less than the *Buffer Ratio* multiplied by the minimum support. This method may be because of too many or too few semi-frequent sequences generated, resulting in problems with the storage space or information loss. In 2012, Ho et al. proposed the IncSpam algorithm [7], which specifies a window size to find the sequential patterns of the customer's recent N transactions, and uses the sliding window to retrieve the stream data and convert it into bit vector array. Although this method can avoid re-scanning the original database, when the transactions exceed the window size, the support and the values in the bit vector array must be recalculated. Furthermore, when the window setting is too large, the computational complexity will be relatively increased, and the customer's recent N transactions may be long time ago, such that it is impossible to find the recent sequential patterns.

Lin et al. [8] proposed a FUSP-tree algorithm, which is based on the FUFP-tree [5] and find the sequential patterns from the whole tree after the transactions are added. FUSP-tree algorithm is extended by FUFP-tree algorithm and IncSpan algorithm [3]. However, FUSP-tree algorithm must spend a lot of time updating and adjusting the structure of the entire tree. Based on the concept of FUFP-tree algorithm, only frequent items are stored in FUSP-tree. If an item is not frequent in the original database, but is frequent in the added transaction database, FUSP-tree algorithm must re-scan the original transaction database. After updating the FUSP-tree, FUSP-tree algorithm uses FP-Growth algorithm [6] to re-discover all the sequential patterns without using the original sequential patterns that already exists.

In this paper, we first propose an algorithm *SPStream*, which retain the sequence structures of the candidate sequences. When a transaction is added, the sequence structure is updated directly, and the updated sequential patterns can be generated without re-scanning the original database and re-generate the existing sequential patterns. However, SPStream may generate a large number of candidate sequences, causing a burden on execution time and memory space, and it cannot handle the situation where the minimum support count will be changed when a new customer sequence is added to the original customer sequence database. Therefore, we also propose another algorithm

SPStream_Ins, which can greatly reduce the number of the candidate sequences, and can handle the situation when the new customer sequences are added.

2 Our Approaches

In this section, we describe our proposed algorithm SPStream and the improved SPStream_Ins algorithm for maintaining the sequential patterns when the transactions are continuously added.

2.1 SPStream Algorithm

For a transaction database, we use Di to represent the transaction date. When the date increases, the value i is also increased. The items in Di are the purchased items on date Di for a certain customer. Table 3 is a customer sequence database with date for each customer.

Table 3. A customer sequence database with date

CID	D1	D2	D3
1	B	C	BCD
2	A	BE	B
3	C	BCD	C

SPStream creates the *sequence structures* for the generated sequences day by day. A sequence structure contains the sequence itself, the support count, and the customer CID followed by a *date string* which is an ordered list of dates on which the customer purchased the sequence. For example, Table 4 is the structure of the sequence <C> in Table 3, in which the customer CID 1 purchased item C on D2 and D3 (i.e., the date string is [2, 3]), and customer CID 3 purchased item C on D1, D2 and D3 (i.e., the date string is [1–3]). Therefore, the support count of item C is 2.

Table 4. The sequence structure for sequence <C> in Table 3

<C>
Count: 2
CID1: [2, 3]
CID3: [1–3]

When the transactions on Di are added, SPStream first creates and updates the sequence structures of 1-sequences. For example, in Table 3, suppose the minimum

Table 5. The sequence structures for 1-sequences on D1 in Table 3

\<A\>	\<B\>	\<C\>
Count: 1	Count: 1	Count: 1
CID2: [1]	CID1: [1]	CID3: [1]

support is 50%, that is the minimum support count is 3 * 50% = 1.5. When all the transactions on D1 are added, the sequence structures are created as shown in Table 5.

If the supports of the items satisfy the minimum support threshold, then these items can be combined each other to generate candidate 2-sequences and their sequence structures are also created. For the above example, all the items on D1 are not frequent items. When the transactions on D2 are added, SPStream inserts CID2: [2] and CID3: [2] in the structure of sequence \<B\>. CID1: [2] is also inserted in the structure of sequence \<C\> which the date string for CID3 is updated, and the structures of sequences \<D\> and \<E\> are also created, which are shown in Table 6.

Table 6. The sequence structures for 1-sequences on D1 and D2 in Table 3

\<A\>	\<B\>	\<C\>	\<D\>	\<E\>
Count: 1	Count: 3	Count: 2	Count: 1	Count: 1
CID2: [1]	CID1: [1]	CID1: [2]	CID3: [2]	CID2: [2]
	CID2: [2]	CID3: [1, 2]		
	CID3: [2]			

Since 1-sequences \<B\> and \<C\> satisfy the minimum support, the candidate 2-sequences \<BC\>, \<CB\>, \<(BC)\>, \<BB\> and \<CC\> can be generated and the sequence structures for these candidates are created. For every two frequent k-sequences ($k \geq 1$), the candidate generation and the sequence structure creation for SPStream are described as follows: for any two frequent k-sequences $X = <x_1, x_2, ..., x_k>$ and $Y = <y_1, y_2, ..., y_k>$, $x_i = y_i$ ($\forall k, 1 \leq i \leq k - 1$) and $x_k \neq y_k$, then the three candidate $(k + 1)$-sequences $Z1 = <x_1, x_2, ..., x_k, y_k>$, $Z2 = <x_1, x_2, ..., x_{k-1}, y_k, x_k>$ and $Z3 = <x_1, x_2, ...x_{k-1}, (x_k, y_k)>$ can be generated, and the candidate sequences $<x_1, x_2, ... x_k, x_k>$ and $<y_1, y_2, ... y_k, y_k>$ also can be generated by the sequence X itself and Y itself, respectively.

If the sequence structures of the two frequent k-sequences X and Y have the same CID, in which the date strings are $[d_1, ..., d_n]$ and $[e_1, ..., e_m]$, respectively, SPStream scans the two date strings from left to right. if $d_1 < e_j$, ($1 \leq j \leq m$), and there is no e_k ($k < j$) such that $d_1 < e_k$, then the date string of the sequence structure of Z1 for the CID is recorded as $[e_j, e_{j+1}, ..., e_m]$; If $e_1 < d_i$ and there is no d_k ($k < i$) such that $e_1 < d_k$, then the date string of the sequence structure of Z2 for the CID is recorded as $[d_i, d_{i+1}, ..., d_n]$, if $d_i = e_j$, ($\forall i, j, 1 \leq i \leq m, 1 \leq j \leq n$), then di is recorded in the date string of this CID for the sequence structure of Z3.

For the two frequent k-sequences $X = <x_1, x_2, ..., x_{i-1}, (x_i, ..., x_k)>$ and $Y = <x_1, x_2, ..., x_{i-1}, (x_i, ..., x_{k-1}, y_k)>$, and $x_k \neq y_k$, SPStream only generates the candidate sequence $<x_1, x_2, ..., x_{i-1}, (x_i, ..., x_k, y_k)>$; For the two frequent k-sequences $X = <(x_1, x_2, ..., x_k)>$ and $Y = <(x_1, x_2, ..., x_{k-1}), y_k)>$, only the candidate $(k + 1)$-sequences $<(x_1, x_2, ..., x_k), y_k>$ can be generated. For the above example, The structures of the candidate sequences $<BC>$, $<CB>$, $<(BC)>$ and $<CC>$ are shown in Table 7.

Table 7. The sequence structures of candidate 2-sequences on D1 and D2 in Table 3

$<BC>$	$<CB>$	$<(BC)>$	$<CC>$
Count: 1	Count: 1	Count: 1	Count: 1
CID1: [2]	CID3: [2]	CID3: [2]	CID3: [2]

When the transactions on D3 in Table 3 are added, SPStream updates the sequence structures of 1-sequences $<B>$, $<C>$ and $<D>$ as follows. D3 is added into the date strings of CID1 and CID2 for 1-sequence $<B>$, and CID1 and CID3 for 1-sequence $<C>$, and CID1 for 1-sequence $<D>$, which are shown in Table 8.

Table 8. The sequence structures for 1-sequences in Table 3

$<A>$	$<B>$	$<C>$	$<D>$	$<E>$
Count: 1	Count: 3	Count: 2	Count: 2	Count: 1
CID2: [1]	CID1: [1, 3]	CID1: [2, 3]	CID1: [3]	CID2: [2]
	CID2:[2, 3]	CID3: [1–3]	CID3:[2]	
	CID3:[2]			

Since the 1-sequences $<B>$, $<C>$, and $<D>$ are frequent sequences, the generated candidate 2-sequences are $<BB>$, $<BC>$, $<CB>$, $<(BC)>$, $<BD>$, $<DB>$, $<(BD)>$, $<CC>$, $<CD>$, $<DC>$, $<DD>$ and $<(CD)>$. If the sequence structure of the candidate sequence has been created, SPStream only needs to update the structure of these sequences. Because customer CID1 purchased items B, C and D on D3, SPStream updates the sequence structures of the candidate sequences which the last items are B, C or D. For example, for sequence $<CC>$, CID1: [3] needs to be inserted into the sequence structure, and the date 3 needs to be inserted into the date string of CID3, because date 3 is larger than the minimum date in the date string of CID3 in the structure of $<C>$. For sequence $<BC>$, the date 3 needs to be inserted into the date string of CID3, because date 3 is larger than the minimum date in the date string of CID1 in the structure of $<B>$. The updated sequence structures for 2-sequences are shown in Table 9. For the sequence structure of the candidate sequence has not been created, SPStream creates the sequence structure for the candidate sequence, which are shown in Table 10.

Because the supports for all the candidate 2-sequences satisfy the minimum support threshold, that is, all the candidate sequences are frequent sequences, the sequence

Table 9. The updated sequence structures for candidate 2-sequences

<BC>	<CB>	<(BC)>	<CC>
Count: 2	Count: 2	Count: 2	Count: 2
CID1: [2, 3]	CID1: [3]	CID1: [3]	CID1: [3]
CID3: [3]	CID3: [2]	CID3: [2]	CID3: [2, 3]

Table 10. The created sequence structures for candidate 2-sequences

<(CD)>	<CD>	<(BD)>	<BB>
Count: 2	Count: 2	Count: 2	Count: 2
CID1: [3]	CID1: [3]	CID1: [3]	CID1: [3]
CID3: [2]	CID3: [2]	CID3: [2]	CID2: [3]

structures for the candidate 3-sequences can be created. The structures for the candidate 3-sequences are shown in Table 11, in which sequence <C(BD)> is generated by the two frequent 2-sequences <CB> and <CD>; sequence <C(BC)> is generated by <CB> and <CC>; sequence <(BCD)> is generated by the sequences <(BC)> and <(BD)>; sequence <C(CD)> is generated by the sequences <(CC)> and <CD>. Finally, the structure of the frequent 4-sequence <C(BCD)> can be generated by the two frequent 3-sequences <C(BD)> and <C(BC)>, which is shown in Table 12.

Table 11. The created sequence structures for candidate 3-sequences

<C(BD)>	<C(BC)>	<(BCD)>	<C(CD)>
Count: 2	Count: 2	Count: 2	Count: 2
CID1: [3]	CID1: [3]	CID1: [3]	CID1: [3]
CID3: [2]	CID3: [2]	CID3: [2]	CID3: [2]

2.2 SPStream_Ins Algorithm

SPStream_Ins algorithm improves SPStream algorithm on candidate generation and the sequence structure as follows. For any two frequent sequences of length k (k $\geq$ 2), only one candidate (k + 1)-sequence needs to be generated. For a k-sequence S, let S.first be the first item of sequence S and S.last be the last item of sequence S. For any two frequent k-sequences S1 and S2, if the sequence generated by removing S1.first from S1 is the same as the sequence generated by removing S2.last from S2, then S1 and S2 can be combined into one candidate k-sequence. There are two cases for the combination: If S2.last is a separate itemset, then the candidate (k + 1)-sequence can be generated by

Table 12. The created sequence structure of the candidate 4-sequence $<$C(BCD)$>$

$<$C(BCD)$>$
Count: 2
CID1: [3]
CID3: [2]

adding S2.last to S1 as the last separate itemset. For example, suppose S1 and S2 are $<(1, 2) (5) (6)>$ and $<(2) (5) (6) (7)>$, respectively, then S1 and S2 can be combined to generate a candidate sequence $<(1, 2) (5) (6) (7)>$. If S2.last is not a separate itemset, then the candidate $(k + 1)$-sequence can be generated by adding S2.last to the last itemset of S1. For example, suppose S1 and S2 are $<(1, 2) (5) (6, 7)>$ and $<(2) (5) (6, 7, 8)>$, respectively, then S1 and S2 can be combined to generate a candidate sequence $<(1, 2)$ $(5) (6, 7, 8)>$. However, for two frequent items x and y, it is necessary to generate five candidate 2-sequences $<(x) (y)>$, $<(y) (x)>$, $<(xy)>$, $<xx>$ and $<yy>$.

Suppose candidate $(k + 1)$-sequence Z is generated by the two frequent k-sequences S1 and S2. If there are the same CIDs in the two sequence structure of S1 and S2, and their date strings are $[d_1, \ldots, d_n]$ and $[e_1, \ldots, e_m]$, respectively, then there are two cases to generate sequence structure for the candidate sequence Z: If S2.last is a separate itemset, $d_1 < e_j$, $(1 \leqq j \leqq m)$, and there is no e_k $(k < j)$ such that $d_1 < e_j$, then the date string of the sequence structure of Z for the CID is recorded as $\left[e_j, e_{j+1}, \ldots, e_m\right]$. If S2.last is not a separate itemset and di = ej ($\forall$ i, j, $1 \leqq i \leqq n$, $1 \leqq j \leqq m$), then di is recorded in the date string of this CID for the sequence structure of Z. By using this way, the number of the generated candidate sequences and the time to generate sequence structures will be reduced.

For example, in Table 3, after processing the transactions on D1, D2 and D3, the frequent 2-sequences are shown in Tables 9 and 10, in which the two sequences $<$BC$>$ and $<$CD$>$ can generate only one candidate 3-sequence $<$BCD$>$ according to the candidate generation method for SPStream_Ins. For frequent 3-sequences $<$C(BC)$>$ and $<$(BCD)$>$, SPStream_Ins generates only one candidate 4-sequence $<$C(BCD)$>$. If the are new customer sequences added, then the minimum support count may be increased, such that the existing frequent sequence S may turn out to be infrequent sequence. If the infrequent sequence S is not a candidate sequence, that is, one of the sequences which generate the infrequent sequence become infrequent, then the infrequent sequence S and its sequence structure would be deleted.

For the items in added transaction, SPStream_Ins first updates the sequence structure for the items. If a k-sequence $(k \geqq 1)$ is not frequent, and becomes frequent after updating, it can be combined with other frequent k-sequences to generate a candidate $(k + 1)$-sequence and construct its sequence structure by using the candidate generation method and the sequence structure construction method described above. If the sequence X is originally frequent and becomes infrequent after the update, it is stored in the deleted set DelItemset. SPStream_Ins deletes the $(k + 1)$-sequence containing X and its sequence structure, since this $(k + 1)$-sequence is not a candidate sequence after the update.

Table 13. The customer sequence database after adding transactions on D4 in Table 3

CID	D1	D2	D3	D4
1	B	C	BCD	
2	A	BE	B	C
3	C	BCD	C	B
4				ABE
5				AC

Table 14. The sequence structures for items after processing the transactions on D4

\<A\>	\<B\>	\<C\>	\<D\>	\<E\>
Count: 3	Count: 4	Count: 4	Count: 2	Count: 2
CID2: [1]	CID1: [1, 3]	CID 1: [2, 3]	CID1: [3]	CID2: [2]
CID4: [4]	CID2: [2–4]	CID2: [4]	CID3: [2]	CID4: [4]
CID5: [4]	CID 3: [2]	CID3: [1–3]		
	CID 4: [4]	CID5: [4]		

For example, Table 13 shows the customer sequences after adding the transactions on D4 into Table 3. After processing the transactions on D4, the sequence structures for all the items can be updated, which is shown in Table 14. Because of the addition of new customers CID4 and CID5, the minimum support count has been increased from the original 1.5 (3 * 50%) to 2.5 (5 * 50%). Therefore, the sequence \<D\> which is originally frequent turns out to be infrequent, because the minimum support count is increased, which indicate that the sequence generated by sequence \<D\> and other sequences is no longer a candidate sequence. Therefore, sequence \<D\> is added to the deleted set DelItemSet. SPStream_Ins deletes the 2-sequence containing the items in the DelItemSet and its sequence structure, so the sequences \<BD\>, \<(BD)\>, \<CD\>, \<(CD)\>, \<DB\>, \<DD\>, \<DC\>, and their sequence structures are deleted.

The two 1-sequences \<B\> and \<C\> are still frequent sequences after the update, so it is only necessary to update the sequence structures of the candidate sequences generated by \<B\> or \<C\>, which are in Table 15. The 1-sequence \<A\> is not a frequent sequence, but it becomes a frequent sequence after updating. Therefore, it needs to be combined with other frequent sequences into a candidate 2-sequences, and their sequence structures can be constructed, which is shown in Table 16.

For the 2-sequences \<(BC)\>, \<CC\> and \<CB\>, they are originally frequent, but turn out to be infrequent after updating, so they are added to the deleted set DelItemSet = { \<D\>, \<(BC)\>, \<CC\>, \<CB\>}. SPStream_Ins then deletes the 3-sequence containing the sequence in the DelItemSet and their sequence structures. After the deletion, the structure of the remaining sequence \<BBC\> can be updated by the structure of the two sequences \<BB\> and \<BC\>. Since there is no frequent 3-sequence, the

Table 15. The updated structures of the candidate 2-sequences after processing the transactions on D4

<BC>	<CB>	<(BC)>	<CC>	<BB>
Count: 3	Count: 2	Count: 2	Count: 2	Count: 3
CID1: [2, 3]	CID1: [3]	CID1: [3]	CID1: [3]	CID1: [3]
CID2: [4]	CID3: [2, 4]	CID3: [2]	CID3: [2, 3]	CID2: [3]
CID3: [3]				CID3: [4]

Table 16. The constructed sequence structures for the candidate 2-sequences after processing the transactions on D4

<AB>	<BA>	<(AB)>	<AC>	<CA>	<(AC)>
Count:1	Count:0	Count:1	Count:1	Count:0	Count:1
CID 1: [2, 3]		CID 4: [4]	CID 2: [4]		CID 5: [4]

4-sequence <C(BCD)> is also deleted. After processing the added transactions on D4, the generated sequential patterns are <B>, <C>, <BC> and <BB>.

3   Experimental Results

In this section, we compare our proposed SPStream and SPStream_Ins algorithms with the recently more efficient algorithm FUSP-tree [8] for execution time and memory usage. We use IBM Data Generator [1, 2] to generate synthetic data set, in which the number of total transactions is 1000, the number of different items is 1000, the average number of a transaction for a customer is 4, and the average number of transactions for a customer is 2.

We increased the transactions from the first day to the 128th day, and set different minimum support thresholds to evaluate the execution time and memory usage for the three algorithms. Figure 1 shows the execution time of the three algorithms for the minimum supports from 1.9% to 1.4% by decreasing 0.1% each time. From this experiment, we can see that the performance gap between FUSP-tree and our algorithms increases as the minimum support decreases. Because FUSP-tree algorithm spends a lot of time to update the FUSP-tree, and finally find all the sequential patterns from the whole tree. It needs to take a lot of time to adjust the tree structure and re-mine the sequential patterns from the updated tree for FUSP-tree algorithm. Our algorithms only needs to update or create the sequence structures for the candidate sequences according to the added transactions, and directly generates sequential patterns from the updated sequence structures without re-mining the sequential patterns. Because SPStream_Ins improves the candidate generation method for SPStream, the number of candidate sequences generated by SPStream_Ins can be reduced and sequence structures can be updated efficiently. Therefore, SPStream_Ins slightly ourperforms SPStream algorithm.

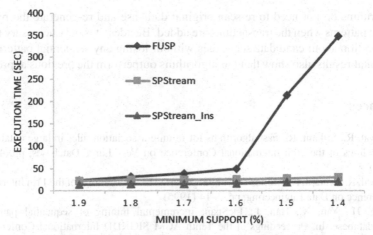

Fig. 1. The execution time for the three algorithms

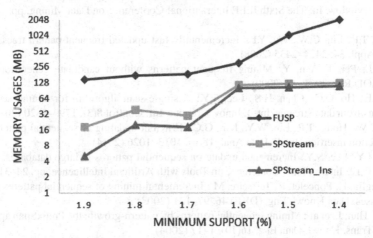

Fig. 2. The memory usages for the three algorithms

Figure 2 shows the memory usage for the three algorithms. Because the smaller the minimum support, the more the frequent itemsets, the larger the structure of the entire FUSP-tree, and the more memory space required. While our algorithms also needs to record CIDs and date strings, it is all digital. In the case of smaller minimum support, compared with FUSP-tree, our algorithms still saves a lot of memory space.

4 Conclusions

In this paper, we propose efficient algorithms for mining sequential patterns in a data stream. Our algorithms record sequence structure for each candidate sequence to generate new sequential patterns, and performs some operations on sequence structures of frequent k-sequences to generate the sequence structure for candidate (k + 1)-sequences.

Our algorithms do not need to re-scan original database and re-generate the previous sequential patterns when the transactions are added. Besides, it is only necessary to store the information about candidate sequences without losing any sequential patterns. The experimental results also show that our algorithms outperform the previous approaches.

References

1. Agrawal, R., Srikant, R.: Fast algorithms for mining association rules in large database. In: Proceedings of the 20th International Conference on Very Large Databases, pp. 487–499 (1994)
2. Agrawal, R., Srikant, R.: Mining sequential patterns. In: Proceedings of the 11th International Conference on Data Engineering, pp. 3–14 (1995)
3. Cheng, H., Yan, X., Han, J.: IncSpan: incremental mining of sequential patterns in large database. In: Proceedings of the Tenth ACM SIGKDD International Conference on Knowledge Discovery and Data Mining, pp. 527–532 (2004)
4. Ho, C.C., Li, H.F., Kuo, F.F., Lee, S.Y.: Incremental mining of sequential patterns over a stream sliding window. In: The Sixth IEEE International Conference on Data Mining, pp. 677–681 (2006)
5. Hong, T.P., Lin, C.W., Wu, Y.L.: Incrementally fast updated frequent pattern trees. Expert Syst. Appl. **34**, 2424–2435 (2008)
6. Han, J., Pei, J., Yin, Y.: Mining frequent patterns without candidate generation. ACM SIGMOD Rec. **29**, 1–12 (2000)
7. Li, H.F., Ho, C.C., Chen, H.S., Lee, S.Y.: A single-scan algorithm for mining sequential patterns from data streams. Int. J. Innov. Comput. Inf. Control **8**(3), 1799–1820 (2012)
8. Lin, C.W., Hong, T.P., Lin, W.Y., Lan, G.C.: Efficient updating of sequential patterns with transaction insertion. Intell. Data Anal. **18**(6), 1013–1026 (2014)
9. Lin, M.Y., Lee, S.Y.: Incremental update on sequential patterns in large databases. In: The Tenth IEEE International Conference on Tools with Artificial Intelligence, pp. 24–31 (1998)
10. Masseglia, F., Poncelet, P., Teisseire, M.: Incremental mining of sequential patterns in large databases. Data Knowl. Eng. (DKE) **46**, 97–121 (2003)
11. Pei, J., Han, J., et al.: Mining sequential patterns by pattern-growth: the PrefixSpan approach. IEEE Trans. Knowl. Data Eng. **16**(10), 1–17 (2004)
12. Pei, J., et al.: PrefixSpan: mining sequential patterns efficiently by prefix-projected pattern growth. In: Proceedings of the 17th International Conference on Data Engineering, pp. 215–224 (2001)
13. Mohammed Zaki, J.: SPADE: an efficient algorithm for mining frequent sequences. Mach. Learn. J. **42**, 31–60 (2001). Special issue on Unsupervised Learning (Doug Fisher, ed.)

Improving Accuracy of Peacock Identification in Deep Learning Model Using Gaussian Mixture Model and Speeded Up Robust Features

Tzu-Ting Chen[1], Ding-Chau Wang[2(✉)], Min-Xiuang Liu[1], Chi-Luen Fu[1],
Lin-Yi Jiang[3], Gwo-Jiun Horng[4(✉)], Kawuu W. Lin[5], Mao-Yuan Pai[6],
Tz-Heng Hsu[4], Yu-Chuan Lin[1], Min-Hsiung Hung[7], and C.-C. Chen[1]

[1] IMIS/CSIE, NCKU, Tainan, Taiwan
{P96084126,P98081077,P96081128}@mail.ncku.edu.tw, duke@imrc.ncku.edu.tw,
chencc@imis.ncku.edu.tw
[2] MIS, STUST, Tainan, Taiwan
dcwang@stust.edu.tw
[3] Star River Design Digital Co, Tainan, Taiwan
chianglin1993@gmail.com
[4] CSIE, STUST, Tainan, Taiwan
grojium@gmail.com, hsuth@stust.edu.tw
[5] CSIE, NKUST, Kaohsiung, Taiwan
linwc@nkust.edu.tw
[6] General Research Service Center,
National Pingtung University of Science and Technology, Neipu, Taiwan
mypai@mail.npust.edu.tw
[7] CSIE, PCCU, Taipei, Taiwan
hmx4@faculty.pccu.edu.tw

Abstract. Data set is a most crucial aspect for the object recognition by the deep learning. The effect of training a deep learning model would be not good with an insufficient quantity of the data set. For example, there are two similar pictures which are captured from a video with a peacock and these two pictures are separated by one second in this video. Results show that the peacock was recognized by the model in the former picture but it failed in the latter picture due to the angle change of the peacock. In order to improve recognition effects of the model, we propose a system based on Gaussian Mixture Model and Speeded Up

This work was supported by Ministry of Science and Technology (MOST) of Taiwan under Grants MOST 107-2221-E-006-017-MY2, 108-2218-E-006-029, 108-2221-E-034-015-MY2, and 107-2221-E-218-024. This work was also supported by the "Intelligent Service Software Research Center" in STUST and the "Allied Advanced Intelligent Biomedical Research Center, STUST" under Higher Education Sprout Project, Ministry of Education, Taiwan. This work was financially supported by the "Intelligent Manufacturing Research Center" (iMRC) in NCKU from The Featured Areas Research Center Program within the framework of the Higher Education Sprout Project by the Ministry of Education in Taiwan.

ⓒ Springer Nature Switzerland AG 2020
N. T. Nguyen et al. (Eds.): ACIIDS 2020, LNAI 12033, pp. 565–574, 2020.
https://doi.org/10.1007/978-3-030-41964-6_49

Robust Features to recognize peacocks in images. We also implement the prototype of this article and conduct a series of experiments to test the proposed solution. Furthermore, experimental results show the scheme did improve the accuracy of the complete training model.

Keywords: Data analysis · Image procession · Deep learning · Iterative adjustment · Avian inspection

1   Introduction

Computer vision and artificial intelligence (AI) are the most exciting areas in recent times. Computer vision can be used in certain recognition with the development of the Convolutional Neural Network (CNN), such as the face recognition. However, it is still a very challenging task for scientists to design these different kinds of calculations.

Data set is a most crucial aspect for the object recognition with the deep learning. However, it can be challenging to collect data for the deep learning. The data collection for the wild life recognition is more difficult specifically because it is supposed to take a lot of time to look for the animal and take photos of the animal from various angles for ensuring the effect of model training.

The training effect of a learning model would be not good with an insufficient quantity of the data set. Furthermore, sometimes there are a lot of difficulties in collecting data set because of time and cost. Therefore, the method of this paper is to correct the identification results on the premise that the model training is not good because of insufficient data sets.

Actually, we use the trained model to test the new data set. Figure 1 shows a comparison chart of some recognition results. Each image caption is the number of seconds showed in the video. It can be observed that the peacock can be identified in the images 11.jpg, 160.jpg and 537.jpg, but it cannot be recognized in the images 12.jpg, 161.jpg and 538.jpg.

In order to improve the identification results of the model, this paper builds an image based feature extraction and the image processing to improve the identification results with the output of the deep learning model. The data set used in this paper is the peacock dataset. And the images of the dataset are captured from a video with peakcock by one second interval. The concept is to use the output results of YOLOv3. First, the system finds the corresponding images according to the time. And the system determines the moving object through the image processing foreground and background separation technique [5,13–15], and determines the position. Then, the system finds out matching features on the image by Speeded Up Robust Features [1,10]. If the count of matching features of the image is conformed with the count of image features of the identified peacock image. The image can be judged to have a peacock.

The rest of the paper is organized as follows. Section 2 gives the related work. Section 3 presents the system architecture. And Sect. 4 describes the design of the system. Then, we conduct experiments and show the results in Sect. 5. Finally, we conclude the paper in Sect. 6.

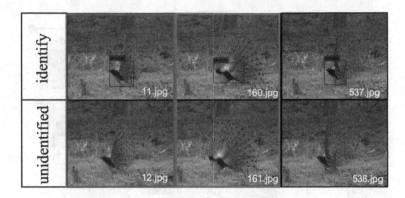

Fig. 1. Deep learning identification results.

2   Related Work

The object detection is a process for detecting certain types of objects (such as people and animals) on the side of images and videos. The majority of works utilizing the deep learning for object detection use the architecture of CNNs, such as [6,7,9] (which proposed new def-pooling layers and new learning strategies). However, others have tried less-used depth models. For example, [2] proposed a rough target localization method based on saliency mechanism combined with DBN for target detection in remote sensing images. [8] proposed a new DBN for 3D object recognition, in which the top-level model is a third-order Boltzmann machine. And the model is trained using a hybrid algorithm which mixes generation and discriminant gradients. [4] uses fused deep learning methods, while [3] explores the representation capabilities of deep models in semi-supervised paradigms.

3   System Architecture

Figure 2 shows the architecture diagram of the whole system. The input of this system is the image. The system is divided into three modules. The first module is the deep learning model. And the paper adopts the state-of-the-art object detection model YOLOv3 [12] as an image recognition method in the system. Figure 3 shows the architecture diagram of YOLO network version 3, YOLOv3 totally includes 108 convolutional and residual layers, where three scales of prediction features are in the tensor forms ($82 \times 82, 96 \times 96, 108 \times 108$). The deep learning model can be arbitrarily replaced with other methods as long as the output of the model is an image and has an bounding box of the identified object.

The second module is Gaussian Mixture Model Get Position (GMM-get position) which uses an algorithm to find the moving object and cuts the range of the object from the image without recognizing the object. The output of the

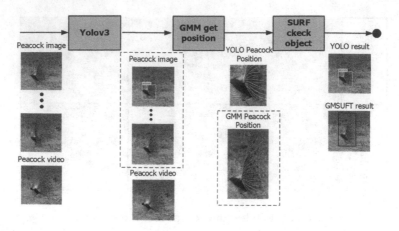

Fig. 2. System architecture.

module is one image per second. The third module is Speeded Up Robust Features Check object (SURF-check object). This module would check whether the peacock image identified by YOLOv3 and the image cut by the GMM-get position module have the same feature in order to confirm that there is a peacock in the image cut by the GMM-get position module.

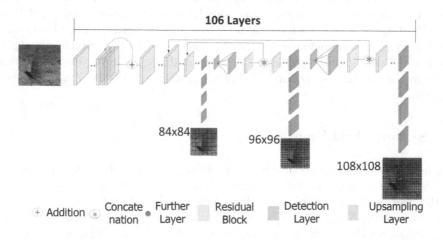

Fig. 3. YOLOv3 system architecture.

4   System Design

Figure 4 shows a flow chart of the GMM-get position module. The GMM-get position module is divided into three parts. The first part of this module would catch images which might be repaired. The second part of the module would extract the peacock position from the bounding box of the image detected by the YOLOv3 model. The third part would select images that need to be repaired from the first part. The system uses the foreground and background separation technique to detect the moving object and cut it down.

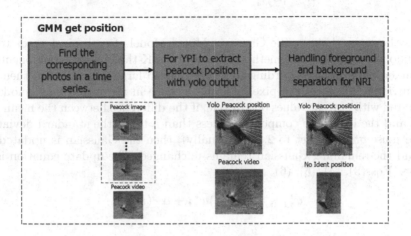

Fig. 4. Gmm model flow chart.

The input data are time-series images so the image file name is labeled in timestamp. The system would find out images that the peacock can be identified in a time series image, but did not recognize the peacock in the image of the next time series. Then these two images would be defined to a set of data. The image with YOLOv3 recognition is called Yolo Peacock Image (YPI), and the image without recognition result is called No Recognize Image (NRI). The output of YOLOv3 has the information including the bounding box. According to the information of the bounding box, the system cuts out the peacock area from YPI and this area is called Yolo Peacock Position (YPP).

The NRI would be cut out the location of the peacock because the set of images is continuous and is the same background. The Gaussian Mixture Model [11] is used to process the foreground and background separation technology. The Gaussian model composed of Gaussian components represents the probability distribution model of the pixel in the time domain.

The probability that the pixel j in the image takes the value x_j at the time t, as in the Eq. (1)

$$P(\mathbf{x}_j) = \Sigma_{i=1}^{K} w_{j,t}^i \cdot \eta(\mathbf{x}_j, \mu_{j,t}^i, \Sigma_{j,t}^i) \tag{1}$$

Where $w_{j,t}^i$ represents the weight of the i-th Gaussian component in the Gaussian Mixed Model of the pixel j at time t. The $\mu_{j,t}^i$ and $\Sigma_{j,t}^i = (\sigma_{j,t}^i)^2$ respectively represent the mean and covariance of the i-th Gaussian component.

The σ is the standard deviation, I is the identity matrix, and η is the Gaussian probability density function, as in Eq. (2), where d is x_j dimension.

$$\eta(\mathbf{x}_j; \mu_{j,t}^i, \Sigma_{j,t}^i) = \frac{1}{(2\pi)^{\frac{d}{2}} |\Sigma_{j,t}^i|^{\frac{1}{2}}} \times \exp\left[-\frac{1}{2}(x_j - \mu_{j,t}^i)^\mathsf{T}(\Sigma_{j,t}^i)^{-1}(x_j - \mu_{j,t}^i) \right] \tag{2}$$

As the scene changes, the Gaussian Mixed Model of each pixel needs to be continuously updated. The method is to sort the K-th Gaussian components in the mixed muse model according to $w_{j,t}^i/\sigma_{j,t}^i$ from large to small, and then use the current value x_j of the pixel. The K-th muse in the mixed muse model is compared with current values one by one, if the difference between the mean $\mu_{j,t}^i$ of x_j and the i-th muse component is less than δ times the standard deviation of the muse $\sigma_{j,t}^i$ (δ is set to 2.5–3.5 usually), then the Gaussian is updated by x_j, and the remaining Gaussian remains unchanged. The update equation is as follows: Eqs. (3), (4), (5), (6)

$$w_{j,t+1}^i = (1 - \alpha)w_{j,t}^i + \alpha \cdot (M_{j,t}^i) \tag{3}$$

$$\mu_{j,t+1}^i = (1 - \rho)\mu_{j,t}^i + \rho \cdot x_j \tag{4}$$

$$(\sigma_{j,t+1}^i)^2 = (1 - \rho)(\sigma_{j,t}^i)^2 + \rho(x_j - \mu_{j,t}^i)^\mathsf{T}(x_j - \mu_{j,t}^i) \tag{5}$$

$$\rho = \frac{\alpha}{w_{j,t}^i} \tag{6}$$

Where α is the learning rate of the model. If the i-th Gaussian component matches x_j, then $M_{j,t}^i$ is 1, otherwise 0. If x_j and pixel j of gaussian ingredient in the K-th Mixture Gaussian model is not match. Then the last Gaussian in the pixel-Gaussian Mixed Model is replaced by a new Gaussian, the new Gaussian mean is x_j, the initial standard deviation and the weight are set to σ_{int} and w_{int}, after the update is completed. The value is normalized so that $\Sigma_{i=1}^K w_{j,t+1}^i = 1$.

The Gaussian Mixed Model of pixel j describes the probability distribution of its eigenvalue x_j in the time domain. In order to determine which Gaussian components in the Gaussian Mixed Model of the pixel to be generated by the background, the order is based on the ratio of the weight of each Gaussian component to its standard deviation. After that, the system would take the B_j Gaussian components as the background distribution, and B_j is calculated according to Eq. (7).

$$B_j = \mathrm{argmin}(\Sigma_{i=1}^b w_{j,t+1}^i > T) \tag{7}$$

The threshold T measures the minimum proportion of the background Gaussian component in its overall probability distribution. When T is small, the background is represented by a single-mode distribution. When T is constant, it can represent the background of the multi-mode distribution. The image removed background is recorded, and the position of the moving object is recorded. The NRI is cut at the position of the moving object, and the new image is called No Recognize Image (NRI). Figure 5 shows the effect after GMM, we can clearly see the position of the moving object on the image which makes this coordinate on the NRI for cutting.

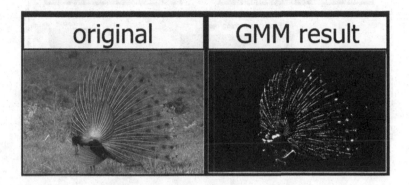

Fig. 5. A result obtained with GMM.

Figure 6 shows a flow chart of the SURF-check object module. The SURF Check object module is divided into two components. The first component would calculate the matching feature number kp for NRI and YPP using Speeded Up Robust Features. The second component would calculate the feature distribution of YPI. If there are matching features, it would be confirmed that it is the same as the object, the coordinate range of the NRI is drawn on the NRI completion mark.

Speeded Up Robust Features have invariances such as scale, translation, rotation, and brightness, so we use Speeded Up Robust Features to detect feature points in the image. The Speeded Up Robust Features algorithm is divided into two steps: (1) feature point detector: use a box filter approximating the Gaussian filter to obtain a simplified Hessian matrix, as Eq. (8). For a point on the image $x = (x, y)$, $\mathcal{H}(x, \sigma)$ is the Hessian matrix of the σ scale at the x point.

Therefore, the characteristics of the integral image can be used to accelerate the computation time of the DoH (Determinant of Hessian) in order to establish the scale space. That the operation efficiency is higher when detecting the feature point. (2) The feature point descriptor: the feature point detected in the previous step is the Harr Wavelet Filter. The system filters the neighboring regions of the feature points to find the main direction of the feature points, which is the gradient direction. In the main direction, the neighboring region is also filtered by the Harr Wavelet Filter to obtain the feature point descriptor. The integral image acceleration operation is also used.

$$\mathcal{H}(x, \sigma) = \begin{bmatrix} L_{xx}(x, \sigma) & L_{xx}(x, \sigma) \\ L_{xx}(x, \sigma) & L_{xx}(x, \sigma) \end{bmatrix} \tag{8}$$

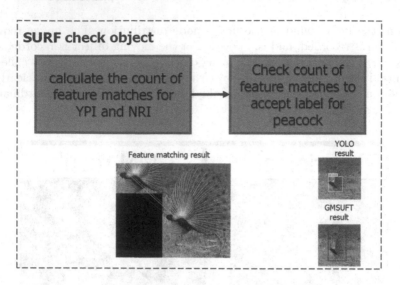

Fig. 6. SURF model flow chart.

5   Case Study

In this chapter we input a peacock data set shot in a fixed background to the system. There are 242 images in the dataset. The time interval between each image is one second. The deep learning model adopts YOLOv3, and the model mainly processes peacock identification training. Figure 7 shows a modification of the YOLOv3 identification results. Initially, the image of the peacock was unrecognizable. But the peacock could be recognized in the image after using the system presented in this paper.

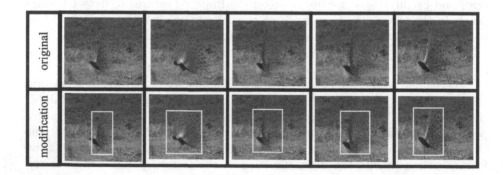

Fig. 7. Modification of YOLOv3 identification results.

Table 1 is a statistical result comparing the output of YOLOv3 with the output of this paper method in accuracy. When only the YOLOv3 identification model is used, 176 pieces of peacock images are identified from 242 images in the data set, the accuracy rate is 73%. After the YOLOv3 output the results, the method of this paper is used to refer to the output of YOLOv3. The method would catch corresponding images and mark them. The position of the mark was similar to that of the identified peacock image, and it would be judged to be the post-peacock mark. As a result, 203 images were identified. The accuracy of the deep learning model identified result was improved to 83%.

Table 1. Comparison of YOLOv3 and the method presented in this paper.

	YOLOv3 result	Paper result
Ident peacock	176	203
Total data	242	242
Accuracy	73%	84%

6 Conclusions

In this paper, we present a system based on the Gaussian Mixture Model and the Speeded Up Robust Feature to identify peacocks in images. This system has three main components. This first component is a deep learning training module. The second component is GMM get position module for finding the moving object, and third component is SURF-check object module for proving that the moving object is a peacock. We also implement the prototype of this article and conduct a series of experiments to test the proposed solution. The Experimental results show that the system improves the accuracy of the complete training model. If the deep laerning model which first module can not identify the object because the object changes its angle, this system can optimize the result that the model can't identify.

References

1. Bay, H., Tuytelaars, T., Van Gool, L.: SURF: speeded up robust features. In: Leonardis, A., Bischof, H., Pinz, A. (eds.) ECCV 2006. LNCS, vol. 3951, pp. 404–417. Springer, Heidelberg (2006). https://doi.org/10.1007/11744023_32
2. Diao, W., Sun, X., Zheng, X., Dou, F., Wang, H., Fu, K.: Efficient saliency-based object detection in remote sensing images using deep belief networks. IEEE Geosci. Remote Sens. Lett. **13**(2), 137–141 (2016)
3. Doulamis, N., Doulamis, A.: Semi-supervised deep learning for object tracking and classification. In: 2014 IEEE International Conference on Image Processing (ICIP), pp. 848–852, October 2014
4. Doulamis, N., Doulamis, A.: Fast and adaptive deep fusion learning for detecting visual objects. In: Fusiello, A., Murino, V., Cucchiara, R. (eds.) ECCV 2012. LNCS, vol. 7585, pp. 345–354. Springer, Heidelberg (2012). https://doi.org/10.1007/978-3-642-33885-4_35

5. Kaewtrakulpong, P., Bowden, R.: An improved adaptive background mixture model for realtime tracking with shadow detection. In: Remagnino, P., Jones, G.A., Paragios, N., Regazzoni, C.S. (eds.) Video-Based Surveillance Systems, pp. 135–144. Springer, Boston (2002). https://doi.org/10.1007/978-1-4615-0913-4_11

6. Liu, J., et al.: Colitis detection on abdominal CT scans by rich feature hierarchies. In: Tourassi, G.D., Armato III, S.G.A. (eds.) Medical Imaging 2016: Computer-Aided Diagnosis. International Society for Optics and Photonics, SPIE, vol. 9785, pp. 423–429 (2016)

7. Luo, G., An, R., Wang, K., Dong, S., Zhang, H.: A deep learning network for right ventricle segmentation in short-axis MRI. In: 2016 Computing in Cardiology Conference (CinC), pp. 485–488, September 2016

8. Nair, V., Hinton, G.E.: 3D object recognition with deep belief nets. In: Bengio, Y., Schuurmans, D., Lafferty, J.D., Williams, C.K.I., Culotta, A. (eds.) Advances in Neural Information Processing Systems 22, pp. 1339–1347 (2009)

9. Ouyang, W., et al.: DeepID-Net: object detection with deformable part based convolutional neural networks. IEEE Trans. Pattern Anal. Mach. Intell. **39**(7), 1320–1334 (2017)

10. Pinto, B., Anurenjan, P.R.: Video stabilization using speeded up robust features. In: 2011 International Conference on Communications and Signal Processing, pp. 527–531, February 2011

11. Rasmussen, C.E.: The infinite gaussian mixture model. In: Advances in Neural Information Processing Systems 12, pp. 554–560 (2000)

12. Redmon, J., Farhadi, A.: YOLOv3: an incremental improvement. CoRR abs/1804.02767 (2018)

13. Reynolds, D.A., Quatieri, T.F., Dunn, R.B.: Speaker verification using adapted Gaussian mixture models. Digit. Signal Proc. **10**(1), 19–41 (2000)

14. Zivkovic, Z.: Improved adaptive Gaussian mixture model for background subtraction, vol. 2, pp. 28–31 (2004)

15. Zivkovic, Z., van der Heijden, F.: Efficient adaptive density estimation per image pixel for the task of background subtraction. Pattern Recogn. Lett. **27**(7), 773–780 (2006)

A Lifelong Sentiment Classification Framework Based on a Close Domain Lifelong Topic Modeling Method

Thi-Cham Nguyen[1,2] ⓘ, Thi-Ngan Pham[1,3(✉)] ⓘ, Minh-Chau Nguyen[1,4] ⓘ,
Tri-Thanh Nguyen[1] ⓘ, and Quang-Thuy Ha[1] ⓘ

[1] Vietnam National University, Hanoi (VNU), VNU-University of Engineering and Technology (UET), No. 144, Xuan Thuy Street, Cau Giay District, Hanoi, Vietnam
nthicham@hpmu.edu.vn, ptngan2012@gmail.com,
chaunguyen.vnu@gmail.com, {ntthanh,thuyhq}@vnu.edu.vn
[2] Hai Phong University of Medicine and Pharmacy, Hai Phong, Vietnam
[3] The Vietnamese People's Police Academy, Hanoi, Vietnam
[4] Japan Advanced Institution of Science and Technology, Nomi, Japan

Abstract. In lifelong machine learning, the determination of the hypotheses related to the current task is very meaningful thanks to the reduction of the space to look for the knowledge patterns supporting for solving the current task. However, there are few studies for this problem. In this paper, we propose the definitions for measuring the "close domains to the current domain", and a lifelong sentiment classification method based on using the close domains for topic modeling the current domain. Experimental results on sentiment datasets of product reviews from Amazon.com show the promising performance of system and the effectiveness of our approach.

Keywords: Close domain · Lifelong topic modeling · Close topic

1 Introduction

The acquisition, representation and transferring domain knowledge, as well as focusing on learning bias approaches are the key scientific concerns that arise in lifelong machine learning (LML) [14]. The determination of only hypotheses related to the current task, instead of selecting all existing ones, is very meaningful thanks to the reduction of the space used to extract the knowledge patterns supporting to solve the current task (as depicted in Fig. 1). The datasets of previous tasks are of base-level hypothesis space. Determining previous tasks related to the current task in LML is very useful, not only efficient (i.e., computation complexity) but also effective (i.e., the performance of problem solving). The problem is also considered in the open-world machine learning (OWML), which is a form of LML. Open world recognition is a kind of OWML [8]. There are some studies for determining the previous tasks related to the current task, i.e., Bendale and Boult [1] proposed a solution for open world image recognition; Fei and Liu [10], Shu et al [12] focused on open world text classification.

© Springer Nature Switzerland AG 2020
N. T. Nguyen et al. (Eds.): ACIIDS 2020, LNAI 12033, pp. 575–585, 2020.
https://doi.org/10.1007/978-3-030-41964-6_50

Lifelong topic modeling (LTM) is a kind of *lifelong unsupervised learning* [8] based on using the hidden topic model [2, 3]. Chen, Z. & Liu, B proposed two algorithms for lifelong topic modeling. While LTM is suitable for the case the dataset of the current task is large enough [4], *Automatically generated Must-links and Cannot-links* (AMC) is suitable for the case the current dataset is small [5]. LML and AMC, in particular lifelong topic modeling, are very suitable for sentiment analysis (opinion mining) [9, 15].

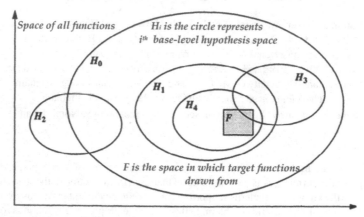

Fig. 1. H_4 is superior to all others: H_0, H_1, and H_4 include F completely, however, H_2 and H_3 do not; H_4 is smaller than H_1 and H_0 ($H_4 \subset H_1 \subset H_0$) [14].

In this paper, we propose a lifelong sentiment classification framework based on a close domain lifelong topic modeling method (CLTM), which uses only the data and knowledge from some of the previous domains (called close domains) for topic modeling the current task. This paper has three main contributions: (i) refine the topic modeling method on the domain level [11], (ii) propose a new definition of closeness measure between domains, (iii) and an application framework for sentiment classification.

The rest of this paper is organized as follows: In next section, two definitions of *close domain* are showed. A lifelong sentiment classification framework and proposed close domain lifelong topic modeling method arc described in Sect. 3. Section 4 shows the experiments and result evaluations. Some related works are analyzed and compared with this paper in Sect. 5. In the last section, we present conclusions and future work.

2   Close Domain

Let S be a text analytic system using lifelong machine learning. Let T_i be history tasks; and D_i be the corresponding dataset of T_i, where $i = 1, 2, .., N$. Let B be the *knowledge base*, which includes all knowledge, information from N previous tasks. B is empty when $N = 0$.

Let T_{N+1} be the current task with its dataset D_{N+1}. The problem is to determine a set D_{close} including previous datasets that are similar to D_{N+1} (called by "close datasets" or "close domains" with D_{N+1}), then apply the knowledge (in B) related to D_{close} for solving the current task T_{N+1}.

A measure of topic-based close domain was defined based on the closeness in three levels of vocabulary, top word and topic [11]. This paper proposes a new classifier-based close domain measure.

Assume that *the ith previous task T_i* be a binary classification problem with the classifier m_i, i.e., m_i is a model to predict whether a new review in i^{th} domain is a positive review or not. The idea for determining the previous domains that are related to the current domain in [1, 8, 10] is stated by two below definitions.

Definition 1. Let x be a document belonging to the current dataset D_{N+1}. The ith *previous dataset D_i* is called close to x with respect to *classifier m_i* iff $m_i(x)$ is *positive*.

Definition 2. (*previous dataset D_i close to the current dataset D_{N+1}*). The ith previous *dataset D_i* is called close to dataset D_{N+1} with respect to *classifier m_i* iff

$$\frac{|x \in D_{N+1} \wedge m_i(x) \, is \, positive|}{|D_{N+1}|} \geq \theta_{classifier}, \tag{1}$$

where $\theta_{classifier}$ is a predefined threshold for deciding whether two datasets are close to each other or not.

3   A Lifelong Sentiment Classification Framework Based on a Close Domain Lifelong Topic Modeling Method

3.1   Problem Formulation

Let S be a sentiment classification system with $T_1, T_2, \ldots, T_N$ be N previously solved sentiment classification tasks in various domains. Let D_i, V_i, and *Topics$_i$* be the dataset, the vocabulary, and the output topic set of T_i, correspondingly.

Let T_{N+1} be a new sentiment classification task (called the *current task*), with a dataset D_{N+1}. The problem is to solve the new sentiment classification task T_{N+1} by exploiting the previous knowledge base in B.

3.2   A Lifelong Sentiment Classification Framework

Figure 2 describes the proposed lifelong sentiment classification framework, in which sentiment documents in D_{N+1} are represented on an enriched topic set $Topics_{N+1}$ by a *Close domain Lifelong Topic Modeling* process (denoted by CLTM) as described in detail in Subsect. 3.3.

The hidden topic set $Topics_{N+1}$ constructed by CLTM is used to build a new feature set for the texts, i.e., a word in a document is replaced by its corresponding topic. For avoiding any exception leaks from the future, the testing Dataset D_{test} is not used in CLTM phase, i.e., for enriching the current topic set $Topics_{N+1}$. This solution has an important meaning in a lifelong machine learning, because the testing data should be

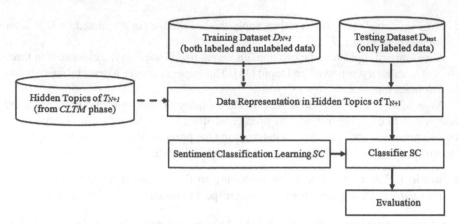

Fig. 2. The proposed lifelong sentiment classification framework

considered as they will be coming in the future. With the support of the knowledge from previous domains, the hidden topic set $Topics_{N+1}$ is adjusted and improved to be better than those extracted from LDA [6, 9].

Finally, a sentiment classifier is built to classify new documents.

3.3   Close Domain Lifelong Topic Modeling Method

Figure 3 describes a five-step procedure CLTM, which refines the procedure in [11].

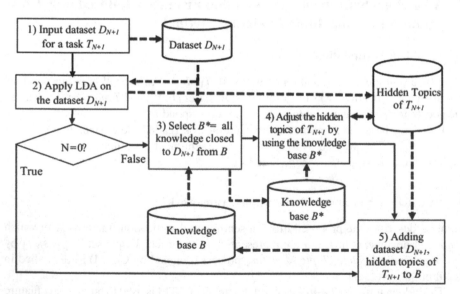

Fig. 3. The close domain lifelong topic modeling method

In the first step, the dataset D_{N+1} for the current task T_{N+1} is received. Step 2, if B is empty (i.e., $N = 0$), build a hidden topic model H on from D_{N+1} by using the

LDA solution, then H is used for solving the current task. After that, B is added all the information and knowledge gained from T_{N+1}.

In the case of N > 0, the knowledge base B is not empty, then hidden topic model H is enriched by exploiting B. Based on approach of learning bias on the domain level, only the information and knowledge of close domains to T_{N+1}, denoted B^*, is used for enrich hidden topic model H.

Step 3, select all the dataset D_i (from N previous datasets) which is close to D_{N+1} (topic-based or classifier-based), and exploit these datasets as well as their corresponding knowledge to form the temporary knowledge base B^* for improving the current hidden topic model H. In case there is not any previous dataset close to D_{N+1}, the whole knowledge base B plays the role of the temporary knowledge base B^*.

In Step (4), the hidden topics $Topics_{N+1}$ are adjusted by the knowledge base as in AMC [5] with two modifications of using the proposed close topic measure and previous topics to mine must-links and cannot-links. The main refining of this framework (as compared to the framework in [11]) includes in two points: (1) It uses new close domain measure to determine whether two domains are close or not. (2) The dataset D_{N+1} is not taken part in the Step (4) to ensure that the hidden topics $Topics_{N+1}$ are adjusted only by the previous knowledge.

After being built and enriched, $Topics_{N+1}$ and dataset D_{N+1}, are added to knowledge base B, and available as the input for other application. In other words, after the T_{N+1} has been solved, all of the knowledge found by it also be added to knowledge base B.

4   Experimental Results

4.1   The Datasets

Table 1. The names of 20 product domains and the proportion of the negative reviews in each domain.

Domain	The proportion negative reviews	Domain	The proportion negative reviews
Alarm Clock	30.51	Flashlight	11.69
Baby	16.45	GPS	19.50
Bag	11.97	Gloves	13.76
CableModem	12.53	GraphicsCard	14.58
Dumbbell	16.04	Headphone	20.99
HomeTheaterSystem	28.84	Projector	20.24
Jewelry	12.21	RiceCooker	18.64
Keyboard	22.66	Sandal	12.11
MagazineSubscriptions	26.88	Vacuum	22.07
MoviesTV	10.86	VideoGames	20.93

We use the dataset of 20 domains including the reviews from different types of products crawled from Amazon.com as in [9] to compare with their approach. The dataset information is listed in Table 1, in which, 19 domains are used as previous knowledge and the remaining domain is used as the current domain.

In [9], the authors spent 1000 reviews each domain and created two kinds of datasets of natural class distribution and balanced class distribution of the positive and negative reviews. For evaluation, they treated each domain as the target domain, and all the 19 remaining domains as the past domains. In our paper, we consider the case of naturally skewed class distribution and we use the same 19 datasets for the prior knowledge, however the dataset for the current domain is treated differently.

When analyzing all the 20 domains, we notice that the minimum and maximum number of positive reviews are 624 and 845 respectively; and the minimum and maximum number of negative reviews are 101 and 274 respectively. Therefore, the proportion of positive reviews per total reviews is around 69% to 89%. In an approximate way, we consider the positive reviews are as four times as the negative ones.

In our experiments, we do not use all 1000 reviews for the current domain. We create 5 different datasets with 20, 40, 60, 80 and 100 reviews for training to illustrate the effectiveness of the proposed approach with small amount of labeled data. In these datasets, the rate of positive is four times negative reviews. The only one testing dataset created from 100 reviews is used in all experiments.

4.2 Experimental Scenarios

For each current training dataset of 20, 40, 60, 80, 100 reviews, we perform three experimental scenarios:

- We extract features of hidden topic distribution from LDA model in the current dataset without using prior knowledge (from previous tasks). This scenario is considered as the baseline for comparison.
- We use all 19 domains as previous domains. This scenario illustrates the use of all prior knowledge without choosing meaningful information, i.e., the close domains.
- We find out the close domains, then use them as the previous knowledge instead of all 19 domains. This scenario is implemented to evaluate the idea of using close domain with valuable information to form knowledge base for the lifelong learning model.

In each scenario, we configured the number of hidden topics for the LDA and AMC models with different values, i.e., 10, 15 and 25. We also built different sentiment classification algorithms, i.e., Decision trees, k Nearest Neighbor (kNN), MultiLayer Perceptrons (MLP), and Naïve Bayes.

In our experiments, the popular measure of F1-scores is used to evaluate the performance of the system of sentiment classification.

4.3 Experimental Results and Discussions

The experimental results are shown in the Tables 2, 3, 4 and 5 in which the first scenario of using only the current domain with LDA model is denoted as LDA with the number of topics, i.e., LDA10, LDA 15, and LDA25; the second scenario of using all 19 domains for Lifelong model is denoted as LL_ALL with the number of topics, i.e., LL_ALL10, LL_ALL15, and LL_ALL25; and the last scenario of using found close domains for Lifelong model is denoted as CLTM with the number of topics, i.e., CLTM10, CLTM15, and CLTM25. The current domain is denoted as C with the number reviews of 20, 40, 60, 80 and 100. The results were collected in groups of scenario, classification algorithm, and the size of the current domain dataset. The best performance of the system when applying the proposed method are formatted in bold. The experiments, in which the baseline scenario using lifelong learning (with previous knowledge) outperform the scenario using only current domain (without previous knowledge), are highlighted with grey color.

Table 2. The experimental results (in F-measure) of the three scenarios with different configurations of Decision Trees classification algorithm and current domain dataset size.

Methods	Current domain dataset size				
	C=20	C=40	C=60	C=80	C=100
LDA10	76.62%	71.90%	74.12%	80.13%	75.09%
LL_ALL10	81.46%	81.41%	83.58%	85.11%	82.49%
CLTM10	75.30%	**81.45%**	**84.69%**	84.03%	**83.48%**
LDA15	76.74%	81.16%	80.26%	75.63%	75.90%
LL_ALL15	78.77%	81.17%	81.95%	82.52%	84.02%
CLTM15	78.33%	**81.31%**	81.40%	**85.99%**	**84.83%**
LDA25	71.41%	79.67%	84.31%	77.79%	79.43%
LL_ALL25	78.80%	81.81%	79.57%	82.42%	80.94%
CLTM25	77.94%	79.59%	83.53%	**84.20%**	**83.11%**

The results show that the two Lifelong models outperform the baseline in lots of experiments. And in many cases, the proposed method using only close domains improves the performance of system in comparison with the approach of using all 19 domains for the Lifelong model. This demonstrates that the prior knowledge from close domains may provide more meaningful contribution for the classification. The close domains may also remove noise, i.e., the data that is not related to current domain, hence, they help to improve the performance of the classification system. Because the reviews of products are often short sentences, and we use the current domain with quite small amount of reviews (from 20 to 100), the number of topics should be not too small or too big. The classification using MLP and Naïve Bayes algorithms in the proposed approach archives the best performance, especially with the number of topics of 15.

Table 3. The experimental results (in F-measure) of the three scenarios with different configurations of Naive Bayes classification algorithm and current domain dataset size.

Methods	Current domain dataset size				
	C=20	C=40	C=60	C=80	C=100
LDA10	63.33%	59.20%	52.73%	53.09%	54.51%
LL_ALL10	82.95%	80.19%	78.86%	79.95%	77.88%
CLTM10	81.28%	79.50%	78.05%	77.56%	**78.57%**
LDA15	67.66%	67.23%	59.99%	50.01%	53.25%
LL_ALL15	82.95%	74.01%	73.88%	76.92%	72.68%
CLTM15	**82.31%**	**78.32%**	**77.04%**	**77.30%**	**79.80%**
LDA25	70.56%	73.51%	69.26%	60.13%	59.02%
LL_ALL25	84.50%	78.96%	78.46%	74.17%	72.61%
CLTM25	**84.75%**	78.36%	75.84%	**76.38%**	**74.11%**

Table 4. The experimental results (in F-measure) of the three scenarios with different configurations of k Nearest Neighbours classification algorithm and current domain dataset size.

Methods	Current domain dataset size				
	C=20	C=40	C=60	C=80	C=100
LDA10	83.87%	81.36%	81.96%	85.17%	78.97%
LL_ALL10	86.91%	84.46%	83.47%	84.97%	83.83%
CLTM10	85.14%	**84.47%**	**84.74%**	84.61%	83.77%
LDA15	76.68%	84.88%	82.03%	85.64%	81.68%
LL_ALL15	86.25%	84.46%	83.30%	85.17%	84.42%
CLTM15	**87.33%**	84.65%	**84.24%**	85.35%	**84.70%**
LDA25	80.54%	86.72%	87.06%	80.42%	82.58%
LL_ALL25	87.06%	87.39%	84.76%	85.20%	84.50%
CLTM25	86.91%	85.14%	84.83%	**85.27%**	**85.23%**

When the current domain dataset is increased (to 80 and 100 reviews), the proposed method may get lower performance than baseline because the bigger amount of training data leads to more useful features than features from other domains. These results also illustrate the effectiveness of using the close domains (i.e., the reasonable amount of previous knowledge) instead of using all 19 domains (all previous knowledge) to support lifelong topic modelling.

Table 5. The experimental results (in F-measure) of the three scenarios with different configurations of MLP classification algorithm and current domain dataset size.

Methods	Current domain dataset size				
	C=20	C=40	C=60	C=80	C=100
LDA10	79.75%	84.46%	83.47%	81.15%	78.91%
LL_ALL10	80.45%	81.36%	83.65%	84.68%	82.63%
CLTM10	78.28%	81.53%	82.45%	**84.78%**	**84.67%**
LDA15	80.31%	84.71%	74.68%	77.09%	72.05%
LL_ALL15	80.30%	76.77%	76.62%	81.58%	80.77%
CLTM15	79.42%	78.75%	**80.53%**	**84.72%**	**84.86%**
LDA25	77.14%	80.77%	79.57%	78.57%	81.72%
LL_ALL25	82.75%	79.00%	79.30%	79.85%	76.80%
CLTM25	**83.00%**	78.58%	78.67%	78.78%	80.22%

5 Related Works

Chen, et al. [6, 9] proposed a lifelong supervised learning approach to sentiment classification (LSC: Lifelong Sentiment Classification) based on Naive Bayes. They exploited probability characteristics for constructing the Past Information Store (PIS) component. Two kinds of knowledge (document-level knowledge and domain-level knowledge) are calculated and used. The concept of domain-level knowledge is a form of close domain. Our paper also focuses on the idea of "close domain" of the current task, however, it uses an unsupervised learning approach for enriching hidden topic model based on learning bias domain-level as [11].

Bendale, Boult [1], Fei, Liu. [10] also considered the previous domains, however, they used the classifiers of previous tasks to determine "close domain" (This paper formalizes their idea in Definition 1 and Definition 2). Some forms of SVM algorithm are used for open-world classification.

Our work is also related to transfer learning [16–18]. However, in [16, 17], transfer learning is used for traditional supervised classification. And [18] using labeled data to build better fitting topic models. We only use previous knowledge to adjust current topic model without using any information of target labeled and unlabeled data as in general transfer learning.

This paper is considered as an upgraded version of [11] by adding a new classifier-based close domain measure. We also implement a lifelong topic modeling for enriching features of topic distribution for the current dataset. In addition, we provide another measure for determining the close domain based on classifier and set of thresholds learnt from past knowledge. The method mines the information of the label set of training data and the relationship among label and features.

6 Conclusions and Future Work

In this paper, we provide a lifelong sentiment classification framework based on a close domain lifelong topic modeling method (CLTM) as *learning bias approach on domain level*. In which, the close domain measure based on classifier is proposed with two definitions of the closeness in classifier between dataset and an element and between two datasets. The experimental results on 20 domains of product reviews crawled from Amazon.com, which are published availably and used in many researches demonstrate the effectiveness of the proposed method.

More experiments should be implemented on balanced class distribution datasets to evaluate the impact of training on the performance of system and compare the two close domain measures. We will take more consideration on designing a more robust solution that can still work well when the training dataset is small.

References

1. Bendale, A., Boult, T.: Towards open world recognition. In: Proceedings of the IEEE Conference on Computer Vision and Pattern Recognition, pp. 1893–1902 (2015)
2. Blei, D.M., Ng, A.Y., Jordan, M.I.: Latent dirichlet allocation. J. Mach. Learn. Res. **3**, 993–1022 (2003)
3. Blei, D.M.: Probabilistic topic models. Commun. ACM **55**(4), 77–84 (2012)
4. Chen, Z., Liu, B.: Topic modeling using topics from many domains, lifelong learning and big data. In: International Conference on Machine Learning, pp. 703–711 (2014)
5. Chen, Z., Liu, B.: Mining topics in documents: standing on the shoulders of big data. In: Proceedings of the 20th ACM SIGKDD International Conference on Knowledge Discovery and Data Mining, pp. 1116–1125 (2014)
6. Chen, Z., Ma, N., Liu, B.: Lifelong learning for sentiment classification. In: Proceedings of the 53rd Annual Meeting of the Association for Computational Linguistics and the 7th International Joint Conference on Natural Language Processing (volume 2: Short Papers), pp. 750–756 (2015)
7. Chen, Z., Liu, B.: Topic models for NLP applications. In: Sammut, C., Webb, G.I. (eds.) Encyclopedia of Machine Learning and Data Mining, 2nd edn, pp. 1276–1280. Springer, Boston (2017). https://doi.org/10.1007/978-1-4899-7687-1
8. Chen, Z., Ma, N., Liu, B.: Lifelong Machine Learning, 2nd edn. Morgan & Claypool Publishers, San Rafael (2018)
9. Chen, Z., Ma, N., Liu, B.: Lifelong Learning for Sentiment Classification. CoRR abs/1801.02808 (2018)
10. Fei, G., Liu, B.: Breaking the closed world assumption in text classification. In: Proceedings of the 2016 Conference of the North American Chapter of the Association for Computational Linguistics: Human Language Technologies, pp. 506–514 (2016)
11. Ha, Q.T., et al.: A new lifelong topic modeling method and its application to vietnamese text multi-label classification. In: Asian Conference on Intelligent Information and Database Systems, pp. 200–210 (2018)
12. Shu, L., Xu, H., Liu, B.: Unseen class discovery in open-world classification. arXiv preprint arXiv:1801.05609 (2018)
13. Thrun, S., Mitchell, T.M.: Lifelong robot learning. Robot. Auton. Syst. **15**(1–2), 25–46 (1995)
14. Thrun, S.: Explanation-Based Neural Network Learning: A Lifelong Learning Approach. Springer, US (1996). https://doi.org/10.1007/978-1-4613-1381-6

15. Wang, S., Chen, Z., Liu, B.: Mining aspect-specific opinion using a holistic lifelong topic model. In: Proceedings of the 25th International Conference on World Wide Web, pp. 167–176 (2016)
16. Xue, G.R., Dai, W., Yang, Q., Yu, Y.: Topic-bridged PLSA for cross-domain text classification. In: SIGIR, pp. 627–634 (2008)
17. Pan, S.J., Yang, Q.: A survey on transfer learning. IEEE Trans. Knowl. Data Eng. **22**(10), 1345–1359 (2010)
18. Kang, J.H., Ma, J., Liu, Y.: Transfer topic modeling with ease and scalability. In: SDM, pp. 564–575 (2012)

Multiple Model Approach to Machine Learning

Data Preprocessing for DES-KNN and Its Application to Imbalanced Medical Data Classification

Maciej Kinal and Michał Woźniak[✉][iD]

Department of Systems and Computer Networks, Faculty of Electronics,
Wrocław University of Science and Technology, Wrocław, Poland
michal.wozniak@pwr.edu.pl

Abstract. Learning from imbalanced data is a vital challenge for pattern classification. We often face the imbalanced data in medical decision tasks where at least one of the classes is represented by only a very small minority of the available data. We propose a novel framework for training base classifiers and preparing the dynamic selection dataset (DSEL) to integrate data preprocessing and dynamic ensemble selection (DES) methods for imbalanced data classification. DES-KNN algorithm has been chosen as the DES method and its modifications base on oversampled training and validations sets using SMOTE are discussed. The proposed modifications have been evaluated based on computer experiments carried out on 15 medical datasets with various imbalance ratios. The results of experiments show that the proposed framework is very useful, especially for tasks characterized by the small imbalance ratio.

Keywords: Dynamic ensemble selection · DES-KNN · Data preprocessing · Imbalanced data · Oversampling

1 Introduction

Learning classifiers from imbalanced data remains the focus of intense research due to its compliance with real-world decision tasks [10], where the most important or interesting classes are usually strongly underrepresented [11]. For example, only 3.59% of the patients suffer from heart disease, then a deterministic classifier which always decides that the patient is not sick may achieve an accuracy of 96.41% while having 0% sensitivity [8].

This work is an attempt to connect two of the important research directions, i.e., classifier ensemble approach [18] as well as imbalanced data analysis. We will concentrate on the classifier ensemble selection methods [1], which employ the idea of *overproduce-and-select*, i.e., for a given classification task we have more classifiers at our disposal and we are trying to exploit their local competences to classify an incoming sample. It needs to propose how these abilities of selected classifiers could be discovered. Basically, two approaches have been proposed:

© Springer Nature Switzerland AG 2020
N. T. Nguyen et al. (Eds.): ACIIDS 2020, LNAI 12033, pp. 589–599, 2020.
https://doi.org/10.1007/978-3-030-41964-6_51

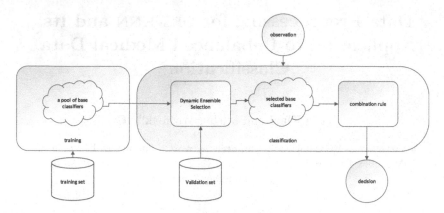

Fig. 1. Idea of DES (dynamic ensemble selection).

- *Static selection*, where a feature space is prior divided into competence areas and the decision about a new instance is made by a classifier assigned to the decision area, which includes the instance. The idea of *Classifier and Selection* algorithm has been formulated by Kuncheva [12]. Then ADASS (*Adaptive Splitting and Selection*) has been proposed by Jackowski et al. [7], who employed an evolutionary algorithm to find the best feature space split and classifier assigning to each created partition.
- *Dynamic Selection* does not divide the feature space before classification [16], but it looks for the best classifier for each incoming instance. Dynamic Ensemble Selection (DES) [5] uses the notion of competence to select the best models to classify a given instance. Usually, the competence of a base classifier is estimated on the basis of *dynamic selection dataset* (DSEL). The idea of DES is presented in Fig. 1. There are several popular strategies for employing DES, as:
  - KNORA-E forms the ensemble based on classifiers which can correctly classify all samples within the local region of competence, while KNORA-U decides based on weighted voting, where the weight assigned to each base classifier equals to the number of correctly classified samples in DESL [9].
  - DES-KNN [17] ranks individual classifiers according to their prediction performance and then the fixed number of the best classifiers are first selected. The final ensemble is formed based on the fixed number of the most diverse preselected individuals.
  - DES-Clustering [17] employs the *K-means* to define DESL, then the most accurate and diverse classifiers ale selected for the ensemble.

This paper also focuses on the data imbalance. One of the most popular and efficient approaches to combat this phenomenon is data preprocessing. One has to mention random replication of minority class examples (*Random Oversampling*) or random removing the majority class example (*Random Undersampling*), but the most popular method is SMOTE (*Synthetic Minority Over-sampling Technique*) [3], which generates new instances from existing minority samples in their neighbourhoods. In this work SMOTE is used to oversample the training and/or validation sets for DES-KNN.

To summarize, this work is focusing on dynamic ensemble selection based on DES-KNN used to mitigate the difficulties related to skewed class distribution using data preprocessing approach. In a nutshell, the main contributions of this work are as follows:

- Presentation of several DES-KNN modifications, which employ oversampled training and/or DESL (*dynamic selection dataset*).
- Experimental evaluation of the discussed approaches based on imbalanced medical datasets.

2   The Proposed Framework

Let's shortly present the proposed modifications of DES-KNN:

- The original version of DES-KNN, where training set used to train the base classifiers and DESL are not preprocessed (denoted as DES-KNN I).
- DES-KNN which uses oversampled training set by SMOTE but DESL is not modified (denoted as DES-KNN II).
- DES-KNN which uses oversampled DESL and original training set. The pseudocode of this modification is presented in Algorithm 1 (denoted as DES-KNN III).
- DES-KNN which uses oversampled DESL as well training set (denoted as DES-KNN IV).

Let's shortly describe pseudocode of DES-KNN III presented in Algorithm 1. The DESL is modified only in the case if at least two minority samples are found in the neighbourhood of a classified instance x. It should protect against homogeneous region of the majority class oversampling. The algorithm checks as many oversampled versions of DSEL as many minority samples are found in the neighbourhood. As the parameter to sampling strategy (oversampling ratio) the exp_IR (presented in line 7) is used (the minimum value 0.75 has been set experimentally). The algorithm chooses the best DSEL according to *gmean* metric [2].

Algorithm 1. Pseudocode for the DES-KNN III (where DESL is overampled)

Input:
   $\mathcal{VS}$ − *validationset*
   x - instance to be classifier
   k - number of neighbours for $DSEL$
1: *gmean* ← 0
2: *label* ← *majority_class*
3: $DSEL$ ← k nearest neighbours of x from $\mathcal{VS}$
4: **if** number of minority class instances in $DSEL$ >1 **then**
5:    **for** each minority instance x_i in $DESL$ **do**
6:       x_j ← nearest neighbour of x_i in $DSEL$ from minority class
7:       $exp\_IR \leftarrow min\left(0.75, \frac{3d(x_i,x_j)}{2(d(x_i,x_j)+1)}\right)$
8:       $DSEL_t$ ← SMOTE($DSEL, exp\_IR$)
9:       $label_t$ ← DES_KNN($x, DSEL_t$)
10:      $gmean_t$ ← gmean of DES_KNN using $DSEL_t$ on $\mathcal{VS}$
11:      **if** $gmean_t > gmean$ **then**
12:         $gmean$ ← $gmean_t$
13:         $label$ ← $label_t$
14:   **end for**
    return *label*

3   Experimental Evaluation

The objective of the experiment is to analyze how the combinations of DES-KNN and oversampling perform in terms of classifying medical imbalanced data with various imbalance ratios.

3.1   Experimental Setup

For training and testing of examined classifiers, we employed a stratified 10-fold cross validation. We evaluate the performance of classifiers on medical imbalanced data using four dedicated skew-insensitive metrics: g-mean (the geometric mean of sensitivity and precision), precision, sensitivity (recall), and F1 Score (the harmonic mean of precision and sensitivity). Additionally, accuracy was calculated for each model.

To offer a more detailed comparison among methods, we have conducted the statistical tests of significance. For pairwise comparison we have employed Wilcoxon signed rank test [6] used confidence level $\alpha = 0.05$.

Experiments were implemented in Python programming language and the following libraries: *scikit-learn* [15], *DESLib* [4], *NumPy* and *Pandas*. The source code is available on *Github*[1].

Neighborhood size for DES-KNN methods is $k = 13$.

---

[1] https://github.com/makonwencjusz/praca_magisterska.

Table 1. Datasets description

Id	Dataset	#instances	#attributes	#min/#maj	IR
1	Cardiovascular disease	70000	10	34979/35021	1,00
2	Diabetic retinopathy debrecen	1151	19	540/611	1,13
3	EEG eye state	14980	14	6723/8257	1,23
4	Heart disease	270	13	120/150	1,25
5	MRI and Alzheimers	336	10	146/190	1,30
6	Breast cancer wisconsin (diagnosis)	569	30	212/357	1,68
7	Breast cancer wisconsin (original)	683	9	239/444	1,86
8	Pima Indians diabetes	768	8	268/500	1,87
9	Yeast1	1484	8	429/1055	2,46
10	Indian liver patient records	579	10	165/414	2,51
11	SPECT heart	267	22	55/212	3,85
12	Thyroid disease (new_thyroid1)	215	5	35/180	5,14
13	Thoracic surgery	470	16	70/400	5,71
14	Ecoli3	336	7	35/301	8,60
15	Cervical cancer risk	725	30	50/675	13,50

To ensure an appropriate diversity of the classifier pool we decided to use the following different models of base classifiers:

- 3 Nearest Neighbours,
- 5 Nearest Neighbours,
- SVM with linear kernel and penalty $C = 0.025$,
- SVM with linear kernel and penalty $C = 1$,
- SVM with RBF kernel and $\gamma = 2$ and penalty $C = 0.025$,
- Gaussian Process Classifier based on Laplace approximation with RBF kernel and $\gamma = 2$,
- Decision Tree (CART) with the maximum depth $= 5$,
- Random Forest with a pool of 10 Decision Trees (CART) with maximum depth $= 5$,
- MultiLayer Perceptron,
- AdaBoost using Decision Tree (CART) with maximum depth $= 5$ as a base classifier.

The proposed framework was evaluated using 15 benchmark data sets form medical domain and their description is presented in Table 1.

The results of experiments are presented in Tables 2, 3, 4, 5 and Fig. 2. The statistical evaluation based on Wicoxon signed-rank test is shown in Table 6.

3.2 Lessons Learned

Based on the statistical analysis, we may see that oversampling can improve the prediction performance of DES-KNN, especially if DESL is preprocessed.

Table 2. Predictive performance qualities of original DES-KNN (ds stands for dataset id according to Table 1)

ds	Accuracy	g-mean	Precision	Sensitivity	F1 score
1	0.63 (± 0.01)	0.62 (± 0.01)	0.66 (± 0.01)	0.52 (± 0.02)	0.58 (± 0.01)
2	0.76 (± 0.06)	0.76 (± 0.06)	0.80 (± 0.08)	0.74 (± 0.06)	0.77 (± 0.06)
3	0.76 (± 0.02)	0.75 (± 0.02)	0.74 (± 0.02)	0.87 (± 0.02)	0.80 (± 0.02)
4	0.86 (± 0.08)	0.85 (± 0.08)	0.85 (± 0.11)	0.91 (± 0.07)	0.87 (± 0.07)
5	0.99 (± 0.01)	0.99 (± 0.01)	0.98 (± 0.03)	1.00 (± 0.00)	0.99 (± 0.02)
6	0.96 (± 0.03)	0.95 (± 0.05)	0.97 (± 0.03)	0.93 (± 0.10)	0.95 (± 0.06)
7	0.97 (± 0.02)	0.97 (± 0.02)	0.96 (± 0.04)	0.97 (± 0.04)	0.96 (± 0.02)
8	0.82 (± 0.04)	0.74 (± 0.08)	0.83 (± 0.08)	0.60 (± 0.14)	0.69 (± 0.10)
9	0.79 (± 0.04)	0.61 (± 0.10)	0.81 (± 0.08)	0.40 (± 0.13)	0.53 (± 0.11)
10	0.74 (± 0.06)	0.85 (± 0.20)	0.74 (± 0.06)	0.97 (± 0.02)	0.84 (± 0.04)
11	0.77 (± 0.10)	0.73 (± 0.13)	0.78 (± 0.16)	0.78 (± 0.24)	0.75 (± 0.15)
12	0.99 (± 0.02)	0.98 (± 0.06)	1.00 (± 0.00)	0.97 (± 0.10)	0.98 (± 0.06)
13	0.85 (± 0.06)	0.14 (± 0.22)	0.30 (± 0.46)	0.07 (± 0.12)	0.11 (± 0.18)
14	0.96 (± 0.05)	0.69 (± 0.38)	0.70 (± 0.40)	0.62 (± 0.40)	0.63 (± 0.37)
15	0.96 (± 0.03)	0.79 (± 0.30)	0.71 (± 0.29)	0.74 (± 0.33)	0.69 (± 0.28)

Table 3. Predictive performance qualities of original DES-KNN II (ds stands for dataset id according to Table 1)

ds	Accuracy	g-mean	Precision	Sensitivity	F1 score
1	0.63 (± 0.01)	0.62 (± 0.01)	0.66 (± 0.01)	0.51 (± 0.02)	0.58 (± 0.01)
2	0.75 (± 0.07)	0.75 (± 0.07)	0.85 (± 0.08)	0.65 (± 0.09)	0.74 (± 0.08)
3	0.77 (± 0.01)	0.78 (± 0.01)	0.83 (± 0.02)	0.73 (± 0.04)	0.78 (± 0.02)
4	0.86 (± 0.07)	0.85 (± 0.08)	0.85 (± 0.11)	0.90 (± 0.07)	0.87 (± 0.08)
5	0.99 (± 0.01)	0.99 (± 0.01)	0.98 (± 0.03)	1.00 (± 0.00)	0.99 (± 0.02)
6	0.96 (± 0.03)	0.95 (± 0.04)	0.95 (± 0.03)	0.95 (± 0.08)	0.95 (± 0.05)
7	0.97 (± 0.02)	0.97 (± 0.02)	0.95 (± 0.03)	0.97 (± 0.03)	0.96 (± 0.03)
8	0.81 (± 0.05)	0.78 (± 0.08)	0.69 (± 0.10)	0.75 (± 0.15)	0.72 (± 0.11)
9	0.77 (± 0.04)	0.76 (± 0.05)	0.61 (± 0.06)	0.73 (± 0.10)	0.66 (± 0.05)
10	0.70 (± 0.05)	0.72 (± 0.05)	0.88 (± 0.06)	0.68 (± 0.07)	0.76 (± 0.04)
11	0.75 (± 0.09)	0.72 (± 0.13)	0.72 (± 0.20)	0.69 (± 0.21)	0.69 (± 0.19)
12	0.99 (± 0.02)	1.00 (± 0.01)	0.97 (± 0.10)	1.00 (± 0.00)	0.98 (± 0.06)
13	0.83 (± 0.06)	0.54 (± 0.14)	0.48 (± 0.18)	0.34 (± 0.15)	0.38 (± 0.16)
14	0.91 (± 0.05)	0.94 (± 0.05)	0.53 (± 0.28)	0.97 (± 0.10)	0.63 (± 0.22)
15	0.97 (± 0.03)	0.90 (± 0.14)	0.75 (± 0.22)	0.85 (± 0.23)	0.77 (± 0.18)

Table 4. Predictive performance qualities of original DES-KNN III (ds stands for dataset id according to Table 1)

ds	Accuracy	g-mean	Precision	Sensitivity	F1 score
1	0.92 (± 0.06)	0.92 (± 0.06)	0.97 (± 0.06)	0.87 (± 0.10)	0.91 (± 0.07)
2	0.92 (± 0.06)	0.92 (± 0.07)	0.92 (± 0.08)	0.93 (± 0.08)	0.92 (± 0.06)
3	1.00 (± 0.00)	1.00 (± 0.00)	1.00 (± 0.00)	1.00 (± 0.00)	1.00 (± 0.00)
4	0.91 (± 0.09)	0.90 (± 0.12)	0.92 (± 0.11)	0.93 (± 0.08)	0.92 (± 0.08)
5	1.00 (± 0.00)	1.00 (± 0.00)	1.00 (± 0.00)	1.00 (± 0.00)	1.00 (± 0.00)
6	0.98 (± 0.03)	0.98 (± 0.03)	0.99 (± 0.03)	0.98 (± 0.04)	0.99 (± 0.02)
7	0.99 (± 0.03)	0.97 (± 0.09)	0.99 (± 0.04)	0.95 (± 0.15)	0.96 (± 0.10)
8	0.92 (± 0.07)	0.92 (± 0.07)	0.97 (± 0.07)	0.87 (± 0.12)	0.91 (± 0.08)
9	0.89 (± 0.07)	0.87 (± 0.08)	0.91 (± 0.12)	0.84 (± 0.13)	0.87 (± 0.09)
10	0.76 (± 0.09)	0.62 (± 0.23)	0.72 (± 0.09)	0.95 (± 0.08)	0.82 (± 0.07)
11	0.79 (± 0.09)	0.76 (± 0.12)	0.84 (± 0.16)	0.68 (± 0.19)	0.74 (± 0.16)
12	0.99 (± 0.02)	0.99 (± 0.02)	1.00 (± 0.00)	0.99 (± 0.04)	0.99 (± 0.02)
13	0.76 (± 0.15)	0.27 (± 0.36)	0.40 (± 0.49)	0.20 (± 0.31)	0.25 (± 0.34)
14	0.91 (± 0.08)	0.58 (± 0.42)	0.67 (± 0.47)	0.52 (± 0.40)	0.57 (± 0.42)
15	0.92 (± 0.08)	0.72 (± 0.37)	0.67 (± 0.38)	0.68 (± 0.39)	0.65 (± 0.35)

Table 5. Predictive performance qualities of original DES-KNN IV (ds stands for dataset id according to Table 1)

ds	Accuracy	g-mean	Precision	Sensitivity	F1 score
1	0.93 (± 0.05)	0.94 (± 0.06)	0.97 (± 0.07)	0.92 (± 0.09)	0.94 (± 0.05)
2	0.91 (± 0.08)	0.90 (± 0.08)	0.96 (± 0.07)	0.86 (± 0.12)	0.90 (± 0.08)
3	1.00 (± 0.00)	1.00 (± 0.00)	1.00 (± 0.00)	1.00 (± 0.00)	1.00 (± 0.00)
4	0.90 (± 0.09)	0.90 (± 0.09)	0.91 (± 0.12)	0.92 (± 0.09)	0.91 (± 0.08)
5	1.00 (± 0.00)	1.00 (± 0.00)	1.00 (± 0.00)	1.00 (± 0.00)	1.00 (± 0.00)
6	0.98 (± 0.05)	0.98 (± 0.05)	0.97 (± 0.08)	0.99 (± 0.04)	0.98 (± 0.06)
7	0.99 (± 0.02)	0.99 (± 0.02)	0.99 (± 0.04)	1.00 (± 0.00)	0.99 (± 0.02)
8	0.91 (± 0.07)	0.91 (± 0.07)	0.92 (± 0.09)	0.91 (± 0.11)	0.91 (± 0.07)
9	0.93 (± 0.06)	0.93 (± 0.07)	0.89 (± 0.13)	0.96 (± 0.06)	0.92 (± 0.08)
10	0.77 (± 0.12)	0.69 (± 0.27)	0.87 (± 0.10)	0.70 (± 0.22)	0.76 (± 0.17)
11	0.78 (± 0.09)	0.73 (± 0.12)	0.81 (± 0.18)	0.73 (± 0.24)	0.73 (± 0.16)
12	0.99 (± 0.02)	1.00 (± 0.01)	0.95 (± 0.15)	1.00 (± 0.00)	0.97 (± 0.10)
13	0.83 (± 0.12)	0.74 (± 0.20)	0.86 (± 0.19)	0.62 (± 0.29)	0.68 (± 0.25)
14	0.81 (± 0.12)	0.78 (± 0.28)	0.52 (± 0.33)	0.87 (± 0.31)	0.60 (± 0.31)
15	0.95 (± 0.07)	0.87 (± 0.29)	0.69 (± 0.34)	0.90 (± 0.30)	0.76 (± 0.30)

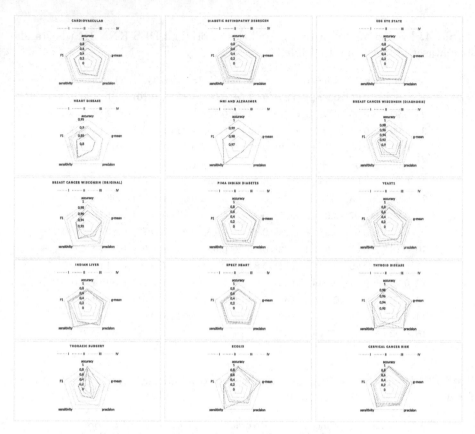

Fig. 2. Comparison of the proposed methods for 5 different metrics and each dataset.

Table 6. Wilcoxon rank test results. Yellow color denotes that the first model statistically significantly outperforms the second one, while green color denotes that the second model is significantly better than the first one, and the red color means that no statistical difference between models is confirmed.

	Accuracy	g-mean	Precision	Sensitivity	F1
DES-KNN I vs DES-KNN II	0.0490	0.0124	0.7835	0.3822	0.4406
DES-KNN I vs DES-KNN III	0.0549	0.0131	0.0109	0.0952	0.0353
DES-KNN I vs DES-KNN IV	0.0300	0.0010	0.0279	0.0156	0.0183
DES-KNN II vs DES-KNN III	0.0213	0.5720	0.0533	0.3304	0.0734
DES-KNN II vs DES-KNN IV	0.0173	0.0411	0.0097	0.0071	0.0062
DES-KNN III vs DES-KNN IV	0.0679	0.4558	0.6888	0.0864	0.2710

Interestingly, there are no statistically differences if only training set or DESL is oversampled (DES-KNN II vs. DES-KNN III and DES-KNN III vs. DES-KNN IV). But, if both training set and DESL are modified then it is possible to outperform methods based on oversampled training set only.

In general, the order of the presented approaches in terms of performance, starting with the worst, is as follows: (1) DES-KNN without the use of any pre-processing methods (DES-KNN I) → (2) DES-KNN using oversampled training set (DES-KNN II) → (3) DES-KNN using oversampled DESL (DES-KNN III i IV). There are no significant differences between DES-KNN III and IV.

Analyzing the radar charts we may observe, that the higher the imbalance ratio, the smaller the differences between approaches. For high imbalanced datasets the predictive performances among tested methods are not clearly visible. Probably, in such cases, the generated synthetic instances cannot allow to better estimate the minority class distributions, because SMOTE uses to small, non-representative sample or instances from the minority class may form small clusters of an unknown structure that are scattered [13].

The experimental results allow us to conclude that using oversampled training set and/or DSEL can improve the predictive performance of DES-KNN, especially for the datasets characterized by not so high imbalance ratio.

4 Conclusions

The main goal of this work was to propose a novel framework for training base classifiers and preparing the *dynamic selection dataset* (DSEL) for the dynamic ensemble selection employing DES-KNN for imbalanced data classification.

The computer experiments confirmed the effectiveness of the proposed framework and based on the statistical analysis we may conclude that DES-KNN coupled with SMOTE is statistically significantly better than the approaches that do not combine both of these concepts.

This work is a step forward towards the use of dynamic ensemble selection using preprocessing techniques for imbalanced data classification tasks. Results obtained in this study encourage us to continue works on alternative approaches to oversampling imbalanced data, with a special focus on the following issues:

- Using different methods of data preprocessing, as SMOTE extensions or method of data undersampling.
- Adapt the proposed concept to different classifier ensemble selection methods as KNORA-U, KNORA-A, etc.
- Employ imbalance distribution metrics (taxonomy based on 5-Nearest Neighbours) to improve the efficiency of the data oversampling [14].
- Evaluate how the proposed framework is robust to label and feature noise.
- Employing proposed concept to non-stationary data stream analysis [19, 20].

Acknowledgement. This work was supported by the Polish National Science Centre under the grant No. 2017/27/B/ST6/01325 as well as by the statutory funds of the Department of Systems and Computer Networks, Faculty of Electronics, Wroclaw University of Science and Technology.

References

1. Burduk, R.: Integration base classifiers based on their decision boundary. In: Rutkowski, L., Korytkowski, M., Scherer, R., Tadeusiewicz, R., Zadeh, L.A., Zurada, J.M. (eds.) ICAISC 2017. LNCS (LNAI), vol. 10246, pp. 13–20. Springer, Cham (2017). https://doi.org/10.1007/978-3-319-59060-8_2
2. Burduk, R., Kasprzak, A.: The use of geometric mean in the process of integration of three base classifiers. In: Saeed, K., Homenda, W. (eds.) CISIM 2018. LNCS, vol. 11127, pp. 246–253. Springer, Cham (2018). https://doi.org/10.1007/978-3-319-99954-8_21
3. Chawla, N.V., Bowyer, K.W., Hall, L.O., Kegelmeyer, W.P.: SMOTE: synthetic minority over-sampling technique. J. Artif. Int. Res. **16**(1), 321–357 (2002)
4. Cruz, R.M.O., Hafemann, L.G., Sabourin, R., Cavalcanti, G.D.C.: DESlib: a dynamic ensemble selection library in Python. arXiv preprint arXiv:1802.04967 (2018)
5. Cruz, R.M., Sabourin, R., Cavalcanti, G.D.: Dynamic classifier selection: recent advances and perspectives. Inf. Fusion **41**(C), 195–216 (2018)
6. García, S., Fernández, A., Luengo, J., Herrera, F.: Advanced nonparametric tests for multiple comparisons in the design of experiments in computational intelligence and data mining: experimental analysis of power. Inf. Sci. **180**(10), 2044–2064 (2010)
7. Jackowski, K., Krawczyk, B., Woźniak, M.: Improved adaptive splitting and selection: the hybrid training method of a classifier based on a feature space partitioning. Int. J. Neural Syst. **24**(3), 1430007 (2014)
8. Khalilia, M., Chakraborty, S., Popescu, M.: Predicting disease risks from highly imbalanced data using random forest. BMC Med. Inform. Decis. Mak. **11**(1), 51 (2011)
9. Ko, A.H., Sabourin, R., Alceu Souza Britto, J.: From dynamic classifier selection to dynamic ensemble selection. Pattern Recogn. **41**(5), 1718–1731 (2008)
10. Krawczyk, B.: Learning from imbalanced data: open challenges and future directions. Prog. Artif. Intell. **5**(4), 221–232 (2016)
11. Ksieniewicz, P., Woźniak, M.: Dealing with the task of imbalanced, multidimensional data classification using ensembles of exposers. In: Torgo, L., Krawczyk, B., Branco, P., Moniz, N. (eds.) Proceedings of the First International Workshop on Learning with Imbalanced Domains: Theory and Applications. Proceedings of Machine Learning Research, vol. 74, pp. 164–175. PMLR, ECML-PKDD, Skopje, Macedonia, 22 September 2017. http://proceedings.mlr.press/v74/ksieniewicz17a.html
12. Kuncheva, L.I.: Clustering-and-selection model for classifier combination. In: Fourth International Conference on Knowledge-Based Intelligent Information Engineering Systems & Allied Technologies, KES 2000, Brighton, UK, 30 August–1 September 2000, Proceedings, 2 Volumes, pp. 185–188 (2000)
13. Napierala, K., Stefanowski, J.: Identification of different types of minority class examples in imbalanced data. In: Corchado, E., Snášel, V., Abraham, A., Woźniak, M., Graña, M., Cho, S.-B. (eds.) HAIS 2012. LNCS (LNAI), vol. 7209, pp. 139–150. Springer, Heidelberg (2012). https://doi.org/10.1007/978-3-642-28931-6_14
14. Napierala, K., Stefanowski, J.: Types of minority class examples and their influence on learning classifiers from imbalanced data. J. Intell. Inf. Syst. **46**, 563–597 (2015)
15. Pedregosa, F., et al.: Scikit-learn: machine learning in Python. J. Mach. Learn. Res. **12**, 2825–2830 (2011)

16. Rejer, I., Burduk, R.: Classifier selection for motor imagery brain computer interface. In: Saeed, K., Homenda, W., Chaki, R. (eds.) CISIM 2017. LNCS, vol. 10244, pp. 122–130. Springer, Cham (2017). https://doi.org/10.1007/978-3-319-59105-6_11

17. Soares, R.G.F., Santana, A., Canuto, A.M.P., de Souto, M.C.P.: Using accuracy and diversity to select classifiers to build ensembles. In: The 2006 IEEE International Joint Conference on Neural Network Proceedings, pp. 1310–1316, July 2006

18. Woźniak, M., Graña, M., Corchado, E.: A survey of multiple classifier systems as hybrid systems. Inf. Fusion **16**, 3–17 (2014)

19. Woźniak, M., Cyganek, B., Kasprzak, A., Ksieniewicz, P., Walkowiak, K.: Active learning classifier for streaming data. In: Martínez-Álvarez, F., Troncoso, A., Quintián, H., Corchado, E. (eds.) HAIS 2016. LNCS (LNAI), vol. 9648, pp. 186–197. Springer, Cham (2016). https://doi.org/10.1007/978-3-319-32034-2_16

20. Zyblewski, P., Ksieniewicz, P., Woźniak, M.: Classifier selection for highly imbalanced data streams with *Minority Driven Ensemble*. In: Rutkowski, L., Scherer, R., Korytkowski, M., Pedrycz, W., Tadeusiewicz, R., Zurada, J.M. (eds.) ICAISC 2019. LNCS (LNAI), vol. 11508, pp. 626–635. Springer, Cham (2019). https://doi.org/10.1007/978-3-030-20912-4_57

Novel Approach to Gentle AdaBoost Algorithm with Linear Weak Classifiers

Robert Burduk$^{(\boxtimes)}$, Wojciech Bożejko$^{(\boxtimes)}$, and Szymon Zacher$^{(\boxtimes)}$

Faculty of Electronic, Wroclaw University of Science and Technology,
Wybrzeze Wyspianskiego 27, 50-370 Wroclaw, Poland
{robert.burduk,wojciech.bozejko,szymon.zacher}@pwr.wroc.pl

Abstract. This paper presents the problem of calculating the value of the scoring function for weak classifiers operating in the sequential structure. An example of such a structure is Gentle AdaBoost algorithm whose modification we propose in this work. In the proposed approach the distance of the object from the decision boundary is scaled in decision regions defined by the weak classifier at first and later transformed by the log-normal function. The described algorithm was tested on sixth public available data sets and compared with Gentle AdaBoost algorithm.

Keywords: Gentle Boost algorithm · Distance to the decision boundary · Score function

1 Introduction

In general, the classifier maps the feature space into a set of class labels. The class label is not only one of the possible types defined as the classifier output. Another type of the classifier output is the classifier score or the support function. In the case of a linear classifier, in particular a linear SVM, the classifier score is proportional to the distance between the instance and the decision boundary. A support function represents a certain value that refers to the degree of membership of the instance in a class label.

Classifier calibration is concerned with the scale on which a classifier's scores are expressed. In general, the calibration converts scores function into a support function, or more precisely transforms classifier outputs into values that can be interpreted as probabilities. The sigmoidal transformation maps the score of a classifier to a calibrated probability output as was proposed by Platt [1].

Ensemble learning (or ensemble of classifiers - EoC) combining several machine learning models into one predictive model. The reasons for the use of EoC include for example, the fact that single classifiers are often unstable (small changes in input data may result in creation of very different decision boundaries).

The three main stages in the process of building EoC include the acquisition of the empirical material (i.e. collection of examples used in the process

© Springer Nature Switzerland AG 2020
N. T. Nguyen et al. (Eds.): ACIIDS 2020, LNAI 12033, pp. 600–611, 2020.
https://doi.org/10.1007/978-3-030-41964-6_52

of creating the system), creation of base classifiers, and their selection for the ensemble [5]. The final stage of the operation of the classifier ensemble is integrating the responses of ensemble's members in order to obtain a clear-cut decision about the membership of an object in a given class label [3]. It should also be emphasized that the selection is optional and is not always considered when building complex classification systems.

The information obtained at the output of a base classifier can be divided into two groups [2]:

- Abstract level (response space)—the output of the base classifier Ψ is the class label assigned to the instance x.
- Support level (support function space)—the base classifier output represents a certain value that refers to the degree of membership of a class label in the input data of the classifier that is the instance x. An example of this type of an output is *a posteriori* probability that determines the membership of an object in a given class label.

With respect to the levels presented here, in the literature there are many proposals of methods for combining the outputs of base classifiers in order to obtain a single class label. In relation to the response space, it is worth to mention, for example, the following works [3–5]. Integration in the support function space is discussed in many works—both of the analytical nature [6–8] and practical nature [9,10].

Another way to categorize EoC is by the structure of how weak learners are combined. In the parallel structure EoC the base learners are generated parallelly and usually heterogeneous weak learners are used. The boosting approach [11] is an example of sequential EoC where a base learner depends on the previous ones. This relationship is represented by a classification error that is measured in the individual iteration of the algorithm. In particular, incorrectly classified objects are assigned greater weights in the next iteration of the algorithm. Various concepts based on the boosting idea were presented in many papers [12–16] also in articles on practical applications [17–19].

In this paper we consider a modification of Gentle AdaBoost algorithm where the proposed modification concerns the calculation of the scoring function. The value of the scoring function depends on the modified distance of the object from the decision boundary defined by the linear base classifier. In addition, no method of transforming the score function to the support function is used and in each iteration of the proposed algorithm each competence region defined by the weak classifier of boosting algorithm is scaled to range $[0, 1]$.

This paper is organized as follows: Sect. 2 presents the details of our modification of Gentle AdaBoost algorithm. In Sect. 3 the experiment results comparing Gentle AdaBoost with our algorithm are presented. Finally, some conclusions and the proposition of future work are presented.

2   Gentle AdaBoost Algorithm with Distance Score Function

Gentle AdaBoost is a type of the boosting algorithm and it calculates weak models by optimizing the weighted least square error iteratively [20]. In this section, we proposed the modification of Gentle AdaBoost algorithm. In particular, the method of calculating the weights of objects as well as the weights of the basic classifiers is not changed. However, base classifiers are not calculated by optimizing the weighted least square error iteratively. We propose that each base classifier has a scoring function, which depends on the distance to the decision boundary defined by this base classifier. Additionally, each competence region defined by the weak classifier in each iteration of boosting algorithm is scaled to range $[0, 1]$. Considering that the features are also normalized the maximum distance of the object x to the decision boundary is equal 1. In order to calculate the value of the scoring function, the distance $d(x)$ of the object x to the decision boundary is transformed by the function

$$SF(x) = \frac{1}{3\sqrt{2\pi}d(x)} \exp\left(-\frac{\ln((d(x)) - \mu)^2}{4}\right) \tag{1}$$

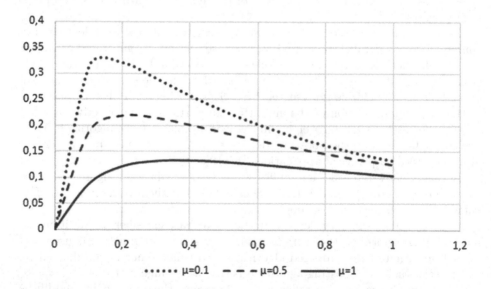

Fig. 1. An example of SF function for three different μ parameter values

The converting function (1) is similar to the lognormal function, however our goal is not to provide a probabilistic interpretation of the scoring function. Additionally, the scoring function is not calibrated to actual probability. The meaning of the function (1) is as follows: an object x very close to the decision boundary has a very small value of the scoring function, the value of SF increases quite quickly as the object's distance $d(x)$ from the decision boundary increases, then the value of SF decreases, but not as fast as it increased. The SF function for three different values of μ parameter is shown in Fig. 1.

Algorithm 1: Gentle AdaBoost algorithm with distance score function

Input: Sequence of N labeled instance, Number of iteration T, Parameter μ of the algorithm

Output: The ensemble classifier decision

1 Scale all features into the range $[0, 1]$
2 Set the initial weights $w_{1,1} = ... = w_{1,n} = 1/N$
3 **for** $t := 1$ **to** T **do**
4   Train a base classifier f_t using weights $w_{t,1}, ..., w_{t,n}$
5   Scale all decision regions defined by f_t into the range $[0, 1]$
6   Compute for each object x_i distance to the decision boundary defined by $d(x_i)$
7   Transform the distance $d(x_i)$ using SF function
8   Update the observations weights: $w_{t+1,i} = \frac{w_{t,i} \exp(-y_t SF(x_i)_t)}{\sum w_{t,i} \exp(-y_t SF(x_i)_t)}$
9 **end**
10 **return** *Output the final ensemble classifier:* $\hat{y}(x_i) = sign\left(\sum_{t=1}^{T} SF(x_i)_t\right)$.

Table 1 presents in detail the variables used by the proposed Algorithm 1.

Table 1. Notation of Gentle AdaBoost algorithm with proposed modification

i	Instance number, $i = 1, ..., N$
t	Stage number, $t = 1, ..., T$
f_t	The weak classifier at the tth stage
$w_{t,i}$	The weight of the ith observation at the tth stage, $\sum_i w_{t,i} = 1$
$d(x_i)$	Distance to the decision boundary for object (instance) x_i
μ	Smoothing parameter of the proposed algorithm
$sign(x)$	$= 1$ if $x \geq 0$ and $= -1$ otherwise

In the paper [21] we present an algorithm in which the transformation function is based on the normal distribution. In the algorithm presented in Algorithm 1 step 5 regarding scaling decision regions defined by each weak classifier has also been added.

3   Experiment Setup

The main aim of the experiments was to compare the proposed modification Gentle AdaBoostM algorithm with the original one Gentle AdaBoost, which is described in [22]. For Gentle BoostM algorithm the value of parameter μ was set to value $\mu = 0.1$, $\mu = 0.5$ or $\mu = 1$. The aim of the experiments was also to examine the influence of parameter μ on the quality of classification. In this experiment, ID3 algorithm with one split is used as the weak classifier.

In the experimental research we use 6 publicly available binary data sets from UCI and KEEL repository.

Table 2. Properties of the data sets used in the experiments

Data set	Attributes	Classes	Examples
Cancer	8	2	699
Parkinsons	23	2	197
Phoneme	5	2	5404
Pima	8	2	768
Sonar	60	2	208
Wdbc	30	2	569

Table 2 presents the properties of the data sets which we used in the experiments. In order to obtain four most informative features the feature selection process [23–25] was performed. Additionally, we use the 10-fold-cross-validation method to obtain a learning and testing sets.

4   Results and Discussion

The results of our experiments are shown in Figs. 2, 3, 4, 5, 6, 7, 8, 9, 10, 11, 12 and 13. The results were obtained for thirty iterations of the boosting algorithm and two classification evaluation metric – classification error and kappa statistic. Figures 2, 3, 4, 5, 6 and 7 show the results for a classification measure which is a classification error, while Figs. 8, 9, 10, 11, 12 and 13 show the results for kappa statistic. Each figure shows the results for Gentle AdaBoost algorithm *GentleB* [22] and the modification of this algorithm proposed in the article *GentleBM*.

Overall, the results obtained depend on the data set used in the experiment. For example, for the *Cancer* data set, the proposed algorithm modification is almost always better than the original algorithm, regardless of the algorithm iteration, parameter value μ and classification evaluation metrics (Figs. 2 and 8). Similar observations apply to *Wdbc* data set, but only in the case of the first 15 iterations of the algorithm and evaluation metric – classification error. In subsequent iterations, there is practically no difference between the algorithms tested

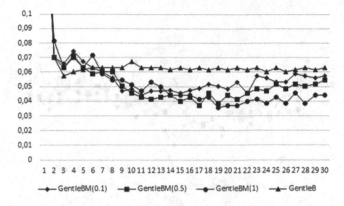

Fig. 2. Cancer – classification error.

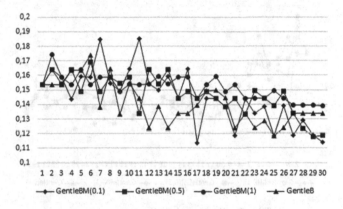

Fig. 3. Parkinsons – classification error.

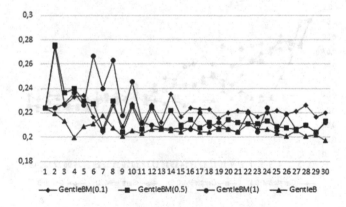

Fig. 4. Phoneme – classification error.

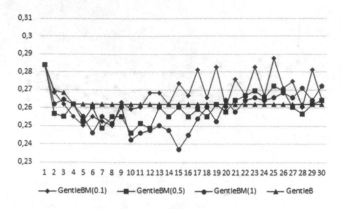

Fig. 5. Pima – classification error.

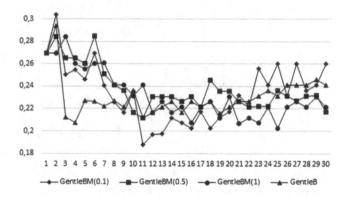

Fig. 6. Sonar – classification error.

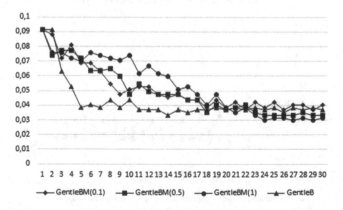

Fig. 7. Wdbc – classification error.

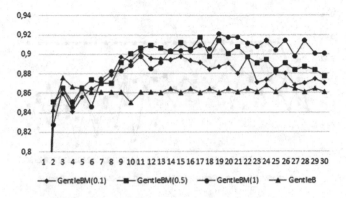

Fig. 8. Cancer – kappa statistics.

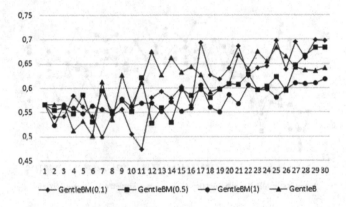

Fig. 9. Parkinsons – kappa statistics.

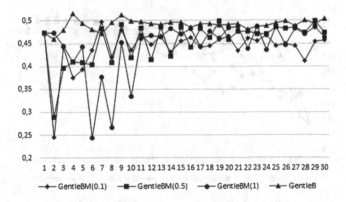

Fig. 10. Phoneme – kappa statistics.

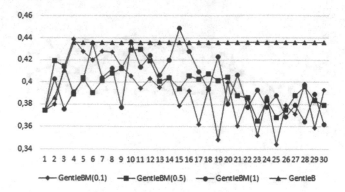

Fig. 11. Pima – kappa statistics.

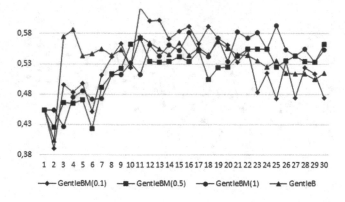

Fig. 12. Sonar – kappa statistics.

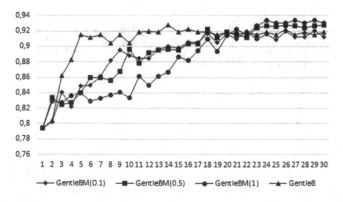

Fig. 13. Wdbc – kappa statistics.

(Figs. 7 and 13). In the case of *Pima* data set, the proposed algorithm proved to be much worse taking into account the kappa statistic measure, regardless of the value of μ parameter (Fig. 11). In the case of a quality measure, which is the classification error, the improvement in classification quality depends on the selection of μ parameter value and the algorithm iteration (Fig. 5). For example, for iteration 14–16, the proposed modification of Gentle AdaBoost algorithm with value $\mu = 1$ improves the quality of the classification by about 1.5%, when in case of $\mu = 0.1$ it worsens by 1.0%. However, in the case of *Sonar* data set, the results depend in particular on the iteration of the algorithm. Figure 8 shows that for the initial iterations the Gentle AdaBoost algorithm is better, while for iterations above 22 better results are for the proposed modification of the algorithm.

For the other two data sets *Parkinsons* and *Phoneme*, the results are no longer so unambiguous. However, Fig. 5 shows that for iteration 13–16 the base algorithm achieves a worse classification result about by 2% than the modification proposed in the work.

5   Conclusions

In this paper we presented the modification of Gentle AdaBoost algorithm. In particular, a method for determining the scoring function has been proposed, which depends on the distance of the object from the decision boundary and the value of the smoothing parameter. The proposed method of calculating the scoring function assumes that objects near the decision boundary and objects far for the decision boundary obtain a lower value of this function than objects between the extreme values. The value of the scoring function depends on the smoothing parameter μ.

The conducted experiments indicate that the proposed modification of Gentle AdaBoost [22] algorithm may improve the quality of this reference algorithm. Generally speaking, the results depend on the data set and the algorithm iteration. Therefore, in future work we want to use real data sets [26, 27] and select the number of the algorithm iterations and μ parameter value for a specific problem.

Acknowledgment. This work was supported by the National Science Centre, Poland under the grant no. 2017/25/B/ST6/01750.

References

1. Platt, J., et al.: Probabilistic outputs for support vector machines and comparisons to regularized likelihood methods. Adv. Large Margin Classif. **10**(3), 61–74 (1999)
2. Kuncheva, L.I.: Combining Pattern Classifiers: Methods and Algorithms, 1st edn. Wiley-Interscience, Hoboken (2004)
3. Lam, L., Suen, S.: Application of majority voting to pattern recognition: an analysis of its behavior and performance. IEEE Trans. Syst. Man Cybern.-Part A: Syst. Hum. **27**(5), 553–568 (1997)

4. Ruta, D., Gabrys, B.: Classifier selection for majority voting. Inf. Fusion **6**(1), 63–81 (2005)
5. Przybyła-Kasperek, M., Wakulicz-Deja, A.: Dispersed decision-making system with fusion methods from the rank level and the measurement level–a comparative study. Inf. Syst. **69**, 124–154 (2017)
6. Fumera, G., Roli, F.: A theoretical and experimental analysis of linear combiners for multiple classifier systems. IEEE Trans. Pattern Anal. Mach. Intell. **6**, 942–956 (2005)
7. Kittler, J., Alkoot, F.M.: Sum versus vote fusion in multiple classifier systems. IEEE Trans. Pattern Anal. Mach. Intell. **25**(1), 110–115 (2003)
8. Kuncheva, L.I., Bezdek, J.C., Duin, R.P.: Decision templates for multiple classifier fusion: an experimental comparison. Pattern Recogn. **34**(2), 299–314 (2001)
9. Woźniak, M., Graña, M., Corchado, E.: A survey of multiple classifier systems as hybrid systems. Inf. Fusion **16**, 3–17 (2014)
10. Xu, L., Krzyzak, A., Suen, C.Y.: Methods of combining multiple classifiers and their applications to handwriting recognition. IEEE Trans. Syst. Man Cybern. **22**(3), 418–435 (1992)
11. Freund, Y., Schapire, R.E.: A decision-theoretic generalization of on-line learning and an application to boosting. J. Comput. Syst. Sci. **55**(1), 119–139 (1997)
12. Burduk, R.: The AdaBoost algorithm with the imprecision determine the weights of the observations. In: Nguyen, N.T., Attachoo, B., Trawiński, B., Somboonviwat, K. (eds.) ACIIDS 2014. LNCS (LNAI), vol. 8398, pp. 110–116. Springer, Cham (2014). https://doi.org/10.1007/978-3-319-05458-2_12
13. Shen, C., Li, H.: On the dual formulation of boosting algorithms. IEEE Trans. Pattern Anal. Mach. Intell. **32**(12), 2216–2231 (2010)
14. Oza, N.C.: Boosting with averaged weight vectors. In: Windeatt, T., Roli, F. (eds.) MCS 2003. LNCS, vol. 2709, pp. 15–24. Springer, Heidelberg (2003). https://doi.org/10.1007/3-540-44938-8_2
15. Freund, Y., Schapire, R.E., et al.: Experiments with a new boosting algorithm. In: ICML, vol. 96, pp. 148–156. Citeseer (1996)
16. Wozniak, M.: Proposition of boosting algorithm for probabilistic decision support system. In: Bubak, M., van Albada, G.D., Sloot, P.M.A., Dongarra, J. (eds.) ICCS 2004. LNCS, vol. 3036, pp. 675–678. Springer, Heidelberg (2004). https://doi.org/10.1007/978-3-540-24685-5_117
17. Frejlichowski, D., Gościewska, K., Forczmański, P., Nowosielski, A., Hofman, R.: Applying image features and AdaBoost classification for vehicle detection in the 'SM4Public' system. In: Choraś, R.S. (ed.) Image Processing and Communications Challenges 7. AISC, vol. 389, pp. 81–88. Springer, Cham (2016). https://doi.org/10.1007/978-3-319-23814-2_10
18. Graczyk, M., Lasota, T., Trawiński, B., Trawiński, K.: Comparison of bagging, boosting and stacking ensembles applied to real estate appraisal. In: Nguyen, N.T., Le, M.T., Świątek, J. (eds.) ACIIDS 2010. LNCS (LNAI), vol. 5991, pp. 340–350. Springer, Heidelberg (2010). https://doi.org/10.1007/978-3-642-12101-2_35
19. Kozik, R., Choraś, M.: The HTTP content segmentation method combined with AdaBoost classifier for web-layer anomaly detection system. In: Graña, M., López-Guede, J.M., Etxaniz, O., Herrero, Á., Quintián, H., Corchado, E. (eds.) SOCO/CISIS/ICEUTE -2016. AISC, vol. 527, pp. 555–563. Springer, Cham (2017). https://doi.org/10.1007/978-3-319-47364-2_54
20. Wu, S., Nagahashi, H.: Analysis of generalization ability for different AdaBoost variants based on classification and regression trees. J. Electrical Comput. Eng. **2015**, 8 (2015)

21. Burduk, R., Bozejko, W.: Gentle AdaBoost algorithm with score function dependent on the distance to decision boundary. In: Saeed, K., Chaki, R., Janev, V. (eds.) CISIM 2019. LNCS, vol. 11703, pp. 303–310. Springer, Cham (2019). https://doi.org/10.1007/978-3-030-28957-7_25
22. Dmitrienko, A., Chuang-Stein, C., D'Agostino, R.B.: Pharmaceutical statisticsusing SAS: a practical guide. SAS Institute (2007)
23. Guyon, I., Elisseeff, A.: An introduction to variable and feature selection. J. Mach. Learn. Res. **3**, 1157–1182 (2003)
24. Rejer, I.: Genetic algorithms for feature selection for brain computer interface. Int. J. Pattern Recogn. Artif. Intell. **29**(5), 1559008 (2015)
25. Szenkovits, A., Meszlényi, R., Buza, K., Gaskó, N., Lung, R.I., Suciu, M.: Feature selection with a genetic algorithm for classification of brain imaging data. In: Stańczyk, U., Zielosko, B., Jain, L.C. (eds.) Advances in Feature Selection for Data and Pattern Recognition. ISRL, vol. 138, pp. 185–202. Springer, Cham (2018). https://doi.org/10.1007/978-3-319-67588-6_10
26. Giełczyk, A., Wawrzyniak, R., Choraś, M.: Evaluation of the existing tools for fake news detection. In: Saeed, K., Chaki, R., Janev, V. (eds.) CISIM 2019. LNCS, vol. 11703, pp. 144–151. Springer, Cham (2019). https://doi.org/10.1007/978-3-030-28957-7_13
27. Topolski, M.: Algorithm of multidimensional analysis of main features of PCA with blurry observation of facility features detection of carcinoma cells multiple myeloma. In: Burduk, R., Kurzynski, M., Wozniak, M. (eds.) CORES 2019. AISC, vol. 977, pp. 286–294. Springer, Cham (2020). https://doi.org/10.1007/978-3-030-19738-4_29

Automatic Natural Image Colorization

Tan-Bao Tran$^{(\boxtimes)}$ and Thai-Son Tran$^{(\boxtimes)}$

University of Science, Vietnam National University, Ho Chi Minh City, Vietnam
`baotrankhtn@gmail.com, ttson@fit.hcmus.edu.vn`

Abstract. We introduce a technique to automatically colorize natural grayscale images that combines both local and global features. Automatic colorization is a hard problem of computer vision and usually requires user interactions such as human-labelled color scribbles or reference images to achieve proper results. Based on Convolutional Neural Networks (CNN), our model is trained in an end-to-end fashion and can process images of any resolution. We improve the model of Iizuka *et al.* [1] by adding edge detection network and adjusting the input of the loss function. We also compare our model with the state of the art and show some improvements. Furthermore, we try colorizing ink wash paintings and achieve a special style.

Keywords: Colorization · Convolutional Neural Network · Self-supervised learning

1 Introduction

Colorizing grayscale images seems impossible because so much information (two out of three dimensions) has been lost. However, the semantics of the image provides meaningful information such as the sky is typically blue, the clouds are typically white. We can not recover the ground truth color, so we try to produce plausible results. Traditional approaches require significant user interaction to produce plausive colorization results while deep learning approaches have provided automatic methods with outstanding results recently. But deep learning models are still facing the problems of color bleeding and desaturation. The deep learning model of Iizuka *et al.* [1] provides good results and is currently one of the best colorization models; however, the colorized images are not

Fig. 1. Colorization results of some natural images

© Springer Nature Switzerland AG 2020
N. T. Nguyen et al. (Eds.): ACIIDS 2020, LNAI 12033, pp. 612–621, 2020.
https://doi.org/10.1007/978-3-030-41964-6_53

vibrant. Beside local features and global features (classification information) like the model of Iizuka *et al.* [1], we add Canny edge information. Local features are computed from small patches, provide information about low level to middle level of the images. Global features include classification information and Canny edge information. Classification information helps the model to know if the image is about forest or sea, indoors or outdoors, etc., while Canny edge information helps the model to reduce color bleeding over object boundaries. We will show that the model with edge information outperforms the model without edge information (Fig. 1).

Our contributions in this paper are in two areas. First, we adjust the model of Iizuka *et al.* [1] at some layers and introduce edge detection network that helps reduce color bleeding. Second, we change the input of the MSE loss function to make the colorization results more vibrant.

2   Related Work

Traditional approaches fall into two categories: scribble-based and transfer-based. Scribble-based methods, introduced by Levin *et al.* [2], require user to input color scribbles then solve a quadratic cost function derived from differences of intensities between a pixel and its neighboring pixels. It is assumed that adjacent pixels with similar luminance should have similar color. Huang *et al.* [3] improve this method by adjusting loss function and images will be processed by edge detection algorithm then each segmented region will be processed by colorization algorithm. Luan *et al.* [4] use the texture similarity to reduce user interactions.

Transfer-based methods require user to provide reference image(s). Welsh *et al.* [5] proposed a technique to colorize grayscale images by creating a set of sample pixels then each pixel in grayscale image will be scanned to find the best matching pixel in the set and transfer chrominance from sample pixel to the target pixel. Finding suitable reference image(s) is difficult for the user so Chia *et al.* [6] proposed a method that allows user to input the semantic label for each object in the scene then automatically download suitable images from the internet and apply colorization algorithm.

Recently, Larson *et al.* [7] proposed automatic colorization method based on deep convolutional neural network. They treat the problem as classification instead of regression and modify VGG-16 model to predict hue and chroma distributions for each pixel. Zhang *et al.* [8] proposed simple CNN architecture but they introduce annealed-mean to keep the vibrancy of the mode while maintaining the spatial coherence of the mean of the distribution. Iizuka *et al.* [1] introduce CNN architecture where global and local features are learned in parallel. They introduce fusion layer to fuse global and local features efficiently to achieve proper colorization results.

In conclusion, traditional methods require significant user interaction for proper colorization results. Automatic methods using CNN do not require any input but still face some problems such as desaturation and color bleeding.

3   Approach

We add edge detection network to the CNN model of Iizuka *et al.* [1], adjust some layers and change the input of the loss function. We use CIE Lab and HSV colorspaces, the input of the model is L and the output is a and b then a and b will be fused with L to form the output image. HSV is used to adjust the saturation of input images. The model is trained on about 200,000 natural images, validated on about 20,000 natural images of **Places** database and tested on natural images database of Massachusetts Institute of Technology. Activation function is ReLU, only the last convolutional layer uses tanh function. To help the model learn complex features and make the computation more effective, we use 3×3 filter.

3.1 Architecture

There are four main parts in the model: a local features network, a classification network, an edge detection network and a colorization network. The local features, classification information and edge information are fused at "fusion layer". Fusion layer is introduced in the paper of Iizuka *et al.* [1] and really efficient in colorization problem. We improve the model of Iizuka *et al.* [1] by adding edge detection network and adjusting the input of the loss function as well as some layers (Fig. 2).

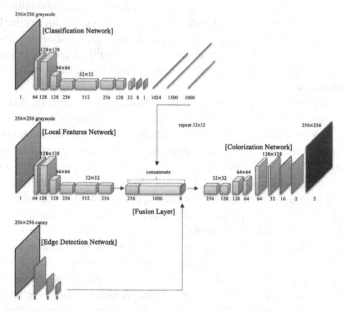

Fig. 2. Overview of our model for automatic colorization of natural grayscale images

Classification Network. The input is L channel and the output is 1000-dimensional vector. Classification information plays an important role in colorization results; for instance, the green colors of leaves will be different between day and night. However, determining the exact class of the image is not necessary in colorization problem. Iizuka *et al.* [1] suppose there are only 256 classes but we suppose there are 1000 classes in the dataset to make the model more adaptive (Table 1).

Table 1. Architecture of classification network

Layer	Filter	Stride	Outputs
Convolution	3×3	2	64
Convolution	3×3	1	128
Convolution	3×3	2	128
Convolution	3×3	2	256
Convolution	3×3	1	512
Convolution	3×3	1	256
Convolution	3×3	1	128
Convolution	3×3	1	32
Convolution	3×3	1	8
Convolution	3×3	1	1
Fully-connected	_	_	1024
Fully-connected	_	_	1500
Fully-connected	_	_	1000

Edge Detection Network. The input is canny edge image extracted from grayscale image, the output is just a small $32 \times 32 \times 8$ matrix because when edge detection network output is bigger it will affect how the model uses colors for objects and make the colorization results desaturated (Table 2).

Table 2. Architecture of edge detection network

Layer	Filter	Stride	Outputs
Convolution	3×3	2	8
Convolution	3×3	2	8
Convolution	3×3	2	8

Canny edge is used because it is the most popular algorithm, strictly defined and provides reliable detection. Furthermore, we can adjust the width of the Gaussian, the low and high threshold for the hysteresis thresholding. Edge detection network can reduce color bleeding and make colorization results more vibrant and realistic (Fig. 3).

Local Features Network. Low level and middle level features will be extracted at this network. The input is L channel and the output is $32 \times 32 \times 256$ matrix. Local features network is responsible for colorizing details in images. This network is similar to the combination of the low-level features network and middle-level features network of Iizuka et al. [1] but we eliminate the last two layers of their low-level features network because we train our model on smaller dataset (Table 3).

Input   Without edge   With edge information   Ground truth
               information

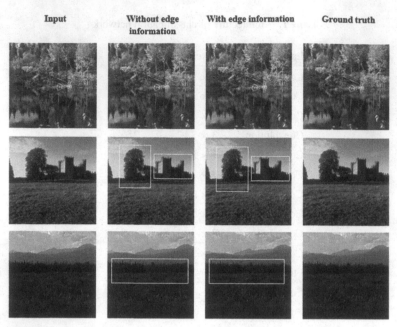

Fig. 3. Colorization results with and without edge information. Color bleeding improvements are in yellow bounding boxes (Color figure online)

Table 3. Architecture of local features network

Layer	Filter	Stride	Outputs
Convolution	3×3	2	64
Convolution	3×3	1	128
Convolution	3×3	2	128
Convolution	3×3	2	256
Convolution	3×3	1	512
Convolution	3×3	1	256

Colorization Network. The input is the output of fusion layer and the output is $256 \times 256 \times 2$ matrix. The output represents a and b channel. To increase the resolution of the output by a factor of 2, upsampling layers consisting of using the nearest neighbor approach are used (Table 4).

Fusion Layer. The output of classification network is 1000-dimensional vector, this will be cloned 32×32 times then arranged to form $32 \times 32 \times 1000$ matrix. Then it will be fused with the output of edge detection network ($32 \times 32 \times 8$ matrix) and local features network ($32 \times 32 \times 256$ matrix) to form $32 \times 32 \times 1264$ matrix.

Table 4. Architecture of colorization network

Layer	Filter	Stride	Outputs
Convolution	3×3	1	256
Convolution	3×3	1	128
Upsampling	_	_	128
Convolution	3×3	1	64
Upsampling	_	_	64
Convolution	3×3	1	32
Convolution	3×3	1	16
Convolution	3×3	1	2
Upsampling	_	_	2

3.2 Loss Function

With N is the number of samples, $Y = \left[y_1; y_2; \ldots; y_N\right]$ is the expected matrix with each row represents (a, b) from a ground truth image, $\widehat{Y} = \left[\hat{y}_1; \hat{y}_2; \ldots; \hat{y}_N\right]$ is the model's output matrix, MSE function used in the paper of Iizuka *et al.* is (1):

$$E = \frac{1}{N} \sum_{i=1}^{N} \left\| y_i - \hat{y}_i \right\|^2 \tag{1}$$

To make colorization results more vibrant, we will use Y' instead of Y. We first convert input images to HSV then increase S by multiplying with a constant T, shown in (2). Then we convert them back to Lab to get Y', so our loss function becomes (3). We found that $T = 1.8$ provides the best results (Fig. 4).

$$S' = T \times S \tag{2}$$

$$E = \frac{1}{N} \sum_{i=1}^{N} \left\| y'_i - \hat{y}_i \right\|^2 \tag{3}$$

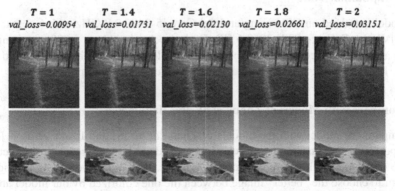

$T = 1$ $T = 1.4$ $T = 1.6$ $T = 1.8$ $T = 2$
val_loss=0.00954   val_loss=0.01731   val_loss=0.02130   val_loss=0.02661   val_loss=0.03151

Fig. 4. Colorization results affected by T

4   Experiments

See Fig. 5.

Fig. 5. Some colorization results of our model

4.1   Computation Time

Our model is trained on CPU using an Intel Core i5-8400 2.8 GHz with 6 cores, GPU using NVIDIA GeForce RTX 2060 6 GB. Each epoch takes about 6,000 s to complete. It takes about 2.5 s to colorize a 256×256 grayscale image and about 3 s to colorize a 1920×1080 grayscale image. We reach the best model after 14 epochs.

4.2   Compare with Models of Zhang *et al.* and Iizuka *et al.*

We compare our model with the models of Zhang *et al.* [8] and Iizuka *et al.* [1]. Our model performs well on natural grayscale images and can not only reduce color bleeding but also make the results more vibrant and realistic. We do a survey on 45 candidates, ask them to choose the "better" image between the one colorized by our model and the one colorized by the model of Iizuka *et al.* [1] in 50 couples of images. The survey's result shows that our model achieves better results on 68% of the couples of images (Fig. 6 and Table 5).

Fig. 6. Compare our model with the models of Zhang et al. [8] and Iizuka et al. [1]

We also do another survey on 30 candidates, ask them to recognize which one is the artificial image between the image colorized by our model and the ground truth image in 10 couples of images. The candidates are classified into 3 groups: **Art** (including people who have artistic jobs such as graphic designer, architect), **Technology** (including people who are familiar with digital images) and **Other**. The result shows that the candidates can recognize 5.6 artificial images in 10 artificial images .

4.3 A Try on Ink Wash Paintings

We also try colorizing ink wash paintings and achieve a special style. Ink wash paintings do not require many colors and typically about natural scenes so our model performs well (Fig. 7).

Table 5. The result of the survey on how good our model is

Group	The number of candidates	The number of right answers (per 10)
Art	3	4.3
Technology	14	6.1
Other	13	5.4
Total	*30*	*5.6*

Fig. 7. A try on ink wash paintings

5   Conclusion

Our model has improved the model of Iizuka *et al.* [1]. Edge detection network helps the model reduce color bleeding while the adjustment in the input of the loss function helps the model achieve more vibrant colorization results. Our deep learning model can process images of any resolution and is trained in an end-to-end fashion. Our model is applied for natural grayscale images only but this can be extended by enlarging the dataset. Automatic colorization is a challenging and an interesting problem that needs more researches to achieve plausible results.

Acknowledgements. This research was supported, in part, by Ngoc Dung Beauty Center. We thank members of Ngoc Dung AI Lab for helpful discussions and Duy-Phu Nguyen for his helpful advice.

References

1. Iizuka, S., Simo-Serra, E., Ishikawa, H.: Let there be color!: joint end-to-end learning of global and local image priors for automatic image colorization with simultaneous classification. ACM Trans. Graph. **35**(4), 1–11 (2016)
2. Levin, A., Lischinski, D., Weiss, Y.: Colorization using optimization. ACM Trans. Graph. **23**(3) (2004)
3. Huang, Y.-C., Tung, Y.-S., Chen, J.-C., Wang, S.-W., Wu, J.-L.: An adaptive edge detection based colorization algorithm. In: ACM International Conference on Multimedia (2005)
4. Luan, Q., Wen, F., Cohen-Or, D., Liang, L., Xu, Y.-Q., Shum, H.-Y.: Natural image colorization. In: Eurographics Conference on Rendering Techniques (2007)
5. Welsh, T., Ashikhmin, M., Mueller, K.: Transferring color to greyscale images. ACM Trans. Graph. **21**(3) (2002)
6. Chia, A.-S., et al.: Semantic colorization with internet images. ACM Trans. Graph. **30**(6), 1–8 (2011)
7. Larsson, G., Maire, M., Shakhnarovich, G.: Learning representations for automatic colorization. In: Leibe, B., Matas, J., Sebe, N., Welling, M. (eds.) ECCV 2016. LNCS, vol. 9908, pp. 577–593. Springer, Cham (2016). https://doi.org/10.1007/978-3-319-46493-0_35
8. Zhang, R., Isola, P., Efros, A.A.: Colorful image colorization. In: Leibe, B., Matas, J., Sebe, N., Welling, M. (eds.) ECCV 2016. LNCS, vol. 9907, pp. 649–666. Springer, Cham (2016). https://doi.org/10.1007/978-3-319-46487-9_40

Remote Usability Testing of Data Input Methods for Web Applications

Krzysztof Osada, Patient Zihisire Muke⬚, Mateusz Piwowarczyk⬚,
Zbigniew Telec⬚, and Bogdan Trawiński(✉)⬚

Department of Applied Informatics,
Wrocław University of Science and Technology, Wrocław, Poland
krzysztof.radoslaw.osada@gmail.com,
{patient.zihisire,mateusz.piwowarczyk,zbigniew.telec,
bogdan.trawinski}@pwr.edu.pl

Abstract. The main purpose of the paper was to conduct a comparative analysis and examine the usability of selected methods and patterns of data entry in web systems and websites. A dedicated web application was developed as an experimental tool for conducting remote unmoderated usability tests. Implemented data entry design patterns and tested with real users included various alignment of labels on forms, entering of logical values, small numbers, dates and time. The metrics collected during the test comprised: time to complete a task, a number of mouse clicks, number of errors committed, the Single Ease Question survey, closed and open questions regarding the subjective assessment of tested patterns. Based on the collected results, recommendations for the best patterns and methods of data entry in the specific context of use were formulated.

Keywords: User experience · Remote usability testing · Data input · Design patterns · Web applications

1 Introduction

Usability testing is one of the most popular methods of assessing usability of web sites and web applications which consists in examining the interface by real users according to a previously developed task scenario. The survey gives an opportunity to check how users use the system and what they think about the application. The method has many variants due to the type of product tested and the purpose of testing. Basically there are two general categories of usability testing methods, namely conventional on-site and remote ones. To perform the traditional usability testing the participants who are selected as the testers and the test evaluator are required to be in the same place at the same time. On the other hand, the remote usability testing method is the opposite of the conventional method [1]. By remote usability testing the participants are separated in space and/or time from the evaluators. We can distinguish between moderated (synchronous) and unmoderated (asynchronous) remote usability testing. With a remote synchronous method the users and evaluators are separated only in space whereas with a remote asynchronous method the users and test evaluators are separated both in space and time [2].

© Springer Nature Switzerland AG 2020
N. T. Nguyen et al. (Eds.): ACIIDS 2020, LNAI 12033, pp. 622–635, 2020.
https://doi.org/10.1007/978-3-030-41964-6_54

Remote tests are most often performed via the Internet and special tools designed for this purpose. All data collected during tests are saved by the software used during the test and sent to the researchers. The researcher does not have to meet the participants and the users do not have to perform tests in a special laboratory. The biggest advantage of remote usability testing is its low cost, because often users voluntarily perform tests or their salary is small. The environment in which the tests are carried out is also important. The user performs tests in his daily environment, in the conditions in which he/she functions at a convenient moment of his choice. The big advantage of testing is the opportunity to acquire a diverse group of respondents in geographical, age or social terms. The disadvantages of remote testing are problems with data confidentiality and security of the site, website and in unmoderated tests, lack of control over users and the inability to observe their behaviour and reactions [3]. According to [4] asynchronous methods do not facilitate the use of observation data and spontaneous verbalization records during remote usability testing sessions. Qualitative data can only be recorded using self-application forms or post-audit questionnaires. However, asynchronous methods allow you to record the activities of a large group of users.

The authors of the paper have conducted several studies in recent years on the usability of responsive applications in laboratory conditions [5, 6] and in real conditions [7]. They also examined the usability of data entry patterns devised for web [8] and mobile applications [9]. In both cases, dedicated applications were developed for the needs of the research and used for moderated on-site testing.

The goal of this paper was to examine the usability of selected data entry patterns in web applications that are commonly used in web forms. In order to conduct experiments, a dedicated web application was prepared that enabled remote unmoderated testing along with data collection and analysis. In his paper we present a part of obtained results covering data entry patterns for label alignment, entering logic values and small numbers as well as date and time input. The second part of the research results, which concerned alternative and auxiliary data entry methods, such as speech recognition, input masking, autocompletion, autosuggestion, and geolocation was published in [10].

2 Related Works

Alghamdi et al. [11] conducted a benchmarking study comparing the two sorts of remote usability testing methods, i.e. synchronous and asynchronous ones. Authors evaluated the usability of a website to compare the effectiveness of the two methods. Three points of comparison were involved: the overall task performance, satisfaction of participants, and the type and number of detected issues.

In turn, Madathil and Greenstein study [12] came up with a new methodology for organizing and carrying out synchronous remote usability studies utilizing a three dimensional virtual usability testing laboratory constructed using the Open Wonderland toolkit. The virtual laboratory methodology was compared with two other frequently applied synchronous usability test methods: WebEx, a web-based screen sharing and conferencing approach as well as normal traditional lab approach. The three methodologies were analyzed taking into account the usability defects identified, severity of these usability defects and time taken to accomplish the tasks.

In addition, Chynal and Sobecki [13] designed a new hybrid usability testing method that facilitated the experimenter to conduct unmoderated remote usability testing of web based systems. The method operated as a software injected into the web pages and collected data from users while doing certain tasks on the website. To evaluate the efficiency of this method, authors designed a benchmark website with intentionally implemented specified number of different usability issues.

Furthermore, Sauer et al. [14] investigated the impact of utilizing non-identical extra laboratorial usability testing procedures. Three experiments were performed employing various artefacts, namely a fully functioning smartphone, computer simulated mobile phone, and website. A comparison was made between various techniques in field testing: asynchronous and synchronous remote testing as well as classical field testing with laboratory testing under various operational conditions. The usability testing typical outcome variables were measured, including workload, perceived user-friendliness, number of clicks, and task execution time.

Yudhoatmojo et al. [1] in their study, concentrated on the remote asynchronous method. Under this method, authors compared and examined the utilization of diary and forum techniques to perform the usability testing. The objective of this research was to determine which one of these two techniques is the most effective in identifying the usability issues, and also to figure out the disadvantages and advantages of the two techniques.

Tullis et al. [15] presented findings of two studies comparing the remote Web based usability testing with traditional lab-based usability testing. Two websites were used for the test: a financial information website and an employee benefits website. An automated technique was employed for the remote tests whereby the users participated in the test from their ordinary work locations using their regular browsers. High correlation between lab and remote tests for the task completion data and the task time data was showed. Both techniques allowed for detection of the most critical usability issues with the web sites.

Wei et al. [16] argued that the traditional lab-based usability testing studies should be complemented by the web-based and remote research studies, testing users on their personal devices at their convenient time. The authors as well outlined the various remote testing tools and methods for empirical experiments and remote usability studies.

Thompson et al. [17] came up with a study that described methodologies for effective and efficient remote testing environments and identified appropriate tools to be utilized during such test. Commercial tools were examined to establish their cost-effectiveness and usefulness for classroom and then an empirical study was carried out to compare remote usability and traditional testing of a web site using one of selected tools.

In addition, Scholtz [18] debated on the approach taken in developing usability testing tools that are rapid, automated, remotely usable and helpful in providing more usability information in a shorter time frame and in a form that can be without hesitation useful to usability professionals.

3   Web Application for Usability Testing

In order to receive a better understanding of the problem that is presented in this paper a series of usability tests with a group of participants was performed. The process was greatly accelerated due to the access to the web application that was designed specifically for this research.

The source code of the application was written with the usage of modern web technologies including HTML, CSS, Sass, JavaScript, React, PHP etc. The program was designed as an *single-page application* thus the interaction with the user was based mainly on rewriting contents of the web page. Though the application development included preparing mostly front-end solutions on the user side, the implementation of the back-end logic played the critical part when it comes to storing the data and controlling the way the program worked.

The main goal of the application was to enable usability testing of the data entry design patterns that might be seen on the Internet. Therefore, the idea of creating a modern web tool that would consist of as many design patterns as possible appeared to be the right direction. In practice, the implemented application contained various input fields that differed in style, the way of working and the possible use case; nonetheless all of them were derived from real web pages and applications. The prepared set of data entry design patterns was extended to the alternative or auxiliary methods of gathering input, such as speech recognition, autosuggestion and word completion.

The usability testing tool worked somewhat similarly to the web questionnaire or the survey: the participant answered simple questions, usually by choosing one of a few available options. Each task performed by the user related to one data entry design pattern or input method and involved transferring (e.g. retyping) the exemplary values stored in the table rows to the corresponding input fields. These tasks, called also exercises, were grouped by context of use in consecutive scenarios. Each scenario was preceded by a short introduction describing the fictional situation in which the given input field or method could be used. All the exercises and scenarios were also summarised by short sets of questions referring to the patterns and methods already tested by the user. In the final stage of the application the participant was asked to fill an additional questionnaire. User satisfaction was likewise measured throughout different self-reported metrics based on miscellaneous usability ratings, e.g. the Single Ease Question, expectation measures and open questions.

When using the application, the participant performed certain actions that were analysed and registered in the background. Among the metrics collected during the test were: numbers of clicks done during each exercise, numbers of errors (i.e. numbers of times when the user tried to validate the data that was not correct) and time on exercise. Every new information was sent to the back-end after completing an exercise or a scenario.

There were two parallel versions of the content for the application; they primarily differed in data entry design patterns and input methods they were using. Both of them consisted of nine scenarios including 18 exercises in total. The tool was fully compatible with Google Chrome; other modern web browsers, i.e. Mozilla Firefox, Opera and Microsoft Edge were having issues with some of the solutions used in the application.

The access to the application was restricted; one could see the interface only after getting a randomized string generated by the system, however the code was valid only for a limited amount of time. Such limitations were made so as to assure that results of the test would be credible and reliable. Last but not least, the order of exercises and scenarios was completely arbitrary, making each instance of the test unique.

4   Setup of Usability Tests

The goal of usability study was to explore various data input techniques and methods developed for web applications. To this end, a dedicated application was developed that was used to collect data on how users fill out web forms depending on the context, the chosen data entry scheme and the data entry method. The collected data was then analyzed in order to evaluate the usability of individual patterns and data entry methods, and to formulate a recommendation of the best solution for a specific application case. A total of 36 different data entry patterns were tested with real users. It this paper we present only part of the results obtained for 20 patterns of 36 including label alignment, input of logic values and small numbers as well as entering dates and time.

The usability testing tool functioned somewhat similarly to the web questionnaire or the survey. The participant answered simple questions, usually by choosing one of a few available options. Each task performed by the user related to one data entry design pattern or input method and involved transferring (e.g. retyping) the exemplary values stored in the table rows to the corresponding input fields. These tasks, called also exercises, were grouped by context of use in consecutive scenarios. Each scenario was preceded by a short introduction describing the fictional situation in which the given input field or method could be applied. All the exercises and scenarios were also summarised by short sets of questions referring to the patterns and methods already tested by the user. In the final stage of the application the participant was asked to fill a personal questionnaire.

While using the application, the participant performed specific activities that were analyzed and recorded in the background. The measures collected during the test included: time spent on performing individual exercises, number of left-clicks made during each exercise, number of errors made, i.e. the number of cases in which the user tried to verify data that was not correct. User satisfaction was also measured by various self-reported indicators based on different usability assessments, e.g. single easy question (SEQ) [19], expectations measures and open questions.

Two parallel versions of the application were developed; they differed mainly in the data entry patterns and the data entry methods applied. Both consisted of five scenarios, including a total of 10 exercises. Access to the application was limited, the interface could only be seen after obtaining a random string generated by the system, but the code was only valid for a limited time. Such restrictions were introduced in order for the test results to be reliable. Finally, the order of exercises and scenarios was completely arbitrary, making each instance of the test unique.

The experiment was conducted with 40 people who were randomly partitioned into two smaller subgroups of 20 people each and assigned to one of two application versions. All tests were carried out remotely. When the users were ready to take part in the study, they received access codes and were able to do exercises on their own personal computers or laptops within one hour. This forced the user to focus on the session and complete all scenarios in this time window.

Most participants, i.e. 67.5%, were under 25 years old, and the remaining part, i.e. 32.5%, were at most 35 years old. The users were relatively well educated because

75% of them had a master's degree and 25% completed high school. The whole group was composed of 23 men (57.5%) and 17 women (42.5%). The participants did a wide variety of jobs - they worked as IT specialists, graphic designers, teachers, advertising agency employees, lawyers, engineers, economists, philologists, office workers, journalists and entrepreneurs. Some users described themselves as students or admitted to being unemployed. Almost all participants claimed to have used internet applications at least once a day. In turn, only two people (5%) said they do it once a week or more. On the other hand, differences can be seen when considering the main reason for using web applications; while for 10 participants (25%) the most important factor was reading news and articles online, slightly less people also mentioned about entertainment, fast searching for information and work. A minority of users said that social contacts, learning or hobbies were the most common reasons for using internet applications. Although the participants were diverse at many levels, they proved to be very similar in terms of the most commonly used pointing devices. The mouse, which was recalled by 28 participants (70%), only seven users (17.5%) mentioned the touchpad. Other answers to this question, i.e. a touch screen, digitizer or other unspecified device, were chosen by one, one and three participants respectively.

5   Results of Usability Tests

Label Alignment. We wanted to examine how the way labels are positioned relative to the text input fields influences the performance of users filling out web forms. We asked the participants to overwrite a fictitious postal address. Each user had to do two exercises based on similar data but varied in the alignment of the input labels. In total we tested four different patterns: right-aligned labels (LA1), left-aligned labels (LA2), placeholders (LA3), and floating labels (LA4).

The efficiency, effectiveness, and satisfaction measures for this scenario are shown in Figs. 1 and 2. It turned out that time of filling out a form depended on the label alignment: while LA1 helped lower time needed, LA2 and LA4 made users act significantly slower. LA3 became the most engaging in terms of the number of mouse clicks and the most error-prone data entry design pattern. On the other hand, the participants were the most satisfied when using LA2 and LA4. It might be possible that such a decrease of efficiency was a result of the way in which LA3 works, i.e. when the field contains a value, its label disappears.

Authors' Recommendations: While filling a form, it is important to ensure that the user sees all the input fields and their description until the end of the process; therefore LA3 ought to be rather avoided. The other data entry design patterns are safer to use; the choice between one of them depends mainly on the context of use, i.e. the type of the form data and the size of the screen.

Entering Logic Values. Sometimes it is necessary to make a decision and choose one of two options – "yes" or "no", zero or one etc. This is a kind of a situation in which the possible values are mutually exclusive and the selection is mandatory. The purpose of this scenario was to determine which data entry design pattern is the best for entering

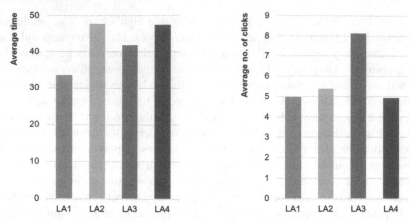

Fig. 1. Average time (left) and number of clicks (right) for label alignment

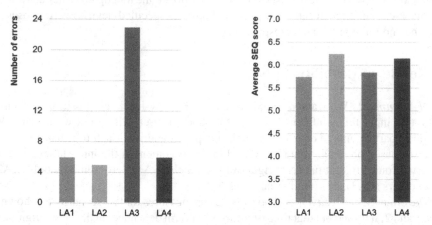

Fig. 2. Number of errors (left) and average SEQ score (right) for label alignment

logic values: a radio button (LV1), toggle switch (LV2), drop-down list (LV3) or toggle button (LV4) and the results are presented in Figs. 3 and 4.

Only a few users needed longer time to accomplish tasks interacting with LV3 and LV4. What is more, LV4 was on average the easiest pattern to use while LV1 and LV3 became the hardest. Surprisingly, many users had a completely contradictory attitude and, according to their comments, they preferred the classic solution (LV1) to the modern one (LV4) due to its misleading behaviour. When considering the average number of mouse clicks, the differences between the patterns tested were not large; nevertheless LV2 turned out to be more effective than LV1. For all patterns the total number of errors made by the participants was very small: only one for LV1 and LV2 and only two for LV3 and LV4.

Authors' Recommendations: Choosing only one of a few values is sometimes critical for the output of the form and has to be done wisely. Seeing that the users like the solutions that they well know, we recommend using LV1 since its way of working is rather obvious. On the other hand, implementing LV2 may be interesting due to its good performance and a broader possibilities when it comes to styling.

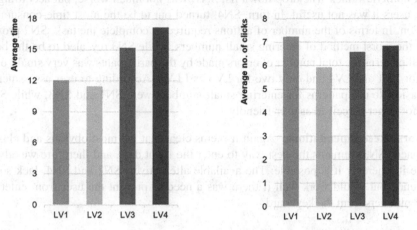

Fig. 3. Average time (left) and number of clicks (right) for entering logic values

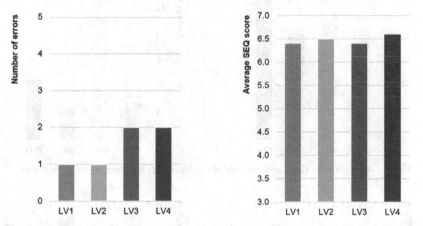

Fig. 4. Number of errors (left) and average SEQ score (right) for entering logic values

Entering Small Numbers. In general, entering numerical values to fill out web forms is not much different from entering standard strings. Inserting numbers and letters using the keyboard on the desktop looks exactly the same There are certainly some design patterns that allow the user to enter numbers more easily (especially smaller ones, e.g., from zero to nine), but they often assume the presence of a mouse or other pointing device, and therefore are usually less effective to use. In this scenario, we employed a drop-down list (SN1), and an input field that cannot be modified directly, but have two separate buttons to increase and decrease the current value (SN2), a conventional text field (SN3), and a range slider (SN4).

The results are illustrated in Figs. 5 and 6. As it turned out, SN3 was not only the simplest data entry design pattern but also the most time-efficient approach for entering small numeric values. The drop-down list SN1, was not much worse, but according to some users it was not useful. In turn, SN4, turned out to be the most time-consuming solution. In terms of the number of actions required to complete the task, SN2 proved to be the worst method of entering small numbers, while SN3 revealed to be the best. For all patterns the total number of errors made by the participants was very small: only one for LV1 and LV4 and only two for LV3 and LV4. According to user assessments, the easiest to use patterns for entering small numbers were SN1 and SN3, while SN4 was not rather perceived as user-friendly.

Authors' Recommendations: Again it seems clear that the most obvious and classic approach (SN3) remains the best way to enter the input data, and therefore we advise to use it wherever it is possible. The available alternatives, SN2 and SN4, lack some efficiency but would work well if there was a need to prevent the user from entering small numbers from the keyboard.

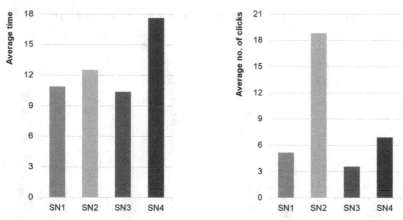

Fig. 5. Average time (left) and number of clicks (right) for entering small numbers

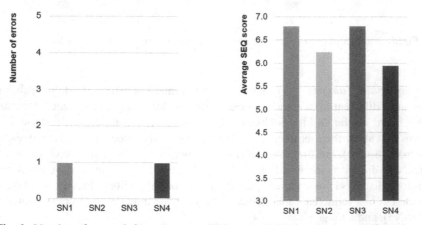

Fig. 6. Number of errors (left) and average SEQ score (right) for entering small numbers

Entering Dates. The ubiquity of dates might be seen on the Internet; everything we plan to do, from booking a hotel room to buying a ticket to the cinema depends on a specific day, a month and a year. One could argue that inputting dates is nontrivial. However, there are many ways to write them and plenty of design patterns that make picking dates possible. We tested four different ways for entering a date into a web form: a group of three text fields (DA1), a calendar picker (DA2), a multi-input field consisting of a text field for a day, a drop-down list for a month and a text field for a year (DA3), and finally an automatically formatted text field (DA4).

The results are depicted in Figs. 7 and 8. The general conclusion is that simpler design patterns worked better than their more sophisticated equivalents. DA1 and DA4 were less time-engaging while DA3 and DA2 worked worse. Moreover, the participants rated DA2 as the most difficult exercise of this scenario; in turn, DA4 was assessed to be the easiest. These findings are supported by the numbers of clicks and errors done during the exercises: DA2 required using the mouse much more frequently than DA4. Only on error was made using DA2, the other design patterns prevented the users from making mistakes.

Authors' Recommendations: The choice of the method of entering the date depends mainly on the context of use. Although DA2 seems to be attractive for choosing days, months and years, it often requires a lot of effort to find the right date, especially when it is distant in time. For most cases DA4 could be the best way to enter dates, however it has to be complemented by an additional label indicating the date format.

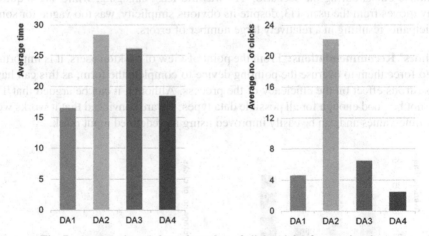

Fig. 7. Average time (left) and number of clicks (right) for entering dates

Entering Time. Time in a web form could be inserted using the keyboard or the mouse. In this scenario, we asked users to enter the time in GG: MM format using two drop-down lists (TI1), a text field with two arrow buttons to set the time 10 min forward or backward (TI2), a text field without any additional facilities (TI3) and a table containing all hours and minutes divided by five (TI4).

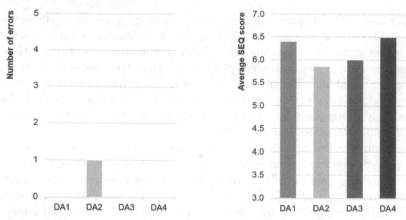

Fig. 8. Number of errors (left) and average SEQ score (right) for entering dates

The results are shown in Figs. 9 and 10. Again, the simple text field (TI3) became the best solution; in this case, it was over 30% better than the design patterns that encouraged the participant to use a pointing device, i.e. TI2 and TI4. However, from the point of view of the difficulty level, the TI4 was as easy to use as the TI3, although the former was unknown to most participants before the test. Moreover, TI4 was often described as difficult to understand. The results from the time metric goes along with the numbers of mouse clicks during the scenario: TI3 was the least engaging, while TI2 required many moves from the user. TI3, despite its obvious simplicity, was too vague for some participants resulting in a relatively large number of errors.

Authors' Recommendations: From the point of view of desktop users, it is important not to force them to overuse the pointing device to complete the form, as this can have a disastrous effect on the efficiency of the process. Although it can be argued that TI3 may not be good enough for all possible data types, we are convinced that it works well with time values and can be easily improved using a predefined input mask.

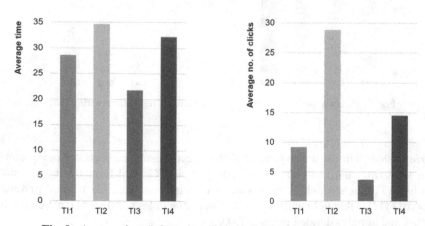

Fig. 9. Average time (left) and number of clicks (right) for entering time

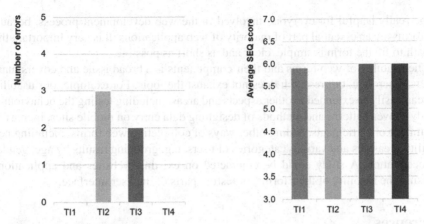

Fig. 10. Number of errors (left) and average SEQ score (right) for entering time

6 Conclusions and Future Works

In this paper, we studied the usability of data entry patterns and techniques that are used in web applications employing the remote usability testing approach. For this purpose we designed and implemented a web application composed of several scenarios presenting sample forms which consisted of various sets of input fields. The application enabled the users to use various data entry patterns and allowed for collecting the results of their activities, analyse and present them. The tool we created was then tested by a group of people invited to take part in remote usability tests. Four different measures of usability: time to complete the task, number of clicks, number of errors made, and SEQ score allowed us to gather useful information about how the given patterns and methods work in practice. Based on the collected results, we proposed recommendations on the selection of forms and methods of data entry design in various application contexts.

Usability tests were carried out by 40 invited participants with different experience, expectations and approach to applications and web forms. We asked participants to do 18 exercises grouped into nine scenarios. In total, we checked how 36 patterns and methods of data entry design work in interaction with real users. The results we collected were not always consistent with our assumptions; As it turned out, sometimes the best solution for a given use case was the simplest. Test participants preferred traditional text fields that can be filled with any input device to complex design patterns that were often more engaging than they should be. We also noticed that some modern methods of data entry did not reduce the time needed to complete the form; on the contrary, the users were often much slower. Among the possible explanations for this contradiction, we found the ambiguity of the method, its lack of usability, low popularity, and even some kind of distrust of users.

The recommendations we presented in our study were based not only on the quantitative data we collected during usability tests, but also on comments and remarks submitted directly by our study participants. For each of the scenarios, which actually reflected the most popular input types, we tried to find the best pattern or design method for entering data for a given application context and justify our choice. We believe that our insights can be really helpful for everyone involved in the web development process, because web forms are an essential part of majority of web applications. It is very important that the path to fill the form is simple, clear and as short as possible.

The usability of web forms and their components is a broad issue and covers many aspects. Therefore, our research does not exhaust the topic. For example, our usability tests can easily be extended to other aspects and areas, including testing the behaviour of already proven patterns and methods of designing data entry on mobile sites, increasing the number of participants, adding other ways of completing web forms, including new usability measures and various categories of users, e.g. grouping results by age, gender, and occupation. A study could be conducted on existing websites and applications, assessing the usability of their forms, or testing parts of the user interface.

References

1. Yudhoatmojo, S.B., Sutendi, S.F.: Empirical study on the use of two remote asynchronous usability testing methods. J. Phys. Conf. Ser. **1193**, 012017 (2019). https://doi.org/10.1088/1742-6596/1193/1/012017
2. Bruun, A., Gull, P., Hofmeister, L., Stage, J.: Let your users do the testing: a comparison of three remote asynchronous usability testing methods. In: Proceedings of the SIGCHI Conference on Human Factors in Computing Systems, CHI 2009, pp. 1619–1628. ACM (2009). https://doi.org/10.1145/1518701.1518948
3. Tylkowska, J., Kurasińska, L.: Moderated and Unmoderated Remote Usability Testing: What Is It, how to Run It, and What Are the Benefits? (2019). https://stxnext.com/blog/2019/05/10/moderated-unmoderated-remote-usability-testing/. Accessed 22 Nov 2019
4. Bastien, J.M.C.: Usability testing: a review of some methodological and technical aspects of the method. Int. J. Med. Informatics **79**(4), e18–e23 (2010). https://doi.org/10.1016/j.ijmedinf.2008.12.004
5. Bernacki, J., Błażejczyk, I., Indyka-Piasecka, A., Kopel, M., Kukla, E., Trawiński, B.: Responsive web design: testing usability of mobile web applications. In: Nguyen, N.T., Trawiński, B., Fujita, H., Hong, T.-P. (eds.) ACIIDS 2016. LNCS (LNAI), vol. 9621, pp. 257–269. Springer, Heidelberg (2016). https://doi.org/10.1007/978-3-662-49381-6_25
6. Błażejczyk, I., Trawiński, B., Indyka-Piasecka, A., Kopel, M., Kukla, E., Bernacki, J.: Usability testing of a mobile friendly web conference service. In: Nguyen, N.-T., Manolopoulos, Y., Iliadis, L., Trawiński, B. (eds.) ICCCI 2016. LNCS (LNAI), vol. 9875, pp. 565–579. Springer, Cham (2016). https://doi.org/10.1007/978-3-319-45243-2_52
7. Krzewińska, J., Indyka-Piasecka, A., Kopel, M., Kukla, E., Telec, Z., Trawiński, B.: Usability testing of a responsive web system for a school for disabled children. In: Nguyen, N.T., Hoang, D.H., Hong, T.-P., Pham, H., Trawiński, B. (eds.) ACIIDS 2018. LNCS (LNAI), vol. 10751, pp. 705–716. Springer, Cham (2018). https://doi.org/10.1007/978-3-319-75417-8_66
8. Drygielski, M., Indyka-Piasecka, A., Piwowarczyk, M., Telec, Z., Trawiński, B., Duong, T.H.: Usability testing of data entry patterns implemented according to material design guidelines for the web. In: Nguyen, N.T., Chbeir, R., Exposito, E., Aniorté, P., Trawiński, B. (eds.) ICCCI 2019. LNCS (LNAI), vol. 11683, pp. 697–711. Springer, Cham (2019). https://doi.org/10.1007/978-3-030-28377-3_58
9. Myka, J., Indyka-Piasecka, A., Telec, Z., Trawiński, B., Dac, H.C.: Comparative analysis of usability of data entry design patterns for mobile applications. In: Nguyen, N.T., Gaol, F.L., Hong, T.-P., Trawiński, B. (eds.) ACIIDS 2019. LNCS (LNAI), vol. 11431, pp. 737–750. Springer, Cham (2019). https://doi.org/10.1007/978-3-030-14799-0_63

10. Osada, K., Zihisire Muke, P., Piwowarczyk, M., Telec, Z., Trawiński, B.: Comparative usability analysis of selected data entry methods for web systems. Cybern. Syst. Int. J. **51**(2), 192–213 (2020). https://doi.org/10.1080/01969722.2019.1705552

11. Alghamdi, A.S., Al-badi, A., Alroobaea, R., Mayhew, P.J.: A comparative study of synchronous and asynchronous remote usability testing methods. Int. Rev. Basic Appl. Sci. **1**(3), 61–97 (2013)

12. Madathil, K.C., Greenstein, J.S.: An investigation of the efficacy of collaborative virtual reality systems for moderated remote usability testing. Appl. Ergon. **65**, 501–514 (2017). https://doi.org/10.1016/j.apergo.2017.02.011

13. Chynal, P., Sobecki, J.: Statistical verification of remote usability testing method. In: Proceedings of the Multimedia, Interaction, Design and Innovation, MIDI 2015, pp. 12:1–12:7 (2015)

14. Sauer, J., Sonderegger, A., Heyden, K., Biller, J., Klotz, J., Uebelbacher, A.: Extra-laboratorial usability tests: an empirical comparison of remote and classical field testing with lab testing. Appl. Ergon. **74**, 85–96 (2019). https://doi.org/10.1016/j.apergo.2018.08.011

15. Tullis, T., Fleischman, S., Mcnulty, M., Cianchette, C., Bergel, M.: An empirical comparison of lab and remote usability testing of Web sites. In: Proceedings of Usability Professionals Conference (2002). http://citeseerx.ist.psu.edu/viewdoc/summary?doi=10.1.1.457.3080. Accessed 15 Dec 2019

16. Wei, C., Barrick, J., Cuddihy, E., Spyridakis, J.: Conducting usability research through the internet: testing users via the WWW. In: Proceedings of the Usability Professional Association, pp. 1–8 (2005)

17. Thompson, K.E., Rozanski, E.P., Haake, A.R.: Here, there, anywhere: remote usability testing that works. In: SIGITE Conference - IT Education - The State of the Art, pp. 132–137 (2004)

18. Scholtz, J.: Adaptation of traditional usability testing methods for remote testing. In: Proceedings of the 34th Annual Hawaii International Conference on System Sciences. IEEE (2002). https://doi.org/10.1109/HICSS.2001.926546

19. Sauro, J., Lewis, J.R.: Quantifying the User Experience: Practical Statistics for User Research, 2nd edn. Morgan Kaufmann, Cambridge (2016)

Machine Learning Models for Real Estate Appraisal Constructed Using Spline Trend Functions

Mateusz Jarosz[1], Marcin Kutrzyński[1] , Tadeusz Lasota[2] ,
Mateusz Piwowarczyk[1] , Zbigniew Telec[1] , and Bogdan Trawiński[1(✉)]

[1] Department of Applied Informatics, Wrocław University of Science and Technology,
Wrocław, Poland
jaroszm91@gmail.com, {marcin.kutrzynski,mateusz.piwowarczyk,
zbigniew.telec,bogdan.trawinski}@pwr.edu.pl
[2] Wroclaw Institute of Spatial Information and Artificial Intelligence, Wrocław, Poland
tadeusz.lasota@wp.pl

Abstract. The paper presents methods of modeling the real estate market using trend functions reflecting changes in real estate prices over time. Real estate transaction prices that are used to create data-driven valuation models must be updated in line with the trend of their change. The primary purpose of the first part of the study was to examine the extent to which splines are suitable for the trend function compared to polynomials of the degree from 1 to 6. In turn, the second part was to compare the performance of prediction models built on the basis of updated data with various trend functions: splines and polynomials. The experiments were conducted using real data on purchase and sale transactions of residential premises concluded in one of the Polish cities. Four machine learning algorithms implemented in the Python environment were used to generate property valuation models. Statistical analysis of the results was carried out using non-parametric Friedman and Wilcoxon tests. The study showed the usefulness of applying splines to model trend functions.

Keywords: Prediction models · Machine learning · Real estate appraisal · Trend functions · Spline functions

1 Introduction

Property valuation is carried out by licensed property appraisers and requires specialized knowledge and sufficient data. One of the most common approaches to determining the market value of real estate is the comparative method. To apply this approach, it is necessary to have transaction prices of real estate sold whose attributes are similar to the assessed one. If good comparable transaction examples are available, it is possible to obtain reliable estimates. Professional standards require data on real estate located close to the evaluated property. In addition, real estate purchase and sale transactions concluded in the recent past are the most reliable, e.g. one or two years before the

© Springer Nature Switzerland AG 2020
N. T. Nguyen et al. (Eds.): ACIIDS 2020, LNAI 12033, pp. 636–648, 2020.
https://doi.org/10.1007/978-3-030-41964-6_55

valuation. Regardless of the period from which the valuation data is downloaded, real estate transaction prices must be updated in line with the trend of their changes.

However, property appraisers often have major problems obtaining adequate data on purchase and sale transactions for similar properties, as they are in most cases scarce or lacking in the market. To solve this problem, the authors proposed methods for determining and combining similar zones to obtain homogeneous areas covering more examples of data. The expert algorithm presented in [1, 2] consisted in searching for zones in the city in which trends in real estate price changes over time were similar and then combining these zones into one homogeneous area. Thanks to this operation, one common price prediction model could be determined for several merged zones. It has been shown that such a model is more effective and reliable because a larger number of transactions was used in its construction. To determine the similarity of trends, the authors divided the period into a number of smaller intervals and for each of them individually determined linear trends of changes in property prices.

To date, many different approaches to property valuation have been proposed; ranging from statistical techniques, through operational research, to computational intelligence methods. In the rich literature on the subject, various techniques can be found, including models created using multiple regression and neural networks [3–5], linear programming [6], decision trees [7], and rough set and fuzzy systems [8, 9].

Our previous research in the field of real estate valuation was inspired by such approaches to machine learning as ensemble learning [10–12], hybrid methods [13], including evolutionary fuzzy systems [14] and evolving fuzzy systems [15]. We developed several models built using various resampling techniques including bagging with decision trees, support vector machines, artificial neural networks, and fuzzy systems implemented in KEEL [16, 17] as well as bagging with genetic fuzzy models [18, 19]. Evolving fuzzy models constructed over cadastral data revealed a relatively good performance [20]. We have also investigated the ensembles of genetic fuzzy systems applied to predict from a data stream of real estate purchase and sale transactions [21].

The main purpose of the study reported in this paper was to examine the extent to which splines are useful to model the trend functions compared to polynomials of the degree from 1 to 6. In turn, the second goal was to compare the performance of models built over updated data on purchase and sale transactions according to the various trend functions. The experiments were carried out using real data on purchase and sale transactions of residential premises concluded in one of the Polish cities. Four machine learning algorithms implemented in the Python environment were used to generate property valuation models. They included linear regression, random forest, bagging of decision trees, and gradient boosting. The obtained results were statistically analyzed using non-parametric Friedman and Wilcoxon tests.

2 Spline Trend Functions for Updating Transactional Prices

There are many concepts for data set interpolation. One of the simple methods is linear interpolation, which divides the domain into intervals and creates straight lines between node points. Node (or knot) points are calculated at the beginning and end of each interval by the method of 2D nearest neighbors [22]. This method is straightforward and fast, but it is not precise between nodes, and the interpolant is not differentiable at nodal points. Achieving greater precision is possible by searching domain cutting points instead of allocating intervals and is known as the NURBS method [23].

The next idea uses higher-order polynomials as interpolants. It is well known that precisely one polynomial of degree n − 1 or less passes through n points [22]. Polynomial interpolation, in contrast to linear interpolation, is differentiable and can estimate local extrema, but errors occur near the endpoints due to the Runge's phenomenon [24]. It is illustrated in Fig. 1.

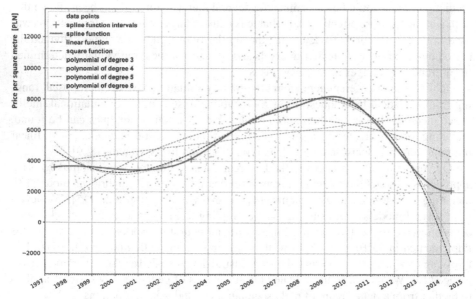

Fig. 1. Illustration of the Runge's phenomenon near the endpoints of the interval

Splines combine both of these ideas. Similar to linear interpolation, the domain is divided into smaller parts, but each part is interpolated by a polynomial [23, 25]. To minimize oscillations of the Runge phenomenon, the degree of a polynomial should be as low as possible. However, it cannot be too small to be able to be differentiated at nodal points. Finally, spline building is an optimization problem to find domain intervals as large as possible, keeping errors (i.e. mean square error) and degree of polynomial at low values. The de Boor's algorithm is an efficient method for evaluating spline curves [23, 26].

Modeling and analysis of the trend functions, especially the spline function, required the development of a tool that enabled both automatic and manual parameter selection, as well as a direct view of the modeled functions. The application has been programmed in Python using the following libraries: SciPy, NumPy, pandas, Matplotlib, scikit-learn, PyYAML and PyQt5. It employed the UnivariateSpline spline function model using the Gauss kernel with an automatically matched sigma parameter of $\sigma = 2\sqrt{n}$, where n is the number of observations in the group.

3 Setup of Evaluation Experiments

The evaluation experiment consisted of two parts. The main purpose of the first part was to examine the extent to which splines are suitable for trend functions compared to polynomials. In turn, the second part of experiments aimed to compare the performance of models built over data updated with various trend functions: splines and polynomials of degrees ranging from 1 to 6.

3.1 Setup for Examination of Matching Trend Functions to Data

The real-world data used in the study comprised real estate purchase and sale transactions derived from a cadastral system in one of the big Polish cities. The initial dataset contained 83,523 records of transaction made in the city within from 1998 to 2015. However, not all transactions were used. First of all, non-market transactions were removed, and only those carried out on the free market and tender were left. In order to obtain segments in which the prices of premises change similarly over time, two partitions of the area of the city were intersected. The first one was the partition into 69 administrative cadastral regions, as shown in Fig. 2. Within the city, cadastral regions are delimited by districts and housing estates, as well as by streets, rivers, railways, etc. The second one was the partition into 250 zones based on maps of the land-use plan worked out by the municipality as illustrated in Fig. 3. The 31 most numerous segments were selected for the experiment, which covered from 314 to 2074 transactions concluded in the same cadastral region and land use zone.

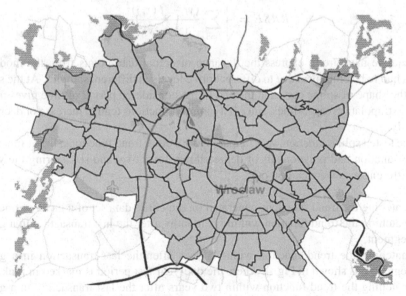

Fig. 2. Area of a city partitioned into cadastral regions

Root relative squared error was used as a fit measure, as shown in Formula 1, where y_i is the i-th data point, i.e. the actual price of a property, $f(x_i)$ – the estimated value for

Fig. 3. Area of a city partitioned into zones based on a land-use plan

the i-th data point, $\bar{y}$ – the arithmetic mean of all prices, and n denotes the number of data points.

$$RRSE = \sqrt{\sum_{i=1}^{n} \frac{(y_i - f(x_i))^2}{(y_i - \overline{y})^2}} \tag{1}$$

The relative measure to express the distance of actual data points from the modeled curves had to be employed due to different price levels in different samples. At the same time, the shape of spline curves at the end of the considered period of time gives much better extrapolation capabilities, since polynomial functions tend to increase or decrease radically.

The tested spline function was compared to six other trend functions: linear function, square function and polynomials of degree three, four, five, and six. During the work four different tests were carried out:

(a) matching the trend functions generated on the entire data set of a given segment,
(b) matching the trend function within six months after the last transaction in a given segment,
(c) matching the trend function within a year after the last transaction in a given segment, as shown in Fig. 4, where the one-year test period is marked in pink,
(d) matching the trend function within two years after the last transaction in a given segment.

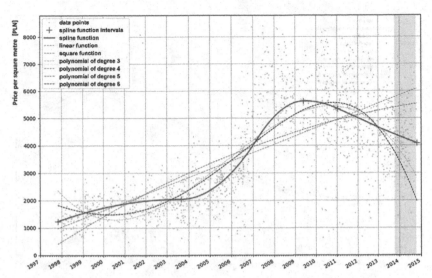

Fig. 4. Illustration of the method of testing the adjustment of the trend function to the data

3.2 Setup for Comparison of Models Built Over Updated Data

The most numerous data segment containing 2074 records was used to conduct comparative testing of machine learning models built over data updated with individual trend functions (see Fig. 5). Machine learning models were examined using a sliding time window, whose width was set at three years and the following year as a test set. This window was moved each time for a year, generating 14 three-year training data sets (from 1998–2000 to 2011–2013) each with a one-year test set (from 2001 to 2014). All transaction prices within each time window were updated on the last day of that time window using individual trend functions generated on the entire data set of the selected segment. The *Delta* method, described in our earlier article [20], was used to update transaction prices. Machine learning models were built over the updated transactional data of each time window. In turn, the accuracy of the models was calculated based on data from one year following a given time window.

Four machine learning algorithms to generate property valuation models were implemented using the *scikit-learn* library:

(1) REG - Model Linear Regression (*sklearn.linear_model.LinearRegression*),
(2) FOR - Random Forest Regressor (*sklearn.ensemble.RandomForestRegressor*),
(3) BAG - Bagging Regressor (*sklearn.ensemble.BaggingRegressor*),
(4) GRA - Gradient Boosting (*sklearn.ensemble.GradientBoostingRegressor*).

Four input features: the usable floor space of the apartment (*Area*), the age of the building structure (*Age*), the number of floors in the building (*Stories*), the distance of the building from the city center (*Center*), and the price per square meter (*Price*) as the input variable were adopted for the construction of models.

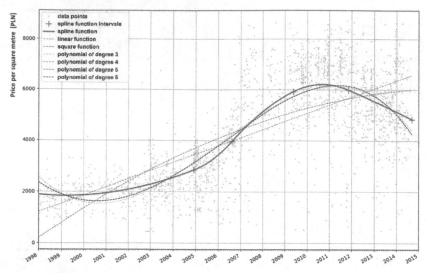

Fig. 5. Trend function graphs generated for the selected data segment

The accuracy of the models was measured with Root Mean Squared Error (RMSE) according to Formula 2, where y_i^a and y_i^p denote the actual and predicted values respectively, and n is the number of transactions in the test set.

$$RMSE = \sqrt{\frac{1}{n} \sum_{i=1}^{n} (y_i^p - y_i^a)^2}$$ (2)

4   Analysis of Experimental Results

4.1   Results of Examination of Matching Trend Functions to Data

The average RRSE values for 31 segments expressing the degree of matching individual trend functions to data within entire segments and a six-months, one-year, and two-year period after the last transaction in individual segments are presented in Table 1. The following denotation was used in Tables 1, 2, 3, 4, 5, 6, 7, 8 and 9: Poly-1, Poly-2, ..., Poly-6 stand for the polynomial trend functions of degree from 1 to 6, respectively.

The Friedman test followed by the paired Wilcoxon signed-ranks test were used to examine the statistically significant differences between individual trend functions in terms of RRSE. The average rank positions of the spline and polynomial functions determined during the Friedman test are shown in Table 2. Ranks generated by the Friedman test mean that the lower the rank value, the more the given trend function is matched to the data.

The results of the non-parametric Wilcoxon signed-rank test for pairwise comparison of the degree of matching individual trend functions to data are given in Table 3. The zero hypothesis stated there were no significant differences in RMSE between individual pairs of functions. In Table 3, + denotes that the trend function in the row revealed significantly better match than the function in the corresponding column, − means that it was significantly worse match, and ≈ signifies that it was statistically equivalent. In turn, / (slashes) separate the results for individual data period. A significance level of 0.05 was set to reject the null hypothesis.

Analyzing the data in Tables 1, 2 and 3, the following general conclusions can be drawn. For entire data from segments, the spline trend function shows significantly better fit to the data only compared to the linear function and the square function. In turn, for a six-month, one-year and two-year period, it significantly surpasses the polynomial functions of the degree four, five, and six.

Table 1. Degree of matching the trend functions to data in terms of the average value of RRSE

	Spline	Poly-1	Poly-2	Poly-3	Poly-4	Poly-5	Poly-6
Entire	0.77	0.83	0.80	0.78	0.77	0.77	0.77
Six-month	1.18	1.47	1.20	1.44	1.62	1.66	1.65
One-year	1.23	1.40	1.27	1.51	1.70	1.71	1.71
Two-years	1.44	1.60	1.59	2.44	3.00	3.58	3.54

Table 2. Average rank positions of tested functions determined during the Friedman test

	Spline	Poly-1	Poly-2	Poly-3	Poly-4	Poly-5	Poly-6
Entire	4.06	6.83	5.64	4.41	2.32	2.32	2.38
Six-month	3.27	4.72	3.51	3.17	4.31	4.51	4.48
One-year	2.93	4.06	3.75	3.13	4.44	4.65	5.00
Two-years	2.58	2.69	3.48	3.24	4.62	5.41	5.96

Table 3. Results of the Wilcoxon test for the Entire/Six-month/One-year/Two-year intervals

	Spline	Poly-1	Poly-2	Poly-3	Poly-4	Poly-5	Poly-6
Spline		+/+/+/≈	+/≈/≈/+	≈/≈/≈/+	≈/+/+/+	≈/+/+/+	≈/+/+/+
Poly-1	−/−/−/≈		−/≈/≈/+	−/−/≈/≈	−/≈/+/+	−/≈/+/+	−/≈/+/+
Poly-2	−/≈/≈/−	+/≈/≈/−		−/≈/≈/≈	−/≈/+/+	−/+/+/+	−/+/+/+
Poly-3	≈/≈/≈/	+/+/≈/≈	+/≈/≈/≈		−/+/+/+	−/+/+/+	−/+/+/+
Poly-4	≈/−/−/−	+/≈/−/−	+/≈/−/−	+/−/−/−		≈/≈/≈/+	≈/≈/≈/+
Poly-5	≈/−/−/−	+/≈/−/−	+/−/−/−	+/−/−/−	≈/≈/≈/−		≈/≈/+/+
Poly-6	≈/−/−/−	+/≈/−/−	+/−/−/−	+/−/−/−	≈/≈/≈/−	≈/≈/−/−	

4.2 Results of Performance Comparison of Models Built Over Updated Data

The results of accuracy of models generated by REG, FOR, BAG, and GRA algorithms in terms of RMSE are presented in Tables 4, 5, 6, and 7, respectively. The values of the RMSE measure differ significantly from year to year due to the different level of property prices in different years.

Table 4. Results of accuracy of REG models in terms of RSME for individual trend functions

Year	Spline	Poly-1	Poly-2	Poly-3	Poly-4	Poly-5	Poly-6
2001	626	626	626	626	626	626	626
2002	1020	1149	1420	1142	1089	1089	1090
2003	763	1010	1314	705	713	713	713
2004	795	1027	1459	795	794	794	794
2005	805	890	1045	853	845	846	846
2006	1799	2035	2425	2176	2121	2121	2121
2007	1972	1971	2172	2174	2136	2136	2137
2008	2813	3541	3055	2688	2772	2771	2771
2009	1452	2588	2130	1547	1595	1595	1595
2010	1066	1829	1658	1105	1107	1107	1107
2011	1872	2458	2417	1975	1966	1966	1966
2012	1263	1142	1150	1212	1238	1238	1238
2013	1425	1440	1390	1507	1535	1535	1535
2014	1285	1360	1314	1294	1294	1294	1294

We employed the Friedman test followed by the paired Wilcoxon signed-ranks test. The average rank positions of the REG, FOR, BAG, and GRA models determined during the Friedman test for RMSE are shown in Table 8. Ranks generated by the Friedman test mean that the lower the rank value, the better the model. The results of nonparametric Wilcoxon signed-rank test to pairwise comparison of the model performance are placed in Table 9. The zero hypothesis stated there were not significant differences in accuracy, in terms of RMSE, between given pairs of models. In Table 9 + denotes that the model in the row worked significantly better than the model in the appropriate column, – means that it was significantly worse, and ≈ signifies that it was statistically equivalent. In turn, / (slashes) separate the results for individual machine learning algorithms. The level of significance taken into account when rejecting the null hypothesis was 0.05.

The models generated using data updated with spline trend functions outperformed significantly the other models only for Model Linear Regression. For other machine learning algorithms no significant differences in model performance were observed despite different rank positions provided by the Friedman test. In turn, the models built over data updated with linear and square trend functions performed generally significantly worse than the models created on data updated with splines and higher order polynomials.

Table 5. Results of accuracy of FOR models in terms of RSME for individual trend functions

Year	Spline	Poly-1	Poly-2	Poly-3	Poly-4	Poly-5	Poly-6
2001	517	518	524	518	527	512	522
2002	963	957	951	961	956	959	958
2003	662	652	671	650	646	652	646
2004	845	1058	1066	836	849	809	809
2005	627	609	641	627	590	615	611
2006	1594	1648	1548	1582	1591	1579	1607
2007	2049	2212	2061	2145	2133	1995	2081
2008	1357	1400	1419	1374	1372	1383	1386
2009	1130	1542	1625	1190	1270	1262	1222
2010	1003	1063	1028	992	1010	953	1044
2011	1550	1605	1552	1510	1524	1507	1518
2012	1248	1326	1257	1284	1322	1274	1252
2013	1271	1248	1260	1326	1307	1305	1276
2014	1174	1319	1184	1143	1157	1126	1134

Table 6. Results of accuracy of BAG models in terms of RSME for individual trend functions

Year	Spline	Poly-1	Poly-2	Poly-3	Poly-4	Poly-5	Poly-6
2001	527	517	538	519	525	526	519
2002	952	972	955	951	961	967	951
2003	647	669	681	681	672	676	676
2004	881	1081	1083	868	823	893	847
2005	590	621	631	605	604	609	610
2006	1599	1649	1570	1564	1597	1617	1594
2007	1992	2215	2082	2101	2046	2077	2039
2008	1391	1382	1319	1379	1385	1381	1403
2009	1103	1485	1460	1197	1268	1207	1207
2010	1032	1057	1018	963	987	976	1007
2011	1540	1650	1533	1525	1474	1529	1482
2012	1259	1282	1300	1325	1216	1205	1287
2013	1296	1254	1228	1334	1303	1330	1291
2014	1170	1240	1169	1161	1158	1156	1084

Table 7. Results of accuracy of GRA models in terms of RSME for individual trend functions

Year	Spline	Poly-1	Poly-2	Poly-3	Poly-4	Poly-5	Poly-6
2001	559	554	557	552	565	575	568
2002	912	921	895	930	925	942	953
2003	656	652	709	738	691	683	686
2004	972	1468	1031	851	952	963	995
2005	679	595	663	638	618	639	630
2006	1538	1558	1653	1527	1508	1563	1520
2007	1988	2025	2010	2063	2235	1956	1968
2008	1352	1372	1463	1355	1442	1400	1374
2009	1119	1342	1528	1264	1409	1332	1325
2010	958	1040	1064	954	951	936	955
2011	1549	1940	1674	1618	1547	1555	1553
2012	1274	1531	1554	1227	1242	1301	1352
2013	1269	1259	1286	1264	1313	1249	1261
2014	1134	1079	1063	1062	1022	1010	1065

Table 8. Average rank positions of compared models determined during the Friedman test

	Spline	Poly-1	Poly-2	Poly-3	Poly-4	Poly-5	Poly-6
REG	2.64	5.00	5.57	3.64	3.64	3.71	3.79
FOR	3.57	5.29	4.86	3.93	4.07	2.71	3.57
BAG	3.79	5.29	4.79	3.57	3.29	4.00	3.29
GRA	3.57	4.21	5.50	3.36	3.71	3.64	4.00

Table 9. Results of the Wilcoxon test for REG/FOR/BAG/GRA models

	Spline	Poly-1	Poly-2	Poly-3	Poly-4	Poly-5	Poly-6
Spline		+/+/+/≈	+/≈/≈/+	≈/≈/≈/≈	+/≈/≈/≈	+/≈/≈/≈	+/≈/≈/≈
Poly-1	−/−/−/≈		≈/≈/−/≈	≈/−/−/≈	≈/−/−/≈	≈/−/−/≈	≈/−/−/≈
Poly-2	−/≈/≈/−	≈/≈/+/≈		−/≈/≈/−	−/≈/≈/≈	−/−/≈/−	−/≈/≈/−
Poly-3	≈/≈/≈/≈	≈/+/+/≈	+/≈/≈/+		≈/≈/≈/≈	≈/−/≈/≈	≈/≈/≈/≈
Poly-4	−/≈/≈/≈	≈/+/+/≈	+/≈/≈/≈	≈/≈/≈/≈		≈/−/≈/≈	≈/≈/≈/≈
Poly-5	−/≈/≈/≈	≈/+/+/≈	+/+/≈/+	≈/+/≈/≈	≈/+/≈/≈		≈/≈/≈/≈
Poly-6	−/≈/≈/≈	≈/+/+/≈	+/≈/≈/+	≈/≈/≈/≈	≈/≈/≈/≈	≈/≈/≈/≈	

5   Conclusions and Future Work

Modeling trend functions brings a lot of benefits. A property appraiser, having at his disposal tools illustrating the trends in the real estate market, is able to estimate their values more accurately. The same applies to models that, thanks to the price update process, are more able to predict the value of valued properties.

The analysis of the obtained results of the function tests does not give an unequivocal answer as to the decisive superiority of the spline functions over the other functions. For data from entire segments, the spline trend functions showed a significantly better fit to the data only compared to the linear function and the quadratic function. In turn, given the intervals of six months, one year and two years at the end of the considered period of time, the spline trend function significantly exceeds only the polynomial functions of the fourth, fifth and sixth degrees.

Models generated using data updated using the spline trend function significantly outperform the other models only for linear regression. For other machine learning algorithms, no significant differences in model performance were observed, despite the different ranking positions given by the Friedman test. In turn, models built on data updated using the linear and quadratic trend functions generally performed much worse than models created on data updated using higher order splines and polynomials.

The study showed the usefulness of using splines to model trend functions. The use of the spline trend function would be beneficial in automatic valuation systems.

References

1. Lasota, T., Sawiłow, E., Trawiński, B., Roman, M., Marczuk, P., Popowicz, P.: A method for merging similar zones to improve intelligent models for real estate appraisal. In: Nguyen, N.T., Trawiński, B., Kosala, R. (eds.) ACIIDS 2015, Part I. LNCS (LNAI), vol. 9011, pp. 472–483. Springer, Cham (2015). https://doi.org/10.1007/978-3-319-15702-3_46
2. Lasota, T., et al.: Enhancing intelligent property valuation models by merging similar cadastral regions of a municipality. In: Núñez, M., Nguyen, N.T., Camacho, D., Trawiński, B. (eds.) ICCCI 2015, Part II. LNCS (LNAI), vol. 9330, pp. 566–577. Springer, Cham (2015). https://doi.org/10.1007/978-3-319-24306-1_55
3. Kontrimas, V., Verikas, A.: The mass appraisal of the real estate by computational intelligence. Appl. Soft Comput. 11(1), 443–448 (2011). https://doi.org/10.1016/j.asoc.2009.12.003
4. Zurada, J., Levitan, A.S., Guan, J.: A comparison of regression and artificial intelligence methods in a mass appraisal context. J. Real Estate Res. 33(3), 349–388 (2011)
5. Peterson, S., Flangan, A.B.: Neural network hedonic pricing models in mass real estate appraisal. J. Real Estate Res. 31(2), 147–164 (2009)
6. Narula, S.C., Wellington, J.F., Lewis, S.A.: Valuating residential real estate using parametric programming. Eur. J. Oper. Res. 217, 120–128 (2012)
7. Antipov, E.A., Pokryshevskaya, E.B.: Mass appraisal of residential apartments: an application of Random forest for valuation and a CART-based approach for model diagnostics. Expert Syst. Appl. 39, 1772–1778 (2012). https://doi.org/10.1016/j.eswa.2011.08.077
8. D'Amato, M.: Comparing rough set theory with multiple regression analysis as automated valuation methodologies. Int. Real Estate Rev. 10(2), 42–65 (2007)
9. Kusan, H., Aytekin, O., Özdemir, I.: The use of fuzzy logic in predicting house selling price. Expert Syst. Appl. 37(3), 1808–1813 (2010). https://doi.org/10.1016/j.eswa.2009.07.031

10. Woźniak, M., Graña, M., Corchado, E.: A survey of multiple classifier systems as hybrid systems. Inf. Fusion **16**, 3–17 (2014). https://doi.org/10.1016/j.inffus.2013.04.006
11. Jędrzejowicz, J., Jędrzejowicz, P.: A family of GEP-induced ensemble classifiers. In: Nguyen, N.T., Kowalczyk, R., Chen, S.-M. (eds.) ICCCI 2009. LNCS (LNAI), vol. 5796, pp. 641–652. Springer, Heidelberg (2009). https://doi.org/10.1007/978-3-642-04441-0_56
12. Burduk, R., Baczyńska, P.: Dynamic ensemble selection using discriminant functions and normalization between class labels – approach to binary classification. In: Rutkowski, L., Korytkowski, M., Scherer, R., Tadeusiewicz, R., Zadeh, L.A., Zurada, J.M. (eds.) ICAISC 2016. LNCS (LNAI), vol. 9692, pp. 563–570. Springer, Cham (2016). https://doi.org/10.1007/978-3-319-39378-0_48
13. Kazienko, P., Lughofer, E., Trawiński, B.: Hybrid and ensemble methods in machine learning. J. Univ. Comput. Sci. **19**(4), 457–461 (2013)
14. Fernández, A., López, V., José del Jesus, M., Herrera, F.: Revisiting evolutionary fuzzy systems: taxonomy, applications, new trends and challenges. Knowl.-Based Syst. **80**, 109–121 (2015). https://doi.org/10.1016/j.knosys.2015.01.013
15. Lughofer, E., Cernuda, C., Kindermann, S., Pratama, M.: Generalized smart evolving fuzzy systems. Evol. Syst. **6**(4), 269–292 (2015). https://doi.org/10.1007/s12530-015-9132-6
16. Lasota, T., Telec, Z., Trawiński, B., Trawiński, K.: A multi-agent system to assist with real estate appraisals using bagging ensembles. In: Nguyen, N.T., Kowalczyk, R., Chen, S.-M. (eds.) ICCCI 2009. LNCS (LNAI), vol. 5796, pp. 813–824. Springer, Heidelberg (2009). https://doi.org/10.1007/978-3-642-04441-0_71
17. Krzystanek, M., Lasota, T., Telec, Z., Trawiński, B.: Analysis of bagging ensembles of fuzzy models for premises valuation. In: Nguyen, N.T., Le, M.T., Świątek, J. (eds.) ACIIDS 2010. LNCS (LNAI), vol. 5991, pp. 330–339. Springer, Heidelberg (2010). https://doi.org/10.1007/978-3-642-12101-2_34
18. Lasota, T., Telec, Z., Trawiński, B., Trawiński, K.: Exploration of bagging ensembles comprising genetic fuzzy models to assist with real estate appraisals. In: Corchado, E., Yin, H. (eds.) IDEAL 2009. LNCS, vol. 5788, pp. 554–561. Springer, Heidelberg (2009). https://doi.org/10.1007/978-3-642-04394-9_67
19. Lasota, T., Telec, Z., Trawiński, G., Trawiński, B.: Empirical comparison of resampling methods using genetic fuzzy systems for a regression problem. In: Yin, H., Wang, W., Rayward-Smith, V. (eds.) IDEAL 2011. LNCS, vol. 6936, pp. 17–24. Springer, Heidelberg (2011). https://doi.org/10.1007/978-3-642-23878-9_3
20. Trawiński, B.: Evolutionary fuzzy system ensemble approach to model real estate market based on data stream exploration. J. Univ. Comput. Sci. **19**(4), 539–562 (2013). https://doi.org/10.3217/jucs-019-04-0539
21. Lughofer, E., Trawiński, B., Trawiński, K., Kempa, O., Lasota, T.: On employing fuzzy modeling algorithms for the valuation of residential premises. Inf. Sci. **181**, 5123–5142 (2011). https://doi.org/10.1016/j.ins.2011.07.012
22. Meijering, E.: A chronology of interpolation: from ancient astronomy to modern signal and image processing. Proc. IEEE **90**(3), 319–342 (2002). https://doi.org/10.1109/5.993400
23. Piegl, L., Tiller, W.: The NURBS Book. Springer, Heidelberg (1995). https://doi.org/10.1007/978-3-642-97385-7
24. Runge, C.: Über empirische Funktionen und die Interpolation zwischen äquidistanten Ordinaten. Zeitschrift für Mathematik und Physik **46**, 224–243 (1901). www.archive.org
25. Dierckx, P.: Curve and Surface Fitting with Splines. Oxford University Press, Oxford (1993)
26. de Boor, C.: A Practical Guide to Splines. Springer, New York (2001)

Author Index

Printed in the United States
By Bookmasters